Handbook of the Tutte Polynomial and Related Topics

Handbook of the Tutte Polynomial and Related Topics

Edited by
Joanna A. Ellis-Monaghan
Korteweg - de Vries Instituut voor Wiskunde, Universiteit van Amsterdam, Netherlands
Iain Moffatt
Royal Holloway, University of London, United Kingdom

CRC Press is an imprint of the
Taylor & Francis Group, an **informa** business
A CHAPMAN & HALL BOOK

First edition published 2022
by CRC Press
6000 Broken Sound Parkway NW, Suite 300, Boca Raton, FL 33487-2742

and by CRC Press
4 Park Square, Milton Park, Abingdon, Oxon, OX14 4RN

Library of Congress Cataloging-in-Publication Data

Names: Ellis-Monaghan, Joanna Anthony, editor. | Moffatt, Iain, editor.
Title: Handbook of the Tutte polynomial and related topics / edited by Joanna A. Ellis-Monaghan, Iain Moffatt.
Description: First edition. | Boca Raton : C&H/CRC Press, 2022. | Includes bibliographical references and index.
Identifiers: LCCN 2021048725 (print) | LCCN 2021048726 (ebook) | ISBN 9781482240627 (hardback) | ISBN 9781032231938 (paperback) | ISBN 9780429161612 (ebook)
Subjects: LCSH: Tutte polynomial. | Graph theory. | Polynomials. | Invariants.
Classification: LCC QA166.249 .H36 2022 (print) | LCC QA166.249 (ebook) | DDC 511/.5--dc23/eng/20211223
LC record available at https://lccn.loc.gov/2021048725
LC ebook record available at https://lccn.loc.gov/2021048726

ISBN: 9781482240627 (hbk)
ISBN: 9781032231938 (pbk)
ISBN: 9780429161612 (ebk)

DOI: 10.1201/9780429161612

Typeset in CMR10
by KnowledgeWorks Global Ltd.

Publisher's note: This book has been prepared from camera-ready copy provided by the authors.

Contents

Preface

The Tutte polynomial is surely the most studied graph polynomial, particularly when all of its many specializations, generalizations, and applications are included. The impact of the Tutte polynomial comes in part from it being a universal invariant. It subsumes graph parameters and invariants that satisfy recursive deletion–contraction relations. Because of this property, the Tutte polynomial appears in various guises throughout mathematics and physics. For example, it encompasses the Potts and Ising models from physics, the Jones and HOMFLYPT polynomials from knot theory, the weight enumerator of linear codes, network reliability, the chromatic polynomial, and the many other applications that appear in this handbook.

This handbook consists of thirty-four chapters on the Tutte polynomial, its applications, and related topics, each written by experts in the field. Each chapter offers a stand-alone account of some aspect of the Tutte polynomial and is written so as to be accessible to non-experts yet comprehensive enough for experts.

The chapters are organized into six parts. Part I describes the fundamental properties of the Tutte polynomial. In particular the first four chapters provide an overview of the Tutte polynomial and the necessary background for the rest of the handbook. Part II is concerned with questions of computation, complexity, and approximation for the Tutte polynomial. Part III covers a selection of graph polynomials that are of interest in their own right, but can be obtained from the Tutte polynomial. Part IV discusses a range of applications of the Tutte polynomial to mathematics and physics. Part V includes various extensions and generalizations of the Tutte polynomial. Part VI provides a history of the development of the Tutte polynomial.

This handbook is not intended to be read linearly. We suggest that readers unfamiliar with the Tutte polynomial begin by reading Chapters 1–4, then dip into the remaining chapters, in any order, according to interest.

Joanna A. Ellis-Monaghan
Iain Moffatt
(March 2021)

Contributors

Federico Ardila
San Francisco State University,
San Francisco, CA, USA
and
Universidad de Los Andes,
Bogotá, Colombia

Spencer Backman
The University of Vermont,
Burlington, VT, USA

Béla Bollobás
University of Cambridge,
Cambridge, UK
and
University of Memphis,
Memphis, TN, USA

Joseph E. Bonin
The George Washington University,
Washington, DC, USA

Magnus Bordewich
Durham University,
Durham, UK

Robert Brijder
Hasselt University,
Diepenbeek, Belgium

Thomas Britz
UNSW Sydney,
Sydney, Australia

Jason I. Brown
Dalhousie University,
Halifax, Nova Scotia, Canada

Jin-Yi Cai
University of Wisconsin–Madison,
Madison, WI, USA

Peter J. Cameron
University of St Andrews,
St Andrews, UK

Sergei Chmutov
Ohio State University at Mansfield,
Mansfield, OH, USA

Timothy Y. Chow
Center for Communications Research
Princeton, NJ, USA

Charles J. Colbourn
Arizona State University,
Tempe, AZ, USA

Anna de Mier
Universitat Politècnica de Catalunya,
Barcelona, Spain

Fengming Dong
Nanyang Technological University,
Singapore

Joanna A. Ellis-Monaghan
University of Amsterdam
Amsterdam, The Netherlands

Michael J. Falk
Northern Arizona University,
Flagstaff, AZ, USA

Graham Farr
Monash University,
Victoria, Australia

Delia Garijo
University of Seville,
Sevilla, Spain

Emeric Gioan
CNRS, Université de Montpellier,
Montpellier, France

Andrew Goodall
Charles University,
Prague, Czech Republic

Gary Gordon
Lafayette College,
Easton, PA, USA

Geoffrey Grimmett
University of Cambridge,
Cambridge, UK

Hendrik Jan Hoogeboom
Leiden University
Leiden, The Netherlands

Stephen Huggett
Cornwall, UK

Bill Jackson
Queen Mary University of London,
London, UK

Khee Meng Koh
National University of Singapore,
Singapore

Tomer Kotek
Berlin, Germany

Joseph P.S. Kung
University of North Texas,
Denton, TX, USA

Johann A. Makowsky
Israel Institute of Technology,
Haifa, Israel

Elizabeth McMahon
Lafayette College,
Easton, PA, USA

Criel Merino
Universidad Nacional Autónoma de México,
Oaxaca, México

Iain Moffatt
Royal Holloway University of London,
Egham, UK

Jaroslav Nešetřil
Charles University,
Prague, Czech Republic

Steven Noble
Birkbeck, University of London,
London, UK

James Oxley
Louisiana State University
Baton Rouge, LA, USA

David Pearce
Victoria University of Wellington,
Wellington, NZ

Guus Regts
University of Amsterdam
Amsterdam, Netherlands

Oliver Riordan
University of Oxford,
Oxford, UK

Gordon F. Royle
University of Western Australia
Crawley, Australia

Adrian Tanasa
University of Bordeaux
Bordeaux, France
and
H. Hulubei Nat. Inst. Phys. Nucl. Engineering,
Magurele, Romania
and
Institut Universitaire de France
Paris, France

Lorenzo Traldi
Lafayette College,
Easton, PA, USA

Tyson Williams
University of Wisconsin–Madison,
Madison, WI, USA

Part I

Fundamentals

1

Graph theory

Joanna A. Ellis-Monaghan • Iain Moffatt

Synopsis

This chapter establishes the graph theoretical conventions used in the handbook.

- Graph theory conventions and notation.
- Graph parameters and properties.
- Standard graph operations.

1.1 Introduction

This chapter establishes the graph theoretical conventions and terminology used throughout this handbook. We assume that the reader is familiar with the fundamentals of graph theory and focus here on concepts essential to the study of the Tutte polynomial. Additional background on graph theory can be found in, for example, [152, 163, 383, 1104, 1147].

1.2 Graph theory conventions

1.2.1 Graphs

Definition 1.1. A *simple graph* G consists of a set V whose elements are called the *vertices* of the graph, together with a set of *edges* $E \subseteq [V]^2$, where

DOI: 10.1201/9780429161612-1

$[V]^2$ is the set of two-element subsets of V. An edge e may be denoted by either (u, v) or uv. The vertices u and v are called the *ends* of $e = uv$.

Definition 1.1 does not allow for the possibility of graphs having more than one edge with the same ends or an edge with both ends being the same vertex. However, for the theory of the Tutte polynomial, it is necessary to allow graphs to have such edges. The following definition provides the necessary formalism for this.

Definition 1.2. A *graph* G is an ordered pair (V, E) consisting of a set of *vertices* V, a set of *edges* E, and an *incidence function* $\psi_G : E \to V^{(2)}$, where $V^{(2)}$ is the set of unordered pairs of (not necessarily distinct) elements of V. With a minor abuse of notation, we write unordered pairs as (u, v). If e is an edge with $\psi(e) = (u, v)$, then we say u and v are the *ends* of e and are *incident* to e. The vertex and edge sets of a graph G are often denoted by $V(G)$ and $E(G)$.

As the notation of Definition 1.2 is somewhat cumbersome, when there is no potential for confusion, it is standard in graph theory to suppress the incidence function. In particular, it is usual to write $G = (V, E)$ for a graph with the implicit understanding that it also has an incidence function ψ_G. We follow this convention here. At times edges will be specified by their ends, for example writing $e = (u, v)$ or $e = uv$ for an edge e such that $\psi(e) = (u, v)$. This notation is ambiguous when there is more than one edge with ends u and v, and will be avoided when there is any potential for confusion.

An edge e is a *loop* if its two ends are the same, i.e. if $\psi(e) = (v, v)$ for some vertex v. If there are edges e and f in G such that $\psi(e) = \psi(f)$, then e and f are said to be *parallel edges*, and G is said to have *multiple edges*. The *multiplicity* of an edge e is the total number of edges that have the same ends as it. Definitions 1.1 and 1.2 are equivalent when we restrict to graphs without loops or multiple edges, and so a simple graph is a graph without loops or multiple edges.

The *degree*, $\deg(v)$, of a vertex v is the number of times it appears as the end of an edge. Thus, a loop edge adds two to the degree of its end vertex. The *maximum degree* (respectively, *minimum degree*) of a graph is the largest (respectively, smallest) degree over all its vertices and is denoted $\Delta(G)$ (respectively, $\delta(G)$). The *order* of a graph is its number of vertices.

Let v be a vertex of $G = (V, E)$. The *(open) neighborhood* of v in G, denoted by $N_G(v)$, is the set $\{w \in V : (v, w) \in E, w \neq v\}$. The *closed neighborhood* of v, denoted by $\bar{N}_G(v)$, is the set $N_G(v) \cup \{v\}$. Note that if v has a loop edge, then v is not in its own neighborhood

1.2.2 Subgraphs

A graph H is a *subgraph* of G if $V(H) \subseteq V(G)$, if $E(H) \subseteq E(G)$, and if the incidence function ϕ_H of H is the restriction of the incidence function ϕ_G of

G to $E(H)$. The subgraph H *induced* by $A \subseteq E(G)$ has edge set $E(H) = A$ and vertex set $V(H)$ consisting of exactly of the vertices in G that are ends of edges in $E(H)$. It is denoted by $G[A]$. The *spanning subgraph* H induced by $A \subseteq E(G)$ has $E(H) = A$ and $V(H) = V(G)$. It is denoted by $G|A$. A *vertex-induced* subgraph $G[S]$ of a graph G induced by $S \subseteq V(G)$ is the subgraph with vertex set S and edge set consisting of exactly those edges in $E(G)$ that have both ends in S. This distinction between the subgraph induced by A and the spanning subgraph induced by A is important. Spanning subgraphs induced by edge sets are the relevant substructures in most, but not all, settings involving the Tutte polynomial.

1.2.3 Isomorphism, invariants, and graph polynomials

Two graphs G and H are *isomorphic*, denoted $G \cong H$, if there are bijections $\theta : V(G) \to V(H)$ and $\varphi : E(G) \to E(H)$ such that $\psi_G(e) = uv$ if and only if $\psi_H(\varphi(e)) = \theta(u)\theta(v)$. A *graph invariant* is a function f on a specified class of graphs such that for all G and H in the class, $f(G) = f(H)$ whenever $G \cong H$. A *graph polynomial* is a graph invariant whose image lies in some polynomial ring.

1.2.4 Walks, trails, and paths; circuits and cycles

A *walk* in a graph $G = (V, E)$ is a sequence $W = v_0e_1v_1 \cdots v_{l-1}e_lv_l$ such that $e_i \in E$ and $e_i = (v_{i-1}, v_i)$, for each i. The number of edges in a walk is its *length*, v_0 is the *initial vertex*, v_l is the *terminal vertex*, and W is a v_0v_l*-walk*. The walk W is *closed* if $v_0 = v_l$, is a *trail* if each edge in it is distinct, a *circuit* if it is a closed trail, a *path* if each vertex and edge is distinct, and a *cycle* if $v_0 = v_l$, but otherwise each vertex and edge is distinct. Walks are often specified by their subsequence of vertices (even when the graph is not simple) if the choice of which parallel edges are taken as the walk is traversed is not important. The *girth* of a graph is the length of a shortest cycle in it. The *distance* between two vertices is the length of a shortest path between them. An edge is a *chord* of a cycle if it is not in the cycle, but its two ends are. A graph that consists of the vertices and edges of a cycle (respectively, path) is also called a *cycle* (respectively, *path*). In this setting a cycle of length m is denoted C_m, and a path of length m is denoted P_m.

An *Euler circuit* is a circuit that visits each edge of a graph exactly once (so every vertex must have even degree), while a *Hamilton cycle* visits each vertex exactly once. A graph is *Eulerian* (respectively, *Hamiltonian*) if it contains an Euler circuit (respectively, Hamilton cycle). By an abuse of terminology, a non-connected graph is often said to be Eulerian if each of its components is Eulerian, or equivalently, if each vertex is of even degree. Such graphs are also sometimes called *even graphs*.

1.2.5 Connected graphs, trees, and forests

A graph $G = (V, E)$ is *connected* if it contains a uv-walk for each $u, v \in V$, otherwise it is *disconnected*. The maximal connected subgraphs of G are called its *components* or *connected components.* A *tree* is a connected graph without cycles. A *forest* is a graph whose connected components are all trees.

1.2.6 Loops, bridges, and ordinary edges

Recall an edge e is a *loop* if its two ends are the same. A *bridge* is an edge whose removal increases the number of components of the graph. A bridge may also be called a *cut edge*, *isthmus*, or *coloop.* An edge that is neither a loop nor a bridge is *ordinary.*

1.2.7 Components, rank, and nullity

We use $k(G)$ to denote the number of connected components of a graph G, $v(G)$ to denote its number of vertices, and $e(G)$ its number of edges. The *rank* of G is $r(G) := v(G) - k(G)$, and the *nullity* of G is $n(G) := e(G) - r(G) = e(G) - v(G) + k(G)$. Given a graph $G = (V, E)$ and $A \subseteq E$, we write $e(A)$, $v(A)$, $k(A)$, $r(A)$, and $n(A)$ each referring to the spanning subgraph $(V(G), A)$ of G. The *corank* is $r(E) - r(A)$. Subscripts, for example $r_G(A)$, indicate the parameters for a specific graph.

1.2.8 Deletion, contraction, and graph minors

Let $G = (V, E)$ be a graph and $e \in E$. The graph $G\backslash e$ is obtained by *deleting* the edge e from G. More formally, $V(G\backslash e) = V(G)$ and $E(G\backslash e) = E(G) \setminus \{e\}$, with incidence function $\psi_{G\backslash e} = \psi_G|_{E\setminus\{e\}}$, i.e. $G\backslash e$ is the spanning subgraph induced by $E(G) \setminus \{e\}$.

If e is a non-loop edge, then the graph G/e is obtained by *contracting* e, that is, identifying the ends of e to create a single new vertex and then removing e. Formally, if $e = (u, v)$, then G/e has vertex set $(V \setminus \{u, v\}) \cup \{w\}$, edge set $E \setminus \{e\}$, and incidence function $\psi_{G/e}$ obtained from $\psi_G|_{E\setminus\{e\}}$ by substituting w for both u and v. If e is a loop, then $G/e := G\backslash e$.

If $A \subseteq E$, we write G/A (respectively $G\backslash A$) to denote the graph obtained by contracting (respectively deleting) all of the edges in A. In this context we often write A^c to denote the set $E \setminus A$. In particular, $G\backslash A^c$ is the spanning subgraph induced by A, which is also denoted by $G|A$. An important observation is that if $e \neq f$, then $(G/e)\backslash f = (G\backslash f)/e$.

If $v \in V$, then $G\backslash v$, the graph obtained from G by *deleting* v, is obtained by removing v and all of its incident edges from G.

A graph H is a *minor* of G if it results from deleting edges, deleting vertices, and contracting edges of G. A class $\mathcal{G}$ of graphs is *minor-closed* if whenever G is in $\mathcal{G}$, then every minor of G is also in $\mathcal{G}$.

1.2.9 Connectivity

A vertex v is a *cut-vertex* if it is not an isolated vertex and $k(G\backslash v) > k(G)$. For a graph $G = (V, E)$, a *vertex cut* (or just *cut*) is a set $X \subset V$ such that $G\backslash X$ has more components than G.

A graph is k*-connected* if there are at least k internally disjoint paths between every pair of distinct vertices, where two paths are internally disjoint if they have no vertices in common except their end points. Note that with this definition, K_2 is 1-connected, and the graph consisting of two vertices joined by n parallel edges is n-connected. By convention any graph on one vertex is 0-connected and 1-connected, but is not k-connected for any $k > 1$.

A *separation* of a connected graph is a decomposition of it into two nonempty connected subgraphs that have exactly one vertex in common. This vertex is a *separating vertex*. A graph is *nonseparable* if it is connected and has no separating vertices and is *separable* otherwise. Maximal nonseparable subgraphs are called *blocks* of the graph.

A *clique* is a set of mutually adjacent vertices in a graph. The maximum size of a clique in a graph G gives its *clique number*, $\omega(G)$.

If $G = (V, E)$ is a graph and $\emptyset \subset X \subset V$, then the set of all edges with one end in X and one end in $V \setminus X$ is called an *edge cut* or *cutset.* An edge cut with k elements is a k*-edge cut.* A *bond* is a minimal nonempty edge cut. A graph is k*-edge-connected* if k is the smallest integer for which it has a k-edge-cut.

A set of edges $A \subseteq E$ is a *flat* if $r(A \cup e) > r(A)$ for each $e \in E \setminus A$, or equivalently if there is no non-loop edge $e \in E \setminus A$ such that e is in a cycle of the spanning subgraph on $e \cup A$.

1.2.10 Graph operations

We use $G \sqcup H$ to denote the *disjoint union* of G and H. This is the graph such that $V(G \sqcup H) = V(G) \sqcup V(H)$, $E(G \sqcup H) = E(G) \sqcup E(H)$, and $\psi_{G \sqcup H}$ is induced by ψ_G and ψ_H. A *one-point join* of two disjoint graphs G and H, denoted by $G * H$, is formed by identifying a vertex u of G and a vertex v of H into a single vertex w of $G * H$. The edge set and incidence function of $G * H$ are induced from those of G and H. The *complete join*, denoted by $G + H$, is obtained starting with the disjoint union of G and H then adding an edge from each vertex of G to each vertex of H.

If G_1 and G_2 are two graphs, each containing a subgraph isomorphic to the complete graph K_r, then we may form two new graphs, the r*-gluing* and the r*-sum.* The r-gluing is formed by identifying the two copies of K_r, while the r-sum is formed by identifying the two copies of K_r and then removing all the identified edges. Note that a 0-gluing and a 0-sum are both equal to the disjoint of G_1 and G_2. Furthermore, a 1-gluing and a 1-sum are both equal to the result of identifying a vertex of G_1 with a vertex of G_2, i.e. forming a one-point join of the two graphs.

The *Cartesian product*, $G \times H$, of two simple graphs G and H is the graph with vertex set $V(G) \times V(H)$ and whose edge set is the set of all pairs $(u_1, v_1)(u_2, v_2)$ such that either $u_1u_2 \in E(G)$ and $v_1 = v_2$, or $v_1v_2 \in E(H)$ and $u_1 = u_2$. If G and H are graphs and H has a distinguished edge d, then the *tensor product*, $G \otimes H$, is the graph obtained by replacing each edge e in the graph G by a copy of H by identifying d and e, then, at the end, all the copies of d are deleted from the graph.

1.2.11 Colorings and matchings

A *k-coloring* of a graph $G = (V, E)$ is mapping $\sigma : V \to \{1, \ldots, k\}$, i.e., it is an assignment of one of k *colors* to each vertex of the graph. A k-coloring is *proper* if $\sigma(u) \neq \sigma(v)$ whenever u and v are adjacent vertices. A graph is *k-colorable* if it has a proper k-coloring, and its *chromatic number* is the minimum k for which it is k-colorable.

A *matching* in a graph is a set of its edges such that no two edges in it have a vertex in common. A matching is *perfect* if every vertex is contained in an edge in the matching.

1.2.12 Common graph families

The path of length m is denoted P_m, the cycle of length m is denoted C_m. A *caterpillar* is a tree with the property that the tree obtained by deleting all its leaves (vertices of degree one) is a path. A *complete graph* on n vertices, K_n, is a simple graph on n vertices in which each pair of vertices is adjacent. A *complete bipartite graph* $K_{p,q}$ is a simple graph whose vertex set can be partitioned into sets X and Y of size p and q, such that for each $x \in X$ and $y \in Y$, there is an edge xy, and every edge has an end in X and an end in Y. Similarly a *complete multipartite graph* $K_{p_1,\ldots,p_r}$ is a graph whose vertex set can be partitioned into sets of size $p_1, \ldots, p_r$ such that for any two vertices u and v in different blocks of the partition there is an edge uv, and every edge has its ends in different blocks of the partition. The *n-spoke wheel* W_n is the graph obtained from the cycle C_n (the *rim*) and another vertex (the *hub*) by adding an edge from each vertex on the rim to the hub. The edges incident with the hub are the *spokes*. A *theta-graph* is the graph on two vertices with three parallel edges between them. A *generalized theta-graph* or *k-theta graph* or *k-bridge graph* is a graph obtained from the graph on two vertices with k parallel edges by subdividing each edge (creating a vertex of degree two in the middle of it) any number of times. A *ladder* is a graph $P_n \times K_2$. A *circular ladder* L_n is a graph $C_n \times K_2$. A *Möbius ladder* M_n is a graph obtained from a cycle C_{2n} by adding an edge between every pair of vertices at distance n.

A *parallel extension* of a graph G is a graph H with an edge that is in a circuit of size two and that when deleted yields the original graph G. Dually, a *series extension* of G is a graph H with an edge that is in a minimal cut set of size two and that when contracted yields the original graph G. A *series-parallel graph* is then a graph that can be constructed from C_2 by iterating the operations of parallel extension and series extension. Equivalently, series-parallel graphs are characterized by being 2-connected and without a K_4 minor. Note that some definitions of series-parallel graphs start with K_1 instead of C_2; starting with C_2 assures 2-connectivity of the class (see [893]).

A *circle graph* is the intersection graph of a chord diagram (i.e., the vertices of a circle graph correspond to the chords, and there is an edge between two vertices if and only if the corresponding chords intersect in the chord diagram).

If $G = (V, E)$ is a graph, then its *line graph* $L(G)$ is the graph with vertex set E and in which two vertices of $L(G)$ are adjacent if and only if the corresponding edges in G are adjacent in G. If G is simple, the *complement* $\overline{G}$ of G is the simple graph on V which has an edge (u, v) if an only if (u, v) is not an edge in G.

1.2.13 Digraphs

Digraphs, or *directed graphs*, are defined in a way similar to graphs, but the incidence function maps into the set V^2 of *ordered* pairs of vertices. A digraph is denoted $\vec{G} = (V, E)$, where E is the set of *(directed) edges*, and where each edge is directed by the order (head, tail) of its ends. We use $e = (u, v)$ or $e = uv$ or $e = \overrightarrow{uv}$ or $e = u \to v$ to specify an edge. Digraph isomorphism is defined in a similar way to graph isomorphism, and other graph terminology is extended to digraphs in the obvious way. In the case of a digraph, the edges of a walk must be consistently oriented so that a walk $W = v_0 e_1 v_1 \cdots v_{l-1} e_l v_l$ in a digraph has $e_i = (v_{i-1}, v_i)$, for each i. A digraph $\vec{G}$ is *strongly connected* if for each pair of vertices $u, v \in V(\vec{G})$ it contains a consistently oriented walk from u to v. It is *weakly connected* if its underlying undirected graph is connected. An *orientation* of a graph is the assignment of a direction to each of its edges so that it forms a digraph. A *source* in a digraph is a vertex with in-degree zero, and a *sink* is a vertex with out-degree zero.

1.2.14 Plane graphs

A connected graph G is a *plane graph* if it is drawn on a sphere S^2 in such a way that edges only intersect at their ends, and such that each connected component of its complement, $S^2 \setminus G$, is homeomorphic to a disc (these discs are called the *faces* of G). A non-connected graph is a plane graph if it is the disjoint union of plane graphs (each component will lie on a different sphere). Plane graphs are usually drawn on the plane, with stereographic projection providing the means to move between the two drawings. A graph is *planar* if it is isomorphic to a plane graph. A plane graph is *outerplane* if all of its vertices lie on the boundary of its outer face, and a graph is *outerplanar* if it has an outerplane embedding.

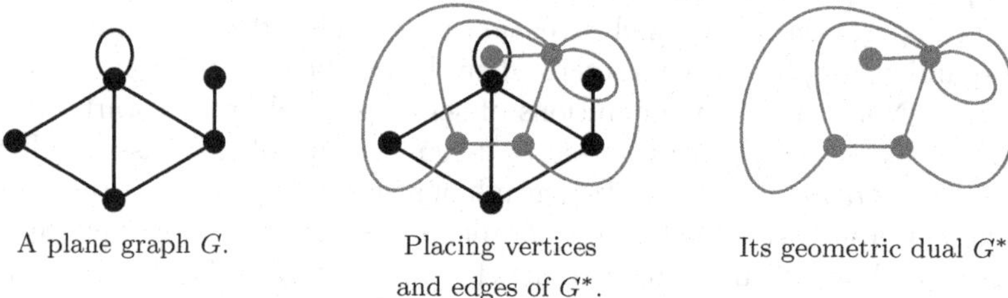

FIGURE 1.1: Forming the geometric dual G^* of a plane graph G.

1.2.15 Embedded graphs

A *cellularly embedded graph* G is a graph drawn on a surface Σ in such a way that edges only intersect at their ends, and such that each connected component of $\Sigma \setminus G$ is homeomorphic to a disc. Thus a plane graph is a graph cellularly embedded in the sphere. Two cellularly embedded graphs $G \subset \Sigma$ and $G' \subset \Sigma'$ are *equivalent* if there is a homeomorphism $\varphi : \Sigma \to \Sigma'$ with the property that $\varphi|_G : G \to G'$ is a graph isomorphism. This homeomorphism should be orientation preserving if Σ is orientable. A *ribbon graph* is a description of a cellularly embedded graph that arises by considering a regular neighborhood of the graph in the surface. It consists of a surface with boundary, decomposed into a number of closed topological discs of two types, *vertex-discs* and *edge-ribbons* such that: (1) the discs of the same type are pairwise disjoint; and (2) the vertex-discs and the edge-discs (which are called ribbons) intersect in disjoint line segments, each such line segment lies on the boundary of precisely one vertex and precisely one edge, and every edge contains exactly two such line segments (see Definition 27.2).

1.2.16 Geometric and Petrie duals, and partial duality

Let G be a plane graph. The *(geometric) dual* G^* is the plane graph obtained from G by placing one vertex in each of its faces and embedding an edge of G^* between two of these vertices whenever the faces of G they lie in are adjacent. Edges of G^* are embedded so that they cross the corresponding face boundary (or edge of G) transversally. An example of the formation of the dual of plane graph is given in Figure 1.1. There is a natural bijection between the edges of G and the edges of G^*, and we use it to identify the edges of a graph and its dual. At times, it is convenient to use e^* to denote the edge in the dual corresponding to an edge e. Note that $(G^*)^* = G$. Geometric duals of cellularly embedded graphs are formed just as in the plane case, by placing vertices in the faces and drawing edges transversally between these vertices when they are on adjacent faces.

There are generalized versions of graph duality. We describe these in the language of ribbon graphs. See [306, 456, 457] for further details of the construction and other formalisms. If $G = (V, E)$ is a ribbon graph, then its dual G^* can be formed by taking one disc for each boundary component of G (these will form the vertices of the dual); for each boundary component of G (which is topologically a circle), identify it with the boundary of one of these discs (resulting in a surface without boundary); finally, in the resulting surface, delete the interiors of the vertex discs in V. This results in the ribbon graph G^*. The discs that were added during the construction for the vertices of G^*, and the edge-ribbons of G form the edge ribbons of G^* but the parts of their boundary that are and are not attached to vertices are switched.

Partial duals arise by modifying this construction of G^* so that the dual is formed with respect to only a subset of edges. Let $G = (V, E)$ and $A \subseteq E$. The *partial dual* of G with respect to A, denoted G^A, is the ribbon graph formed as follows. Consider the spanning ribbon subgraph (V, A) as a subset of G. The boundary of (V, A) defines a set of closed curves on G. For each of these closed curves, take a disc (which will form a vertex of G^A) and identify the curve and the boundary of this disc. Finally, delete the interior of each vertex disc in V. The resulting ribbon graph is G^A. Observe that $G^E = G^*$ and $G^{\emptyset} = G$.

The *partial Petrial*, $G^{\tau(e)}$, with respect to an edge of a ribbon graph e results from detaching one end of the edge-ribbon from the vertex-disc, giving the ribbon a half-twist, and reattaching it. For $A \subseteq E(G)$, the partial Petrial $G^{\tau(A)}$ is the result of forming the partial Petrial with respect to every element in A (in any order). The Petrie dual $G^{\times}$ of a G is $G^{\tau(E)}$. This construction is equivalent to keeping the vertices and edges of an embedded graph but replacing the faces by left-right walks, also known as Petrie polygons. See, for example, [1165]. A *twisted dual* of G is the result of applying any sequence of partial Petrial and partial dual operations to G (see [456]).

1.2.17 Medial graphs

Let G be a graph embedded in a surface Σ. Its *medial graph* G_m is the 4-regular graph embedded in Σ obtained placing a vertex of degree four on each edge of G, and then drawing the edges of the medial graph by following the face boundaries of G. See Figure 1.2.

There are two common conventions for constructing medial graphs when G has isolated vertices. The first is that the medial graph of an isolated vertex is the empty graph. The second is that it is a *free loop*, which is an edge that forms a closed loop and is incident to no vertex. Both conventions are used in this handbook, but where there is ambiguity it is clearly stated which convention is being used.

A *checkerboard coloring* of a cellularly embedded graph is an assignment of the color gray or white to each face such that adjacent faces receive different colors (i.e., it is a face 2-coloring). A medial graph is always 4-regular and is always checkerboard colorable. One way to do this is to color the faces corre-

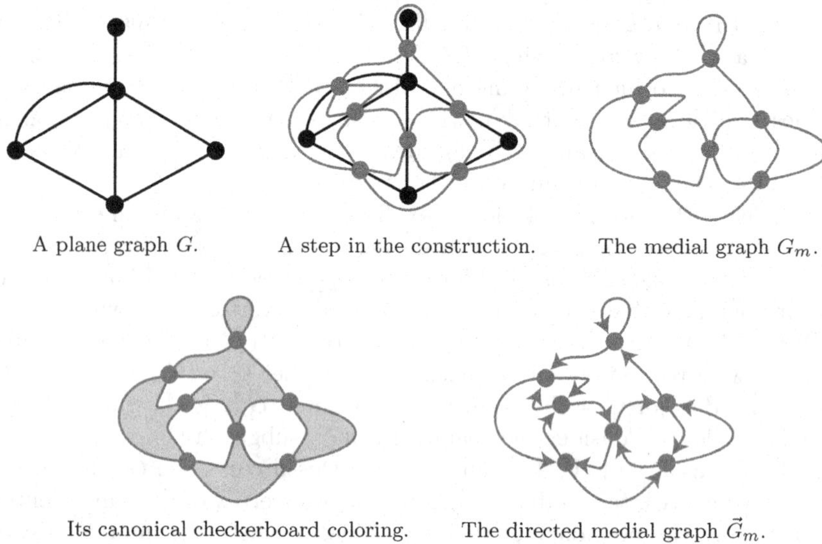

A plane graph G.
A step in the construction.
The medial graph G_m.
Its canonical checkerboard coloring.
The directed medial graph $\vec{G}_m$.

FIGURE 1.2: Forming the medial graph G_m of a plane graph G.

sponding to a vertex of the original graph gray, and the remaining faces white. We call this face 2-coloring of a medial graph G_m the *canonical checkerboard coloring*. See Figure 1.2. for an example.

If G is embedded in an oriented surface, its *directed medial graph* $\vec{G}_m$ is the directed graph obtained by orienting each edge of the medial graph G_m by traveling round the gray faces of its canonical checkerboard coloring in a direction consistent with the orientation of the surface. See Figure 1.2.

Tait graphs provide a way to recover a graph from a canonically checkerboard colored medial graph. Let F be a 4-regular checkerboard colored cellularly embedded graph. Its *Tait graph* F_{gr} is the embedded graph constructed

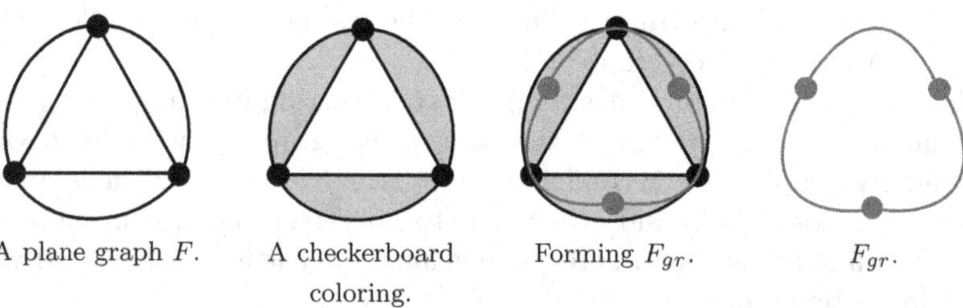

A plane graph F.
A checkerboard coloring.
Forming F_{gr}.
F_{gr}.

FIGURE 1.3: Forming a Tait graph.

by placing one vertex in each gray face and adding an edge between two of these vertices whenever the corresponding regions meet at a vertex of F. If F is a 4-regular graph in a plane (rather than a sphere), its *Tait graph* is obtained by choosing the face 2-coloring in which the unbounded region is white and forming F_{gr}. See Figure 1.3.

2

The Tutte polynomial for graphs

Joanna A. Ellis-Monaghan • Iain Moffatt

Synopsis

This chapter establishes several equivalent definitions of the Tutte polynomial for graphs and concludes with the universality property of the Tutte polynomial.

- The dichromatic polynomial.
- The state sum or spanning subgraph definition of the Tutte polynomial.
- The deletion–contraction or recursive definition of the Tutte polynomial.
- The activities or spanning forest definition of the Tutte polynomial.
- A universality and recipe theorem for the Tutte polynomial.

2.1 Introduction

We give several standard definitions of the Tutte polynomial, all of which are equivalent, but very different in appearance. This variety of expression accounts for the versatility and wide range of applications of the Tutte polynomial, since, as subsequent chapters in this handbook will illustrate, each facilitates different proof techniques and combinatorial insights. Furthermore, the Tutte polynomial is the universal object for essentially all multiplicative deletion–contraction graph invariants, which is one of its most powerful properties. Thus, any multiplicative graph invariant that has a deletion–contraction reduction must be an evaluation of the Tutte polynomial. We give this property in the form of a "recipe theorem" that provides the specific evaluation of the Tutte polynomial needed to recover a given deletion–contraction invariant.

DOI: 10.1201/9780429161612-2

A	Contribution to $Z(G)$
$\{a,b,c\}$	u^1v^3
$\{a,b\}$	u^2v^2
$\{a,c\}$	u^1v^2
$\{b,c\}$	u^2v^2

A	Contribution to $Z(G)$
$\{a\}$	u^2v^1
$\{b\}$	u^3v^1
$\{c\}$	u^2v^1
$\emptyset$	u^3v^0

TABLE 2.1: For the computations in Example 2.2.

2.2 The standard definitions of the Tutte polynomial

The Tutte polynomial has its origins in work by H. Whitney and by W. T. Tutte. A discussion of its history can be found in Chapter 34. Here we present several standard definitions of the Tutte polynomial, and see three quite different approaches and also some minor variations in formulation for each of them. We include several of the different names under which these definitions appear in the literature. The three main formulations of the Tutte polynomial given here are the most ubiquitous in the literature and will appear repeatedly throughout the handbook. However, there are others that are more specialized and require significant description to establish their settings, so we leave them to their respective chapters. Chapter 34 gives the history of the Tutte polynomial and details the development of its various definitions.

2.2.1 The dichromatic polynomial and rank generating formulation of the Tutte polynomial

The dichromatic polynomial is a two variable extension of the single variable chromatic polynomial (see Chapter 11), hence its name. It is simply a generating function for the number of components and edges in the spanning subgraphs of G. Although the dichromatic polynomial and the Tutte polynomial are equivalent, often one or the other form is easier to work with in a given setting.

Definition 2.1. If $G = (V, E)$ is a graph, then the *dichromatic polynomial*, $Z(G; u, v) \in \mathbb{Z}[u, v]$ is

$$Z(G; u, v) = \sum_{A \subseteq E(G)} u^{k(A)} v^{|A|}. \tag{2.1}$$

Example 2.2. For the graph G_1 in Figure 2.1a, the spanning subgraphs and their contributions to $Z(G_1)$ are given in the Table 2.1. These give that

$$Z(G_1; u, v) = u^3(v + 1) + u^2(2v^2 + 2v) + u(v^3 + v^2).$$

The Tutte polynomial, originally called the *dichromate* by Tutte, has the following expression as a *state sum expansion*, or *spanning subgraph expansion*.

(a) A graph G_1.

(b) A graph G_2.

FIGURE 2.1: Graphs used in Examples 2.2 and 2.5.

Definition 2.3. If $G = (V, E)$ is a graph, then the *Tutte polynomial* of G, $T(G; x, y) \in \mathbb{Z}[x, y]$, is defined by

$$T(G; x, y) = \sum_{A \subseteq E} (x-1)^{r(E)-r(A)}(y-1)^{n(A)}, \tag{2.2}$$

where the rank $r(A)$ and nullity $n(A)$ are as defined in Section 1.2.7.

The polynomial $W(G; x, y) = T(G; x+1, y+1)$ is known as the *Whitney rank generating function*, or the *corank–nullity polynomial.* We often refer to $T(G; x, y)$ as the *classical Tutte polynomial* of a graph G. The Whitney rank generating function predates the Tutte polynomial, and the Tutte polynomial is often called the *Tutte–Whitney polynomial* to reflect this.

Remark 2.4. The use of the name "dichromatic polynomial" in Definition 2.1 reflects modern usage of the term, and differs from its use in Tutte's papers. In his PhD thesis, Tutte used the term dichromate for what is now known as the Whitney rank generating function. Outside of his thesis he is generally consistent with the use of "dichromatic polynomial" for the for Whitney rank generating function, and "dichromate" for what is now called the Tutte polynomial. See Chapter 34 for historical details.

Example 2.5. For the graph G_2 in Figure 2.1b, the spanning subgraphs and their contributions to $T(G_2)$ are given in Table 2.2. Thus we see that

$$T(G_2; x, y) = (x-1)^3+4(x-1)^2+6(x-1)+(x-1)(y-1)+(y-1)+3 = x^3+x^2+xy.$$

Theorem 2.6. *The dichromatic polynomial and the Tutte polynomial are related through*

$$T(G; x, y) = (x-1)^{-k(G)}(y-1)^{-v(G)} Z\left(G; (x-1)(y-1), (y-1)\right), \tag{2.3}$$

and

$$Z\left(G; u, v\right) = u^{k(G)} v^{r(G)} T\left(G; \frac{u+v}{v}, v+1\right). \tag{2.4}$$

The equivalences given in Theorem 2.6 may be proved by using the definitions of rank and nullity to expand the exponents, and then collecting terms to move the prefactor out of the sum.

A	Contribution to $T(G)$
$\{a,b,c,d\}$	$(x-1)^0(y-1)^1$
$\{a,b,c\}$	$(x-1)^1(y-1)^1$
$\{a,b,d\}$	$(x-1)^0(y-1)^0$
$\{a,c,d\}$	$(x-1)^0(y-1)^0$
$\{b,c,d\}$	$(x-1)^0(y-1)^0$
$\{a,b\}$	$(x-1)^1(y-1)^0$
$\{a,c\}$	$(x-1)^1(y-1)^0$
$\{a,d\}$	$(x-1)^1(y-1)^0$

A	Contribution to $T(G)$
$\{b,d\}$	$(x-1)^1(y-1)^0$
$\{b,c\}$	$(x-1)^1(y-1)^0$
$\{c,d\}$	$(x-1)^1(y-1)^0$
$\{a\}$	$(x-1)^2(y-1)^0$
$\{b\}$	$(x-1)^2(y-1)^0$
$\{c\}$	$(x-1)^2(y-1)^0$
$\{d\}$	$(x-1)^2(y-1)^0$
$\emptyset$	$(x-1)^3(y-1)^0$

TABLE 2.2: For the computations in Example 2.5.

2.2.2 Linear recursion definitions

The Tutte polynomial has a linear recursion reduction that expresses the Tutte polynomial of a graph as the sum of the Tutte polynomials of the two graphs that result from deleting or contracting any of its ordinary edges. This central property of the Tutte polynomial not only facilitates inductive arguments but is also at the heart of the universality statement in Section 2.4. The linear recursion reduction leads to the following formulations of the Tutte polynomial, known as the *linear recursion* or *deletion–contraction* definitions. Work is required to verify these definitions, in particular to show that the deletion–contraction operation may be applied to the edges in any order so it results in a well-defined function. This point is discussed in more detail at the end of this section.

The formulation of Definition 2.7 successively deletes and contracts ordinary edges until only forests with loops remain, and these are evaluated by counting the number of loops and bridges. This terminal form of the recursion, which often provides the base case for induction arguments, is variously called the *terminal form*, the *base case*, or the *boundary condition.*

Definition 2.7. Let $G = (V, E)$ be a graph. Then the *Tutte polynomial*, $T(G; x, y) \in \mathbb{Z}[x, y]$ is the graph polynomial defined recursively by the relations

$$T(G;x,y) = \begin{cases} T(G\backslash e;x,y) + T(G/e;x,y) & \text{if } e \text{ is ordinary edge,} \\ x^b y^l & \text{if } G \text{ consists of exactly } b \text{ bridges and } l \text{ loops.} \end{cases} \tag{2.5}$$

When the variables are clear from context, it is common to suppress them and write $T(G)$ for $T(G; x, y)$.

Example 2.8. An example of the computation of the Tutte polynomial using Equation (2.5) is shown in Figure 2.2. The edge being deleted or contracted at each step is shown in bold.

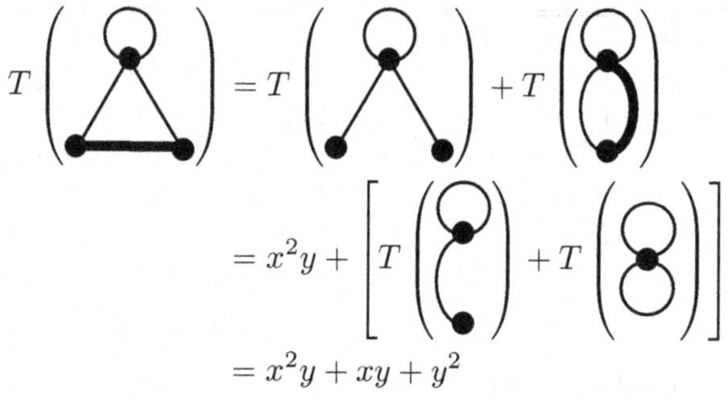

$$T(\ldots) = T(\ldots) + T(\ldots)$$
$$= x^2y + \left[T(\ldots) + T(\ldots)\right]$$
$$= x^2y + xy + y^2$$

FIGURE 2.2: A computation of the Tutte polynomial of a graph.

Example 2.9. A *computation tree* (also sometimes called a *resolution tree*) for the Tutte polynomial of a graph is a diagram showing the successive deletion–contractions steps. A computation tree for the Tutte polynomial using the terminal forms from Equation (2.5) is shown in Figure 2.3, giving that

$$T(\ldots) = x^2 + x + xy + y + y^2.$$

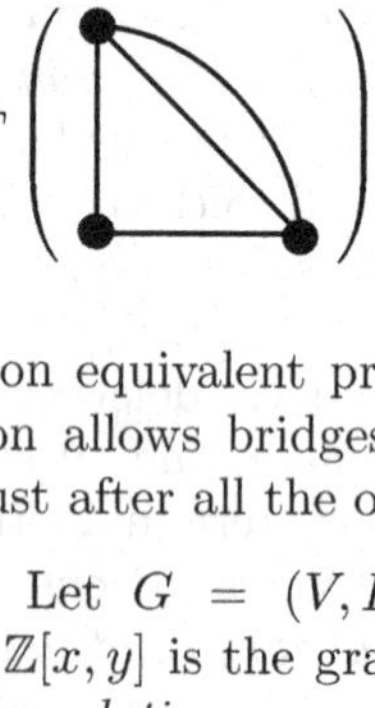

Another common equivalent presentation of the Tutte polynomial using deletion–contraction allows bridges and loops to be "factored out" at any point, instead of just after all the ordinary edges have been removed.

Definition 2.10. Let $G = (V, E)$ be a graph. Then the *Tutte polynomial*, $T(G; x, y) \in \mathbb{Z}[x, y]$ is the graph polynomial defined recursively by the *deletion–contraction relations*

$$T(G;x,y) = \begin{cases} x\,T(G/e;x,y) & \text{if } e \text{ is a bridge,} \\ y\,T(G\backslash e;x,y) & \text{if } e \text{ is a loop,} \\ T(G\backslash e;x,y) + T(G/e;x,y) & \text{if } e \text{ is ordinary edge,} \\ 1 & \text{if } E(G) = \emptyset. \end{cases} \tag{2.6}$$

The identities in Equation (2.6) are known as the *deletion–contraction relations* for the Tutte polynomial.

Note that since the Tutte polynomial is multiplicative on disjoint unions and one-point joins (see Section 2.3), if e is a bridge, then $x\,T(G/e; x, y)$ can be replaced by $x\,T(G\backslash e; x, y)$ in Equation (2.6).

Example 2.11. An example of computing the Tutte polynomial using Equation (2.6) is shown in Figure 2.4. The edge being deleted or contracted at each step is shown in bold.

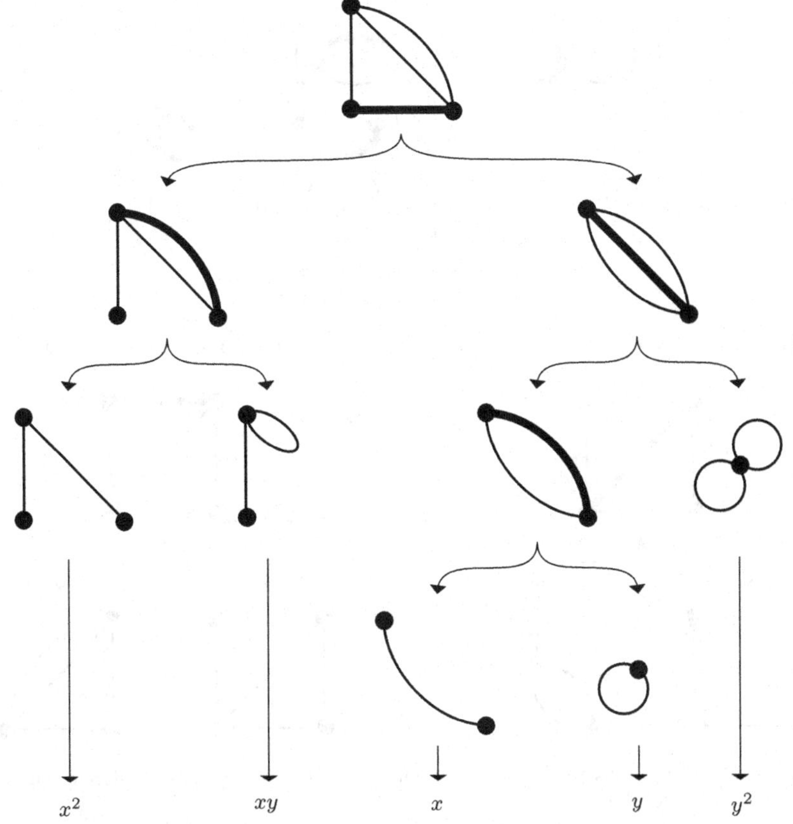

FIGURE 2.3: A computation tree for the Tutte polynomial of a graph.

The dichromatic polynomial $Z(G)$, since equivalent to the Tutte polynomial, must also have a deletion–contraction reduction. However, since $Z(G)$ is a shift of the Tutte polynomial, the form of its deletion–contraction reduction is a little different, in particular in its terminal form.

Definition 2.12. If $G = (V, E)$ is a graph, then the *dichromatic polynomial*, $Z(G; u, v) \in \mathbb{Z}[u, v]$ is given recursively by

$$Z(G; u, v) = \begin{cases} Z(G\backslash e; u, v) + vZ(G/e; u, v) & \text{if } e \text{ is any edge of } G, \\ Z(G) = u^n & \text{if } E(G) = \emptyset \text{ and } v(G) = n. \end{cases} \tag{2.7}$$

In general, if e and f are ordinary edges, then the two graphs that result from deleting and contracting e are not isomorphic to the two graphs that result from deleting and contracting f, as in Figure 2.5. Thus, it is necessary to prove that the Tutte polynomial is independent of the order in which the

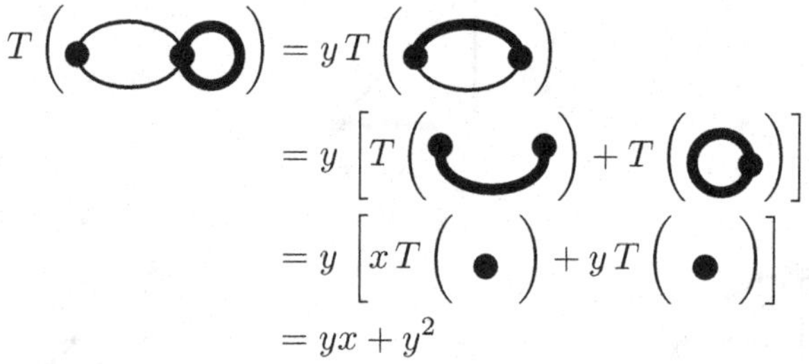

$$
\begin{aligned}
T\Big(\quad\Big) &= yT\Big(\quad\Big)\\
&= y\left[T\Big(\quad\Big) + T\Big(\quad\Big)\right]\\
&= y\left[xT\Big(\quad\Big) + yT\Big(\quad\Big)\right]\\
&= yx + y^2
\end{aligned}
$$

FIGURE 2.4: A computation of the Tutte polynomial of a graph using Equation (2.6).

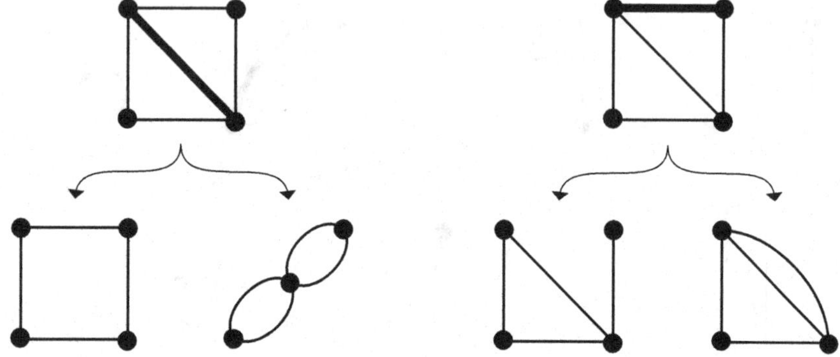

FIGURE 2.5: The graphs that result from deleting and contracting two different edges are generally not isomorphic.

deletion–contraction relation is applied to the edges of a graph. There are various approaches to this. One is to show that switching the order of deletion and contraction of two edges does not change the outcome. This appears for matroids in [249], and for graphs follow, for example, the proof of Proposition 4.1 of [885]. The proof depends on the observation that $(G\backslash e)/f = (G/f)\backslash e$. Another approach is to show by induction that the deletion–contraction formulation is equivalent to another formulation, usually the state sum expansion of Definition 2.3, which is clearly independent of any ordering of the edges. A proof of this may be found in [152]. Alternatively, if the state sum expansion of Definition 2.3 is taken to be the definition of the Tutte polynomial, then the universality of the Tutte polynomial given by Theorem 2.24 in the next section may be applied; however, first proving the universality requires as much work as a direct proof.

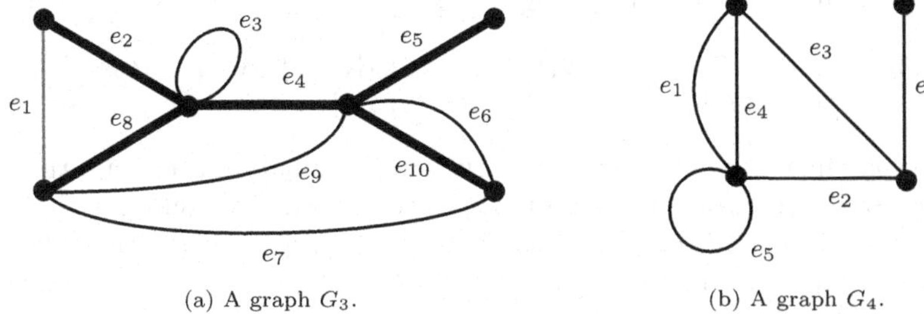

(a) A graph G_3. (b) A graph G_4.

FIGURE 2.6: Graphs used in Examples 2.13 and 2.18.

2.2.3 Spanning tree or activities expansions

The formulations of the Tutte polynomial in this subsection are sums over maximal spanning forests. Note that although the nomenclature in the literature often uses "spanning tree expansion" to describe this formulation of the Tutte polynomial, the sums are actually over maximal spanning forests when the graphs are not connected. The expressions for the Tutte polynomial use a specified ordering of the edges but can be shown to be independent of the choice of order. The equivalence to the deletion–contraction formulation, and also a direct proof of the independence of edge order, may be found in [152]. Other "activity expansions" for the Tutte polynomial can be found in Chapter 5.

We begin with the concepts necessary to define the internal and external activities of a maximal spanning forest with respect to a given order of the edges of a graph G. Let $G = (V, E)$ be a graph and F be a maximal spanning forest of G. The maximality of F means that if G is connected then F is a spanning tree, and otherwise restricting F to any component of G gives a spanning tree of that component. For each edge $e \in E\setminus E(F)$ there is a unique cycle in G consisting of e and a subset of edges in F. This is the *cycle defined by* e, denoted $Z_F(e)$ so

$$Z_F(e) = (\text{the edge set of the unique cycle in } F \cup e).$$

Also, for each edge e in F, the *cut defined by* e, denoted $U_F(e)$, is the set

$$U_F(e) = \{f \in E : (F\setminus e) \cup f \text{ is a maximal spanning forest of } G\}.$$

Thus, an edge f is in $U_F(e)$ if and only if replacing e with f in F results in another maximal spanning forest.

Example 2.13. If G_3 is the graph in Figure 2.6a with maximal spanning forest F with edges $\{e_2, e_4, e_5, e_8, e_{10}\}$ shown in bold, then the cycles determined by each of the edges not in F are:

$$Z_F(e_1) = \{e_1, e_2, e_8\}, \quad Z_F(e_3) = \{e_3\}, \quad Z_F(e_6) = \{e_6, e_{10}\},$$
$$Z_F(e_7) = \{e_4, e_7, e_8, e_{10}\}, \quad Z_F(e_9) = \{e_4, e_8, e_9\}.$$

The cuts determined by each of the edges in F are:

$$\begin{aligned} &U_F(e_2) = \{e_1, e_2\}, && U_F(e_4) = \{e_4, e_7, e_9\}, && U_F(e_5) = \{e_5\}, \\ &U_F(e_8) = \{e_1, e_7, e_8, e_9\}, && U_F(e_{10}) = \{e_6, e_7, e_{10}\}. \end{aligned}$$

For the spanning tree definition of the Tutte polynomial we need an (arbitrary) linear ordering $\prec$ of the edges of G. For convenience, we suppose that the edges are indexed by this order, so that if $E = \{e_1, \ldots, e_m\}$, then the order is given by $e_i \prec e_j$ if and only if $i < j$.

Definition 2.14. If $E = \{e_1, \ldots, e_m\}$ gives a linear order on the edges of G, and if F is a maximal spanning forest of G, then we say that an edge e of G is:

1. *internally active* with respect to F if e is in F, and it is the smallest edge in $U_F(e)$, the cut defined by e (otherwise it is *internally inactive*);
2. *externally active* with respect to F if e is not in F, and it is the smallest edge in $Z_F(e)$, the cycle defined by e (otherwise it is *externally inactive*);
3. F is an (i,j)*-forest* if it is a is a maximal spanning forest of G, and G has exactly i internally active edges and exactly j externally active edges with respect to F.

Example 2.15. If G_3 is the graph in Figure 2.6a with maximal spanning forest F on edges $e_2, e_4, e_5, e_8, e_{10}$, and edge order given by the indices, then the internally active edges of G with respect to F and this ordering are e_4, e_5, and the externally active edges are e_1, e_3, e_6.

We can now give the *spanning tree expansion* or *activities expansion* of the Tutte polynomial.

Definition 2.16. Let $G = (V, E)$ be a graph with a fixed linear order on E. The *Tutte polynomial*, $T(G; x, y) \in \mathbb{Z}[x, y]$, of G is

$$T(G; x, y) = \sum_{\substack{F \text{ a max.} \\ \text{span. forest}}} x^{|\mathrm{IA}(F)|} y^{|\mathrm{EA}(F)|}, \tag{2.8}$$

where the sum is over all maximal spanning forests of G and where $\mathrm{IA}(F)$ (respectively $\mathrm{EA}(F)$) denotes the set of internally active (respectively, externally active) edges of G with respect to F and the linear ordering of $E(G)$.

Collecting like terms gives the other common formulation of the spanning tree expansion of the Tutte polynomial.

Definition 2.17. Let $G = (V, E)$ be a graph with a fixed linear order on E. The *Tutte polynomial*, $T(G; x, y) \in \mathbb{Z}[x, y]$, of G is

$$T(G; x, y) = \sum_{i,j \geq 0} t_{ij} x^i y^j, \tag{2.9}$$

Spanning tree	Internally active	Externally active	Contribution to $T(G)$
e_1, e_2, e_6	e_1, e_2, e_6	e_5	x^3y
e_1, e_3, e_6	e_1, e_6	e_5	x^2y
e_2, e_3, e_6	e_6	e_1, e_5	xy^2
e_2, e_4, e_6	e_2, e_6	e_1, e_5	x^2y^2
e_3, e_4, e_6	e_6	e_1, e_2, e_5	xy^3

TABLE 2.3: Activities for Example 2.18.

where, t_{ij} (which is also written $t_{i,j}$) is the number of maximal spanning forests of G with internal activity i and external activity j, i.e., t_{ij} is the number of (i, j)-forests in G.

Example 2.18. For the graph G_4 from Figure 2.6b, and with edge order given by the indices, the internally active and externally active edges with respect to each spanning tree are given in Table 2.3. It follows that the Tutte polynomial of G_4 is

$$T(G_4) = x^3y + x^2y^2 + x^2y + xy^2 + xy^3.$$

2.2.4 Equivalence of the various definitions

Any of the definitions given for $T(G)$ or $Z(G)$ in Sections 2.2.1–2.2.3 may be taken as *the* definition of the Tutte polynomial and the rest derived from it. As noted above, if either the deletion–contraction definition or spanning tree expansion is taken as the definition, it must first be shown independent of the choice of edge order. If the state sum expansion is taken to be the definition, then it requires some work to show that it has a deletion–contraction reduction. Showing that the deletion–contraction and state sum definitions, Definitions 2.10 and 2.3, are equivalent is a fairy routine induction argument. It involves specifying an edge e and splitting the sum in (2.3) into to parts, one where e is in A and one where it is not, and considering the three possible edge types, loop, bridge, and ordinary, for e. Showing that the spanning tree definition, Definition 2.16, satisfies the deletion–contraction relations (2.6) is more involved. The book [152] gives lucid proofs of equivalence of the three definitions.

2.3 Multiplicativity

The Tutte polynomial is multiplicative on the disjoint union and one-point joins of graphs.

Proposition 2.19. *If G and H are graphs then*

$$T(G \sqcup H) = T(G) \cdot T(H), \tag{2.10}$$

and

$$T(G * H) = T(G) \cdot T(H). \tag{2.11}$$

This follows readily from Definitions 2.10 or 2.16 by induction on the number of ordinary edges in $G * H$ or $G \sqcup H$.

2.4 Universality of the Tutte polynomial

The universality property of the Tutte polynomial is one of its most important features. Essentially, it says that any graph invariant that is multiplicative on disjoint unions and one-point joins of graphs and that has a deletion–contraction reduction must be an evaluation of the Tutte polynomial.

2.4.1 Tutte–Grothendieck invariant

Definition 2.20. Let $\mathcal{G}$ be a minor-closed class of graphs, $\mathfrak{R}$ be a commutative ring with unity and $a, b \in \mathfrak{R}$. A graph invariant $f : \mathcal{G} \to \mathfrak{R}$ is a *(generalized) Tutte–Grothendieck invariant*, or *T–G invariant*, if

$$f(G) = af(G\backslash e) + bf(G/e) \quad \text{if } e \text{ is ordinary,} \tag{2.12}$$

$$f(\bullet) = 1, \tag{2.13}$$

$$f(G \sqcup H) = f(G) \cdot f(H), \tag{2.14}$$

$$f(G * H) = f(G) \cdot f(H). \tag{2.15}$$

2.4.2 Universality

The Tutte polynomial is a Tutte–Grothendieck invariant, and, in fact, since the following two results give both universal and unique extension properties, it is essentially the *only* Tutte–Grothendieck invariant, in that any other must be an evaluation of it. This universality property of the Tutte polynomial can be used to identify various graph polynomials as specializations of the Tutte polynomial (see Chapter 3). Theorem 2.21 is known as a *recipe theorem* since it specifies how to recover a Tutte–Grothendieck invariant as an evaluation of the Tutte polynomial.

Theorem 2.21 (Universality Property of the Tutte Polynomial). *Let $\mathcal{G}$ be a minor-closed class of graphs, let $\mathfrak{R}$ be a commutative ring with unity, and let $f : \mathcal{G} \to \mathfrak{R}$. If there exists $a, b \in \mathfrak{R}$ such that f is a Tutte–Grothendieck* *invariant, then*

$$f(G) = a^{n(G)} b^{r(G)} T\left(G; \frac{x}{b}, \frac{y}{a}\right), \tag{2.16}$$

where $f(\bullet\!\!-\!\!\!-\!\!\bullet) = x$ *and* $f\left(\bullet\!\!\!\bigcirc\right) = y$.

Example 2.22. As an illustration of the application of Theorem 2.21, let $f(G)$ denote the number of maximal spanning forests of a graph G. If e is an ordinary edge of G, then $f(G)$ can be written as the sum of the number of maximal spanning forests of G that contain e and the number of maximal spanning forests of G that do not contain e. It is readily verified that the first term of this sum equals $f(G/e)$ and the second equals $f(G\backslash e)$, so $f(G) = f(G\backslash e) + f(G/e)$ if e is ordinary. It is easily checked that Equations (2.13)–(2.15) hold, and so f is a Tutte–Grothendieck invariant. Since $f(\bullet\!\!-\!\!\!-\!\!\bullet) = 1$ and $f\left(\bullet\!\!\!\bigcirc\right) = 1$, it follows that $T(G; 1, 1)$ equals the number of maximal spanning forests of G.

We have the following unique extension property, which says that if we specify any four elements $a, b, x, y \in \mathfrak{R}$, then there is a unique well-defined Tutte–Grothendieck invariant on these four elements.

Theorem 2.23. *Let $\mathcal{G}$ be a minor-closed class of graphs, let $\mathfrak{R}$ be a commutative ring with unity, and let $a, b, x, y \in \mathfrak{R}$. Then there is a unique Tutte–Grothendieck invariant $f : \mathcal{G} \to \mathfrak{R}$ with $f(\bullet\!\!-\!\!\!-\!\!\bullet) = x$ and $f\left(\bullet\!\!\!\bigcirc\right) = y$. Furthermore, this function f is given by*

$$f(G) = a^{n(G)} b^{r(G)} T\left(G; \frac{x}{b}, \frac{y}{a}\right). \tag{2.17}$$

If a or b are not units of $\mathfrak{R}$, then (2.16) and (2.17) are interpreted to mean using expansion (2.2) of Definition 2.3, and cancelling before evaluating.

As we shall see in subsequent chapters, a common occurrence is that a graph invariant may not quite satisfy the conditions of Theorem 2.18, but a minor reformulation of it will, for example with the introduction of a prefactor. However, the following slight generalization from [152] may sometimes streamline a proof.

Theorem 2.24. *Let $\mathcal{G}$ be a minor-closed class of graphs. Then there is a unique map $U : \mathcal{G} \to \mathbb{Z}[x, y, a, b, \alpha]$ such that*

$$U(G) = \begin{cases} x\,U(G\backslash e) & \text{if } e \text{ is a bridge,} \\ y\,U(G\backslash e) & \text{if } e \text{ is a loop,} \\ a\,U(G\backslash e) + b\,U(G/e) & \text{if } e \text{ is ordinary edge,} \\ \alpha^n & \text{if } E(G) = \emptyset \text{ and } v(G) = n. \end{cases} \tag{2.18}$$

Moreover

$$U(G) = \alpha^{k(G)} a^{n(G)} b^{r(G)} T\left(G; \frac{\alpha x}{b}, \frac{y}{a}\right). \tag{2.19}$$

Here, if $\alpha \neq 1$ then U is not multiplicative on disjoint unions, so unlike the case for $T(G)$ in Definition 2.10 in general $U(G\backslash e) \neq U(G/e)$.

Example 2.25. If we begin with the deletion–contraction relations for the dichromatic polynomial in Equation (2.7), then taking $(x, y, a, b, \alpha) = (1 + v/u, v+1, 1, v, u)$ and applying Theorem 2.24 gives Equation (2.4). On the other hand, by starting with Equation (2.4), we can recover Equation (2.7) via Theorem 2.24.

The results in this section can be proven by induction on the number of ordinary edges from the deletion–contraction definition of the Tutte polynomial. See, for example, Brylawski [251], Oxley and Welsh [904], Brylawski and Oxley [247], Welsh [1139], and Bollobás [152] for detailed discussions of these theorems and their consequences.

2.5 Duality

For plane graphs, duality interchanges the role of deletion and contraction: $(G^*/e^*) = (G\backslash e)^*$ and $(G^*\backslash e^*) = (G/e)^*$, where e^* is the edge in G^* corresponding to an edge e of G. It also changes loops to bridges and bridges to loops. These observations can be used to write the deletion–contraction formulation of the Tutte polynomial of the dual G^* in terms of the primal graph G. An application of universality then gives the *duality relation* for the Tutte polynomial:

Theorem 2.26. *If G is a plane graph and G^* its dual, then*

$$T(G^*; x, y) = T(G; y, x). \tag{2.20}$$

Theorem 2.26 can be rephrased in terms of planar graphs by using algebraic duals, and also holds for matroids (See Theorem 4.125). However it does not hold, in general, for non-plane graphs.

3

Essential properties of the Tutte polynomial

Béla Bollobás • Oliver Riordan

Synopsis

The impact of the Tutte polynomial derives from the richness of its properties and the wealth of combinatorial information it encodes. We illustrate the importance of the Tutte polynomial by discussing briefly some central topics in its theory.

- Evaluations of the Tutte polynomial at special integer points.
- Basic properties of the coefficients of the Tutte polynomial, including Crapo's β-invariant.
- Evaluations of the Tutte polynomial along particular curves: the chromatic, flow and reliability polynomials.
- Connections between the Tutte polynomial and the Ising, Potts and random-cluster models.
- Duality and the Tutte polynomial.
- Medial graphs and the Martin and circuit partition polynomials; the interlace polynomial.
- Connections to knot theory, in particular the Jones polynomial.
- Signed, colored and topological Tutte polynomials.

DOI: 10.1201/9780429161612-3

(x, y)	Interpretation	Discussed in
$(1, 1)$	spanning trees	Proposition 3.1
$(2, 1)$	spanning forests	Proposition 3.1
$(1, 2)$	connected spanning subgraphs	Proposition 3.1
$(2, 2)$	$2^{e(G)}$	Proposition 3.1
$(2, 0)$	acyclic orientations	Proposition 3.2
$(1, 0)$	... with a given unique source	Proposition 3.3
$(0, 2)$	totally cyclic orientations	Section 3.2.2
$(0, 1)$	... with no clockwise cycle	Section 3.2.2
$(0, -2)$	ice configurations	Section 3.2.2
$(-1, -1)$	size of bicycle space	Section 3.2.3
$(1, 2)$	score sequences	Section 3.2.3
$(3, 3)$	tetromino tilings	Section 3.2.3

TABLE 3.1: Evaluations of the Tutte polynomial.

3.1 Introduction

Our aim in this chapter is to outline the importance of the Tutte polynomial from a number of different points of view. The Tutte polynomial of a graph contains an enormous amount of information about the graph; we illustrate this with a number of simple and not-so-simple examples. We also outline some of the most successful applications of the Tutte polynomial, in particular in knot theory. Most of the topics we touch upon here are treated in depth in later chapters; the brief details here are intended to whet the reader's appetite for what is to come in the rest of the book.

3.2 Evaluations at special points

Throughout we work with the formulation of the Tutte polynomial given in Definition 2.3, or any of the equivalent definitions in Chapter 2.

A striking feature of the Tutte polynomial $T(G; x, y)$ is that its evaluations at various specific points count a wide variety of combinatorial objects associated to the graph G. We include a table (Table 3.1) of such points discussed in this section for ease of reference. This table is not supposed to make sense on its own, but simply directs the reader to the relevant proposition or subsection for the interpretation of each entry, and in particular the (sometimes limited) class of graphs to which it applies.

More often than not, the achievement is noticing that the Tutte polynomial counts a specific type of object; proving that this is the case tends to be easy, either from the deletion–contraction relation (see below) or directly from an expansion. We start with a very simple proposition whose proof is immediate from an expansion: in this case, the rank–nullity expansion (2.2).

Proposition 3.1. *Let G be a connected graph. Then $T(G;1,1)$ counts the number of spanning trees of G, while $T(G;2,1)$ counts the number of spanning forests, and $T(G;1,2)$ the number of connected spanning subgraphs. Moreover, $T(G;2,2) = 2^{e(G)}$ is the number of all spanning subgraphs of G.*

To see this, note, for example, that a set $A \subseteq E$ of edges of $G = (V,E)$ describes a spanning tree if and only if the spanning subgraph $H = (V,A)$ is connected, i.e., $r(A) = r(E) = |G| - 1$, and H is a forest, i.e., A has nullity 0. With $x = y = 1$, the formula (2.2) counts 1 for each such set A, and 0 for all other $A \subseteq E$. The other statements follow similarly.

Proposition 3.1 extends to disconnected graphs G by replacing "spanning tree" by "maximal spanning forest" (i.e., a subgraph consisting of a spanning tree of each component), and "connected spanning subgraph" by "spanning subgraph with no more components than G".

3.2.1 Acyclic orientations

Much of the time, to prove that a certain "count" $c(G)$ is given by an evaluation of the Tutte polynomial, one uses the recursive definition (2.6) or some variant, for example (2.5). To show that $c(G) = T(G;x,y)$ for specific x and y, it suffices to establish the relevant boundary condition, together with the deletion–contraction relation for $c(G)$. We illustrate this with a well-known example due to Stanley [1027], concerning the number of acyclic orientations (orientations without cyclically oriented cycles). We give proofs of this and the next result since the basic method is essential to understanding the significance of the Tutte polynomial.

Proposition 3.2. *For any graph G, $T(G;2,0)$ is the number of acyclic orientations of G.*

Proof. Let $A(G)$ denote the set of acyclic orientations of G, and set $a(G) = |A(G)|$. If G contains a loop, then clearly $a(G) = 0$; if G is a forest, then $a(G) = 2^{e(G)}$. This shows that the boundary condition in (2.5) is satisfied. It thus suffices to show that for any ordinary edge $e = uv$ of G we have

$$a(G) = a(G\backslash e) + a(G/e); \tag{3.1}$$

then we have the same recurrence relation and the same boundary condition as for $T(G;2,0)$. Identifying the edge-sets of $G\backslash e$ and G/e in the natural way, we have $A(G/e) \subseteq A(G\backslash e)$, so $A(G\backslash e)$ is the disjoint union of $A_+ := A(G/e)$ and $A_- := A(G\backslash e) \setminus A(G/e)$. Furthermore, an acyclic orientation O of $G\backslash e$ is invalid for G/e (i.e., is not acyclic) if and only if it contains either a directed uv-path, or a directed vu-path. Since O is acyclic, it cannot contain paths of both types. Considering the possible orientations of $e = uv$, we see that each $O \in A_+$ extends to two acyclic orientations of G, and each $O \in A_-$ extends to exactly one. Hence

$$a(G) = 2|A_+| + |A_-| = |A_+| + (|A_+| + |A_-|) = a(G/e) + a(G\backslash e),$$

proving (3.1). □

Stanley stated the result above in the context of the chromatic rather than the Tutte polynomial (see Section 3.4 below for the chromatic polynomial, and also Chapters 11 and 31 for further details); in addition, he gave a related interpretation of $T(G; n, 0)$ for any integer $n \geq 2$.

We next give a proof of a slightly more involved evaluation, due to Greene and Zaslavsky [580]. A *source* in a directed graph is a vertex with in-degree zero.

Proposition 3.3. *For $u \in V(G)$ let $a_u(G)$ denote the number of acyclic orientations of G in which u is the only source. If G is connected, then*

$$a_u(G) = T(G; 1, 0). \tag{3.2}$$

In particular, $a_u(G)$ is independent of the choice of the vertex u.

Proof. Note that every acyclic orientation of a connected graph has at least one source. Applying this to each component, we see that if G is not connected, then $a_u(G) = 0$.

Write $A_u(G)$ for the set of orientations counted by $a_u(G)$. The proof of (3.2) proceeds by induction on $e(G)$. The base case $e(G) = 0$ is trivial, so suppose that G is connected and $e(G) \geq 1$, and let $e = uv$ be any edge incident with u. We may assume that e is not a loop; otherwise $a_u(G) = 0 = T(G; 1, 0)$. Let w be the vertex of G/e formed by contracting the edge e (i.e., identifying u and v). Then, it is straightforward to check that each orientation $O \in A_u(G)$ belongs to precisely one of the sets $A_u(G\backslash e)$ and $A_w(G/e)$; the key point is that this orientation O is not in $A_w(G/e)$ if and only if in O there is some edge other than e oriented towards v. Conversely, every $O \in A_u(G\backslash e) \cup A_w(G/e)$ extends to a unique $O \in A_u(G)$, orienting the edge e from u to v. This establishes that $a_u(G) = a_w(G/e) + a_u(G\backslash e)$.

If e is an ordinary edge, then $T(G; 1, 0) = T(G/e; 1, 0) + T(G\backslash e; 1, 0)$ and so we are done by induction. If not, e is a bridge, so $G\backslash e$ is disconnected and $a_u(G\backslash e) = 0$. Thus $a_u(G) = a_w(G/e) = T(G/e; 1, 0) = T(G; 1, 0)$. □

The argument above illustrates an important point concerning the deletion–contraction relations. Depending on the context, to show that a certain function is an evaluation of the Tutte polynomial, it often suffices to establish the relevant deletion–contraction relation for one particular ordinary edge e, together with the appropriate boundary condition, which is usually trivial. The power of the Tutte polynomial is that it then tells us that the relation holds for *any* ordinary edge. Of course, in many contexts, it is just as easy to establish the relation directly for an arbitrary ordinary edge.

In the examples above, the quantity considered satisfied $c(G) = c(G/e) + c(G\backslash e)$ for each ordinary edge e; when the "count" $c(G)$ satisfies such a relation with coefficients other than 1, one can of course renormalize, or appeal to the "recipe theorem", Theorem 2.24.

3.2.2 Further types of orientations

Let us note some further types of orientation counted by the Tutte polynomial, taken from a list given by Welsh and Merino [1145]. First, as shown by Las Vergnas [745], $T(G; 0, 2)$ counts the number of "totally cyclic" orientations of G, i.e., orientations in which every edge is contained in a directed cycle.

If G is planar, $T(G; 0, 1)$ counts the number of totally cyclic orientations of G containing no clockwise cycle; see Green and Zaslavsky [580] or [1145]. In particular, this number is independent of the way the planar graph G is drawn in the plane.

If G is 4-regular, $T(G; 0, -2)$ counts the number of *ice configurations* of G, i.e., orientations of the edges in which each vertex has two edges directed towards it and two away from it. These configurations were introduced and much studied by physicists; see, e.g., Pauling [907, 908], Lieb [773] and Temperley and Lieb [1056]. In a graph theoretical setting, such configurations are also called *2-in 2-out digraphs.* If G is 4-regular, it is trivial that the number of ice configurations of G is equal to the number nowhere-zero $\mathbb{Z}_3$-flows in G; see Section 3.4.

3.2.3 Further values at special points

For the next evaluation of the Tutte polynomial, recall that the *edge space* of a graph $G = (V, E)$ is the power set $\mathcal{P}(E)$, viewed as a vector space over $\mathbb{Z}_2$. The *cycle space* of G is the subspace of this space generated by (the edge sets of) the cycles; the orthogonal complement of the cycle space is called the *cocycle space.* The intersection of the cycle space and the cocycle space is the *bicycle space* B of the graph G. In connection with their *principal edge tripartition theorem,* Rosenstiehl and Read [969] proved that $T(G; -1, -1) = (-1)^{|E|}(-2)^{\dim(B)}$.

A *score sequence* or *score vector* of a graph $G = (V, E)$ is any function $s : V \to \mathbb{N}$ that is the out-degree sequence of some orientation of G. Stanley [1030] noted that any graph G has $T(G; 2, 1)$ score sequences. Thus, the number of such sequences is equal to the number of spanning forests. A bijection between the corresponding sets was found by Kleitman and Winston [694], but this bijection is far from simple. This illustrates a general phenomenon: sometimes the deletion–contraction relations allow one to easily show that two sets associated to a graph have the same size, without giving any explicit bijection.

For certain graphs G, namely (roughly speaking), "simply connected" induced subgraphs of the square grid, Korn and Pak [706] gave a very surprising combinatorial interpretation of $T(G; 3, 3)$: if R is the region in the plane formed by replacing each vertex of G by a square of side-length 4 (so if G is a rectangular grid, for example, then R is a rectangle), then the number of tilings of R by T-tetrominos is exactly $2T(G; 3, 3)$.

The examples listed above are not the only specific points at which $T(G; x, y)$ has a combinatorial interpretation. Indeed, we shall give some further examples later, in the context of interpretations of the Tutte polynomial along certain curves in the x-y plane.

3.3 Coefficient properties and irreducibility

The following properties of the coefficients of $T(G; x, y)$ are immediate from, for example, the deletion–contraction relations. Recall that the *rank* of a graph G is $r(G) = v(G) - k(G)$, where $k(G)$ is the number of components of G, and that the *nullity* is $n(G) = e(G) - v(G) + k(G)$, i.e., the minimal number of edges that must be deleted to turn G into a forest.

Proposition 3.4. *Let G be any graph and let $T(G; x, y) = \sum_{i,j} t_{i,j} x^i y^j$ be its Tutte polynomial. Then*

1. *$t_{i,j} \geq 0$ for all $i, j \geq 0$, and $t_{0,0} = 0$ unless $e(G) = 0$;*
2. *the highest power of x in $T(G; x, y)$ is $x^{r(G)}$, and its coefficient is y to the power of the number of loops of G;*
3. *the highest power of y in $T(G; x, y)$ is $y^{n(G)}$, and its coefficient is x to the power of the number of bridges in G;*
4. *the number of non-trivial components of G is $\min\{i + j : t_{i,j} > 0\}$; and*
5. *if G is a 2-connected loopless graph which is neither a cycle nor its dual (a 2-vertex graph consisting only of parallel edges), then $t_{1,1} > 0$.*

Brylawski [251] established (in the more general context of matroids; see Chapter 4) the following monotonicity property of the non-zero coefficients.

Theorem 3.5. *Suppose that G is 2-connected and that $t_{i,j} > 0$. Then $t_{i',j'} > 0$ for all $(i', j') \neq (0, 0)$ with $i' \leq i$ and $j' \leq j$.*

In the same paper, Brylawski established an infinite sequence of coefficient identities for graphs that are not too small.

Theorem 3.6. *Let $k \geq 1$ be an integer and let G be a graph with $e(G) \geq k$. Then*

$$\sum_{i+j<k} (-1)^j \binom{k-i-1}{j} t_{i,j} = 0.$$

3.3.1 Factorizing Tutte polynomials

As noted in Section 2.3, it is immediate from the definition that the Tutte polynomial is multiplicative over disjoint unions and one-point joins. Hence if G has blocks $B_1, \ldots, B_r$ then

$$T(G) = \prod_{i=1}^{r} T(B_i).$$

Thus, for many purposes, it suffices to study the Tutte polynomial restricted to 2-connected graphs.

One might wonder whether any further factorization is possible. The answer is no.

Theorem 3.7. *If G is 2-connected, then $T(G; x, y)$ is irreducible in $\mathbb{C}[x, y]$.*

This result, which settles a conjecture of Brylawski [251], was proved by Merino, de Mier and Noy [840] using Theorem 3.6 as the main tool.

3.3.2 Crapo's β-invariant

In [347] Crapo introduced a matroid invariant, called the *β-invariant*, (that we here describe in the context of graphs) defined as follows: for any graph G,

$$\beta(G) = (-1)^{r(G)} \sum_{A \subseteq E(G)} (-1)^{|A|} r(A).$$

This function is a deletion–contraction invariant, in the sense that if e is neither a bridge nor a loop, then $\beta(G) = \beta(G \backslash e) + \beta(G/e)$. Trivially $\beta(G) = 0$ if G contains a loop. If G contains a bridge and at least one other edge then $\beta(G) = 0$. This invariant is sometimes known as the *chromatic invariant* of G, since it can be expressed as

$$\beta(G) = (-1)^{r(G)+1} \chi'(G, \lambda)|_{\lambda=1}. \tag{3.3}$$

The connection to the Tutte polynomial is the following result, easily established using deletion–contraction relations, or (3.3).

Theorem 3.8. *Let G be a graph with at least two edges, with Tutte polynomial $\sum_{i,j} t_{i,j} x^i y^j$. Then $t_{1,0} = t_{0,1} = \beta(G)$.*

The β invariant has many interesting properties. Firstly, it is a "structural" invariant, in the sense that if G has at least two edges, then $\beta(G)$ is unchanged by series or parallel connections. It is then immediate that if G is a series-parallel graph, then $\beta(G) = 1$ (see Section 1.2.12). The reverse is much harder, and was proved by Brylawski [251].

Theorem 3.9. *A graph G has $\beta(G) = 1$ if and only if $G = K_2$, or G is series-parallel.*

For further properties of the β-invariant see, for example, Ellis-Monaghan and Merino [454] and the references therein.

3.4 Evaluations along curves

3.4.1 The chromatic polynomial

Given a graph G and an integer $r \geq 1$, let us (temporarily) write $N_r(G)$ for the number of proper vertex colorings of G with r colors, i.e., maps $c : V \to \{1, 2, \ldots, r\}$ with the property that if u and v are adjacent, then $c(u) \neq c(v)$. Clearly, if G has no edges then $N_r(G) = r^{v(G)}$. Also, if $e = uv$ is any edge of G, then considering colorings of $G \backslash e$ in which u and v do, or do not, have the same color, we see that $N_r(G \backslash e) = N_r(G) + N_r(G/e)$. It follows easily that for each graph G there is a polynomial $\chi(G; \lambda)$, the *chromatic polynomial* of G, such that for every positive integer r we have $\chi(G; r) = N_r(G)$. Moreover, we have the recurrence relation $\chi(G; \lambda) = \chi(G \backslash e; \lambda) - \chi(G/e; \lambda)$. It should be no surprise that the Tutte polynomial has the chromatic polynomial as a specialization. More precisely, for any graph G we have

$$\chi(G; \lambda) = (-1)^{r(G)} \lambda^{k(G)} T(G; 1 - \lambda, 0). \tag{3.4}$$

This is perhaps most easily seen by taking as boundary condition the value $\chi(G; \lambda) = \lambda^{k(G)}(\lambda - 1)^{r(G)}$ for a forest G. Historically, Tutte developed his polynomial as a generalization of the chromatic polynomial; hence the original name *dichromate*. For more details on the chromatic polynomial see Chapter 11, and for the history of the Tutte polynomial see Chapter 34.

3.4.2 The flow polynomial

Let A be an abelian group, and let G be a graph with the edges oriented arbitrarily. A *nowhere-zero* A*-flow* on G is an assignment of a non-zero element $f(e)$ of A to each edge e of G such that the net flow into a vertex (the sum of the weights of the incoming edges) is equal to the net flow out. By replacing $f(e)$ by $-f(e)$ whenever we reverse the orientation of an edge e, we see readily that the number of such flows is independent of the orientation chosen. More surprisingly, this number depends only on the order $n = |\mathsf{A}|$ of the abelian group A, not on its structure. Indeed, there is a polynomial $F(G; \lambda)$, the *flow polynomial* of G, such that, for any abelian group of order n, the number of nowhere-zero A-flows on G is exactly $F(G; n)$. Moreover,

$$F(G; \lambda) = (-1)^{n(G)} T(G; 0, 1 - \lambda). \tag{3.5}$$

For further information concerning the flow polynomial, see Chapter 12.

3.4.3 The reliability polynomial

Another simply defined polynomial associated to G is the *(all terminal) reliability polynomial*. Let $H = (V, A)$ be the random subgraph of $G = (V, E)$

obtained by keeping each edge of G with probability p, independently of the others. (One can think of the links as failing with probability $1-p$, independently.) Then, if G is a connected graph, $\mathrm{Rel}(G;p)$ is the probability that at the end the random subgraph H is connected. In general, it is the probability that H is "as connected as possible", i.e., has exactly as many components as G. It is easy to see that $\mathrm{Rel}(G;p)$ is indeed a polynomial, and that

$$\mathrm{Rel}(G;p) = (1-p)^{n(G)}p^{r(G)}T(G;1,1/(1-p)).$$

The reliability polynomial is discussed in detail in Chapter 15.

3.4.4 The Ising, Potts and random-cluster models

Let $G=(V,E)$ be a graph, and $q \geq 1$ an integer. The q-state *Potts model* on G, introduced by Potts in his thesis [927], is defined as follows. A *state* or *configuration* of the model is a function $\omega : V \to \{1,2,\ldots,q\}$. The (Potts) *Hamiltonian* of ω is simply the number $H_{\mathrm{P}}(\omega)$ of edges $ab \in E$ with $\omega(a)=\omega(b)$. If $\beta > 0$ is a positive number (typically the inverse temperature scaled by the Boltzmann constant), then each state has probability $e^{-\beta H_{\mathrm{P}}(\omega)}/Z_{\mathrm{P}}(G;\beta,q)$ of occurring, where

$$Z_{\mathrm{P}}(G;\beta,q) = \sum_{\omega} e^{-\beta H_{\mathrm{P}}(\omega)}$$

is the *partition function* of the model. The Potts model represents a physical system in which particles have spins that may take one of q values, and neighboring particles like to have the same spin (the *ferromagnetic* Potts model). As so often in statistical physics, the partition function is a key quantity that encodes much of the behavior of the system. It may be calculated easily from the Tutte polynomial:

$$Z_{\mathrm{P}}(G;\beta,q) = e^{-\beta e(G)}q^{k(G)}(e^{\beta}-1)^{r(G)}T\left(G;1+\frac{q}{e^{\beta}-1},e^{\beta}\right); \qquad (3.6)$$

see Fortuin and Kasteleyn [498]. In other words, along the part of the hyperbola $(x-1)(y-1)=q$ with $x,y>1$, the Tutte polynomial and the partition function of the q-state Potts model are simple transformations of each other. An important special case of the Potts model is the case $q=2$, the *Ising model*, defined by Lenz in 1920 [762].

When $q=1$, the q-state Potts model is trivial, and so is the Tutte polynomial: on $(x-1)(y-1)=1$, from (2.2) and a simple application of the binomial theorem, it is easy to see that $T(G;x,1+1/(x-1)) = x^{e(G)}(x-1)^{-n(G)}$.

In 1972, Fortuin and Kasteleyn [498] generalized and unified both the Potts model and the Tutte polynomial, defining a new polynomial associated to a graph $G=(V,E)$ with weights $(p_e)_{e\in E}$ on the edges. In the special case where all weights p_e are between 0 and 1 this can be thought of as a probabilistic model, the *random-cluster model*: a configuration is now a subset A of the edges, and the probability of a configuration is proportional to

$q^{k(A)} \prod_{e \in A} p_e \prod_{e \in E \setminus A}(1 - p_e)$, where, as usual, $k(A)$ is the number of components of the graph (V, A). If $q = 1$ this is simply the standard random subgraph model. When $q < 1$ the model is biased towards more connected configurations, and when $q > 1$ towards less connected ones. Taking all p_e to be equal to p, the partition function of this model is a polynomial in p and q, and is given by

$$Z_{\mathrm{RC}}(G; p, q) = (1-p)^{n(G)} p^{r(G)} q^{k(G)} T\left(1 + \frac{q(1-p)}{p}, 1 + \frac{p}{1-p}\right).$$

Thus the polynomials $Z_{\mathrm{RC}}(G; p, q)$ and $T(G; x, y)$ are simple transformations of each other. However, $Z_{\mathrm{RC}}(G; p, q)$ only makes sense as the partition function of a statistical mechanical model if $q > 0$ and $0 \leq p \leq 1$, corresponding to the region $x, y > 1$ in the *Tutte plane*, i.e., the set $\{(x, y)\}$ of inputs to the Tutte polynomial $T(G; x, y)$. For more details concerning the Potts and random-cluster models, see Chapter 20.

We shall describe evaluations of the Tutte polynomial along two further curves in later sections: on $x = y$, the Tutte polynomial of a plane graph G corresponds to the Martin polynomial of the medial graph of G (see Section 3.5.1), and on $xy = 1$ the Tutte polynomial of a plane graph G corresponds to the Jones polynomial of the associated alternating link (see Section 3.6).

3.5 Dual graphs and medial graphs

Although the Tutte polynomial is defined for general graphs, in the special case of planar graphs it has many additional beautiful properties. Let G be a connected plane graph with dual G^*, and for each edge e of G let e^* denote the *dual edge*, joining the vertices of G^* corresponding to the faces on the two sides of e. Note that e^* is a loop if and only if these two faces are the same, i.e., if and only if e is a bridge. Similarly, e^* is a bridge if and only if e is a loop (so the drawing of e separates the plane). Moreover, if e is an ordinary edge, then deleting e merges the two faces bounded by e, corresponding to contracting the edge e^* of G^*, and contracting e corresponds to deleting e^*. In other words,

$$(G \backslash e)^* = G^*/e^* \quad \text{and} \quad (G/e)^* = G^* \backslash e^*.$$

From these facts we immediately have the following *duality relation* for connected plane graphs G:

$$T(G^*; x, y) = T(G; y, x).$$

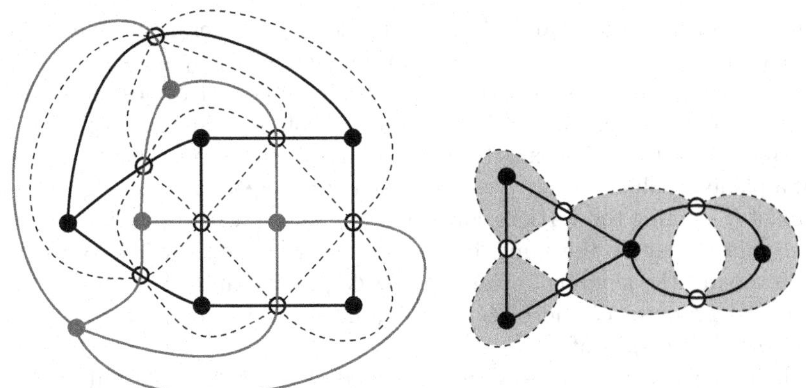

FIGURE 3.1: On the left, a plane graph G (black), its dual G^* (gray), and their common medial graph $G_m = (G^*)_m$ (dashed). The right figure shows a shading of a 4-regular plane graph H (dashed) and the associated graph G (black) with $H = G_m$.

This relation and the formulas (3.4) and (3.5) give a simple relationship between the chromatic polynomial of G and the flow polynomial of G^* for a connected plane graph G:

$$\chi(G; \lambda) = \lambda F(G^*; \lambda).$$

Of course, this relationship is easy to establish directly. In the light of this connection, it is perhaps not too surprising that, *for planar graphs*, there is a single two-variable polynomial encoding both the chromatic and the flow polynomials. What is remarkable is that such a polynomial, the Tutte polynomial, exists for *all* graphs.

3.5.1 Medial graphs

If G is a connected plane graph, there is a very nice way to define a graph G_m that sits "between" G and its dual G^*: take a vertex $m(e)$ for each edge e of G and, for each vertex v of G, if $e_1, \ldots, e_r$ are the edges of G meeting v in cyclic order, then join $m(e_i)$ to $m(e_{i+1})$ for $i = 1, \ldots, r$, as shown in Figure 3.1. Of course, here we take $e_{r+1} = e_1$. If some face of G is bounded by only two edges, the corresponding vertices will be joined by a double edge in G_m. The graph G_m is called the *medial graph* associated to G.

The construction of G_m may be described in a dual way: rather than draw cycles around the vertices of G (as above), we may draw cycles around the faces. This leads to the same result, and it follows that

$$(G^*)_m = G_m.$$

There is a very simple way to reverse the construction just described. The medial graph is always connected and 4-regular. Starting with a connected 4-regular plane graph H, one can shade its regions gray and white, so that each edge borders two regions of different colors. (In general this is possible if and only if all degrees are even—this can be seen either by a local argument, or by writing such a graph as an edge-disjoint union of cycles, and adding the cycles one-by-one, each time reversing all colors inside the cycle.) When H is connected, there are exactly two shadings of this type. Given a shading, form a plane graph G with a vertex for each gray region, and an edge for each vertex of H, joining the vertices corresponding to the two gray regions meeting there. (These may be the same, in which case e is a loop.) It is easy to check that reversing the shading replaces G by G^*, and that H is simultaneously the medial graph of G and of G^*; see Figure 3.1. The graphs G and G^* are called the *Tait graphs* of H.

There is a natural way to orient the medial graph G_m, obtaining the *directed medial graph* $\vec{G}_m$. This is the directed graph obtained as above by orienting the cycle around each vertex anticlockwise (say).

In his thesis, Martin [824] introduced two polynomials, one associated to undirected graphs and one to directed 2-in 2-out graphs; the latter satisfies

$$m(\vec{G}_m; x) = T(G; x, x).$$

Las Vergnas [748] generalized this *Martin polynomial*, defining a version that (up to a very simple reparameterization) is the *circuit partition polynomial* of [153]. This polynomial is defined for any Eulerian graph $\vec{G}$, i.e., any directed graph in which each vertex has as many edges directed into it as out of it, and is given by

$$j(\vec{G}; x) = \sum_{k \geq 0} f_k(\vec{G}) x^k,$$

where $f_k(\vec{G})$ is the number of partitions of the edge set of $\vec{G}$ into exactly k directed circuits. The Martin polynomial satisfies $xm(\vec{G}; x+1) = j(\vec{G}; x)$.[1]

Martin [824, 825] used the connection outlined above to give a combinatorial interpretation of $T(G; -1, -1)$ for any planar graph G: it is simply $(-1)^{e(G)}(-2)^{s-1}$, where s is the number of *strings* (or anti-circuits) in the medial graph G_m of G, i.e., the number of components formed when we connect each pair of opposite edges in G_m. This quantity is also known as the number of *crossing circuits* or the number of *anti-circuits*. The latter name comes from viewing G_m as a directed graph: there is then a unique decomposition into cycles along which the orientations of the edges alternate; these are exactly the strings. Still for planar graphs, Las Vergnas [751] showed that $T(G; 3, 3) = q_G 2^{s(G)-1}$ for some odd integer q_G depending on G. For more on the Tutte–Martin connection and evaluations of $T(G; x, y)$ when $x = y$, see Ellis-Monaghan [450]. For a detailed discussion of connections between the Tutte polynomial and medial graphs, see Chapter 13.

[1] In [153], Bollobás describes the relationship between the Martin and circuit partition polynomials incorrectly; the formulas for the directed and undirected cases are swapped.

Let D be a 2-in 2-out directed graph (for example the medial graph of some planar graph). Surprisingly, there is another, very different, connection between the Martin polynomial of D and a polynomial defined on arbitrary undirected graphs. If D is connected, let C be any Euler circuit of D. Then C visits each vertex twice, and we say that two vertices a and b are *interlaced* if these visits occur as $\ldots a \ldots b \ldots a \ldots b \ldots$. The *interlace graph* $H = H(D, C)$ is the (undirected) graph on $V(D)$ in which a and b are joined if and only if they are interlaced on C. Arratia, Bollobás and Sorkin [49] defined a polynomial $q(G; x)$, the *interlace polynomial*, such that under these conditions $q(H; x) = m(D; x)$. However, $q(G; x)$ is defined for *all* graphs, not just graphs that arise as interlace graphs. Later, Arratia, Bollobás and Sorkin [50] further generalized the interlace polynomial to a two-variable polynomial defined on arbitrary undirected graphs. For more information on interlace polynomials, see Chapter 14.

3.6 Knots

One of the most important, and most surprising, applications of the Tutte polynomial is that to knot theory. For the reader unfamiliar with this area, who might well wonder what the two topics have to do with each other, we attempt in this section to give a brief but self-contained outline of the connection, and a very brief historical discussion of the resulting applications. This topic is discussed more fully in Chapter 18, with many more details and more of a focus on recent results.

In topology, a *knot* is a simple closed curve in $\mathbb{R}^3$, and a *link* is a finite set of pairwise disjoint simple closed curves in $\mathbb{R}^3$. Two knots or links are *equivalent* if one can be deformed into the other in the natural way. The *unknot* is the "unknotted" knot, for example, a geometric circle. Often, it is easy to see that two links are equivalent—simply exhibit a deformation. But how can one tell that two links are *not* equivalent, or that a given knot is not the unknot? Sometimes there is an easy answer based on a simple invariant, such as the number of components in a link, or the "linking number" (closely related to winding number). But in general this is a very hard problem.

The most common way of representing a link is via a *link diagram*, obtained by projecting the link into the plane in the natural way (avoiding any degenerate points etc.). Rather than give a formal definition, we draw some examples in Figure 3.2. Of course, link diagrams can be viewed as 4-regular plane graphs with some extra information at each "crossing", namely which strand passes on top. Reidemeister [957], and independently Alexander and Briggs [19], showed that two links are equivalent if and only if their link diagrams can be transformed into each other by a sequence of "Reidemeister moves", certain simple local operations on the diagram (see Theorem 18.5). Unfortunately, there is no obvious bound on the length of the sequence that

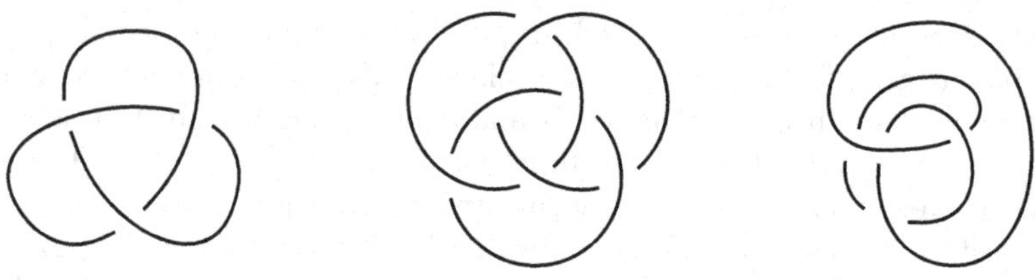

FIGURE 3.2: Three link diagrams; the first two are alternating. These represent, respectively, the trefoil knot, the Borromean rings, and the unknot.

FIGURE 3.3: A crossing and its two resolutions (the "A-resolution" and the "B-resolution").

may be needed, since Reidemeister moves can increase as well as decrease the complexity of a diagram. Such bounds are a topic of ongoing research; see in particular the paper of Lackenby [739] giving a polynomial bound in the case that the diagram can be reduced to the unknot, as well as a discussion of what is known in other cases.

V. Jones [668] started a revolution in knot theory when he introduced the *Jones polynomial*, a link invariant defined via von Neumann algebras that is extremely powerful—it distinguishes many pairs of knots/links. However, the real breakthrough came only a little later, when Kauffman [681] found a combinatorial way of using link diagrams to calculate (and re-prove the existence of) the Jones polynomial, based on an expansion (or, equivalently, a recurrence) involving "resolving" a crossing in two ways, as in Figure 3.3. Thistlethwaite [1061] noticed that this *Kauffman bracket* is extremely closely related to the Tutte polynomial. Indeed, for *alternating links*, where the crossings alternate over, under, over, under, etc. as one moves along a strand of the link, the Kauffman bracket, and hence the Jones polynomial, is given by an evaluation of the Tutte polynomial.

To state this connection, it is convenient to view a link diagram as, in essence, the medial graph of a *signed* plane graph: starting with a gray/white shading of the regions formed by the link diagram, as in Section 3.5.1, form a graph G with a vertex for each gray region. Take an edge for each crossing, joining the vertices corresponding to the two gray regions which touch there, but signed + or − according to the sense of the crossing, as in Figure 3.4. Then we obtain a signed plane graph G that encodes the link diagram; the diagram is formed from the medial graph G_m of G by taking appropriate under/over crossings according to the signs on the edges of G. See Figure 3.5. Of course, this correspondence is 2-to-1, since replacing G by G^* with all signs reversed gives the same link diagram. In the special case of an alternating link, one choice for the corresponding graph G has all edges +, and so G may be viewed as a standard (unsigned) plane graph.

FIGURE 3.4: The sign convention when representing a shaded link diagram as a signed graph.

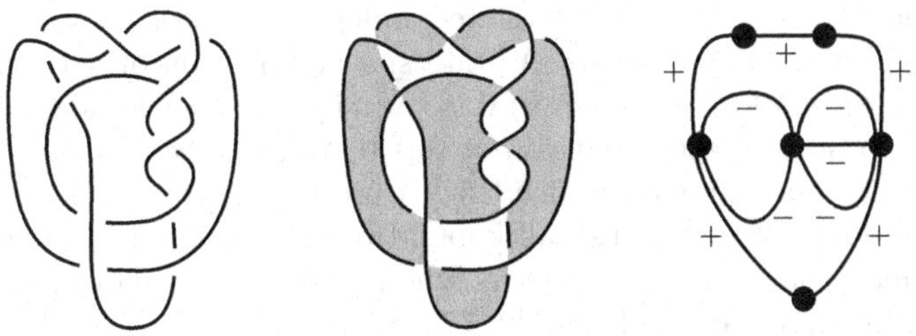

FIGURE 3.5: A non-alternating knot and one of its Tait graphs.

Let D be any alternating diagram of an (alternating) link L, and let G be the corresponding planar graph as defined above. Then the Jones polynomial of L is equal to

$$V_L(t) = (-A)^{-3w(D)} \langle D \rangle \big|_{A=t^{-1/4}},$$

where the Kauffman bracket $\langle D \rangle$ satisfies

$$\langle D \rangle = A^{2v(G)-e(G)-2} T(G; -A^{-4}, -A^{4})$$

and $w(D)$, the *writhe* of D, is a quantity easily read off from D. (See Chapter 18 for details.) For general link diagrams, the Jones polynomial can be derived in a similar way from Kauffman's *signed Tutte polynomial* [680], an analogue of the Tutte polynomial defined for graphs with signs on the edges.

Using Kauffman's "bracket polynomial", Kauffman himself [681], Murasugi [870, 871] and Thistlethwaite [1061] gave remarkably simple proofs of a number of conjectures attributed to Tait that had been open for well over a century. The most basic of these is that any alternating knot diagram that cannot obviously be simplified (by a simple "twist") cannot be transformed in *any* way into a simpler diagram, i.e., one with fewer crossings. In particular, any non-trivial alternating knot diagram does not represent the unknot. Thistlethwaite's proof of this uses ideas of Tutte, in particular the notion of the "activity" of an edge in the spanning tree expansion of the Tutte polynomial. For more details see, for example, the accessible survey of Lickorish [771] written near the time. Similar ideas were also an important ingredient in Menasco and Thistlethwaite's proof [836, 837] of Tait's "flyping conjecture" from [1052]. (A special case of this conjecture was proved around the same time by Schrijver [989] using different methods.) Unfortunately the Jones polynomial, or

Kauffman bracket, does not distinguish all knots; for example Eliahou, Kauffman and Thistlethwaite constructed an infinite family of links with the same Jones polynomial as the unknot [446]; again Kauffman's Tutte-like bracket is a key tool in their method.

For a different proof that any reduced (twist-free) alternating knot diagram represents a non-trivial link, based on a connection between the interlace polynomial and the Alexander polynomial (a knot invariant preceding the Jones polynomial), see Balister, Bollobás, Riordan and Scott [68]. A generalization of the Jones polynomial was introduced by Freyd, Yetter, Hoste, Lickorish, Millett and Ocneanu [503], and related independent work was undertaken by Przytycki and Traczyk [939]. A state space expansion of this HOMFLYPT *polynomial* (or HOMFLY *polynomial*) was given by Bollobás, Pebody and Weinreich [157]. A state space expansion of this HOMFLYPT *polynomial* was given by Bollobás, Pebody and Weinreich [157]. Unfortunately, even the HOMFLYPT polynomial does not distinguish all knots; two knots related by *mutation* are known to have the same HOMFLYPT polynomial (see [503]). For more on these topics see, for example, the books [2, 772, 815].

3.7 Signed, colored and topological Tutte polynomials

We have seen above that in the context of link diagrams, it is very natural to consider graphs with two "types" or "colors" of edges, and a Tutte-like polynomial that obeys deletion–contraction relations analogous to (2.6) or (2.18), but with coefficients that depend on the type of the edge being deleted/contracted. What about graphs with a more general set of colors (or weights) on the edges? One such *colored Tutte polynomial* is in fact the (polynomial associated to) the random-cluster model of Fortuin and Kasteleyn [498]. For the most general invariant of this type, see [158]; and for a detailed discussion of multivariable and parameterized Tutte polynomials, see Chapter 24.

In a different direction, the connection between the Tutte polynomial and knot theory was one motivation behind the development in [160, 161] of a Tutte polynomial for graphs embedded into surfaces. This 3-variable *topological Tutte polynomial* or *Bollobás–Riordan polynomial* has been studied extensively by Chmutov, Ellis-Monaghan, Moffatt, Pak and others, and was recently generalized to a 4-variable polynomial by Krushkal [724]. For more on this topic see Chapter 27.

3.8 Open problems

One of the best known open questions relating to specific values of the Tutte polynomial is Tutte's *5-flow conjecture* from [1092]. Tutte showed that if a graph G has a nowhere-zero n-flow, then it has a nowhere-zero $(n+1)$-flow. He noted that the four color theorem (conjecture at that time) is equivalent to the statement that every bridgeless *planar* graph has a nowhere-zero 4-flow. Without the condition of planarity this is not true (the Petersen graph is a counterexample). However, Tutte conjectured that every bridgeless graph G, planar or not, has a nowhere-zero 5-flow, i.e., has $T(G; 0, -4) \neq 0$. This conjecture has remained open for over 60 years, despite substantial progress by Jaeger [644, 645] and Seymour [999], who proved that the answer for 8-flows and 6-flows respectively is yes. In Section 7 of [1094], Tutte also conjectured (in algebraic language) that if G is bridgeless and does not have the Petersen graph as a minor, then G has a nowhere-zero 4-flow; this is also open.

Another longstanding open question concerning values of the Tutte polynomial is the *Merino–Welsh conjecture* [844] that

$$\max\{T(G; 2, 0), T(G; 0, 2)\} \geq T(G; 1, 1).$$

A number of special cases have been proved, for example, the series-parallel case by Noble and Royle [884], and there are a number of related results, for example the result of Jackson [639] that if $a > 0$ and $b \geq a(a+2)$, then $T(G; 0, b)T(G; b, 0) \geq T(G; a, a)^2$. Despite considerable attention, the original conjecture still remains open.

Although the Tutte polynomial encodes an enormous quantity of information about a graph G, it does not capture everything: there are non-isomorphic graphs with the same Tutte polynomial. Graphs with the same cycle matroid (see Chapter 4) trivially have the same Tutte polynomial. Much less trivially, Tutte [1097] showed in 1974 that there are non-isomorphic 5-connected graphs with the same Tutte polynomial, and asked whether this can be extended to arbitrarily high connectivity, a question answered positively by Brylawski [244] in 1981. Still, it seems likely, and was conjectured by Bollobás, Pebody and Riordan [156], that almost all graphs are Tutte-unique: if G is a graph chosen uniformly at random from all (labelled) graphs on $\{1, 2, \ldots, n\}$, then as $n \to \infty$ the probability that there exists any graph H not isomorphic to G with $T(G; x, y) = T(H; x, y)$ tends to 0. There are number of variants of this conjecture: is the same true for the chromatic polynomial? Or for the (stronger) deletion–contraction invariant defined in [156]? For more information about Tutte uniqueness, see Chapter 6.

Turning to knot theory, although there are examples of distinct knots with the same Jones polynomial, it is open whether there is any non-trivial *knot* with the same Jones polynomial as the unknot. (There are such links, as mentioned above.) This may be viewed as a question about the Tutte polynomial, in the form of Kauffman's extension to signed graphs.

4

Matroid theory

James Oxley

Synopsis

This chapter introduces matroids, gives several basic examples of them, describes the fundamental constructions for matroids, and defines the Tutte polynomial for matroids.

- The matroid theory conventions throughout the handbook.
- Circuits, independent sets, bases, rank, closure, flats, and hyperplanes.
- Examples of matroids including those from graphs, matrices, and linear codes.
- Deletion, contraction, duality, direct sum, 2-sum.
- The Tutte polynomial of a matroid and its basic properties.

4.1 Introduction

"If a theorem about graphs can be expressed in terms of edges and circuits only it probably exemplifies a more general theorem about matroids." These words of Tutte [1101] have guided much of the research in matroid theory and particularly that associated with the polynomial that bears his name. That polynomial was introduced for graphs in Tutte's 1947 paper [1090] "A ring in graph theory". In his 1948 Cambridge Ph.D. thesis [1091], Tutte extended much of the theory of this polynomial to what he called "nets" but are now known as representable matroids. Tutte never published this work for matroids and it was not until 1969 that H.H. Crapo [349] published a fully general

DOI: 10.1201/9780429161612-4

matroid form of this theory. (See Chapter 34 for the history of the Tutte polynomial.)

This chapter introduces matroids by beginning with examples of such structures that arise from graphs and from matrices. Both in the context of these basic examples and in general, it then looks at the many different ways that matroids can be defined. The operations of deletion and contraction of an edge in a graph are special cases of matroid operations with the same name. This leads to a theory of matroid minors. Each matroid has a dual, which is also a matroid, and the dual of the matroid associated with a plane graph is the matroid associated with the planar dual of the graph. Importantly, matroid theory provides a context for a meaningful dual of a non-planar graph. The dual of the matroid associated with a linear code is the matroid associated with the dual code. The operations of 1-sum and 2-sum for graphs have analogs for all matroids, and 3-sum has an analog for the matroids arising from matrices over the 2-element field.

The various different definitions of the Tutte polynomial of a graph extend quite easily to matroids. Much of this chapter is devoted to indicating how other properties of the graph polynomial are just special cases of properties of the matroid polynomial. In particular, the universality property and the recipe theorem both extend, and many of the elementary combinatorial evaluations noted for graphs generalize to matroids.

The proofs of most of the results in this chapter have been omitted. Proofs of the basic matroid results may be found in Oxley [894], while many of the Tutte-polynomial results are proved in Brylawski and Oxley [247].

4.2 Fundamental examples and definitions

Matroids were introduced by Whitney [1159] in 1935 to provide a common framework for the notions of dependence found in graph theory and linear algebra.

Example 4.1. For the graph G in Figure 4.1, the edge set E is $\{1,2,\ldots,8\}$ and the set $\mathcal{C}$ of edge sets of cycles is $\{\{8\},\{2,3\},\{2,4,5\},\{3,4,5\},\{4,6,7\},\{2,5,6,7\},\{3,5,6,7\}\}$. The pair $(E,\mathcal{C})$ is an example of a matroid. One way to begin to get intuition for matroids is to consider them as consisting of a finite set and a set of special subsets of that set that behave somewhat like the edge sets of cycles in a graph. Like most initial approximations, this notion will need considerable refinement.

Definition 4.2. A *matroid* M is an ordered pair $(E,\mathcal{C})$ consisting of a finite set E, called the *ground set*, and a set $\mathcal{C}$ of subsets of E, called *circuits*, that obey the following three conditions.

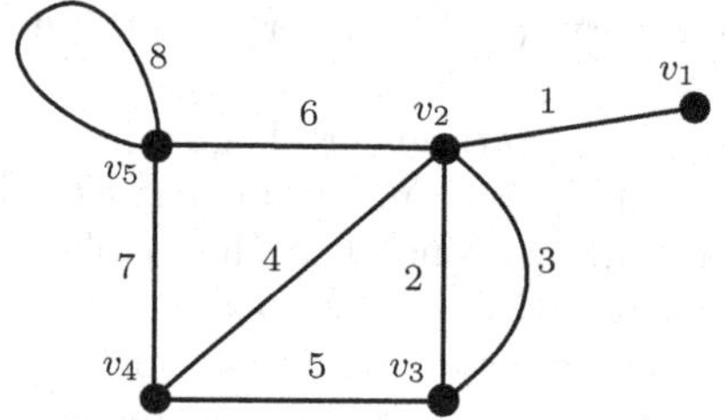

FIGURE 4.1: The graph G in Example 4.1.

(C1) $\emptyset \notin \mathcal{C}$.

(C2) If C_1 and C_2 are in $\mathcal{C}$ and $C_1 \subseteq C_2$, then $C_1 = C_2$.

(C3) If C_1 and C_2 are distinct members of $\mathcal{C}$ and $e \in C_1 \cap C_2$, then $\mathcal{C}$ contains a member C_3 such that $C_3 \subseteq (C_1 \cup C_2) \setminus \{e\}$.

In a matroid M, we often write $E(M)$ for the ground set and $\mathcal{C}(M)$ for the set of circuits, especially when several matroids are being considered.

Theorem 4.3. *Let G be a graph with edge set E and $\mathcal{C}$ be the set of edge sets of cycles of G. Then $(E, \mathcal{C})$ is a matroid.*

Definition 4.4. The matroid whose existence is asserted in the last theorem is called the *cycle matroid* of the graph G and is denoted by $M(G)$.

In Example 4.1, both the loop 8 and the pair $\{2, 3\}$ of parallel edges correspond to circuits in $M(G)$. This leads us to the first of numerous examples in which graph-theory terminology is carried over into matroid theory.

Definition 4.5. Let M be the matroid $(E, \mathcal{C})$. An element e of E is a *loop* if $\{e\}$ is in $\mathcal{C}$; distinct elements f and g of E are *parallel* if $\{f, g\}$ is in $\mathcal{C}$. A *parallel class* of M is a maximal subset X of E such that any two distinct members of X are parallel and no member of X is a loop. A parallel class is *trivial* if it contains just one element. If M contains no loops and no non-trivial parallel classes, then M is called a *simple matroid* or sometimes a *geometry.*

In Example 4.1, the sets $\{1, 5, 7\}$ and $\{1, 4, 5, 6\}$ are both edge sets of forests of G. Indeed, a set X of edges in a graph H is the edge set of a forest if and only if no cycle of H has its edge set contained in X. This idea is generalized as follows.

Definition 4.6. Let M be the matroid $(E, \mathcal{C})$. A subset I of E is *independent* in M if no circuit of M is contained in I. A set that is not independent is called *dependent.* The set of independent sets of M is denoted by $\mathcal{I}(M)$.

Clearly a set C is a circuit in a matroid M if and only if C is a minimal dependent set of M. While the set $\mathcal{C}(M)$ of circuits of M certainly determines the set of independent sets of M, if we know the set of independent sets, then we know the set of dependent sets and the minimal ones of those are the circuits. Thus matroids are often described by listing the independent sets rather than the circuits. The next theorem characterizes precisely which sets of subsets of a set can be the set of independent sets of a matroid.

Theorem 4.7. *Let $\mathcal{I}$ be a set of subsets of a finite set E. Then $\mathcal{I}$ is the set of independent sets of a matroid on E if and only if $\mathcal{I}$ satisfies the following conditions.*

(I1) *$\mathcal{I}$ is non-empty.*

(I2) *Every subset of a member of $\mathcal{I}$ is also in $\mathcal{I}$.*

(I3) *If I_1 and I_2 are in $\mathcal{I}$ and $|I_1| < |I_2|$, then there is an element e of $I_2 \setminus I_1$ such that $I_1 \cup \{e\}$ is in $\mathcal{I}$.*

Whitney [1159] modeled the matroid notion of independence on the familiar notion of linear independence. In the next example, we consider how the cycle matroid of a graph can be derived from its vertex-edge incidence matrix before extrapolating from this to define a matroid from an arbitrary matrix.

Example 4.8. Let A be the following matrix over the field $\mathbb{R}$ of real numbers.

$$\begin{array}{c} \\ v_1 \\ v_2 \\ v_3 \\ v_4 \\ v_5 \end{array}\begin{array}{c} \begin{array}{cccccccc} 1 & 2 & 3 & 4 & 5 & 6 & 7 & 8 \end{array} \\ \left[\begin{array}{cccccccc} 1 & 0 & 0 & 0 & 0 & 0 & 0 & 0 \\ -1 & 1 & 1 & 1 & 0 & 1 & 0 & 0 \\ 0 & -1 & -1 & 0 & 1 & 0 & 0 & 0 \\ 0 & 0 & 0 & -1 & -1 & 0 & 1 & 0 \\ 0 & 0 & 0 & 0 & 0 & -1 & -1 & 0 \end{array}\right] \end{array}$$

Evidently, A is the vertex-edge incidence matrix for the directed graph that is obtained from the graph G in Example 4.1 by directing each edge v_iv_j with $i \le j$ from v_i to v_j. Thus the column corresponding to the loop 8 is the zero vector. Now let $E = \{1, 2, \ldots, 8\}$ and let $\mathcal{C}$ be the set of subsets X of E such that the multiset of columns labelled by X is not a linearly independent set but, for every proper subset X' of X, the multiset of columns labelled by X' is a linearly independent set. Then $\{2, 3\} \in \mathcal{C}$ since the columns labelled by 2 and 3 are equal. Noting that $\{8\}$ is in $\mathcal{C}$ but that no member of $\mathcal{C}$ contains 1, it is straightforward to check that $\mathcal{C} = \{\{8\}, \{2, 3\}, \{2, 4, 5\}, \{3, 4, 5\}, \{4, 6, 7\}, \{2, 5, 6, 7\}, \{3, 5, 6, 7\}\}$. Thus the pair $(E, \mathcal{C})$ is precisely the matroid we considered in Example 4.1. Hence the particular matrix A chosen above gives rise to a matroid. The next theorem notes that every matrix over every field yields a matroid in precisely this way.

Expressing the following theorem in terms of independent sets rather than circuits yields a somewhat cleaner statement.

Theorem 4.9. *Let E be the set of column labels of an $m \times n$ matrix over a field $\mathbb{F}$ and let $\mathcal{I}$ be the set of subsets X of E for which the multiset of columns labelled by X is a set that is linearly independent over $\mathbb{F}$. Then $\mathcal{I}$ is the set of independent sets of a matroid on E.*

Definition 4.10. The matroid whose existence is asserted by the last theorem is called the *vector matroid* of the matrix A and is denoted by $M[A]$.

We observed that the cycle matroid $M(G)$ of the graph G in Example 4.1 and the vector matroid of the matrix A in Example 4.8 are equal since they have the same ground sets and the same sets of circuits. More generally, two matroids are isomorphic if they have the same structure.

Definition 4.11. Let M_1 and M_2 be the matroids $(E_1, \mathcal{C}_1)$ and $(E_2, \mathcal{C}_2)$. We say that M_1 and M_2 are *isomorphic* and write $M_1 \cong M_2$ if there is a bijection $\varphi : E_1 \to E_2$ such that a subset C of E_1 is a circuit of M_1 if and only if $\varphi(C)$ is a circuit of M_2.

Definition 4.12. A matroid M is *graphic* if $M \cong M(G)$ for some graph G.

Definition 4.13. A matroid M is $\mathbb{F}$-*representable* if $M \cong M[A]$ for some matrix A over the field $\mathbb{F}$. When the latter occurs, the matrix A is called an $\mathbb{F}$-*representation* of M. We call M *binary* if it is $\mathbb{F}_2$-representable; M is *ternary* if it is $\mathbb{F}_3$-representable.

In Example 4.8, a specific orientation of the graph G was used to derive the matrix A. In fact, any orientation of G could have been used. To see this, note that if we multiply some collection of columns of A by -1, then we do not alter which sets of column labels correspond to linearly independent sets of vectors. Hence we do not change the matroid. But multiplying a column of A by -1 has the effect of reversing the direction on the corresponding edge of G. In general, for any matrix A over a field $\mathbb{F}$, if we multiply some column of A by a non-zero member of $\mathbb{F}$ to give the matrix A', then $M[A] = M[A']$. This is one of several ways in which one can alter a matrix without changing the associated vector matroid. Some other allowable moves one can perform will be identified in Lemma 4.57.

Whenever the vertex-edge incidence matrix for an orientation of a graph is used to produce a real matrix as in Example 4.8, this matrix has the following special property.

Definition 4.14. A matrix A over $\mathbb{R}$ is *totally unimodular* if every square submatrix of A has its determinant in $\{0, 1, -1\}$.

Such matrices give rise to a special class of matroids.

Definition 4.15. A matroid M is *regular* if M has an $\mathbb{R}$-representation that is totally unimodular.

It is easy to see that if a matroid is regular, then it is representable over every field. The converse is true but is non-trivial to prove.

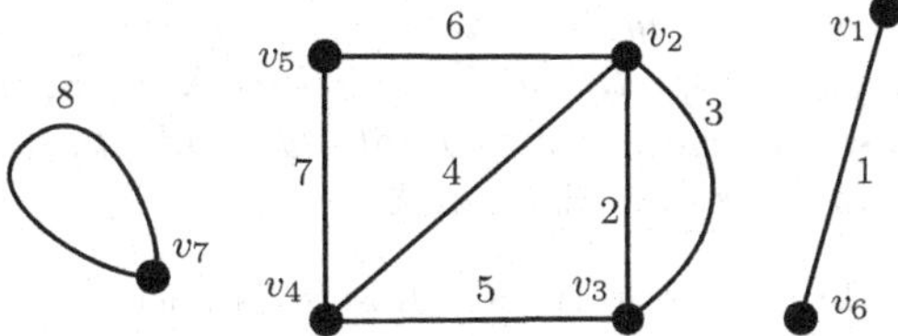

FIGURE 4.2: A graph G' having the same cycle matroid as the graph G in Example 4.1.

Theorem 4.16. *A matroid is regular if and only if it is representable over every field.*

The technique used in Example 4.8 can be used to prove the following.

Theorem 4.17. *Every graphic matroid is regular.*

Let G' be the graph in Figure 4.2. Clearly G' and G are non-isomorphic graphs. However, the edge sets of these two graphs are equal as are the edge sets of the cycles of the two graphs. Hence $M(G) = M(G')$. Similarly, the addition of arbitrarily many isolated vertices to G will not change the collection of edge sets of cycles and so will not alter the cycle matroid. These ideas easily generalize to give the next theorem. Whitney determined the precise relationship between two graphs that have the same cycle matroids. His result is stated in Theorem 4.102.

Theorem 4.18. *For every graph G, there is a connected graph G' such that $M(G) = M(G')$.*

Evidently, every subset of an independent set is also independent. Hence, to describe a matroid, it suffices to list the maximal independent sets. We have already seen an example of graph-theory terminology carrying over into matroid theory. The next definition exemplifies how linear algebra is also a frequent source for matroid terminology.

Definition 4.19. A maximal independent set in a matroid M is called a *base* or a *basis* of M. The set of bases of M is denoted by $\mathcal{B}(M)$.

Example 4.20. For the graph G in Figure 4.1, the set $\{1, 2, 4, 7\}$ is a basis of $M(G)$. It is also the set of edges of a spanning tree in G. In general, for a connected graph H, the bases of $M(H)$ are precisely the edge sets of spanning trees in H. For an $m \times n$ matrix A over a field $\mathbb{F}$, the columns of A are vectors in $V(m, \mathbb{F})$, the m-dimensional vector space over $\mathbb{F}$. Let W be the subspace of $V(m, \mathbb{F})$ that is spanned by the columns of A. If E is the set of column labels of A, then the bases of $M[A]$ are precisely the bases of W all of whose elements are in E.

It is well-known that all the bases of a vector space are equicardinal. An immediate consequence of **(I3)** is that all bases of a matroid M are equicardinal. We can say even more along these lines using the following observation.

Theorem 4.21. *Let M be a matroid having E and $\mathcal{I}(M)$ as its ground set and set of independent sets, respectively. For a subset X of E, there is a matroid $M|X$ with ground set X whose collection of independent sets consists of those members of $\mathcal{I}(M)$ that are contained in X.*

Definition 4.22. The matroid $M|X$ whose existence is asserted in the last theorem is called the *restriction* of M to X. This matroid is also denoted $M\backslash(E \setminus X)$ and is called the *deletion* of $E \setminus X$ from M.

The last theorem enables us to make the following:

Definition 4.23. Let M be a matroid with ground set E. The *rank*, $r(M)$, of M is the common cardinality of the bases of M. For each subset X of E, the *rank*, $r(X)$, of X is $r(M|X)$. In particular, $r(E) = r(M)$. When more than one matroid is being considered, it is common to write the rank of X in M as $r_M(X)$.

Example 4.24. In Example 4.8, the rank in $M[A]$ of $\{2,3,4,5\}$ is 2. More generally, the rank of a set X of column labels is the rank of the matrix obtained from A be removing all of the columns except those labelled by members of X. In Example 4.1, the rank in $M(G)$ of $\{1,5,6,8\}$ is 3 since $\{1,5,6\}$ is a maximal independent subset of $\{1,5,6,8\}$.

For a set X of edges in a graph H, let $H[X]$ be the subgraph of H induced by X, that is, X is the edge set of $H[X]$ while the vertex set is $H[X]$ is the set of vertices that meet some edge in X.

Proposition 4.25. *For a set X of edges of a graph H, the rank of X in $M(H)$ is the number of vertices of $H[X]$ minus the number of components of $H[X]$.*

Definition 4.26. Let E be an n-element set and r be an integer with $0 \leq r \leq n$. The set of subsets of E with at most r elements is the set of independent sets of a matroid $U_{r,n}$ with ground set E. This matroid has rank r and is called the *uniform matroid* of rank r on an n-element set.

Proposition 4.27. *For the uniform matroid $U_{r,n}$ with ground set E,*

$$\mathcal{C}(U_{r,n}) = \{C \subseteq E : |C| = r+1\} \text{ and}$$

$$\mathcal{B}(U_{r,n}) = \{B \subseteq E : |B| = r\}.$$

Definition 4.28. The matroid $U_{0,0}$, whose ground set is the empty set and whose set of circuits is empty, is called the *empty matroid.*

Example 4.29. There are exactly two matroids on a 1-element set $\{e\}$. These are $U_{0,1}$ and $U_{1,1}$, which have $\{\emptyset\}$ and $\{\{e\}\}$ as their sets of bases. More generally, it is easy to check that the matroids $U_{0,n}$, $U_{1,n}$, $U_{n-1,n}$, $U_{n,n}$ are isomorphic to the cycle matroids of, respectively, an n-edge graph in which all edges are loops, a graph with two vertices that are joined by exactly n parallel edges, an n-edge cycle, and an n-edge tree. However, it can be shown that no other uniform matroid is graphic. The reader can check this in the case of $U_{2,4}$ by trying to build a 4-edge graph in which every set of three edges is the edge set of a cycle.

Let C_1 and C_2 be distinct circuits in a matroid M. If M is graphic, then C_1 and C_2 are edge sets of cycles, and it is easily checked that $(C_1 \cup C_2) \setminus (C_1 \cap C_2)$ contains the edge set of a cycle. But, if $M \cong U_{2,4}$, then $(C_1 \cup C_2) \setminus (C_1 \cap C_2)$ contains only two elements and so does not contain a circuit. In fact, $U_{2,4}$ is the unique smallest non-graphic matroid.

Although an arbitrary matroid does not satisfy the condition on circuits considered at the end of the last example, all matroids do satisfy the following property. It is called the *strong circuit elimination axiom* to contrast it with **(C3)**, which is called just the *circuit elimination axiom* or, sometimes, the *weak circuit elimination axiom*.

Theorem 4.30. *Let C_1 and C_2 be circuits in a matroid M. If $f \in C_1 \setminus C_2$ and $e \in C_1 \cap C_2$, then M has a circuit that contains f and is contained in $(C_1 \cup C_2) \setminus \{e\}$.*

4.3 The many faces of a matroid

In the last section, we defined matroids in terms of their circuits and saw how matroids arise naturally from graphs and from matrices. We also characterized which collections of subsets of a set can be the set of independent sets of a matroid. In addition, we defined the bases of a matroid and the rank of each set in a matroid. One of the attractive features of matroid theory is that there are numerous equivalent ways in which these structures can be defined. In this section, we describe three others and identify some additional key structures within a matroid.

We begin by characterizing when a collection of subsets of a finite set E can be the set of bases of a matroid on E and when a function on the set 2^E of subsets of E can be the rank function of a matroid on E.

Theorem 4.31. *Let E be a finite set and $\mathcal{B}$ be a set of subsets of E. Then $\mathcal{B}$ is the set of bases of a matroid on E if and only if $\mathcal{B}$ satisfies the following conditions.*

(B1) *$\mathcal{B}$ is non-empty.*

(B2) *If B_1 and B_2 are in $\mathcal{B}$, and $x \in B_1 \setminus B_2$, then there is an element y of $B_2 \setminus B_1$ such that $(B_1 \setminus \{x\}) \cup \{y\}$ is in $\mathcal{B}$.*

Axiom **(B2)** is called a *basis exchange axiom.* There are a number of such axioms for matroids some of which can be used in place of **(B2)** in the last theorem. The reader is warned that there are also several conditions that look similar to **(B2)** that cannot be used in place of **(B2)**.

Theorem 4.32. *Let E be a finite set. A function $r : 2^E \to \mathbb{Z}$ is the rank function of a matroid on E if and only if it satisfies the following conditions.*

(R1) *If $X \subseteq E$, then $0 \leq r(X) \leq |X|$.*

(R2) *If $X \subseteq Y \subseteq E$, then $r(X) \leq r(Y)$.*

(R3) *If X and Y are subsets of E, then*

$$r(X) + r(Y) \geq r(X \cup Y) + r(X \cap Y).$$

The inequality in axiom **(R3)** is called the *submodular inequality* or the *semimodular inequality* for the rank function.

We can characterize independent sets, circuits, and bases of a matroid in terms of the rank function.

Theorem 4.33. *Let M be a matroid with rank function r and let X be a subset of $E(M)$. Then*

1. *X is independent in M if and only if $r(X) = |X|$;*
2. *X is a circuit of M if and only if $r(X) < |X|$ and $r(X \setminus \{x\}) = |X| - 1$ for all x in X; and*
3. *X is a basis of M if and only if $r(X) = |X| = r(E)$.*

The rank function in a matroid behaves rather like the dimension function in a vector space V. Given a set X of vectors in V, the span of X can be thought of as the set of vectors x of V such that the subspaces of V generated by X and by $X \cup \{x\}$ have the same dimension. This notion carries through to matroids.

Definition 4.34. Let M be a matroid with rank function r. For a subset X of $E(M)$, define the *closure*, $\mathrm{cl}(X)$, of X by

$$\mathrm{cl}(X) = \{x \in E(M) : r(X \cup \{x\}) = r(X)\}.$$

Sometimes $\mathrm{cl}(X)$ will be written as $\mathrm{cl}_M(X)$ to clarify the matroid in which the closure is being taken.

In Example 4.1, $\mathrm{cl}(\{1,4,5\}) = \{1,2,3,4,5,8\}$ and $\mathrm{cl}(\{1,2,4,6,7\}) = \{1,2,3,4,5,6,7,8\}$. Moreover, $\mathrm{cl}(\emptyset) = \{8\}$. In general, the closure of the empty set in a matroid M consists of the set of loops of M.

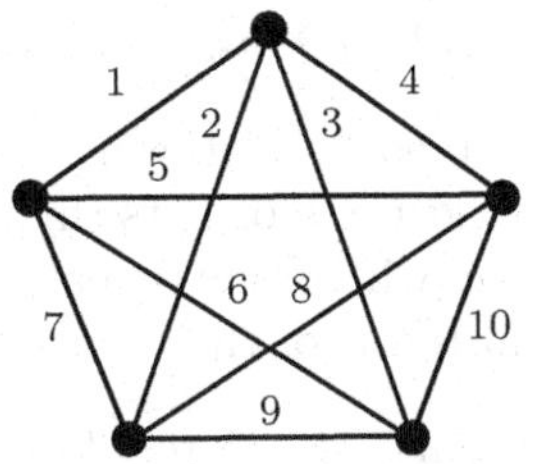

FIGURE 4.3: K_5.

Definition 4.35. Let M be a matroid and X and Y be subsets of $E(M)$. If $\mathrm{cl}(X) = X$, then X is a *closed set* or a *flat* of M. If $\mathrm{cl}(Y) = E$, then Y is a *spanning set* of M. A *hyperplane* of M is a flat of M of rank $r(M) - 1$.

Example 4.36. In an arbitrary matroid M, the unique flat of rank 0 is the set of loops of M. Thus $\{8\}$ is the only rank-0 flat in the cycle matroid, $M(G)$, of the graph G in Figure 4.1. Indeed, as 8 is a loop, it is contained in every flat of $M(G)$. The rank-1 flats of $M(G)$ consist of $\{2,3,8\}$ and all sets of the form $\{i,8\}$ where $i \in \{1,4,5,6,7\}$. The sets $\{1,7,8\}$, $\{5,6,8\}$, and $\{2,3,4,5,8\}$ are examples of rank-2 flats, while $\{1,2,3,7,8\}$ and $\{2,3,4,5,6,7,8\}$ are examples of hyperplanes.

Example 4.37. For the cycle matroid, $M(K_5)$, of the graph K_5 labelled as in Figure 4.3, we see that $\{5,6,7,8,9,10\}$ and $\{1,2,7,10\}$ are hyperplanes. By symmetry, there are five hyperplanes consisting of all of the edges that avoid some vertex of K_5, and there are ten hyperplanes that consist of an edge e along with the edges of the K_3 on the three vertices avoiding e. These fifteen hyperplanes are all of the hyperplanes of $M(K_5)$. Thus a set of edges is a hyperplane of $M(K_5)$ if and only if it is the complement of a minimal edge-cut (a *bond*) in K_5. This is true in general.

Theorem 4.38. *Let G be a graph. A set X of edges of G is a hyperplane in $M(G)$ if and only if $E(G) \setminus X$ is a minimal set of edges of G whose removal from G produces a graph with more connected components than G.*

The bonds in a graph turn out to be very significant. We shall see in the next section why they behave like the edge sets of cycles.

Like Theorem 4.33, the next result ties together some of the many concepts introduced above.

Theorem 4.39. *Let M be a matroid and X be a subset of $E(M)$. Then*

1. *X is spanning if and only if $r(X) = r(M)$;*

2. *X is a basis if and only if X is both spanning and independent;*

3. *X is a basis if and only if X is a minimal spanning set; and*

4. X is a hyperplane if and only if X is a maximal non-spanning set.

There are numerous other equivalent ways to axiomatize matroids, for example, in terms of their collections of flats, hyperplanes, or spanning sets. These can be found in [894] and will not be repeated here. We close this section by characterizing matroids in terms of their closure operators.

Theorem 4.40. *Let E be a finite set. A function $\mathrm{cl} : 2^E \to 2^E$ is the closure operator of a matroid on E if and only if it satisfies the following conditions.*

(CL1) *If $X \subseteq E$, then $X \subseteq \mathrm{cl}(X)$.*

(CL2) *If $X \subseteq Y \subseteq E$, then $\mathrm{cl}(X) \subseteq \mathrm{cl}(Y)$.*

(CL3) *If $X \subseteq E$, then $\mathrm{cl}(\mathrm{cl}(X)) = \mathrm{cl}(X)$.*

(CL4) *If $X \subseteq E$ and $x \in E$, and $y \in \mathrm{cl}(X \cup \{x\}) \backslash \mathrm{cl}(X)$, then $x \in \mathrm{cl}(X \cup \{y\})$.*

4.4 Duality

Example 4.41. Figure 4.4 shows the construction of a planar dual G^* for the graph G in Figure 4.1 where the latter is viewed as a plane graph. The figure also shows a redrawing of G^*. Observe that $\{2, 4, 5\}$ is the edge set of a cycle in G and is a bond in G^*. Similarly, $\{2, 3, 5\}$ is a bond in G and the edge set of a cycle in G^*. Indeed, it is well known and easily verified that, for a plane graph G and a planar dual G^* of it, if $\{G, G^*\} = \{G_1, G_2\}$, a subset X of $E(G)$ is the edge set of a cycle in G_1 if and only if X is the bond in G_2. We know that the cycle matroid $M(G)$ is a matroid on $E(G)$ in which the circuits are the edge sets of cycles. When G is a planar graph, we now have a second way to define a matroid on $E(G)$, namely, we construct a planar dual G^* of a plane embedding of G and look at its cycle matroid $M(G^*)$. Our observations above tell us that the edge sets of the cycles of G^* are the bonds of G. Hence $M(G^*)$ is a matroid on $E(G)$ whose circuits are the bonds of G.

The construction in the last example can be generalized to all graphs, planar or otherwise.

Theorem 4.42. *Let G be a graph and $\mathcal{C}^*$ be the collection of bonds of G. Then $(E(G), \mathcal{C}^*)$ is a matroid.*

Definition 4.43. The matroid whose existence is asserted in the last theorem is called the *bond matroid* of the graph G. It is denoted by $M^*(G)$. A matroid is *cographic* if it is isomorphic to the bond matroid of some graph.

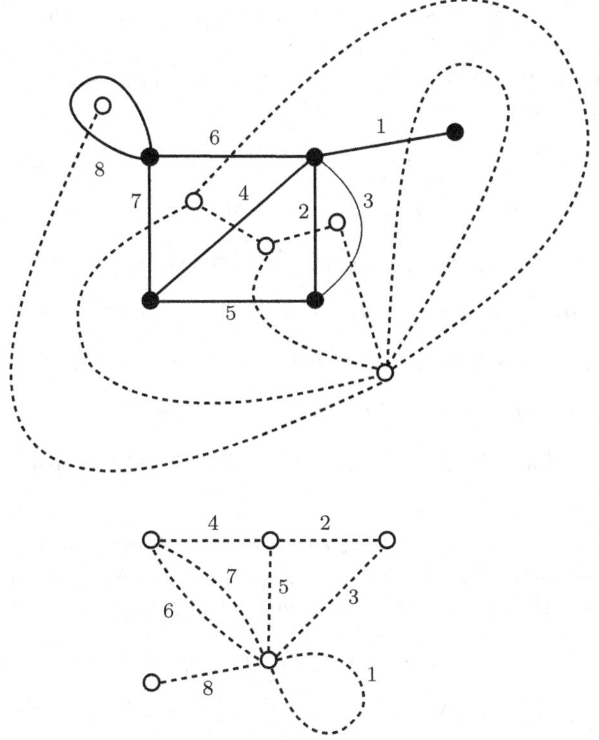

FIGURE 4.4: Construction of a planar dual of the graph G in Figure 4.1 and a redrawing G^* of this planar dual.

The bond matroid of a graph G is the dual of the cycle matroid of G. A fundamental part of matroid theory is the existence of a dual for every matroid. A simple way to define duals directly is in terms of bases. To verify the validity of this definition, we use the fact that matroids obey the following deceptively similar variant of the basis exchange axiom **(B2)**.

Lemma 4.44. *Let B_1 and B_2 be bases of a matroid M. If $x \in B_2 \setminus B_1$, then there is an element y of $B_1 \setminus B_2$ such that $(B_1 \setminus \{y\}) \cup \{x\}$ is a basis of M.*

Theorem 4.45. *Let M be a matroid and $\mathcal{B}^*(M)$ be $\{E(M) \setminus B : B \in \mathcal{B}(M)\}$. Then $\mathcal{B}^*(M)$ is the set of bases of a matroid on $E(M)$.*

Definition 4.46. The matroid in the last theorem whose ground set is $E(M)$ and whose set of bases is $\mathcal{B}^*(M)$ is called the *dual matroid* of M and is denoted by M^*.

Theorem 4.47. *If M is a matroid, then $(M^*)^* = M$.*

Example 4.48. Since the bases of $U_{r,n}$ are the r-element subsets of an n-element set E, the bases of the dual matroid are the $(n-r)$-element subsets of E. Hence $U^*_{r,n} = U_{n-r,n}$.

Definition 4.49. The bases, circuits, independent sets, spanning sets, hyperplanes, and loops of M^* are called *cobases, cocircuits, coindependent sets, cospanning sets, cohyperplanes*, and *coloops* of M. The rank function and closure operator of M^* are denoted by r^* and cl^*.

The next result gives some relationships between these sets.

Theorem 4.50. *Let M be a matroid on a set E and suppose $X \subseteq E$. Then*

1. *X is independent if and only if $E \setminus X$ is cospanning;*
2. *X is spanning if and only if $E \setminus X$ is coindependent;*
3. *X is a hyperplane if and only if $E \setminus X$ is a cocircuit; and*
4. *X is a circuit if and only if $E \setminus X$ is a cohyperplane.*

To derive a formula for the rank function of M^*, we use the following result.

Lemma 4.51. *In a matroid M, let I and I^* be disjoint subsets of $E(M)$ such that I is independent and I^* is coindependent. Then M has a basis B and a cobasis B^* such that B and B^* are disjoint, $I \subseteq B$, and $I^* \subseteq B^*$.*

Theorem 4.52. *For all subsets X of the ground set E of a matroid M,*

$$r^*(X) = |X| + r(E \setminus X) - r(M).$$

The next three theorems give some important links between a matroid and its dual. It is easily checked that a cycle and a bond in a graph must meet in an even number of edges. The corresponding result does not hold for arbitrary matroids. However, we do have the following.

Theorem 4.53. *A circuit and a cocircuit in a matroid do not have exactly one common element.*

Theorem 4.54. *Let $(X, Y, \{z\})$ be a partition of the ground set E of a matroid M where X or Y may be empty. Then $z \in \mathrm{cl}(X)$ if and only if $z \notin \mathrm{cl}^*(Y)$.*

Example 4.55. Let G be a connected graph. Two players, Constructor and Destroyer, take turns tagging the edges of G. An edge that is tagged by Destroyer is destroyed, that is, deleted, while an edge that is tagged by Constructor is made invulnerable to destruction. Destroyer's goal in the game is to produce a disconnected graph while Constructor's goal is to maintain a connected graph. Hence Destroyer's goal is to tag the edges of some bond while Constructor's goal is to tag the edges of a spanning tree. Evidently, the bonds of G are the minimal subsets of $E(G)$ that meet every spanning tree of G; and the spanning trees of G are the minimal subsets of $E(G)$ that meet every bond. It can be proved that, with Destroyer playing first, Constructor can win against all possible strategies of Destroyer if and only if G has two edge-disjoint spanning trees [438].

This example demonstrates the following.

Theorem 4.56. *Let X be a subset of the ground set of a matroid M. Then*

1. *X is a cocircuit if and only if X is a minimal set that meets every basis;*
2. *X is a basis if and only if X is a minimal set that meets every cocircuit.*

We have already seen that matroid duality generalizes duality for planar graphs. It also generalizes vector space orthogonality. To show this, we first revisit the issue of how a matrix can be altered without changing the associated vector matroid.

Lemma 4.57. *Let A be a matrix over a field $\mathbb{F}$. Let E be the set of column labels on A. Let $M = M[A]$. The matroid M is unchanged if any of the following operations is performed on A.*

1. *Interchange two rows.*
2. *Multiply a row by a non-zero member of $\mathbb{F}$.*
3. *Replace a row by the sum of that row and another.*
4. *Adjoin or remove a zero row.*
5. *Interchange two columns (the labels moving with the columns).*
6. *Multiply a column by a non-zero member of $\mathbb{F}$.*
7. *Replace each matrix entry by its image under some automorphism of $\mathbb{F}$.*

Example 4.58. While each of the operations in the last lemma change the matrix without altering the associated vector matroid, two matrices may yield the same matroid and yet not be related by these operations. As an example, consider the matrix

$$A = \begin{array}{c} \begin{matrix} 1 & 2 & 3 & 4 & 5 \end{matrix} \\ \begin{bmatrix} 1 & 0 & 0 & 1 & 1 \\ 0 & 1 & 0 & 1 & \alpha \\ 0 & 0 & 1 & 1 & \beta \end{bmatrix} \end{array}$$

over the 5-element field $\mathbb{F}_5$ where α and β are distinct members of $\mathbb{F}_5 \setminus \{0,1\}$. The reader can check that there are six choices for (α, β) and, for each of these, the matroid $M[A]$ is isomorphic to $U_{3,5}$. However, none of the six choices for A can be obtained from another of these choices by a combination of operations in Lemma 4.57.

The following is a straightforward consequence of the last lemma. The exclusion of zero matrices from the next lemma is not significant since the vector matroid of a zero matrix with n columns is isomorphic to $U_{0,n}$. The $r \times r$ identity matrix is denoted by I_r.

Lemma 4.59. *Let A be a non-zero matrix with n columns. Then $M[A] = M[I_r|D]$ where r is the rank of the matrix A, and D is some $r \times (n-r)$ matrix.*

Definition 4.60. A matrix in the form $[I_r|D]$ is called a *standard representative matrix* for the associated vector matroid.

The matrix A in Example 4.58 is a standard representative matrix for $U_{3,5}$ over $\mathbb{F}_5$. Sometimes, the identity matrix is suppressed in such a representation with the rows being labelled in the natural way. Hence

$$\begin{array}{c} \\ 1 \\ 2 \\ 3 \end{array}\begin{array}{c} \begin{array}{cc} 4 & 5 \end{array} \\ \begin{bmatrix} 1 & 1 \\ 1 & \alpha \\ 1 & \beta \end{bmatrix} \end{array}$$

also represents $U_{3,5}$. Such a representation is known as a *reduced standard representative matrix* or sometimes as just a standard representative matrix.

The next result implies that the class of $\mathbb{F}$-representable matroids is closed under taking duals.

Theorem 4.61. *Let M be the vector matroid of the matrix $[I_r|D]$ where the columns of this matrix are labelled, in order, $e_1, e_2, \ldots, e_n$ and $1 \le r < n$. Then M^* is the vector matroid of the matrix $[-D^T|I_{n-r}]$ where its columns are also labelled $e_1, e_2, \ldots, e_n$ in that order.*

The dual pair of matroids $U_{0,n}$ and $U_{n,n}$ are not covered by the last theorem. Each is representable over all fields, the first by a zero matrix with n columns, the second by the matrix I_n. Combining this observation with the last theorem gives the following.

Corollary 4.62. *If a matroid is representable over a field $\mathbb{F}$, then M^* is also representable over $\mathbb{F}$.*

By Lemma 4.57, the vector matroids of $[-D^T|I_{n-r}]$ and $[D^T|I_{n-r}]$ are equal. The significance of using $[-D^T|I_{n-r}]$ in Theorem 4.61 is that it highlights the link between matroid duality and vector space orthogonality.

Definition 4.63. Let W be a subspace of $V(n,\mathbb{F})$, the n-dimensional vector space over the field $\mathbb{F}$. Let $\mathsf{W}^\perp$ be the set of vectors of $V(n,\mathbb{F})$ whose inner product with every vector in W is zero. Then $\mathsf{W}^\perp$ is a subspace of $V(n,\mathbb{F})$ called the *orthogonal subspace* of W. If A is an $m \times n$ matrix over $\mathbb{F}$, the *row space* $\mathcal{R}(A)$ of A is the subspace of $V(n,\mathbb{F})$ generated by the rows of A.

Theorem 4.64. *Let $[I_r|D]$ be an $r \times n$ matrix over a field $\mathbb{F}$ where r is in $\{1, 2, \ldots, n-1\}$. Then the orthogonal subspace of $\mathcal{R}[I_r|D]$ is $\mathcal{R}[-D^T|I_{n-r}]$.*

Definition 4.65. An $[n,r]$ *linear code over* $\mathbb{F}_q$, or an $[n,r]$*-code over* $\mathbb{F}_q$, is an r-dimensional subspace C of the n-dimensional vector space over $\mathbb{F}_q$. We call r and n, the *dimension* and *length*, respectively, of C.

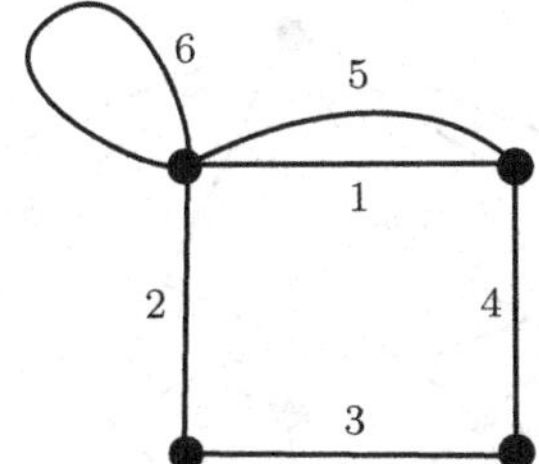

FIGURE 4.5: $M(G) = M[A]$ for the matrix A in Example 4.69.

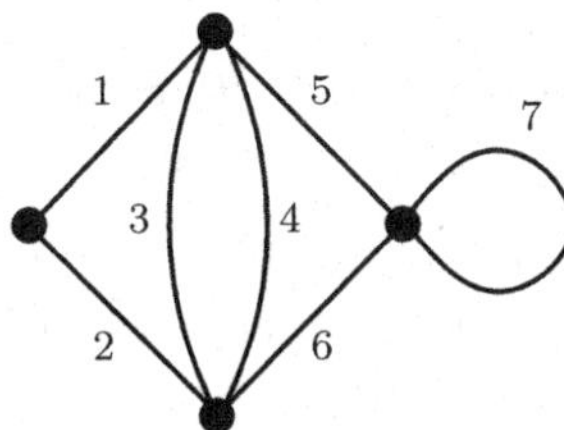

FIGURE 4.6: The graph H in Example 4.70.

Example 4.66. When $\mathbb{F} = \mathbb{F}_q$, the $r \times n$ matrix $[I_r|D]$ is a *generator matrix in standard form* for the $[n, r]$-code $\mathcal{R}[I_r|D]$, while $[-D^T|I_{n-r}]$ is the *parity-check matrix* for the code. (See Chapter 16 for more on codes.)

Lemma 4.67. *Let B be a basis of a matroid M. If $e \in E(M)\backslash B$, then $B \cup \{e\}$ contains a unique circuit, $C(e, B)$. Moreover, $e \in C(e, B)$.*

Definition 4.68. The circuit $C(e, B)$ is called the *fundamental circuit of e with respect to B.*

Example 4.69. Consider the matrix

$$A = \begin{array}{c} \begin{matrix} 1 & 2 & 3 & 4 & 5 & 6 \end{matrix} \\ \begin{bmatrix} 1 & 0 & 0 & 3 & -2 & 0 \\ 0 & 1 & 0 & -1 & 0 & 0 \\ 0 & 0 & 1 & 7 & 0 & 0 \end{bmatrix} \end{array}$$

over $\mathbb{R}$. Then $\{1, 2, 3\}$ is a basis of $M[A]$. Moreover, $C(4, B) = \{1, 2, 3, 4\}$ while $C(5, B) = \{1, 5\}$, and $C(6, B) = \{6\}$. If we view the rows of A as being labelled, in order, by $1, 2$, and 3, we observe that an element j of B is in $C(e, B)$ if and only if the entry in row j and column e of A is non-zero.

We observe that $M[A] = M(G)$ for the graph G in Figure 4.5.

Example 4.70. Figure 4.7 shows all of the flats of $M(H)$ where H is the graph in Figure 4.6 and the flat $\{3, 4, 5, 6, 7\}$, for example, has been abbreviated as 34567. These flats have been arranged in layers corresponding to their ranks.

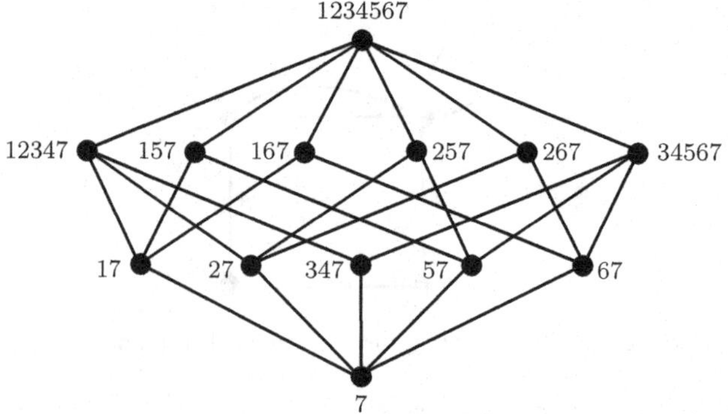

FIGURE 4.7: The lattice of flats of $M(H)$ for the graph H in Figure 4.6.

The lines between consecutive layers indicate inclusion. This picture is called the Hasse diagram for the partially ordered set of flats of $M(H)$ ordered by inclusion.

In general, we have the following.

Lemma 4.71. *The set of flats of a matroid M, ordered by inclusion, is a partially ordered set. Moreover, if X and Y are flats of M, then*

1. *$X \cap Y$ is the unique maximal flat of M that is contained in both X and Y; and*

2. *$\mathrm{cl}(X \cup Y)$ is the unique minimal flat of M that contains both X and Y.*

A consequence of this lemma is that the flats of a matroid form a lattice.

Definition 4.72. For a matroid M, the set of flats of M, ordered by inclusion, forms a lattice, $\mathcal{L}(M)$, called the *lattice of flats* of M.

Those lattices that arise as the lattice of flats of some matroid were characterized by Birkhoff [119]. Such lattices are now called geometric lattices. They are discussed, for example, in Section 1.7 of Oxley [894]. One noteworthy feature of $\mathcal{L}(M)$ is that, in the absence of labels on the elements, one cannot detect loops or parallel elements in M. For instance, in Example 4.70, the lattices $\mathcal{L}(M(H))$ and $\mathcal{L}(M(H)\backslash\{4,7\})$ are isomorphic.

4.5 Basic constructions

The graph operations of deletion and contraction play a fundamental role in one of the ways that the Tutte polynomial for graphs is defined. In Section 4.3, we looked briefly at the matroid operation of deletion. In this section, we will

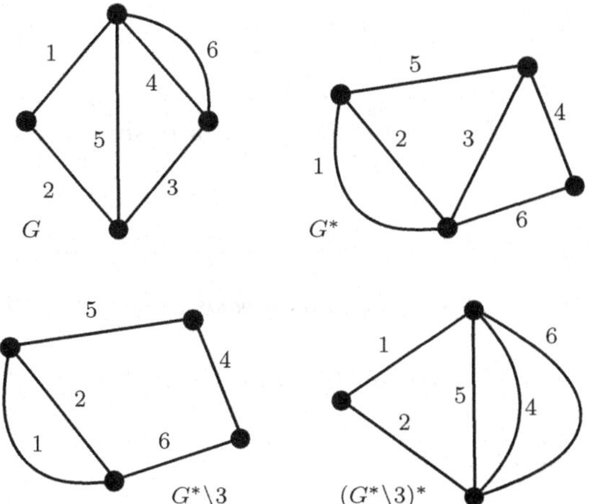

FIGURE 4.8: Deletion and contraction are dual operations for graphs.

define contraction for matroids as the dual operation of deletion. We will also consider some ways of gluing matroids together.

Example 4.73. Let $M = M(G)$ where G is the plane graph shown in Figure 4.8. In that figure, we also see G^*, a planar dual of G, along with the graphs $G^*\backslash 3$ and its dual $(G^*\backslash 3)^*$. Note that the last graph could have been obtained directly from G by contracting the edge 3.

Recall from Definition 4.22 that, for a subset Z of the ground set E of a matroid M, the deletion $M\backslash Z$ of Z from M is the matroid with ground set $E\backslash Z$ whose independent sets are the independent sets of M that are contained in $E \setminus Z$.

Definition 4.74. Let Z be a subset of the ground set E of a matroid M. The *contraction* M/Z *of* Z *from* M is given by

$$M/Z = (M^*\backslash Z)^*.$$

Sometimes M/Z is denoted by $M.(E\backslash Z)$ and called the *contraction of* M *onto* $E \setminus Z$. Moreover, if $Z = \{e_1, e_2, \ldots, e_n\}$, then M/Z and $M\backslash Z$ are sometimes written as $M/e_1, e_2, \ldots, e_n$ and $M\backslash e_1, e_2, \ldots, e_n$, respectively.

Applying Definition 4.74 to calculate a particular contraction is too convoluted. The next result gives several more-direct ways in which the contraction can be found.

Theorem 4.75. *Let Z be a subset of the ground set E of a matroid M.*

1. *For all $X \subseteq E \setminus Z$,*

$$r_{M/Z}(X) = r_M(X \cup Z) - r_M(Z).$$

2. *Let B_Z be a basis of $M|Z$. Then*

$$\begin{aligned}\mathcal{I}(M/Z) &= \{I \subseteq E \setminus Z : I \cup B_Z \in \mathcal{I}(M)\} \\ &= \{I \subseteq E \setminus Z : M|Z \text{ has a basis } B \text{ such that } I \cup B \in \mathcal{I}(M)\}\end{aligned}$$

and

$$\begin{aligned}\mathcal{B}(M/Z) &= \{B' \subseteq E \setminus Z : B' \cup B_Z \in \mathcal{B}(M)\} \\ &= \{B' \subseteq E \setminus Z : M|Z \text{ has a basis } B \text{ such that } B' \cup B \in \mathcal{B}(M)\}.\end{aligned}$$

3. *The circuits of M/Z consist of the minimal non-empty members of $\{C\backslash Z : C \in \mathcal{C}(M)\}$.*

4. *For all $X \subseteq E \setminus Z$,*

$$\mathrm{cl}_{M/Z}(X) = \mathrm{cl}_M(X \cup Z) \setminus Z.$$

The corresponding result for deletion is much simpler than the last result.

Theorem 4.76. *Let Z be a subset of the ground set E of a matroid M.*

1. *For all $X \subseteq E \setminus Z$,*

$$r_{M\backslash Z}(X) = r_M(X).$$

2. $\mathcal{I}(M\backslash Z) = \{I \subseteq E \setminus Z : I \in \mathcal{I}(M)\}$.

3. *$\mathcal{B}(M\backslash Z)$ is the set of maximal members of $\{B \setminus Z : B \in \mathcal{B}(M)\}$.*

4. $\mathcal{C}(M\backslash Z) = \{C \subseteq E \setminus Z : C \in \mathcal{C}(M)\}$.

5. *For all $X \subseteq E \setminus Z$,*

$$\mathrm{cl}_{M\backslash Z}(X) = \mathrm{cl}_M(X) \setminus Z.$$

The operations of deletion and contraction for graphic matroids behave completely predictably.

Theorem 4.77. *Let Z be a subset of the edge set of a graph G. Then $M(G\backslash Z) = M(G)\backslash Z$ and $M(G/Z) = M(G)/Z$.*

The operations of deletion and contraction commute with each other and with themselves.

Theorem 4.78. *Let X and Y be disjoint subsets of the ground set of a matroid M. Then*

1. $M\backslash X\backslash Y = M\backslash(X \cup Y) = M\backslash Y\backslash X$;

2. $M/X/Y = M/(X \cup Y) = M/Y/X$;

3. $M\backslash X/Y = M/Y\backslash X$.

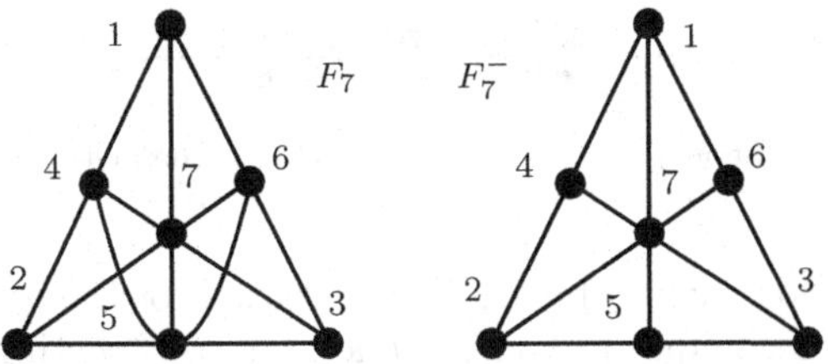

FIGURE 4.9: The Fano and non-Fano matroids.

Example 4.79. For a field $\mathbb{F}$, consider the matrix

$$A_{\mathbb{F}} = \begin{array}{c} \begin{matrix} 1 & 2 & 3 & 4 & 5 & 6 & 7 \end{matrix} \\ \begin{bmatrix} 1 & 0 & 0 & 1 & 0 & 1 & 1 \\ 0 & 1 & 0 & 1 & 1 & 0 & 1 \\ 0 & 0 & 1 & 0 & 1 & 1 & 1 \end{bmatrix} \end{array}$$

viewed over the field $\mathbb{F}$. In $M[A_{\mathbb{F}_2}]$, the set $\{4, 5, 6\}$ is a circuit since the vectors labelled by $4, 5$, and 6 add to the zero vector over $\mathbb{F}_2$. But, in $M[A_{\mathbb{F}_3}]$, the set $\{4, 5, 6\}$ is independent. The matroids $M[A_{\mathbb{F}_2}]$ and $M[A_{\mathbb{F}_3}]$ are called the *Fano matroid* and the *non-Fano matroid*, respectively, and are denoted by F_7 and F_7^-. Figure 4.9 shows *geometric representations* for F_7 and F_7^-. Such diagrams are useful for drawing matroids of rank at most four. When the matroid is simple, each element is denoted by a point, and a set of points is dependent if it contains three collinear points or four coplanar points or five points in space. Since F_7 and F_7^- are both simple of rank three, the circuits consist of all sets of three collinear points together with all sets of four points no three of which are collinear. Notice that the curved line $\{4, 5, 6\}$ in the representation of F_7 is to indicate that this set is a circuit.

For all subsets Z of $\{1, 2, \ldots, 7\}$, it is clear that $M[A_{\mathbb{F}}] \backslash Z$ is $\mathbb{F}$-representable since, by deleting the columns of $A_{\mathbb{F}}$ that are labelled by elements of Z, we obtain an $\mathbb{F}$-representation. Combining this observation with Corollary 4.62, we see that $M[A_{\mathbb{F}}]/Z$ is also $\mathbb{F}$-representable. We would like a direct way to construct an $\mathbb{F}$-representation for this contraction. By Theorem 4.78, it suffices to describe this when $|Z| = 1$. The matroid $M[A_{\mathbb{F}}]/1$, that is, $M[A_{\mathbb{F}}]/\{1\}$, is represented by the following matrix

$$\begin{array}{c} \begin{matrix} 2 & 3 & 4 & 5 & 6 & 7 \end{matrix} \\ \begin{bmatrix} 1 & 0 & 1 & 1 & 0 & 1 \\ 0 & 1 & 0 & 1 & 1 & 1 \end{bmatrix} \end{array},$$

which is obtained by deleting the column labelled by 1 together with the row containing the unique non-zero entry in column 1. To contract an element like 4 that does not have a single non-zero entry in its column in $A_{\mathbb{F}}$, we first subtract row 2 from row 1 in $A_{\mathbb{F}}$. This does not alter the corresponding matroid but means that 4 now labels a column with a single non-zero entry,

FIGURE 4.10: A common geometric representation for the matroids $F_7/1$ and $F_7^-/1$.

and we already know how to find a representation for the contraction of such an element. The technique just exemplified for constructing representations of $M[A]\backslash Z$ and $M[A]/Z$ is easily seen to be valid in general.

Because the matroid $M[A_{\mathbb{F}}]/1$ is the same irrespective of the choice of the field $\mathbb{F}$, the matroids $F_7/1$ and $F_7^-/1$ are equal. Geometrically, this rank-2 matroid is represented as in Figure 4.10 with the pairs of touching points indicating pairs of parallel elements. Now suppose we contract 2 from $F_7/1$. In the resulting matroid, $F_7/\{1,2\}$, the element 4 is a loop, while $\{3,5,6,7\}$ is a parallel class. Geometrically, this rank-1 matroid is represented by four touching points labelled by $3,5,6$, and 7. The element 4 is labelled as a loop and put in an inset box.

The matroid $F_7/1$ is clearly not simple. One can obtain a simple matroid from it by deleting all but one element of each parallel class. This matroid is denoted by $\mathrm{si}(F_7/1)$. Note that the last matroid is only defined up to isomorphism since, in general, there is no satisfactory way to specify which element from each parallel class is retained.

Definition 4.80. For a matroid M, the *simplification* of M, or the *simple matroid associated with* M, is the matroid $\mathrm{si}(M)$ that is obtained from M by deleting all of the loops of M along with all but one element from each parallel class.

To define the dual operation directly, we will need a preliminary definition:

Definition 4.81. In a matroid M, elements e and f are *series elements* if $\{e,f\}$ is a cocircuit. A *series class* of M is a maximal set X of elements of M none of which is a coloop such that every two distinct elements of X are series elements.

Definition 4.82. For a matroid M, the *cosimplification* of M is the matroid $\mathrm{co}(M)$ that is obtained from M by contracting all of the coloops of M along with all but one element from each series class.

Every minor of a graph can be produced by deleting and contracting edges and deleting isolated vertices. Since isolated vertices have no effect on the cycle matroid of a graph, the definition of a minor of a matroid is quite predictable.

Definition 4.83. A *minor* of a matroid M is any matroid N that can be written in the form $M\backslash X/Y$ for some disjoint subsets X and Y of $E(M)$. The matroid N is a *proper minor* of M if $X \cup Y \neq \emptyset$. A class $\mathcal{M}$ of matroids is *closed under minors* or is *minor-closed* if every minor of a member of the class is also in the class.

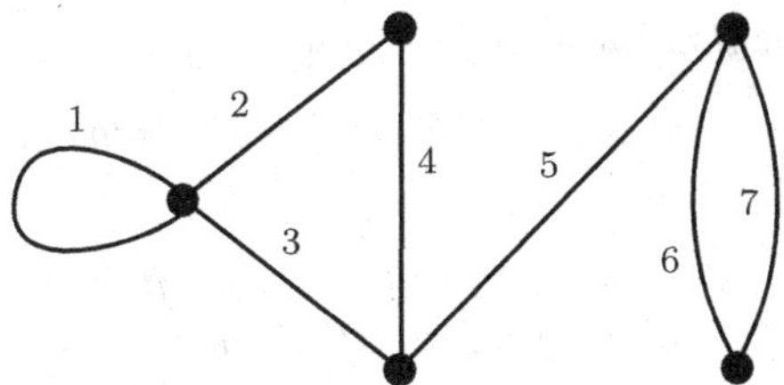

FIGURE 4.11: A graph H.

Example 4.84. The classes of graphic, $\mathbb{F}$-representable, and uniform matroids are minor-closed. The class of cycle matroids of bipartite graphs is closed under deletion but is not closed under contraction, so it is not minor-closed. The class $\{U_{0,0}\} \cup \{U_{n,n+1} : n \geq 0\}$ is closed under contraction but not under deletion, so it is not minor-closed either. Just as for classes of graphs, when a class of matroids is minor-closed, it can be characterized by listing the *excluded minors*, that is, the minor-minimal matroids that are not in the class. For instance, the minor-closed class $\{U_{n,n} : n \geq 0\}$ has $U_{0,1}$ as its unique excluded minor. In general, finding the excluded minors for a minor-closed class can be very difficult. Indeed, even determining whether or not the number of excluded minors is finite or infinite can be a major challenge. This contrasts with the situation for graphs where, by Robertson and Seymour's excluded minors theorem [965], every minor-closed class has a finite collection of excluded minors.

Example 4.85. For the graph H in Figure 4.11, by comparing their sets of circuits, we see that $M(H)\backslash 1 = M(H)/1$, and $M(H)\backslash 5 = M(H)/5$. The element 1 is a loop of $M(H)$ while 5 is a coloop. One can easily check that 1 and 5 are the only elements e of $M(H)$ for which $M(H)\backslash e = M(H)/e$. However, there are other subsets Z of $E(H)$ for which $M(H)\backslash Z = M(H)/Z$. In particular, the minimal non-empty sets Z for which the last equation holds are $\{1\}, \{2,3,4\}, \{5\}$ and $\{6,7\}$. Indeed, the sets Z for which $M(H)\backslash Z = M(H)/Z$ are precisely the unions of arbitrarily many of the four sets just noted.

This example illustrates the following results.

Theorem 4.86. *Let e be an element of a matroid M. Then $M\backslash e = M/e$ if and only if e is a loop or a coloop of M.*

Theorem 4.87. *Let Z be a subset of the ground set E of a matroid M. Then the following statements are equivalent.*

1. $M\backslash Z = M/Z$;
2. $r(Z) + r(E \setminus Z) = r(M)$;
3. $r(Z) + r^*(Z) = |Z|$;

4. *M has no circuit that meets both Z and $E \setminus Z$;*

5. *a subset I of E is independent if and only if both $I \cap Z$ and $I \cap (E \setminus Z)$ are independent;*

6. *a subset B of E is a basis of M if and only if $B \setminus (E \setminus Z)$ is a basis of $M \backslash (E \setminus Z)$ and $B \setminus Z$ is a basis of $M \backslash Z$; and*

7. *a subset C of E is a circuit of M if and only if C is a circuit of $M \backslash (E \setminus Z)$ or of $M \backslash Z$.*

Definition 4.88. A set Z in a matroid M that satisfies the equivalent conditions in the last theorem is called a *separator* of M. It is a *non-trivial separator* if $Z \notin \{\emptyset, E(M)\}$. A minimal non-empty separator of M is called a *component* of M.

In Example 4.85, the components of $M(H)$ are $\{1\}, \{2, 3, 4\}, \{5\}$ and $\{6, 7\}$ while the separators of $M(H)$ are unions of arbitrarily many of these components. In Theorem 4.87, we began with a matroid and identified certain subsets of it. Focusing on Part 5 of that theorem suggests a way to stick matroids together.

Theorem 4.89. *Let M_1 and M_2 be matroids with disjoint ground sets E_1 and E_2. Then there is a matroid with ground set $E_1 \cup E_2$ whose set of independent sets is $\{I_1 \cup I_2 : I_1 \in \mathcal{I}_1$ and $I_2 \in \mathcal{I}_2\}$.*

Definition 4.90. The matroid whose existence is established by the last theorem is denoted by $M_1 \oplus M_2$ and is called the *direct sum*, or sometimes the 1-*sum*, of M_1 and M_2.

In particular, taking the direct sum of a matroid M and a copy of $U_{0,1}$ or $U_{1,1}$ having ground set $\{e\}$ produces a matroid with ground set $E(M) \cup \{e\}$ in which e is a loop or a coloop, respectively.

The direct sum has numerous attractive properties, many of which can be deduced from Theorem 4.87.

Corollary 4.91. *Let M_1 and M_2 be matroids with disjoint ground sets E_1 and E_2. Suppose X is a subset of $E_1 \cup E_2$. Then*

1. $M_2 \oplus M_1 = M_1 \oplus M_2$;

2. $(M_1 \oplus M_2)^* = M_1^* \oplus M_2^*$;

3. $(M_1 \oplus M_2)\backslash X = (M_1\backslash(X \cap E_1)) \oplus (M_2\backslash(X \cap E_2))$;

4. $(M_1 \oplus M_2)/X = (M_1/(X \cap E_1)) \oplus (M_2/(X \cap E_2))$;

5. $r_{M_1 \oplus M_2}(X) = r_{M_1}(X \cap E_1) + r_{M_2}(X \cap E_2)$;

6. *a set B is a basis of $M_1 \oplus M_2$ if and only if B is the union of a basis of M_1 and a basis of M_2; and*

7. *a set C is a circuit of $M_1 \oplus M_2$ if and only if C is a circuit of M_1 or C is a circuit of M_2.*

In the Fano matroid F_7 in Example 4.79, if we wish to contract the set $\{1, 2, 4\}$, then, once we have contracted $\{1, 2\}$, the element 4 is a loop and so, by Theorem 4.86, we can delete 4 rather than contracting it; that is, $F_7/\{1, 2, 4\} = F_7/\{1, 2\}\backslash 4$. We can think of contracting elements until the rank is correct and then deleting elements without changing the rank. The next result comes from generalizing this idea.

Theorem 4.92. *Every minor of a matroid M can be written in the form $M/I\backslash I^*$ where I is independent and I^* is coindependent. Moreover, every loopless minor of M can be written in the form $M/F\backslash D$ where F is a flat of M.*

Example 4.93. When G_1 and G_2 are disjoint graphs, $M(G_1) \oplus M(G_2)$ is a graphic matroid. Indeed, $M(G_1) \oplus M(G_2) = M(G)$ where G is the disjoint union of G_1 and G_2, that is, G is the 0-sum of G_1 and G_2. In addition, we see that $M(G_1) \oplus M(G_2) = M(H)$ for any 1-sum H of G_1 and G_2, that is, for any graph that is obtained by identifying a single vertex in G_1 with a single vertex in G_2.

Suppose M_1 and M_2 are $\mathbb{F}$-representable matroids on disjoint ground sets E_1 and E_2. If A_i is an $\mathbb{F}$-representation of M_i for each i, then $M_1 \oplus M_2$ is represented over $\mathbb{F}$ by the matrix

$$\begin{bmatrix} A_1 & 0 \\ 0 & A_2 \end{bmatrix}.$$

In the last example, we saw that both 0-sum and 1-sum of graphs correspond to the operation of direct sum for matroids. In particular, there is no matroid analog of the graph property of connectedness. There is, however, a matroid analog of the graph property of 2-connectedness. The following lemma gives a way of characterizing 2-connected graphs without referring to vertex deletions and so readies us to apply the maxim of Tutte with which this chapter began.

Lemma 4.94. *Let G be a loopless graph without isolated vertices and suppose that $|V(G)| \geq 3$. Then G is 2-connected if and only if, for every two distinct edges e and f, there is a cycle that contains $\{e, f\}$.*

Definition 4.95. Let M be a matroid in which, for every two distinct elements e and f, there is a circuit containing $\{e, f\}$. Then M is called *2-connected* or *connected.* A matroid that does not have this property is called *disconnected.*

It is slightly disconcerting that the terms "connected" and "2-connected" actually have the same meaning for matroids but that is the situation. Indeed, historically, the term "connected" has probably been more commonly used. Clearly a non-empty matroid M is connected if and only if $E(M)$ is a component of M. Moreover, M is disconnected if and only if, for some proper non-empty subset Z of $E(M)$, each of the equivalent set of conditions in Theorem 4.87 holds.

The following very attractive property of connected matroids is ideal for use in inductive arguments.

Theorem 4.96. *Let e be an element of a connected matroid M. Then $M\backslash e$ or M/e is connected.*

In the 2-sum of graphs G_1 and G_2, it is straightforward to specify the (edge sets of the) cycles of the 2-sum in terms of the cycles of G_1 and G_2. Again, Tutte's maxim suggests that the operation can be generalized to matroids. Recall from Theorem 4.86 that a single element is a separator in a matroid if it is a loop or a coloop.

Theorem 4.97. *Let M_1 and M_2 matroids on disjoint ground sets. For each i in $\{1,2\}$, let p_i be an element of M_i that is not a separator. Then there is a matroid $M_1 \oplus_2 M_2$ with ground set $(E(M_1)\setminus\{p_1\})\cup(E(M_2)\setminus\{p_2\})$ whose set of circuits is the union of $\mathcal{C}(M_1\backslash p_1)\cup\mathcal{C}(M_2\backslash p_2)$ with $\{(C_1\setminus p_1)\cup(C_2\setminus p_2) : p_i \in C_i \in \mathcal{C}(M_i)$ for each $i \in \{1,2\}\}$.*

Definition 4.98. The matroid whose existence is asserted by the last theorem is called the 2-*sum* of M_1 and M_2 with respect to the *basepoints* p_1 and p_2, respectively. Sometimes, $M_1 \oplus_2 M_2$ is defined for matroids M_1 and M_2 whose ground sets meet in a single element p that is not a separator of either matroid. In that case, p_1 and p_2 are replaced by p in the last theorem.

Theorem 4.99. *Let M_1 and M_2 be matroids whose ground sets meet in a single element p that is not a separator of either matroid. Then*

1. *$M_1 \oplus_2 M_2 = M_2 \oplus_2 M_1$; and*
2. *$(M_1 \oplus_2 M_2)^* = M_1^* \oplus_2 M_2^*$.*

Example 4.100. Consider the graphs G_1 and G_2 in Figure 4.12. Both G_3 and G_4 are 2-sums of G_1 and G_2 with respect to the basepoint p. The first is obtained by identifying the vertices u_1 and v_1 with the vertices u_2 and v_2, respectively; the second is obtained by, instead, identifying u_1 with v_2 and v_1 with u_2. Clearly $M(G_3) = M(G_1) \oplus_2 M(G_2) = M(G_4)$ even though G_3 and G_4 are different graphs. We saw earlier how the 0-sum and 1-sum of two disjoint graphs have the same cycle matroid. The "twist" that produces G_4 from G_3 is another way to alter a graph without changing its cycle matroid. The next theorem [1156] shows that two graphs have the same cycle matroids if and only if one can be obtained from the other by a combination of these moves.

Definition 4.101. A graph G is 2-*isomorphic* to a graph H if H can be transformed into G by a sequence of the following operations:

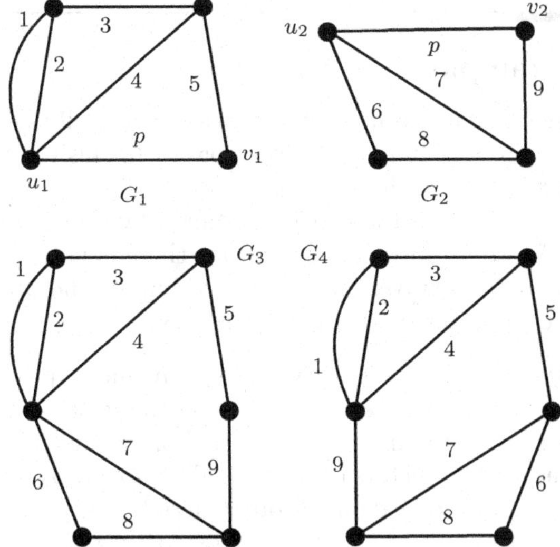

FIGURE 4.12: Two 2-sums of G_1 and G_2.

1. identifying a vertex of one component with a vertex of a different component;
2. the reverse of the previous operation; and
3. when one graph is obtained from disjoint graphs G_1 and G_2 by identifying the vertices u_1 and v_1 of G_1 with the vertices u_2 and v_2, respectively, of G_2, the other graph is obtained from G_1 and G_2 by identifying, instead, the vertices u_1 and v_1 with the vertices v_2 and u_2, respectively. The two graphs thus constructed are said to be related by a *Whitney flip* or a *Whitney twist.*

Theorem 4.102 (Whitney's 2-isomorphism Theorem). *The cycle matroids of graphs G and H are isomorphic if and only if the graphs G and H are 2-isomorphic.*

If one of the operations in Definition 4.101 is applied to a graph G to produce a graph H, then G or H is disconnected or has a 1- or 2-vertex cut. Thus the following is an immediate consequence of the last theorem.

Corollary 4.103. *Let G and H be loopless graphs without isolated vertices. If G is 3-connected and $M(H) \cong M(G)$, then $H \cong G$.*

4.6 More examples

In earlier sections, we have seen uniform matroids, as well as examples of matroids arising from graphs and from vector spaces. In this section, we consider some additional ways in which matroids arise.

Every n-vertex simple graph can be obtained by deleting edges from the complete graph K_n. It is not hard to see that there is no corresponding universal rank-r matroid. However, the classes of binary and ternary matroids, for example, do contain universal rank-r members.

Definition 4.104. For a nonnegative integer n and a prime power q, let $V(n,q)$ be the n-dimensional vector space over the q-element field $\mathbb{F}_q$. View $V(n,q)$ as a matroid by taking the vector matroid of an $n \times q^n$ matrix over $\mathbb{F}_q$ whose columns are the vectors of $V(n,q)$. The *projective geometry* $PG(n-1,q)$ is the simplification of $V(n,q)$. More generally, for an arbitrary field $\mathbb{F}$, the projective geometry $PG(n-1,\mathbb{F})$ is the structure that is obtained from $V(n,\mathbb{F})$, the n-dimensional vector space over $\mathbb{F}$, by deleting the zero vector and then deleting all but one non-zero element from every one-dimensional subspace of $V(n,\mathbb{F})$.

Example 4.105. From Example 4.79, we see that the projective geometry $PG(2,2)$ is equal to the Fano matroid. In general, it is clear that every rank-n simple matroid that is representable over $\mathbb{F}_q$ can be viewed as a restriction of $PG(n-1,q)$. The corresponding result holds for every rank-n simple matroid that is representable over an arbitrary field $\mathbb{F}$ with the caveat that, although each $PG(n-1,q)$ can be viewed as a matroid, when $\mathbb{F}$ is infinite, $PG(n-1,\mathbb{F})$ has infinitely many elements and so cannot be a matroid. We observe, however, that if E is the set of elements of $PG(n-1,\mathbb{F})$ and $\mathcal{C}$ is the set of minimal linearly dependent subsets of E, then $(E,\mathcal{C})$ satisfies **(C1)**–**(C3)**. In this way, we can view $PG(n-1,\mathbb{F})$ as an "infinite matroid".

Example 4.106. The theorem of Robertson and Seymour [965] noted in Example 4.84 establishes that there is no infinite set of graphs none of which is isomorphic to a minor of another. By contrast, no matroid in the infinite set $\{PG(2,p) : p \text{ is prime}\}$ of projective planes of prime order is a minor of another.

Next we extend the geometric representations of matroids shown in Figures 4.9 and 4.10 to produce a matroid that is not representable over any field.

Example 4.107. The geometric configuration shown in Figure 4.13 is known as the *non-Pappus matroid.* The matroid has ground set $\{1,2,\ldots,9\}$ and has rank 3. Its circuits are all sets of three collinear points along with all sets of four points no three of which are collinear. In view of the next theorem, this is indeed a matroid. However, it is not a representable matroid since a

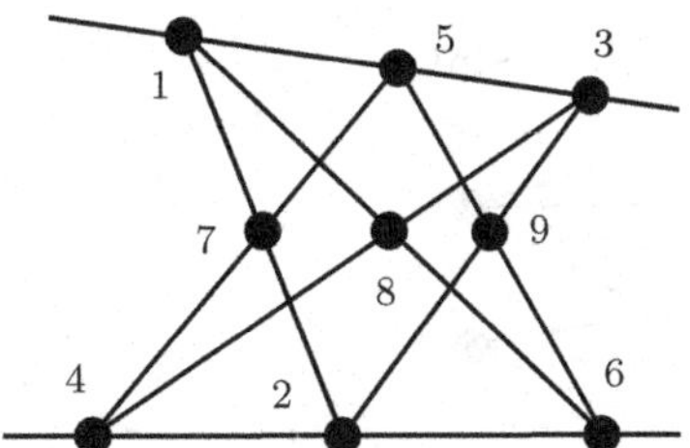

FIGURE 4.13: The non-Pappus matroid, which is nonrepresentable.

theorem of Pappus asserts that, in a projective geometry $PG(2, \mathbb{F})$, when such a configuration arises with the vertices of a hexagon occurring on two lines as shown, the intersections of opposite sides of the hexagon must be collinear. That means that, in order for the configuration shown to correspond to a representable matroid, there should be an extra line through the points 7, 8, and 9. When this line exists, the resulting matroid is called the *Pappus matroid* and it is, for example, $\mathbb{R}$-representable although it is neither binary nor ternary.

As the next theorem shows, in order for a configuration such as that shown in Figure 4.13 to represent a matroid, when two lines meet, they do so in at most one element.

Theorem 4.108. *Let E be a finite set and Λ be a collection of subsets of E called lines each having at least three elements such that every two distinct lines meet in at most one element. Then $(E, \mathcal{C})$ is a simple matroid where $\mathcal{C}$ is composed of all 3-element subsets of E that are contained in some common line along with all four-element subsets of E no three of which lie on a common line. Moreover, every simple matroid of rank at most three arises in this way.*

Next we observe that a matroid of rank r is uniform if and only if all of its circuits have at least $r+1$ elements. Since no circuit can have more than $r+1$ elements, the latter condition means that all of the circuits have exactly $r+1$ elements. The next class of matroids is so fundamental that it is believed that most matroids belong to the class; that is, it is conjectured [829] that, as $n \to \infty$, the fraction of matroids on $\{1, 2, \ldots, n\}$ that are in the class approaches 1.

Definition 4.109. A matroid M is *paving* if M has no circuits with fewer than $r(M)$ elements.

The next result characterizes paving matroids of rank at least two. Every matroid M of rank at most one is clearly paving. Indeed, if $n = |E(M)|$, then $M \cong U_{0,n}$ or $M \cong U_{0,k} \oplus U_{1,n-k}$ for some k with $0 \le k < n$.

Theorem 4.110. *Let E be a finite set and suppose $r \ge 2$. The set $\mathcal{H}$ of hyperplanes of every rank-r paving matroid on E has the properties that*

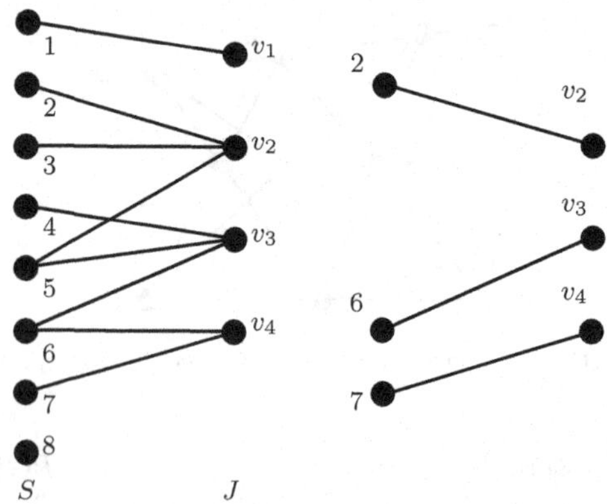

FIGURE 4.14: A bipartite graph Δ and a matching in it.

1. *every member of $\mathcal{H}$ has at least $r-1$ elements; and*
2. *every $(r-1)$-element subset of E is in a unique member of $\mathcal{H}$.*

Moreover, every collection $\mathcal{H}$ of subsets of E that obeys Items 1 and 2 is the set of hyperplanes of a (unique) rank-r paving matroid on E.

Example 4.111. Consider the bipartite graph Δ shown on the left in Figure 4.14, letting the vertex classes be labelled S and J as shown. Then $\{2v_2, 6v_3, 7v_4\}$ is a matching in H. Thus $\{2,6,7\}$ is a subset of S for which there is a matching in Δ whose set of endpoints in S is $\{2,6,7\}$. Another such subset of S is $\{1,3,4,6\}$. Making a list of all such subsets of E and comparing it with the list of cycle-free subsets of the edge set of the graph in Figure 4.1, we see that these sets coincide. This means that we have a matroid on the set S. Indeed, this construction can be used for all bipartite graphs.

An alternative way to view this construction is to consider the family (A_1, A_2, A_3, A_4) of subsets of S where $A_i = \{v \in S : vv_i \in E(\Delta)\}$. Picking the elements $1, 2, 4$, and 7 from A_1, A_2, A_3, and A_4 gives a system of distinct representatives (a *transversal*) for (A_1, A_2, A_3, A_4). This transversal corresponds to the set of edges of the matching $\{1v_1, 2v_2, 4v_3, 7v_4\}$.

Theorem 4.112. *Let Δ be a bipartite graph with vertex classes S and J. Let $\mathcal{I}$ be the set of subsets X of S such that there is a matching in Δ whose set of endpoints in S equals X. Then $\mathcal{I}$ is the set of independent sets of a matroid whose ground set is S.*

Definition 4.113. The matroid whose existence is asserted by the last theorem is called a *transversal matroid* since its collection of independent sets coincides with the set of systems of distinct representatives for subfamilies of the family $\{A_j : j \in J\}$ of subsets of S where $A_j = \{v \in S : vj \in E(\Delta)\}$.

The dual of a transversal matroid is not necessarily transversal. But there is a way in which such duals arise from directed graphs as we now observe.

Definition 4.114. Let G be a directed graph and X and Y be subsets of the vertex set V of G. We say X is *linked into* Y if there are $|X|$ disjoint directed paths in G whose initial vertex is in X and whose final vertex is in Y.

Theorem 4.115. *Let G be a directed graph having vertex set V and let B_0 be a subset of V. Then the set of subsets of V that are linked into B_0 is the set of independent sets of a matroid on V.*

Definition 4.116. The matroid whose existence is asserted in the last theorem is called a *strict gammoid.* A *gammoid* is a matroid that is isomorphic to a restriction of a strict gammoid.

Theorem 4.117. *A matroid M is transversal if and only if M^* is a strict gammoid.*

The class of transversal matroids is closed under deletion but not under contraction. Dually, the class of strict gammoids is closed under contraction but not under deletion.

Theorem 4.118. *The smallest minor-closed class of matroids that contains all transversal matroids is the class of gammoids. Moreover, the class of gammoids is closed under taking duals.*

Definition 4.119. Let G be a graph. For all v in $V(G)$, let A_v be the set of edges of G that meet v. The transversal matroid $\mathrm{Bi}(G)$ of the family $\{A_v : v \in V(G)\}$ of subsets of $E(G)$ is called the *bicircular matroid* of G.

Bicircular matroids were introduced by Simões-Pereira [1008]. Item 2 of Theorem 4.121 makes the reason for this name apparent.

Example 4.120. Let H_1 be the graph with two vertices joined by four parallel edges. Then $\mathrm{Bi}(H_1) \cong U_{2,4}$. Note that, in addition, $\mathrm{Bi}(H_2) \cong U_{2,4}$ where H_2 is the two-vertex graph with a single loop at each vertex and two edges joining these vertices. Perhaps surprisingly, $\mathrm{Bi}(K_4) \cong U_{4,6}$. Lest these examples give a false impression, $\mathrm{Bi}(K_5)$ is not uniform.

Theorem 4.121. *Let G be a graph.*

1. *The independent sets of* $\mathrm{Bi}(G)$ *are those subsets X of $E(G)$ for which each component of $G[X]$ has at most one cycle.*

2. *The circuits of* $\mathrm{Bi}(G)$ *are those subsets Y of $E(G)$ for which $G[Y]$ is either a Θ-graph or consists of two cycles along with a path of possibly zero length that shares exactly one vertex with each cycle.*

4.7 The Tutte polynomial of a matroid

This section defines the Tutte polynomial of a matroid and gives a number of its basic properties.

Definition 4.122. Let M be a matroid with ground set E and rank function r. The *Tutte polynomial* $T(M;x,y)$ of M is defined by

$$T(M;x,y) = \sum_{A\subseteq E} (x-1)^{r(E)-r(A)}(y-1)^{|A|-r(A)}.$$

To distinguish it from several alternative definitions of the Tutte polynomial for a matroid, we call the last definition the *rank definition* of the Tutte polynomial. It is also commonly referred to as the *corank–nullity definition* or the *state sum definition* of the Tutte polynomial. In this definition, if $r(E)-r(A)=0$, then we always interpret $(x-1)^{r(E)-r(A)}$ as having the value 1 even when $x=1$. Likewise, when $|A|-r(A)=0$, we always interpret $(y-1)^{|A|-r(A)}$ as being 1. This means that, for the empty matroid $U_{0,0}$, we have

$$T(U_{0,0};x,y) = 1.$$

Example 4.123. Applying the rank definition, we see that, for $U_{0,1}$ and $U_{1,1}$, the two matroids on a 1-element set,

$$T(U_{1,1};x,y) = (x-1)+1 = x \quad\text{and}\quad T(U_{0,1};x,y) = 1+(y-1) = y.$$

Moreover,

$$\begin{aligned} T(U_{2,4};x,y) &= (x-1)^2 + \binom{4}{1}(x-1) + \binom{4}{2} + \binom{4}{3}(y-1) + (y-1)^2 \\ &= x^2+2x+2y+y^2. \end{aligned}$$

The next theorem generalizes the results in the last example.

Theorem 4.124. *For the rank-r uniform matroid on an n-element set,*

$$T(U_{r,n};x,y) = \begin{cases} x^n & \text{if } r=n; \\ y^n & \text{if } r=0; \end{cases}$$

and, otherwise,

$$T(U_{r,n};x,y) = \sum_{i=0}^{r-1}\binom{n-r-1+i}{i}x^{r-i} + \sum_{j=0}^{n-r-1}\binom{r-1+j}{j}y^{n-r-j}.$$

Observe that the Tutte polynomial of $U_{2,4}$ is symmetric with respect to x and y. This is a special case of the next result, which is an easy consequence of Theorem 4.52.

Theorem 4.125. *For all matroids M,*

$$T(M^*;x,y) = T(M;y,x). \tag{4.1}$$

The next lemma states the deletion–contraction formula for Tutte polynomials that, in a sense to be made precise by the subsequent theorem, characterizes such polynomials.

Lemma 4.126. *Let e be an element of a matroid M. Then*

$$T(M;x,y) = \begin{cases} T(M\backslash e;x,y) + T(M/e;x,y) & \text{if } e \text{ is not a loop or a coloop;} \\ xT(M\backslash e;x,y) & \text{if } e \text{ is a coloop;} \\ yT(M\backslash e;x,y) & \text{if } e \text{ is a loop.} \end{cases}$$

The apparent lack of symmetry under duality between the last two cases in the previous lemma is explained by the fact that $T(M\backslash e;x,y) = T(M/e;x,y)$ when e is a loop or a coloop (see Theorem 4.86). Indeed, the second and third cases of that lemma can both be stated as $T(M;x,y) = T(M|Z;x,y)T(M\backslash Z;x,y)$ when Z is a one-element separator of M. In fact, this assertion holds for all separators.

Corollary 4.127. *Let M_1 and M_2 be matroids on disjoint sets. Then*

$$T(M_1 \oplus M_2;x,y) = T(M_1;x,y)T(M_2;x,y).$$

The next assertion, that every coefficient of the Tutte polynomial is nonnegative, is not immediately apparent from the definition, but follows easily from Lemma 4.126.

Corollary 4.128. *For every matroid M, the Tutte polynomial can be written as*

$$\sum_{i\geq 0}\sum_{j\geq 0} t_{ij}x^i y^j$$

where $t_{ij} \geq 0$ for all i and j. Moreover, $t_{ij} = 0$ unless $i \leq r(M)$ and $j \leq r^*(M)$.

The double summation in the last result will usually be abbreviated to

$$T(M;x,y) = \sum t_{ij}x^i y^j. \tag{4.2}$$

Sometimes, to avoid ambiguity, we write $t_{i,j}$ for t_{ij}.

The Tutte polynomials of two matroids are equal when the matroids are isomorphic.

Definition 4.129. A matroid *isomorphism invariant* is a function f on the class of all matroids such that $f(M) = f(N)$ whenever $M \cong N$.

Definition 4.130. Let $\mathcal{K}$ be a class of matroids that is closed under isomorphism and the taking of minors. Let $\mathfrak{R}$ be a commutative ring with unity. An isomorphism invariant f from $\mathcal{K}$ into $\mathfrak{R}$ is called a *Tutte–Grothendieck invariant* or *T–G invariant* if it satisfies the following conditions for every matroid M in $\mathcal{K}$ and every element e of M.

1. (Deletion–contraction) $f(M) = f(M\backslash e) + f(M/e)$ if e is neither a loop nor a coloop of M; and

2. (Direct sum) $f(M) = f(M|\{e\})f(M\backslash e)$ if e is a loop or a coloop.

By Lemma 4.126, the Tutte polynomial is a Tutte–Grothendieck invariant. Indeed, as the next theorem establishes, it is, in effect, the universal Tutte–Grothendieck invariant. While this result was first published by Brylawski [251], it is essentially in Tutte's Ph.D. thesis [1091] (see Chapter 34).

Theorem 4.131. *Let f be a function from the set of matroids into the polynomial ring $\mathbb{Z}[x,y]$ satisfying the following conditions.*

1. *$f(M) = f(N)$ if $M \cong N$.*

2. *$f(U_{1,1}) = x$ and $f(U_{0,1}) = y$.*

3. *If e is an element of a matroid M and e is not a loop or a coloop, then*

$$f(M) = f(M\backslash e) + f(M/e).$$

4. *If e is a loop or a coloop of a matroid M, then*

$$f(M) = f(M|\{e\})f(M\backslash e).$$

Then, for all matroids M,

$$f(M) = T(M; x, y).$$

Furthermore, let $\mathfrak{R}$ be a commutative ring with unity and let g be any function from the class of non-empty matroids into $\mathfrak{R}$. If g satisfies Items 1, 3, and 4, whenever $|E(M)| \geq 2$, then, for all non-empty matroids M,

$$g(M) = T(M; g(U_{1,1}), g(U_{0,1})).$$

If the deletion–contraction formula is replaced by a more general recursion, then a straightforward modification of the last theorem holds. Welsh [1133] called this result [904] the *recipe theorem* and that name has persisted.

Corollary 4.132. *Let σ and τ be elements of a field $\mathbb{F}$. Let $\mathcal{K}$ be a class of matroids that is closed under isomorphism and the taking of minors. Let f be an isomorphism invariant mapping $\mathcal{K}$ into $\mathbb{F}$. Suppose f satisfies the following recursions for all matroids M and all elements e of M.*

1. *$f(M) = \sigma f(M\backslash e) + \tau f(M/e)$ if e is neither a loop nor a coloop of M; and*

2. *$f(M) = f(M|\{e\})f(M\backslash e)$ if e is a loop or a coloop of M.*

Then

$$f(M) = \sigma^{|E(M)|-r(M)}\tau^{r(M)}T(M; \tau^{-1}f(U_{1,1}), \sigma^{-1}f(U_{0,1})).$$

To interpret the last result when, for example, $\tau = 0$, one does the calculation initially ignoring this fact. The multiplicative term $\tau^{r(M)}$ will cancel out all of the negative powers of τ. Only after this is done should one make the substitution for τ.

Equation (4.2) prompts a question as to what the coefficients t_{ij} actually count. To answer this, we need the following.

Definition 4.133. Let B be a basis of a matroid M and assume that there is a linear order on $E(M)$. An element e of $E(M) \setminus B$ is *externally active* in B if e is the least element in $C(e, B)$, the unique circuit of M contained in $B \cup \{e\}$; otherwise e is *externally inactive.* Dually, an element f of B is *internally active* if f is the least element in $C_{M^*}(f, E(M) \setminus B)$; otherwise f is *internally inactive.* Let $\mathrm{IA}(B)$ and $\mathrm{EA}(B)$ denote the sets of internally active and externally active elements of B.

Example 4.134. Consider the Fano matroid F_7 labelled as in Figure 4.9. Then every 3-element subset of $\{1, 2, \ldots, 7\}$ is a basis except for the seven sets of three collinear points. Since $\binom{7}{3} - 7 = 28$, this is the number of bases in F_7. Clearly $\{1, 2, 3\}$ is a basis. With respect to this basis, since $1, 2$, and 3 are the three lowest element labels, all of the elements of $\{4, 5, 6, 7\}$ are externally inactive, while all of the elements of $\{1, 2, 3\}$ are internally active. By comparison, for the basis $\{1, 3, 5\}$, the element 2 is externally active, while 4, 6, and 7 are externally inactive. To find the internally active and inactive elements with respect to the basis $\{1, 3, 5\}$, we consider the basis $\{2, 4, 6, 7\}$ of F_7^* that is the complement of $\{1, 3, 5\}$. When we add an element f from $\{1, 3, 5\}$ to this set, we find a unique cocircuit C_f^* of F_7 contained in $\{2, 4, 6, 7\} \cup \{f\}$. Equivalently, F_7 has a unique hyperplane that contains $\{1, 3, 5\} \setminus \{f\}$. The complement of this hyperplane is C_f^*. Thus 1 is internally active while 3 and 5 are internally inactive. The Tutte polynomial of F_7, calculated by Lemma 4.126, is

$$T(F_7; x, y) = x^3 + 4x^2 + 3x + 7xy + 3y + 6y^2 + 3y^3 + y^4. \tag{4.3}$$

The coefficient of the x^3-term here is 1. It corresponds to the basis $\{1, 2, 3\}$, the unique basis with 3 internally active elements and 0 externally active elements. The basis $\{1, 3, 5\}$, which has one internally active element and one externally active element contributes 1 to the coefficient of xy.

The general principle being exemplified above is captured in the next result.

Theorem 4.135. *Let M be a matroid. Put a linear order on the elements of $E(M)$. Then*

$$T(M; x, y) = \sum_{B \in \mathcal{B}(M)} x^{|\mathrm{IA}(B)|} y^{|\mathrm{EA}(B)|}.$$

Thus the coefficient t_{ij} of $x^i y^j$ is the number of bases of M with i internally active elements and j externally active elements.

Observe that this theorem implies that a count of the number of bases of M with i internally active elements and j externally active elements will not depend on the particular ordering imposed on the elements of M.

Definition 4.136. The *characteristic polynomial* $\chi(M;\lambda)$ of a matroid M is

$$\chi(M;\lambda) = \sum_{A \subseteq E(M)} (-1)^{|A|}\lambda^{r(M)-r(A)}. \tag{4.4}$$

This polynomial is also known as the *chromatic polynomial* of M.

The next theorem follows directly by substituting into Definition 4.122.

Theorem 4.137. *For a matroid M,*

$$\chi(M;\lambda) = (-1)^{r(M)}T(M;1-\lambda,0).$$

In particular, if e is neither a loop nor a coloop of M, then

$$\chi(M;\lambda) = \chi(M\backslash e;\lambda) - \chi(M/e;\lambda).$$

From (4.3), we find that

$$\chi(F_7;\lambda) = (\lambda-1)(\lambda-2)(\lambda-4). \tag{4.5}$$

Corollary 4.138. *If a matroid M has a loop, then $\chi(M;\lambda) = 0$; otherwise $\chi(M;\lambda) = \chi(\mathrm{si}(M);\lambda)$.*

Crapo and Rota [350] proved the following remarkable theorem. In view of the last corollary, this theorem is stated for simple matroids. Recall that a linear functional on a vector space is a linear map from the vector space into its underlying field.

Theorem 4.139. *Let M be a simple $\mathbb{F}_q$-representable matroid and A be an $r \times n$ matrix over $\mathbb{F}_q$ for which $M = M[A]$. Let $N(M[A],q,k)$ be the number of k-tuples $(f_1, f_2, \ldots, f_k)$ of linear functionals on $V(r,\mathbb{F}_q)$ such that there is no element e of $E(M[A])$ for which $f_i(e) = 0$ for all i. Then*

$$N(M[A],q,k) = \chi(M;q^k).$$

The striking feature of the last theorem is that $N(M[A],q,k)$ is independent of the matrix A that represents M. As noted in Example 4.105, the Fano matroid F_7 equals the projective geometry $PG(2,2)$. By using the last theorem and the fact that the kernel of a non-trivial linear functional on $V(r,\mathbb{F}_q)$ is a hyperplane of $V(r,\mathbb{F}_q)$, one can show that the pattern that is apparent in (4.5) holds for all projective geometries. Specifically,

$$\chi(PG(r-1,q);\lambda) = (\lambda-1)(\lambda-q)(\lambda-q^2)\ldots(\lambda-q^{r-1}). \tag{4.6}$$

As the next result notes, the chromatic polynomial $\chi(G;\lambda)$ of a graph G and the characteristic polynomial $\chi(M(G);\lambda)$ of its cycle matroid differ only by a multiplicative factor.

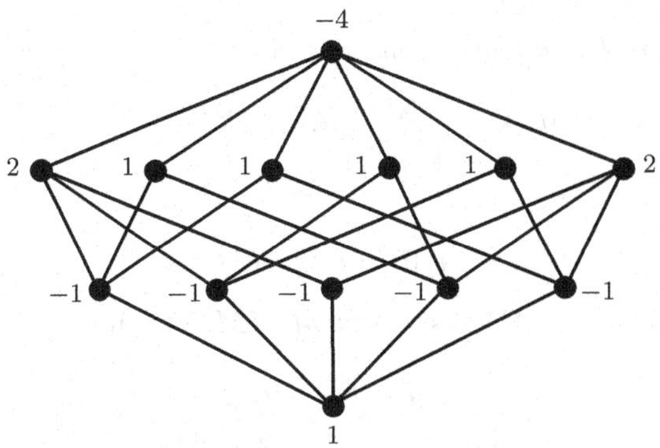

FIGURE 4.15: Each flat F is labelled by $\mu(\mathrm{cl}(\emptyset), F)$.

Theorem 4.140. *For a graph G having $k(G)$ connected components,*

$$\chi(G; \lambda) = \lambda^{k(G)} \chi(M(G); \lambda).$$

An immediate consequence of this theorem is that, for all $n \geq 2$,

$$\chi(M(K_n); \lambda) = (\lambda - 1)(\lambda - 2)(\lambda - 3) \ldots (\lambda - (n-1)). \tag{4.7}$$

The Möbius function, originally introduced in 1832 as a tool in number theory, has been generalized to partially ordered sets. This generalization was examined in detail by Rota [970]. We shall focus here on the partially ordered sets that are the lattice of flats of some matroid.

Definition 4.141. For a matroid M having lattice of flats $\mathcal{L}(M)$, the *Möbius function* is the map μ from $\mathcal{L}(M) \times \mathcal{L}(M)$ into $\mathbb{Z}$ that is defined recursively by setting $\mu(X, X) = 1$ for all X and

$$\mu(X, Z) = -\sum_{X \subseteq Y \subsetneq Z} \mu(X, Y) \text{ whenever } X \subseteq Z.$$

Example 4.142. For $M(H)$ where H is the graph in Figure 4.6, the lattice of flats is shown in Figure 4.7. In Figure 4.15, each flat F of $M(H)$ has been labelled by the value of $\mu(\mathrm{cl}(\emptyset), F)$. Because the definition of the Möbius function of a matroid M depends only on the lattice of flats of M, its value is unchanged if we replace M by $\mathrm{si}(M)$ and perform the natural relabelling of the flats. In particular, after accounting for this labelling change, the Möbius functions of $M(H)$ and $M(H)\backslash\{4,7\}$ coincide. As one can easily check, $\chi(M(H)\backslash\{4,7\}; \lambda) = (\lambda - 1)(\lambda - 2)^2$ so, letting $E = E(M(H)\backslash\{4,7\})$, we have $\mu(\emptyset, E) = -4 = \chi(M(H)\backslash\{4,7\}; 0)$.

As the next theorem notes, the last identity in this example holds in general.

Theorem 4.143. *For a loopless matroid M,*

$$\chi(M;\lambda) = \sum_{F\in\mathcal{L}(M)} \mu(\emptyset, F)\lambda^{r(M)-r(F)}.$$

In particular,

$$\chi(M;0) = \mu(\emptyset, E(M)).$$

Definition 4.144. The *Möbius invariant* $\mu(M)$ of a matroid M is

$$\mu(M) = \begin{cases} \mu(\emptyset, E(M)), & \text{if } M \text{ is loopless;} \\ 0, & \text{otherwise.} \end{cases} \tag{4.8}$$

It follows immediately from this definition and Theorem 4.143 that, for all matroids M,

$$\mu(M) = \chi(M;0) = (-1)^{r(M)}T(M;1,0). \tag{4.9}$$

Next we define another polynomial that is equivalent to the Tutte polynomial. The coboundary polynomial was introduced by Crapo [349] and can be used [173] to lift some results on characteristic polynomials to Tutte polynomials.

Definition 4.145. The *coboundary polynomial* $\overline{\chi}(M;x,y)$ of a matroid M is

$$\overline{\chi}(M;x,y) = \sum_{A\subseteq E(M)} y^{|A|}\chi(M/A;x) = \sum_{F\in\mathcal{L}(M)} y^{|F|}\chi(M/F;x). \tag{4.10}$$

The following formula, which is well-known and easy to prove, establishes the equivalence between the Tutte and coboundary polynomials.

$$T(M;x,y) = \frac{\overline{\chi}(M;(x-1)(y-1),y)}{(y-1)^{r(M)}} \tag{4.11}$$

From Example 4.79, we know that the non-Fano matroid F_7^- is obtained from the Fano matroid F_7 by taking the set $\{4,5,6\}$ that is both a circuit and a hyperplane of F_7 and declaring it to be a basis; that is, $\mathcal{B}(F_7^-) = \mathcal{B}(F_7)\cup\{\{4,5,6\}\}$. This process always produces a matroid.

Proposition 4.146. *Let M be a matroid and C be a subset of $E(M)$ that is both a circuit and a hyperplane of M. Then there is a matroid M' with ground set $E(M)$ whose set of bases is $\mathcal{B}(M)\cup\{C\}$.*

Definition 4.147. The matroid M' in the last proposition is said to have been obtained from M by *relaxing the circuit-hyperplane C*.

For comparison with (4.3), we note that the Tutte polynomial of the non-Fano matroid is given by

$$T(F_7^-;x,y) = x^3 + 4x^2 + 4x + 6xy + 4y + 6y^2 + 3y^3 + y^4. \tag{4.12}$$

Thus $T(F_7^-;x,y) = T(F_7;x,y) - xy + x + y$. Indeed, this fact generalizes.

Proposition 4.148. *If the matroid M' is obtained from a matroid M by relaxing a circuit-hyperplane, then*

$$T(M';x,y) = T(M;x,y) - xy + x + y.$$

The next theorem gives a formula for the Tutte polynomial of the 2-sum of two matroids where, for notational convenience, we have omitted the variables x and y. This formula appears explicitly in Oxley and Welsh [905] but is an easy consequence of formulas derived by Brylawski [250].

Theorem 4.149. *Let the matroid M be the 2-sum of matroids M_1 and M_2 with respect to the basepoint p. Then*

$$T(M) = \frac{1}{xy-x-y}[(x-1)T(M_1/p)T(M_2/p) + (y-1)T(M_1\backslash p)T(M_2\backslash p) - T(M_1\backslash p)T(M_2/p) - T(M_1/p)T(M_2\backslash p)].$$

The next theorem is a result of Kook, Reiner, and Stanton [704] that they call a *convolution formula*. This theorem was implicit in Etienne and Las Vergnas [469].

Theorem 4.150. *For a matroid M with ground set E,*

$$T(M;x,y) = \sum_{A\subseteq E} T(M|A;0,y)T(M/A;x,0).$$

4.8 Some particular evaluations

The recipe theorem gives us a way of identifying certain statistics associated with a matroid as being evaluations of its Tutte polynomial. We begin this section with an example of how this theorem can be used. The rest of the section interprets evaluations of the Tutte polynomial at a number of points in the (x,y)-plane.

Theorem 4.151. *The number of bases of a matroid M is $T(M;1,1)$.*

Proof. Take e to be an element of M. Partition the set $\mathcal{B}(M)$ of bases of M into $\mathcal{B}'$, the set of bases that avoid e, and $\mathcal{B}''$, the set of bases that contain e. Clearly

$$|\mathcal{B}(M)| = |\mathcal{B}'| + |\mathcal{B}''|.$$

First suppose that e is not a coloop of M. Then $r(M\backslash e) = r(M)$ and

$$\mathcal{B}' = \mathcal{B}(M\backslash e).$$

Now suppose, instead, that e is not a loop. Then $\{e\}$ is a basis of $M|\{e\}$. Thus, by Theorem 4.75, $\mathcal{B}(M/e) = \{B' \subseteq E\setminus\{e\} : B' \cup \{e\} \in \mathcal{B}(M)\}$. Hence

$$|\mathcal{B}(M/e)| = |\mathcal{B}''|.$$

We conclude that, when e is neither a loop nor a coloop,

$$|\mathcal{B}(M)| = |\mathcal{B}(M\backslash e)| + |\mathcal{B}(M/e)|. \tag{4.13}$$

Evidently,

$$|\mathcal{B}(U_{1,1})| = 1 = |\mathcal{B}(U_{0,1})|. \tag{4.14}$$

When e is a loop, no basis of M contains it, so $|\mathcal{B}(M)| = |\mathcal{B}(M\backslash e)|$. Dually, when e is a coloop, every basis of M contains it and, again, $|\mathcal{B}(M)| = \mathcal{B}(M\backslash e)|$. Thus, when e is a loop or a coloop of M,

$$|\mathcal{B}(M)| = |\mathcal{B}(M|\{e\})| \cdot |\mathcal{B}(M\backslash e)|. \tag{4.15}$$

By (4.13)–(4.15) and Corollary 4.132, we immediately obtain the theorem. □

The last theorem can also be proved directly from Definition 4.122 as follows. When the sum for $T(M;1,1)$ is evaluated, the only subsets A of E that make a contribution are those for which $r(E) = r(A)$ and $|A| = r(A)$. Evidently such sets A are precisely the bases of M. Moreover, recalling that we interpret 0^0 as 1, each such set contributes exactly one to the sum, and the theorem follows.

Each part in the next theorem can be proved by either of the techniques used to prove the last theorem. But the proof in Example 4.153 depends crucially on the recipe theorem.

Theorem 4.152. *For a matroid M with ground set E,*

1. *$T(M;2,1)$ is the number of independent sets of M;*
2. *$T(M;1,2)$ is the number of spanning sets of M; and*
3. *$T(M;2,2) = 2^{|E|}$.*

Example 4.153. Let M be a matroid with ground set E. Suppose that every element of M has, independently of all other elements, a probability $1-p$ of being deleted from M, and assume that $0 < p < 1$. The resulting restriction $\omega(M)$ of M is called a *random submatroid* of M. When M is the cycle matroid of the complete graph K_n, this corresponds in the obvious way to construct a random graph on n vertices. Let $\Pr(M)$ denote the probability that $\omega(M)$ has the same rank as M. Clearly $\Pr(U_{1,1}) = p$ and $\Pr(U_{0,1}) = 1$. Moreover,

$$\Pr(M) = \begin{cases} (1-p)\Pr(M\backslash e) + p\Pr(M/e) & \text{if } e \text{ is neither a loop nor a coloop;} \\ \Pr(M|\{e\})\Pr(M\backslash e) & \text{otherwise.} \end{cases}$$

It follows by Corollary 4.132 that

$$\Pr(M) = (1-p)^{|E|-r(M)} p^{r(M)} T(M; 1, \tfrac{1}{1-p}).$$

In the next example, we consider an isomorphism invariant that obeys the deletion–contraction recursion for elements that are not loops or coloops but does not obey the multiplicative formula for elements that are loops or coloops.

Example 4.154. For a matroid M, let i_{r-k} be the number of independent sets of M having $r(M)-k$ elements. Suppose e is an element of M that is neither a loop nor a coloop. Partition the set of $(r(M)-k)$-element independent sets of M into the set $\mathcal{I}'_{r(M)-k}$, consisting of those such sets avoiding e, and $\mathcal{I}''_{r(M)-k}$, those such sets that contain e. Clearly $|\mathcal{I}'_{r(M)-k}| = i_{r(M\backslash e)-k}$. Moreover, for a subset I of $E(M) \setminus \{e\}$, we see that $I \cup \{e\} \in \mathcal{I}''_{r(M)-k}$ if and only if I is an $(r(M/e)-k)$-element independent set of M/e. Thus $|\mathcal{I}''_{r(M)-k}| = i_{r(M/e)-k}$. We deduce that

$$i_{r-k}(M) = i_{r-k}(M\backslash e) + i_{r-k}(M/e).$$

But, for $k = 2$, say, $U_{1,1}$ and $U_{0,1}$ have no $(r(M)-k)$-element independent sets, so the multiplicative recursion does not hold.

Although the invariant arising in the last example is not an evaluation of the Tutte polynomial itself, we can obtain it as a function of the coefficients of that polynomial.

Theorem 4.155. *Let* A *be an abelian group. A function* g *from the class of non-empty matroids into* A *is an isomorphism invariant for which*

$$g(M) = g(M\backslash e) + g(M/e) \text{ provided } e \text{ is not a loop or a coloop of } M$$

if and only if

$$g(M) = \sum_i \sum_j t_{ij}\alpha_{ij}$$

where $T(M;x,y) = \sum_i \sum_j t_{ij}x^i y^j$ *and* $g(U_{i,i} \oplus U_{0,j}) = \alpha_{ij}$ *for all* i *and* j *such that* $i + j > 0$.

Corollary 4.156. *For a matroid* M*, the number* i_{r-k} *of independent sets of* M *having* $r(M)-k$ *elements is given by*

$$i_{r-k} = \sum_i \sum_j t_{ij}\binom{i}{k}.$$

The next result lists some more of the properties of a matroid that one can determine from its Tutte polynomial. Following that is an example that indicates some of the features of a matroid that one cannot determine from its Tutte polynomial.

Theorem 4.157. *For a matroid* M*, let* $T(M;x,y) = \sum t_{ij}x^i y^j$*. Then*

1. $r(M)$ *is the highest power of* x *in* $T(M;x,y)$*;*

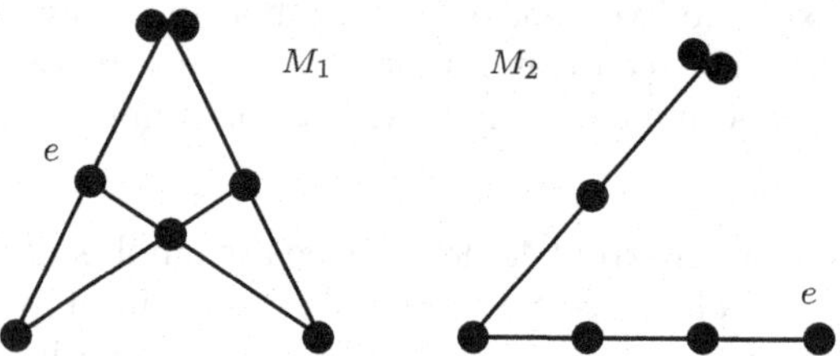

FIGURE 4.16: Two non-isomorphic matroids with the same Tutte polynomial.

2. *$r^*(M)$ is the highest power of y in $T(M; x, y)$;*

3. *if $t_{r(M),t} > 0$, then t is the number of loops of M and $t_{r(M),t} = 1$;*

4. *if $t_{s,r^*(M)} > 0$, then s is the number of coloops of M and $t_{s,r^*(M)} = 1$;*

5. *the number of rank-1 flats of M is $r(M)+t_{(r(M)-1),t}$ where t is the number of loops of M; and*

6. *provided M has no loops and no coloops, the number of connected components of M is $\min\{j : t_{0j} > 0\} = \min\{i : t_{i0} > 0\}$. In general, provided M is non-empty, the number of connected components of M is $\min\{i + j : t_{ij} > 0\}$.*

Example 4.158. Consider the 7-element rank-3 matroids M_1 and M_2 for which geometric representations are shown in Figure 4.16. Clearly $M_1\backslash e = M_2\backslash e$ and $M_1/e = M_2/e$. Thus, by Lemma 4.126, $T(M_1; x, y) = T(M_2; x, y)$. Observe, however, that these two matroids differ in the following ways.

- M_1 is isomorphic to the cycle matroid of a planar graph so M_1 is graphic, cographic, regular, and binary. However, M_2 has none of these properties.

- M_2 is transversal but M_1 is not even a gammoid.

- Although M_1 and M_2 have the same number of rank-1 flats and the same number of 4-element and 3-element rank-2 flats, M_1 has two 2-element rank-2 flats, whereas M_2 has three. Hence M_1 and M_2 have different numbers of flats and different numbers of hyperplanes.

- M_1 and M_2 each have one 2-element circuit and six 3-element circuits, but M_2 has five 4-element circuits while M_2 has six such circuits. Thus M_1 and M_2 have different numbers of circuits.

By comparing the known lower bound on the number of non-isomorphic n-element matroids with the crude upper bound of $n^2 2^{n^3}$ on the number of possible Tutte polynomials of such matroids, we get the following result [245].

Proposition 4.159. *For each positive integer k, there are at least k non-isomorphic matroids that all have the same Tutte polynomial.*

4.9 Some basic identities

In this section, we make some basic observations about the coefficients t_{ij} of the Tutte polynomial. In the Tutte polynomials of $U_{2,4}$, F_7, and F_7^- in Example 4.123, (4.3), and (4.12), we observe that $t_{1,0} = t_{0,1}$. This is not true for $U_{0,1}$ or $U_{1,1}$ but these are the only exceptions.

Proposition 4.160. *For a matroid M with more than one element,*

$$t_{10} = t_{01}.$$

Moreover, M is connected if and only if t_{10} is positive.

In the next proposition of Brylawski [251], we use $t_{ij}(M)$ to denote the coefficient t_{ij} of $x^i y^j$ in the Tutte polynomial of M.

Proposition 4.161. *Let M be a connected matroid.*

1. *For a minor N of M and all nonnegative integers i and j with $i + j > 0$,*

$$t_{ij}(N) \leq t_{ij}(M).$$

2. *If $t_{ij}(M) > 0$, then $t_{i'j'}(M) > 0$ for all $(i', j') \neq (0, 0)$ with $i' \leq i$ and $j' \leq j$.*

Example 4.162. To see that the last proposition can fail when M is not connected, we observe that $T(U_{1,1} \oplus U_{0,1}; x, y) = xy$. Thus $t_{10}(U_{1,1} \oplus U_{0,1}) = 0$. But $U_{1,1}$ is a minor of $U_{1,1} \oplus U_{0,1}$, yet $t_{10}(U_{1,1}) = 1$.

The identity in Proposition 4.160 is a special case of the following general identity involving coefficients of the Tutte polynomial, which was proved by Brylawski [251].

Theorem 4.163. *For all matroids M with $|E(M)| > k$,*

$$\sum_{i=0}^{k} \sum_{j=0}^{k-i} (-1)^j \binom{k-i}{j} t_{ij} = 0.$$

In particular,

1. $t_{00} = 0$ *if* $|E(M)| \geq 1$;
2. $t_{10} = t_{01}$ *if* $|E(M)| \geq 2$;
3. $t_{20} - t_{11} + t_{02} = t_{10} = 0$ *if* $|E(M)| \geq 3$; *and*
4. $t_{30} - t_{21} + t_{12} - t_{03} = t_{11} - 2t_{02} + t_{10}$ *if* $|E(M)| \geq 4$.

5

Tutte polynomial activities

Spencer Backman

Synopsis

Activities are certain statistics associated to spanning forests and more general objects which can be used for defining the Tutte polynomial. This chapter is intended to serve as an introduction to activities for graphs and matroids. We describe various activity expansions of the Tutte polynomial.

- Tutte's original spanning forest activities.
- Gordon and Traldi's subgraph activities.
- Gessel and Sagan's depth-first search activities.
- Bernardi's embedding activities.
- Gordon and McMahon's generalized subgraph activities.
- Las Vergnas' orientation activities.
- Etienne and Las Vergnas' activity bipartition.
- Crapo's activity interval decomposition.
- Las Vergnas' active orders.
- Shellability and the algebraic combinatorics of activities.

5.1 Introduction

Unlike Whitney's definition of the corank–nullity generating function (Definition 2.3) Tutte's definition (Definition 2.16) of $T(G; x, y)$ requires a total

DOI: 10.1201/9780429161612-5

order on the edges of which the polynomial is a posteriori independent. Tutte presented his definition in terms of *internal and external activities* of maximal spanning forests. Although this definition may appear somewhat ad hoc upon first inspection, subsequent work by various researchers has demonstrated that activity is a deep combinatorial concept. In this chapter, we provide an introduction to activities for graphs and matroids. We describe several notions of activity for graphs which admit expansions of the Tutte polynomial. Additionally, we describe some fundamental structural theorems, and outline connections to the topological notion of shellability as well as several topics in algebraic combinatorics.

We use the language of graphs except in Sections 5.9 and 5.10 where matroid terminology is employed, although Sections 5.2–5.4, and 5.7 apply equally well to matroids, and Section 5.8 applies to oriented matroids.

5.2 Activities for maximal spanning forests

We recall Whitney's original definition of the Tutte polynomial [1154]. (See Chapter 34 for the history of the Tutte polynomial.) The rank of a subset of edges A of a graph G, written $r(A)$, is the maximum cardinality of a forest contained in A.

Definition 5.1. If $G = (V, E)$ is a graph, then the Tutte polynomial of G is

$$T(G; x, y) = \sum_{A \subseteq E} (x-1)^{r(E)-r(A)} (y-1)^{|A|-r(A)}. \tag{5.1}$$

Let $G = (V, E)$ be a graph and F be a maximal spanning forest of G. The maximality of F means that if G is connected, then F is a spanning tree, and if G is not connected, restricting F to any component of G gives a spanning tree of that component.

Definition 5.2. Let F be a maximal spanning forest of G, $f \in F$, and $e \in E \setminus F$. The *fundamental cut* associated with F and e is

$$U_F(e) = \{\text{the edges of the unique cut in } (E \setminus F) \cup e\}.$$

Similarly, the *fundamental cycle* associated with F and e is

$$Z_F(e) = \{\text{the edges of the unique cycle in } F \cup e\}.$$

This definition is modified for matroids by replacing "cut" with "cocircuit" in the definition of $U_F(e)$, and "cycle" with "circuit" in the definition of $Z_F(e)$. We now describe Tutte's activities using fundamental cuts and cycles.

Definition 5.3. Let $G = (V, E)$ be a graph with a total order on E, and F be a maximal spanning forest of G. We say that an edge $e \in E$ is

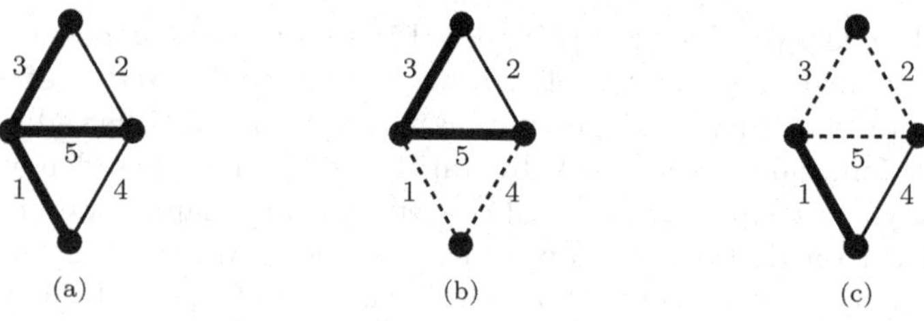

FIGURE 5.1: Graphs used in Example 5.4.

1. *internally active* ($e \in \mathrm{IA}(F)$) with respect to F if $e \in F$, and it is the smallest edge in $U_F(e)$,
2. *externally active* ($e \in \mathrm{EA}(F)$) with respect to F if $e \notin F$, and it is the smallest edge in $Z_F(e)$.

Observe that all bridges are internally active and all loops are externally active.

Example 5.4. Figure 5.1a shows a graph with a total order on the edges and a spanning tree T (given by the bold edges). The edge 1 is the only internally active edge of T, and its fundamental cut is shown as the dashed edges in Figure 5.1b. The edge 2 is the only externally active edge of T, and its fundamental cycle is shown as the dashed edges in Figure 5.1c.

We can now give the *spanning tree (maximal spanning forest) activities expansion* of the Tutte polynomial, which is due to Tutte [1092].

Definition 5.5. If $G = (V, E)$ is a graph with a fixed total order of E, then

$$T(G; x, y) = \sum_F x^{|\mathrm{IA}(F)|} y^{|\mathrm{EA}(F)|}, \tag{5.2}$$

where the sum is over all maximal spanning forests of G.

Tutte demonstrated that this polynomial is well defined, i.e., it is independent of the total order on the edges. An example of using Definition 5.5 to compute a Tutte polynomial can be found in Example 2.13.

5.3 Activity bipartition

Etienne and Las Vergnas showed that the activities of a maximal spanning forest induce a canonical bipartition of the edge set of a graph or more generally a matroid. First we recall the definition of a flat.

Definition 5.6. Let $G = (V, E)$ be a graph and $\mathcal{F} \subseteq E$. If there exists no $e \in E \setminus \mathcal{F}$ such that e is contained in a cycle in $e \cup \mathcal{F}$, we say that $\mathcal{F}$ is a *flat* of G. A flat is *cyclic* if it is a union of cycles.

Theorem 5.7. *Given a maximal spanning forest F of G, there exists a unique cyclic flat $\mathcal{F}$ of G such that $\mathcal{F} \cap F$ is a maximal spanning forest of $G|\mathcal{F}$ with no internal activity, and $\mathcal{F}^c \cap F$ is a maximal spanning forest of $G/\mathcal{F}$ with no external activity.*

As an application of this decomposition, due to Etienne and Las Vergnas [469], one may obtain a convolution formula for the Tutte polynomial which was independently discovered by Kook, Reiner, and Stanton via incidence algebra methods in [704], and has since been substantially refined and generalized [62, 424, 532, 735, 959].

Theorem 5.8. *Given a graph G, then*

$$T(G; x, y) = \sum_{\mathcal{F}} T(G/\mathcal{F}; x, 0)\, T(G|\mathcal{F}; 0, y),$$

where the sum is over all cyclic flats of G.

5.4 Activities for subgraphs

Gordon and Traldi [573] introduced a notion of activities for arbitrary subgraphs, and used this to provide a 4-variable expansion of the Tutte polynomial which naturally specializes to both Whitney's and Tutte's original expansions.

Definition 5.9. Let $G = (V, E)$ be a graph with a total order on E, let $S \subseteq E$ and $S^c := E \setminus S$. Then an edge e is:

1. *internally active present* with respect to S, written $e \in \mathrm{IAP}(S)$, if $e \in S$ and e is the smallest edge in some cut in $S^c \cup e$;

2. *internally active absent* with respect to S, written $e \in \mathrm{IAA}(S)$, if $e \notin S$ and e is the smallest edge in some cut in S^c;

3. *externally active present* with respect to S, written $e \in \mathrm{EAP}(S)$, if $e \in S$ and e is the smallest edge in some cycle in S;

4. *externally active absent* with respect to S, written $e \in EAA(S)$, if $e \notin S$ and e is the smallest edge in some cycle in $S \cup e$.

Theorem 5.10. *If $G = (V, E)$ is a graph with a fixed total order of E, then*

$$T(G; x + w, y + z) = \sum_{S \subseteq E} x^{|\mathrm{IAP}(S)|} w^{|\mathrm{IAA}(S)|} y^{|\mathrm{EAP}(S)|} z^{|\mathrm{EAA}(S)|}. \tag{5.3}$$

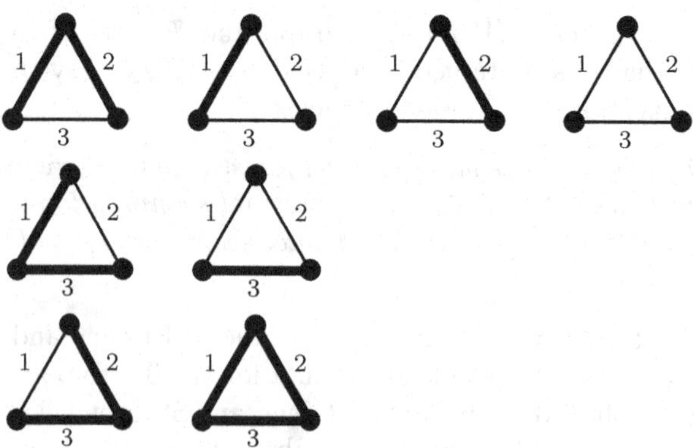

FIGURE 5.2: A Crapo decomposition of the subgraphs of K_3.

By setting $x = 1$ and $z = 1$, we recover Whitney's definition, and by setting $w = 0$ and $y = 0$, we recover Tutte's definition.

While Gordon and Traldi's expansion is proven recursively via deletion–contraction, from which they obtain more general formulas, the 4-variable expansion is equivalent to an earlier theorem of Crapo from [349].

Theorem 5.11. *Suppose $G = (V, E)$ is a graph with a fixed total order of E. Let $P(E)$ be the Boolean lattice of subgraphs of G ordered by containment. Given a spanning forest F, define an interval in this lattice $[F\backslash \mathrm{IA}(F), F \cup \mathrm{EA}(F)]$. Then*

$$P(E) = \bigsqcup_{F} [F\backslash \mathrm{IA}(F), F \cup \mathrm{EA}(F)]$$

where the disjoint union is over all maximal spanning forests.

Figure 5.2 illustrates the this decomposition. In the figure, each row collects the subgraphs (given by the bold edges) corresponding to the spanning tree on the left side of the row.

In Sections 5.5–5.7 we describe other notions of activity for maximal spanning forests which employ input data different from a total order on the edges, but still allow for expansions of the Tutte polynomial and an analogue of Crapo's interval decomposition.

5.5 Depth-first search external activity

In [520], Gessel and Sagan introduced a notion of external activity for maximal spanning forests based on depth-first search (DFS). For simplicity sake, we will assume that our graph is connected and has no parallel edges. In what

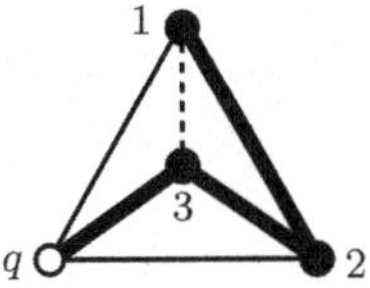

FIGURE 5.3: K_4 with a root q and a total order on its vertices. A spanning tree is shown in bold, and a DFS externally active edge is shown as dashed.

follows, we view a tree rooted at a vertex q to be oriented so that every vertex is reachable from q by a directed path. DFS externally activity, defined below, is illustrated in Figure 5.3.

Definition 5.12. Let $<$ be a total order on the vertices of G, and F be a spanning tree of G rooted at the smallest vertex q. Let $e = (u, v)$ be an edge of $G \backslash F$. We say that e is *depth-first search externally active (DFS externally active)*, and write $e \in \mathrm{EA}_{\mathrm{DFS}}(F)$, if either $u = v$, or (u, w) is an oriented edge in F belonging to the unique cycle in $F \cup (u, v)$, and $w > v$.

The name DFS externally active is justified by the following observation. Given a spanning subgraph of G, we can produce a spanning forest F by performing a DFS search which favors larger labeled vertices. Then (u, v) is DFS externally active if when we apply DFS search to the graph $F \cup (u, v)$, we obtain F. Gessel and Sagan [520] showed that DFS external activity when combined with Tutte's notion of internal activity allows for an expansion of the Tutte polynomial.

Theorem 5.13. *If G is a connected graph with a total order of its vertices, then*

$$T(G; x, y) = \sum_{T} x^{|\mathrm{IA}(T)|} y^{|\mathrm{EA}_{\mathrm{DFS}}(T)|}, \tag{5.4}$$

where the sum is over all spanning trees T of G.

We note that in the same article, Gessel and Sagan introduced a notion of neighbors-first search activity.

5.6 Activities via combinatorial maps

Bernardi, in [99], defined an activity induced by a rooted combinatorial map, which is essentially an embedding of a graph $G = (V, E)$ with a distinguished half-edge h into an oriented surface. In what follows, we assume that G is connected and loopless, although these restrictions are not essential.

Informally, given a spanning tree T of G, we can use a rooted combinatorial map to tour the edges of G by starting at h and traveling against the

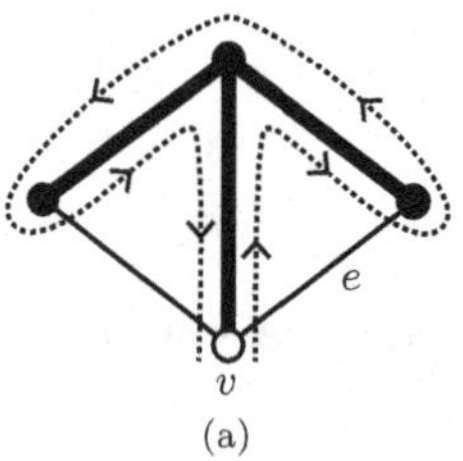

(a)

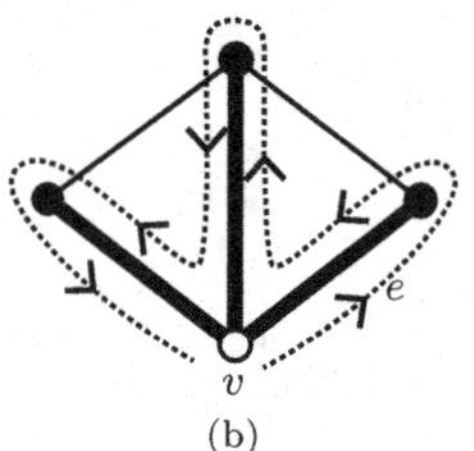

(b)

FIGURE 5.4: Tours of two different spanning trees induced by the motion operator associated to the same rooted combinatorial map.

orientation around the outside of T. We then declare an edge $e \notin T$ to be externally active if it is the first edge in its associated fundamental cycle which we meet during the tour. We similarly say an edge $e \in T$ is internally active if it is the first edge in its associated fundamental cut which we meet during the tour.

We now describe Cori's maps [331] and Bernardi's activities more formally. We define a *half-edge* to be an edge and an incident vertex, e.g., if $e \in E$ with $e = (u, v)$ such that $u, v \in V$, then (e, u) and (e, v) are the two associated half-edges. Let σ be a permutation of the half-edges of G such that $\sigma((e, u)) = (e', u)$ some other half-edge incident to u, and for any two half-edges (e, u) and (e', u) with the same endpoint, there exists some $k > 0$ such that $\sigma^k((e, u)) = (e', u)$. Let h be a distinguished half-edge. We define a *rooted combinatorial map* to be a triple (G, σ, h). Let α be the involution on the set of half-edges such that for all $e = (u, v) \in E$, $\alpha((e, u)) = (e, v)$. Given a half-edge i and a spanning tree T, we define the *motion operator*

$$t(i) := \begin{cases} \sigma(i) & \text{if } i \notin T, \\ \sigma \circ \alpha(i) & \text{if } i \in T. \end{cases}$$

It is easy to check that the iterated motion operator defines a tour of the half-edges of G. This tour induces a total order $<_T$ on the edges given by the first time one of its half-edges is visited. We use this total order to define internally and externally active edges.

Example 5.14. Figure 5.4 shows two copies of the same rooted combinatorial map, and tours of two different spanning trees (shown in bold) induced by the motion operator. In the figure, the root is the vertex v, and (e, v) is the distinguished half-edge.

Definition 5.15. Let (G, σ, h) be a rooted combinatorial map, T a spanning tree of G, and e an edge of G.

1. The edge e is *embedding-internally active*, written $e \in \mathrm{IA}_B(T)$, if $e \in T$ and e is the minimum edge with respect to $<_T$ in its associated fundamental cut.

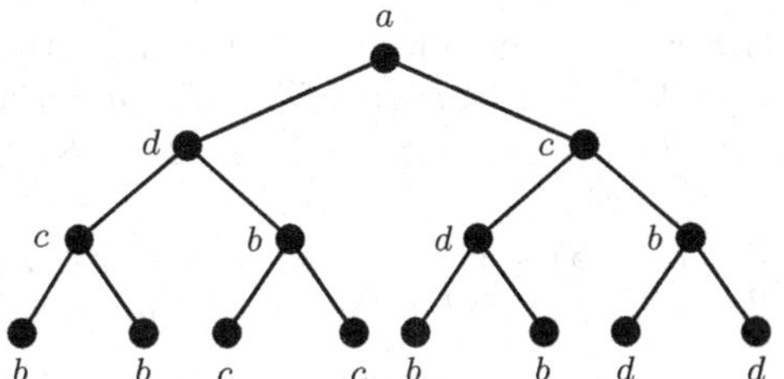

FIGURE 5.5: An example of a decision tree for a graph on 4 edges.

2. The edge e is *embedding-externally active*, written $e \in \mathrm{EA}_B(T))$, if $e \notin T$ and e is the minimum edge with respect to $<_T$ in its associated fundamental cycle.

Bernardi [99] showed that this definition of activity admits an expansion of the Tutte polynomial.

Theorem 5.16. *Let (G, σ, h) be a rooted combinatorial map, then*

$$T(G; x, y) = \sum_{T} x^{|\mathrm{IA}_B(T)|} y^{|\mathrm{EA}_B(T)|}, \tag{5.5}$$

where the sum is over all spanning trees.

We remark that Courtiel recently introduced a different notion of activity via combinatorial maps which he calls the "blossom activity" in [341].

5.7 Unified activities for subgraphs via decision trees

The Gordon–Traldi activities were further generalized by Gordon and McMahon [571] in a way which also applies to greedoids. The Gordon–McMahon notion of activity was rediscovered by Courtiel [341] who showed that it unifies all of the aforementioned types of activity. We describe these activities following Courtiel and using the language of decision trees. (Figure 5.5 provides an example of a decision tree for a graph on four edges.)

Definition 5.17. A *decision tree* D for G consists of a perfect binary tree (i.e., a rooted tree such that all non-leaf nodes have exactly two descendants) and a labeling of each node of the tree by elements of E such that the labels along any particular branch give a permutation of E.

Given a decision tree D and a subgraph $S \subseteq E$, we can use D to partition S into four sets: $IA_D(S)$, $EA_D(S)$, S_E, S_I. We describe the recursive algorithm for producing this partition informally and refer the reader to [341] for a pseudocode presentation.

Algorithm 5.18 (Recursive generalized activities algorithm).
Input: A graph $G = (E, V)$*, a decision tree* D*, and a collection of edges* $S \subseteq E$*.*
Output: a partition of the edge set into four sets $IA_D(S)$*,* $EA_D(S)$*,* S_I*, and* S_E*.*
Initialize with $X = E, \mathrm{IA}_D(S) = \mathrm{EA}_D(S) = S_I = S_E = \emptyset$*, and* e *corresponding to the label of the root of* D*. While* $X \neq \emptyset$*, do the following:*

1. *If* e *is a bridge, add* e *to* $\mathrm{IA}_D(S)$*, contract* e *in* X*, and move to the right descendant of* e *in* D*.*

2. *If* $e \in S$ *is neither a bridge nor a loop, add* e *to* S_I*, contract* e *in* X*, and move to the right descendant of* e *in* D*.*

3. *If* e *is a loop, add* e *to* $\mathrm{EA}_D(S)$*, delete* e *in* X*, and move to the left descendant of* e *in* D*.*

4. *If* $e \notin S$ *is neither a bridge nor a loop, add* e *to* S_E*, delete* e *in* X*, and move to the left descendant of* e *in* D*.*

   After we move to a descendant, we update e *to be the label of the new node and recurse.*

Theorem 5.19. *Let* G *be a graph,* D *a decision tree for* G*, and* $\mathrm{IA}_D(S)$ *and* $\mathrm{EA}_D(S)$ *be the output of Algorithm 5.18, then the Tutte polynomial has the following expansion*

$$T(G; x + w, y + z) = \sum_{S \subseteq E} x^{|S \cap \mathrm{IA}_D(S)|} w^{|S^c \cap \mathrm{IA}_D(S)|} y^{|S^c \cap \mathrm{EA}_D(S)|} z^{|S \cap \mathrm{EA}_D(S)|}.$$

5.8 Orientation activities

A famous result of Stanley states that $T(G; 2, 0)$ (equivalently, the absolute value of the chromatic polynomial evaluated at -1) counts the number of acyclic orientations of G [1027]. This result was generalized to hyperplane arrangements by Zaslavsky [1181], and to oriented matroids by Las Vergnas [745].

Las Vergnas [750] introduced orientation activities, which parallel those of subsets, and allow for an orientation expansion of the Tutte polynomial which recovers Stanley's result. He later introduced refined orientation activities [755], which we now describe. Similar to the way that Tutte's activities are defined in terms of fundamental cuts and cycles, Las Vergnas' orientation activities are defined in terms of directed cuts and cycles. (Additional information on this topic can be found in Chapter 31.)

FIGURE 5.6: A total order and reference orientation for Example 5.22.

Definition 5.20. Let $\mathcal{O}$ be an orientation of the edges of G. Let Z be a cycle in G, and $U = (X, V \backslash X)$, be a cut in G. We say that Z is a *directed cycle* if we can walk around the cycle traveling in the direction of the edge orientations. We similarly define U to be a *directed cut* if all of its edges are, without loss of generality, oriented from X to $V \setminus X$.

Following Las Vergnas, we use directed cuts and cycles to define orientation activities.

Definition 5.21. Let $G = (V, E)$ be a graph with a total order on E, and $\mathcal{O}_{\text{ref}}$ be a reference orientation of the edges of G. If $\mathcal{O}$ is an orientation of G and $e \in E$, then we say e is:

1. *positive cut active,* written $e \in \text{CUA}(\mathcal{O})^+$, if e is the smallest edge in some directed cut and is oriented in agreement with $\mathcal{O}_{\text{ref}}$;

2. *negative cut active,* written $e \in \text{CUA}(\mathcal{O})^-$, if e is the smallest edge in some directed cut and is oriented in disagreement with $\mathcal{O}_{\text{ref}}$;

3. *positive cycle active,* written $e \in \text{CYA}(\mathcal{O})^+$, if e is the smallest edge in some directed cycle and is oriented in agreement with $\mathcal{O}_{\text{ref}}$;

4. *negative cycle active,* written $e \in \text{CYA}(\mathcal{O})^-$, if e is the smallest edge in some directed cycle and is oriented in disagreement with $\mathcal{O}_{\text{ref}}$.

Example 5.22. Figure 5.6a shows a total order and reference orientation of the edges of a graph G. Using this data, Figure 5.6b shows an orientation $\mathcal{O}$ of G with $\text{CUA}(\mathcal{O})^+ = \emptyset, \text{CUA}(\mathcal{O})^- = \{1\}, \text{CYA}(\mathcal{O})^+ = \{2\}$, and $\text{CYA}(\mathcal{O})^- = \emptyset$.

Theorem 5.23. *Let G be a graph with a fixed total order and reference orientation, and $\text{CUA}(\mathcal{O})^+, \text{CUA}(\mathcal{O})^-, \text{CYA}(\mathcal{O})^+$, and $\text{CYA}(\mathcal{O})^-$ be as in Definition 5.21. Then*

$$T(G; x + w, y + z) = \sum_{\mathcal{O}} x^{|\text{CUA}(\mathcal{O})^+|} w^{|\text{CUA}(\mathcal{O})^-|} y^{|\text{CYA}(\mathcal{O})^+|} z^{|\text{CYA}(\mathcal{O})^-|}, \quad (5.6)$$

where the sum is over all orientations of E.

Las Vergnas' orientation expansion above (which is from [755]) holds for all oriented matroids (see Chapter 31). By specializing variables $x = w = u/2$ and $y = z = v/2$, we recover Las Vergnas' following earlier expansion (from [750]) which does not make use of a reference orientation.

Corollary 5.24. *Let G be a graph with a fixed total order on its edge set E, and let* $\mathrm{CUA}(\mathcal{O})$ *and* $\mathrm{CYA}(\mathcal{O})$ *to be the set of edges which are minimum is some directed cut or cycle, respectively, then*

$$T(G; x, y) = \sum_{\mathcal{O}} \left(\frac{u}{2}\right)^{|\mathrm{CUA}(\mathcal{O})|} \left(\frac{v}{2}\right)^{|\mathrm{CYA}(\mathcal{O})|}, \tag{5.7}$$

where the sum is over all orientations of E.

We remark that Berman [95] was the first to propose an orientation expansion of the Tutte polynomial, although his definition was not correct. There are natural notions of orientation activity classes which parallel Crapo's subset intervals. The refined orientation expansion of the Tutte polynomial also follows as a direct consequence of "the active bijection" of Gioan and Las Vergnas which gives a bijection between orientation activity classes and Crapo subset intervals, which respects the four different activities [525, 534, 538]. See Chapter 31 for further details.

A *fourientation* of a graph is a choice for each edge of the graph whether to orient that edge in either direction, leave it unoriented, or biorient it. One may view fourientations as a mixture of orientations and subgraphs where absent and present edges correspond to unoriented and bioriented edges, respectively. Backman, Hopkins, and Traldi [61] introduced activities for fourientations which provide a common refinement of the Gordon and Traldi subgraph activities and Las Vergnas' orientation activities.

5.9 Active orders

Further deepening the theory of activities, Las Vergnas introduced three *active orders* on the bases of a matroid. We describe these partial orders and Las Vergnas' key result that they induce lattice structures on bases.

Let B_1 and B_2 be bases of a matroid M with fixed order on its ground set. We say that B_1 is obtained from B_2 by an *externally active pivoting* if $B_2 = (B_1 \setminus e) \cup f$, where e is the minimum element in $Z_{B_1}(f)$, and we write $B_1 \leftarrow_M B_2$. Dually, we say that B_1 is obtained from B_2 by an *internally active pivoting* if $B_1 = (B_2 \setminus e) \cup f$, where e is the smallest element in $U_{B_1}(f)$, and we write $B_1 \leftarrow *_M B_2$. We let $<_{\mathrm{Ext}}$ and $<_{\mathrm{Int}}$ denote the partial orders on the bases obtained by taking the transitive closures of the relations $\leftarrow_M$ and $\leftarrow *_M$, respectively. We refer to $<_{\mathrm{Ext}}$ as the *external order*, and $<_{\mathrm{Int}}$ as the *internal order*.

Las Vergnas also defined the following join of the external and internal orders. Let B_1 and B_2 be bases, then we say that $B_1 <_{\text{Ext/Int}} B_2$ if there exists bases $C_1, \ldots, C_k$ such that $B_1 = C_1$, $B_2 = C_k$, and for each i either $C_i \leftarrow_M C_{i+1}$ or $C_i \leftarrow *_M C_{i+1}$.

Recall that a lattice is a poset such that every pair of elements have a join and meet. Las Vergnas [753] proved the following.

Theorem 5.25. *Let $\mathcal{B}(M)$ be the set of bases of a matroid M with a total order on its ground set. Then the posets $(\mathcal{B}(M) \cup \{\mathbf{0}\}, <_{\text{Ext}})$, $(\mathcal{B}(M) \cup \{\mathbf{1}\}, <_{\text{Int}})$, and $(\mathcal{B}(M), <_{\text{Ext/Int}})$ are lattices.*

Las Vergnas observed that the lattice associated to the external order is not distributive, although it is atomistic. This appears to have been remedied in the recent thesis of Gillespie [521] where it is shown that by extending the external order to all independent sets, a supersolvable join-distributive lattice is obtained.

5.10 Shellability and activity

There are important connections between activity and combinatorial topology. We refer the reader to Björner [131] for an excellent introduction to this topic. Let $[n]$ denote the finite set $\{1, \ldots, n\}$. An *abstract simplicial complex* (often just called a *simplicial complex*) Δ on n elements is a collection of subsets (faces) of $[n]$ which is closed under taking subsets. Informally, a simplicial complex Δ is *shellable* if there is an ordering of the maximal faces (*facets*) of Δ so that each facet can be added to the previous ones by glueing along codimension 1 faces.

The *f-polynomial* of a simplicial complex Δ is $f_\Delta(x) = \sum_{i=0}^{n} f_i x^{d-i}$ where f_i is the number of faces of Δ of size i. The *h-polynomial* of Δ is the $h_\Delta(x) = f_\Delta(x-1)$. The f-vector and h-vector of Δ are the vectors whose entries are the coefficients of the f-polynomial and h-polynomial, respectively. A shellable complex is homotopy equivalent to a wedge of spheres, and its h-vector is nonnegative.

The following complexes are shellable, and the proofs of shellability are related to activities.

1. The *independence complex*, $\text{IN}(M)$, whose faces are the independent sets of a matroid M [932, 935]. Its h-polynomial is $T(M; x, 1)$. Matroids are characterized by the fact that the $\text{IN}(M)$ are the pure simplicial complexes which are lexicographically shellable with respect to any order on the ground set [131].

2. The *no broken-circuit complex*, NBC(M), whose faces are the independent sets of a matroid with no external activity in the sense of Tutte [932, 935]. Its h-polynomial is $T(M; x, 0)$.

3. The *external activity complex* defined on $E(M) \times E(M)$, whose facets are given by $B \cup \mathrm{EA}(B) \times B \cup (B \cup \mathrm{EA}(B))^c$, where B ranges over the bases of M [39]. The shelling makes use of Las Vergnas' order $<_{\mathrm{Ext/Int}}$. It has the same h-polynomial as IN(M) [39].

4. The *order complex of* IN(M) modulo Las Vergnas' external active order $<_{\mathrm{Ext}}$ [521].

5. The *order complex of the lattice of flats* [129]. It has Euler characteristic $T(M; 1, 0)$.

These complexes admit many interesting connections with algebraic and geometric combinatorics. We briefly mention a few. Orlik and Solomon [890] introduced a certain graded algebra which is isomorphic to the cohomology ring of the complement of a complex hyperplane arrangement, and showed that the monomials corresponding to faces of NBC(M) give a basis for this algebra. (See Chapter 30 for an introduction to Orlik–Solomon algebras.)

The external activity complex was introduced by Ardila and Boocher in their investigation of commutative algebraic aspects of the closure of a linear space in a product of projective lines [37]. The shellability of this complex is closely related to Las Vergnas' active orders. The ideals they consider are homogenizations of ones considered earlier by Proudfoot and Speyer [931], and Terao [1060]. A slight variation of these objects play an important role in Huh and Wang's proof of the Dowling–Wilson conjecture for realizable matroids [632].

Tropical geometry is a certain piecewise-linear version of algebraic geometry. Roughly speaking, the tropicalization of a variety over $\mathbb{C}$ is a balanced polyhedral fan which occurs as a logarithmic degeneration of the variety. Sturmfels observed that the tropicalization of a linear space L retains precisely the information provided by the matroid associated to L. Bergman fans are a natural generalization of such fans to arbitrary matroids, and Ardila and Klivans [41] showed that they are unimodularily triangulated by the fan over the order complex of the lattice of flats. This description has lead researchers to uncover interesting connections between algebraic geometry and matroids, most notably the proof of the Heron–Rota–Welsh conjecture that the f-vector of NBC(M) is log-concave by Adirprasito, Huh, and Katz [3] building on earlier works of Huh [629], and Huh and Katz [631]. Recently, Fink, Speyer, and Woo [487] introduced an *extended* NBC *complex* in order to shed some light on these different manifestations of the f-vector of NBC(M).

5.11 Open problems

Open Question 5.26. Fink and Speyer proved that Whitney's corank-nullity expression for the Tutte polynomial can be recovered geometrically from the K-theory of the Grassmannian [18]. Elucidate to what extent Tutte activities play a natural role in their setting.

Open Question 5.27. Backman, Hopkins, and Traldi provided a 12 variable activities expansion of the Tutte polynomial of a graph in terms of fourientations, i.e. generalized orientations where each edge of the graph can be oriented in either directions, bioriented, or left unoriented [61]. There are 8 types of activities in that setting and 2 of these activities $I^b(O)$ and $L^u(O)$, those which generalize Tutte's original activities, have yet to be defined in a non-recursive manner; see Remark 4.3 in [61]. Thus we propose finding a direct description of $I^b(O)$ and $L^u(O)$.

Open Question 5.28. A *multicomplex* Δ is a finite collection of points in $\mathbb{Z}^n_{\geq 0}$ which is downward closed, i.e. if $p \in \Delta$ and $q \in \mathbb{Z}^n_{\geq 0}$ with $q_i \leq p_i$ for all $1 \leq i \leq n$, then $p \in \Delta$. A multipcomplex Δ is *pure* if there exists some positive integer k such that each maximal point p in Δ has $\sum_{i=1}^n p_i = k$. Stanley's famous matroid h-vector conjecture states that for each matroid M there exists a pure multicomplex Δ such that the f-polynomial of Δ is $T(x,1)$ [1028].

Open Question 5.29. Gillespie's work extending Las Vergnas' external active order to independent sets is one of the most interesting developments in the combinatorics of activities in recent years [521]. We refer the reader to Gillespie's article for some open questions in this direction; see the arXiv version of this chapter for updated questions.

6

Tutte uniqueness and Tutte equivalence

Joseph E. Bonin • Anna de Mier

Synopsis

This chapter considers graphs or matroids that have the same Tutte polynomial, as well as graphs or matroids that, up to isomorphism, are distinguished from all others by their Tutte polynomial. We call the former Tutte equivalent, and the latter Tutte unique.

- Tutte invariants (data that the Tutte polynomial contains).
- Operations that preserve Tutte uniqueness or equivalence.
- Connections between the graph and matroid Tutte-uniqueness problems.
- Constructions of large families of Tutte-equivalent graphs and matroids.
- Tutte-unique graphs and matroids, and characterizations of some of these graphs and matroids by Tutte invariants.
- Related lines of research.

6.1 Introduction

The Tutte polynomial contains a wealth of information about a graph or a matroid. From the Tutte polynomial of a graph, we can determine the number of edges, the girth, the chromatic number, and much more. Likewise, the Tutte polynomial of a matroid gives us the number of independent sets of each size, the number of flats of certain sizes and ranks, and far more. When does all this information suffice to determine a particular graph or matroid, up to isomorphism? That is, in the terminology introduced below, which graphs or

DOI: 10.1201/9780429161612-6

matroids are Tutte unique? How can we construct graphs or matroids that have the same Tutte polynomial, and so share all this information, and yet are not isomorphic? That is, which non-isomorphic graphs or matroids are Tutte equivalent? These are the questions behind the results that we discuss in this chapter.

The answers to such questions for graphs can differ from the answers to their matroid counterparts. For instance, it is possible for a graph to be Tutte unique while its cycle matroid is not, and vice versa. Moreover, while the conjecture that almost all graphs are Tutte unique is open, it is known that Tutte uniqueness of matroids is relatively rare.

We discuss techniques to construct large families of non-isomorphic Tutte-equivalent graphs or matroids, often with certain additional properties, such as high connectivity, or, for matroids, representability over a given finite field.

Our survey of graphs and matroids that have been shown to be Tutte unique highlights common themes in the proofs of such results. In particular, we discuss characterizations of graphs or matroids by Tutte invariants since proofs of Tutte uniqueness are built on such results.

6.2 Basic notions and results, and initial examples

A basic question for any invariant is how strong it is: how effectively does it distinguish between non-isomorphic structures? The study of Tutte uniqueness and Tutte equivalence explores this issue for the Tutte polynomial. While Section 6.3 surveys this topic for graphs, and Section 6.4 does likewise for matroids, in this section we discuss graphs and matroids together so that we can efficiently treat properties that are common to the two settings, as well as explain how the graph and matroid problems are related.

We will use the following concepts throughout this chapter.

Definition 6.1. Two graphs (or matroids) are *Tutte equivalent* if they have the same Tutte polynomial. The equivalence classes of this equivalence relation are called *Tutte classes.* A graph (or matroid) is *Tutte unique* if any other graph (or matroid) with the same Tutte polynomial is isomorphic to it; in other words, all members of its Tutte class are isomorphic.

We do not consider graphs that have isolated vertices since adding isolated vertices to a graph does not change the Tutte polynomial; if isolated vertices were allowed, no graph would be Tutte unique. On the other hand, adding loops or elements parallel to existing ones changes the Tutte polynomial, so both loops and parallel elements are allowed in graphs and matroids.

The well-known formula

$$T(G \sqcup H; x, y) = T(G * H; x, y) = T(G; x, y) \cdot T(H; x, y) \qquad (6.1)$$

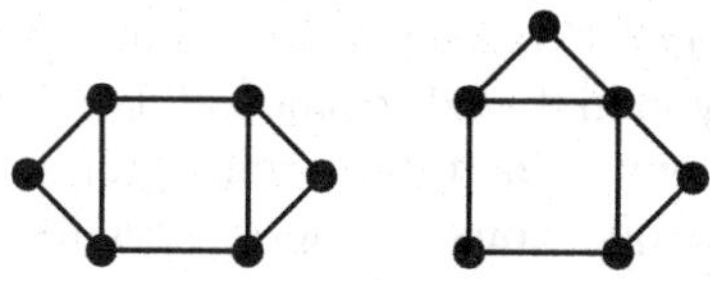

FIGURE 6.1: Two non-isomorphic graphs that are 2-isomorphic, and thus Tutte equivalent.

shows that the disjoint union, $G \sqcup H$, and the one-point join, $G * H$, of two graphs are Tutte equivalent. Thus, Tutte-unique graphs are connected and have no cut-vertices. By slightly abusing the usual definition, we call these 2-connected, so that K_2 and the graph consisting of a single loop are considered 2-connected in this chapter. (This definition agrees with 2-connectivity, or connectivity, for matroids.) From Equation (6.1), it also follows that the Tutte polynomial of a forest on n edges is x^n, and it is easy to check that the set of all such forests is a Tutte class.

A number of graphs have been shown to be chromatically unique, that is, determined up to isomorphism by their chromatic polynomials (see Chapter 11). The chromatic polynomial is related to the Tutte polynomial by the formula

$$\chi(G;x) = (-1)^{|V(G)|-k(G)} x^{k(G)} T(G; 1-x, 0), \tag{6.2}$$

where $k(G)$ is the number of components of G. The following result is an easy consequence of Theorems 6.3 and 6.4 below.

Theorem 6.2. *Chromatically-unique 2-connected graphs are Tutte unique.*

A graph G and its cycle matroid $M(G)$ have the same Tutte polynomial, so graphs with isomorphic cycle matroids are Tutte equivalent. Whitney [1156] introduced 2-isomorphism and showed that two graphs have isomorphic cycle matroids if and only if the graphs are 2-isomorphic. (See Definition 4.101 and Theorem 4.102. The graphs $G \sqcup H$ and $G * H$ in Equation (6.1) are 2-isomorphic.) Thus, Whitney gave a sufficient (but not necessary) condition for graphs to be Tutte equivalent. Figure 6.1 gives an example of two graphs that are 2-isomorphic and so are Tutte equivalent.

The first known pair of non-2-isomorphic, Tutte-equivalent graphs were the Gray graphs (see [1097]). Each has six vertices and a pair of parallel edges. A computer search shows that there are no simple, non-2-isomorphic, Tutte-equivalent graphs on six or fewer vertices, and exactly one such pair (shown in Figure 6.2) on seven vertices. Only one of these Tutte-equivalent graphs is planar, and 2-isomorphism preserves planarity, so these graphs are not 2-isomorphic.

Figure 6.3 shows the only non-isomorphic, Tutte-equivalent matroids on six elements; all matroids with fewer elements are Tutte unique.

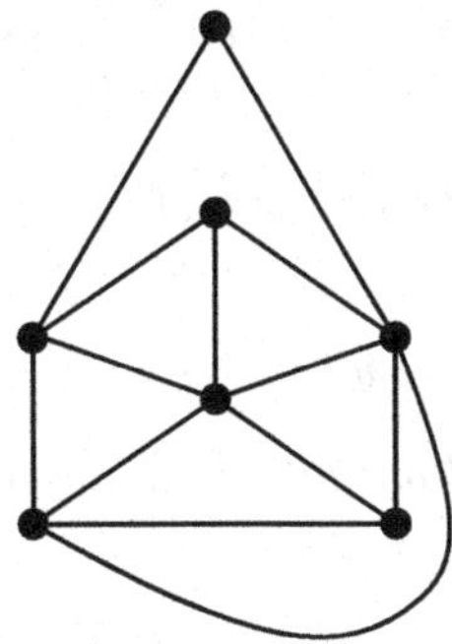
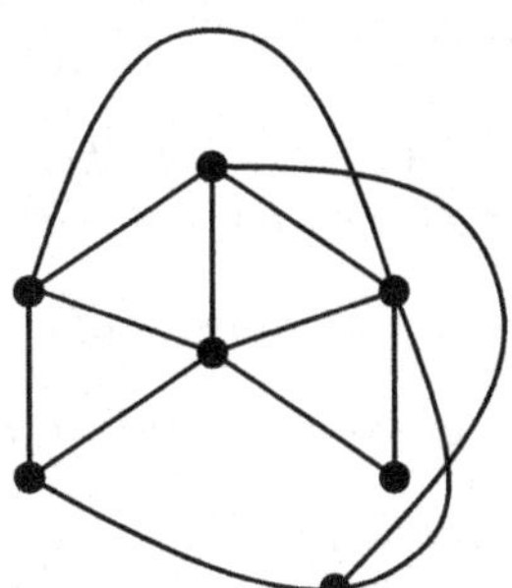

FIGURE 6.2: The unique pair of non-2-isomorphic Tutte-equivalent simple graphs on seven or fewer vertices.

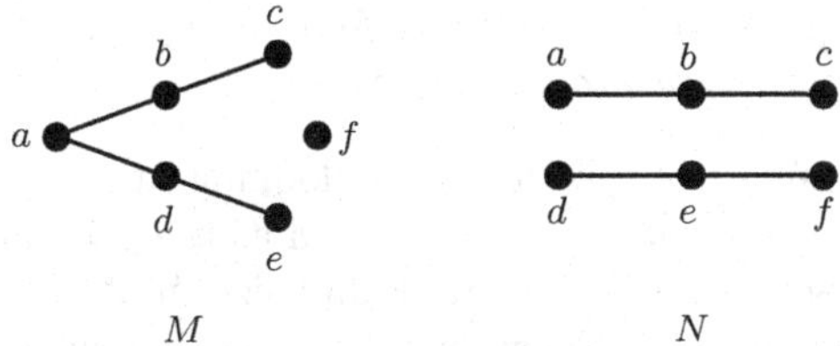

FIGURE 6.3: The pair of non-isomorphic Tutte-equivalent matroids on fewest elements.

6.2.1 Tutte invariants

A prerequisite for showing that a given graph or matroid is Tutte unique is knowing what information is contained in the Tutte polynomial. In this section, we summarize some of the many Tutte invariants, that is, properties (such as being 2-connected) or numerical quantities (such as the number of edges or elements) that are shared by all Tutte-equivalent graphs or matroids.

Given formula (6.1), from $T(G; x, y)$ we generally cannot deduce whether G is connected. However, by the following result of Brylawski [251], we know whether G is 2-connected.

Theorem 6.3. *A graph or matroid is* 2*-connected if and only if the coefficient of either x or y in its Tutte polynomial is non-zero.*

(Except for single-edge graphs and single-element matroids, the coefficients of x and y in the Tutte polynomial are equal.)

The next result gives some of the graph parameters that can be deduced from the Tutte polynomial.

Theorem 6.4. *Let G be a* 2*-connected graph. The parameters of G below are determined by $T(G; x, y)$:*

1. *the number of vertices and the number of edges;*

2. *for integers i and j with $0 \leq i \leq r(G)$ and $0 \leq j \leq |E(G)|$, the number of subsets A of $E(G)$ with $r_G(A) = i$ and $|A| = j$;*

3. *the girth $g(G)$ (so, whether G is simple), the number of cycles C with $|C| = g(G)$, and the number with $|C| = g(G) + 1$;*

4. *for every k, the number of edges with multiplicity k;*

5. *the edge-connectivity (which is a lower bound on the minimum degree);*

6. *the chromatic number.*

If G is simple, then we can also deduce:

7. *the number of cliques of each order and the clique-number;*

8. *the number of cycles of length three, four, and five, and the number of cycles of length four with exactly one chord.*

Most of the statements above follow by looking at certain coefficients of the Tutte polynomial or from its interpretation as the generating function for the number of edge-sets according to rank and size. Statement 5 follows from Theorem 6.5 and formula (6.3) below by observing that the edge-connectivity of G is the girth of the dual of the cycle matroid of G.

In contrast, the length of the longest cycle and, in particular, whether a graph is Hamiltonian, cannot be determined from the Tutte polynomial (see Schwärzler [991]). The degree sequence is also not a Tutte invariant, as witnessed by the pairs in Figures 6.1 and 6.4.

The pair in Figure 6.2 shows that planarity is not a Tutte invariant for graphs. The Tutte polynomial of a graph G can be symmetric in x and y without G being self-dual, as witnessed by the non-planar graph in [1152] that has a symmetric Tutte polynomial. (This example was rediscovered by computer search in [846], where one can also find a (planar) non-self-dual 3-connected graph Tutte equivalent to its dual.)

It is routine to prove the following similar theorem for matroids.

Theorem 6.5. *The following invariants of a matroid M can be deduced from $T(M; x, y)$: the rank, $r(M)$, the size, $|E(M)|$, the number of loops, the number of coloops, and, with the obvious adaptations (e.g., circuit replacing cycle), the counterparts of Items 2–4 of Theorem 6.4.*

The following result is due to Brylawski [245].

Theorem 6.6. *For a rank-r matroid M and integer i with $0 \leq i \leq r$, let c_i be the largest cardinality among the rank-i flats of M. For each j with $c_{i-1} < j \leq c_i$, the number of flats of M of rank i and cardinality j can be deduced from $T(M; x, y)$.*

As Example 4.158 shows, many basic properties (including being graphic, or binary, or transversal) need not be shared by Tutte-equivalent matroids.

The members of some special classes of matroids can be identified by their Tutte polynomials, even if they are not Tutte unique. The following result of Brylawski [251] illustrates this, and we will see more examples later. A *series-parallel graph* is a graph that is 2-connected (that is, connected with no cut-vertex) and that can be obtained from a single-edge graph by iterating the operations of parallel extension and series extension. Equivalently, it is 2-connected and has no minor isomorphic to K_4. Brylawski's result below states that one can tell whether a matroid is a direct sum of cycle matroids of series-parallel graphs by looking at certain coefficients of its Tutte polynomial. In particular, a connected matroid with at least two elements is a cycle matroid of a series-parallel graph if and only if both the coefficient of x and that of y are equal to one.

Theorem 6.7. *A matroid M with $T(M;x,y) = \sum_{i,j\geq 0} t_{i,j}x^i y^j$ having no loops and no coloops is a direct sum of n cycle matroids of series-parallel graphs if and only if*

1. $t_{i,j} = \binom{n}{i}$ *whenever* $i + j = n$, *and*

2. $t_{i,j} = 0$ *whenever* $i + j < n$.

Furthermore, condition 1 can be replaced by the condition:

1′. $t_{i,j} = \binom{n}{i}$ *for some pair* i, j *with* $i + j = n$.

We cite a second result of this type. A graph is *outerplanar* if it has a planar embedding in which all vertices lie in the outer face. Outerplanar graphs are those that have no minor isomorphic to K_4 or $K_{2,3}$, so in particular 2-connected outerplanar graphs are series-parallel graphs. Goodall, de Mier, Noble, and Noy [559] showed that from the coefficients of the Tutte polynomial one can tell whether a graph is a simple outerplanar graph.

Theorem 6.8. *Let G be a simple series-parallel graph with $G \neq K_2$. Then $t_{2,0} \geq t_{0,2} + 1$, with equality if and only if G is outerplanar.*

As mentioned above, while the Tutte polynomial identifies series-parallel graphs and simple outerplanar graphs, not all such graphs are Tutte unique. Omer Giménez [522] found the pair of Tutte-equivalent non-2-isomorphic outerplanar graphs in Figure 6.4.

6.2.2 Some constructions

From the well-known equation $T(G;x,y) = T(G^*;y,x)$, relating the Tutte polynomials of a planar graph and its dual, we get the following result.

Theorem 6.9. *Two planar graphs G and H are Tutte equivalent if and only if their duals G^* and H^* are Tutte equivalent.*

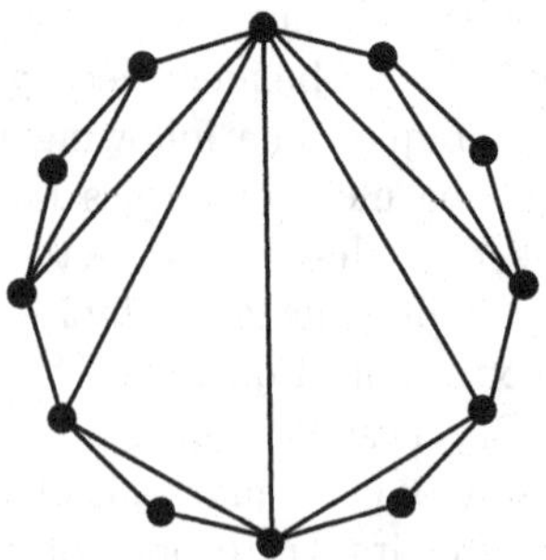

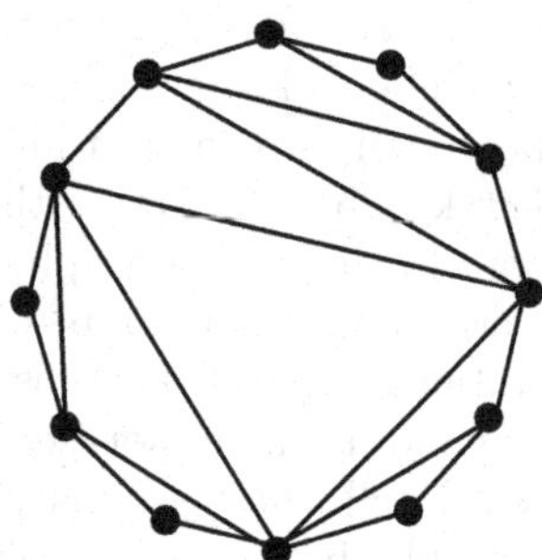

FIGURE 6.4: A pair of Tutte-equivalent non-2-isomorphic outerplanar graphs.

The dual of a Tutte-unique planar graph may fail to be Tutte unique. For instance, while the planar graph G in Figure 6.2 is Tutte equivalent to the non-planar graph also shown there, a computer search shows that G^* is Tutte unique.

While graph duality is limited to planar graphs, every matroid has a dual. Thus, we get the following stronger result for matroids as a consequence of the well-known formula

$$T(M; x, y) = T(M^*; y, x). \tag{6.3}$$

Theorem 6.10. *Two matroids M_1 and M_2 are Tutte equivalent if and only if their duals M_1^* and M_2^* are Tutte equivalent. A matroid is Tutte unique if and only if its dual is Tutte unique.*

As discussed above, with the product formula (6.1), it is easy to construct non-isomorphic Tutte-equivalent graphs. This does not carry over to matroids because the two graph operations, disjoint union and one-point join, have the same matroid counterpart, the direct sum. From the formula

$$T(M_1 \oplus M_2; x, y) = T(M_1; x, y) \cdot T(M_2; x, y), \tag{6.4}$$

it follows that direct sum preserves Tutte equivalence.

Theorem 6.11. *If M_1 and M_1' are Tutte equivalent, and likewise for M_2 and M_2', then $M_1 \oplus M_2$ and $M_1' \oplus M_2'$ are Tutte equivalent.*

In [840], Merino, de Mier, and Noy proved the next theorem, which settled a conjecture of Brylawski [251].

Theorem 6.12. *If M is a connected matroid, then $T(M; x, y)$ is irreducible in $\mathbb{C}[x, y]$.*

From Theorem 6.12, it follows that Tutte equivalence and uniqueness for disconnected matroids reduces to that of the connected components.

Corollary 6.13. *For connected matroids M_1 and M_2, if $M_1 \oplus M_2$ and $M_1' \oplus M_2'$ are Tutte equivalent, then, up to switching M_1' and M_2', the matroids M_1 and M_1' are Tutte equivalent, as are M_2 and M_2'.*

Two matroids are Tutte unique if and only if their direct sum is.

Examples show that there are no counterparts of Theorem 6.11 and Corollary 6.13 for parallel connections, 2-sums, and their generalizations. While Equation (6.4) has counterparts for these operations (see Bonin and de Mier [167] and Chapter 7 of this handbook), they involve the Tutte polynomials of certain minors of M_1 and M_2.

Later in this chapter we give some less straightforward constructions of Tutte-equivalent graphs or matroids (see in particular Sections 6.3.1, 6.4.1, 6.4.5, and 6.4.6). Some of these constructions use the deletion–contraction formula to show they indeed produce Tutte-equivalent graphs or matroids.

In this direction, Brylawski [245, Section 3] defined a relation $\sim$ recursively on the set of all matroids by setting $M \sim M'$ whenever

- M is isomorphic to M', or
- there are elements $p \in E(M)$ and $p' \in E(M')$, neither of which is a loop or coloop, for which $M\backslash p \sim M'\backslash p'$ and $M/p \sim M'/p'$.

By the deletion–contraction formula, if $M \sim M'$, then M and M' are Tutte equivalent. For instance, the non-isomorphic matroids in Figure 6.3 are related by taking p and p' above to be the points labeled c. Brylawski asked whether the relation $\sim$ is the same as Tutte equivalence. It is not. This is because the set of Tutte polynomials of the single-edge deletions of the first graph in Figure 6.4 is disjoint from the set of Tutte polynomials of the single-edge deletions of the second graph.

6.2.3 The connection between Tutte uniqueness for graphs and matroids

It follows from Whitney's 2-isomorphism theorem (Theorem 4.102) that if graphs G and G' have isomorphic cycle matroids and G is 3-connected, then G and G' are isomorphic. Thus we have the following result.

Theorem 6.14. *If G and H are Tutte-equivalent graphs and $M(G)$ is Tutte unique, then G and H are 2-isomorphic. A 3-connected graph is Tutte unique if its cycle matroid is Tutte unique.*

Forests are not Tutte unique; however, their cycle matroids are Tutte unique since the cycle matroid of a forest with n edges is isomorphic to the uniform matroid $U_{n,n}$. The graphs in Figure 6.1 give a 2-connected example of this phenomenon, as one can show with Theorem 6.5 and basic arguments.

A graph can be Tutte unique while its cycle matroid is not. For example, the graph obtained by adding one edge parallel to any edge of K_4 is Tutte unique (this follows from Theorem 6.4), but its cycle matroid, the matroid M_1 in Figure 4.16, is Tutte equivalent to the matroid M_2 in that figure.

As the examples above show, knowing that a graph G is Tutte unique usually gives no insight into whether its cycle matroid $M(G)$ is also Tutte unique. Even if $M(G)$ is Tutte unique, it typically takes much more effort to prove it; for instance, the Tutte uniqueness of K_n follows from Theorem 6.4 since K_n is the only simple graph on n vertices with $\binom{n}{2}$ edges, but there are many more conditions in the characterization of $M(K_n)$ that is used to establish its Tutte uniqueness (see Theorem 6.61). One reason for this is that, as Example 4.158 shows, the Tutte polynomial does not distinguish graphic from non-graphic matroids. Thus, to show that $M(G)$ is Tutte unique one must compare it to an arbitrary matroid, and the number of matroids on n elements is much larger than that of graphs on n edges. In Section 6.4.2 we sketch a counting argument that shows that the proportion of Tutte-unique matroids is bound to be very small. For graphs the situation is thought to be just the opposite, as conjectured by Bollobás, Pebody, and Riordan [156].

Conjecture 6.15. Asymptotically, almost all simple graphs are Tutte unique.

That is, among simple graphs with vertex set $[n]$, the proportion of those that are Tutte unique is conjectured to tend to 1 as n tends to infinity.

6.3 Tutte uniqueness and equivalence for graphs

In the first section we review some constructions that yield Tutte-equivalent graphs, while in the next four sections we survey families of graphs that are known to be Tutte unique and we briefly describe the techniques used in the proofs. In three of those sections, we focus on graphs with special properties: edge subdivisions (Section 6.3.3), local grid-like structure (Section 6.3.4), and line graphs (Section 6.3.5). In Section 6.3.6, we present a family of outerplanar graphs whose cycle matroids yield exponentially many Tutte-unique matroids.

6.3.1 Constructions of Tutte-equivalent graphs

The first construction technique in this area was developed by Tutte [1097], who introduced rotors to produce pairs of Tutte-equivalent, k-connected, non-2-isomorphic graphs for $k \leq 5$. Brylawski [244] adapted this construction to give such pairs with arbitrarily high connectivity.

Bollobás, Pebody, and Riordan [156] gave a simpler construction of highly-connected, non-isomorphic, Tutte-equivalent graphs. They used the pair G_1, G_2 of graphs shown in Figure 6.5, and the following properties of those graphs:

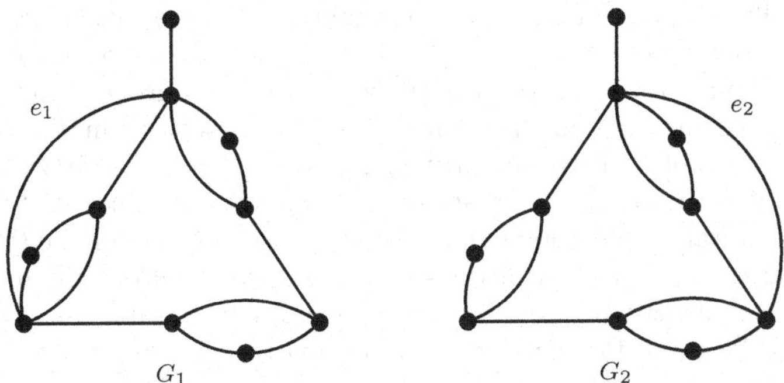

FIGURE 6.5: Two Tutte-equivalent graphs with $G_1\backslash e_1 \cong G_2\backslash e_2$ and $G_1/e_1 \cong G_2/e_2$.

$G_1\backslash e_1 \cong G_2\backslash e_2$ and there is an isomorphism from G_1/e_1 onto G_2/e_2 that maps the vertex obtained by contracting e_1 to that obtained by contracting e_2. Now take the disjoint union of G_1 and K_{k-1}, and add all edges uv with $u \in V(G_1)$ and $v \in V(K_{k-1})$, and construct a second graph similarly from G_2 and K_{k-1}. Clearly these two graphs are k-connected and non-isomorphic, and by the deletion–contraction formula they are Tutte equivalent.

The following variation gives exponentially many such graphs. Take a highly-connected graph H with vertex set $\{1, 2, \ldots, n/10\}$ and with trivial automorphism group. Replace each vertex i of H by a graph H_i that is isomorphic to either G_1 or G_2, and then, whenever ij is an edge of H, add all edges uv with $u \in V(H_i)$ and $v \in V(H_j)$. The resulting $2^{n/10}$ graphs on n vertices are highly-connected, Tutte-equivalent, and non-isomorphic.

Another way of proving that two graphs have the same Tutte polynomial is by ensuring that they have the same number of edge-subsets of the same rank and size. In turn, this follows if the two graphs have the same number of edge-subsets with the same size that induce the same number of components of each order; in other words, if both graphs have the same U polynomial (see Chapter 26). Azarija [58] exploited this idea to construct graphs that are Tutte equivalent to their complements. Starting with a non-self-complementary graph G having the same U polynomial as its complement (found by computer search), he showed that joining all vertices of G to the endpoints of a 4-vertex path yields another non-self-complementary graph that has the same U polynomial as its complement. By iterating the construction one obtains infinitely many pairs of non-self-complementary graphs that are Tutte equivalent to their complements.

6.3.2 Families of Tutte-unique graphs

As Theorem 6.2 states, 2-connected chromatically-unique graphs are also Tutte-unique, so research naturally started by looking at graphs that play prominent roles in graph theory and either are not, or are not known to be, chromatically unique. For many of the families below, chromatic uniqueness has been settled only when the number of vertices is small (see Chapter 11).

A typical proof that a graph G is Tutte unique shows that G is determined, up to isomorphism, by some Tutte invariants, such as those in Theorem 6.4. Thus, from a proof of Tutte uniqueness, one obtains a characterization of G in terms of such invariants; we state such characterizations in selected cases.

Complete graphs and complete bipartite graphs (except for the tree $K_{1,p}$) are chromatically unique, but little is known about the chromatic uniqueness of complete multipartite graphs with more than two stable sets. De Mier and Noy [847] proved that all complete multipartite graphs are Tutte unique (again, except $K_{1,p}$). The following lemma is the core of their proof.

Lemma 6.16. *Let $r \geq 3$, let $p_1, p_2, \ldots, p_r$ be positive integers, and let H be a simple r-partite graph. For each k with $1 \leq k \leq r$, suppose that the number of k-cliques of H is*

$$\sum_{1 \leq i_1 < i_2 < \cdots < i_k \leq r} p_{i_1} p_{i_2} \cdots p_{i_k}.$$

Suppose also that the number of subgraphs of H isomorphic to K_{r+1}^-, the single-edge deletion of K_{r+1}, is $p_1 \cdots p_r \sum_i (p_i - 1)/2$. Then H is isomorphic to the complete multipartite graph $K_{p_1,\ldots,p_r}$.

This lemma, along with Theorem 6.4, gives the next result.

Theorem 6.17. *Every complete multipartite graph $K_{p_1,\ldots,p_r}$ other than $K_{1,p}$ is Tutte unique.*

The *r-spoke wheel* W_r is the graph obtained from the cycle C_r (the *rim*) and another vertex (the *hub*) by making every rim vertex adjacent to the hub. The edges incident with the hub are the *spokes*. If r is even, then W_r is chromatically unique. For odd r, what is known is that W_3 (which is K_4) and W_9 are chromatically unique and that W_5 and W_7 are not. Wheels were proved to be Tutte unique by de Mier and Noy [847] with an argument independent of the parity of r. The core of the proof is a characterization of wheels by some subgraph counts.

Lemma 6.18. *For $r \geq 5$, let H be a simple graph with $r+1$ vertices, $2r$ edges, r cycles of length 3, and r cycles of length 4. Suppose also that each 4-cycle has exactly one chord, that no edge belongs to more than two cycles of length 3, and that H has no subgraph isomorphic to W_p for $p < r$. Then H is isomorphic to the r-spoke wheel W_r.*

Theorem 6.19. *The r-spoke wheel W_r is Tutte unique for all $r \geq 3$.*

The square of a cycle, C_n^2, is obtained from the cycle C_n by adding all edges between vertices at distance 2. The theorem below appears in [847] but the proof, which is very similar to that of Theorem 6.19, can be found in [846].

Theorem 6.20. *The graph C_n^2 is Tutte unique for all $n \geq 3$.*

A variation on wheels is the *twisted wheel* W_{k_1,k_2}, which is constructed from two vertex-disjoint paths $u_1,\ldots,u_{k_1}$ and $v_1,\ldots,v_{k_2}$ by adding the edges $\{v_1u_i : 1 \le i \le k_1\} \cup \{u_{k_1}v_j : 1 \le j \le k_2\} \cup \{u_1v_{k_2}\}$. Duan, Wu, and Yu [422] proved that twisted wheels are Tutte unique. It is clear that any graph that is Tutte equivalent to W_{k_1,k_2} has k_1+k_2 vertices, $2(k_1+k_2)-2$ edges, and k_1+k_2-2 cycles of length 3. The difficulty of the proof lies in showing that these cycles of length 3 are glued to each other as in W_{k_1,k_2}.

Theorem 6.21. *Twisted wheels, W_{k_1,k_2}, are Tutte unique for all $k_1,k_2 \ge 4$.*

Now we turn to two families of graphs that are formed by gluing cycles of length 4. The *circular ladder* L_n is the graph $C_n \times K_2$. The *Möbius ladder* M_n is the graph obtained from a cycle C_{2n} by adding an edge between every pair of vertices at distance n. If a graph is Tutte equivalent to either ladder, it can be shown that it is 3-regular and contains n cycles of length 4 forming a path-like structure $P_n \times K_2$; the chromatic number then distinguishes L_n from M_n. The details of this argument can be found in de Mier and Noy [847].

Theorem 6.22. *Circular ladders, L_n, are Tutte unique for all $n \ge 3$, and Möbius ladders, M_n, are Tutte unique for all $n \ge 2$.*

For positive integers m and n with $m \ge 2n+1$, the *generalized Petersen graph* $P(m,n)$ has vertex set $\{u_1,\ldots,u_m,v_1,\ldots,v_m\}$ and edge set

$$\{u_iu_{i+1} : 1 \le i < m\} \cup \{u_1u_m\} \cup \{u_iv_i : 1 \le i \le m\} \cup \{v_iv_j : |j-i| \in \{n, m-n\}\}.$$

The Petersen graph is the case $m=5$ and $n=2$. Kuhl [727] proved that the generalized Petersen graph $P(m,2)$ is Tutte unique. The role played by cycles of length 3 or 4 in the previous families is now played by cycles of length 5. To analyze how these cycles are incident to each other, it is necessary to look at configurations including cycles of length 8. In general the Tutte polynomial does not give much information about these cycles since many other subgraphs have rank 7 and size 8, but as the girth of $P(m,2)$ is 5, the number of options is small enough to extract the needed information. See [727] for further details and also for an indication of how to prove the Tutte uniqueness of $P(m,3)$.

Theorem 6.23. *The generalized Petersen graph $P(m,2)$ is Tutte unique for all $m \ge 5$.*

Another well-studied graph is the n-cube, Q_n, which is the cartesian product, $K_2 \times \cdots \times K_2$, of n copies of K_2. Its Tutte uniqueness was proved in [847]. The proof follows easily from the fact that Q_n is the only n-regular connected graph with 2^n vertices in which each pair of vertices at distance two have exactly two common neighbors.

Theorem 6.24. *The n-cube is Tutte unique for all $n \ge 2$.*

6.3.3 Edge subdivisions and chains

A *chain* in a graph G is a path in G in which all vertices, except possibly the end-vertices, have degree two in G. The *length* of a chain is its number of edges. If G is a plane graph, then the edges of a chain in G give multiple edges in the dual graph G^* (in general this need not account for all multiple edges in G^*). From this perspective, the following result of Read and Whitehead [952] can be seen as a dual of Item 4 in Theorem 6.4. (See also Section 24.4.1.) To state the result, we need one more notion. Chains can be introduced by repeatedly subdividing edges, that is, replacing an edge by a path of length two. When, to a graph G, the reverse of subdividing edges is applied to eliminate all vertices of degree two, we obtain the *reduced graph* of G.

Theorem 6.25. *If the reduced graph of G is 3-edge-connected, then, for any integer $k > 0$, the number of chains of length k can be deduced from $T(G; x, y)$.*

Unlike in the following applications of this theorem to Tutte uniqueness, it may not be possible to check, from $T(G; x, y)$, that the reduced graph of G is 3-edge-connected.

An *s-theta graph* (or *graph s-bridge*) is a graph obtained from the graph on two vertices with s multiple edges by subdividing each edge any number of times. Similarly, a *homeomorph of K_4* is obtained from K_4 by subdividing edges. Read and Whitehead [952] proved the following result.

Theorem 6.26. *All s-theta graphs, with $s \geq 2$, are Tutte unique, as are all homeomorphs of K_4.*

The question of which s-theta graphs are chromatically unique is far from being settled. Many subfamilies are known to be, and there are s-theta graphs that are not. Likewise, it is known that almost all homeomorphs of K_4 are chromatically unique, but it is not known exactly which ones are. (See Chapter 11.)

6.3.4 Graphs with a grid-like structure

The graph $C_p \times C_q$ can be naturally embedded on a torus, resulting in a tiling of the surface by squares; we will refer to this graph as a *toroidal grid*. Locally grid graphs are defined so that around each vertex they look like a toroidal grid. Up to small technical details, they correspond to the quadrilateral tilings studied by Thomassen [1062].

Definition 6.27. A 4-regular graph G is a *locally grid graph* if, for each vertex x of G, the graph induced on x, its four neighbors, and the vertices that are adjacent to at least two neighbors of x is isomorphic to the grid $P_2 \times P_2$.

By this definition, the toroidal grid $C_p \times C_q$ is a locally grid graph if and only if both p and q are at least five; this fits the technical requirements of the results below.

The project of proving that locally grid graphs are Tutte unique was initiated by Márquez, de Mier, Noy, and Revuelta [820], and continued by Garijo, Márquez, and Revuelta [513]. The first step is to show that the Tutte polynomial captures the locally grid condition.

Theorem 6.28. *If G and H are Tutte-equivalent graphs and G is locally grid with no cycle of length* 5*, then H is also locally grid and has no cycle of length* 5.

Locally grid graphs fall into three families, one of which is embeddable in the torus and the other two in the Klein bottle. One can show that all locally grid graphs contain a spanning subgraph isomorphic to $P_p \times P_q$; the different families account for the various options for the edges between the boundary vertices (see [820, 1062] for details). A cycle in a locally grid graph is called *essential* if, in the corresponding surface, it is not contractible to a point; the *essential girth* is the length of the shortest essential cycle. For instance, for the toroidal grid $C_p \times C_q$ the essential girth is $\min\{p, q\}$. The main tool for distinguishing locally grid graphs by their Tutte polynomials is the following result from [820].

Theorem 6.29. *If G and G' are Tutte-equivalent locally grid graphs, then they have the same essential girth g and the same number of essential cycles of length g.*

Thus, one needs to compute the essential girth and the number of shortest essential cycles for all locally grid graphs, and then use ad hoc arguments to prove that the Tutte polynomial distinguishes those pairs where both quantities coincide. This is done in [820, 513], leaving just a few cases undecided.

Theorem 6.30. *Most locally grid graphs without cycles of length five are Tutte unique. In particular, $C_p \times C_q$ is Tutte unique for all $p, q \geq 6$.*

The same program as the one described above for locally grid graphs has been carried out by Garijo, Márquez, and Revuelta for another family of graphs with a grid-like structure, hexagonal tilings. These are graphs that locally resemble a tiling of the torus by hexagons; thus, they are 3-regular and each vertex belongs to three cycles of length six. See [514] for a detailed description of the method and the results.

6.3.5 Line graphs

Recall that the *line graph* $L(G)$ of a graph G has as vertices the edges of G, and two vertices of $L(G)$ are adjacent if the corresponding edges of G have a common endpoint. Thus, the set of edges of G that are incident with a given vertex become a clique in $L(G)$. In fact, line graphs are precisely those graphs whose edge set can be partitioned into cliques such that no vertex lies in more than two of these cliques. The only pair of non-isomorphic graphs with isomorphic line graphs are K_3 and $K_{1,3}$.

In [849], de Mier and Noy proved that line graphs of some families of complete multipartite graphs are Tutte unique. The next result is one of the main tools they developed to accomplish this.

Theorem 6.31. *Let G be a simple d-regular d-edge-connected graph on n vertices for some $d \geq 3$. If $d = 3$, assume also that G is triangle-free. If a graph H is Tutte equivalent to $L(G)$, then $H = L(G_0)$ for some simple d-regular connected graph on n vertices. Moreover, G_0 has the same number of triangles as G.*

This theorem easily gives the first and third parts of the next result, and, with a triangle-counting argument, it yields the second assertion.

Theorem 6.32. *The line graphs of the following graphs are Tutte unique: K_n, for $n \geq 2$; the t-partite graph $K_{p,\ldots,p}$, for all $p \geq 2$ and $t \geq 1$; and $K_{2n} - nK_2$ (a complete graph with a perfect matching removed), for $n \geq 2$.*

Theorem 6.31 does not apply to $K_{p,q}$ if $p \neq q$, but $L(K_{p,q})$ is also Tutte unique, as shown in [846].

Theorem 6.33. *The graph $L(K_{p,q})$ is Tutte unique for all $p, q \geq 1$.*

Kuhl [727] also relied on Theorem 6.31 to prove the uniqueness of the line graph of a generalized Petersen graph.

Theorem 6.34. *For $m \geq 13$, the graph $L(P(m, 2))$ is Tutte unique.*

Given Theorem 6.31, one may wonder if there are non-isomorphic Tutte-equivalent line graphs of d-regular graphs. Such graphs are constructed in [849], using Tutte's rotor construction.

6.3.6 An exponentially large family of Tutte-unique graphic matroids

As we saw in Theorems 6.7 and 6.8, from $T(G; x, y)$, we can tell whether the graph G is a series-parallel graph or a simple outerplanar graph. However, among such graphs Tutte uniqueness is rare since 2-isomorphism gives many Tutte-equivalent pairs; also, as Figure 6.4 illustrates, not all cycle matroids of series-parallel graphs are Tutte-unique matroids. We next describe a class of outerplanar graphs that de Mier and Noy [848] defined and showed have Tutte-unique cycle matroids; this class contains exponentially many matroids with a given rank.

A simple 2-connected outerplanar graph has an embedding in the plane as a polygon with some diagonals. The maximum number of edges in such a graph on n vertices is $2n - 3$; those with $2n - 3$ edges correspond to triangulations of an n-gon and are called *maximal simple outerplanar graphs* since they have the maximum number of edges among simple outerplanar graphs.

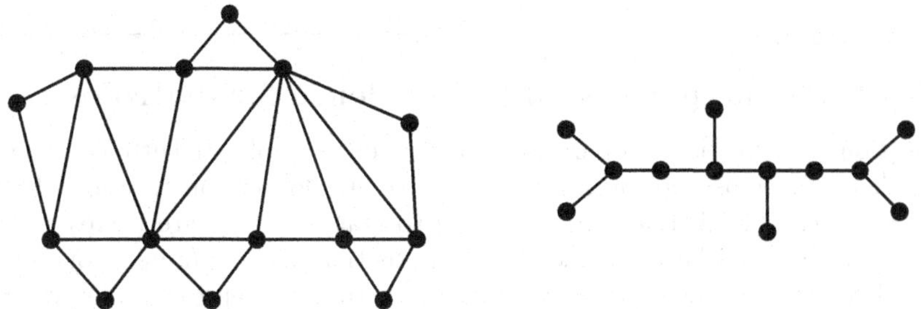

FIGURE 6.6: A maximal simple outerplanar graph and its dual tree.

For a maximal simple outerplanar graph G, if, from the dual graph of G, we delete the vertex that corresponds to the outer face, we get the incidence graph $\tau(G)$ of the triangles in the embedding of G as a triangulation. Note that $\tau(G)$ is a tree with maximum degree at most three. We call $\tau(G)$ the *dual tree* of G (see Figure 6.6 for an example). Two maximal simple outerplanar graphs have isomorphic dual trees if and only if they are 2-isomorphic.

Let the class $\mathcal{C}$ consist of the maximal simple outerplanar graphs G that satisfy the following two conditions.

1. The dual tree $\tau(G)$ is a caterpillar, that is, removing all leaves from $\tau(G)$ produces a path, called the *spine*.

2. The two end points of the spine have degree three in $\tau(G)$.

The graph in Figure 6.6 belongs to $\mathcal{C}$. Note that the graph on the left-hand side of Figure 6.4 satisfies the first but not the second of the conditions.

Any matroid that is Tutte equivalent to the cycle matroid of a maximal simple outerplanar graph is the cycle matroid of such a graph. The key to proving the following counterpart for $\mathcal{C}$ is counting certain subgraphs in the dual graph.

Lemma 6.35. *For any graph G in $\mathcal{C}$, any matroid that is Tutte equivalent to $M(G)$ is $M(H)$ for some H in $\mathcal{C}$.*

For a graph G in $\mathcal{C}$ and a choice of an initial vertex of the spine of $\tau(G)$, the sequence of degrees in $\tau(G)$ of the successive vertices in the spine is a word in the alphabet $\{2, 3\}$; we let $\omega(G)$ denote this word. For instance, for the graph in Figure 6.6, both possible initial vertices yield the palindrome $\omega(G) = 3, 2, 3, 3, 2, 3$. In [848], de Mier and Noy proved the following result about the graphs G in $\mathcal{C}$ for which $\omega(G)$ is a palindrome.

Theorem 6.36. *For a graph G in $\mathcal{C}$, if $\omega(G)$ is a palindrome, then its cycle matroid is Tutte unique. For odd r, such graphs yield $\Omega(\alpha^r)$ Tutte-unique graphic matroids of rank r, where α is the square root of the golden ratio.*

It is not known whether $M(G)$ is Tutte unique for all G in $\mathcal{C}$.

6.4 Tutte uniqueness and equivalence for matroids

In Section 6.4.1 we discuss operations that, given examples of Tutte-equivalent matroids, can be used to obtain more. We treat another construction of this type in Section 6.4.5, but it applies only to matroids that are representable over $\mathbb{F}_q$. In Section 6.4.6, we sketch a construction of large families of Tutte-equivalent, non-graphic, 3-connected matroids that are representable over $\mathbb{F}_q$.

Paving and sparse paving matroids, which we treat in Section 6.4.2, are important in part because they are so numerous; indeed, it is conjectured that almost all matroids are sparse paving. We characterize Tutte-equivalent paving matroids, and do the same for the somewhat larger class of near designs.

In many of the other sections we treat the Tutte uniqueness of matroids that play major roles in matroid theory: projective and affine geometries, Bose–Burton geometries, cycle matroids of complete graphs and complete bipartite graphs, wheels and whirls, and certain spikes. For other important matroids, such as Dowling geometries over groups, we identify their Tutte classes.

A matroid can be described relatively compactly by giving its cyclic flats and the rank of each. Section 6.4.11 contains several Tutte-uniqueness and equivalence results that relate to cyclic flats.

6.4.1 Matroid operations that preserve Tutte equivalence

We start by treating operations that, from a collection of Tutte-equivalent matroids, produce more such matroids. Section 6.2.2 covered direct sums and duality; here we focus on operations that are not related to graph operations.

Crapo [346] introduced the following two operations, which have long played important roles in matroid theory.

Definition 6.37. The *free extension of M by e*, denoted $M+e$, is the matroid on $E(M) \cup e$ whose bases are the subsets X of $E(M) \cup e$ with $|X| = r(M)$ for which $X \setminus e$ is independent in M. The *free coextension of M by e*, denoted $M \times e$, is the dual operation, that is, $M \times e = (M^* + e)^*$.

Geometrically, $M + e$ extends M by putting e as freely as possible into M without increasing the rank. The Tutte polynomial of $M + e$ is given by

$$T(M + e; x, y) = \frac{x}{x-1} T(M; x, y) + \left(y - \frac{x}{x-1}\right) T(M; 1, y). \qquad (6.5)$$

Combining free extension and contraction gives another important operation.

Definition 6.38. Fix an element e not in $E(M)$. The matroid $(M + e)/e$ is the *truncation*, $\mathrm{Tr}(M)$, of M. Dually, $(M \times e)\backslash e$ is the *elongation* or *free lift* of M. Iterating truncation k times gives the *k-fold truncation*, denoted $\mathrm{Tr}^k(M)$.

The bases of $\mathrm{Tr}(M)$ are the bases of the hyperplanes of M. Thus, $\mathrm{Tr}(M)$ is the freest way to reduce the rank of M by one.

The next result follows from Equations (6.3) and (6.5), and the deletion–contraction formula.

Theorem 6.39. *Free extension, free coextension, truncation, and elongation each preserve Tutte equivalence.*

Qin [940] observed that if M and N are connected, non-isomorphic, Tutte-equivalent matroids that are not self-dual, then $M \oplus N^*$ is not self-dual and its Tutte polynomial is symmetric in x and y, so $M \oplus N^*$ is Tutte equivalent to its dual. He proved that the same conclusion holds for $(M \oplus N^*) + e_1 \times e_2$, and that by iterating this construction, following each free extension by a free coextension, one obtains non-self-dual matroids that are Tutte equivalent to their duals and have arbitrarily high (Tutte and vertical) connectivity.

We turn to a non-commutative operation that Crapo and Schmitt [343, 345] introduced.

Definition 6.40. Let M and N be matroids on disjoint sets. The *free product*, $M \,\square\, N$, of M and N is the matroid on $E(M) \cup E(N)$ whose bases are the subsets B of $E(M) \cup E(N)$ with $|B| = r(M) + r(N)$ for which $B \cap E(M)$ is independent in M and $B \cap E(N)$ spans N.

Free extension and free coextension are special cases of the free product: if N is the rank-0 matroid on $\{e\}$, then $M \,\square\, N$ is $M + e$; dually, if M is the rank-1 matroid on $\{e\}$, then $M \,\square\, N$ is $N \times e$.

This operation has many attractive properties, among which we cite two and refer readers to [343, 345] for others. The free product $M \square N$ is the freest matroid whose restriction to $E(M)$ is M and whose contraction to $E(N)$ is N (in contrast, $M \oplus N$ is the least free such matroid). Also, $(M \square N)^* = N^* \square M^*$.

Bonin and de Mier [169] proved the following result.

Theorem 6.41. *The Tutte polynomial of $M \,\square\, N$ can be computed from just the Tutte polynomials of M and N. Thus, if M_1 and M_1' are Tutte equivalent, and likewise for M_2 and M_2', then $M_1 \square M_2$ and $M_1' \square M_2'$ are Tutte equivalent.*

In the same way one can get more Tutte-equivalent matroids from known Tutte-equivalent matroids using Brylawski's tensor product of matroids [245]. (See Section 1.2.10 for the corresponding graph operation.)

Definition 6.42. In a matroid N, fix an element e that is neither a loop nor a coloop. For each element p in M, take the 2-sum of M, at p, and a copy of N, at e, thereby replacing each element of M by a copy of $N \backslash e$, with these copies disjoint. The result is the *tensor product*, $M \otimes N_e$, of M and N.

Brylawski proved the following result. (See Diao, Hetyei, and Hinson [382] for an attractive proof using basis activities.)

Theorem 6.43. *With M, N, and e as above,*

$$T(M \otimes N_e) = T_L(N,e)^{r(M)} T_C(N,e)^{|E(M)|-r(M)} T\Big(M; \frac{T(N\backslash e)}{T_L(N,e)}, \frac{T(N/e)}{T_C(N,e)}\Big),$$

where $T_L(N,e)$ and $T_C(N,e)$ are uniquely determined by the equations

$$\begin{aligned} T(N/e) &= T_C(N,e) + (y-1)T_L(N,e) \\ T(N\backslash e) &= (x-1)T_C(N,e) + T_L(N,e). \end{aligned}$$

6.4.2 Paving and sparse paving matroids, perfect matroid designs, and near designs

A matroid M is *paving* if $|C|$ is either $r(M)$ or $r(M)+1$ for all circuits C of M. It is *sparse paving* if each non-spanning circuit of M is a hyperplane. Thus, sparse paving matroids are paving.

The matroids in Figure 6.3 are sparse paving; adding a fourth point freely to any three-point line yields a paving matroid that is not sparse paving.

Note that M is paving if and only if each of its flats of rank $r(M)-2$ or less is independent. Also, M is sparse paving if and only if both M and M^* are paving. Theorem 6.5 implies that any matroid that is Tutte equivalent to a paving matroid is paving, and likewise for sparse paving.

When M is paving, from $r(M)$, $|E(M)|$, and the number of hyperplanes of each size, a counting argument gives the multiset $\{(|A|, r(A)) : A \subseteq E(M)\}$, which is equivalent to $T(M; x, y)$. Indeed, we get the following characterization of the Tutte classes of paving matroids.

Theorem 6.44. *A matroid M is Tutte equivalent to a paving matroid N if and only if M is paving, they have the same rank, say r, the same number of elements, say n, and, for each j with $r \le j < n$, the same number of hyperplanes of size j.*

Paving matroids with at most one dependent hyperplane are of course Tutte unique, but even with just two dependent hyperplanes, say H and H', there can be multiple options for $|H \cap H'|$ (as in Figure 6.3), so such matroids are typically not Tutte unique. Tutte classes of paving matroids are commonly quite large, as the following remarks, based on Brylawski [245], show. He showed that at most $2^{(n+1)^3/4}$ polynomials are Tutte polynomials of matroids on n elements. Knuth [697] showed that there are at least

$$\frac{1}{n!} 2^{\binom{n}{\lfloor n/2 \rfloor}/2n} \tag{6.6}$$

non-isomorphic sparse paving matroids on n elements. Thus, some Tutte classes of these matroids contain a huge number of non-isomorphic matroids.

Note that the truncation of a paving matroid is a uniform matroid, and all flats of the same rank in a uniform matroid have the same size. These examples illustrate the next definition.

Definition 6.45. A *perfect matroid design* is a matroid in which all flats of the same rank have the same cardinality. A *near design* is a matroid M whose truncation $\mathrm{Tr}(M)$ is a perfect matroid design.

Other basic examples of perfect matroid designs include finite projective and affine geometries (see the next section), as well as their truncations.

With a near design M of rank r on an n-element set, we associate a triple $p_M = (\mathbf{f}, \mathbf{h}, n)$ where $\mathbf{f} = (f_0, f_1, \ldots, f_{r-2})$ and f_i is the common cardinality of the flats of rank i, and where $\mathbf{h} = (h_{r-1}, h_r, \ldots, h_{n-1})$ and h_j is the number of hyperplanes of M of cardinality j. The following result of Brylawski [245] significantly strengthens Theorem 6.44.

Theorem 6.46. *Whether a matroid is a near design can be determined from its Tutte polynomial, as can the triple p_M of a near design M. Also, two near designs M and M' are Tutte equivalent if and only if $p_M = p_{M'}$.*

6.4.3 Projective and affine geometries

All projective and affine geometries that we consider are finite. The projective geometry $PG(r-1,q)$ is the simplification of the vector matroid on the r-dimensional vector space $\mathbb{F}_q^r$ over the field $\mathbb{F}_q$. The *affine geometry* $AG(r-1,q)$ is the deletion $PG(r-1,q)\backslash H$, where H is a hyperplane of $PG(r-1,q)$; all such deletions are isomorphic. There are many finite projective and affine planes besides $PG(2,q)$ and $AG(2,q)$; see [628] for many constructions of these non-desarguesian planes.

All lines in a finite projective or affine geometry have the same number of points; see [103, 628]. The *order* of a finite projective geometry is q if its lines have $q+1$ points. The *order* of a finite affine geometry is q if its lines have q points. In both cases, $q \geq 2$. Thus, $PG(r-1,q)$ and $AG(r-1,q)$ have order q. One can show that projective planes of order q are exactly simple matroids with q^2+q+1 points in which all lines have $q+1$ points. Also, affine planes of order q are exactly rank-3 simple matroids with q^2 points in which all lines have q points. Thus, Theorem 6.44 gives the next result.

Theorem 6.47. *The projective planes of order q form a Tutte class, as do the affine planes of order q.*

Brylawski observed in [245] that, due to this result, answering the question of which polynomials are Tutte polynomials would require knowing which integers are the orders of projective planes. It is unknown whether the order of a non-desarguesian plane must be a prime power.

For $r > 3$, there is only one rank-r projective geometry of order q, namely $PG(r-1,q)$, and similarly for $AG(r-1,q)$. Both are Tutte unique.

Theorem 6.48. *For $r > 3$, the projective geometry $PG(r-1,q)$ and the affine geometry $AG(r-1,q)$ are Tutte unique.*

This result can be shown in many ways. Behind some of the simplest proofs is the following extremal result, the first part of which is due to Kung [730], with the extension to affine geometries observed by Bonin [164].

Theorem 6.49. *Let M be a simple matroid of rank r. If M has no minor isomorphic to the uniform matroid $U_{2,q+2}$, then*

$$|E(M)| \leq \frac{q^r - 1}{q - 1},$$

and this upper bound is attained only by projective geometries of order q.

If, in addition, M has at most q points on each line, then $|E(M)| \leq q^{r-1}$. This upper bound is attained only by affine geometries of order q.

The next result is one of several characterizations of projective and affine geometries, using Tutte invariants, that Bonin and Miller [172] deduced from Theorem 6.49.

Theorem 6.50.

1. *Any rank-r simple matroid with $(q^r - 1)/(q - 1)$ elements in which each line has at least $q + 1$ points is a projective geometry of order q.*
2. *Any rank-r simple matroid with q^{r-1} elements in which all lines have q points and all planes have at least q^2 points is an affine geometry of order q.*

Kung gave a number of subtler characterizations of these matroids in [733], all using Tutte invariants. For instance, he showed that $PG(r - 1, q)$, with $r > 3$, is characterized by any three of the following four numbers: (i) the size of its ground set, (ii) the number of lines, (iii) the number of sets of size three and rank two, and (iv) the coefficient of x^{r-2} in its characteristic polynomial, which we define next.

Definition 6.51. The *characteristic polynomial* $\chi(M; x)$ of M is

$$\chi(M; x) = (-1)^{r(M)} T(M; 1 - x, 0).$$

Comparing this definition to Equation (6.2) shows that the characteristic polynomial is the matroid counterpart of the chromatic polynomial.

6.4.4 The Bose–Burton geometries

Each simple graph on r vertices is a restriction of K_r; likewise, each rank-r simple matroid that is representable over $\mathbb{F}_q$ is a restriction of $PG(r - 1, q)$. Thus, a graph with no K_m-subgraph is like an $\mathbb{F}_q$-representable matroid with no $PG(m - 1, q)$-restriction. In this spirit, the following theorem, due to Bose and Burton [182], is a counterpart of Turán's theorem in extremal graph theory.

Theorem 6.52. *Let M be a restriction of $PG(r-1,q)$. If no restriction of M is isomorphic to $PG(m-1,q)$, then*

$$|E(M)| \leq \frac{q^r - q^{r-m+1}}{q-1}.$$

Moreover, equality holds if and only if M is $PG(r-1,q)\backslash F$ where F is a flat of rank $r-m+1$ of $PG(r-1,q)$.

Up to isomorphism, the deletion $PG(r-1,q)\backslash F$ depends only on the rank of the flat F. These counterparts of Turán graphs are called *Bose–Burton geometries.* They generalize affine geometries $AG(r-1,q)$ (the case $m=2$). The characterization of Bose–Burton geometries below was given by Ankney and Bonin [26], and was applied to establish their Tutte uniqueness.

Theorem 6.53. *Fix integers r and k with $r \geq 4$ and $1 \leq k \leq r-3$. A simple matroid M of rank r is isomorphic to $PG(r-1,q)\backslash F$, where F is a rank-k flat of $PG(r-1,q)$, if and only if the following four conditions hold.*

1. $|E(M)| = (q^r - q^k)/(q-1)$
2. *No rank-$(r-k+1)$ flat of M contains $(q^{r-k+1}-1)/(q-1)$ points.*
3. *Each hyperplane of M has cardinality*

$$\frac{q^{r-1}-q^k}{q-1} \quad \text{or} \quad \frac{q^{r-1}-q^{k-1}}{q-1}.$$

4. *Each flat of rank $r-2$ of M has cardinality*

$$\frac{q^{r-2}-q^k}{q-1}, \quad \frac{q^{r-2}-q^{k-1}}{q-1}, \quad \text{or, if } k>1, \quad \frac{q^{r-2}-q^{k-2}}{q-1}.$$

Theorem 6.54.

1. *The set of single-element deletions of projective planes of order q is a Tutte class.*
2. *For $r \geq 4$ and $1 \leq k \leq r-2$, the Bose–Burton geometry $PG(r-1,q)\backslash F$, where F is a rank-k flat of $PG(r-1,q)$, is Tutte unique.*

6.4.5 q-cones

In this section, we discuss a way to produce Tutte-equivalent matroids that are representable over a finite field.

Whittle [1161] introduced q-cones (or q-lifts) in his work on tangential blocks. To get a q-cone of a simple matroid M that is representable over $\mathbb{F}_q$, embed M in $PG(r(M),q)$, fix a point a (the apex) outside of the hyperplane

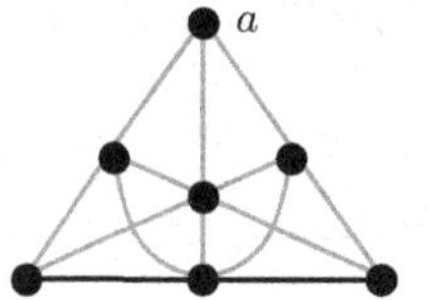

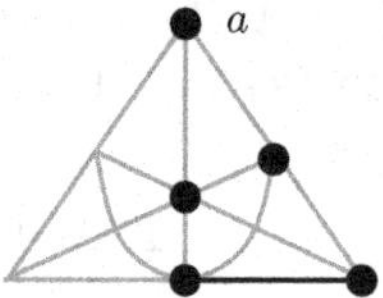

FIGURE 6.7: On the left, the Fano plane, $PG(2,2)$, is a 2-cone of $PG(1,2)$ (the darker line). On the right, a 2-cone of a 2-point line (the darker line).

that the image of M spans, and restrict $PG(r(M),q)$ to the union of all lines spanned by a and, in turn, the different elements in the image of M. For instance, each q-cone of $PG(r-1,q)$ is $PG(r,q)$, and each q-cone of $AG(r-1,q)$ is formed from $PG(r,q)$ by deleting all but one element in some hyperplane of $PG(r,q)$. See Figure 6.7 for examples.

We say "a q-cone" because the result can depend on the embedding: using inequivalent representations, Oxley and Whittle [900] showed that a matroid can have non-isomorphic q-cones. However, Bonin and Qin [173] showed that all of its q-cones are Tutte equivalent.

Theorem 6.55. *Let M be a rank-r simple matroid that is representable over $\mathbb{F}_q$. If M' is a q-cone of M, then $T(M';x,y)$ is given by*

$$\frac{y(y^q-1)^r}{(y-1)^{r+1}}T\Big(M;\frac{(x-1)(y-1)}{y^q-1}+1,y^q\Big)+\frac{q^r(xy-x-y)}{y-1}T\Big(M;\frac{x-1}{q}+1,y\Big).$$

Thus, any two q-cones of M are Tutte equivalent. Also, if M and M' are non-isomorphic Tutte-equivalent matroids that are representable over $\mathbb{F}_q$, then the same is true of their q-cones.

Behind the proof is the following result of Kung [731, Theorem 8.23].

Theorem 6.56. *The characteristic polynomial of any q-cone M' of a rank-r simple matroid M is given by $\chi(M';\lambda)=(\lambda-1)q^r\chi(M;\lambda/q)$.*

Another element in the proof is the coboundary polynomial, which Crapo [349] defined and which is equivalent to the Tutte polynomial. As in [173], it can be used to lift some results on characteristic polynomials to Tutte polynomials.

Definition 6.57. The *coboundary polynomial* $\overline{\chi}(M;x,y)$ is

$$\overline{\chi}(M;x,y)=\sum_{A\subseteq E(M)}y^{|A|}\chi(M/A;x)=\sum_{F\in\mathcal{F}(M)}y^{|F|}\chi(M/F;x), \qquad (6.7)$$

where $\mathcal{F}(M)$ is the set of flats of M.

The following formula, which is well-known and easy to prove, shows that $T(M;x,y)$ and $\overline{\chi}(M;x,y)$ are equivalent.

$$T(M;x,y)=\frac{\overline{\chi}(M;(x-1)(y-1),y)}{(y-1)^{r(M)}} \qquad (6.8)$$

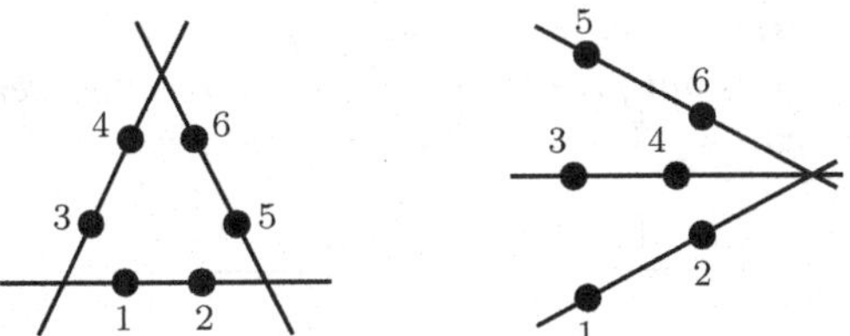

FIGURE 6.8: In an embedding of the uniform matroid $U_{3,6}$ in a projective plane $PG(2,q)$ with q sufficiently large, the lines spanned by $\{1,2\}$, $\{3,4\}$, and $\{5,6\}$ may intersect in three different points of $PG(2,q)$ or in one point of $PG(2,q)$.

6.4.6 Complements in $PG(r-1,q)$

In this section we sketch an application of inequivalent representations to construct large families of non-isomorphic Tutte-equivalent $\mathbb{F}_q$-representable matroids. The construction, from Bonin [165], uses the following special case of a result of Brylawski [242, Theorem 5.9].

Theorem 6.58. *For any sets S and T in $PG(r-1,q)$, if the restrictions $PG(r-1,q)|S$ and $PG(r-1,q)|T$ are Tutte equivalent, then so are the deletions $PG(r-1,q)\backslash S$ and $PG(r-1,q)\backslash T$.*

For example, both matroids in Figure 6.3 are $\mathbb{F}_q$-representable for all prime powers $q \geq 4$, and if $PG(r-1,q)|S$ and $PG(r-1,q)|T$ are isomorphic to M and N, respectively, then $PG(r-1,q)\backslash S$ and $PG(r-1,q)\backslash T$ are Tutte equivalent. These deletions are not isomorphic since each has exactly two lines with exactly $q-2$ elements, but only in the second case do those lines intersect. In general, no conclusion about whether $PG(r-1,q)\backslash S$ and $PG(r-1,q)\backslash T$ are isomorphic can be drawn solely from knowing whether $PG(r-1,q)|S$ and $PG(r-1,q)|T$ are isomorphic: an argument is needed.

We call $PG(r-1,q)\backslash S$ a *complement* of M if $PG(r-1,q)|S$ is isomorphic to M. By Theorem 6.58, the complements of M are Tutte equivalent.

A representation of a simple matroid M by an $r \times n$ matrix over $\mathbb{F}_q$ is an embedding of M in $PG(r-1,q)$. As Figure 6.8 suggests, two such embeddings can be geometrically different due to information unrelated to M that comes with an embedding. In matrix terms, for $q > 3$ there can be pairs of matrix representations of M that cannot be obtained from each other by matrix operations that preserve the matroid. See [894, Section 14.6] for an in-depth discussion of inequivalent representations and, in particular, for the results of Oxley, Vertigan, and Whittle [896] about r-swirls and free r-spikes (defined in Sections 6.4.8 and 6.4.9) that are key to the constructions behind the results that we cite below. The idea in [165] is to start with a single matroid, such as an r-swirl or free r-spike, having numerous inequivalent representations, and extend it in a certain way so that these representations yield non-isomorphic Tutte-equivalent complements. The following two results are obtained this way. (Similar results hold using embeddings in affine geometries.)

Theorem 6.59. *Assume that q is p^t for some prime p, and $q > 5$; if $p = 2$, assume also that $q-1$ is not prime. Let d be the largest proper divisor of $q-1$. For each integer r with $r \geq 3$, there is a set of at least*

$$\frac{(d-1)^{r-1}(q-1-d)(q-2-d)}{t}$$

non-isomorphic, Tutte-equivalent, 3-connected matroids of rank $2r+2$ that are representable over $\mathbb{F}_q$.

Theorem 6.60. *Let $q = p^t$ with p prime, $t > 1$, and $q > 4$. For each integer r with $r > 4$, there is a set of at least*

$$\frac{(p^{t-1}-1)^{r-2}(p^t-p^{t-1})(p^t-p^{t-1}-1)}{t}$$

non-isomorphic, Tutte-equivalent, 3-connected matroids of rank $r(r+3)/2$ that are representable over $\mathbb{F}_q$.

6.4.7 The matroid $M(K_{r+1})$, its truncations, and $M(K_{m,n})$

By proving the following characterization of the cycle matroid of the complete graph K_{r+1}, Bonin and Miller [172] proved that $M(K_{r+1})$ is Tutte unique.

Theorem 6.61. *Fix a positive integer r. Let M be a simple matroid with*

1. *$\binom{r+1}{2}$ points,*
2. *$\binom{r+1}{3}$ lines with three points,*
3. *no 5-point planes, $\binom{r+1}{4}$ planes with six points, no planes with more than six points, and*
4. *no rank-4 flats with more than ten points.*

If $r(M) = r$, then $M \cong M(K_{r+1})$. Thus, $M(K_{r+1})$ is Tutte unique.

The following extension of Theorem 6.61 is from de Mier and Noy [848]. Recall from Section 6.4.1 that $\mathrm{Tr}^k(M(K_{r+1}))$ is the k-fold truncation of $M(K_{r+1})$ and so has rank $r-k$.

Theorem 6.62. *Fix positive integers r and k with $r-k \geq 5$. Assume that a simple matroid M satisfies conditions 1–4 of Theorem 6.61 and*

5. *for i with $4 \leq i \leq r-k$, M has the same number of i-element independent sets as $\mathrm{Tr}^k(M(K_{r+1}))$.*

If $r(M) = r-k$, then $M \cong \mathrm{Tr}^k(M(K_{r+1}))$. Thus, $\mathrm{Tr}^k(M(K_{r+1}))$ is Tutte unique.

The proof builds on that of Theorem 6.61 and uses the theory of k-chordal matroids from Bonin and de Mier [166], which we now sketch. For the following, $g(M)$ denotes the *girth* of a matroid M, which is the minimum circuit size of M; if M has no circuits, then $g(M) = \infty$.

Definition 6.63. A circuit C of a matroid M is *chordal* if there are circuits C_1 and C_2 of M and an element e in $C_1 \cap C_2$ such that $|C_1|, |C_2| < |C|$ and $C = (C_1 \cup C_2) \setminus e$. For an integer k with $g(M) < k \leq r(M) + 1$, a matroid M is k-*chordal* if each circuit C of M with $g(M) < |C| \leq k$ is chordal.

By the following result from [166], knowing that one in a pair of rank-r Tutte-equivalent matroids is r-chordal greatly simplifies what must be checked in order to show that the two matroids are isomorphic.

Theorem 6.64. *Assume that M is $r(M)$-chordal, and that M and N have*

1. *the same rank and the same girth,*
2. *the same number of circuits of cardinality $g(M)$, and*
3. *the same number of i-element independent sets for each i with $g(M) < i \leq r(M)$.*

If $\phi : E(M) \to E(N)$ is a bijection such that for each circuit C of M with $|C| = g(M)$, its image $\phi(C)$ is a circuit of N, then ϕ is an isomorphism of M onto N.

Note that conditions 1–3 hold if $T(M; x, y) = T(N; x, y)$. See [166] for a similar result that applies to the truncations of M and N to rank k when M is known to be k-chordal. Theorem 6.64 is also one of the tools that de Mier and Noy [848] used to prove the following result.

Theorem 6.65. *For all positive integers m and n, the cycle matroid $M(K_{m,n})$ is Tutte unique.*

The heart of the proof is a characterization of $M(K_{m,n})$ using the following invariants: the size of the ground set, the rank, the girth, the maximum size of flats of each rank, the number of four-element planes, the numbers of rank-4 sets having five and six elements, the numbers of rank-5 flats with seven, eight, and nine elements, and the numbers of independent sets of each size four and greater.

6.4.8 Wheels, whirls, and some generalizations

The cycle matroid $\mathcal{W}_r$ of the r-spoke wheel W_r is also called the rank-r *wheel*. Wheels and whirls (defined below) play major roles in structural results about graphs and matroids (see [894]). Label the spokes of W_r as $s_1, s_2, \ldots, s_r$ and the rim edges as $c_1, c_2, \ldots, c_r$ so that $\{s_i, c_i, s_{i+1}\}$ is a line of $\mathcal{W}_r$ (interpret subscripts modulo r). Thus, $\{s_1, s_2, \ldots, s_r\}$ is a basis of $\mathcal{W}_r$, and $\{c_1, c_2, \ldots, c_r\}$ is a circuit-hyperplane (i.e., a circuit and a hyperplane). Below we use circuit-hyperplane relaxation, as defined in Chapter 4.

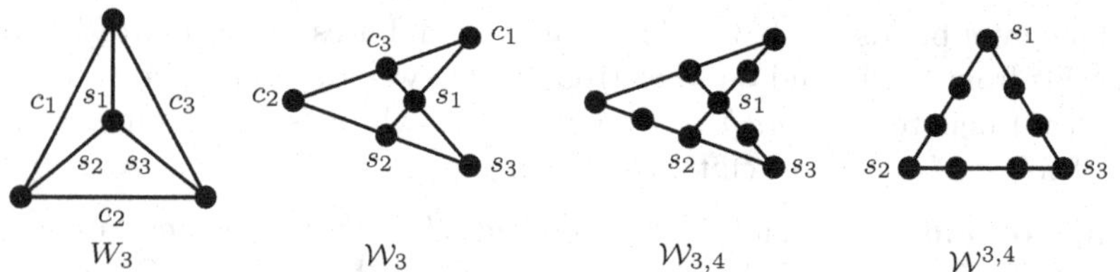

FIGURE 6.9: The 3-wheel W_3 (which is K_4), its cycle matroid $\mathcal{W}_3$, its extension $\mathcal{W}_{3,4}$, and the relaxation, the $(3,4)$-whirl $\mathcal{W}^{3,4}$.

Definition 6.66. For an integer $t \geq 3$, let $X_1, X_2, \ldots, X_r$ be sets of size $t-3$ that are disjoint from each other and from $E(\mathcal{W}_r)$. The *(r,t)-wheel* $\mathcal{W}_{r,t}$ is formed from $\mathcal{W}_r$ by, for each i with $1 \leq i \leq r$, adding the points in X_i freely to the line $\{s_i, c_i, s_{i+1}\}$. To get the *(r,t)-whirl* $\mathcal{W}^{r,t}$, relax the circuit-hyperplane $\{c_1, c_2, \ldots, c_r\}$ of $\mathcal{W}_{r,t}$. In particular, relaxing this circuit-hyperplane in $\mathcal{W}_r$ gives the *r-whirl* $\mathcal{W}^r$.

Adding points freely to a flat, illustrated in Figure 6.9, is made precise by iterating the operation of principal extension. The (r,t)-whirl $\mathcal{W}^{r,t}$ is also obtained from the whirl $\mathcal{W}^r$ by, for each i with $1 \leq i \leq r$, adding the points in X_i freely to the line $\{s_i, c_i, s_{i+1}\}$. Note that $\mathcal{W}_{r,3} = \mathcal{W}_r$ and $\mathcal{W}^{r,3} = \mathcal{W}^r$. The $(r,4)$-whirls are called *swirls*.

The Tutte uniqueness of $\mathcal{W}_{r,t}$ and $\mathcal{W}^{r,t}$ follows from the characterization of these matroids, due to Bonin and de Mier [166], stated next; the proof uses the theory of k-chordal matroids (see Section 6.4.7). Theorem 6.67 does not cover $\mathcal{W}_3$ and $\mathcal{W}^3$, which are also Tutte unique.

Theorem 6.67. *Let r and t be integers with $r, t \geq 3$ and $r+t > 6$. Let M be a simple rank-r matroid with $|E(M)| = (t-1)r$ for which*

1. *the number of t-point lines is r, and*
2. *$|F| \leq (r(F)-1)(t-1)+1$ for all flats F with $2 \leq r(F) \leq r-1$.*

If, for each $s \leq r$, the matroid M has the same number of independent sets of s elements as $\mathcal{W}_{r,t}$ (respectively, $\mathcal{W}^{r,t}$), then M is isomorphic to $\mathcal{W}_{r,t}$ (respectively, $\mathcal{W}^{r,t}$). In particular, (r,t)-wheels and (r,t)-whirls are Tutte unique.

6.4.9 Spikes and some generalizations

Spikes, which play prominent roles in matroid structure theory (see [894] and [516, Section 3]), are the special case of $(r,r,3)$-spikes in the next definition.

Definition 6.68. Let r, s, and t be integers with $r \geq 3$, $s \geq r-1$, and $t \geq 3$. An *(r,s,t)-spike with tip a* is a rank-r simple matroid on the union of s lines $\ell_1, \ell_2, \ldots, \ell_s$, each having t elements and containing a, such that each union of $r-2$ of these lines is a hyperplane.

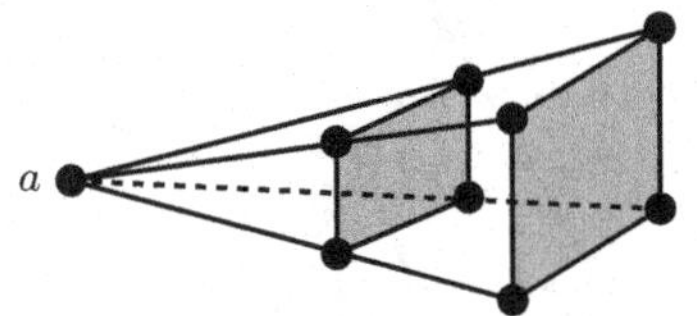

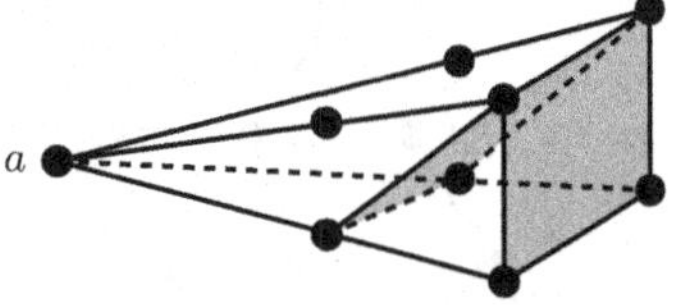

FIGURE 6.10: Two Tutte-equivalent, non-isomorphic $(4, 4, 3)$-spikes, each having two circuit-hyperplanes (shaded).

For example, $PG(2, q)$ is a $(3, q+1, q+1)$-spike; any point can serve as the tip (a phenomenon that can happen only in rank 3).

The hyperplanes of an (r, s, t)-spike M that contain the tip a are those given in the definition; a hyperplane H that does not contain a contains at most one point from each of $\ell_1, \ell_2, \ldots, \ell_s$, so $r - 1 \leq |H| \leq s$. For i with $r - 1 \leq i \leq s$, let h_i^M be the number of hyperplanes H of M with $a \notin H$ and $|H| = i$. The following result of Bonin and de Mier [166] is analogous to Theorem 6.44.

Theorem 6.69. *Two (r, s, t)-spikes M and N are Tutte equivalent if and only if $h_i^M = h_i^N$ for all i with $r - 1 \leq i \leq s$.*

Note that $h_r^M \leq 2^{r-1}$ for any $(r, r, 3)$-spike. One can show that there is a unique $(r, r, 3)$-spike with $h_r^M = 2^{r-1}$; furthermore, it is binary. For example, the only $(3, 3, 3)$-spike with $h_3^M = 4$ is $PG(2, 2)$. There is also a unique (r, s, t)-spike in which each hyperplane that does not contain a is independent, so $h_{r-1}^M = \binom{s}{r-1}(t-1)^{r-1}$ and $h_i^M = 0$ for $r \leq i \leq s$.

Definition 6.70. The *binary r-spike* is the $(r, r, 3)$-spike with $h_r^M = 2^{r-1}$. The *free (r, s, t)-spike* is the (r, s, t)-spike M with $h_{r-1}^M = \binom{s}{r-1}(t-1)^{r-1}$.

The following result of Bonin and de Mier [166] shows that in many cases, matroids that are Tutte equivalent to an (r, s, t)-spike are (r, s, t)-spikes. The third condition mildly limits the scope of this result.

Theorem 6.71. *Let r, s, and t be integers with $r \geq 5$, $s \geq r - 1$, and $t \geq 3$. Let M be a rank-r simple matroid with $|E(M)| = s(t-1) + 1$ that has*

1. *s lines with t points, and $s\binom{t}{3}$ circuits with three elements,*
2. *$\binom{s}{2}$ planes with $2t - 1$ points, and $\binom{s}{2}\binom{t-1}{2}^2$ circuits with four elements, and*
3. *for each j with $j \geq r - 1$, no hyperplane with $j(t-1) + 1$ points.*

Then M is an (r, s, t)-spike. If in addition there are no r-circuits, then M is the free (r, s, t)-spike. If $s = r$, $t = 3$, and M has 2^{r-1} circuits with r elements, then M is the binary r-spike. Thus, the free (r, s, t)-spike and the binary r-spike are Tutte unique.

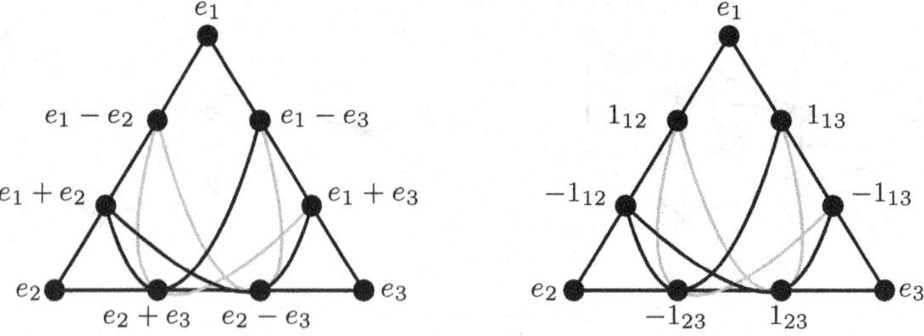

FIGURE 6.11: The Dowling geometry $Q_r(3)$, labeled, on the left, by points in $PG(2,3)$, and, on the right, as in Theorem 6.72 using the group $\mathsf{G} = \{1, -1\}$. One can see such diagrams as geometric encodings of the group table.

Together Theorems 6.69 and 6.71 identify many Tutte classes that contain only spikes.

6.4.10 Dowling geometries

The matroids Dowling introduced in [414], which were later named Dowling geometries (or lattices), are, in many ways, to groups as projective geometries are to fields (see, e.g., [90]).

We start with the special case that Dowling explored in [413, 415]. Since $PG(r-1,q)$ is the simplification of the matroid on the vector space $\mathbb{F}_q^r$, we can take its elements to be vectors if we identify vectors that span the same 1-dimensional subspace. Thus, the set $\{e_1, e_2, \ldots, e_r\}$ of standard basis vectors is a basis of $PG(r-1,q)$. Let S be the union of the lines of $PG(r-1,q)$ that are spanned by pairs of elements in this basis, that is,

$$S = \bigcup_{1 \leq i < j \leq r} \operatorname{cl}(\{e_i, e_j\}).$$

The restriction $PG(r-1,q)|S$, which we denote by $Q_r(q)$, is a representable Dowling geometry. We can write each element in $\operatorname{cl}(\{e_i, e_j\}) \setminus \{e_i, e_j\}$ in the form $e_i - a\,e_j$ for a unique element a in the multiplicative group $\mathbb{F}_q^*$ of non-zero elements of $\mathbb{F}_q$. Note that for any $a, b \in \mathbb{F}_q^*$, the set

$$\{e_i - a\,e_j,\ e_j - b\,e_k,\ e_i - ab\,e_k\}$$

is a line of $Q_r(q)$, and the collection of such lines encodes the group $\mathbb{F}_q^*$. To generalize $Q_r(q)$, we use an arbitrary group in place of $\mathbb{F}_q^*$, as follows.

Theorem 6.72. *Let* G *be a group, written multiplicatively. There is a unique rank-*r *matroid* $Q_r(\mathsf{G})$*, called the* rank-r Dowling geometry over G*, with*

$$E(Q_r(\mathsf{G})) = \{p_1, p_2, \ldots, p_r\} \cup \{a_{ij} : a \in \mathsf{G} \text{ and } 1 \leq i < j \leq r\}$$

that has the following lines:

1. $\text{cl}(\{p_i, p_j\}) = \{p_i, p_j\} \cup \{a_{ij} : a \in \mathsf{G}\}$ *for* $1 \le i < j \le r$*;*

2. $\{a_{ij}, b_{jk}, (ab)_{ik}\}$ *for* $a, b \in \mathsf{G}$ *and* $1 \le i < j < k \le r$*; and*

3. $\{p_i, a_{jk}\}$ *and* $\{a_{hi}, b_{jk}\}$*, where* h, i, j, k *are distinct.*

If $r = 3$, then the same is true for any quasigroup (Latin square) G. We call $p_1, p_2, \ldots, p_r$ the *joints* of $Q_r(\mathsf{G})$. We set $a_{ji} = (a^{-1})_{ij}$. When G is finite (as assumed in the rest of this section), $|E(Q_r(\mathsf{G}))| = r + \binom{r}{2}|\mathsf{G}|$. If G is the trivial group, then $Q_r(\mathsf{G})$ is isomorphic to $M(K_{r+1})$ (identify $p_1, p_2, \ldots, p_r$ with the edges of a $K_{1,r}$-subgraph). (All Dowling geometries can be seen via group-theoretic expansions of complete graphs; see [894, Section 6.10].)

Definition 6.73. A flat X of a matroid M is *modular* if

$$r(X) + r(Y) = r(X \cup Y) + r(X \cap Y)$$

for every flat Y of M.

One can show that the closure of any set of joints of $Q_r(\mathsf{G})$ is a modular flat. Stanley [1026] defined a matroid M to be *supersolvable* if it has a chain of flats, $F_0 \subsetneq F_1 \subsetneq \cdots \subsetneq F_{r(M)}$ where each F_i is modular and $r(F_i) = i$. Thus, Dowling geometries are supersolvable. Stanley [1026] proved the following factorization of $\chi(M; x)$ when M is supersolvable. (See Björner and Ziegler [136] for a related factorization of a broken-circuit complex.)

Theorem 6.74. *If* $F_0 \subsetneq F_1 \subsetneq \cdots \subsetneq F_r$ *is a chain of modular flats in a simple rank-*r *supersolvable matroid* M*, then its characteristic polynomial is*

$$\chi(M; x) = \prod_{i=1}^{r} (x - |F_i - F_{i-1}|).$$

With this, one can show that for a Dowling geometry,

$$\chi(Q_r(\mathsf{G}); x) = \prod_{i=0}^{r-1} (x - i|\mathsf{G}| - 1).$$

Thus, $\chi(Q_r(\mathsf{G}); x)$ depends only on r and $|\mathsf{G}|$. Dowling [414] showed that for each rank-$(r-i)$ flat X of $Q_r(\mathsf{G})$, the simplification of the contraction $Q_r(\mathsf{G})/X$ is isomorphic to $Q_i(\mathsf{G})$. The description of the flats in [414] implies that the number of flats of each rank i and cardinality j also depends only upon $|\mathsf{G}|$, r, i, and j. Combining these results, Definition 6.57, and Equation (6.8) gives the next theorem.

Theorem 6.75. *If* $|\mathsf{G}| = |\mathsf{G}'|$*, then* $Q_r(\mathsf{G})$ *and* $Q_r(\mathsf{G}')$ *are Tutte equivalent.*

By this and the next result, from Bonin and Miller [172], the set of all Dowling geometries of rank r based on groups (or quasigroups, if $r = 3$) of a fixed order is a Tutte class. (Theorem 6.61 treats $g = 1$.)

Theorem 6.76. *Fix an integer $g > 1$. If, for a rank-r simple matroid M,*

1. $|E(M)| = \binom{r}{2}g + r$;
2. *for $2 \leq k \leq 6$, flats of rank k have at most $\binom{k}{2}g + k$ points, and $\binom{r}{k}$ of them have $\binom{k}{2}g + k$ points;*
3. *$\binom{r}{3}g^2$ lines of M have three points;*
4. *$\binom{r}{4}\binom{r-4}{2}g$ rank-5 flats have $6g + 5$ points;*
5. *no rank-6 flat F satisfies $14g + 6 \leq |F| < 15g + 6$; and*
6. *all rank-7 flats (if any) have fewer than $22g + 8$ points;*

then $M \cong Q_r(\mathsf{G})$ for some group (or quasigroup, if $r = 3$) G of order g. Thus, any matroid that is Tutte equivalent to $Q_r(\mathsf{G})$ is a Dowling geometry over a group (or quasigroup, if $r = 3$) of order $|\mathsf{G}|$. Also, for a prime p and integer $r > 3$, the Dowling geometry $Q_r(\mathbb{Z}_p)$ is Tutte unique.

We now turn to two variations on these results. The matroids defined next play important roles in the critical problem (see [1160] and [731, Section 8]).

Definition 6.77. The *rank-r jointless Dowling geometry over* G, denoted $Q'_r(\mathsf{G})$, is $Q_r(\mathsf{G})\backslash\{p_1, p_2, \ldots, p_r\}$ where $p_1, p_2, \ldots, p_r$ are the joints of $Q_r(\mathsf{G})$.

Theorem 6.75 has the following counterpart.

Theorem 6.78. *If $|\mathsf{G}| = |\mathsf{G}'|$, then $Q'_r(\mathsf{G})$ and $Q'_r(\mathsf{G}')$ are Tutte equivalent.*

Sarmiento [983] proved a counterpart of the first part of Theorem 6.76 for rank four and higher (with similar conditions on the flats of rank seven or less), and obtained the following result.

Theorem 6.79. *For $r \geq 4$ and $g \geq 4$, jointless Dowling geometries of rank r over groups of order g form a Tutte class. This also applies to $r = 3$, using quasigroups, if, in addition, g is even.*

Via the operation of complete principal truncation at a modular flat of rank k in $Q_{r+k-1}(\mathsf{G})$, we obtain a variation on $Q_r(\mathsf{G})$, denoted $Q_{r,k}(\mathsf{G})$, that is also supersolvable and in which the lines $\mathrm{cl}(p_i, p_r)$ contain multiple copies of G. The following result of Qin [941] is a counterpart of Theorem 6.72.

Theorem 6.80. *Let G be a group. There is a unique matroid $Q_{r,k}(\mathsf{G})$ of rank r with*

$$E(Q_{r,k}(\mathsf{G})) = \{p_1, p_2, \ldots, p_r\} \cup \{a_{ij} : a \in \mathsf{G} \text{ and } 1 \leq i < j < r + k\}$$

in which the lines that have more than two points are:

1. $\mathrm{cl}(\{p_i, p_j\}) = \{p_i, p_j\} \cup \{a_{ij} : a \in \mathsf{G}\}$ *for* $1 \leq i < j < r$,

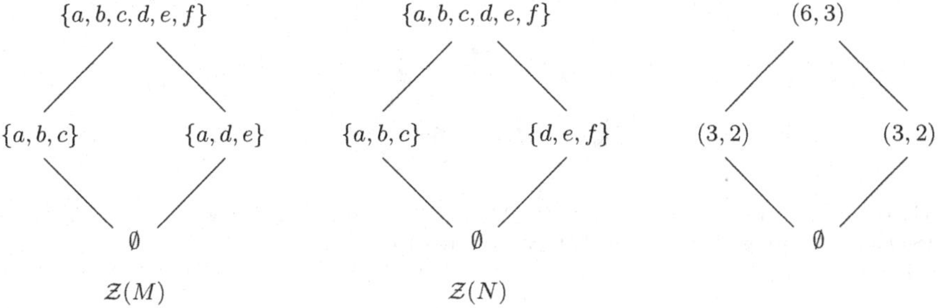

FIGURE 6.12: The lattices of cyclic flats of the matroids in Figure 6.3, along with their common configuration.

2. $\text{cl}(\{p_i, p_r\}) = \{p_i, p_r\} \cup \{a_{ij} : a \in \mathsf{G}, r \leq j < r + k\}$, *for* $1 \leq i < r$,
3. $\{a_{ij}, b_{jk}, (ab)_{ik}\}$ *for* $a, b \in \mathsf{G}$ *and* $1 \leq i < j < k < r$,
4. $\{a_{ij}, b_{jk}, (ab)_{ik}\}$ *for* $a, b \in \mathsf{G}$, $1 \leq i < j < r$ *and* $r \leq k < r + k$.

Qin [941] gave a counterpart, for $Q_{r,k}(\mathsf{G})$ with $r \geq 4$, of Theorem 6.76, using similar information about flats of rank seven or less.

6.4.11 Cyclic flats and two classes of transversal matroids

Cyclic flats play several roles in this section.

Definition 6.81. A flat in a matroid is *cyclic* if it is a union of circuits.

A matroid M is determined by its cyclic flats and the rank of each. The set of cyclic flats of M, ordered by inclusion, is a lattice, which we denote by $\mathcal{Z}(M)$. Two examples are shown in Figure 6.12. The meet, in $\mathcal{Z}(M)$, of cyclic flats X and Y is the union of the circuits that are contained in $X \cap Y$; their join is $\text{cl}(X \cup Y)$. (See [169] for more on this topic.)

Definition 6.82. A matroid is *nested* if it can be obtained from a matroid on one element by repeatedly (i) adding a coloop and (ii) taking a free extension.

See Figure 6.13 for an example. Nested matroids have been rediscovered several times in different settings since Crapo [346] introduced them (see [168, Section 4]). Oxley, Prendergast, and Row [895] showed that a matroid M is nested if and only if $\mathcal{Z}(M)$ is a chain. That property plays a role in the proof of the following result from de Mier [846], which identifies $\binom{n}{r}$ Tutte-unique matroids of rank r on n elements.

Theorem 6.83. *Each nested matroid is Tutte unique.*

Omer Giménez constructed a family of transversal matroids that were later shown to be Tutte equivalent; the family grows super-exponentially in the size of the ground set. An account of his work is given in [169]. (The perspective in the next definition differs from that in [169].)

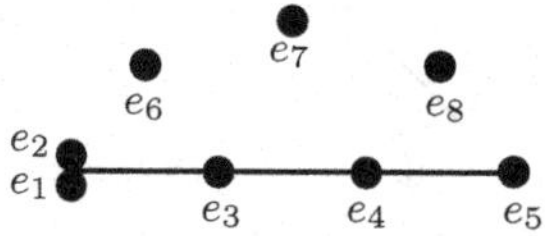

FIGURE 6.13: A rank-3 nested matroid. The labels $e_2, e_3, \ldots, e_8$ show the order in which these elements are added to extend the rank-1 matroid on e_1.

Definition 6.84. Fix an integer $n \geq 2$. Let E be the union of four disjoint sets,

$$X = \{x_1, x_2, \ldots, x_n\}, \quad Y = \{y_1, y_2, \ldots, y_n\}, \quad S_0, \quad \text{and} \quad T_0,$$

where $|S_0| = n+2$ and $|T_0| = n+3$. Fix a permutation σ of $[n]$. For an integer i with $1 \leq i \leq n$, let

$$S_i = S_{i-1} \cup \{x_i, y_i\} \quad \text{and} \quad T_i^\sigma = T_{i-1}^\sigma \cup \{x_i, y_{\sigma(i)}\},$$

where $T_0^\sigma = T_0$. Set $\mathcal{A}^\sigma = \{S_i : 0 \leq i \leq n\} \cup \{T_i^\sigma : 0 \leq i \leq n\}$. Let Γ^σ be the bipartite graph on the vertex set $E \cup \mathcal{A}^\sigma$ whose edge set is

$$\{(e, Z) : e \in E,\ Z \in \mathcal{A}^\sigma, \text{ and } e \in Z\}.$$

Let M^σ be the transversal matroid on E given by Γ^σ, that is, the bases of M^σ are the subsets B of E for which there is a matching of B onto $\mathcal{A}^\sigma$.

Thus, this defines a collection of $n!$ transversal matroids of rank $2n+2$ on a set of $4n+5$ elements. Giménez proved the following result.

Theorem 6.85. *For different permutations σ and τ of $[n]$, the transversal matroids M^σ and M^τ are not isomorphic, but their (abstract) lattices of cyclic flats are isomorphic.*

Shoda [1005] proved the following result.

Theorem 6.86. *For any integer $n \geq 2$, the transversal matroids M^σ, for all permutations σ of $[n]$, are Tutte equivalent.*

As noted in [169], Giménez's construction can be generalized by using a k-tuple of permutations of $[n]$ (the generalization need not be transversal; also, the size of the sets on which such matroids are defined is a function of both n and k). Shoda [1005] also extended Theorem 6.86 to that setting.

Eberhardt [435] developed a broad generalization of these results that yields many Tutte-equivalent matroids and gives us a better idea of how much data is needed to compute the Tutte polynomial. He defined the *configuration* of a matroid M to be the isomorphism class of the lattice $\mathcal{Z}(M)$, along with the size and rank of each cyclic flat. For example, the matroids in

Figure 6.3 have the same configuration, which is shown in Figure 6.12. He used the following reformulation of the Tutte polynomial, due to Plesken and Bächler [922]:

$$T(M;x,y) = \sum_{Z\in\mathcal{Z}(M)} c(M,Z;x-1)\, f(M,Z;y-1),$$

where, using the map e that takes a flat X of M to the union of the circuits that are contained in X (equivalently, e removes the coloops of $M|X$ from X),

$$c(M,Z;x) = \sum_{\substack{\text{flats } A\\ e(A)=Z}} x^{r(M)-r(A)},$$

and

$$f(M,Z;y) = \sum_{\substack{\text{sets } B\\ \mathrm{cl}(B)=Z}} y^{|B|-r(B)}.$$

He showed how to recursively compute the polynomials $c(M,Z;x)$ and $f(M,Z;y)$, and from that obtained the following result.

Theorem 6.87. *The Tutte polynomial of a matroid can be computed from its configuration and its number of elements. Thus, matroids with the same configuration and the same number of elements are Tutte equivalent.*

He also identified potentially smaller sets of data (e.g., when the matroid has a lot of symmetry) that suffice to determine the Tutte polynomial.

6.5 Related results

We first discuss the implications of Tutte uniqueness for reconstruction, both for graphs and for matroids, and then survey some results akin to Tutte uniqueness for polynomials that are related to the Tutte polynomial. We omit results that appear in other chapters (see Chapter 11 for the chromatic polynomial, Chapter 26 for a discussion of the question of whether trees are determined by the U polynomial, and Chapter 33 for results about antimatroids and greedoids).

6.5.1 Reconstruction

The *deck of deletions* of a graph G is the multiset of its unlabeled, single-vertex deletions; thus, for each graph H on $|V(G)|-1$ vertices, the deck tells us how many vertices $v \in V(G)$ have $G\backslash v$ isomorphic to H. We say G is *reconstructible* if all graphs that have the same deck as G are isomorphic to G.

The deck of single-edge deletions and edge reconstructible graphs are defined similarly. The graph reconstruction conjecture of Kelly and Ulam (in Kelly [687]) asserts that every graph with at least three vertices is reconstructible. The edge reconstruction conjecture asserts that every graph with at least four edges is edge reconstructible. Both conjectures are still open.

Greenwell [581] showed that reconstructible graphs without isolated vertices are edge reconstructible. Tutte [1100] proved that $T(G; x, y)$ can be computed from the deck of deletions of G. Thus, the next result follows.

Theorem 6.88. *Any graph that is Tutte unique is reconstructible and so edge reconstructible.*

The cocircuits (complements of the hyperplanes) of a matroid M can play a role like that of the vertices of a graph, so a counterpart of the deck of deletions of a graph is the *deck of hyperplanes*, which is the multiset of the unlabeled restrictions of M to its hyperplanes. Thus, for each matroid N of rank $r(M) - 1$, the deck tells us the number of hyperplanes H of M for which $M|H$ is isomorphic to N. We call M *hyperplane reconstructible* if matroids that have the same deck of hyperplanes as M are isomorphic to M. Many matroids are not hyperplane reconstructible. For instance, a matroid M is paving if and only if all matroids in its deck of hyperplanes are uniform; also, the deck gives us the sizes of the hyperplanes, as well as the number of elements in M (see Miller [850]); thus, by Theorem 6.44, a paving matroid is hyperplane reconstructible if and only if it is Tutte unique.

For a matroid M, its *deck of deletions* is the multiset of its unlabeled single-element deletions. This is a counterpart of the deck of single-edge deletions of a graph. A matroid M is *deletion reconstructible* if all matroids that have the same deck of deletions as M are isomorphic to M.

Brylawski [243] extended Tutte's result by showing that $T(M; x, y)$ can be computed from the deck of hyperplanes of M. Also, Miller [850] extended Greenwell's work by showing that the deck of hyperplanes can be deduced from the deck of deletions, so hyperplane reconstructible matroids are also deletion reconstructible. This gives the following counterpart of Theorem 6.88.

Theorem 6.89. *Each Tutte-unique matroid is hyperplane reconstructible and so deletion reconstructible.*

6.5.2 The chromatic and flow polynomial

The flow polynomial $F(G; y)$ of a graph G is related to its Tutte polynomial by the equation

$$F(G; y) = (-1)^{|V(G)|-|E(G)|+k(G)} T(G; 0, 1 - y).$$

Adapting Definition 6.1 in the obvious way defines flow-unique graphs. When studying flow-unique graphs, one considers only graphs with no bridges (otherwise $F(G; y) = 0$) and no degree-two vertices (since G and its reduced graph have the same flow polynomial). Duan, Wu, and Yu [421] showed that K_5, $K_{3,3}$, and the dual graphs of 3-theta graphs are flow-unique. They also initiated the study of graphs G that are determined by the pair $(\chi(G;x), F(G;y))$, where now bridges and degree-two vertices are allowed. This yields a notion between chromatic and Tutte uniqueness. In [421], they obtained the following result.

Theorem 6.90. *The following graphs G are determined up to isomorphism by the pair $(\chi(G;x), F(G;y))$:*

1. *the wheel W_5;*
2. *the circular ladder $C_n \times K_2$, with $n \geq 3$;*
3. *the Möbius ladder M_n, with $n \geq 2$;*
4. *the square C_n^2 of the n-cycle, with $n \geq 3$.*

Parts 2–4 strengthen the results in Theorems 6.20 and 6.22. Note that W_5 is neither chromatically unique nor flow unique. Read's conjecture in [945] that C_n^2 is chromatically unique is still open.

6.5.3 k-state Potts uniqueness

The evaluation of the Tutte polynomial along the hyperbola

$$H_k = \{(x, y) : (x - 1)(y - 1) = k\},$$

for a positive integer k, plays an important role in statistical physics; see Chapter 4 of Welsh [1139] and Chapter 20 of this handbook. Garijo, Goodall, and Nešetřil [511] initiated the study of graphs determined by this evaluation of the Tutte polynomial (together with a monomial prefactor, as in the case of chromatically-unique graphs). They considered the *k-state Potts partition function* $Z_{\mathrm{PP}}(G; k, y)$ of a graph G defined by

$$Z_{\mathrm{PP}}(G; k, y) = k^{k(G)}(y - 1)^{r(G)} T(G; \tfrac{k}{y-1} + 1, y)$$

(see Equation (3.6) of this handbook). For a positive integer k, the coefficient of y^m is the number of (not necessarily proper) colorings of G with k colors with exactly m monochromatic edges, so in the literature it is also known as the monochrome or the bad coloring polynomial. It corresponds also to the evaluation $Z(G; k, y - 1)$ of the dichromatic polynomial and, up to the factor $k^{k(G)}$, to the coboundary polynomial of its cycle matroid. See Chapter 22 for the connection between the k-state Potts partition function and homomorphism counting.

For a fixed integer $k > 0$, graphs that are determined by $Z_{\mathrm{PP}}(G; k, y)$ are called *k-state Potts unique*. In [511], Garijo, Goodall, and Nešetřil presented a result, akin to Theorem 6.4, that gives graph invariants that are determined by $Z_{\mathrm{PP}}(G; k, y)$. With that, they proved the next result, mostly by adapting the proofs from [421].

Theorem 6.91. *The following graphs are k-state Potts unique for $k \geq 2$:*

1. *the complete graph K_n, with $n \geq 1$;*
2. *the cycle C_n, with $n \geq 3$;*
3. *the complete bipartite graph $K_{3,3}$ and the wheel W_5;*
4. *the circular ladder $C_n \times K_2$, with $n \geq 3$;*
5. *the Möbius ladder M_n, with $n \geq 2$;*
6. *the square C_n^2 of the n-cycle, with $n \geq 3$;*
7. *all 3-theta graphs.*

Actually, for Item 1 above, only the cases $n = 4, 5$ are addressed in [511], but the argument used applies in general.

In contrast to Theorem 6.26, not all s-theta graphs are k-state Potts unique. Indeed, the 4-theta graph with path lengths 1, 1, 1, 3 has the same 2-state Potts partition function as a path of length 3 with all edges doubled.

Naturally, since k-state Potts uniqueness has just begun to be studied, many basic problems are still unanswered. For instance, are there pairs of non-Tutte-equivalent graphs that share the same k-state-Potts partition function for arbitrarily many values of k? Markström [819] gives an example of a pair of graphs G, G' with $Z_{\mathrm{PP}}(G;3,y) = Z_{\mathrm{PP}}(G';3,y)$ but $Z_{\mathrm{PP}}(G;k,y) \neq Z_{\mathrm{PP}}(G';k,y)$ for $k = 2, 4$ (and possibly for many other values of k). It can be shown that for $k > 2^{e(G)}$ one can deduce from $Z_{\mathrm{PP}}(G;k,y)$ the number of sets of edges with a given size and rank, and hence the Tutte polynomial of G. Thus, as k increases, the size of two graphs with the same k-state Potts partition function but different Tutte polynomial must also increase.

6.5.4 The characteristic polynomial of a matroid

For the same reason that the theory of chromatically-unique graphs considers only simple graphs, here we consider only simple matroids. It is rare for a simple matroid to be determined by its characteristic polynomial. The case of equality in the following extremal result of Björner [130, 131] is an exception.

Theorem 6.92. *For a simple matroid M of rank r and girth g on a set with n elements, the absolute value of the coefficient of x^i in $\chi(M;x)$ is at least*

$$\sum_{j=0}^{g-2} \binom{n-r+j-1}{j} \binom{r-j}{i-j}.$$

Furthermore, equality holds for some i with $i \geq g-1$, or equivalently for all i with $0 \leq i \leq r-1$, if and only if M is isomorphic to the direct sum of the uniform matroid $U_{g-1,n-r+g-1}$ and the free matroid $U_{r-g+1,r-g+1}$.

Earlier Brylawski [251, Corollary 7.14] gave a related characterization of uniform matroids, and Dowling and Wilson [416] addressed the case of $g = 3$. Rota [970] developed the theory of the broken-circuit complex of a matroid, generalizing that of Whitney [1152, 1154] for a graph; the characteristic polynomial is a generating function for the face numbers of this complex. Björner's work in [130, 131] applies more broadly to shellable simplicial complexes.

An immediate corollary of Theorem 6.92 is that any simple matroid that has the same characteristic polynomial as $U_{g-1,n-r+g-1} \oplus U_{r-g+1,r-g+1}$ is isomorphic to it. This result is atypical, as the discussion in Kung [731, Sections 5.5 and 5.6] shows. Only by greatly strengthening the hypotheses are other results of this sort obtained. For instance, [731, Theorem 5.16] states that for a simple matroid M of rank r, if, for each rank-t flat F of M with t in $\{0, r-3, r-2\}$, the contraction M/F has the same characteristic polynomial as $M(K_{r-t+1})$, then M is isomorphic to $M(K_{r+1})$. A similar result is given for projective geometries.

6.6 Open problems

A large part of the motivation for investigating Tutte-equivalent and Tutte-unique graphs and matroids is to learn how strong an invariant the Tutte polynomial is. Thus, it is of interest to have further enlightening examples of Tutte-equivalent and Tutte-unique graphs or matroids. In particular, the proofs of the Tutte-uniqueness results we have cited use a relatively limited amount of the information that the Tutte polynomial contains; having examples that use more of, or other parts of, this data would help advance the theory. Likewise, there are many properties for which it is not known whether they can can be deduced from the Tutte polynomial, such as being outerplanar or 3-connected, for graphs.

In the spirit of the results in Section 6.5.2, it would be interesting to know which Tutte-unique matroids M are determined by less information, specifically, by the characteristic polynomials of M and M^* together. Similarly, one can investigate what information in addition to its Tutte polynomial suffices to determine a given graph or matroid. For instance, [171, Proposition 7.4] shows that a Dowling geometry of rank four or more is determined by its configuration.

As noted above, Bollobás, Pebody, and Riordan [156] made the following conjecture.

Conjecture 6.93. Asymptotically, almost all simple graphs are Tutte unique.

While not addressing this conjecture, the next result, by Loebl, Matoušek, and Pangrać [782], establishes a sense in which Tutte equivalence is rare.

Theorem 6.94. *The probability that two independently chosen random simple graphs on n vertices have the same Tutte polynomial is of order $\mathcal{O}(1/\log n)$.*

The analogue to Conjecture 6.93 for the chromatic polynomial has also been posed.

Also as discussed above, the counterpart, for matroids, of Conjecture 6.93 is known to be false. However, the justification uses sparse paving matroids, which are far less prevalent in many commonly-studied classes of matroids (e.g., those that are representable over a fixed finite field, or transversal matroids) than in the class of all matroids. Thus, studying such questions in special classes of matroids would be of interest. For instance, it remains to be explored whether almost all, or almost no, $\mathbb{F}_q$-representable matroids are Tutte unique, and likewise for transversal matroids.

In [452, Section 4.4], Farr cast the following problem, due to Whitney [1152], in terms of the Tutte polynomial (the problem predates the Tutte polynomial). If G is a graph for which there is a graph H with $T(M(G)^*; x, y) = T(H; x, y)$, must G be 4-colorable? This includes in particular the question of whether having a symmetric Tutte polynomial implies 4-colorability. As noted in Section 6.2.1 there are non-planar graphs with this symmetry property, and the ones known to us are indeed 4-colorable.

Part II

Computation

7

Computational techniques

Criel Merino

Synopsis

This chapters covers various techniques for finding general formulas for the Tutte polynomial for various families of graphs and matroids.

- Direct computation using the basic definitions of the Tutte polynomial and duality.
- Formulas for operations on graphs and matroids.
- Formulas for polynomials equivalent to the Tutte polynomial.
- Applying the transfer-matrix method.
- Using 1-, 2-sums and parallel and series connections to decompose a graph or matroid.
- Using counting arguments to find the coefficients in the Tutte polynomial.
- A list of known Tutte polynomials of common graphs and matroids.

7.1 Introduction

The Tutte polynomial, being an invariant, is an important tool to discern, identify, and extract information from a graph or matroid. Thus, many times we face the task of computing the Tutte polynomial, or evaluating it, for a family of graphs or matroids. This is not always an easy task, and is often time-consuming. For example, computing the Tutte polynomial for complete graphs has been done many times. The most famous strategy is due to Tutte [1095],

DOI: 10.1201/9780429161612-7

and even though the different strategies are not exceedingly difficult to discover, neither are they straightforward.

Here we survey some of the best-known formulas for some interesting families of graphs and matroids, including complete and complete bipartite graphs, and projective and affine geometries. Some of these families are quite general, such as the abundant paving matroids. Note that the complexity of computing the Tutte polynomial is addressed in Chapter 9, of approximating it in Chapter 10, and computer packages for computing the Tutte polynomial are discussed in Chapter 8.

We have grouped together the formulas by the method used to obtain them. This goes from the obvious direct computation, using the state sum expansion definition of the Tutte polynomial, to using equivalent polynomials when the structure of the graphs or matroids to be considered has a good leverage point to use such polynomials.

7.2 Direct computation

The deletion and contraction formula for computing the Tutte polynomial for a graph is given in Equation (2.6), and for matroids in Lemma 4.126. This formula reduces the problem of computing the Tutte polynomial of a graph to the same problem for (potentially) two new smaller graphs or matroids. For graphs or matroids with very few elements this is often the most effective method of computation. There is another common way to compute the Tutte polynomial which is called the state sum expansion, and it appears in Definition 2.3. The corresponding expansion for matroids is given in Definition 4.122. Examples of its use in graphs and matroids are in the corresponding chapters, in particular, Theorem 4.124, which gives a formula for the Tutte polynomial of uniform matroids.

7.2.1 Deletion and contraction

Computing the Tutte polynomial by deletion and contraction leads naturally to linear recurrence relations. As a first example we compute the Tutte polynomial of the n-cycle C_n.

By deleting an edge e from C_n we obtained a path whose Tutte polynomial is x^{n-1}, while if we contract e we get a smaller cycle C_{n-1}. Thus, we obtain the non-homogeneous linear recurrence relation $x^{n-1} = T(C_n; x, y) - T(C_{n-1}; x, y)$. As the Tutte polynomial of the 2-cycle is $x + y$, we obtain, by solving the recurrence relation, the following formula:

$$T(C_n; x, y) = \sum_{i=1}^{n-1} x^i + y. \tag{7.1}$$

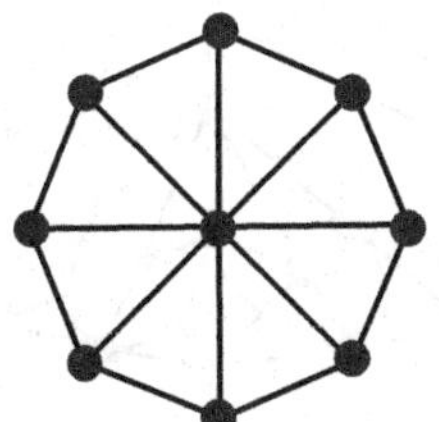

FIGURE 7.1: The 8-wheel, W_8.

For our second example, let us consider the *wheel* W_n. This graph consists of an n-cycle plus an apex vertex, that is an extra vertex that is adjacent to all the other vertices, see Figure 7.1. The vertices and edges on the n-cycle are said to be on the *rim* of W_n and the apex vertex is called the *hub*.

Using a sequence of deletions and contractions and algebraic manipulation the linear homogeneous recurrence relation $T_{n+3} - (w+1)T_{n+2} + (w + xy)T_{n+1} - (xy)T_n = 0$ is obtained, see [598]. Here we write $w := x + y + 1$ and $T(W_n; x, y) =: T_n$ to simplify the equation. Standard methods for solving homogeneous recurrence relations give the following formula for the Tutte polynomial of the wheel graphs:

$$T_n = \frac{1}{2^n}[w + (w^2 - 4xy)^{1/2}]^n + \frac{1}{2^n}[w - (w^2 - 4xy)^{1/2}]^n + xy - w. \quad (7.2)$$

For example,

$$T(W_4; x, y) = x^4 + 4x^3 + 6x^2 + 3x + 4x^2y + 9xy + 4xy^2 + 3y + 6y^2 + 4y^3 + y^4.$$

By deleting an edge from the rim of W_n, the *fan* F_n, is obtained. The graph F_n is well-known for having the $2n$-th Fibonacci number as its number of spanning trees, see [615]. A formula for its Tutte polynomial can be obtained by the same method as in [598]. However, standard methods for generating functions also lead to formulas for the Tutte polynomial of W_n and F_n, as in [197]. The following formula for F_n was obtained in this way by extracting the appropriate coefficient in a generating function.

$$T(F_{n+2}) = \sum_{i=1}^{n}\sum_{j=0}^{n-i}\binom{i+j}{j}\left[\binom{n-j}{i-1}x + \binom{n-j}{i}\right]x^j y^{n+1-i-j} + \sum_{j=0}^{n} x^j y^{n+1-j} + \sum_{j=0}^{n+1}\binom{n+1}{j}x^{j+1}. \quad (7.3)$$

For example,

$$T(F_4; x, y) = x^4 + 3x^3 + 3x^2 + x + 3x^2y + 2xy^2 + 4xy + y + 2y^2 + y^3.$$

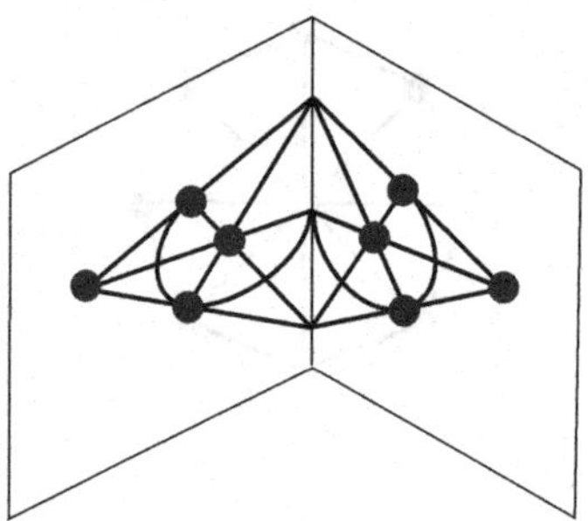

FIGURE 7.2: The geometric representation of $AG(3,2)$.

7.2.2 State sum expansion

As an example of the use of the state sum expansion, Equation (2.2) and Definition 4.122, we consider a generalization of uniform matroids. A rank-r *paving matroid* M is a matroid whose circuits all have a size of at least r (see Definition 4.109). As part of this family we have *sparse paving matroids* which were introduced in [660, 830] (see Section 6.4.2). We say that a rank-r matroid M is sparse paving if M is paving and if for every pair of circuits C_1 and C_2 of size r we have $|C_1 \triangle C_2| > 2$.

In [842], it is shown that when M is a paving matroid of rank $r \geq 1$, the matroid M is sparse paving if and only if all the hyperplanes of M have size r or $r-1$. Also, all hyperplanes of size r are necessarily circuits. Using the state sum expansion, it is easy to prove that the Tutte polynomial of a rank-r sparse paving matroid M with n elements and λ circuit–hyperplanes is given by:

$$T(M) = \sum_{i=0}^{r-1} \binom{n}{i} (x-1)^{r-i} + \binom{n}{r} + \lambda(xy-x-y) + \sum_{i=r+1}^{n} \binom{n}{i} (y-1)^{i-r}. \quad (7.4)$$

Examples of sparse paving matroids include uniform matroids, the Fano matroid F_7, the non-Fano matroid F_7^-, and $AG(3,2)$. The Tutte polynomials for F_7 and F_7^- have already been considered in Chapter 4. The binary affine cube $AG(3,2)$ has the geometric representation given in Figure 7.2. As this matroid has 14 circuit-hyperplanes, we have

$$T(AG(3,2); x, y) = x^4 + 4x^3 + 10x^2 + 6x + 14xy + 6y + 10y^2 + 4y^3 + y^4.$$

7.3 Duality and matroid operations

Duality can be a useful tool in computing Tutte polynomials. For a plane graph G we have the duality relation $T(G;x,y) = T(G^*;y,x)$ from Equation (2.20). Thus if we know the Tutte polynomial of a planar graph, we know the Tutte polynomial of its planar duals. Care must be taken since this relation may only be used for planar graphs. However, for any matroid M, we have $T(M;x,y) = T(M^*;y,x)$ (see Theorem 4.125), and so we can always deduce $T(M)$ from $T(M^*)$. These observations are particularly useful when they are used in conjunction with other computational results.

For example, the n-cycle, C_n, has as a dual the graph consisting of two vertices and n parallel edges between them. By duality, the Tutte polynomial of the graph C_n^* is

$$T(C_n^*;x,y) = \sum_{i=1}^{n-1} y^i + x.$$

7.3.1 Relaxation

Matroid relaxation is defined in Definition 4.147 and the formula for the Tutte polynomial of M', the relaxation of M, was given in Proposition 4.148: $T(M';x,y) = T(M;x,y) - xy + x + y$.

For example, the rim of the wheel graph W_n is a circuit-hyperplane of the corresponding graphic matroid $\mathcal{W}_n$. The relaxation of this circuit-hyperplane gives the matroid $\mathcal{W}^n$, called the *whirl matroid.* Whirl matroids are important in matroid theory, see [894]. (See also Section 6.4.8 for more on whirls.) From the expression in Equation (7.2), it is easy to compute an expression for the Tutte polynomial of whirls.

$$\begin{aligned} T(\mathcal{W}^n;x,y) &= \frac{1}{2^n}[(1+x+y) + ((1+x+y)^2 - 4xy)^{1/2}]^n \\ &+\frac{1}{2^n}[(1+x+y) - ((1+x+y)^2 - 4xy)^{1/2}]^n - 1. \end{aligned}$$

7.3.2 Free extension and coextension

For a matroid M with ground set E and an element $e \notin E$, the free extension $M + e$ and the free coextension $M \times e$ are defined in Definition 6.37. Equation (6.5) gives a formula for the Tutte polynomial of $M + e$ in terms of the Tutte polynomial of M. The formula for the free coextension can be obtained by duality:

$$T(M \times e;x,y) = \frac{y}{y-1}(T(M;x,y) - T(M;x,1)) + xT(M;x,1).$$

One class of matroids in which we can apply the extension and coextension formulas are *nested matroids*. Nested matroids were defined in Definition 6.82 as the matroids obtained from the empty matroid by iterating the operations of adding coloops and taking free extensions. In [170], *generalized Catalan matroids* were introduced and proved to be the same as nested matroids. Thus, as noted in [170], the formula for the Tutte polynomial of a nested matroid can be computed in polynomial time. Also, in [245] Brylawski obtains a formula for the Tutte polynomial of a nested matroid by using the cardinality-corank polynomial.

When a nested matroid is obtained from the empty matroid by a sequence of length $2n$ of first making a free extension, then alternately adding a coloop and doing a free extension, and ending by adding a coloop, the matroid is called a *Catalan matroid*. In [170] the following formula for the Tutte polynomial of the Catalan matroid M_n, for $n > 1$ is given.

$$T(M_n; x, y) = \sum_{i,j>0} \frac{i+j-2}{n-1} \binom{2n-i-j-1}{n-i-j+1} x^i y^j.$$

This results also appear in [30], and an interpretation in terms of Dyck paths can be found in [30, 170].

7.3.3 Truncation and free lift

An operation related to free extension is the truncation $\mathrm{Tr}(M)$ of a matroid M of rank $r \geq 1$ (see Definition 6.38). Because $\mathrm{Tr}(M) = (M+p)/p$, we have, by deletion and contraction, $T(\mathrm{Tr}(M); x, y) = T(M+p; x, y) - T((M+p)\backslash p; x, y)$. But $(M+p)\backslash p \cong M$. Thus we obtain the following formula:

$$T(\mathrm{Tr}(M); x, y) = \frac{1}{x-1}\left(T(M; x, y) - xT(M; 1, y)\right) + yT(M; 1, y).$$

This formula is from [212]. However, formulas for the Tutte polynomial of free extension and truncation were given much earlier by Brylawski in [245].

The dual operation of truncation is the free lift or elongation, which is $(M \times p)\backslash p$, for a matroid M (see Definition 6.38). Thus, by duality we get the following formula:

$$T((M \times p)\backslash p; x, y) = \frac{1}{y-1}\left(T(M; x, y) - yT(M; x, 1)\right) + xT(M; x, 1).$$

7.3.4 Double extension

The *double extension* of a matroid M by p and q is $(M+p)\times q$. It is not difficult to prove that the order of the operations is irrelevant, that is, $(M \times q) + p =$

$(M+p)\times q$. Using the formulas for free extension and coextension of the Tutte polynomial we get the following result that appears in [940]:

$$T(M+p\times q;x,y) = \frac{xy}{(x-1)(y-1)}T(M;x,y) - \frac{xy-x-y}{(x-1)(y-1)}T(M;1,1) + \frac{xy-x-y}{(x-1)(y-1)}(xT(M;x,1)+yT(M;1,y)).$$

7.4 Using equivalent polynomials

The Tutte polynomial is equivalent to many other polynomials that have appeared throughout the history of combinatorics. Polynomials like the q-state Potts model partition function (see [1139] and Chapter 20), the coboundary polynomial (see [247], Definitions 4.145 and 6.57) or the weight enumerator of a linear code (see [247], and Chapter 16). In this context, two polynomials are *equivalent* if computing one polynomial gives the other polynomial after substituting some variables and multiplying by an easily computable factor. A computational approach is not computing the Tutte polynomial directly, but rather computing an equivalent polynomial with the hope that this computation is easier because of the structure of the family of graphs or matroids being studied.

7.4.1 The bad coloring polynomial

Computing the Tutte polynomial of complete graphs using the state sum expansion definition seems hopeless. Using deletion–contraction also seems very complicated, but there is some hope of computing the Tutte polynomial for a given K_n, and a dynamic programing algorithm, see [27]. However, there is another way to get a general expression for the Tutte polynomials of complete graphs, and it is by using the exponential formula to compute the bad coloring polynomial.

For $n > 0$, we let $[n]$ be the vertex set of K_n, and we let $B_n(\lambda, t)$ denote the bad coloring polynomial (defined in Section 12.3) of K_n. We set $B_0(\lambda, t) = 1$. To compute $B_n(\lambda, t)$, observe that any λ-coloring of K_n partitions the vertex set $[n]$ into λ color classes. For $i = 1 \ldots \lambda$ let V_i denote the set of vertices in the color class i, and let n_i denote its cardinality, So, we have that $n_1 + \cdots + n_\lambda = n$. The number of bad edges with both ends in the set V_i is $\binom{n_i}{2}$, and the contribution of these edges to the bad coloring for this λ-coloring will be $t^{\binom{n_i}{2}}$. The first part in the partition can be colored with λ colors, the second one with $\lambda - 1$ and so on, until the k-th part that can be colored with any of the $\lambda - k + 1$ colors left. Using the "composition formula" from [1038] we get:

$$1 + \sum_{n \geq 1} B_n(\lambda, t) \frac{u^n}{n!} = \left(\sum_{n \in \mathbb{N}_0} t^{\binom{n}{2}} \frac{u^n}{n!} \right)^{\lambda}. \tag{7.5}$$

Let $T_n(x, y)$ be the Tutte polynomial of K_n. Tutte in [1095] and Welsh in [1142] give the following exponential generating function for $T_n(x, y)$ that follows from the Equation (7.5) and the relation between the bad coloring and Tutte polynomials given in Theorem 12.18.

$$1 + (x-1) \sum_{n \geq 1} (y-1)^n T_n(x, y) \frac{u^n}{n!} = \left(\sum_{n \geq 0} y^{\binom{n}{2}} \frac{u^n}{n!} \right)^{(x-1)(y-1)}. \tag{7.6}$$

Another approach for computing the Tutte polynomial of K_n is to use the finite field method as describe in Section 29.6. Also, in [244], Brylawski computes the cardinality-corank polynomial (which is equivalent to the Tutte polynomial) of K_n.

Even if the previous formulas seem difficult to handle, computing $B_n(\lambda, t)$ for a fixed value of n using Equation (7.5) is straightforward in `Maple` or `Mathematica`. For the calculation, only the first $n + 1$ terms in the sum inside the parentheses on the right-hand side of the equation are required. Then, the right-hand side is differentiated n times with respect to u and evaluated at $u = 0$ to obtain the value of $B_n(\lambda, t)$. By making the appropriate substitution, $T_n(x, y)$ is obtained. For example, for $n = 5$ the result is

$$\begin{aligned} T_5(x, y) = {} & x^4 + 6\,x^3 + 11\,x^2 + 6\,x + 10\,x^2 y + 5\,x y^3 + 20\,xy \\ & + 15\,x y^2 + 6\,y + 15\,y^2 + 15\,y^3 + 10\,y^4 + 4\,y^5 + y^6. \end{aligned}$$

It is natural to apply this technique to complete bipartite graphs as well. For details (but in the language of the coboundary polynomial), see [822]. For the resulting expression, let $T_{n,m} := T(K_{n,m}; x, y)$ be the Tutte polynomial of the complete bipartite graph $K_{n,m}$. Then

$$\begin{aligned} 1 + \sum_{\substack{(n,m) \in \mathbb{N}_0^2 \\ (n,m) \neq (0,0)}} (x-1)(y-1)^{nm} T_{n,m} \frac{u^n}{n!} \frac{v^m}{m!} = & \\ \left(\sum_{(n,m) \in \mathbb{N}_0^2} y^{nm} \frac{u^n}{n!} \frac{v^m}{m!} \right)^{(x-1)(y-1)} & . \end{aligned} \tag{7.7}$$

As before, it is straightforward to use, for example, `Maple` to compute $T_{n,m}(x, y)$ for small values of n and m. In this way, we get the following:

$$\begin{aligned} T_{3,3}(x, y) = {} & x^5 + 4\,x^4 + 10\,x^3 + 11\,x^2 + 5\,x + 9\,x^2 y \\ & + 15\,xy + 6\,x y^2 + 5\,y + 9\,y^2 + 5\,y^3 + y^4. \end{aligned}$$

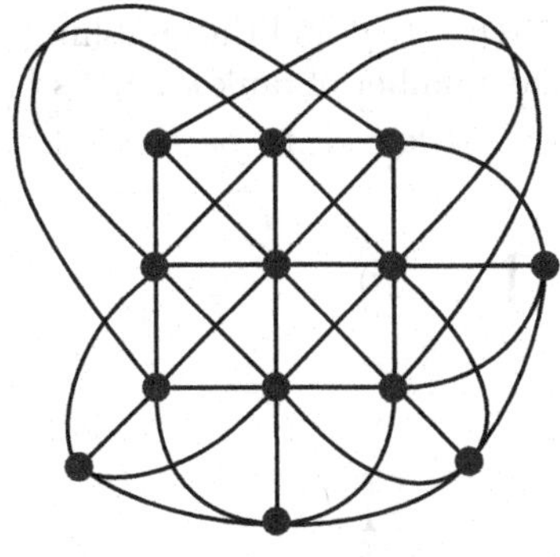

(a) The projective plane $PG(2,3)$.

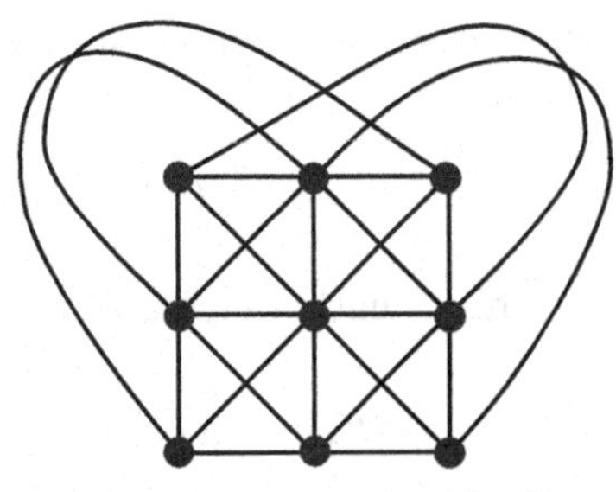

(b) The affine plane $AG(2,3)$.

FIGURE 7.3: Projective and affine planes.

7.4.2 The coboundary polynomial

We now consider Crapo's coboundary polynomial $\overline{\chi}(M)$ (see Definition 4.145). For the following computations we follow Mphako [867] and Brylawski [245]. A matroid is a *near-design* if all closed sets of rank m have the same cardinality $k(m)$ for $m = 0, 1, 2, \ldots, r-2$ (see Definition 6.45). Examples of near-designs are paving matroids, where $k(m) = m$ for all $m \leq r-2$, and *perfect matroid designs*, like projective and affine geometries, where all the flats of a fixed rank, including the hyperplanes, are equicardinal. Brylawski [242] showed that, for any perfect matroid design, all minors obtained by contracting a k-element flat have the same characteristic polynomial $\chi(M)$. Hence, if we denote by n_k the number of flats of size k in a perfect matroid design $M = (E, r)$, and let X_k be a flat of size k, then

$$\overline{\chi}(M;\lambda,t) = \sum_{k=0}^{|E|} n_k\, \chi(M/X_k;\lambda)t^k. \tag{7.8}$$

Brylawski also computed the coboundary polynomial, which he called the *Poincaré polynomial*, of a perfect matroid design in [245]. An expression for the Tutte polynomial of a near-design matroid is also given on page 208 of [245].

For all nonnegative integers m and k the *Gaussian coefficients* are

$$\begin{bmatrix} m \\ k \end{bmatrix}_q := \frac{(q^m-1)(q^m-q)\cdots(q^m-q^{k-1})}{(q^k-1)(q^k-q)\cdots(q^k-q^{k-1})}. \tag{7.9}$$

Note that $\begin{bmatrix} m \\ 0 \end{bmatrix}_q = 1$. Gaussian coefficients are also called *q-binomial coefficients* and they count the number of k-dimensional subspaces of an m-dimensional vector space over a field with q elements.

Let $PG(r-1,q)$ be the $(r-1)$ dimensional projective geometry over $\mathbb{F}_q$ (see Definition 4.104). Then, as a matroid, it has rank r and $\begin{bmatrix} r \\ 1 \end{bmatrix}_q$ elements.

Also, every rank-k flat X is isomorphic to $PG(k-1,q)$ and the simplification of M/X is isomorphic to $PG(r-k-1,q)$. The number of rank-k flats is $\left[{r \atop k}\right]_q$. The characteristic polynomial of $PG(r-1,q)$ is known to be the following (see [247]):

$$\chi(PG(r-1,q);\lambda) = \prod_{i=0}^{r-1}(\lambda - q^i).$$

Thus, using Equation (7.8), we get

$$\overline{\chi}(PG(r-1,q);\lambda,t) = \sum_{k=0}^{r} t^{\left[{k \atop 1}\right]_q} \left[{r \atop k}\right]_q \prod_{i=0}^{r-k-1}(\lambda - q^i). \tag{7.10}$$

The matroid $PG(2,3)$, has a geometric representation given in Figure 7.3a. To compute its Tutte polynomial we could use the above formula to get $\overline{\chi}(PG(2,3);\lambda,t) = (\lambda-1)(\lambda-3)(\lambda-9)+13t(\lambda-1)(\lambda-3)+13t^4(\lambda-1)+t^{13}$. The Tutte polynomial of $PG(2,3)$ is obtained from the coboundary by substitutions $\lambda = (x-1)(y-1)$ and $t = y$, and by multiplying by the factor $1/(y-1)^3$. This gives

$$\begin{aligned} T(PG(2,3);x,y) &= x^3 + 10x^2 + 16x + 13xy^2 + 26xy + 16y + 32y^2 \\ &\quad + 36y^3 + 28y^4 + 21y^5 + 15y^6 + 10y^7 + 6y^8 + 3y^9 + y^{10}. \end{aligned}$$

A similar computation of the Tutte polynomial of $AG(r-1,q)$, the $r-1$ dimensional affine geometry over $\mathbb{F}_q$, (see Section 6.4.3) can be obtained. This matroid has rank r and q^{r-1} elements. Any rank-k flat X is isomorphic to $AG(k-1,q)$ and the simplification of $AG(r,q)/X$ is isomorphic to $PG(r-k-1,q)$. The number of rank-k flats is $q^{r-k}\left[{r-1 \atop k-1}\right]_q$, for $1 \le k \le r$, and there is just one rank-0 flat. From [247], the characteristic polynomial of $AG(r-1,q)$ is

$$\chi(AG(r-1,q);\lambda) = (\lambda-1)\sum_{k=0}^{r-1}(-1)^k\lambda^{r-k-1}\prod_{i=0}^{k-1}(q^{r-i-1}-1).$$

Using Equation (7.8) we get

$$\begin{aligned} &\overline{\chi}(AG(r-1,q);\lambda,t) \\ &\quad = \chi(AG(r-1,q);\lambda) + \sum_{k=1}^{r} t^{q^{k-1}} q^{r-k} \begin{bmatrix} r-1 \\ k-1 \end{bmatrix}_q \prod_{i=0}^{r-k-1}(\lambda - q^i). \end{aligned} \tag{7.11}$$

A geometric representation of the affine plane $AG(2,3)$ is obtained by deleting all the points in a line from the projective plane $PG(2,3)$, as in Figure 7.3b. From Equation (7.11) we have

$$\overline{\chi}(AG(2,3);\lambda,t) = (\lambda-1)(\lambda^2-8\lambda+16)+9t(\lambda-1)(\lambda-3)+12t^3(\lambda-1)+t^9,$$

from which it follows that

$$T(AG(2,3);x,y) = x^3+6x^2+9x+12xy+9y+15y^2+10y^3+6y^4+3y^5+y^6.$$

7.4.3 The q-cone

The q-cone has been considered in considerable depth in Section 6.4.5. Theorem 6.55 there gives the formula for the Tutte polynomial of the q-cone. As an example, the 2-cone of the Fano matroid is the matroid $PG(3,2)$. The Tutte polynomial of the Fano matroid is $x^3+4x^2+3x+7xy+3y+6y^2+3y^3+y^4$, so applying Theorem 6.55 gives

$$\begin{aligned}T(PG(3,2);x,y) &= x^4+11x^3+31x^2+21x+35x^2y+80xy\\ &\quad +90xy^2+45xy^3+15xy^4+21y+70y^2+105y^3+105y^4\\ &\quad +84y^5+56y^6+35y^7+20y^8+10y^9+4y^{10}+y^{11}.\end{aligned}$$

As the q-cone of $PG(r-1,q)$ is $PG(r,q)$, this method can also be used to compute the Tutte polynomial of any $PG(r,q)$.

7.5 Transfer matrix method

Using, for example, the deletion–contraction or state sum definition to compute Tutte polynomials can quickly become prohibitive as the number of subgraphs to consider grows exponentially with the size of the graph. However, when the family of graphs being studied is constructed using a "non-complicated" graph that is repeated in a "path-like" fashion, a computation can be managed—the bookkeeping of the contribution of each repetition can be done with a matrix, and the update can be done by matrix multiplication after the graph grows. This gives rise to the transfer-matrix method. The theoretical background for this method can be found in [1040]. Here we apply it to rectangular lattices and bracelets.

7.5.1 No boundary conditions

The *rectangular lattice* $L_{m,n}$ is the graph having vertex set $\{1,\dots,m\}\times\{1,\dots,n\}$ and where two vertices (i,j) and (i',j') are adjacent if $|i-i'|+|j-j'|=1$. The transfer-matrix method gives a way to compute $T(L_{m,n};x,y)$,which is described in Calkin et. al. [261].

Theorem 7.1. *For integers $n,m\geq 2$, with m fixed, we have*

$$T(L_{m,n};x+1,y+1)=x^{nm-1}X_m^T\cdot(\Lambda_m)^{n-1}\cdot\mathbf{1},$$

where X_m is a vector of length c_m, and Λ_m is a $c_m\times c_m$ matrix, both depending on x,y and m but not n; and where $\mathbf{1}$ is the vector of length c_m with all entries equal to 1.

The quantity c_m is the m–th Catalan number, so the method is only practical for small values of m. In practice, computing the vectors X_m and the matrix Λ_m can easily be done by computer.

For example, for $L_{2,n}$ we get that $T(L_{2,n}; x+1, y+1)$ equals

$$x^{2n-1}(x^{-1}\ 1)\begin{pmatrix} x^{-1}+3x^{-2}+yx^{-2} & 1+2x^{-1} \\ x^{-1}+2x^{-2}+x^{-3} & 1+2x^{-1}+x^{-2} \end{pmatrix}^{n-1}\begin{pmatrix} 1 \\ 1 \end{pmatrix}. \tag{7.12}$$

7.5.2 Boundary conditions

The transfer matrix method can be combined with the idea of boundary conditions. In this situation additional linear algebra is required but the method is still suitable to use in a computer algebra package.

In [281], Chang and Shrock use results from [1006] to compute the Tutte polynomial of the wheel W_n by using the Potts model partition function together with the transfer matrix method. As previously mentioned, the Tutte polynomial and the Potts model partition function are equivalent. (See Section 3.4.4 or Chapter 20 for background on the Potts model and its relation to the Tutte polynomial.)

In the work of Biggs [109], the transfer matrix method is used to compute the coboundary polynomial of a family of graphs that are called *bracelet* graphs. A bracelet G_n is formed by taking n copies of a graph G and joining each copy to the next by a set of edges L (with $n+1=1$ by convention). For each choice of G and L, we get a family of graphs for which the transfer matrix method is appropriate. For example if G is K_2 and L is the pair of edges joining corresponding vertices in each copy, we get that G_n is the *cubic planar ladder* CPL_n. In particular, CPL_4 is the *cube*.

Using boundary conditions requires computing the trace of a large matrix. In [109] representation theory is used to get a simple expression for bracelet graphs, and we refer the reader to the reference for the formula, as it involves aspects of representation theory that we are not defining here. The expression for the cubic planar ladder is the following:

$$(x-1)\,T(CPL_n; x, y) =$$

$$\operatorname{Tr}\begin{pmatrix} x^2 & x^2 \\ x+1 & x+y+1 \end{pmatrix}^n + w_1 \operatorname{Tr}\begin{pmatrix} x & 0 & x \\ 0 & x & x \\ 1 & 1 & y+2 \end{pmatrix}^n + w_2,$$

where Tr denotes the trace of the matrix, $w_1 = xy - x - y$ and $w_2 = x^2y^2 - 2x^2y - 2xy^2 + x^2 + xy + y^2 + x + y - 1$. Thus, for $n = 4$, we get the Tutte polynomial of the cube.

$$T(CPL_4; x, y) = x^7 + 5x^6 + 15x^5 + 29x^4 + 40x^3 + 32x^2 \\ + 11x + 6x^4y + 24x^3y + 12x^2y^2 + 52x^2y + 8xy^3 \\ + 39xy^2 + 46xy + 11y + 25y^2 + 20y^3 + 7y^4 + y^5.$$

7.6 Decomposition

A standard strategy for tackling any problem is to decompose it into smaller subproblems. This strategy can be applied to computing the Tutte polynomial by using deletion–contraction, as we have shown previously. However, it can also be applied by decomposing a graph or matroid into smaller graphs or matroids where computing the Tutte polynomial becomes relatively easy. The direct sum and the 2-sum of graphs or matroids are usually employed in this strategy, but we described the less common operations of parallel and series connection, and the tensor product. However, this restricts the applicability of the strategy to graphs or matroids with low connectivity. The definition of disjoint union and one-point joins of graphs is given in Section 1.2.9, and the formula for the Tutte polynomial under these operations is given in Proposition 2.19. For matroids, the definition of direct sum is given in Definition 4.90 and the formula for the Tutte polynomial of direct sums is given in Corollary 4.127. The definition of 2-sum of graphs and matroids is given in Section 1.2.10 and Definition 4.98, respectively, and the formula for the Tutte polynomial of 2-sums of matroids is given in Theorem 4.149.

As an example, consider R_6, which is the 2-sum of $U_{2,4}$ with itself. By symmetry, the choice of base point is irrelevant. Theorem 4.149 gives

$$T(R_6; x, y) = x^3 + 3x^2 + 4x + 2xy + 4y + 3y^2 + y^3.$$

For the 3-sum of matroids there are formulas for the Tutte polynomial although they are rather involved. We refer the reader to the work of [24]. There is a generalization of k-sums for matroids and a corresponding formula. However, the formula is quite intricate, so we refer the reader to the original paper of Bonin and de Mier [167] for the results.

7.6.1 Parallel and series connection

Let $G_1 = (V_1, E_1)$ and $G_2 = (V_2, E_2)$ be two graphs and let $p_1 = (u_1, v_1) \in E_1$ and $p_2 = (u_2, v_2) \in E_2$ be two edges that are not loops or bridges. We assign an orientation to p_i from u_i to v_i, for $i = 1, 2$. The parallel connection of G_1 and G_2 is the graph obtained by identifying p_1 and p_2 so that their directions agree. The *series connection* of G_1 and G_2 is the graph obtained by, first, deleting p_1 and p_2, then identifying the vertices u_1 and u_2 and finally add a new edge joining v_1 and v_2. For example, when both graphs are the 3-cycle C_3, their *parallel connection* is the graph K_4 minus an edge, while their series connection is the 5-cycle C_5. Although the resulting graph depends on the chosen orientation in general, the cycle matroid of the resulting graph does not. These definitions can be found in [894].

These operations extend to matroids. A *pointed matroid* N_d is a matroid on a ground set which includes a distinguished element, the *point* d, which will be assumed to be neither a loop nor coloop. Given two pointed matroids $M_d = (E_1, \mathcal{C}_1)$ and $N_d = (E_2, \mathcal{C}_2)$, where $E_1 \cap E_2 = \{d\}$, the *parallel connection* of M_d and N_d, denoted $P(M_d, N_d)$ is the matroid with family of circuits given by $\mathcal{C}(P(M_d, N_d)) := \{C_1 \cup C_2 \setminus d : C_i \text{ is a circuit of } M_i, \text{ with } d \in C_i,\ i = 1, 2\} \cup \mathcal{C}_1 \cup \mathcal{C}_2$.

The series connection is the dual operation and can be defined as $S(M_d, N_d) := (P(M_d^*, N_d^*))^*$. Observe that the 2-sum $M_d \oplus_2 N_d$ equals the parallel connection $P(M_d, N_d)$ with the element d deleted. Alternatively, $M_d \oplus_2 N_d$ equals the series connection $S(M_d, N_d)$ with the element d contracted. If $M \cong M(G_1)$ and $N \cong M(G_2)$, with p_1 and p_2, respectively, as the distinguished element d in each matroid, then the matroids $S(M_d, N_d)$ and $P(M_d, N_d)$ are isomorphic to the cycle matroids of the graphs that are series and parallel connections, respectively, of G_1 and G_2.

Identities for computing the Tutte polynomial of the parallel and series connections of the matroids M_d and N_d are given in [24, 167, 245, 250, 1075]. The following formula is from [167], for brevity writing $T(M)$ for $T(M; x, y)$ and $w = xy - x - y$.

$$T(P(M_d, N_d)) = \frac{1}{w} \left[T(M_d) \quad T(M_d/d) \right] \begin{bmatrix} y-1 & -y \\ -y & xy \end{bmatrix} \begin{bmatrix} T(N_d) \\ T(N_d/d) \end{bmatrix}, \tag{7.13}$$

For example, the parallel connection of $U_{2,4}$ with itself is isomorphic to the 3-cone of a 2-point line and has Tutte polynomial

$$T(P(U_{2,4}, U_{2,4}); x, y) = x^3 + 4x^2 + 4x + 4xy + 2xy^2 + 4y + 4y^2 + 3y^3 + y^4.$$

The equation for the series connection is similar and can be obtained through the duality formula for the Tutte polynomial.

There is also a *generalized parallel connection* of the matroids M and N along a common subset T (which has to satisfy a number of technical requirements). A formula for the Tutte polynomial of generalized parallel connections can be found in [167].

7.6.2 Thickening, stretch, and tensor product

Given a graph G and a positive integer k, we define the *k-thickening* of G, $G^{(k)}$, as the graph obtained by replacing every edge by k parallel edges and every loop by k loops. Similarly, if we replace every edge in G by k edges

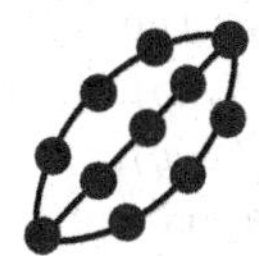

(a) A 3-theta graph.

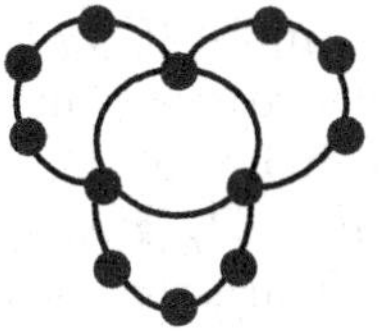

(b) A flower graph.

FIGURE 7.4: Examples of the tensor product of graphs.

in series (i.e., a path with k edges) this new graph is called the *k-stretch*, $G_{(k)}$. These two operations are examples of the more general operation called the *tensor product.* For two graphs G and H and an edge d of H, the tensor product $G \otimes H_d$ is the graph obtained by replacing every edge e in the graph G by a fixed graph H with a marked edge d that is identified with e, and at the end all the copies of d are deleted from the graph. As in the previous subsection, the resulting graph depends on the chosen way to identify the edges, but the cycle matroid of the resulting graph does not.

Thus, for example, the 3-thickening of K_2 is 3 edges in parallel, and the 3-stretch of the resulting graph is shown in Figure 7.4a. The graph is a *3-theta graph*, that is a graphs that consists of three internally disjoint paths of any length. The theta graph of the example can also be obtained from the tensor product of a 3-thickening of K_2 and a 5-cycle.

We present the formula for the Tutte polynomial of these three graph operations in the more general framework of matroids. For a simple matroid M with no coloops and a pointed matroid N_d, where d is not a loop nor a coloop, the tensor product $M \otimes N_d$ is the matroid obtained by taking simultaneously, at each point e of M, the 2-sum of M with N_d, where we identify e with d for each 2-sum. If $M \cong M(G)$ and $N \cong M(H)$, then the matroid $M \otimes N_d$ is isomorphic to the graphic matroid $M(G \otimes H_d)$.

The Tutte polynomial of $M \otimes N_d$, is then given by

$$T(M \otimes N_d; x, y) = f^{|E|-r(M)} g^{r(M)} T\Big(M; \frac{(x-1)f+g}{g}, \frac{f+(y-1)g}{f}\Big), \quad (7.14)$$

where

$$\begin{aligned} f &= f(x,y) &= \frac{(y-1)T(N_d \backslash d; x,y) - T(N_d/d; x,y)}{xy-x-y}, \text{ and} \\ g &= g(x,y) &= \frac{(x-1)T(N_d/d; x,y) - T(N_d \backslash d; x,y)}{xy-x-y}. \end{aligned}$$

The proof of the formula uses a generalization of the Tutte polynomial to pointed matroids, called the *pointed Tutte polynomial*, that can be found in [245]. Another proof can be found in [382], but here we follow the exposition in [247, 653].

As an example, if N_d is the graphic matroid of an m-cycle with one edge replaced by two edges in parallel, one of which is d, then $M(C_n) \otimes N_d$ is the graphic matroid of a *flower graph* (see Figure 7.4b). In this example, $T(N_d \backslash d; x, y) = P_m(x, y)$ and $T(N_d/d; x, y) = yP_{m-1}(x, y)$, where $P_n(x, y) = T(C_n; x, y)$. Thus, the Tutte polynomial of flower graphs can be obtained by substituting $f = P_{m-1}(x, y) + 1 - y$ and $g = P_{m-1}(x, y) + 1$ in (7.14). An alternative method using a linear recurrence may be found in [869].

When a matroid N has a transitive automorphism group the choice of the distinguished point d is immaterial. Thus, if N is $U_{1,k+1}$, for $k \geq 1$, we get the *k-thickening* of M, denoted $M^{(k)}$. By duality, we get the *k-stretch* of M, denoted $M_{(k)}$. If $M \cong M(G)$, then the matroids $M^{(k)}$ and $M_{(k)}$ are isomorphic to the cycle matroids of the graphs $G^{(k)}$ and $G_{(k)}$, respectively. For the special case of (7.14), for thickening we have

$$T(M^{(k)}; x, y) = \left(\frac{y^k - 1}{y - 1}\right)^{r(M)} T\Big(M; 1 + \frac{(x-1)(y-1)}{y^k - 1}, y^k\Big).$$

A direct proof of this may be obtained by using the fact that any flat of $M^{(k)}$ is the k-thickening of a flat of M. Thus, by using the definition of the coboundary polynomial we get that

$$\overline{\chi}(M^{(k)}; q, t) = \overline{\chi}(M; q, t^k).$$

A similar formula for the k-stretch can be obtained by duality.

7.7 Counting arguments

In [245], Brylawski wrote "Knowing the Tutte polynomial of M is equivalent to knowing, for all parameters, the number of all sets in M of fixed cardinality and corank". To illustrate our final method for computing Tutte polynomials we present two counting arguments. In both cases we count subsets of certain types in the matroid. With this information, it is possible to recover the Tutte polynomial or an equivalent polynomial.

7.7.1 Computing coefficients

We present a formula for the Tutte polynomial of paving matroids. For this we start with a general result of Brylawski about how to count flats of corank k and nullity j in a matroid using the coefficients of the Tutte polynomial (see Theorem 6.3 in [251]).

Given integers $k > 1$ and $m > 0$, a collection $\mathcal{T} = \{T_1, \ldots, T_k\}$ of subsets of a set E, such that each member of $\mathcal{T}$ has at least m elements and each m-element subset of E is contained in a unique member of $\mathcal{T}$, is called an *m-partition* of E. The elements of the partition are called *blocks* of the partition.

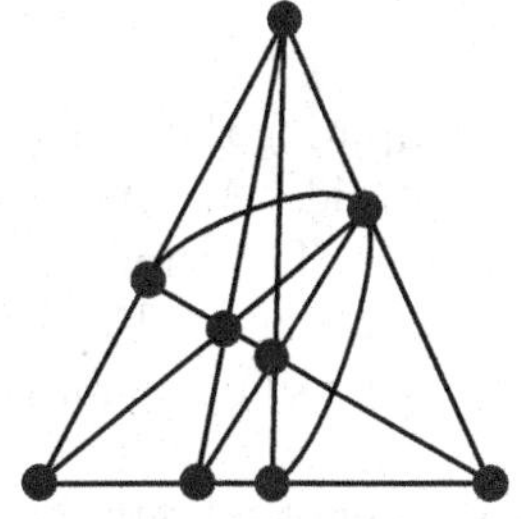

(a) The geometric representation of R_9.

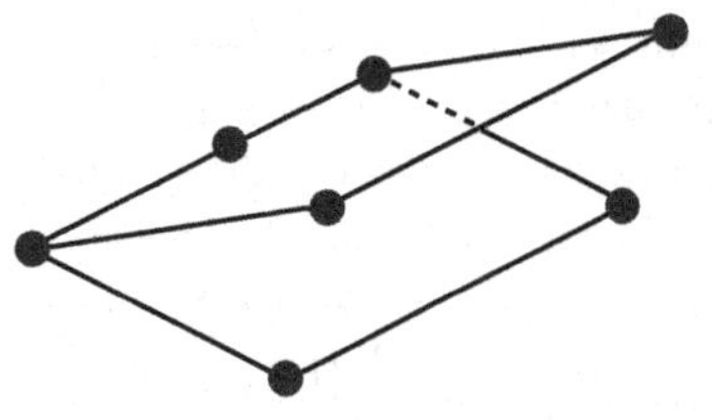

(b) The geometric representation of $U_{2,3}\Box U_{1,2}^{(2)}$.

FIGURE 7.5: Two examples of matroids.

The set of hyperplanes of a rank-r paving matroids defines an $(r-1)$-partition, for $r \geq 2$. This property characterizes paving matroids because an m-partition defines the set of hyperplanes of a paving matroid. Thus, for example, all the matroids coming from Steiner systems are paving, see [1138]. These results were applied in [251] to obtain the following.

Proposition 7.2. *Let M be a rank-r matroid with n elements and Tutte polynomial $\sum_{i,j} t_{ij}x^iy^j$. Then, M is paving if and only if $t_{ij} = 0$ for all $(i,j) \geq (2,1)$ (in lexicographic order). In addition if the $(r-1)$-partition of E has b_k blocks of cardinality k, for $k = r-1, \dots n$, then*

$$\begin{aligned} t_{i0} &= \binom{n-i-1}{r-i}, \qquad \text{for all } i \geq 2; \\ t_{10} &= \sum_{k=0}^{\infty} \binom{r-2+k}{r-2} b_{k+r-1} + \binom{n-2}{r-1} - \binom{n}{r-1}; \end{aligned}$$

and for all $j > 0$,

$$\begin{aligned} t_{1j} &= \sum_{k=0}^{\infty} \binom{r-2+k}{r-2} b_{k+j+r-1}, \text{ and} \\ t_{0j} &= \binom{n-j-1}{r-1} - \sum_{k=0}^{\infty} \binom{r-1+k}{r-1} b_{k+j+r-1}. \end{aligned}$$

The expression for the Tutte polynomial of near-design matroids in [245] (see Subsection 7.4) can be applied to paving matroids and the result is, not surprisingly, the same as above.

As an example, consider the matroid R_9 with geometric representation given in Figure 7.5a. This matroid is paving but not sparse paving. The blocks correspond to lines in the representation. There are 7 blocks of size 3, and 2 blocks of size 4, and then 3 additional blocks of size 2, corresponding to trivial lines that do not appear in Figure 7.5a. Using the above formulas, the Tutte polynomial of R_9 is

$$T(R_9; x, y) = x^3 + 6x^2 + 8x + 11xy + 2xy^2 + 8y + 13y^2 + 10y^3 + 6y^4 + 3y^5 + y^6.$$

7.7.2 Free product

As our second example, we present a formula for computing the Whitney rank generating function of the free product of two matroids, $M\Box N$, a matroid operation that was considered in detail in Section 6.4.1. The papers [343, 169] contain the following expression for the rank function of $M\Box N$. For $X \subseteq E_1$ and $Y \subseteq E_2$, we have $r_{M\Box N}(X \cup Y) = r_M(X) + r_N(Y) + \min\{r(M) - r_M(X), |Y| - r_N(Y)\}$.

With this formula, it is possible to count the number of subsets A of $M\Box N$ with corank i and nullity j given the corresponding sets in M and N. Thus, we are really computing the Whitney rank generating function $W(M\Box N; x, y)$, which is equivalent to the Tutte polynomial as $W(M\Box N; x, y) = T(M\Box N; x+1, y+1)$ (see Chapter 2). The following formula is from [169]. Let $W(M; x, y) = \sum_{i,j} a_{ij}x^iy^j$ and $W(N; x, y) = \sum_{k,l} b_{kl}x^ky^l$. Then, the coefficient of x^py^q in $W(M\Box N; x, y)$ is

$$\sum_{\substack{i,j,k,l:\\ i+k-\min\{i,l\}=p\\ j+l-\min\{i,l\}=q}} a_{ij}b_{kl}. \tag{7.15}$$

For example, if M is a 3-point line (which is the cycle matroid of a complete graph on 3 vertices) and N is two double points (which is the cycle matroid of two pairs of parallel edges) the matroid $M\Box N$ is the matroid in Figure 7.5b. The Tutte polynomial of M is x^2+x+y and the corresponding polynomial of N is $(x+y)^2$. Thus, the Whitney rank generating functions are, respectively, $x^2+3x+y+3$ and $x^2+y^2+4+4x+4y+2xy$. By using (7.15) we obtain $W(M\Box N; x, y) = x^4+7x^3+21x^2+34x+25+10xy+x^2y+2xy^2+19y+7y^2+y^3$. Thus, the Tutte polynomial of $M\Box N$ is

$$T(U_{2,3}\Box U_{1,2}^{(2)}; x, y) = x^4+3x^3+5x^2+3x+4xy+x^2y+2xy^2+3y+2y^2+y^3. \tag{7.16}$$

7.8 Other strategies

There are least other two techniques that we did not survey in this chapter. One is the finite field method that is covered in Chapter 29. The other computes the expansion of the Tutte polynomial in terms of basis activities. This last method has less applicability but it has been used, for example in computing the Tutte polynomial of Catalan matroids, see [30, 170].

By deletion and contraction, the Tutte polynomial of open and closed uniform polygon chain graphs was obtained in [1007]. Other examples of the use of recurrence relations to compute the Tutte polynomial include benzenoid graphs [483], some outerplanar graphs [770], Sierpiński and Schreier graphs [411], $C_2 \times P_n$ and $\overline{K_2} \vee P_n$ [664] and modified Koch curve and Austria graphs [557].

Generating function techniques were used to compute the Tutte polynomial of the 4-uniform book graph, the k-uniform polygonal chain [412] and the bicircular matroid $B(K_n)$ associated to the complete graph [524]. Computations for the coboundary polynomial the Coxeter arrangements of type A_n, B_n and D_n and be found in [32]; and also for E_6, E_7 and E_8 in [518]. The Tutte polynomial was computed for all exceptional root systems in [365]. Also, the Tutte polynomials of the complete r-partite graphs $K_{n_1,\ldots,n_r}$ can be obtained with this method [994].

Other examples of the use of the transfer matrix method include quartic planar ladder [109, 1006], homogeneous clan graphs [284], Möbius strips [283], strips of the triangular lattice [281], strips of the honeycomb lattice [285], strips of the rectangular lattice [280, 282], and Archimedean tilings [510]. For the use of parallel connections to compute the Tutte polynomial of twisted wheel graphs see [598].

The method of computing the coboundary polynomial can be viewed as computing the Tutte polynomial by using the lattice of flats of the matroid. In [435], a new method for computing the rank generating polynomial from the lattice of cyclic flats of the matroid is demonstrated and it can be used to compute the Tutte polynomial of near design matroids.

7.9 Computation for common graphs and matroids

Table 7.1 shows the Tutte polynomials for some common graphs and matroids, also references to the corresponding formulas to compute the expression are provided.

7.10 Open problems

Among the most studied Tutte polynomials come from *square lattices* $P_n \times P_n$ and lattice graphs in general, like the cylindrical, $P_n \times C_m$, and toroidal, $C_n \times C_m$, lattices. However, thus far, formulas for these polynomials have eluded any approach. The importance of this problem resides in the connection of the Tutte polynomial and statistical physics (see Subsection 8.2.2 and Chapter 20). Another related important problem is to find expressions for the Tutte polynomials of the cubic lattices $P_n \times P_n \times P_n$.

Graph	Tutte polynomial	Family	Formula
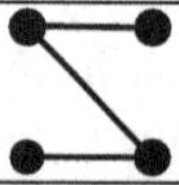	x^3	paths	p. 142
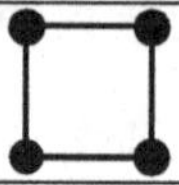	$x^3 + x^2 + x + y$	cycles	(7.1)
	$y^2 + y + x$	3-theta graphs	p. 155
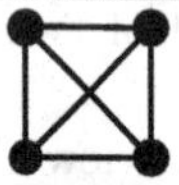	$x^3 + 3x^2 + 2x + 4xy$ $+2y + 3y^2 + y^3$	complete graphs	(7.6)
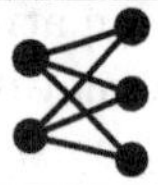	$x^4 + 2x^3 + 3x^2 + x$ $+3xy + y + y^2$	complete bipartite graphs	(7.7)
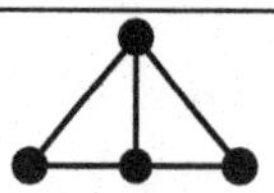	$x^3 + 2x^2 + x$ $+2xy + y + y^2$	fan graphs	(7.3)
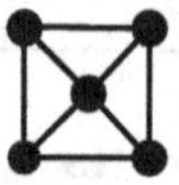	$x^4 + 4x^3 + 6x^2 + 3x$ $+4x^2y + 4xy^2 + 9xy$ $+3y + 6y^2 + 4y^3 + y^4$	wheel graphs	(7.2)
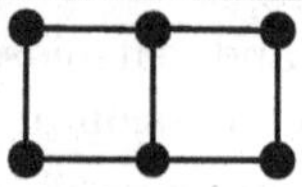	$x^5 + 2x^4 + 3x^3$ $+2x^2 + x + 2x^2y$ $+2xy + y + y^2$	ladder graphs	(7.12)
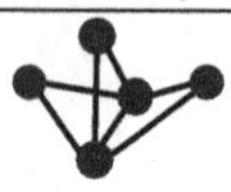	$x^4 + 3x^3 + 3x^2$ $+x + 3x^2y + 3xy^2$ $+3xy + y + y^2 + y^3$	book graphs	(7.13)
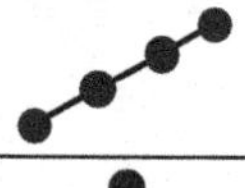	$x^2 + 2x + 2y + y^2$	uniform matroids	(7.4)
	$x^3 + 4x^2 + 3x + 7xy$ $+3y + 6y^2 + 3y^3 + y^4$	projective geometries	(7.10)

TABLE 7.1: Tutte polynomial for common graphs and matroids.

8

Computational resources

David Pearce • Gordon F. Royle

Synopsis

This chapter describes a range of currently available implementations (i.e., runnable computer programs rather than theoretical algorithms) for the computation of Tutte polynomials of general graphs. For each of the following implementations, we describe the underlying algorithm and very briefly examine their empirical performance. The programs considered are:

- The program `tutte` by Haggard, Pearce and Royle
- The program `tutte` by Bedini and Jacobsen.
- The program `tutte_bhkk`.
- The C++ implementation of the program `tuttepol`
- The function `tutte_polynomial()` in the `sage.matroids` package of `SageMath`
- The built-in functions for computing Tutte polynomials `Maple`, `Mathematica` and `SageMath`.

8.1 Introduction

Almost every researcher studying the theory or one of the applications of the Tutte polynomial (or any of its specializations) will—at least occasionally—wish to explicitly compute the Tutte polynomial of a specific (necessarily-small) graph or matroid. In this chapter we survey the main options that are readily available to such a researcher (although these options are very limited indeed for the matroid-theorist working on non-binary matroids).

DOI: 10.1201/9780429161612-8

The choices range from using built-in functions from standard computer algebra systems, both free (`SageMath`) and licensed (`Mathematica`, `Maple`), to downloading source code from individual researchers. These latter programs need compiling, mostly for a Unix system (Linux, MacOS X, etc.), which is not always straightforward, especially if various auxiliary packages and libraries need to be installed first. Thus there is a possibly large time-cost in the initial set-up of these programs.

The selection of implementations was made on the basis of whether the code was widely available (i.e., publicly posted on `GitHub` or a webpage etc.) and whether the implemented algorithm is applicable to graphs in general—at least in principle. We did evaluate implementations with minor restrictions that are straightforward to overcome (i.e., programs that require the input graphs to be simple and/or connected), but we did not consider programs that are fundamentally designed for particular subclasses of graphs.

Thus, the researcher's choice of how to proceed will depend on the number, size, and nature of the graphs they wish to process, their ability and willingness to install and compile software, and how easy it is to work around minor restrictions in the software. It is our intention in this chapter to help such a researcher make an informed choice.

Finally we observe that there is continual progress in this area, that researchers frequently write their own programs for personal use that may or may not be generally available, and that stand-alone programs that prove valuable are often incorporated into general purpose computer algebra systems. In summary, caveat emptor!

8.2 Implementations

In this section, we describe the algorithms underlying each of the implementations we analyze, along with some indication of how the programs manage input/output and limitations (if any) on the types of graph or matroid that the program can handle.

8.2.1 Haggard, Pearce and Royle (HPR)

First we consider the program `tutte` by Haggard, Pearce and Royle [596, 909], which is available (at the time of writing) from `http://homepages.ecs.vuw.ac.nz/~djp/tutte/`. The input to this program is a comma-separated list of the edges of the graph, in the format `1--2,2--3,1--3` and the output is a text representation of the Tutte polynomial that can be directly read by almost any computer algebra system. The input graph may have loops, multiple edges and/or be disconnected. The program is based on the standard deletion–contraction relation

$$T(G;x,y) = \begin{cases} 1 & \text{if } G \text{ has no edges,} \\ T(G\backslash e;x,y) + T(G/e;x,y) & e \text{ not a loop or bridge,} \\ xT(G/e,x,y) & e \text{ is a bridge,} \\ yT(G\backslash e,x,y) & e \text{ is a loop,} \end{cases}$$

where G/e and $G\backslash e$ are the graphs arising by contracting e or deleting e, respectively, retaining any loops or multiple edges that may be created.

The program performs a depth-first search on the search tree implied by the deletion–contraction algorithm. During this search graphs may be encountered that are isomorphic to graphs that have previously been processed to completion, and so a naive implementation of deletion–contraction potentially repeats computations unnecessarily. To avoid this, the HPR algorithm maintains a cache storing the canonically-labelled isomorphs of the most recently-encountered intermediate graphs along with their Tutte polynomials. At every node of the search tree, the graph currently being processed is canonically labelled (using Brendan McKay's program `nauty` [833]) and compared to those in the cache. If the graph is already in the cache, then the stored Tutte polynomial is re-used, rather than being re-computed.

The cache dramatically improves the performance, but in a rather unpredictable manner. The success of this strategy is determined by the number of beneficial collisions in the cache, which in turn depends on a variety of user-selectable parameters such as the cache size, the cache-removal mechanism, the edge-selection heuristic and the initial labelling of the graph.

8.2.2 Bedini and Jacobsen (BJ)

Now we consider the program `tutte` by Bedini and Jacobsen [85] which is available from `https://github.com/andreabedini/tutte`. This uses the same edge-list input format as HPR (although vertices must be indexed from 0), and produces as output a polynomial in Q and v. This output uses juxtaposition for product, so each term has a form similar to `5 Q v^2`; this can be read directly by `Mathematica` but needs reformatting (to something like `5*Q*v^2`) for other systems. The input graph may have multiple edges, but must be connected, and (currently) loops are not processed correctly.

The program BJ actually computes the partition function $Z(G;q,v)$ of the q-state Potts model which, in the Fortuin–Kastelyn expansion, can be defined as follows:

$$Z(G;q,v) = \sum_{A \subseteq E(G)} v^{|A|} q^{k(A)} \tag{8.1}$$

where $k(A)$ denotes the number of connected components of the graph $(V(G),A)$. This partition function is the same as the dichromatic polynomial of Definition 2.1, and thus equivalent to the Tutte polynomial by a change of

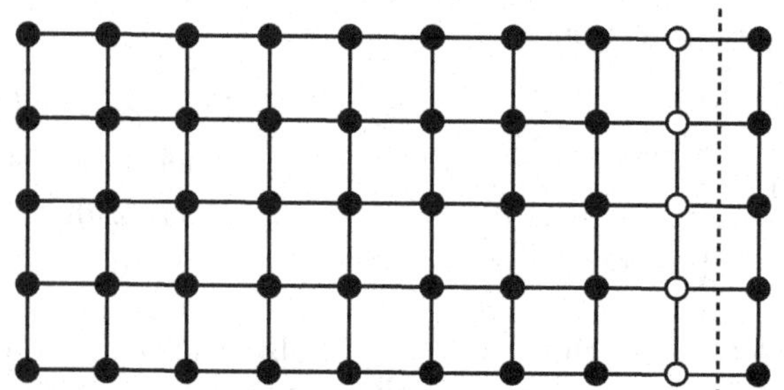

FIGURE 8.1: Adding a layer to a section of the square lattice.

variables (see Theorem 2.6):

$$T(G; x, y) = (x-1)^{-k(G)}(y-1)^{-v(G)} Z(G; (x-1)(y-1), y-1).$$

Statistical physicists are particularly concerned with the behavior of $Z(G; q, v)$ on various structured families of graphs of increasing size, in particular various types of lattice, where one dimension is fixed while the other dimension increases (see Figure 8.1). (See Chapter 20 for details about the connections between statistical physics and the Tutte polynomial.)

In these cases, the lattice is viewed as being built up layer-by-layer, with each new layer attaching to the boundary vertices of the previous graph. By keeping some auxiliary information, it is possible to incrementally update the partition function as each new layer is added. Each subset A of edges in the expression (8.1) determines a partition of the boundary vertices where two vertices are in the same cell if they are connected by edges in A. For each partition π of the boundary vertices, the partial partition function $Z^{\pi}(G)$ is the contribution to $Z(G; q, v)$ made by edge subsets with boundary partition π, and it is this list that is maintained through each stage of the algorithm.

Now suppose that a new layer is added to the graph G forming a larger graph G'. An arbitrary subset of $E(G')$ has the form $A \cup A'$, where $A \subseteq E(G)$ and A' is a subset of the newly added edges. The graph $(V(G), A)$ has a particular number of edges and connected components, and determines a particular partition π of the old boundary vertices. After adding the edges A', the number of edges and connected components of $(V(G'), A \cup A')$ changes, and a partition π' of the new boundary vertices is created. The key point is that these changes are determined only by the partition π and the subset A', and not by the exact subset A. Therefore for each boundary partition π' of G', the partial partition function $Z^{\pi}(G')$ is a linear combination (whose coefficients are polynomials in q, v) of the partial partition functions $Z^{\pi}(G)$. The *transfer matrix* T is a matrix whose rows and columns are indexed by partitions of the old boundary vertices and new boundary vertices respectively, and whose entries contain these polynomials (see Section 7.5). If v is a (column) vector indexed by partitions of the boundary vertices, such that v_{π} is the partial partition function associated with π, then Tv is the same vector for the graph with one layer added. If each layer has the same structure, as is the case for regular lattices, then this process can be iterated, in which case $T^m v$ is the result of adding m layers. In this simple situation, the limiting behavior is determined by the spectral properties (eigenvalues and eigenvectors) of T.

Bedini and Jacobsen's algorithm is based on applying the underlying principles of the transfer matrix method to graphs with no a priori layered structure. In its simplest form, their algorithm processes a graph vertex-by-vertex, maintaining a "boundary" between the processed and unprocessed vertices, along with the partial partition functions indexed by the boundary partitions. The boundary evolves as the algorithm runs, and it will grow and shrink as each vertex is included, processed, and then removed. As every boundary partition must be analyzed, it is important to keep the boundary as small as possible.

A *tree decomposition* of a graph G is a tree T whose vertices are labelled by subsets of $V(G)$ called *bags*, such that the two vertices of each edge are contained in at least one bag, and for each vertex v of G, the subgraph of T induced by the bags containing v is a connected subtree of T. The *tree-width* of a graph is one less than the minimum, over all tree decompositions, of the maximum bag size in that decomposition. The tree-width can be viewed as measuring how "tree-like" the graph is; the graphs of tree-width one are actually trees, while graphs of tree-width two are series-parallel graphs. The algorithm starts by heuristically finding a "good" tree decomposition for the graph (finding an optimal tree decomposition is NP-hard), and then using that tree decomposition to determine the order in which the vertices are processed, with the maximum bag size determining the maximum boundary size that will be encountered during this process. (Details of how the algorithm deals with gluing together the recursively calculated partial partition functions at a tree vertex of degree more than two are omitted here, but may be found in [85].)

If a graph is qualitatively tree-like ("long and skinny"), then a good tree decomposition is found, and the program runs fast even for quite large graphs. However, if the graph has large tree-width, then every tree decomposition will have some large bags and both the time and space requirements will increase accordingly. It is important to note that if a particular input is "easy" (low tree-width), then the program will automatically detect and exploit this fact with no user intervention.

Overall, this algorithm appears to be extremely effective, particularly for relatively sparse graphs (such as planar graphs).

8.2.3 Björklund, Husfeldt, Kaski and Koivisto (BHKK)

In this section we consider the program `tutte_bhkk` by Björklund, Husfeldt, Kaski and Koivisto [127] that is available from `https://github.com/thorehusfeldt/tutte_bhkk`. The input to this program is the 01-adjacency matrix of the graph and the output is the array of coefficients of the Tutte polynomial. The input graph must be simple, having no loops or multiple edges, but it may have any number of connected components.

This algorithm works quite differently to the others in that it processes every subset of the vertices of the graph. Their paper [127] is largely concerned with proving that the Tutte polynomial $T(G;x,y)$ can be calculated in vertex-exponential time, and they describe two algorithms to achieve this. As a "proof of concept", they have implemented one of these algorithms, and it is this implementation (which we call BHKK) that we analyze in this section.

The starting point is to consider another expression for the partition function of the q-state Potts model (see Chapter 20), where q is taken to be a fixed, but otherwise arbitrary, positive integer. Define

$$Z_{\text{Potts}}(G;q,v) = \sum_{\sigma:V\to\{1,2,\ldots,q\}} \prod_{e=xy} (1+v\delta(\sigma(x),\sigma(y))), \tag{8.2}$$

where the sum is taken over all q-colorings σ (proper or improper) of $V(G)$, the product is taken over all edges $e \in E(G)$, and δ is the Kronecker delta function. In other words, a q-coloring with m monochromatic edges contributes $(1+v)^m$ to the sum. This formulation reflects the physical basis of the partition function of the q-state Potts model as a sum over all possible states of a graph when each "site" (vertex) has one of q possible "spins". Fortuin and Kastelyn [498] first proved the equivalence of (8.1) and (8.2).

Each coloring σ determines a partition of $V(G)$ into color classes, and each term of the sum depends on the number of edges contained in the induced subgraphs on each color class. For $X \subseteq V(G)$, let $G[X]$ denote the subgraph of G induced by the vertices in X, let $m(X)$ denotes the number of edges of $G[X]$, and let $f(X) = (1+v)^{m(X)}$. Then

$$Z(G;q,v) = \sum_{(X_1,X_2,\ldots,X_q)} \prod_{i=1}^{q} f(X_i), \tag{8.3}$$

where the summation is over all ordered partitions $(X_1, X_2, \ldots, X_q)$ of V, and where an ordered partition is a sequence of pairwise-disjoint subsets whose union is V.

The BHKK algorithm proceeds by numerically evaluating this expression for $1 \le v \le m-n+2$ and $2 \le q \le n$ (where G has n vertices and m edges) and then using Lagrange interpolation to determine the coefficients of the Tutte polynomial. To avoid manipulation of arbitrary length integers as much as possible, the calculations are done in $\mathbb{Z}_p$ where p is a large prime, and then repeated for as many additional primes as necessary to use the Chinese remainder theorem to determine the coefficients unambiguously.

We need some more terminology before explaining exactly how (8.3) is evaluated. Suppose that $f: 2^V \to \mathfrak{R}$ is a function from subsets of V to a ring $\mathfrak{R}$ (think of $\mathfrak{R}$ as some $\mathbb{Z}_p$). Then the *ζ-transform* of f, denoted $f\zeta$, is the function $f\zeta : 2^V \to \mathfrak{R}$ given by

$$(f\zeta)(X) = \sum_{X' \subseteq X} f(X'),$$

or in other words, the sum of f over all subsets of X (including $\emptyset$ and X itself). The *Möbius transform* of f, denoted $(f\mu)$, is the function $f\mu : 2^V \to \mathfrak{R}$ such that

$$(f\mu)(X) = \sum_{X' \subseteq X} (-1)^{|X \setminus X'|} f(X'),$$

and it is the inverse of the ζ-transform, i.e., $f\zeta\mu = f\mu\zeta = f$ for all f. The rationale for introducing these functions is that it is known that

$$((f\zeta)^q \mu)(V) = \sum_{(X_1, X_2, \ldots, X_q)} \prod_{i=1}^{q} f(X_i), \tag{8.4}$$

where the summation is over all ordered covers $(X_1, X_2, \ldots, X_q)$ of V and $(f\zeta)^q$ means the point-wise q-th power of $(f\zeta)$. Importantly, note that an ordered cover is a sequence of subsets—not necessarily pairwise-disjoint—whose union is V.

Comparing (8.4) and (8.3) we see (8.4) includes all the terms of (8.3), along with many extra terms. However, BHKK found a clever way to extract precisely the required information from (8.4). They introduced a new indeterminate z and a new function $f_z : 2^V \to \mathfrak{R}[z]$ defined by

$$f_z(X) = (1+v)^{m(X)} z^{|n(X)|}$$

which is a monomial (in z) in the polynomial ring $\mathfrak{R}[z]$. In other words, the value $f_z(X)$ now carries information about the size of X in addition to the value $f(X)$. Therefore the expression $((f\zeta)^q\mu)(V)$ is a polynomial in z. An ordered cover $(X_1, X_2, \ldots, X_q)$ contributes to the coefficient of z^n if and only if $n(X_1) + n(X_2) + \cdots + n(X_q) = n$, in which case the cover is necessarily a partition. Therefore the value we are seeking is *precisely* the coefficient of z^n in $((f_z\zeta)^k\mu)(V)$.

Thus a simplified overview of the overall strategy of the BHKK algorithm is that it first fixes a specific value for v, then computes the (polynomial-valued) function $(f_z\zeta)$ (that is, it is calculated for each subset of $V(G)$). Then by polynomial multiplication, each of $(f_z\zeta)^2, (f_z\zeta)^3, \ldots, (f_z\zeta)^n$ is calculated. During this procedure, the contribution that each subset will ultimately make to the z^n coefficient of $((f_z\zeta)^k\mu)(V)$ is calculated and accumulated. There are some obvious optimizations that can be made—for example, every polynomial can be truncated after the z^n term, because no higher order terms can arise from ordered partitions. In the actual implementation, some additional clever optimizations speed up various aspects of the overall strategy, but we omit the details here.

The BHKK algorithm is competitive for relatively small, but dense, graphs, but as the number of vertices increases, it becomes unusably slow. The reason for this is clear from the nature of the algorithm—each of the functions f_z and its various transforms and powers is stored in image form, i.e, as an array of 2^n polynomials in z, and each evaluation uses the entire array.

8.2.4 Tutte polynomials of matroids

In this section, we consider two programs that can compute the Tutte polynomial of non-graphic matroids. The first of these, `tuttepol`, was developed by Thomas Britz, coded in C++ by the mathematical historian and philosopher Henrik Kragh Sørensen (`henrikkragh.dk`), and is available at `https://github.com/hksorensen/TuttePol`. The second is the function `tutte_polynomial()`, coded by Rudi Pendavingh, available in the `sage.matroids` package of the free open-source computer algebra system `SageMath` (`sagemath.org`).

Britz and Sorensøn's `tuttepol` is for binary matroids only, and is a highly optimized program based on deletion–contraction that exploits the fact that binary vectors can be represented as bit patterns and manipulated by the bitwise operators of C/C++, along with numerous other coding optimizations. The user must first obtain a binary matrix representation of the matroid of the form $[I \mid D]$, and it is the matrix D that forms the input to `tuttepol`. The output is a list of each of the individual terms of the Tutte polynomial written in a form that would require slight reformatting before entering into a computer algebra system. There are no restrictions on the matroid input, which can have loops and/or parallel elements.

Rudi Pendavingh's implementation of `tutte_polynomial()` in `SageMath` is the only program currently readily available that can compute the Tutte polynomial of an arbitrary matroid. It works directly from the definition of Tutte polynomial as a sum over the bases of the matroid (see Theorem 4.135) where each base B contributes a single term of the form x^iy^j where i, j are the internal and external activities of B respectively. The program is very convenient to use as it is fully integrated into `SageMath` and so requires no compilation or installation and the polynomial that is returned can immediately be used. (`SageMath` itself is available in ready-to-use binaries.) The time it takes is dependent on the number of bases of the matroid, and this can grow very rapidly as the size of the matroid increases, and so it is only practical for really quite small matroids. However, it is the only available program that can compute the Tutte polynomial for non-binary matroids, and as it is simple and robust, we have a high degree of confidence in its correctness, making it a valuable tool for the matroid theorist.

Of course a graph is also a binary matroid, and a binary matroid is also an arbitrary matroid, and so the researcher examining a graph may wonder if there could be any advantage in treating the graph as a binary matroid and using the more general program `tuttepol` (and similarly treating a binary matroid as a general matroid and using `tutte_polynomial()`). Our empirical investigations show that there is never an advantage in doing this, and so the researcher with a graph should always choose one of the programs designed explicitly for graphs, and similarly for a binary matroid.

8.2.5 Computer algebra systems

In this section we consider computer algebra systems that have functions for computing the Tutte polynomial of a graph, namely `Maple`, `Mathematica` and `SageMath`, which are available from `www.maplesoft.com`, `www.wolfram.com` and `www.sagemath.org`. The first two systems are commercial products with significant license costs and hence somewhat limited availability, while the latter is a free open-source program.

`Mathematica` uses the same method (and some of the same code) as HPR, but with no user-access to the various parameters regarding cache-size, edge-selection heuristics and so on. Therefore the performance of `Mathematica` is approximately the same as the untuned performance of HPR (though the empirical results reported below seem to indicate that `Mathematica` has some additional overhead for very small graphs).

`Maple` uses an algorithm due to Monagan [858], the source code for which is available at `http://www.cecm.sfu.ca/~mmonagan/tutte`. As with HPR, the underlying idea is that of a deletion–contraction method with caching of the polynomials of completed graphs. However, in this variation, the cache lookup is based on equality rather than isomorphism. This avoids the considerable expense of canonically labelling every intermediate graph, but of course reduces the chances of a beneficial collision in the cache. However, it turns out that many of the cache collisions occur with pairs of graphs that are created very close to each other in the computation tree, and so are already identical or almost identical. In empirical testing, Monagan finds that when using a judiciously chosen ordering of the edges, along with a strategy for relabelling the vertices of the graph (contractions remove a vertex), as many as 95% of the cache isomorphisms are actually equalities. Therefore the algorithm captures many of the benefits of a caching strategy, but sheds many of the costs. The trade-off between the reduction in cost and the reduction in benefit determines its overall performance.

On some occasions, the gains are quite spectacular, as illustrated by the graph of the truncated icosahedron. This is a planar cubic graph with 60 vertices and 90 edges, more commonly known as buckminsterfullerene, that was adopted as a "challenge graph" by the first author of HPR. Eventually, its Tutte polynomial was found with HPR, but at the cost of hundreds of hours of cpu time, whereas the `Maple` algorithm can perform the same task in minutes.

While the performance gains are not always of this magnitude, the implementation found in `Maple` performs very well overall, and—for those with access to `Maple`—would be the obvious first choice. For those without access to `Maple`, we note that Monagan's source code is freely available, and although it is itself written in the Maple language, it would be relatively straightforward to implement it for another computer algebra system or indeed as a standalone program.

SageMath uses the same algorithm as HPR, but as it is written in Python and uses its own canonical labelling routine, it is considerably slower than the compiled versions of the same program. However for very small graphs, the convenience of having the routine integrated into a computer algebra system that can then immediately manipulate the resulting polynomial makes this a useful tool.

8.3 Comparative performance

In this section, we report briefly on experiments comparing Maple, Mathematica, HPR and BJ. (We omitted SageMath altogether and BHKK for all except the small graphs test as they cannot complete more than a fraction of the input graphs before reaching our timeout.)

Given the enormous variety of possible input graphs, the number of user-selectable parameters in some programs, and the variation that comes with different labelings of the same graph, it is impossible to be comprehensive, and so the results in this section should be taken as giving just a very rough qualitative impression of the expected performance.

8.3.1 Experimental procedure

We consider just a few types of graph that occur frequently in the literature on graph polynomials, and are commonly used as benchmarks, namely small graphs on a fixed number of vertices and edges, regular graphs of degrees three and four, planar cubic graphs and (dually) planar triangulations.

To generate random connected graphs, we employed the tool genrang (supplied with Brendan McKay's package nauty [833]) to construct random graphs with a given number of edges; from these, we selected connected graphs until there were 100 for each value of $|E|$ or $|V|$ (depending upon experiment). The genrang tool constructs a random graph by generating a random edge, adding it to the graph (if not already present), and then repeating this until enough edges have been added. We also used genrang to generate random simple regular graphs—this essentially works by generating a random regular multigraph and then eliminating it if it contains loops or multiple edges. Random planar triangulations and the dual class of random cubic planar graphs were generated by a variant of Brinkmann and McKay's [207] program plantri.

Each set of 100 test graphs were then randomly relabelled and the total time taken by each of Maple, Mathematica, HPR and BJ to find the Tutte

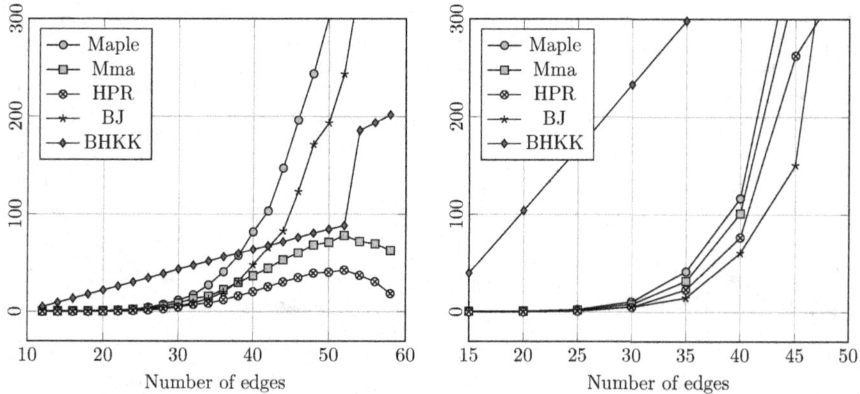

FIGURE 8.2: Graphs on 12 and 14 vertices.

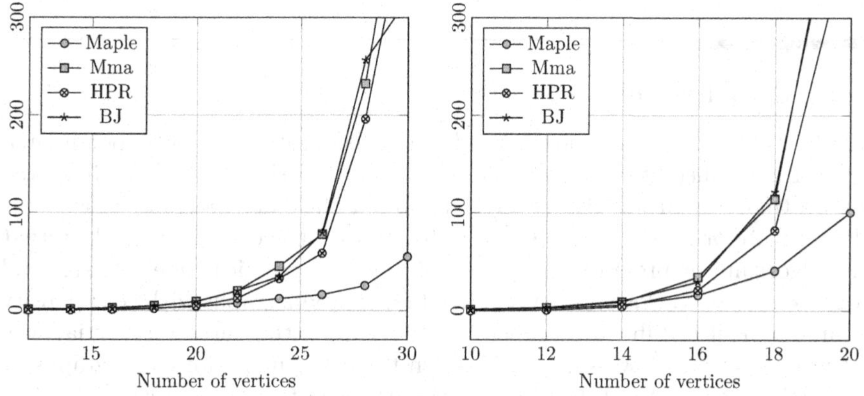

FIGURE 8.3: Cubic and quartic graphs.

polynomials of all 100 graphs was computed. Each of the figures shows this total time in seconds on the vertical axis.

Each of the programs was executed without any additional command-line parameters being provided or any tuning performed. This gives an "out-of-the-box" comparison of the tools (reflecting our expectation that most researchers will not perform extensive tuning and parameter selection). All experiments were performed on an Intel Core i7 4.2GHz with 64GB of memory, running macOS 10.13.2. As `Mathematica` and `Maple` may silently use multiple computer cores on their default settings, the programs were explicitly constrained to use a single core for these experiments.

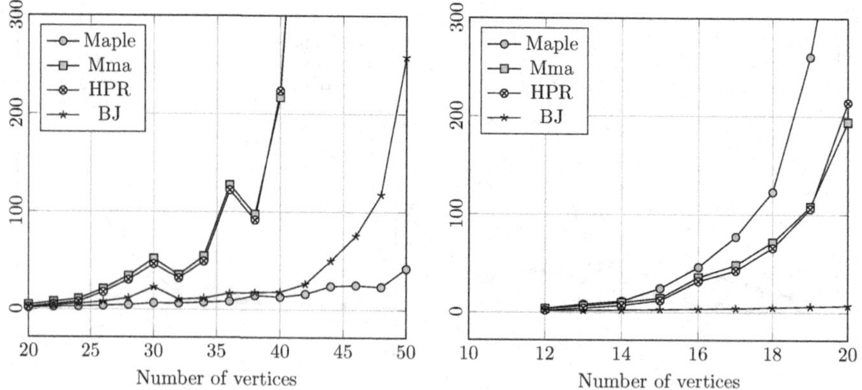

FIGURE 8.4: Cubic planar graphs and planar triangulations.

8.4 Conclusions

The first conclusion that one can reach is that computing the Tutte polynomial is hard both theoretically and in practice. On the theoretical side, it is well known that computing almost anything about the Tutte polynomial is either ♯P-complete or NP-hard (see Chapter 9), and so it is extremely unlikely that any algorithm or program will scale well. On the practical side, the size and edge-density of the graph are the crucial factors. For graphs with edge density around 50%, it is difficult or impossible to compute the Tutte polynomial even for graphs with as few as 20 vertices. On the other hand, for cubic graphs, it is possible to compute the Tutte polynomial even for graphs with more than 60 vertices, as many graphs of this size can be processed in a few minutes. Unfortunately, some take far longer, and we know of no way of distinguishing the difficult cases in advance.

Overall, the situation has improved considerably over the last few years, and researchers have easier access to a wider range of computational tools than ever before. The emergence of computational tools for computing Tutte polynomials that are *not* based on deletion–contraction makes it feasible to compute the Tutte polynomials of whole new families of graphs. Drawing firm conclusions about which program or algorithm is "the best" is impossible because the empirical performance of each of the programs varies dramatically, but not predictably, according to a numerous different factors. Obviously the size (numbers of vertices and edges) and structure (planar, non-planar etc.) of the graph are important, and some of the programs have user-selectable parameters that can dramatically influence performance. Worst of all, just altering the initial labelling of the graph can in the worst case make several orders of magnitude difference.

Compounding this problem is that most graph generators produce graphs with some sort of canonical labelling (for example, the vertex 0 is often adjacent to the vertices 1, 2, 3, ...) while the Tutte polynomial algorithms will—after exhausting any applicable vertex-selection heuristics—process the vertices in some arbitrary order, usually just by increasing vertex number. In practice, the unpredictable interaction between these two sorts of "incidental structure" often seems to mean that graph generating programs produce graphs that are non-typical.

Our experiments have shown that there are frequently used classes of graphs on which each of `Maple`, `Mathematica`/HPR and BJ outperform the other two programs, sometimes significantly. However, subject to our earlier caveats about problem dependence and variability, our experiments show that `Maple` is the best overall performer and should be carefully considered by any researcher needing to compute a Tutte polynomial.

8.5 Open problems

Although researchers now have access to more and better tools than ever before, the results in this chapter demonstrate that computing Tutte polynomials is still more of a black art than a science, involving numerous choices of unpredictable effect. The following list describes some future directions for research that might ameliorate this situation and provide new computational tools.

1. Performance analysis:

   In the timing experiments described above, the averages are often skewed by the presence of one or two pathologically difficult graphs taking several thousand times longer than the other—supposedly similar—graphs in the same experiment.

   Understanding why some graphs are so much easier or harder than others would be of great use in helping to avoid the pathological cases and/or actively exploit the beneficial ones (through relabelling the graph, or reordering the search, or in any other way).

2. Concurrency:

   Modern computers have multiple computing cores and so are inherently capable of parallel computing, while the deletion/contraction recurrence for Tutte polynomials provides an obvious way to distribute the computation. Although the total cpu time (summed over all cores) will be at least as high, programs that can transparently exploit multiple cores will be particularly effective in interactive use.

In this respect, the commercial computer algebra systems have an advantage as they already have some facilities for general purpose concurrency that benefits any computation. In our experiments described above, the `Maple` program ran faster (in real time) when the restriction to a single core was lifted, because other cores were automatically used to perform tasks such as memory garbage collection, etc.

Implementing a truly concurrent version of a Tutte polynomial program, where separate cores communicate usefully during the computation, rather than independently performing disjoint portions of the computation is likely to be very difficult.

3. Standard benchmarks:

   In many fields, computational progress is measured by reference to performance on a benchmark collection of problems; for example, many papers on clique finding and graph coloring refer to the DIMACS graphs ([667]). This collection contains a range of problem instances from easy to unsolved, and authors can report their progress on solving the problems in this collection. In addition to providing a collection of "challenge problems", this ameliorates the problem of authors (often unconsciously) primarily reporting performance data for instances on which their own implementation excels.

   It would be useful to develop a similar library of benchmark instances for the computation of the Tutte polynomial and its specializations.

4. Special-purpose programs:

   Individual researchers often focus on a (small, even tiny) subclass of graphs relevant to their own research, and it is sometimes possible to exploit the very specific nature of those graphs to write a very fast bespoke program to compute their Tutte polynomials.

   While some of these special-purpose programs are too specialized to be of general use, it is sometimes possible to implement the program so that it runs (though perhaps not very fast) on any graph. Bedini and Jacobsen's program BJ is a particularly interesting example. Motivated by techniques tailored for graphs of low tree-width, the program runs on all graphs, but just more slowly if the graph does not have low tree-width.

   A suite of different special-purpose programs, each catering primarily to graphs of a particular type would be a useful adjunct to the general-purpose programs described above.

9

The exact complexity of the Tutte polynomial

Tomer Kotek • Johann A. Makowsky

Synopsis

In this chapter we explore the complexity of exactly computing the Tutte polynomial and its evaluations for graphs and matroids in various models of computation.

- The Turing complexity of evaluating the Tutte polynomial exactly and the resulting dichotomy theorem.
- The special case of planar and bipartite planar graphs.
- The impact of various kinds of graph width (tree-width, clique-width, branch-width) on computing the Tutte polynomial.
- The role of encoding matroids as inputs for computing the Tutte polynomial: Turing complexity for succinct presentations of matroids versus matroid oracles.
- The complexity in algebraic models of computation: Valiant's uniform families of algebraic circuits and the computational model of Blum–Shub–Smale (BSS).

9.1 Introduction

When evaluating the Tutte polynomial we look at the problem of computing the exact value of $T(G; x_0, y_0)$ where the input is a graph G and two elements $x_0, y_0 \in \mathbb{C}$. One could, alternatively, also ask for the list of all coefficients of

DOI: 10.1201/9780429161612-9

$T(G;x_0,y_0)$, or, more modestly, for the sign of $T(G;x_0,y_0)$, when $x_0,y_0 \in \mathbb{R}$. If we can evaluate efficiently, we can use interpolation to compute the coefficients, see [653] for a detailed discussion. Surprisingly, M. Jerrum and L. Goldberg showed in [548] that even computing the sign of $T(G;x_0,y_0)$ can be very hard. In this chapter we concentrate on the complexity of exactly evaluating the Tutte polynomial and its relatives as a function of the order of the graph G. Approximate computation of the Tutte polynomial is treated in Chapter 10.

The landmark paper of F. Jaeger, D. Vertigan, and D. Welsh [653] initiated the study of the complexity of the Tutte polynomial. This paper sets the paradigm for further investigations. It studies the complexity of the problem, given a graph G as input, to compute the value of all the coefficients of the polynomial $T(G;x,y)$. The question is stated and answered in the Turing model of computation and the complexity class suitable to capture the most difficult cases is $\sharp$P, which was introduced by Valiant in [1110]. In the course of the chapter we see that the results can be extended to any subfield of $\mathbb{C}$, the elements of which can be represented as finite binary strings. The main result of [653] states that for almost all pairs $(x_0,y_0) \in \mathbb{C}$ the problem of evaluating $T(G;x_0,y_0)$ is $\sharp$P-complete, and the set for which it is not $\sharp$P-complete (the exception set) is given explicitly as a semialgebraic set of lower dimension. We shall discuss this result in detail in Section 9.3.1.

9.2 Complexity classes and graph width

In this section we collect some background material needed for discussing the complexity of the Tutte polynomial for graphs. Further background for the case of matroids follows in Section 9.4.

9.2.1 Complexity classes

Background on complexity in the Turing model of computation can be found in [47, 556, 666, 906, 1013], and for Valiant-style algebraic complexity in [254]. For complexity over the reals and complex numbers we refer to [144], and for quantum computing we refer to [590, 880]. For parameterized complexity we refer the reader to [355, 417, 418, 490, 494].

Informally, a *decision problem* is a computational problem for which the answer is either "yes" or "no". An example of a decision problem is: Given a graph G, does it have a proper r-coloring? A *counting problem* is a computational problem for which the answer is the number of satisfying configurations. An example of a counting problem is: given a graph G, how many proper r-colorings does it have? We denote by P, respectively FP, the set of decision problems and counting problems which can be solved in time polynomial in the size of the input. Problems which belong to P or to FP are called *tractable*, and are considered to be efficiently computable.

We denote by NP the set of decision problems for which it is possible to *verify* whether a given configuration is correct in polynomial time in the size of the input. We denote by ♯P (pronounced "number P" or "sharp P") the set of counting problems which count configurations whose correctness can be verified in polynomial time. For example the problem of deciding whether a graph is r-colorable is in NP, and computing the number of r-colorings is in ♯P. Whether P = NP and whether FP = ♯P are famous and notoriously difficult open questions in theoretical computer science. A problem X is *NP-hard* (respectively, *♯P-hard*) if using it as an oracle allows one to solve any problem in NP (respectively in ♯P) in polynomial time. Problems which are NP-hard (or ♯P-hard) are provably the hardest problems in NP (respectively ♯P), and are are considered *intractable.* They are not considered effectively computable, unless NP = P, respectively ♯P = FP.

So far the complexity of computational problems was measured with respect to one parameter of the input only: the size of the input (e.g., the number of vertices or edges of the input graph). *Parameterized complexity* measures the complexity of problems in terms of additional parameters of the graph. A problem is *fixed-parameter tractable* with respect to a parameter k if there is a computable function f, a polynomial p, and an algorithm solving the problem in time $f(k) \cdot p(n)$, where n is the size of the problem, and the function $f(k)$ only depends on k. However, f may grow arbitrarily in k, and indeed is often exponential in k. The set of fixed-parameter tractable decision problems is denoted by FPT.

9.2.2 Structural graph parameters

Tree-width and clique-width are graph parameters which measure the compositionality of a graph. Tree-width is usually defined as the minimum width of a certain map of a graph into a tree called a *tree-decomposition* (see, e.g., [383]). In contrast, graphs of clique-width k are defined inductively. For uniformity of presentation, we also give an inductive definition of graphs of tree-width k.

The various notions of graph width have had a large impact on the study of efficient algorithms for graph problems, [617]. Due to the inductive definitions of graphs of bounded tree-width and clique-width, problems which are NP-hard or ♯P-hard (on general graphs) are often fixed parameter tractable with respect to tree-width or clique-width [337, 339].

Let $[k] = \{1, \ldots, k\}$. A *k-graph* is a graph $G = (V, E)$ together with a partition $\bar{R} = (R_1, \ldots, R_k)$ of V. The sets R_i are called *labels* of G and $\bar{R}$ is a *labeling.* The classes $TW(k)$ and $CW(k)$ of k-graphs are defined inductively below.

Definition 9.1. The class $TW(k)$ of k-graphs of *tree-width* at most k is defined inductively as follows:

1. All $(k+1)$-graphs of order at most $k+1$ belong to $TW(k)$.
2. $TW(k)$ is closed under the following operations:
   (a) Disjoint union.
   (b) Renaming of labels $\rho_{i \to j}$: all vertices in R_i are moved to R_j.
   (c) Fusion $fuse_i$: all vertices in R_i are identified into a single vertex.

A graph has *tree-width* k, written $\mathrm{tw}(G)$, if k is the minimal value such that there is a labeling $\bar{R}$ of G for which $(G, \bar{R}) \in TW(k)$.

All trees have tree-width 1 and all cycles have tree-width 2. A clique of size k has tree-width $k-1$.

Definition 9.2. The class $CW(k)$ of k-graphs of *clique-width* at most k is defined inductively as follows:

1. All k-graphs of order 1 belong to $CW(k)$.
2. $CW(k)$ is closed under the following operations:
   (a) Disjoint union.
   (b) Renaming of labels $\rho_{i \to j}$.
   (c) Edge addition: $\eta_{i,j}$: all possible edges are added between R_i and R_j.

A graph has *clique-width* k, written $\mathrm{cw}(G)$, if k is the minimal value such that there is a labeling $\bar{R}$ of G for which $(G, \bar{R}) \in CW(k)$.

Every graph of tree-width at most k has clique-width at most $2^{k+1}+1$ (cf. [340]) while there are classes of graphs of bounded clique-width which have unbounded tree-width, such as cliques (clique-width 1) or complete bipartite graphs (clique-width 2). *Logarithmic clique-width* is defined similarly to clique-width, with the exception that $R_1, \ldots, R_k$ are no longer required to be disjoint. Every graph of tree-width at most k has logarithmic clique-width at most $k+2$. A graph $G=(V,E)$ of tree-width k has at most $k|V|$ edges, while a graph of clique-width k can have the maximal number of edges of a loop-free undirected graph, $\binom{|V|}{2}$.

A $TW(k)$-expression is an expression t consisting of base elements and operations as given in Definition 9.1. A $CW(k)$-expression is defined correspondingly using Definition 9.2 instead. Computing the exact tree-width or the exact clique-width of a graph is NP-hard. However, computing a $TW(k)$-expression is fixed parameter tractable in k due to the fixed parameter tractability of computing a tree decomposition [146]. For clique-width, the computation of an exponential upper bound k' on the clique-width k of a graph and a $CW(k')$-expression for the graph is fixed parameter tractable (see [892]). Other notions of widths of graphs are studied in the literature (one survey is [617]). Among these we have $\mathrm{bw}(G)$, the *branch-width* of G, which is related to tree-width;

and rw(G), the *rank-width* of G, which is related to clique-width. The following theorem makes this precise. Its first result is from [964], and second from [892].

Theorem 9.3. *For a graph G,*

1. $\mathrm{bw}(G) \leq \mathrm{tw}(G) + 1 \leq \lfloor \frac{3}{2}\mathrm{bw}(G) \rfloor$,
2. $\mathrm{rw}(G) \leq \mathrm{cw}(G) \leq 2^{1+\mathrm{rw}(G)} - 1$.

Among these notions of width, only branch-width generalizes to matroids (see Section 9.4.6).

9.3 Exact evaluations on graphs

9.3.1 The main paradigm: a dichotomy theorem

The essential computational task associated with the Tutte polynomial is to compute, for a given graph G, the table of coefficients of $T(G; x, y)$. However, this task is $\sharp$P-hard, since the number $(-1)^{r(E)} 3^{-k(G)} T(G; -2, 0)$ of proper 3-colorings of G is $\sharp$P-hard, and the prefactor $(-1)^{r(E)} 3^{-k(G)}$ is polynomial-time computable. In particular, the evaluation $(-2, 0)$ of the Tutte polynomial is $\sharp$P-hard.

F. Jaeger, D. Vertigan and D. Welsh began the study of the complexity of the Tutte polynomial in their classic paper [653]. They studied the problem of computing the Tutte polynomial on input a graph G at a fixed point (a, b) in the complex plane. The main result of [653], stated below, is a *dichotomy theorem*, a theorem which classifies the complexity of evaluations (a, b) of the Tutte polynomial into *tractable* or *$\sharp$P-hard*. Technically, the complexity is measured in an computable extension field of the rational numbers containing a and b.

Theorem 9.4. *Let G be a graph, let H_1 be the hyperbola*

$$\{(x, y) \in \mathbb{C}^2 : (x - 1)(y - 1) = 1\},$$

and let H be the union of H_1 with

$$\{(1, 1), (-1, -1), (0, -1), (-1, 0), (i, -i), (-i, i), (j, j^2), (j^2, j)\},$$

where $j = e^{2\pi i/3}$. Then, for every $(a, b) \in \mathbb{C}^2$:

1. *if $(a, b) \notin H$, then $T(G; a, b)$ is $\sharp$P-hard;*
2. *if $(a, b) \in H$, then $T(G; a, b)$ is computable in polynomial-time.*

See Figure 9.1 for a plot of the $\mathbb{R}^2$-part of the "Tutte plane".

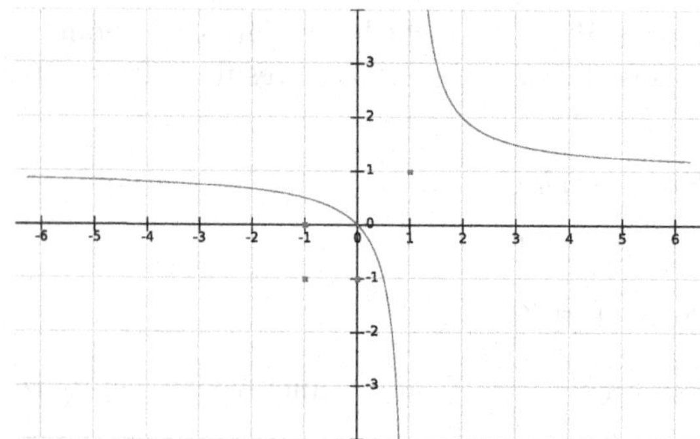

FIGURE 9.1: $T(G;x,y)$ is polynomial-time computable on the hyperbola $(x-1)(y-1)=1$ and the points $(1,1),(-1,-1),(0,-1),(-1,0)$. Otherwise it is $\sharp$P-hard.

9.3.2 Planar and bipartite planar graphs

Because of the connection between the Potts and Ising models of statistical mechanics on one hand and the Tutte polynomial on the other (see Chapter 20), results in one area inform the other. For example, the method of Kasteleyn [678] for tractable computation of the Ising partition function on planar graphs carries over to the Tutte polynomial.

Theorem 9.5. *For planar graphs evaluating the Tutte polynomial on the hyperbola*

$$H_2 = \{(x,y) \in \mathbb{C}^2 : (x-1)(y-1) = 2\}$$

is in P. *Otherwise, points which are $\sharp$P-hard for general graphs remain $\sharp$P-hard even for bipartite planar graphs.*

Vertigan and Welsh, [1118, 1119], studied the complexity of evaluating the Tutte polynomial on the class of graphs which are both bipartite and planar.

Theorem 9.6. *Let H_{bp} be the union of $\{(1,1),(-1,-1),(j,j),(j,j)\}$, H_1, and H_2. For every $(a,b) \in \mathbb{C}^2$,*

1. *if $(a,b) \notin H_{bp}$, then $T(G;a,b)$ is $\sharp$P-hard for bipartite planar graphs;*
2. *if $(a,b) \in H_{bp}$, then $T(G;a,b)$ is computable in polynomial-time for bipartite planar graphs.*

See Figure 9.2 for a plot of the Tutte plane for bipartite planar graphs.

9.3.3 Graphs of bounded tree-width and clique-width

The Tutte polynomial and variants of it are efficiently computable on graphs of bounded tree-width or clique-width. J. Oxley and D. Welsh initiated the study of the complexity of evaluating the Tutte polynomial on graph classes of bounded width, [905]. However, their notion of width is more restricted

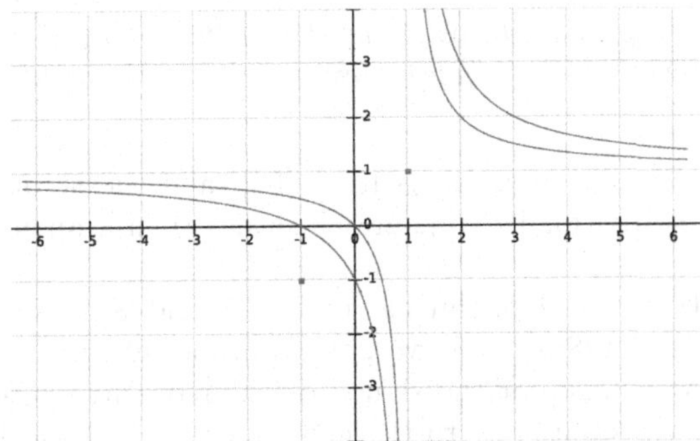

FIGURE 9.2: $T(G;x,y)$ is polynomial-time computable on the hyperbolas $(x-1)(y-1)=1$ or 2 and the points $(1,1),(-1,-1)$ for bipartite planar graphs. Otherwise it is ♯P-hard.

than tree-width, but does include the series–parallel graphs, which are of tree-width 2. The first general results on the computation of the Tutte polynomial on graphs of bounded tree-width were obtained independently by A. Andrzejak in [25] and S. Noble in [882]. They proved that for every fixed evaluation (a,b) of $T(G;x,y)$, there an algorithm which computes $T(G;a,b)$ using a linear number of arithmetic operations.

Theorem 9.7. *$T(G;x,y)$ can be evaluated in linear time in $|V|$ on graphs with bounded tree-width. In fact, $T(G;x,y)$ is fixed-parameter tractable with respect to tree-width.*

Analogues of Theorem 9.7 for variants of the Tutte polynomial are given below. The coefficients of the Tutte polynomial and its variants can be computed in polynomial-time on graphs of bounded tree-width. In the case of $T(G;x,y)$, the list of coefficients can be computed in time $O(|V|^3)$ by evaluating $T(G;x,y)$ at sufficiently many points to interpolate it. There is a matching cubic lower bound for the computation of all the coefficients [882]. Theorem 9.7 has a wide generalization based on logical techniques which covers colored versions of the Tutte polynomial, the Jones polynomial in knot theory, and many other graph polynomials which can be defined in monadic second order logic.

L. Kauffman [680] defined a generalization of the Tutte polynomial for signed graphs called the signed Tutte polynomial. This was shown by B. Bollobás and O. Riordan in [158] to be a substitution instance of their colored Tutte polynomial. (See Section 24.5 for the discussion of colored Tutte polynomials.) T. Zaslavsky [1183] introduced another variation of the Tutte polynomial, the *normal function* of the colored matroid, hence also for colored graphs. It is a substitution instance of the extension of the bivariate chromatic polynomial introduced in Remark 11 of [55].

Theorem 9.8.

1. *Evaluating the normal function of a colored graph is fixed parameter tractable with respect to tree-width.*

2. *Evaluating the signed Tutte polynomial and the colored Tutte polynomial with a finite set of colors is fixed parameter tractable with respect to tree-width.*

The first item in Theorem 9.8 is from [55], and the second from [811]. Similar theorems stated for knot diagrams rather than graphs can be found in [810].

The complexity of the algorithms given in [55] and [811] is asymptotically the same as in [25] and [882]. However, the runtimes of the algorithms given in [55] and [811] work also for the signed or colored Tutte polynomials, but involve very large constants due to the generality of the logical methods involved which make them impractical. The algorithms given in [25] and [882] give workable upper bounds, but do not work for the signed or colored Tutte polynomials. A more efficient algorithm for the colored Tutte polynomial was given in [1076] based on [25].

S. Noble in [883] extended his approach to a further generalization of the Tutte polynomial, the weighted graph polynomial U (see Chapter 26). He showed that evaluating $U(G;\mathbf{x},y)$ is also fixed-parameter tractable with respect to tree-width (see Section 26.5).

On graphs of bounded clique-width the Tutte polynomial can be computed in subexponential time. However, unlike the case of tree-width, $T(G;x,y)$ is unlikely to be fixed-parameter tractable with respect to clique-width.

Theorem 9.9. *$T(G;x,y)$ can be computed in time $\exp\left(O(|V|^{1-1/(k+2)})\right)$ on graphs of clique-width at most k.*

Theorem 9.10. *Given a graph G and $r \in \mathbb{N}$, deciding whether the chromatic polynomial $\chi(G;r)$ is non-zero is not fixed-parameter tractable with respect to clique-width (under the complexity-theoretic assumption $\mathrm{FPT} \neq \sharp\mathrm{W}[1]$). In particular, this shows that the same applies to evaluating the chromatic polynomial.*

Theorem 9.9 is from [523], and Theorem 9.10 from [492, 493].

The chromatic polynomial, however, is polynomial-time computable if the clique-width is fixed, as shown in [814]:

Theorem 9.11. *The chromatic polynomial $\chi(G,\lambda)$ can be computed in time $O(|V|^{f(k)})$ on graphs of clique-width at most k, where $f(k) = O(2^k)$ is a function depending only on k.*

It is an open question whether Theorem 9.9 can be improved to match the bound for the chromatic polynomial given in Theorem 9.11.

We summarize the results of Sections 9.3.1–9.3.3 in Table 9.1.

9.3.4 Exact computation of the $T(G;x,y)$ on small graphs

Computing the Tutte polynomial on large graphs is a major challenge. The deletion–contraction reduction formula of the Tutte polynomial gives a naive

Graph class	♯P-hard	Subexponential	FPT	P	Theorem
All graphs	$\mathbb{C}^2 - H$	H	H	H	9.4
Planar	$\mathbb{C}^2 - H_2$	H_2	H_2	H_2	9.5
Bipartite planar	$\mathbb{C}^2 - H_{bp}$	H_{bp}	H_{bp}	H_{bp}	9.6
$TW(k)$	$\emptyset$	$\mathbb{C}^2$	$\mathbb{C}^2$	H	9.7
$CW(k)$	$\emptyset$	$\mathbb{C}^2$	H	H	9.9

TABLE 9.1: The complexity of evaluating the Tutte polynomial for various graph classes.

algorithm for computing the list of coefficients of the Tutte polynomial whose runtime is exponential in the number of edges of the graph. Sekine, Imai and Tani [996] gave an algorithm computing the Tutte polynomial of a planar graph of order n whose runtime is $O(2^{O(\sqrt{n})})$. Björklund et al. [127] gave an algorithm for general graphs which is exponential in the number of vertices and uses polynomial space. Haggard, Pearce and Royle [596] implemented an algorithm which exploits isomorphisms in the computation tree to allow more efficient computation of the Tutte polynomial and used it disprove a conjecture by Welsh on the roots of the Tutte polynomial. See Chapters 7 and 8 for more information on computing the Tutte polynomial.

9.3.5 An exponential-time dichotomy theorem

An algorithm has *subexponential running time* if it runs in time $2^{n^{o(1)}}$. While there are subexponential algorithms for the Tutte polynomial on restricted classes of graphs [523, 996], a subexponential algorithm for the class of all graphs is unlikely to exist, since the problem 3-SAT of whether propositional 3-CNF formulas are satisfiable is reducible via a linear-size reduction to the chromatic polynomial. Indeed, Dell et al. [370] gave a dichotomy theorem under the exponential time hypothesis (♯ETH), which is the assumption that there is no sub-exponential algorithm for the number of proper 3-colorings. The lower bounds in this dichotomy theorem are tight under the assumption of ♯ETH, due to the vertex-exponential algorithm of [127].

Theorem 9.12. *Let $(a, b) \in \mathbb{Q}^2$. Under ♯ETH:*

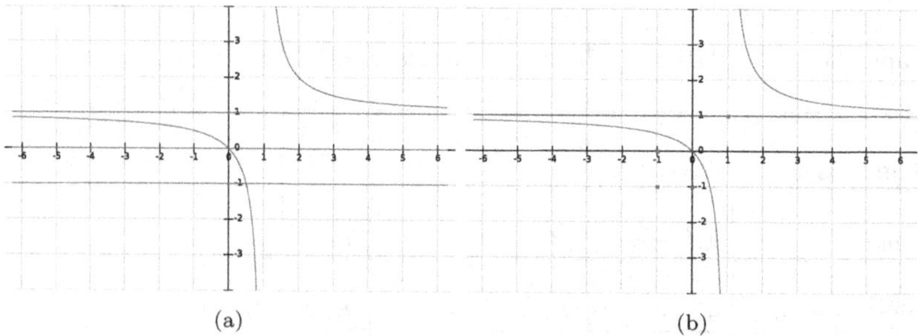

(a) (b)

FIGURE 9.3: Plots of subexponential lower bounds for $T(G;x,y)$.

1. *$T(G;a,b)$ cannot be computed in subexponential time in $|V|$ if $(a,b) \notin H_1$ and $b \notin \{-1,0,1\}$;*

2. *$T(G;a,b)$ cannot be computed in subexponential time in $|V|$ if $y = 0$ and $a \notin \{-1,0,1\}$ on simple graphs;*

3. *$T(G;a,b)$ cannot be computed in subexponential time in $(|E|/\log^2|E|))$ if $a = 1$ and $b \neq 1$ on simple graphs;*

4. *$T(G;a,b)$ cannot be computed in subexponential time in $(|E|/\log^3|E|))$ if $(a-1)(b-1) \neq \{0,1\}$ and $(a,b) \notin \{(-1,-1),(-1,0),(0,-1)\}$ on simple graphs.*

The line $y = 1$ is still open (♯P-hardness is the best known).

Using results of R. Curticapean [354] items (3) and (4) of Theorem 9.12 can be improved by eliminating the logarithmic factor.

See Figure 9.3 for plots of subexponential lower bounds for $T(G;x,y)$. Figure 9.3a shows that except for the hyperbola $(x-1)(y-1) = 1$ and the lines $y \in \{0,-1,1\}$, $T(G;x,y)$ cannot be computed in subexponential vertex-time. Figure 9.3a shows that except for the hyperbola $(x-1)(y-1) = 1$, the line $y = 1$ and the points $(-1,-1),(-1,0),(0,-1)$, $T(G;x,y)$ cannot be computed in time which is subexponential in $|E|/\log^3|E|$ on simple graphs.

9.4 Exact evaluation on matroids

9.4.1 Presentation of matroids

Background on matroids can be found in Chapter 4. Recall a matroid is an ordered pair $(E,\mathcal{C})$ consisting of a finite set E, called the *ground set* and a set $\mathcal{C}$ of subsets of E, called the *circuits*. Matroids have many equivalent definitions. Instead of $\mathcal{C}$ one can consider $\mathcal{I}$, the *independent sets*, $\mathcal{S}$, the *spanning sets*, $\mathcal{F}$, the *flats*, $\mathcal{B}$, the *bases*, $\mathcal{H}$, the *hyperplanes*. Other definitions use a *closure operator* on the subsets of E, a *rank function* or a *girth function.* All these definitions of matroids are equivalent from an axiomatic point of view, but differ from an algorithmic point of view (cf. [602, 828]).

In all the definitions of matroids there is an exponential gap between the size of the underlying set E, and the information needed to capture the properties of the family of subsets, the closure operation, the rank function or the girth function defining the matroid. In order to develop a complexity theory for matroid problems three approaches are used: matroid *oracle computations*, the restriction to matroids with *succinct presentations*, and using the full listing of the family of subsets (circuits, independent sets, etc.) or the functions on the subsets of the ground sets (rank function, closure operation, etc.). Pioneering papers for the first two approaches are [154, 439, 441, 602, 657, 967]. More recent discussions can be found in [828, 1019].

9.4.2 Full presentations of matroids

A matroid is essentially a finite set with a structured family of subsets, which can be described explicitly. This approach has originally not been studied because the input for a Turing machine was considered too large if the size of the matroid should be polynomial in the size of the cardinality of the ground set E. It was suspected that this would make most, if not all, computational matroid problems solvable in polynomial time in the size of its true input. However, D. Mayhew analyzed the situation carefully in [828] by comparing the various input sizes of RANK, INDEPENDENT, SPANNING, BASIS, CIRCUIT, FLAT and HYPERPLANE.

Suppose that INPUT_1 and INPUT_2 are two methods for describing a matroid M. Then $\text{INPUT}_1 \leq \text{INPUT}_2$ if and only if there exists a polynomial-time Turing machine which will produce for every matroid M the presentation (M, INPUT_2) given the presentation (M, INPUT_1). We write $\text{INPUT}_1 < \text{INPUT}_2$ if and only if it is not the case that $\text{INPUT}_2 \leq \text{INPUT}_1$.

Comparing input modes gives the following picture from [828].

Theorem 9.13. *Under the partial ordering $\leq$ of comparing input modes we have*

1. RANK $<$ INDEPENDENT, *and* RANK $<$ SPANNING. *Furthermore* INDEPENDENT *and* SPANNING *are incomparable;*

2. SPANNING $<$ BASIS $<$ CIRCUIT *and* BASIS $<$ HYPERPLANES*; Furthermore,* CIRCUITS *and* HYPERPLANES *are incomparable;*

3. INDEPENDENT $<$ FLATS $<$ HYPERPLANES *and* INDEPENDENT $<$ BASIS. *Furthermore,* BASIS *and* FLATS *are incomparable.*

9.4.3 Succinct presentation of matroids

Let $\mathcal{A}$ be a class of matroids closed under matroid isomorphisms, and Σ^* be the finite words over a finite alphabet Σ. Let $\ell(w)$ denote the length of $w \in \Sigma^*$, and let $\mathcal{A}_n$ be the class of matroids in $\mathcal{A}$ with ground set $[n] = \{1, \ldots, n\}$. A class of matroids has a *succinct description*, or is *succinct*, if there is an injective mapping $e : \mathcal{A} \to \Sigma^*$ and a polynomial $p \in \mathbb{N}[X]$ such that for each $M \in \mathcal{A}_n$ we have that $\ell(e(M)) \leq p(n)$. It is easy to see that $\mathcal{A}$ is succinct if and only if $|\mathcal{A}_n| = O(2^{q(n)})$ for some $q \in \mathbb{N}[X]$. In this section complexity results are always formulated for succinct classes of matroids.

Proposition 9.14.

1. *If a matroid M is representable over a finite field, then it is succinct (see [827]).*

2. *Graphic matroids, regular matroids and binary matroids are succinct, as they are representable over the field $\mathbb{F}_2$ (see Chapter 4).*

3. *Transversal matroids are representable over sufficiently large (finite) fields, hence they form a succinct class of matroids (see [920]). (Special cases of transversal matroids are bicircular matroids and lattice-path matroids.)*

Let M be an arbitrary matroid. The k-*stretch* $s_k(M)$ is defined by $s_k(M) = M \otimes U_{k,k+1}$, and the k-*thickening* $t_k(M)$ is $t_k(M) = M \otimes U_{1,k+1}$. (Recall the tensor product of matroids from Definition 6.42.)

Definition 9.15.

1. A class of matroids $\mathcal{C}$ is *closed under expansions* if it is closed under $s_k(M)$ and $t_s(M)$ for every $k \in \mathbb{N}$.

2. A class of matroids $\mathcal{C}$ is *expand-succinct* or a JVW-*class* if it is succinct and $s_k(M)$ and $t_k(M)$ can be constructed in polynomial time for each $M \in \mathcal{C}$.

These classes of matroids were defined in [653, 1116]. Typical JVW classes are graphic matroids, regular matroids, and matroids representable over a fixed finite field. The class of transversal matroids is not a JVW-class as it is closed under deletion, but not under contraction.

9.4.4 Complexity for succinct presentations

We let $\mathbf{A}$ denote the set of algebraic numbers. The paper [653] was the first to analyze the complexity of evaluating the Tutte polynomial of a matroid for pairs of algebraic numbers $x_0, y_0 \in \mathbf{A}$. The Tutte polynomial of a graphic matroid $M(G)$ is the same as the Tutte polynomial of G, so here we only discuss classes of non-graphic matroids.

Theorem 9.16. *Let $\mathcal{C}$ be* JVW*-class of matroids and $M \in \mathcal{C}$. Then evaluating $T(M; a, b)$ is* ♯P*-hard on all points $(a, b) \notin H$, where H is as in Theorem 9.4. If $(a, b) \in H$, then $T(M; a, b)$ is computable in polynomial time.*

Although the class of transversal matroids is not a JVW-class, it is succinct (see Proposition 9.14), and we still have the following dichotomy from [322].

Theorem 9.17. *For the class of transversal matroids, and M in this class, the evaluation the Tutte polynomial $T(M, x_0, y_0)$ is* ♯P*-hard for all $(x_0, y_0) \in \mathbf{A}$, unless $(x_0 - 1)(y_0 - 1) = 0$, in which case it is in* P.

To prove this theorem, the authors replace the operations k-stretching and k-thickening used in the proof of Theorem 9.16 by operations k-expansion and k-augmentation which preserve transversality.

The class of bicircular matroids (see Definition 4.119) is contained in the class of transversal matroids, which, unlike transversal matroids, is closed under minors. A bicircular matroid $B = \mathrm{Bi}(G)$ is determined by its underlying graph G, but it need not be graphic. The k-stretching of a bicircular matroid B, denoted by $s_k(B)$ can be shown to be $\mathrm{Bi}(s_k(G))$. Hence, bicircular matroids are closed under k-stretching. However, they are not closed under k-thickening. The following result is from [524].

Theorem 9.18. *Let $\mathcal{B}$ denote the class of bicircular matroids.*

1. *For a matroid $M \in \mathcal{B}$, the evaluation of the Tutte polynomial $T(M; x_0, y_0)$ is* ♯P*-hard for all $(x_0, y_0) \in \mathbf{A}$, unless $(x_0 - 1)(y_0 - 1) = 0$. (For $x_0 = 0$ and $x_0 = -1$ the complexity has not yet been determined.)*

2. *For the complete graph K_n and the bicircular matroid $M = B(K_n) \in \mathcal{B}$, the evaluation of the Tutte polynomial $T(M; x_0, y_0)$ is in* P.

A further special case of transversal matroids is the class of the *lattice-path matroids.* They are obtained from the lattice with vertices in $(i, j) \in \mathbb{N}^2$ with $0 \leq i \leq m$ and $0 \leq j \leq r$ with steps EAST and NORTH. Let P and Q be two paths in this lattice. The lattice paths that go from $(0, 0)$ to (m, r) and that remain in the region bounded by P and Q can be identified with the bases of a particular type of transversal matroid, [168]. The following theorem is from [865].

Theorem 9.19. *Let $\mathcal{LP}$ denote the class of lattice-path matroids. For $M \in \mathcal{LP}$, the evaluation of the Tutte polynomial $T(M; x_0, y_0)$ is in* P.

9.4.5 Matroid oracles

The situation for matroids which have no succinct presentation is less well understood. Matroid oracles are used to show that most matroid properties are hard to compute for general matroids (see [657, 967]). This is largely due to the fact that the number of matroids with underlying set E of size

n is doubly exponential in n (see [657]). Oracles may vary (cf. [602]) but the most widely used is the *independence oracle*, which takes as its input a set of matroid elements, and returns as output a Boolean value, true if the given set is independent and false otherwise. Naturally, every one of the nine axiomatic definitions of matroids, INDEPENDENT, BASIS, CIRCUIT, SPANNING, FLAT, HYPERPLANE, RANK, GIRTH and CLOSURE, gives rise to an oracle. The oracles return Boolean values also for basis sets, circuits, spanning sets, flats and hyperplanes. In the case of the rank or girth function, the oracle returns an element of $\{0, 1, \ldots, |E|, \infty\}$, and in the case of the closure, a subset of E. The computational strength of the different oracles can be compared by polynomial time simulation. Let O_1, O_2 be two oracles of matroids on a ground set E. We say that O_1 is *polynomially reducible to* O_2, denoted by $O_1 \to O_2$, if and only if one call in O_1 can be simulated by at most polynomially many calls on O_2. In the case, we say that O_1 is *weaker* than O_2. If $O_1 \to O_2$ and $O_2 \to O_1$, we say O_1 and O_2 are *bireducible.*

Surprisingly, comparing input modes gives a different picture than comparing matroid oracles. From [602]:

Theorem 9.20. *Under the partial order of polynomial reducibility between oracles we have:*

1. BASIS, CIRCUIT, FLAT, HYPERPLANE *are incomparable and weaker as oracles than* RANK*;*
2. RANK, INDEPENDENT, SPANNING *and* CLOSURE *polynomially bireducible;*
3. GIRTH *is strictly stronger than all the other oracles.*

Many natural matroid parameters are hard to compute, [657].

Theorem 9.21. *The following are not polynomial-time computable using the* INDEPENDENT*-oracle.*

1. *Decide whether M is uniform, self-dual, orientable, bipartite, Eulerian or representable.*
2. *Compute the girth or the connectivity of M.*
3. *Count the number of circuits, bases, hyperplanes or flats of M.*

Information on the computability of the Tutte polynomial when the matroid is given by the oracle INDEPENDENT is at the end of Subsection 9.4.6, and a detailed comparison of the computational power of various oracles may be found in [828].

9.4.6 Matroid width

J. Oxley and D. Welsh in [905] introduced a very restricted definition of matroid width. A class of matroids $\mathcal{M}$ has *OW-width* k if a largest 3-connected member of $\mathcal{M}$ has a ground set of k elements. They show that, under some stronger assumption than succinct representability, evaluating the Tutte polynomial on matroid classes of bounded OW-width is in FPT. We shall not pursue this further.

A *branch decomposition* of a matroid M with ground set E and rank function r is a tree T such that all inner nodes of T have degree three, and the leaves of T are in one-one correspondence with E. An edge e of T splits T into two subtrees with leaves corresponding to $E_1(e)$ and $E_2(e)$ which partition E. The width of e is defined by $r(E_1(e)) + r(E_2(e)) - r(E) + 1$. The *width* of the branch decomposition T is the maximum width of an edge of T. Finally the *branch-width of* M is the minimum width over all branch decompositions of M, and is denoted by $\text{bw}(M)$.

We collect some basic facts listed in the survey paper [617].

Theorem 9.22.

1. *The branch-width of a bridgeless graph equals the branch-width of its cycle matroid.*
2. *Given a matroid M with n elements, and a positive integer k, it is possible to test, using an* INDEPENDENT*-oracle, in polynomial-time (with the degree depending on k), whether* $\text{bw}(M) \leq k$.
3. *For matroids M representable over a fixed finite field, deciding whether* $\text{bw}(M) \leq k$ *is in* FPT. *Furthermore, the same algorithm also outputs a branch-decomposition of M of width at most $3k$.*

Although testing whether $\text{bw}(M) \leq k$ is easy in the oracle framework, applications usually require that the matroid is representable, which is hard by Theorem 9.21. To remedy this, other definitions of matroid width were recently introduced: *decomposition width* in [721] and *amalgam width* in [805]. Without going into details, we note that matroids of decomposition width $\text{dw}(M) = k$ (respectively, amalgam width $\text{aw}(M) = k$) can be represented by a *decomposition tree* (respectively, *amalgam tree*), although this tree is not known to be computable in polynomial time. However, if the decomposition or amalgam tree is given together with M and is of width k, then many parameters can be computed in polynomial time using the INDEPENDENT-oracle, [721, 805]. We also note that for matroids M representable by a finite field $\mathbb{F}$ and branch-width $\text{bw}(M) \leq k$ both $\text{dw}(M)$ and $\text{aw}(M)$ are bounded by functions which depend on k and the size of $\mathbb{F}$.

We can now summarize here what is known about the computability of the Tutte polynomial when the matroid is given by the oracle INDEPENDENT.

Theorem 9.23. *Let M be a matroid of size n and rank r. Using the* INDEPENDENT*-oracle we have:*

1. *The Tutte polynomial of M is not computable in polynomial time.*
2. *If the matroid M has branch-width* $\mathrm{bw}(M) \leq k$ *and is representable over a finite field $\mathbb{F}$, then $T(M;x,y)$ is computable in time $O(n^6 \log(n) \log\log(n))$ even without the oracle.*
3. *If M has decomposition-width* $\mathrm{dw}(M) \leq k$ *and is given with its decomposition tree, then $T(M;x,y)$ is computable in time $O(k^2n^3r^2)$ and evaluated in time $O(k^2n)$ (under the assumption of unit cost of the arithmetic operations).*
4. *If M has amalgam-width* $\mathrm{aw}(M) \leq k$ *and is given with its amalgam tree, then $T(M;x,y)$ is computable in time $O(n^d)$ for some constant d independent of k.*

The first item in the theorem is due to P. Jensen and B. Korte [657], the second P. Hliněný [616], the third D. Král, [721], and the fourth L. Mach and T. Toufar, [805].

9.5 Algebraic models of computation

The complexity of the Tutte polynomial was also studied in various alternative models of computation.

9.5.1 Valiant's algebraic circuit model

In [1111], Valiant introduced an algebraic model of computation based on uniformly defined families of algebraic circuits and a notion of reducibility based on substitutions called *p-projections* (for background see, for example, [254]). In this framework, it is natural to consider multivariate versions of graph polynomials (see Chapter 24). In this model there is analogue for polynomial time called VP and for non-deterministic polynomial time called VNP. Computing the family of determinants DET_n of an $(n \times n)$ matrix where the entries are treated as indeterminates is in VP, computing the corresponding permanent PER_n is VNP-complete. The major problem in Valiant's framework is the question whether VP $=$ VNP. A detailed development of Valiant's theory can be found in [254].

Typical families of functions in VNP are generating functions of graph properties. These are polynomials depending on graphs with indeterminate weights on the edges. Let $\mathcal{P}$ be a graph property, i.e., a class of finite graphs closed under isomorphisms. For a graph $G = (V, E)$ the generating function of a graph property is

$$GF(G, \mathcal{P}) = \sum \prod_{e \in A} X_e,$$

where the sum is over spanning subgraphs $A \subseteq E$ with $A \in \mathcal{P}$, and the X_e's are indeterminates indexed by E.

The complexity of generating functions of graph properties in Valiant's model was first studied by M. Jerrum in [659] and further developed by P. Bürgisser in [254].

The Tutte polynomial of a graph has only two variables, therefore it cannot be VNP-complete in Valiant's model. However, adding variable weights to the edges of a graph leads to a VNP-complete family of Tutte polynomials in two ways, one via the multivariate Tutte polynomial (see Chapters 20 and 24) and the other via the colored Tutte polynomial (see Chapter 24) defined independently by Zaslavsky [1183], and Bollobás and Riordan [158].

The multivariate Tutte polynomial uses independent variables v_e indexed by the edges of a graph, and is given by

$$Z(G; q, \mathbf{v}) = \sum_{A \subseteq E} q^{k(A)} \prod_{e \in A} v_e.$$

Subject to relations among the variables to ensure that the function is well-defined (again see Chapter 24), the colored Tutte polynomial may be written as

$$f(G) = \alpha \cdot \sum_{T} \Big(\prod_{e \in \mathrm{IA}(T)} X_e \Big) \Big(\prod_{e \in T \setminus \mathrm{IA}(T)} a_e \Big) \Big(\prod_{e \in \mathrm{EA}(T)} Y_e \Big) \Big(\prod_{e \notin T \cup \mathrm{EA}(T)} b_e \Big),$$

where $G = (V, E)$ is a connected graph with a linear order on its edges, X_e, Y_e, a_e, b_e are indeterminates indexed by $e \in E$, and the sum is over all spanning trees T. For graphs that are not connected we add additional indeterminates.

The complexity of the Tutte polynomial in Valiant's model was studied in [788]. To get a meaningful complexity analysis, the notion of p-projections is modified to allow also a wider class of substitutions which the authors call *polynomial oracle reductions.* Not surprisingly, they show that in this (slightly) modified framework of Valiant's model the multivariate Tutte polynomial is VNP-complete.

Theorem 9.24.

1. *The multivariate dichromatic polynomial* $Z(G; q, \mathbf{v})$ *is* VNP*-complete via polynomial oracle reductions.*

2. *There is a family of graphs* G_n *such that the family of colored Tutte polynomials* $f(G_n)$ *is* VNP*-complete under p-projections.*

For details the reader is referred to [618, 788].

9.5.2 The Blum–Shub–Smale model

We close by remarking briefly on another approach to understanding the complexity of the Tutte polynomial. In the early 1970s computability over arbitrary first order structures was introduced independently in [465, 504] by

E. Engeler and H. Friedman. (See, for example, [434] for background on first order structures.) What they proposed was, roughly speaking, to use register machines over first order structures $\mathfrak{A}$, where the registers contain elements of $\mathfrak{A}$ and the tests and operations are defined by the relations and functions of $\mathfrak{A}$. Their framework was rediscovered by L. Blum, M. Shub and S. Smale in [145] (see also [144] and [505] for a comparison between [145] and [504]). The main advantage of [145] is the introduction of a complexity theory for computability over the real and complex numbers which includes complexity classes $\mathrm{P}_{\mathfrak{R}}$ for deterministic polynomial-time computable problems and $\mathrm{NP}_{\mathfrak{R}}$ for non-deterministic polynomial-time computable problems, and the existence of $\mathrm{NP}_{\mathfrak{R}}$-complete problems, where $\mathfrak{R}$ is an arbitrary (possibly ordered) ring. We call this model of computation the BSS *model of computation.* Later, K. Meer, [835], also introduced a complexity class $\sharp\mathrm{P}_{\mathfrak{R}}$ for the BSS model of computation.

As it is customary to consider the Tutte polynomial $T(G;x,y)$ as a polynomial in $\mathbb{R}[x,y]$ or $\mathbb{C}[x,y]$, and graphs can be viewed as 0–1-matrices, the complexity of the Tutte polynomial is most naturally discussed in the BSS model of computation, [812]. However, the theory of counting complexity in the BSS model of computation has various drawbacks. In particular, there is no satisfactory theory for $\sharp\mathrm{P}_{\mathfrak{R}}$-complete problems. Furthermore, it is not clear whether, and it seems unlikely that, evaluating the chromatic polynomial at the point $x = 3$ is $\mathrm{NP}_{\mathfrak{R}}$-complete. A detailed description of these problems can be found in [711, 713], where also an abstract algebraic version of Theorem 9.4 is formulated. In [44] the descriptive complexity of counting problems in the BSS model of computation over $\mathbb{R}$ is also discussed using essentially the same framework as in [711, 713].

9.6 Open problems

The complexity of computing or evaluating the Tutte polynomial in the Turing model of computation in a ring $\mathfrak{R}[x,y]$ is well understood for graphs without restrictions on their presentations and for matroids (provided the matroids have succinct presentations) provided the arithmetic operations in the ring $\mathfrak{R}$ are polynomial-time computable in some standard presentation of $\mathfrak{R}$. The same is true for special classes of graphs (planar, bipartite, etc.) and special succinct classes of matroids (transversal, bicircular, etc.). For graph classes of bounded tree-width, evaluating the Tutte polynomial is fixed parameter tractable. However, for graph classes of bounded clique-width, the situation is not completely understood.

Problem 9.25. Determine the complexity of evaluating the Tutte polynomial for graphs of fixed clique-width.

Planar graphs are a special case of a minor-closed class of graphs for which the complexity of the Tutte polynomial is known, but little is known for other classes.

Problem 9.26. Determine the complexity of evaluating the Tutte polynomial for other minor-closed graph classes.

The JVW theorem, 9.4, and its variations show a dichotomy, the *difficult point property* (DPP). This states that the evaluation of the Tutte polynomial at fixed evaluation points is $\sharp$P hard on all points in $\mathfrak{R}^2 \setminus A$ where $A \subset \mathfrak{R}^2$ is a semialgebraic (quasialgebraic) set of dimension 1. Versions of DPP have been proven also for many other graph polynomials, including the colored Tutte polynomial, [141], the interlace polynomial, [142], the cover polynomial for directed graphs, [139, 140]. the bivariate matching polynomial for multigraphs in [56, 619], and many more which are surveyed in [713].

Problem 9.27. Characterize the graph polynomials for which DPP holds.

If one studies the complexity in the Turing model of computation, the real or complex numbers have no recursive presentation. Moreover, the recursive presentations of the countable subrings which do have a recursive presentation are not uniform. This makes the statement about the complexity over the reals or complex numbers somehow artificial. In the BSS model of computation (see [713]) this problem of non-uniformity does not arise.

Problem 9.28. Refine the complexity analysis of the Tutte polynomial in the BSS model of computation.

In the case of matroids without succinct presentation, it is known that the Tutte polynomial is not polynomial-time computable using the INDEPENDENT-oracle, Theorem 9.23. But to the best of our knowledge, the analogues of NP-completeness, or higher levels of the complexity hierarchy within exponential time oracle computability of matroids has not been developed.

Problem 9.29. Develop a coherent theory of complexity for computations with matroid oracles in general.

10

Approximating the Tutte polynomial

Magnus Bordewich

Synopsis

This chapter explores the approximability of the Tutte Polynomial. What is the complexity of approximating the value of $T(G; x, y)$? How does this depend on the values of x and y? What can be done when G is restricted to some special class of graphs?

- Fully polynomial randomized approximation schemes.
- Approximation complexity and the role of ♯BIS.
- Approximation of the Ising model and its relation to the Tutte plane.
- Approximating the Tutte polynomial on dense graphs.
- Approximating for bounded degree graphs, e.g., approximately counting the number of proper k-colorings of a graph, given by $T(G; 1-k, 0)$.
- Inapproximability results for regions of the Tutte plane.
- ♯BIS-hardness of approximating at most points in the positive quadrant of the Tutte plane.
- Links between quantum computation and additive approximations of the Jones polynomial.

10.1 Introduction

Chapter 9 has already discussed the complexity of exactly evaluating $T(G; x, y)$ for various classes of graph G and at various points (x, y): in sum-

DOI: 10.1201/9780429161612-10

mary it is ♯P-hard in almost all cases. In this chapter we consider the complexity of *approximately* evaluating $T(G; x, y)$. In Section 10.3 we formally define the type of approximation we are concerned with: *fully polynomial randomized approximation schemes* (FPRAS) [677], and we introduce the concept of an *approximation preserving reduction* [431].

In Section 10.4 we present positive approximation results, starting with an FPRAS for the hyperbola corresponding to the partition function of the ferromagnetic Ising model. For further discussion of the relationship between the Tutte polynomial and the Ising and other models in statistical physics, see Chapter 20. While there are no further positive results to report for general graphs, for restricted classes of graphs approximations can be made at various additional points. Our attention first turns to approximations when the underlying graphs satisfy various kinds of density. For graphs with minimum degree $\Omega(n)$, there is an FPRAS for points in the positive quadrant $x > 1, y > 1$. Using an alternative definition of dense graph, that the minimum edge cut has size at least $c \log n$ for some constant c, an FPRAS exists for the upper half plane $y > 1$, including the all-terminal network reliability polynomial along $x = 1, y > 1$ (for more on the connections between reliability and the Tutte polynomial, see Chapter 15). Next we consider degree bounded graphs: there are Markov chain Monte Carlo based approximation schemes for approximating the number of proper graph colorings given sufficiently many colors. This corresponds to approximating the Tutte polynomial at points on the line $y = 0, x < 0$. We describe these results and extensions to points off the mentioned line.

On the negative side (Section 10.5), it is easily seen that some points, for example the point which counts the number of proper 3-colorings of G, cannot be approximated unless NP=RP, since even determining if the number of 3-colorings of a graph is non-zero is NP-complete. (For more on complexity classes see for example [906].) In Section 10.5 such results are extended to give a large region of the Tutte plane at which there can be no FPRAS unless there is a randomized polynomial-time algorithm for exactly solving every problem in ♯P. Much of the remaining Tutte plane is shown to be ♯BIS-hard to approximate, that is at least as hard as the problem of counting the number of independent sets in bipartite graphs.

Finally, in Section 10.6, we briefly summarize links between quantum computation and the Jones polynomial, and introduce additive approximations.

Many of the theorems stated in this chapter were originally stated in different settings, for example in terms of the partition function of the Ising or Potts models, or in terms of approximately counting the number of proper colorings of a graph. We have attempted to rephrase all results in terms of the Tutte polynomial under the usual parametrization wherever possible.

10.2 Notation

As with other topics on the Tutte polynomial, the hyperbolas on which $(x-1)(y-1)$ is constant will play a significant role. We therefore define the curves H_q, H_q^+, H_q^- as follows to facilitate the discussions.

Notation 10.1. For any $q \in \mathbb{Q}$ define the hyperbola H_q and its positive and negative branches H_q^+ and H_q^- as follows:

- $H_q = \{(x,y) \in \mathbb{Q} \times \mathbb{Q} : (x-1)(y-1) = q\}$,
- $H_q^+ = \{(x,y) \in \mathbb{Q} \times \mathbb{Q} : (x-1)(y-1) = q, x \geq 1, y \geq 1\}$, and
- $H_q^- = \{(x,y) \in \mathbb{Q} \times \mathbb{Q} : (x-1)(y-1) = q, x \leq 1, y \leq 1\}$.

10.3 Randomized approximation schemes

Chapter 9 considers the complexity of exactly evaluating $T(G;x,y)$ for various classes of graph G and at various points $(x,y) \in \mathbb{C}^2$. In this chapter we will restrict attention to real rational points (x,y): issues of representation arise when considering complexity questions at additional points. For a discussion of such issues relating to the Tutte polynomial we refer the reader to [653] and for a detailed handling of complexity issues arising from computing with real numbers to [144]. In summary the main result is: for $(x,y) \in \mathbb{Q}^2$, $T(G;x,y)$ is ♯P-hard to evaluate for except when (x,y) is one of the four special points $(1,1), (-1,-1), (-1,0), (0,-1)$, or lies on H_1. The existence of a polynomial time algorithm for evaluating a ♯P-hard function would not only imply that P=NP, but indeed that the entire polynomial hierarchy collapses [906, 1046]. It is therefore widely believed that no such algorithm can exist.

If we do not know how to evaluate some function f on some input x in time polynomial in the size of that input ($|x|$), we are left with two choices: compute it exactly and take more than polynomial time, or compute only an approximate value in polynomial time. This chapter is concerned with the second approach. Work on approximating ♯P functions was initiated by Karp and Luby [676] in 1985. In 1989 Karp, Luby and Madras [677] extended the ideas to give a *fully polynomial randomized approximation scheme* (FPRAS) for the ♯P-complete counting problem ♯DNF, that of counting the number of satisfying assignments to a Boolean formula in disjunctive normal form. We formally define an FPRAS next.

Definition 10.2. A *randomized approximation scheme* (RAS) for a function $f(x)$ is a randomized approximation algorithm taking input x along with a rational error parameter ϵ, such that for any given $\epsilon > 0$, with probability at least 3/4 the output $\hat{f}(x,\epsilon)$ is within a relative error of $1 \pm \epsilon$. That is: $\mathbf{Pr}[(1-\epsilon)f(x) < \hat{f}(x,\epsilon) < (1+\epsilon)f(x)] \geq 3/4$.

Definition 10.3. A RAS is described as a *fully polynomial randomized approximation scheme* (FPRAS), if the running time of the algorithm is bounded by a polynomial in $|x|$ and $1/\epsilon$.

Note that the constant 3/4 in the above definition could be replaced by any constant strictly greater than 1/2, as with repeated trials this probability can be amplified. With this definition in hand, the fundamental question we are addressing in this chapter may be stated as follows.

Open Question 10.4. For what rational values of x and y does there exist an FPRAS for the function taking input a graph G and having output the value $T(G;x,y)$?

When, for fixed x, y, the function f taking input a graph G restricted to be in some class $\mathcal{G}$, and having output the value $T(G;x,y)$, has an FPRAS, we shall simply say there is an FPRAS for $T(G;x,y)$ for $G \in \mathcal{G}$.

The existence of an FPRAS for a function is a strong result and corresponds to the idea of (randomized) tractability. Thus the existence of an FPRAS for a $\sharp$P (counting) problem could be considered the analogue of a BPP algorithm (bounded-error probabilistic polynomial time) for a decision problem. An important feature of an FPRAS is that it reliably differentiates between zero and non-zero: if $\hat{f}$ is an FPRAS for a function f, then for any $\epsilon \in (0,1)$

$$\begin{aligned} &\mathbf{Pr}[\hat{f}(x,\epsilon) = 0] > 3/4 && \text{if } f(x) = 0, \\ &\mathbf{Pr}[\hat{f}(x,\epsilon) = 0] < 1/4 && \text{if } f(x) \neq 0. \end{aligned}$$

Thus, on the assumption that NP $\not\subseteq$ BPP, there can be no FPRAS for any counting function for which the associated decision problem is NP-complete. Indeed the assumption RP $\neq$ NP is sufficient, where RP is Randomized Polynomial time, since it is known that if NP $\subseteq$ BPP, then NP $\subseteq$ RP (see [906]). Since the number of (proper) k-colorings of a graph G is given by $T(G;1-k,0)$, and determining if there exists a proper k-coloring of a graph is NP-complete for integer $k > 2$, the following proposition easily follows:

Proposition 10.5. *There can be no FPRAS for $T(G;x,0)$, for any integer $x \leq -2$, unless* RP=NP.

In some other regions of the (x,y) plane the decision problem of whether $T(G;x,y)$ is non-zero is not NP-complete, indeed it is often trivial: every non-empty graph has an acyclic orientation so $T(G;2,0) \neq 0$. At these points the assumption that NP$\neq$RP does not immediately rule out an FPRAS. Nevertheless, we shall see the above result extended in Section 10.5 below.

In 2003, Dyer et al. [179] turned their attention to the relative complexity of approximate counting problems, introducing the idea of approximation preserving reductions. The essence of an ordinary Turing reduction is that if $A \leq_T B$, then the existence of a polynomial time algorithm for problem B implies the existence of a polynomial time algorithm for problem A. The idea behind an *approximation preserving (AP) reduction* is the same: $f \leq_{AP} g$ roughly means that an FPRAS for function g would imply the existence of an FPRAS for f. Formally:

Definition 10.6. A $\sharp$P function f is *AP-reducible* to g (denoted $f \leq_{AP} g$) if there is a probabilistic oracle Turing machine M such that:

1. if each oracle call made by M is to a valid FPRAS for g, then on any input (x, ϵ) the machine M returns a value in $[(1-\epsilon)f(x), (1+\epsilon)f(x)]$ with probability at least 3/4, i.e., the output satisfies the condition for being a RAS for f;
2. each oracle call $g(y, \epsilon')$ made by M satisfies $|y|, 1/\epsilon' \leq poly(|x|, 1/\epsilon)$; and
3. M runs in time polynomial in $|x|$ and $1/\epsilon$, i.e., M operates as an FPRAS for f.

Definition 10.7. If $f \leq_{AP} g$ and $g \leq_{AP} f$ we say f and g are *AP-interreducible.*

Dyer et al. [179] showed that there appear to be three distinct classes of problems relative to such reductions, which are defined with reference to the following two counting problems.

Problem: $\sharp$SAT
Input: A Boolean formula F in conjunctive normal form.
Output: The number of satisfying assignments for F.

Problem: $\sharp$BIS
Input: A bipartite graph G.
Output: The number of independent sets of G, that is subsets of the vertices inducing no edges.

The three classes of problems in the approximation hierarchy are then: those that can be approximated in polynomial time (admit an FPRAS), those that are as hard to approximate as $\sharp$SAT, and the class of counting problems AP-interreducible with $\sharp$BIS. This last class may be logically defined, see [179] for details. Despite significant progress in the area of approximation algorithms over the last decade, it is still the case that most natural problems fall into one of these three classes and that it is open as to whether $\sharp$BIS admits an FPRAS, is as hard to approximate as $\sharp$SAT (i.e., $\sharp\text{SAT} \leq_{AP} \sharp\text{BIS}$), or is genuinely in an intermediate class.

10.4 Positive approximation results

In this section we shall outline the main results showing the existence of FPRASs for evaluating $T(G; x, y)$ at various points (x, y) and for G in various classes of graphs.

10.4.1 The ferromagnetic Ising model

There is only one positive result for approximating any point in the Tutte plane that is valid for the class of all graphs: that of Jerrum and Sinclair [662]. They presented an FPRAS for the partition function of the Ising model of ferromagnetism, which is equivalent (in the case with no external field) to approximating $T(G; x, y)$ for (x, y) on H_2^+, the positive branch of the hyperbola $(x-1)(y-1) = 2$ (for details of the equivalence see Chapter 20).

Theorem 10.8. *There is an FPRAS for the partition function of a ferromagnetic Ising system.*

Corollary 10.9. *There is an FPRAS for the function taking input a graph G and $(x, y) \in H_2^+$ and outputting $T(G; x, y)$.*

Note that here we have the point (x, y) specified as input to the FPRAS, not fixed in advance: this is a stronger result than we initially sought. Figure 10.1 shows this curve in the Tutte plane, augmenting the depiction of the complexity of the Tutte plane given in Chapter 9 to include these approximable points. In the figure, the points and solid lines are those at which $T(G; x, y)$ may be computed exactly in polynomial time for all graphs. The points on the dashed line are those on H_2^+ for which there is an FPRAS for $T(G; x, y)$ for all graphs.

The approach taken was the *Markov chain Monte Carlo* (MCMC) method. Jerrum and Sinclair first expressed the partition function of the Ising model for the graph G as a sum of weights over all subgraphs of G, which is known as the high temperature expansion in statistical physics. They then showed that this sum can be efficiently approximated by sampling subgraphs with probability approximately proportional to their weight using MCMC. The reader is referred to [1012] for further information on this approach.

10.4.2 The dense case

Despite Proposition 10.5, not all hope is lost even at the points in the Tutte plane corresponding to counting proper k-colorings. For graphs with very large minimum degree the problem of determining whether the number of k-colorings is non-zero is not NP-hard anymore. Indeed, Edwards [442] showed, in terms of proper colorings, the following proposition which obtains *exact* evaluation in polynomial time.

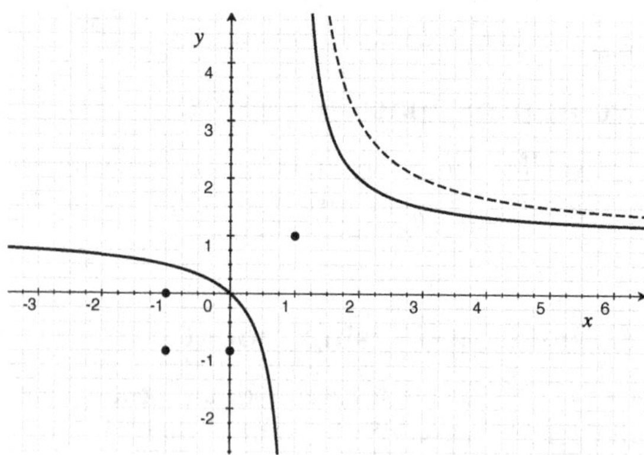

FIGURE 10.1: The real Tutte plane showing the complexity of evaluating $T(G; x, y)$.

Proposition 10.10. *Fix $k \in \mathbb{Z}^+$ and let $\alpha = \frac{(k-2)}{(k-1)}$. Then, $T(G; 1-k, 0)$ may be exactly evaluated in polynomial time for G in the class of graphs with minimum degree αn.*

In this area, graphs with minimum degree $\Omega(n)$ are known as dense graphs [22, 28, 176]. The first Tutte approximation result to make use of denseness was that of Annan [28], who gave an FPRAS for the number of forests of a dense graph, which is given by $T(G; 2, 1)$, and some further Tutte evaluations.

Definition 10.11. Let $\mathcal{G}_\alpha$ denote the class of graphs with minimum degree at least αn. A graph $G \in \mathcal{G}_\alpha$ is an *α-dense* graph. A graph in $\mathcal{G}_{1/2}$ is called strongly-dense.

Theorem 10.12. *Fix $\alpha > 0$ and $x \in \mathbb{N}$. Then there exists an FPRAS for $T(G; x, 1)$ on the class of α-dense graphs.*

Alon, Frieze and Welsh [22] went further and showed that an FPRAS exists for the entire region $x \geq 1$, $y \geq 1$ for dense graphs, and for all $y \geq 1$ for strongly-dense graphs.

Theorem 10.13. *Fix constants α, x and y. Then there exists an FPRAS for $T(G; x, y)$ on the class of α-dense graphs when:*

1. *$\alpha > 0$, $x \geq 1$ and $y \geq 1$; or*

2. *$\alpha > \frac{1}{2}$ and $y \geq 1$.*

The approach of Alon, Frieze and Welsh is as follows. For any graph G and $p \in [0, 1]$ let G_p denote a random spanning subgraph of G obtained by selecting each edge independently with probability p. Assume $x, y > 1$ and let $p = (y-1)/y$ and $Q = (x-1)(y-1)$. Let $k = k(G_p)$ denote the number of connected components of G_p; thus k is a random variable. Then we can express the Tutte polynomial in the following form:

$$T(G; x, y) = \frac{y^m}{(x-1)(y-1)^n}\mathbf{E}(Q^k).$$

With this, approximating $T(G; x, y)$ reduces to approximating $\mathbf{E}(Q^k)$. For dense graphs, it is likely that the number of components of a random subgraph is small, and so $\mathbf{E}(Q^k)$ can be accurately approximated by a sample average after sampling only a polynomial number of random subgraphs. Note that if x and y are regarded as inputs to the function, then the running time is bounded by $\text{poly}(|G|, 1/\epsilon, \max\{Q, 1/Q\})$.

Karger [674] proved the existence of a similar scheme for graphs under a different definition of denseness: for those graphs with no small cutset. The paper gives an FPRAS (for all graphs) for the probability that a random spanning subgraph is disconnected. For dense graphs, it is shown that most spanning subgraphs are connected, and that small cutsets (that have a non-negligible probability of being absent in a random subgraph) can be efficiently enumerated. From this the following theorem is obtained.

Theorem 10.14. *Let $\mathcal{G}'_\alpha$ be the class of graphs with edge connectivity at least $\alpha \log n$. Fix x, y such that $y > 1$. Then there is an α (depending on x, y) such that there is an FPRAS for $T(G; x, y)$ for graphs in the class $\mathcal{G}'_\alpha$.*

Indeed Karger shows something stronger: for fixed x, y such that $y > 1$, and with α as given in Theorem 10.14,

$$T(G; x, y) = \frac{y^m}{(y-1)^{n-1}}(1 + O(1/n))$$

and moreover there is an FPRAS for the second order term.

Theorem 10.15. *Let $\Delta T(G; x, y) = \frac{y^m}{(y-1)^{n-1}} - T(G; x, y)$. Fix x, y with $y > 1$. Then there is an α (depending on x, y) such that there is an FPRAS for $\Delta T(G; x, y)$ for graphs in the class $\mathcal{G}'_\alpha$.*

The success of these results lead Welsh to the following conjecture [1141].

Conjecture 10.16. For any $x, y \geq 1$, there exists an FPRAS for $T(G; x, y)$ for all graphs.

Although this conjecture is still open, progress has been made suggesting the conjecture may not hold, as we shall see in Theorem 10.31.

In related work, Bordewich [175, 176] obtained an FPRAS for a class of sparse graphs. The approach was essentially a dualization (in the matroid sense) of Karger's approach, and restricted the graphs to have no small cycles. Although this is a less natural class than dense graphs, it does include, for example, standard constructions of large expanders (see [175] for details). For the following theorem, recall that $T(G; 2, 0)$ counts the number of acyclic orientations of the graph G.

Theorem 10.17. *Let $\delta > 0, x > 1, y \geq 1$ be fixed, such that $(x-1)(y-1) \leq 1$, and let $\mathcal{G}_{\delta,x}$ be the class of graphs with girth $g \geq (5+\delta)\log_x(n)$. Then*

1. *there is an FPRAS for $T(G;x,y)$ for all $G \in \mathcal{G}_{\delta,x}$,*
2. *there is an FPRAS for $T(G;2,0)$ for all $G \in \mathcal{G}_{\delta,2}$.*

10.4.3 Bounded degree graphs

As noted above, the most obvious difficulty in approximating the Tutte polynomial is at the points which correspond to counting colorings of the underlying graph, where even determining if there is a proper coloring is NP-hard. However if we restrict ourselves to bounded degree graphs, then there is no such obstacle and progress can be made. Jerrum [661] showed the first result of this kind.

Definition 10.18. For $\Delta \in \mathbb{N}$, let $\mathcal{G}_\Delta$ be the class of graphs with maximum degree at most Δ.

Theorem 10.19. *Fix $\Delta \in \mathbb{N}$. For integer $q > 2\Delta$, there is an FPRAS for $T(G;1-q,0)$ for G in $\mathcal{G}_\Delta$.*

The approach is again via Markov chain Monte Carlo estimation. This relies on the fact that for all self-reducible problems (essentially reducing the search problem to the decision problem), approximate counting can be achieved by almost uniform sampling (see [663] for details). In practice most natural $\sharp$P problems turn out to be self-reducible, and evaluations of the Tutte polynomial are self-reducible because the deletion–contraction formulae reduce the evaluation to smaller instances in polynomial time. Recall that $T(G;1-q,0)$ gives the number of proper q-colorings of G. First an arbitrary proper q-coloring of G is constructed. A Markov chain on the set of all proper colorings of G, the transitions of which involve randomly recoloring a single vertex, is then used to sample almost uniformly at random from the set of proper colorings of G. Provided the number of colors is more than twice the maximum degree of the graph, this Markov chain will mix rapidly, yielding a sufficiently close-to-uniform sample in polynomial time.

This result has been followed by numerous refinements, notably Vigoda [1123] who analyzed a different Markov chain, and obtained the following.

Theorem 10.20. *Fix $\Delta \in \mathbb{N}$. For integer $q > 11\Delta/6$, there is an FPRAS for $T(G;1-q,0)$ for G in $\mathcal{G}_\Delta$.*

This is the best result to date applying to *all* graphs of a given minimum degree; other results have gradually reduced the ratio of colors to degree required, but introduced additional restrictions on the minimum girth: see [506] for a fairly recent survey.

These approximation results hold only at isolated points in the Tutte plane.

However the following result of Salas and Sokal [979], covers half-curves corresponding to evaluations of the partition function of the antiferromagnetic Potts model.

Theorem 10.21. *Fix $\Delta \in \mathbb{N}$, and $(x, y) \in H_q^- \cap \{y \in [0,1]\}$ for some integer $q > 2\Delta(1-y)$. There is an FPRAS for $T(G; x, y)$ for G in $\mathcal{G}_\Delta$.*

For the restricted case of grid graphs (G is a finite portion of $\mathbb{Z}^2$), the above result gives an FPRAS for $q \geq 8, y \in [0,1]$; Goldberg et al. [550] have extended this result.

Theorem 10.22. *Fix $(x, y) \in H_q^-$ for some integer q. There is an FPRAS for $T(G; x, y)$ for G a finite portion of $\mathbb{Z}^2$ in the following cases:*

1. *$q = 6$ and $y \in [0, 1]$,*
2. *$q = 5$ and $y \in [0.127, 1]$,*
3. *$q = 4$ and $y \in [0.262, 1]$,*
4. *$q = 3$ and $y \in [0.393, 1]$.*

In the positive quadrant, or ferromagnetic region when viewing the Tutte polynomial as the partition function of the Potts model, Bordewich, Greenhill and Patel [178] obtained similar results relating approximability for bounded degree graphs to sufficiently high q.

Theorem 10.23. *For any $\nu \in (0, 1)$, there is a constant $c = c(\nu)$ such that the following statements hold.*

1. *Fix $\Delta \in \mathbb{N}$, and $(x, y) \in H_q^+$ such that $q \geq \Delta y^\Delta + 1$. Then there is an FPRAS for $T(G, x, y)$ for G in $\mathcal{G}_\Delta$.*
2. *Fix $(x, y) \in H_q^+$ such that $q \geq cy^{2+\nu}$. Then there is an FPRAS for $T(G, x, y)$, for G in the class of toroidal grid graphs.*

10.5 Negative results: inapproximability

Proposition 10.5 was extended first to points on the negative branch of the hyperbola H_2 by Jerrum and Sinclair [662], who used a reduction to MAXCUT, and then to the negative branches of the hyperbolas H_q for integer $q \geq 2$ and more by Welsh [1140], using the approach of thickening and stretching (see Section 7.6.2) to relate points in the Tutte plane.

Theorem 10.24. *On the assumption that* RP$\neq$NP *the following statements are true:*

1. *There is no FPRAS for the function taking input a graph G and $(x, y) \in H_3^-$ and outputting $T(G; x, y)$, even if G is restricted to the class of planar graphs.*

2. *Let q be an integer $q \in \{2, 3, 4, 5, \ldots\}$. There is no FPRAS for the function taking input a graph G and $(x, y) \in H_q^- \cap \{x < 0\}$ and outputting $T(G; x, y)$.*

Note that the theorem above rules out an FPRAS on a curve or region where the coordinates (x, y) are considered part of the input to the problem, but it does not rule out an FPRAS at a specific point in the region. For example, the point $(-1, 0)$ is in H_2^-, and yet $T(G; -1, 0)$ is exactly computable in polynomial time. Goldberg and Jerrum [551] vastly clarified the picture with a string of related results, giving stronger results ruling out an FPRAS at specific points, in the sense of Question 10.4.

Theorem 10.25. *Under the assumption that* RP≠NP, *the function taking input G and outputting $T(G; x, y)$ has no FPRAS whenever*

1. *$x < -1$ and (x, y) is not on H_0 or H_1,*

2. *$y < -1$ and (x, y) is not on H_1 or H_2,*

3. *(x, y) is not on H_1, $|x| < 1$, $|y| < 1$ and (x, y) is in the region given by $\{y < -1 - 2x\} \cup \{x < -1 - 2y\}$,*

4. *$|x| < 1$, $|y| < 1$ and $q = (x - 1)(y - 1) > 1.5$ (excluding the special points at which exact computation is possible).*

These results are depicted in Figure 10.2. In the figure, the points and solid lines are those at which $T(G; x, y)$ may be computed exactly in polynomial time for all graphs. The points on the dashed line are those on H_2^+ for which there is an FPRAS for $T(G; x, y)$ for all graphs. The light gray regions are points at which $T(G; x, y)$ has no FPRAS unless RP=NP.

In related work, Bläser, Dell and Makowsky [141] have extended the results of Goldberg and Jerrum to the colored Tutte Polynomial of Bollobás and Riordan. Goldberg and Jerrum have also shown that even if the graph is restricted to be planar, much of the inapproximability result still holds [552].

Theorem 10.26. *Under the assumption that* RP≠NP, *the function taking input a* planar *graph G and outputting $T(G; x, y)$ has no FPRAS whenever*

1. *$x < 0, y < 0$ and (x, y) is on $H_q, q > 5$,*

2. *$x > 1, y < -1$,*

3. *$x < -1, y > 1$,*

4. *$(x, y) \in H_3^-$.*

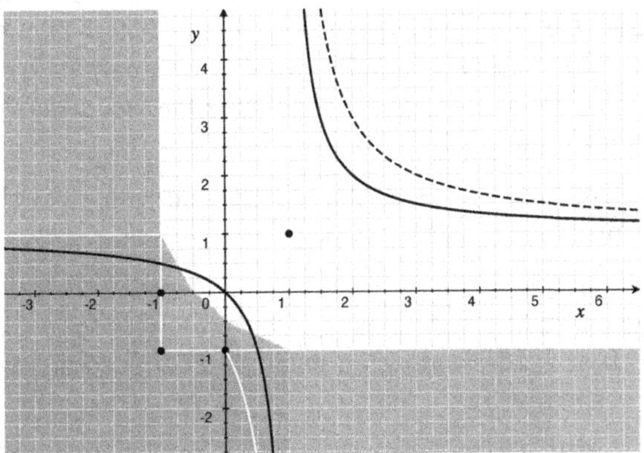

FIGURE 10.2: The real Tutte plane showing the complexity of evaluating $T(G; x, y)$.

Two further results in [551] highlighted some particular curves on which more can be said about the approximability of $T(G; x, y)$. The first is for the curve $H_2^- \cap \{y < -1\}$, which, apart from the easy-to-evaluate H_1, contains the only points in the half-plane $y < -1$ excluded from the theorem above. (See Figure 10.2.) Although it is unknown whether there is an FPRAS for points on this curve, the resolution of the question is closely connected with the long standing problem ♯PerfectMatching.

Problem: ♯PerfectMatching
Input: A graph G.
Output: The number of perfect matchings of G.

Theorem 10.27. *Fix (x, y) in $H_2 \cap \{y < -1\}$. Then the function taking input G and outputting $T(G; x, y)$ is* AP*-interreducible (equivalent in difficulty) to* ♯PerfectMatching, *the function taking as input a graph, and outputting the number of perfect matchings of that graph.*

The second result shows that even where there is no FPRAS, some points may be "more intractable" than others.

Theorem 10.28. *Fix (x, y) in $H_4 \cap \{-1 < y < 0\}$. Then the function taking input G and outputting $T(G; x, y)$ has no FPRAS unless there is a randomized polynomial-time algorithm for exactly solving every problem in* ♯P.

Thus approximating $T(G; x, y)$ at these points is essentially as hard as exact counting (♯P), whereas for most counting problems, including all those in ♯P and evaluating $T(G; x, y)$ for (x, y) in the positive quadrant $x, y > 0$, approximation is only as hard as solving an NP decision problem. To be precise: such problems may be efficiently approximated using calls to an NP oracle, by the bisection technique of Valiant and Vazirani [1113]. Exploring links with quantum computation (see Section 10.6), Kuperberg [738] strengthened this result to the following.

Theorem 10.29. *Let $(x, y) \in \mathbb{Q}^2$ such that $q = (x-1)(y-1) > 4$ and $x, y < 0$. Then the function taking input a planar graph G and outputting $T(G; x, y)$ has no FPRAS unless there is a randomized polynomial-time algorithm for exactly solving every problem in* $\sharp$P.

In a subsequent paper [548] Goldberg and Jerrum take this further by considering the problem SIGNTUTTE(x, y):

Problem: SIGNTUTTE(x, y)
Input: A graph G
Output: $+, 0$ or $-$, depending on whether $T(G; x, y)$ is positive, zero or negative respectively.

Observe that an FPRAS for $T(G; x, y)$ at a point (x, y) would be able to reliably answer SIGNTUTTE(x, y). Thus if for some fixed (x, y) the problem SIGNTUTTE(x, y) is shown to be $\sharp$P hard, then there can be no FPRAS for evaluating $T(G; x, y)$ unless there is a randomized polynomial-time algorithm for exactly solving every problem in $\sharp$P. By classifying first the complexity of determining the sign of the chromatic polynomial, and then using thickening and stretching, and more complex gadgets, to relate points in the Tutte plane, they obtained the following.

Theorem 10.30. *For all x, y let $q = (x-1)(y-1)$. Under the assumption that there is no randomized polynomial-time algorithm for exactly solving any $\sharp$P-complete problem, the function taking input G and outputting $T(G; x, y)$ has no FPRAS whenever*

1. $\min\{x, y\} \leq -1$ *and* $\max\{x, y\} < 0$, *except the point* $(-1, -1)$;
2. $x < -1$ *and* $y > 1$;
3. $x < -1$ *and* $0 < y < 1$, *except for when* $q \in \mathbb{Z}$;
4. $y < -1$ *and* $x > 1$;
5. $y < -1$ *and* $0 < x < 1$, *except for when* $q \in \mathbb{Z}$ *and all* $q > 4$;
6. $\max\{|x|, |y|\} < 1$ *and* $q > 32/27$;
7. $\max\{|x|, |y|\} < 1, x < -2y - 1$ *and* $q \leq 32/27, q \neq 1$;
8. $\max\{|x|, |y|\} < 1, y < -2x - 1$ *and* $q \leq 32/27, q \neq 1$.

The full theorem of Goldberg and Jerrum in [548] contains details of the status at boundaries between these regions, points at which SIGNTUTTE(x, y) may be efficiently computed, and which points in the Tutte plane remain unresolved.

Although the results above rule out an FPRAS for a large portion of the Tutte plane, they do not impinge on the region $x, y \geq 1$. In a further paper [547], Jerrum and Goldberg presented strong evidence that even in this region an FPRAS may be unlikely.

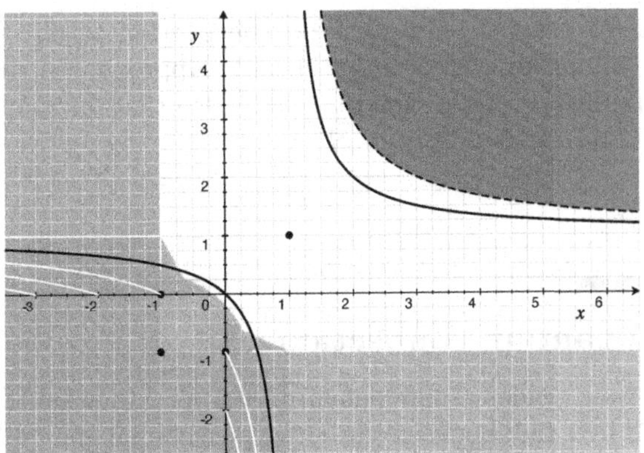

FIGURE 10.3: The real Tutte plane showing the complexity of evaluating the sign of $T(G; x, y)$.

Theorem 10.31. *Fix $q > 2$ and $(x, y) \in H_q^+$. Let f be the function taking input a graph G and outputting $T(G; x, y)$. Then $\sharp$BIS $\leq_{AP} f$.*

These results are depicted in Figure 10.3. In the figure, the points and solid curve are those at which $T(G; x, y)$ may be computed exactly in polynomial time for all graphs, and hence the sign also. The points on the dashed line are those on H_2^+ for which there is an FPRAS for $T(G; x, y)$ for all graphs. The light gray regions are points at which $T(G; x, y)$ has no FPRAS unless there is a randomized polynomial-time algorithm for exactly solving every problem in $\sharp$P. The darker gray region in the positive quadrant covers points at which $T(G; x, y)$ has no FPRAS unless there is also an FPRAS for $\sharp$BIS. For full details of the boundary points, see [548]. Thus the existence of an FPRAS for any point on $H_q^+, q > 2$ would imply the existence of an FPRAS for $\sharp$BIS. The latter seems unlikely, and Goldberg and Jerrum [547] conjecture that no FPRAS exists.

Conjecture 10.32. The counting problem $\sharp$BIS does not admit an FPRAS.

In a final extension, Goldberg and Jerrum [549] have also considered the problem of evaluating $T(M; x, y)$ for a binary matroid M. The previously stated hardness results for graphs carry over to binary matroids, but there are additional results on the hyperbola H_2.

Theorem 10.33. *The following statements hold.*

1. *For (x, y) on H_2^-, there can be no FPRAS for $T(M; x, y)$, where the input M is a binary matroid, unless* NP=RP.
2. *For (x, y) on H_2^+,* $\sharp$BIS *is* AP*-reducible to the problem of evaluating $T(M; x, y)$, where the input M is a binary matroid.*

Recall that for graphs, if (x, y) is on the portion of H_2^- with $y < -1$, then $T(G; x, y)$ is not known to be NP-hard to approximate, only as hard as ♯PERFECTMATCHING to approximate. Recall also that for graphs there is an FPRAS for $T(G, x, y)$ when $(x, y) \in H_2^+$.

10.6 The quantum connection

In this section we consider a different form of approximation to the standard RAS, motivated by a link between the Tutte polynomial and quantum computation. Freedman, Kitaev, Larson and Wang [501] showed that the "quantum part" of any quantum computation can be replaced by an approximate evaluation of the Jones polynomial of a related braid. (Recall that the Jones polynomial is a specialization of the Tutte polynomial, see Chapter 18.) A classical polynomial time algorithm can convert a quantum circuit into a braid, such that the probability that the output of the quantum computation is zero is a simple (polynomial time) function of the Jones polynomial of the braid at a 5-th root of unity. For an exact statement of this see Freedman, Kitaev, Larsen and Wang [501]. Bordewich, Freedman, Lovász, and Welsh [177] then showed that in fact only an approximate evaluation of the Jones polynomial at this point is required to simulate the quantum computation, and that a weaker form of approximation than a RAS, an additive approximation, is sufficient.

Definition 10.34. Let $\mathcal{I}$ be a set of valid instances of some problem encoded as finite binary strings. Given any function $f : \mathcal{I} \mapsto \mathbb{C}$ and a normalization $u : \mathbb{N} \mapsto \mathbb{R}^+$, an additive approximation for (f, u) is a probabilistic algorithm which given any $I \in \mathcal{I}$ and $\epsilon > 0$ produces an output $\hat{f}(I)$, such that

$$\mathbf{Pr}[|f(I) - \hat{f}(I)| > \epsilon u(|I|)] < 1/4,$$

in time polynomial in $|I|$ and ϵ^{-1}.

Recall that a RAS for a function f with error parameter ϵ allows a *multiplicative* error. That is, the output of the RAS $\hat{f}(I, \epsilon)$ on input (I, ϵ) should be at most $\epsilon f(I)$ (with probability at least 3/4). In an additive approximation the error is allowed to be up to $\epsilon u(I)$, where the normalization function u is typically taken to be natural bound on the range of f. For example, if we were to consider an additive approximation for $T(G; 2, 0)$, which counts the number of acyclic orientations for G, then u might be taken to be $u(G) := 2^{|E(G)|}$ (the total number of orientations).

It follows that an ability to additively approximate $T(G; e^{-2\pi i/5}, e^{-2\pi i/5})$, with a suitable normalization, is sufficient to simulate any quantum computation. Bordewich et al. [177] gave a number of elementary additive approximations for Tutte evaluations, and asked whether the evaluation of

$T(G; e^{-2\pi i/5}, e^{-2\pi i/5})$ had a suitable additive approximation, which would amount to a simulation of a BQP-complete problem. For a precise statement see [177]. BQP (bounded error quantum polynomial time) is the complexity class of decision problems solvable by a quantum computer in polynomial time with bounded error. It is analogous to the classical complexity class BPP; for further information see [100].

A number of papers have built upon the connection between BQP and the Jones polynomial (e.g. [1167]) to show first that approximating the Jones polynomial is in BQP at a number of points [8], and then that it is BQP-hard [6]. More recent work by Aharonov, Arad, Eban and Landau [7] shows that this extends to the (multivariate) Tutte polynomial. Kuperberg [738] gave an enlightening analysis of links between additive approximations and multiplicative approximations, along the way obtaining Theorem 10.29.

10.7 Open problems

This chapter has gone some way to answering Open Question 10.4, although it is far from completely resolved. In particular resolving any of the following questions would represent significant progress.

Open Question 10.35. Determine whether approximating $T(G; x, y)$ for fixed x, y on $H_q^+, q > 2$ is AP-interreducible with ♯BIS or harder, likely AP-interreducible with ♯SAT.

Open Question 10.36. Resolve the approximation complexity of ♯BIS. That is, decide if ♯BIS admits an FPRAS, is AP-interreducible with ♯SAT or is genuinely of intermediate complexity.

Open Question 10.37. For any $(x, y) \in \mathbb{C}^2$, for what classes of graphs is there an FPRAS for $T(G; x, y)$?

Open Question 10.38. What is the complexity of approximating $T(G; x, y)$ for fixed x, y in any of the following regions:

1. $x, y \geq 1$ and $0 < (x-1)(y-1) < 2$, except $(1, 1)$ and the curve H_1^+, which are exactly computable in polynomial time;

2. $-1 < x < 1$ and $y \geq 1$;

3. $-1 < y < 1$ and $x \geq 1$;

4. the remaining points near the origin ($|x|, |y| < 1$) not already determined in Theorem 10.30. In particular the region $|x|, |y| < 1$ and $x + y \geq 0$.

Part III

Specializations

11

Foundations of the chromatic polynomial

Fengming Dong • Khee Meng Koh

Synopsis

The chromatic polynomial of a graph evaluated at λ gives the number of ways to properly color the graph with λ colors. It arose from the four color conjecture and in turn gave rise to the Tutte polynomial, which can be viewed as a two variable generalization of the chromatic polynomial. This chapter gives an overview of some of the central topics in the study of the chromatic polynomial.

- Computing the chromatic polynomial, including classes of graphs for which it can be computed in polynomial time.
- Properties of chromatic polynomials, such as interpretations of coefficients, factorizations, combinatorial interpretations, unimodality of coefficients, inequalities of chromatic polynomials, and encoding of graph connectivity.
- Identifying chromatically equivalent graphs, that is, graphs with the same chromatic polynomial.
- Significance and locations of the roots of chromatic polynomials.

11.1 Introduction

The beginnings of the Tutte polynomial trace back to the chromatic polynomial, a graph polynomial that is of independent interest. The chromatic polynomial arises from questions in graph coloring. Given a positive integer λ, a mapping $f : V \to \{1, 2, \cdots, \lambda\}$ is called a *proper λ-coloring* of a graph $G = (V, E)$ if $f(u) \neq f(v)$ whenever the vertices u and v are adjacent. We say

DOI: 10.1201/9780429161612-11

that two proper λ-colorings f and g of G are *distinct* if $f(v) \neq g(v)$ for some vertex v in G, and denote the number of distinct proper λ-colorings of G by $\chi(G;\lambda)$. Since each set of vertices of a given color in a proper coloring of a graph forms an independent set, $\chi(G;\lambda)$ also counts the number of ways to partition the vertices of G into λ independent sets.

Remarkably, $\chi(G;\lambda)$ turns out to be a polynomial in λ. There are many ways to show this, and to reveal the relation between the chromatic polynomial and the Tutte polynomial as given in the following sections. Although the chromatic polynomial is a specialization of the Tutte polynomial, it preceded the Tutte polynomial, and indeed expanding the properties and formulation of the chromatic polynomial led to the development of the Tutte polynomial (see Chapter 34 for details of this history).

The chromatic polynomial was introduced by Birkhoff in 1912 [120] for planar graphs. He hoped that the chromatic polynomial could be used as a tool to prove the four color conjecture, which holds if and only if $\chi(G;4) > 0$ for any planar graph G. The chromatic polynomial was then extended to general graphs by Whitney [1152, 1154] in 1932.

So far, using the chromatic polynomial as a tool to prove the four color conjecture has not been successful. However, we know of no compelling argument to suggest that this approach cannot yield a theoretical proof of this famous conjecture. Many mathematicians, including both professional researchers and amateur investigators, still seek to prove the four color conjecture by understanding the properties of the chromatic polynomial. Although proving the four color conjecture using chromatic polynomial as a tool has not succeeded, the function $\chi(G;\lambda)$ itself has attracted extensive study by many researchers as a fascinating object in its own right, with substantive surveys including [301, 392, 404, 638, 640, 950]. Attributes of the chromatic polynomial of particular interest that we discuss here include its computability, expressions, coefficients, chromatically equivalent classes, and locations of its roots.

11.2 Computing chromatic polynomials

11.2.1 Definition of the chromatic polynomial

Definition 11.1. *The chromatic polynomial* of a graph G is the polynomial $\chi(G;\lambda)$ such that whenever λ is a positive integer, $\chi(G;\lambda)$ is the number of distinct proper λ-colorings of G.

An inclusion-exclusion argument for counting the number of proper λ-colorings of a graph readily yields the following result, giving an alternative definition of the chromatic polynomial. It also makes clear that $\chi(G;\lambda)$ is in fact a polynomial, thus validating Definition 11.1. Furthermore, the relation between the chromatic polynomial and the dichromatic polynomial

$Z(G; u, v) = \sum_{A \subseteq E(G)} u^{k(A)} v^{|A|}$ (see Definition 2.1) is now immediately apparent (as is the name "dichromatic").

Proposition 11.2. *For a graph $G = (V, E)$,*

$$\chi(G; \lambda) = \sum_{A \subseteq E(G)} (-1)^{|A|} \lambda^{k(A)}, \tag{11.1}$$

where $k(A)$ is the number of components of the spanning subgraph (V, A).

The chromatic polynomial may also be viewed as a special case of the chromatic symmetric function $X(G; x_1, x_2, \cdots)$ introduced by Stanley [1033].

Definition 11.3. For any graph $G = (V, E)$ with vertices $v_1, v_2, \cdots, v_n$, the *chromatic symmetric function* is defined by

$$X(G; x_1, x_2, \ldots) = \sum_{p} x_{p(v_1)} x_{p(v_2)} \cdots x_{p(v_n)},$$

where $x_1, x_2, \cdots$ are variables and the sum runs over all mappings $p : V \to \mathbb{N}$ such that $p(v_i) \neq p(v_j)$ whenever v_i and v_j are adjacent in G.

Clearly, for any positive integer λ, $\chi(G; \lambda) = X(G; x_1, x_2, \cdots)$, where $x_1 = x_2 = \cdots = x_\lambda = 1$ and $x_i = 0$ for all $i \geq \lambda + 1$.

Example 11.4. The following chromatic polynomials can be easily derived from the definition.

1. For $\overline{K}_n$, the edgeless graph of order n, we have $\chi(\overline{K}_n; \lambda) = \lambda^n$.

2. If G has a loop, then $\chi(G; \lambda) = 0$.

3. If G is a graph with at least one edge, and G' is a graph formed by adding an edge in parallel to any existing edge of G, then $\chi(G'; \lambda) = \chi(G; \lambda)$.

4. For K_n, the complete graph of order n, the chromatic polynomial is $\chi(K_n; \lambda) = \lambda(\lambda - 1) \cdots (\lambda - n + 1)$.

5. If G is a tree of order n, then $\chi(G; \lambda) = \lambda(\lambda - 1)^{n-1}$.

6. If G is a disconnected graph with components $G_1, G_2, \cdots, G_k$, then

$$\chi(G; \lambda) = \prod_{i=1}^{k} \chi(G_i; \lambda).$$

7. For a graph $G = (V, E)$ and nonnegative integers m and n,

$$\chi(G; m + n) = \sum_{X \subseteq V} \chi(G[X]; m)\, \chi(G[V \setminus X]; n).$$

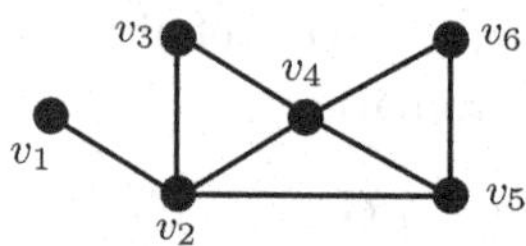

FIGURE 11.1: A chordal graph G with a perfect elimination ordering $(v_1, v_2, \cdots, v_6)$.

For any vertex v in a graph G, let $N_G(v)$ (or simply $N(v)$) denote the neighborhood of v in G, that is, all the vertices adjacent to v, but not v itself. A vertex v is a *simplicial vertex* if it has no loops and either $N(v) = \emptyset$ or the subgraph of G induced by $N(v)$ is a clique.

By Definition 11.1, if v is a simplicial vertex of G, then we have the following expression for $\chi(G; \lambda)$.

Proposition 11.5. *If v is a simplicial vertex of a graph G, then*

$$\chi(G; \lambda) = (\lambda - \deg(v))\, \chi(G \backslash v; \lambda).$$

Proposition 11.5 can be applied to determine the chromatic polynomials of chordal graphs. A graph G is called a *chordal graph* if for any cycle C in G with $|V(C)| \geq 4$, the subgraph of G induced by $V(C)$ always contains more than $|V(C)|$ edges. Dirac [385] showed that a graph G of order n is chordal if and only if it has a *perfect elimination ordering*, that is, an ordering $(v_1, v_2, \cdots, v_n)$ of its vertices such that each v_i is a simplicial vertex of the subgraph of G induced by $\{v_1, v_2, \cdots, v_i\}$.

Corollary 11.6. *If G is a chordal graph of order n with a perfect elimination ordering $(v_1, v_2, \cdots, v_n)$, then*

$$\chi(G; \lambda) = \prod_{i=1}^{n} (\lambda - d_i),$$

where $d_i = |N(v_i) \cap \{v_1, v_2, \cdots, v_i\}|$.

Example 11.7. The graph shown in Figure 11.1 is a chordal graph with a perfect elimination ordering $(v_1, v_2, \cdots, v_6)$. By Corollary 11.6, we have

$$\chi(G; \lambda) = \lambda(\lambda - 1)^2(\lambda - 2)^3,$$

as $d_1 = 0, d_2 = d_3 = 1, d_4 = d_5 = d_6 = 2$.

11.2.2 Chromatic polynomials for graphs in general

In general, given any graph G, the chromatic polynomial $\chi(G; \lambda)$ can be determined by applying a recursive formula repeatedly until all new graphs obtained are complete graphs or edgeless graphs.

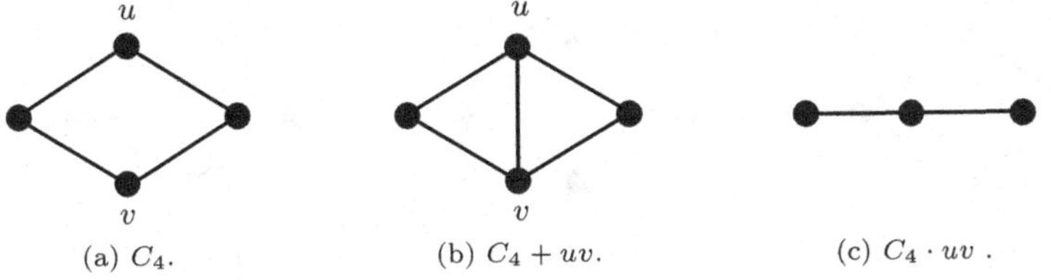

(a) C_4. (b) $C_4 + uv$. (c) $C_4 \cdot uv$.

FIGURE 11.2: The graphs C_4, $C_4 + uv$ and $C_4 \cdot uv$.

For non-adjacent vertices u and v in a graph G, let $G + uv$ denote the graph obtained from G by adding a new edge joining u and v, and let $G \cdot uv$ denote the graph obtained from G by identifying u and v and removing all but one edge from any set of parallel edges that might result. Observe that $\chi(G + uv; \lambda)$ is equal to the number of proper λ-colorings of G when u and v are assigned different colors while $\chi(G \cdot uv; \lambda)$ is equal to the number of proper λ-colorings of G when u and v are assigned the same color. This leads to the following fundamental reduction theorem for the chromatic polynomial.

Theorem 11.8. *If u and v are non-adjacent vertices in a graph G, then*

$$\chi(G; \lambda) = \chi(G + uv; \lambda) + \chi(G \cdot uv; \lambda). \tag{11.2}$$

Example 11.9. Let u and v be two non-adjacent vertices in C_4 as shown in Figure 11.2. Applying Theorem 11.8, $\chi(C_4; \lambda)$ is obtained as follows:

$$\begin{aligned} \chi(C_4; \lambda) &= \chi(C_4 + uv; \lambda) + \chi(C_4 \cdot uv; \lambda) \\ &= \lambda(\lambda-1)(\lambda-2)^2 + \lambda(\lambda-1)^2 \\ &= (\lambda-1)^4 + (\lambda-1). \end{aligned}$$

Theorem 11.8 also leads to another form of recursive expression for the chromatic polynomial. If $e = uv$ is an edge of a graph G that is not a loop nor a multiple edge, then we can let $H = G\backslash e$, and thus $G = H + uv$. Since, by Item 3 of Example 11.4, $\chi(G/e; \lambda) = \chi(H \cdot uv; \lambda)$, applying Theorem 11.8 to $H = G\backslash e$ yields the recursion in Theorem 11.10 for this type of edge. The fact that the chromatic polynomial of any graph with a loop is 0 then extends the recursion to an arbitrary edge e of G.

Theorem 11.10. *If e is an edge in G, then*

$$\chi(G; \lambda) = \chi(G\backslash e; \lambda) - \chi(G/e; \lambda). \tag{11.3}$$

Both (11.2) and (11.3) are referred to as the *fundamental reduction theorem* for the chromatic polynomial.

Example 11.11. Let e be an edge in C_4 as shown in Figure 11.3. Applying Theorem 11.10, $\chi(C_4; \lambda)$ can be obtained as follows:

$$\begin{aligned} \chi(C_4; \lambda) &= \chi(C_4\backslash e; \lambda) - \chi(C_4/e; \lambda) \\ &= \lambda(\lambda-1)^3 - \lambda(\lambda-1)(\lambda-2) \\ &= (\lambda-1)^4 + (\lambda-1). \end{aligned}$$

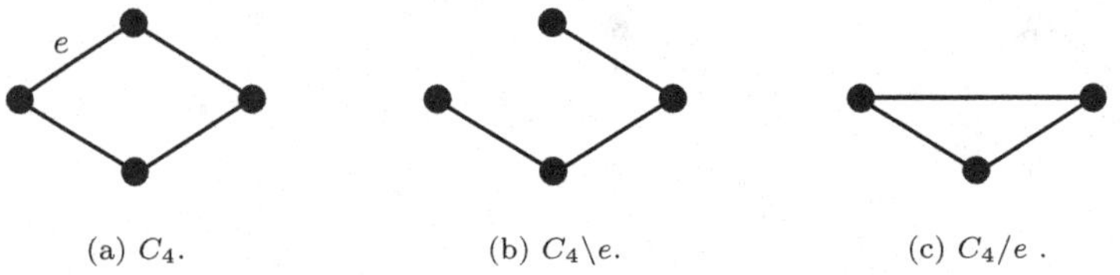

(a) C_4. (b) $C_4\backslash e$. (c) C_4/e .

FIGURE 11.3: The graphs C_4, $C_4\backslash e$ and C_4/e.

Example 11.12. For any integer $n \geq 3$, by applying Theorem 11.10 it can be shown that

$$\chi(C_n;\lambda) = (\lambda-1)^n + (-1)^n(\lambda-1).$$

Applying Theorem 2.21 to the function $(-1)^{r(G)}\lambda^{k(G)}\chi(G;\lambda)$ and using the recursion in Theorem 11.10 give the following relation between the chromatic polynomial and the Tutte polynomial.

Theorem 11.13. *For a graph G,*

$$\chi(G;\lambda) = (-1)^{r(G)}\lambda^{k(G)}\,T(G;1-\lambda,0).$$

Remark 11.14. Although computing the chromatic polynomial of a graph in general is a NP-hard problem (see Chapter 9 and [897]), the preceding theorems yield the following theoretically useful heuristics. Let G be any graph of order n.

1. If G is a chordal graph, then $\chi(G;\lambda)$ can be obtained by Corollary 11.6.

2. If G is not a chordal graph, $\chi(G;\lambda)$ can be obtained by applying Theorem 11.8 (or Theorem 11.10) repeatedly until all new graphs obtained are chordal graphs.

3. If G is connected, $\chi(G;\lambda)$ can be obtained by applying Theorem 11.10 repeatedly until all new graphs obtained are trees, and thus $\chi(G;\lambda)$ can be expressed as

$$\chi(G;\lambda) = \sum_{i=1}^{n} t_i\lambda(\lambda-1)^{i-1},$$

where t_i is an integer.

4. $\chi(G;\lambda)$ can also be obtained by applying Theorem 11.8 until all new graphs obtained are complete graphs, and thus $\chi(G;\lambda)$ can be expressed as

$$\chi(G;\lambda) = \sum_{i=1}^{n} a_i(\lambda)_i,$$

where a_i is a nonnegative integer and $(\lambda)_i = \lambda(\lambda-1)\cdots(\lambda-i+1)$ for any nonnegative integer i.

5. $\chi(G;\lambda)$ can also be obtained by applying Theorem 11.10 until all new graphs obtained are edgeless, and thus $\chi(G;\lambda)$ can be expressed as

$$\chi(G;\lambda) = \sum_{i=1}^{n} b_i \lambda^i$$

where b_i is an integer for all i.

11.2.3 Computing chromatic polynomials by matrices

In this subsection, we introduce a matrix method, due to Biggs [108], to compute chromatic polynomials of graphs from some special families.

A *graph scheme* is a pair (H, J), where H is a graph and J is a subset of the set $\{(a, b) : a, b \in V(H)\}$. Let Q_k denote the set of proper k-colorings of H. Thus $\chi(H, k) = |Q_k|$. Any two proper colorings α and β in Q_k are called *compatible* with respect to J if $\alpha(u) \neq \beta(v)$ for all $(u, v) \in J$.

Definition 11.15. For any integer $n \geq 2$, let $G_n(H, J)$ denote the graph obtained from n disjoint copies of H by adding edges joining vertices u in the i-th copy to vertices v in the $(i + 1)$-st copy whenever $(u, v) \in J$ for all $i = 1, 2, \cdots, n$, where the $(n + 1)$-th copy is the first copy by convention.

Definition 11.16. For a given graph scheme (H, J) and a given positive integer k, the *compatible matrix*, denoted by $M_k(H, J)$, is the $|Q_k| \times |Q_k|$ matrix $(m_{i,j})$ where

$$m_{i,j} = \begin{cases} 1 & \text{if } \alpha_i \text{ and } \alpha_j \text{ are compatible,} \\ 0 & \text{otherwise} \end{cases}$$

and where α_i is the i-th member in Q_k for $i = 1, 2, \cdots, |Q_k|$.

Theorem 11.17. *For any given graph scheme (H, J) and positive integers k and n, where $n \geq 2$,*

$$\chi(G_n(H, J); k) = \mathrm{Tr}(M_k(H, J)^n).$$

Corollary 11.18. *If the matrix $M_k(H, J)$ has eigenvalues $\lambda_1, \lambda_2, \cdots, \lambda_s$, with multiplicities $c_1, c_2, \cdots, c_s$, respectively, then*

$$\chi(G_n(H, J); k) = \sum_{i=1}^{s} c_i \lambda_i^n.$$

In what follows, we apply Corollary 11.18 to find the chromatic polynomial of the circular ladder L_n of order $2n$, as shown in Figure 11.4. Note that L_n is the Cartesian product $C_n \times P_2$.

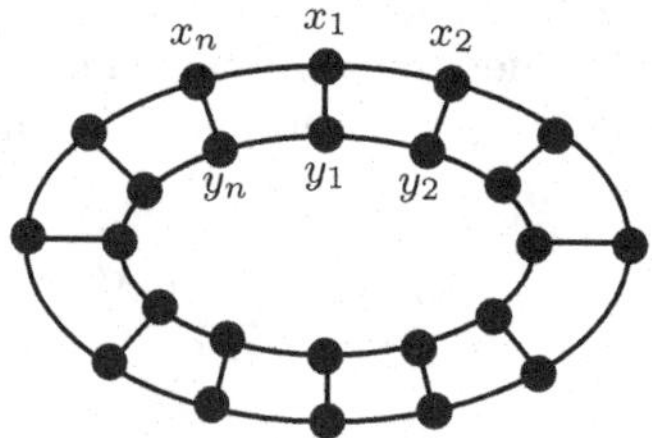

FIGURE 11.4: Circular ladders L_n of order $2n$, where $n \geq 2$.

Eigenvalue	$k^2 - 3k + 3$	$-k + 1$	$3 - k$	1
Multiplicity	1	$k - 1$	$k - 1$	$(k-1)(k-2) - 1$

TABLE 11.1: Eigenvalues of $M_k(H, J)$, from Example 11.19, and their multiplicities.

Example 11.19. The circular ladder L_n of order $2n$ is the graph $G_n(H, J)$, where H is the complete graph K_2 with vertex set $\{a, b\}$ and $J = \{(a, a), (b, b)\}$. For any positive integer k, H has exactly $k(k-1)$ different proper k-colorings: the coloring $\alpha_{i,j}$ assigns color i to vertex a and color j to vertex b for all $i = 1, 2, \cdots, k$ and $j = 1, 2, \cdots, k$ with $j \neq i$. We may assume that the matrix $M_k(H, J)$ is obtained with respect to the following ordering of $\alpha_{i,j}$'s:

$$\alpha_{1,2}, \alpha_{1,3}, \cdots, \alpha_{1,k}, \alpha_{2,1}, \alpha_{2,3}, \cdots, \alpha_{2,k}, \cdots, \alpha_{k,1}, \alpha_{k,2}, \cdots, \alpha_{k,k-1}.$$

The eigenvalues of $M_k(H, J)$ are shown in Table 11.1. By Corollary 11.18, we find an expression for $\chi(L_n, k)$, which was first published [114]:

$$\chi(L_n, k) = (k^2 - 3k + 3)^n - (1 - k)^{n+1} + (k - 1)(3 - k)^n + (k^2 - 3k + 1).$$

11.3 Properties of chromatic polynomials

This section introduces properties of chromatic polynomials, including expressions of coefficients, factorizations, combinatorial interpretations, unimodal property of coefficients, inequalities of chromatic polynomials and the connectivities of graphs which can be determined by chromatic polynomials.

11.3.1 Interpretation of coefficients

The interpretations of the coefficients of $\chi(G; \lambda)$ have a long history of study, beginning with Whitney [1154], followed by, among others, Bari and Hall [80], Read [945], Eisenberg [444], Hong [624] and, Dong, Koh and Soh [403].

Notice that Proposition 11.2 gives an interpretation of the coefficients of $\chi(G;\lambda)$, which was first provided by Whitney in [1154].

Theorem 11.20. *For a graph G on n vertices and m edges,*

$$\chi(G;\lambda) = \sum_{i=1}^{n}\left(\sum_{r=0}^{m}(-1)^r N(i,r)\right)\lambda^i, \tag{11.4}$$

where $N(i,r)$ is the number of spanning subgraphs of G with exactly i components and r edges.

Whitney substantially simplified Equation (11.4) by introducing broken-cycles.

Definition 11.21. Let $G=(V,E)$ be a graph on n vertices and m edges, and let $\beta : E \to \{1,2,\cdots,m\}$ be a bijection giving an ordering of its edges. For any cycle C in G, if e is the edge on C such that $\beta(e) > \beta(e')$ for any edge e' in $E(C)\backslash\{e\}$, the path $C\backslash e$ is called a *broken-cycle* in G with respect to the ordering β.

The following celebrated result from [1154] is known as *Whitney's broken-cycle theorem.*

Theorem 11.22. *Let $G=(V,E)$ be a graph on n vertices and m edges, and $\beta : E \to \{1,2,\cdots,m\}$ be any bijection. Then*

$$\chi(G;\lambda) = \sum_{i=1}^{n}(-1)^{n-i}h_i(G)\lambda^i,$$

where $h_i(G)$ is the number of spanning subgraphs of G that have exactly $n-i$ edges and that contain no broken-cycles with respect to β.

Bari and Hall [80] gave another interpretation for $h_i(G)$, which is now known as Bari–Hall's broken-cycle formula, by introducing another concept, called a minimal spanning subgraph of G containing r broken-cycles.

Definition 11.23. Let $\beta : E \to \{1,2,\cdots,m\}$ be a bijection ordering the edges of a graph G on n vertices and m edges. Suppose that there are q broken-cycles in G induced by β. For any r of these q broken-cycles, where $1 \le r \le q$, any spanning subgraph of G whose edge set consists precisely of those edges in the given r broken-cycles is called a *minimal spanning subgraph of G containing r broken-cycles.*

The *Bari–Hall broken-cycle formula,* from [80], is then as follows.

Theorem 11.24. *Let G be a graph on n vertices and m edges, and $\beta : E \to \{1,2,\cdots,m\}$ be a bijection ordering the edges of G. Then*

$$h_{n-i}(G) = \binom{m}{i} + \sum_{j=2}^{i}\sum_{r=1}^{q}(-1)^r\binom{m-j}{i-j}n_{j,r}, \tag{11.5}$$

where $n_{j,r}$ is the number of minimal spanning subgraphs of G with exactly j edges containing r broken-cycles.

A broken-cycle in G is said to be *proper* if it contains no other broken-cycles. Bari and Hall [80] pointed out that, to apply (11.5), it suffices to consider those spanning subgraphs of G containing proper broken-cycles only, yielding the following corollary.

Corollary 11.25. *Let G be a graph on n vertices and m edges, and $\beta : E \to \{1, 2, \cdots, m\}$ be a bijection ordering the edges of G. Then*

$$h_{n-i}(G) = \binom{m}{i} + \sum_{j=2}^{i} \sum_{r=1}^{q} (-1)^r \binom{m-j}{i-j} n^*_{j,r}, \tag{11.6}$$

where $n^*_{j,r}$ is the number of minimal spanning subgraphs of G with exactly j edges containing r proper broken-cycles.

The coefficient h_1 of λ in $\chi(G;\lambda)$ reveals a number of properties of the graph G.

Theorem 11.26. *Let G be a graph of order $n \geq 2$. Then*

1. *G is connected if and only if $h_1 \geq 1$;*
2. *G is a tree if and only if $h_1 = 1$;*
3. *$(\chi(G) - 1)!|h_1$, recalling that $\chi(G)$ is the chromatic number of G;*
4. *G is connected and bipartite if and only if h_1 is odd;*
5. *if G is uniquely k-colorable, where $k \geq 2$, then h_1 is divisible by $(k-1)!$ but not by $k!$.*

In Theorem 11.26, Items 1 and 2 are from [444, 945], Items 3 and 4 from [624], and Item 5 from [403].

11.3.2 Explicit expressions for some coefficients

Explicit expressions for some coefficients of chromatic polynomials have been found. For the following two results, we let $s_G(H)$ denote the number of subgraphs of a graph G which are isomorphic to H, and $i_G(H)$ denote the number of induced subgraphs of G which are isomorphic to H.

As an immediate consequence of Theorem 11.20 or 11.22, we have the following result.

Corollary 11.27. *Let G be a graph on n vertices and m edges and with girth g. Then*

$$\chi(G;\lambda) = \sum_{i=1}^{n} (-1)^{n-i} h_i \lambda^i$$

is a polynomial in λ such that

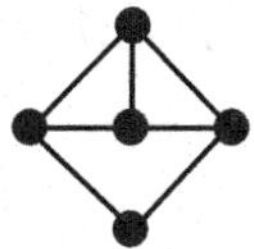

FIGURE 11.5: The graph H in Theorem 11.28.

1. *the degree of* $\chi(G;\lambda)$ *is* n*;*
2. *for* $0 \leq i \leq g-2$, $h_{n-i} = \binom{m}{i}$ *(in particular,* $h_n = 1$ *and* $h_{n-1} = m$*);*
3. *the coefficients are integers and alternate in sign;*
4. *the constant term is zero;*
5. $h_{n-(g-1)} = \binom{m}{g-1} - s_G(C_g)$ *(in particular,* $h_{n-2} = \binom{m}{2} - s_G(C_3)$*).*

Recall that W_n is the wheel of order $n+1$, so $W_n = C_n + K_1$, the complete join of C_n and K_1.

Theorem 11.28. *Let H be the graph in Figure 11.5, and let G be a graph on n vertices and m edges and with girth g and chromatic polynomial* $\chi(G;\lambda) = \sum_{i=1}^{n}(-1)^{n-i}h_i\lambda^i$. *Then*

1. $h_{n-3} = \binom{m}{3} - (m-2)s_G(K_3) - i_G(C_4) + 2s_G(K_4)$*;*
2. $h_{n-4} = \binom{m}{4} - \binom{m-2}{2}s_G(K_3) + \binom{s_G(K_3)}{2} + (2m-9)s_G(K_4) - 6s_G(K_5)$ $-(m-3)i_G(C_4) - i_G(C_5) + i_G(K_{2,3}) + 2i_G(H) + 3i_G(W_4)$*;*
3. *for an edge e in* $K_{3,3}$,
$$\begin{aligned} h_{n-5} &= \binom{m}{5} - \binom{m-3}{2}s_G(C_4) - i_G(C_6) + (m-3)s_G(K_{2,3}) \\ &\qquad - s_G(K_{2,4}) + i_G(K_{3,3}\backslash e) + 4s_G(K_{3,3})); \end{aligned}$$
4. *for any integer k with* $g \leq k \leq \lceil \frac{3}{2}g \rceil - 3$,
$$h_{n-k} = \binom{m}{k} - \sum_{r=g}^{k+1} s_G(C_r)\binom{m-r+1}{k-r+1};$$
5. *for* $1 \leq i \leq n-1$,
$$h_i \leq \binom{m}{n-i} - \binom{m-g+2}{n-i-g+2} + \binom{m-s_G(C_g)-g+2}{n-i-g+2};$$
6. *for* $1 \leq i \leq n-1$ *and l with* $3 \leq k \leq l \leq n$,
$$h_i \geq \sum_{j=0}^{l-1}\binom{m-n+j}{j}\binom{n-1-j}{i-1} - \sum_{k=3}^{l}\left(s_G(C_k)\sum_{j=0}^{l-k}\binom{m-n+j}{j}\binom{n-k-j}{i-1}\right).$$

In the theorem, the first two items are from [481, 915], the third is from [915], the fourth from [914, 1059], and the final two items are from [768].

11.3.3 The chromatic polynomial at negative integers

Stanley [1027] found combinatorial interpretations for the values $\chi(G;\lambda)$ and $(-1)^{|V|}\chi(G;-\lambda)$ when λ is a positive integer. In particular, $(-1)^{|V|}\chi(G;-1)$ is the number of acyclic orientations of G, i.e., those orientations of G in which there are no directed cycles. (See Chapter 31 for more on the Tutte polynomial and orientations.)

Proposition 11.29. *If $G=(V,E)$ is a graph and λ is a nonnegative integer, then $\chi(G;\lambda)$ is equal to the number of pairs (σ,ρ), where σ is any map: $V\to\{1,2,\cdots,\lambda\}$ and ρ is any orientation of G, subject to the condition that if $u\to v$ in the orientation ρ, then $\sigma(u)>\sigma(v)$.*

For a positive integer λ, let $\bar{\chi}(G;\lambda)$ denote the number of pairs (σ,ρ), where σ is any map: $V\to\{1,2,\cdots,\lambda\}$ and ρ is an acyclic orientation of G such that if $u\to v$ in the orientation, then $\sigma(u)\geq\sigma(v)$.

Theorem 11.30. *For any positive integer λ and any graph G on n vertices,*

$$\bar{\chi}(G;\lambda)=(-1)^n\chi(G;-\lambda).$$

Corollary 11.31. *For any graph G of order n, $(-1)^n\chi(G;-1)$ is equal to the number of acyclic orientations of G.*

11.3.4 Unimodal properties

In this subsection we introduce the celebrated result about the coefficients of the chromatic polynomial due to Huh [629] and Huh and Katz [631], who in 2012 proved the unimodal conjecture proposed by Read [945] in 1968. The unimodal conjecture was acknowledged to be one of the most renowned outstanding open problems in the theory of chromatic polynomials, and was proved in the context of the strong results given in Theorem 11.36.

Read [945] noticed that for any graph G it appears that the coefficients of $\chi(G;\lambda)$ always increase in absolute value first, and decrease eventually. As pointed out by Read and Tutte [950], it was Nijenhuis and Wilf [881] who first formulated this observation into a conjecture, which is now known as the unimodal conjecture

Conjecture 11.32 (Unimodal Conjecture)**.** For a graph G of order n, if $\chi(G;\lambda)=\sum_{i=1}^{n}(-1)^{n-i}h_i\lambda^i$, then there always exists an integer k with $2\leq k\leq n-1$ such that

$$h_1\leq h_2\leq\cdots\leq h_{k-1}\leq h_k\geq h_{k+1}\geq\cdots\geq h_n.$$

In connection with the unimodal conjecture, there is the following result (e.g., see [404]).

Proposition 11.33. *Let G be a connected graph of order n. Then for any integer i with $1 \leq i \leq n-1$, we have that $h_{i+1} \leq \left(\frac{n}{i} - 1\right) h_i$, where equality holds if and only if G is a tree.*

Corollary 11.34, due to Lovász [792], follows as a corollary of Proposition 11.33.

Corollary 11.34. *Let G be a connected graph of order n.*

1. *If n is odd, then $h_n < h_{n-1} < \cdots < h_{\frac{n+1}{2}}$; and*

2. *if n is even, then $h_n < h_{n-1} < \cdots < h_{\frac{n}{2}+1} \leq h_{\frac{n}{2}}$, and $h_{\frac{n}{2}+1} = h_{\frac{n}{2}}$ if and only if G is a tree.*

A sequence $a_1, a_2, \cdots, a_n$ is said to be *log-concave* if the inequality $a_{i-1}a_{i+1} \leq a_i^2$ holds for all $i = 2, 3, \cdots, n-1$, and it is said to have *no internal zero* if the indices of the non-zero elements are consecutive integers.

Hoggar [620] proposed considering, instead of the unimodal conjecture, the following conjecture, known as the *strong logarithmic concavity conjecture*. Since log concavity implies unimodality, this conjecture implies the unimodal conjecture.

Conjecture 11.35 (Strong Logarithmic Concavity Conjecture)**.** If k is any positive integer with $k \leq n-3$, then $h_k h_{k+2} < h_{k+1}^2$.

This conjecture, and hence the unimodal conjecture, have been resolved. In 2012, Huh [629] obtained the more general result for the characteristic polynomial of a matroid, stated here in Theorem 11.36. See Definition 4.136 for the definition of the characteristic polynomial, and Theorem 4.140 for its connection with the chromatic polynomial.

Theorem 11.36. *If M is a matroid representable over a field of characteristic zero, then the coefficients of its characteristic polynomial of $\chi(M;\lambda)$ form a sign-alternating log-concave sequence of integers with no internal zeros.*

Huh and Katz extended the above result to all representable matroids in [631].

11.3.5 Factorizations of chromatic polynomials

The polynomial $\chi(G;\lambda)$ can be factorized when G has a clique whose removal disconnects G or when G is the complete join of two other graphs. Morgan and Farr [861, 862, 863] showed that there are some other graphs whose chromatic polynomials can also be factorized by introducing the process of a certificate of factorization.

If G is disconnected, $\chi(G;\lambda)$ can be expressed as the product of $\chi(G_i;\lambda)$ over all its components G_i (see Example 11.4). A graph G is called a *K_r-gluing* of two graphs G_1 and G_2 if G is obtained from G_1 and G_2 by identifying a clique K_r in G_1 with a clique K_r in G_2 in an arbitrary manner. Note that $G\backslash S$ is disconnected for some clique S of G if and only if G is a K_r-gluing of two graphs for some $r \geq 1$. Zykov [1191] proved the following.

Theorem 11.37. *Let G_1 and G_2 be any two graphs and r be an integer with $0 \leq r \leq \min\{\omega(G_1), \omega(G_2)\}$, where $\omega(G_i)$ is the clique number of G_i. If G is a K_r-gluing of G_1 and G_2, then*

$$\chi(G;\lambda) = \frac{\chi(G_1;\lambda)\chi(G_2;\lambda)}{\chi(K_r;\lambda)}. \tag{11.7}$$

Note that Proposition 11.5 follows directly from Theorem 11.37. By Theorem 11.37, when G is connected but has more than one block, $\chi(G;\lambda)$ has a factorization similar to Item 6 of Example 11.4.

Corollary 11.38. *If G is connected with blocks $G_1, G_2, \cdots, G_b$, then*

$$\chi(G;\lambda) = \frac{1}{\lambda^{b-1}} \prod_{i=1}^{b} \chi(G_i;\lambda).$$

An expression for the chromatic number of the complete join of two graphs, $\chi(G+H;\lambda)$, in terms of $\chi(G;\lambda)$ and $\chi(H;\lambda)$ was provided by Zykov [1191].

Theorem 11.39. *For any two graphs G and H, if*

$$\chi(G;\lambda) = \sum_{i=1}^{n_1} a_i(\lambda)_i \quad \textit{and} \quad \chi(H;\lambda) = \sum_{j=1}^{n_2} b_j(\lambda)_j,$$

then

$$\chi(G+H;\lambda) = \sum_{i=1}^{n_1}\sum_{j=1}^{n_2} a_i b_j (\lambda)_{i+j},$$

where $(\lambda)_i = \lambda(\lambda-1)\cdots(\lambda-i+1)$.

For integers n and k with $0 \leq k \leq n$ the *Stirling number of the second kind*, $S(n,k)$, counts the number of ways to partition a set of n objects into k nonempty subsets. For any $1 \leq k \leq n$, the function $S(n,k)$ satisfies the recursion

$$S(n,k) = S(n-1,k-1) + k\,S(n-1,k).$$

The definitions of $S(n,k)$ and $\chi(\overline{K}_r;\lambda)$ imply that

$$\chi(\overline{K}_r;\lambda) = \lambda^r = \sum_{i=1}^{r} S(r,i)(\lambda)_i.$$

If G is any complete k-partite graph for $k \geq 2$, Theorem 11.39 gives an explicit expression for $\chi(G;\lambda)$.

Corollary 11.40. *For the complete k-partite graph $K_{p_1,p_2,\cdots,p_k}$,*

$$\chi(K_{p_1,p_2,\ldots,p_k};\lambda) = \sum_{r_k=1}^{p_k} \cdots \sum_{r_1=1}^{p_1} \left(\prod_{i=1}^{k} S(p_i, r_i) \right) (\lambda)_{r_1+\cdots+r_k}.$$

The following corollary gives a useful special case of Theorem 11.39 that occurs when $H = K_1$, i.e. when we have a graph on n vertices with a vertex of degree $n-1$.

Corollary 11.41. *If $G = (V, E)$ is a graph with a vertex w such that* $\deg(w) = |V| - 1$, *then*

$$\chi(G;\lambda) = \lambda\,\chi(G\backslash w;\lambda - 1).$$

Corollary 11.41 can be strengthened to the following proposition from [402].

Proposition 11.42. *Let w be any vertex in a graph G. Then*

$$\chi(G;\lambda) = \lambda \sum_{A \subseteq \mathcal{I}_w} \chi(G\backslash A;\lambda - 1),$$

where $\mathcal{I}_w$ is the family of independent sets in G which contain w.

Theorem 11.37 shows that $\chi(G;\lambda)$ can be factorized if G is clique-separable (i.e., G is either disconnected or a K_r-gluing of some pair of graphs). However, Morgan and Farr [861, 862, 863] showed the existence of non-clique-separable graphs whose chromatic polynomials can also be factorized in the form of (11.7). They introduced certificates of factorization, which are sequences of algebraic transformations based on identities for the chromatic polynomials.

Definition 11.43. A *certificate of factorization* of $\chi(G_0;\lambda)$ for a graph G_0 is a sequence of polynomials $(\chi_0, \chi_1, \cdots, \chi_k)$ such that $\chi_0 = \chi(G_0;\lambda)$ and $\chi_k = \chi(H_1;\lambda)\chi(H_2;\lambda)/\chi(K_r;\lambda)$ for some graphs H_1 and H_2 and some positive integer r. Furthermore, for $1 \le i \le k$, each χ_i is obtained from χ_{i-1} by one of the following steps:

(CS1) $\chi(G';\lambda)$ becomes $\chi(G'\backslash e;\lambda) - \chi(G'/e;\lambda)$ for some $e \in E(G')$;

(CS2) $\chi(G_1;\lambda) - \chi(G_2;\lambda)$ becomes $\chi(G';\lambda)$ where G' is isomorphic to $G_1 + uv$, and G_2 is isomorphic to $G_1 \cdot uv$ for some non-adjacent vertices u and v in G_1;

(CS3) $\chi(G';\lambda)$ becomes $\chi(G' + uv;\lambda) + \chi(G' \cdot uv;\lambda)$ for non-adjacent vertices u and v in G';

(CS4) $\chi(G_1;\lambda) + \chi(G_2;\lambda)$ becomes $\chi(G';\lambda)$, where G' is isomorphic to $G_1\backslash e$ and G_2 is isomorphic to G_1/e for some $e \in E(G_1)$;

(CS5) $\chi(G_1;\lambda) - \chi(G_2;\lambda)$ becomes $\chi(G';\lambda)$, where G' is isomorphic to G_2/e and G_1 is isomorphic to $G_2\backslash e$ for some $e \in E(G_2)$;

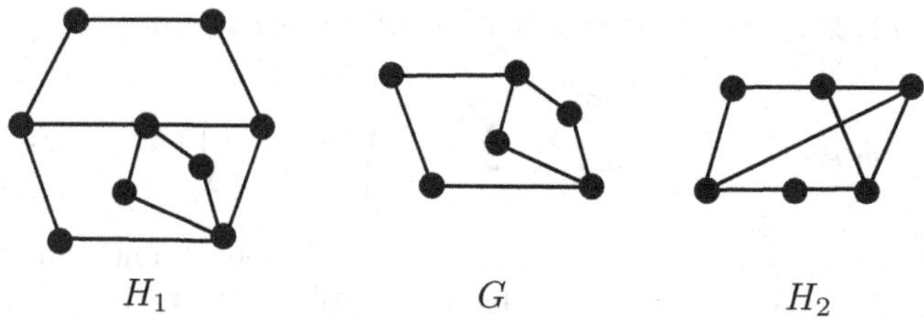

FIGURE 11.6: The graphs G, H_1 and H_2 for Example 11.44.

(CS6) $\chi(G';\lambda)$ becomes $\chi(G_1;\lambda)\chi(G_2;\lambda)/\chi(K_r;\lambda)$, where G' is isomorphic to the graph obtained by a K_r-gluing of G_1 and G_2;

(CS7) $\chi(G_1;\lambda)\chi(G_2;\lambda)/\chi(K_r;\lambda)$ becomes $\chi(G';\lambda)$, where G' is isomorphic to the graph obtained by a K_r-gluing of G_1 and G_2;

(CS8) By applying the field axioms, for the field $\mathbb{Q}(p_0, p_1, \cdots)$ of rational functions in indeterminates $p_0, p_1, \cdots$, a finite number of times, so as to produce a different expression for the same field element;

(CS9) $\chi(G';\lambda)$ becomes $\chi(G'';\lambda)$ where G' and G'' have the same chromatic polynomial.

Example 11.44. Let G, H_1 and H_2 be graphs shown in Figure 11.6. Although G is non-clique-separable we have

$$\chi(G;\lambda) = \frac{\chi(H_1;\lambda)\chi(H_2;\lambda)}{\chi(K_3;\lambda)}.$$

This follows by applying certificates for factorizations of chromatic polynomials in the order CS1, CS6 (twice), CS8, CS7 (twice), and CS2, as shown in Figure 11.7, where we represent $\chi(G;\lambda)$ of a graph G by a drawing of G.

Morgan and Farr [862] showed that there exist infinitely many non-clique-separable graphs whose chromatic polynomials have factorizations in the form of Equation (11.7) as follows.

Proposition 11.45. *For any integer $n \geq 2$, let G be the graph of order $4n$ and H be the graph of order $2n+2$ shown in Figure 11.8. Then G is non-clique-separable but*

$$\chi(G;\lambda) = \frac{\chi(H;\lambda)\,\chi(C_{2n+1};\lambda)}{\chi(K_3;\lambda)}.$$

11.3.6 Connectivities and chromatic polynomials

Theorem 11.26 and Example 11.4 imply that the number of components of G can be determined by the multiplicity of its chromatic root 0. Woodall [1170]

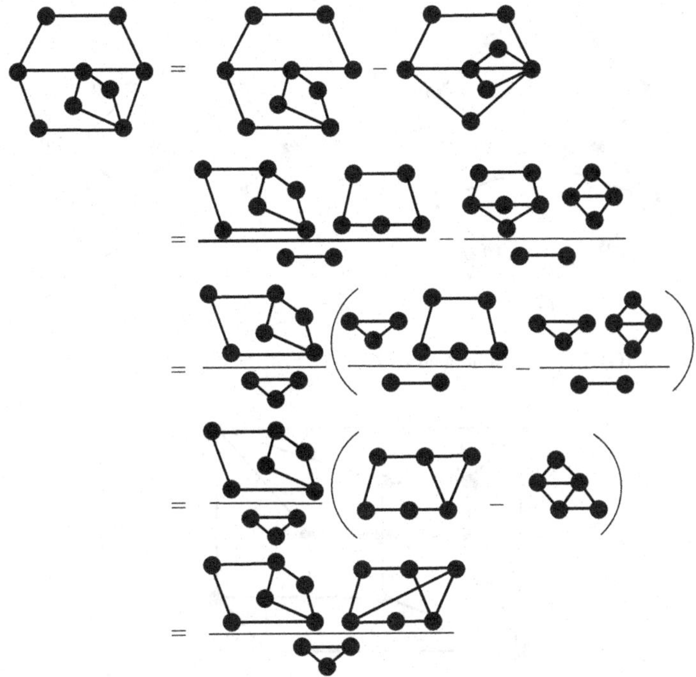

FIGURE 11.7: Example of applying certificates of factorizations.

and Whitehead and Zhao [1150] independently showed that the number of blocks of G can be determined by the multiplicity of its chromatic root 1 (See Theorem 11.46, Item 3). For some cases, $\chi(G; \lambda)$ can also tell if G is 3-connected.

Theorem 11.46. *For any graph G of order n,*

1. *the number of components of G is equal to the multiplicity of the root* 0 *of* $\chi(G; \lambda)$;
2. *the coefficient of λ^i in $\chi(G; \lambda)$ is non-zero if and only if $k(G) \leq i \leq n$;*
3. *if G is connected with $|V(G)| \geq 2$, the multiplicity of the chromatic root* 1 *of G is equal to the number of blocks in G.*

By Theorem 11.46, if G is connected, then λ^2 is not a factor of $\chi(G; \lambda)$; if G is 2-connected, then $(\lambda - 1)^2$ is not a factor of $\chi(G; \lambda)$. However, it is not true that if G is 3-connected, then $(\lambda - 2)^2$ is not a factor of $\chi(G; \lambda)$. The graph G in Figure 11.9, found by Jackson (unpublished see [402]), is 3-connected but $(\lambda - 2)^2$ is indeed a factor of $\chi(G; \lambda) = \lambda(\lambda - 1)(\lambda - 2)^2 f(\lambda)$, where $f(\lambda)$ is a polynomial of degree nine. An infinite family of such 3-connected graphs was found by Dong and Koh [402]. For any $r \geq 2$, both G'_r and G''_r in Figure 11.10 are 3-connected, but $(\lambda - 2)^2$ is a factor of both $\chi(G'_r; \lambda)$ and $\chi(G''_r; \lambda)$ when r

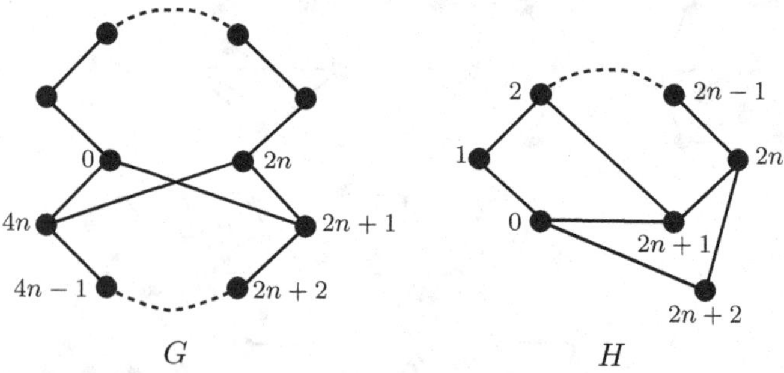

FIGURE 11.8: The graphs G and H of Proposition 11.45.

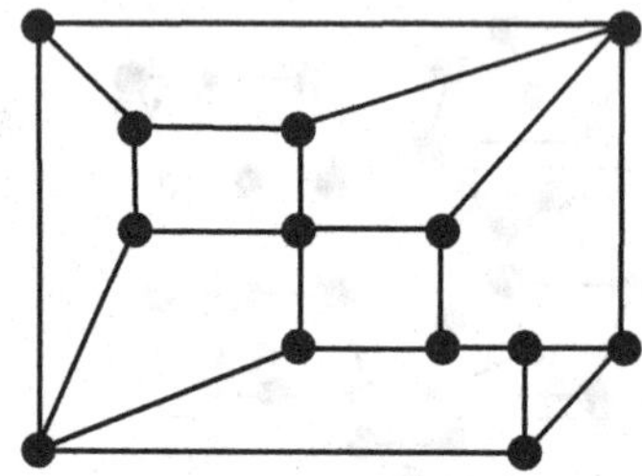

FIGURE 11.9: A 3-connected graph G for which $(\lambda-2)^2|\chi(G;\lambda)$.

is odd. They also showed that $(\lambda-2)^2$ is not a factor $\chi(G;\lambda)$ for 3-connected graphs G when $\Delta(G)+\delta(G)$ is large enough.

Theorem 11.47. *For any 3-connected graph G of order n with non-decreasing degree sequence $(d_1, d_2, \cdots, d_n)$, if $d_1 + d_n \geq n$ and $(d_3, d_n) \neq (3, n-3)$, then $(\lambda-2)^2$ is not a factor of $\chi(G;\lambda)$.*

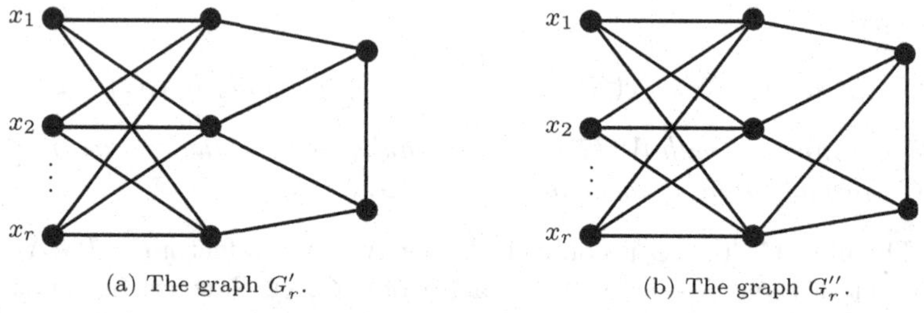

FIGURE 11.10: Two families of 3-connected graphs.

11.3.7 Inequalities on chromatic polynomials

This subsection is concerned with inequalities of chromatic polynomials related to the problem of, given a set $\mathcal{S}$ of graphs and a positive integer λ, determining the maximum value of $\chi(G;\lambda)$ over all graphs G in $\mathcal{S}$.

Proposition 11.48. *Let G be a graph of order n and λ be a nonnegative integer. Then*

1. *$\chi(G;\lambda) \le \lambda^n$, where equality holds for $\lambda \ge 1$ if and only if $G \cong \overline{K}_n$.*
2. *When G is connected, $\chi(G;\lambda) \le \lambda(\lambda-1)^{n-1}$, where equality holds for $\lambda \ge 3$ if and only if G is a tree of order n.*

When $\mathcal{S}$ is the set of graphs of order n and chromatic number k, the problem has been studied by Tomescu [1069, 1070] and Brown and Erey [223].

Theorem 11.49. *Let G be a graph of order n and chromatic number k. Then the following hold.*

1. *For every integer $\lambda \ge 0$, the inequality $\chi(G;\lambda) \le (\lambda)_k\,\lambda^{n-k}$ holds. Moreover, for every integer $\lambda \ge k$, it holds if and only if G has exactly one component isomorphic to K_k and $n-k$ components isomorphic to K_1.*
2. *For every integer $\lambda \ge 0$,*
$$\chi(G;\lambda) \le (\lambda)_k(\lambda-1)^{\Delta(G)-(k-1)}\lambda^{n-1-\Delta(G)}.$$
3. *If $k=3$, then $\chi(G;\lambda) \le (\lambda-1)^n-(\lambda-1)^{2-p(n)}$ holds for every integer $\lambda \ge 3$, where $p(n)=0$ when n is even and $p(n)=1$ otherwise. Furthermore, for every integer $\lambda \ge 3$, the above is an equality if and only if G is isomorphic to C_n for odd n and to the one-point join $C_{n-1} * K_2$ for even n.*
4. *If $k=4$ and G is planar, then $\chi(G;\lambda) \le (\lambda)_4(\lambda-1)^{n-4}$ holds for very integer $\lambda \ge 4$. Moreover, for every integer $\lambda \ge 4$, the above is an equality if and only if G has exactly one block isomorphic to K_4 and $n-4$ blocks isomorphic to K_2.*

The first result in this theorem is from [1069], the second from [223], and the final two from [1070]. In this last reference, Tomescu conjectured that $(\lambda)_k(\lambda-1)^{n-k}$ is the maximum chromatic polynomial over all connected graphs of order n and chromatic number $k \ge 4$. While the full conjecture remains open, there are the following partial results, the first due to Brown and Erey [223], and the second to Tomescu [1071].

Theorem 11.50. *Let G be a connected graph of order n and chromatic number k, where k is an integer with $k \ge 4$. Then*
$$\chi(G;\lambda) \le (\lambda)_k(\lambda-1)^{n-k}$$
holds for all real numbers $\lambda > n-2+\left(\binom{n}{2}-\binom{k}{2}-n+k\right)^2$.

Tomescu [1071] determined the maximum chromatic polynomial for the family of 2-connected graphs. Let D_n ($n \geq 5$) be the graph consisting of a 4-cycle in which two nonadjacent vertices are connected by a newly added path of length $n-3$. Thus D_5 is isomorphic to $K_{2,3}$. If "nonadjacent" is replaced by "adjacent", the resulting graph is denoted F_n. Thus F_n is a K_2-gluing of C_4 and C_{n-2}.

Theorem 11.51. *Let G be a 2-connected graph of order n, where $n \geq 3$. Then the following hold.*

1. *For any integer $\lambda \geq 3$, the inequality $\chi(G;\lambda) \leq (\lambda-1)^n + (-1)^n(\lambda-1)$, holds, with equality if and only if $G \cong C_n$ (or $G \cong K_{2,3}$ for the case that $n=5$ and $\lambda = 3$).*
2. *If $G \ncong C_n$ and $\lambda \geq 4$, then $\chi(G;\lambda) \leq \max\{\chi(D_n;\lambda), \chi(F_n;\lambda)\}$, with equality if and only if $G \cong D_n$ for odd n and $G \cong F_n$ for even n.*

Now we introduce some other inequalities for chromatic polynomials. Bartels and Welsh [82] showed that $\chi(G;\lambda)$ is convex on the positive integers greater than the chromatic number of G:

Theorem 11.52. *For any G and any integer k greater than the chromatic number of G, $\chi(G;k+1) + \chi(G;k-1) \geq 2\chi(G;k)$.*

Brenti [198] and Bartels and Welsh [82] conjectured that the inequality $\chi(G;k+1)\chi(G;k-1) \leq \chi(G;k)^2$ holds for all integers $k \geq 1$. This conjecture was disproved by Seymour [998] who showed that it does not hold for $k=6$ and some graphs G with chromatic number $\chi(G) = 3$. However, in [404] Dong, Koh and Teo conjectured that if G has order n, then the inequality holds for all $\lambda \geq n$, and made the following progress towards the conjecture.

Theorem 11.53. *Let G be a connected graph on n vertices and m edges. Then $\chi(G;\lambda+1)\chi(G;\lambda-1) \leq \chi(G;\lambda)^2$ holds for all real $\lambda \geq \max\{n-1, \sqrt{2}(m-n+2.5)\}$.*

Bartels and Welsh [82] proposed an inequality known as *the shameful conjecture*, that if G has order n, then the ratio of $\chi(G;n)$ to $\chi(G;n-1)$ exceeds $\frac{n^n}{(n-1)^n}$ (which approaches the number e as n grows). Seymour [998] showed that $\chi(G;n)/\chi(G;n-1) \geq 685/252$ ($= 2.71825\cdots$), and the conjecture was proved by Dong [394]:

Theorem 11.54. *For any graph G of order n and any real number $\lambda \geq n$, $(\lambda-1)^n\chi(G;\lambda) \geq \lambda^n\chi(G;\lambda-1)$.*

11.4 Chromatically equivalent graphs

In 1968, Read [945] asked what a necessary and sufficient condition is for two graphs G and H to have the same chromatic polynomial. This question actually asks, given a graph G, what is the set of graphs H with $\chi(H;\lambda) = \chi(G;\lambda)$?

If $\chi(H;\lambda) = \chi(G;\lambda)$, then G and H are called *chromatically equivalent*, or, simply, χ-*equivalent*, written as $G \sim H$. Given a graph G, the set $\mathcal{G}[G] := \{H \in \mathcal{G} : H \sim G\}$ is called the *chromatic equivalence class* or χ-*equivalence class* determined by G. If $\mathcal{G}[G] = \{G\}$, G is said to be *chromatically unique* or, simply, χ-*unique* (see [288]).

11.4.1 Invariants for chromatically equivalent graphs

In this subsection, we introduce some χ-invariants for graphs, where a χ-*invariant* is a graph-function f such that $f(G) = f(H)$ holds whenever $G \sim H$. Such invariants are very important and useful in the study of *chromaticity* (i.e., determining the set $\mathcal{G}[G]$ for a graph G or determining the set $\cup_{G\in\mathcal{S}}\mathcal{G}[G]$ for a set $\mathcal{S}$ of graphs). The following χ-invariants follow immediately from the definitions, Corollary 11.27, Theorems 11.28 and 11.46 Item 3, and a result due to Liu [779] (the final item in the following proposition). For the proposition below, $\alpha(G;k)$ denotes the number of ways of partitioning $V(G)$ into k non-empty independent sets, and $s_G(H)$ (respectively, $i_G(H)$) is number of subgraphs (respectively, induced subgraphs) in G which are isomorphic to H.

Proposition 11.55. *Let G and H be any graphs with $G \sim H$. Then*

1. $|V(G)| = |V(H)|$;
2. $|E(G)| = |E(H)|$;
3. $\chi(G) = \chi(H)$;
4. $s_G(K_3) = s_H(K_3)$;
5. $i_G(C_4) - 2s_G(K_4) = i_H(C_4) - 2s_H(K_4)$;
6. $k(G) = k(H)$;
7. $b(G) = b(H)$, *where $b(G)$ is the number of non-trivial blocks in G;*
8. *G is connected if and only if H is connected;*
9. *G is 2-connected if and only if H is 2-connected;*
10. $g(G) = g(H)$, *where $g(G)$ is the girth of G;*
11. $s_G(C_k) = s_H(C_k)$, *where $g \le k \le \lceil 3g/2 \rceil - 2$ and g is the girth of G;*
12. $\chi(G;k) = \chi(H;k)$ *for all integers k with $\chi(G) \le k \le |V(G)|$;*
13. $\alpha(G,k) = \alpha(H,k)$ *for all integers k with $\chi(G) \le k \le |V(G)|$;*

14. $s_{\bar{G}}(K_3) - \sum_{u\in V(G)} \deg_{\bar{G}}(u)^2 = s_{\bar{H}}(K_3) - \sum_{v\in V(H)} \deg_{\bar{H}}(v)^2$, *where* $\overline{G}$ *is the complement of* G.

Items 1 and 2 in Proposition 11.55 directly imply that both K_n and $\overline{K}_n$ are χ-unique. Items 1, 2 and 9 yield that C_n is χ-unique for $n \geq 3$. Items 1, 2, 7 and 8 imply that if T is a tree of order $n \geq 1$, then $\mathcal{G}[T]$ is the set of trees of order n. Thus any tree of order at least 4 is not χ-unique.

We end this subsection by introducing a χ-invariant found by Lundow and Markström [798], although it has been seldom applied in the study of chromaticity. For a graph $G = (V, E)$ of order n, assume that

$$\chi(G;\lambda) = \sum_{i=1}^{n} (-1)^{n-i} h_i \lambda^i.$$

Let β be an ordering of the edges, i.e. a bijection from E to $\{1, 2, \cdots, |E|\}$. A subgraph H of G is said to be *broken-cycle-free* if it does not contain any broken-cycle of G with respect to β. Let $\epsilon(G)$ denote the mean size of broken-cycle-free spanning subgraphs of G. Lundow and Markström [798] showed that

$$\epsilon(G) = \frac{h_{n-1} + 2h_{n-2} + \cdots + (n-1)h_1}{h_1 + h_2 + \cdots + h_n} = n + \frac{\chi'(G;-1)}{\chi(G;-1)}. \tag{11.8}$$

Obviously $\epsilon(G)$ is a χ-invariant of graphs. It is not difficult to verify that $\epsilon(T) < \epsilon(G)$ holds for any tree T of order n and any connected graph G of order n which is not a tree.

11.4.2 The χ-unique graphs which are not 2-connected

The χ-unique graphs which are not 2-connected have been completely determined by Chia [298], Giudici [540], Read [947] and Xu [1175]. A graph G is said to be *vertex-transitive* if for any two vertices u and v in G, there is an automorphism φ of G such that $\varphi(u) = v$.

Theorem 11.56. *Let* G *be a graph which is not 2-connected.*

1. *If* G *is disconnected, then* G *is* χ*-unique if and only if* $G \cong H \sqcup \overline{K}_n$ *for some positive integer* n*, where* H *is a connected* χ*-unique graph without cut-vertices (i.e.,* H *is 2-connected or* $H \cong K_i$*,* $i = 1, 2$*).*

2. *If* G *is connected, then* G *is* χ*-unique if and only if* $G = H * K_2$*, where* H *is a vertex-transitive and* χ*-unique graph.*

The first item of this theorem is from [540, 947, 1175], and the second from [298, 947, 1175].

11.4.3 The χ-unique bipartite graphs

C.Y. Chao in 1978 (unpublished) conjectured that the complete bipartite graph $K_{p,p+k}$ is χ-unique if $p \geq 2$ and $0 \leq k \leq 2$. Chao's conjecture was confirmed by Salzberg, López and Giudici [982] who further conjectured that $K_{p,q}$ is χ-unique whenever $p \geq q \geq 2$. This conjecture was verified by Teo and Koh [1057]. A couple more alternative proofs also appeared in Dong [393] and Teo and Koh [1058].

Theorem 11.57. *The complete bipartite graph $K_{p,q}$ is χ-unique for all integers p, q with $q \geq p \geq 2$.*

Dong, Koh, Teo, Little and Hendy [405] further showed that some bipartite graphs obtained from $K_{p,q}$ by deleting some edges are also χ-unique.

11.4.4 Chromaticity of complete multipartite graphs

Some complete tripartite graphs were confirmed to be χ-unique by Chia, Goh and Koh [303], Liu, Zhao and Ye [778], Chia and Ho [304], Zou [1189] and Zou and Shi [1190].

Theorem 11.58. *For integers p_1, p_2, p_3 with $2 \leq p_1 \leq p_2 \leq p_3$, K_{p_1,p_2,p_3} is χ-unique for the following cases:*

1. *$p_1 = p_2 = p_3 - k$, where $k \geq 2$ and $p_1 > \frac{k+k^2}{3}$ when $k \geq 4$;*
2. *$p_1 = q - k$, $p_2 = q$ and $p_3 = q + k$, where either $q \geq 5$ when $k = 2$ or $q > \frac{2\sqrt{3}}{3}k + k^2$ when $k \geq 3$;*
3. *$p_2 = p_3$;*
4. *$p_1 + 2 \leq p_3 \leq 2p_1$ and $p_2 = p_3 - 1$;*
5. *$p_1 \geq 3$ and $p_3 \leq p_1 + 2$.*

The first two items in Theorem 11.58 are from [303, 1189, 1190], the third from [303, 304, 778], the fourth from [778], and fifth from [1189, 1190].

Note that the graph $K_{1,n,n+k}$ is isomorphic to both $\bar{K}_n + K_{1,n+k}$ and $\bar{K}_{n+k} + K_{1,n}$. As both $K_{1,n+k}$ and $K_{1,n}$ are trees, by Theorem 11.39, we have

$$\mathcal{G}[K_{1,n,n+k}] \supseteq \{\bar{K}_n + T : T \in \mathcal{T}_{n+k+1}\} \cup \{\bar{K}_{n+k} + T' : T' \in \mathcal{T}_{n+1}\},$$

where $\mathcal{T}_n$ is the set of trees on n vertices. This has ben shown to be an equality for $k = 0$ by Chia and Ho [304] and for $1 \leq k \leq 2$ by Ng and Dong [879]:

Theorem 11.59. *Let n be any positive integer.*

1. *$\mathcal{G}[K_{1,n,n}] = \{\bar{K}_n + T : T \in \mathcal{T}_{n+1}\}$;*
2. *For $k = 1, 2$ and $n \geq 2$,*

$$\mathcal{G}[K_{1,n,n+k}] = \{\bar{K}_n + T : T \in \mathcal{T}_{n+k+1}\} \cup \{\bar{K}_{n+k} + T' : T' \in \mathcal{T}_{n+1}\}.$$

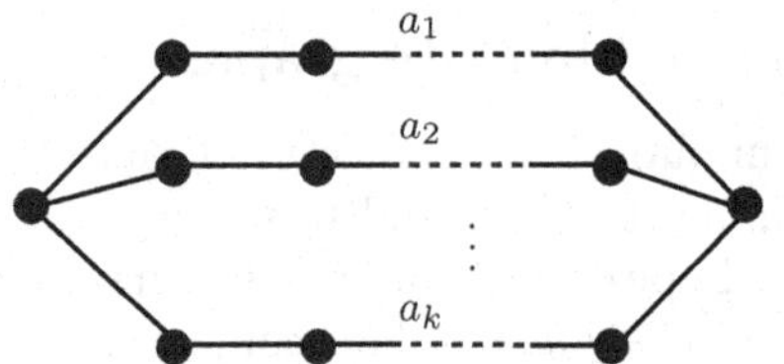

FIGURE 11.11: A generalized theta-graph $\theta(a_1, a_2, \cdots, a_k)$.

Some χ-unique complete t-partite graphs have been confirmed by Chao and Novacky Jr. [287], Chia [302], Li and Liu [765] (the first item of the following theorem); Giudici and López [541] (the second item of the following theorem); and Zhao, Li, Liu and Ye [1187] (the third and fourth items of the following theorem).

Theorem 11.60. *For integers $t \geq 2$ and $1 \leq p_1 \leq p_2 \leq \cdots \leq p_t$, the multipartite graph $K_{p_1,p_2,\cdots,p_t}$ is χ-unique for the following cases:*

1. $p_t \leq p_1 + 1$;

2. $p_2 = p_{t-1} = p_1 + 1 \geq 3$ *and* $p_t = p_1 + 2$;

3. $p_1 \geq e(K[t; |V(G)|]) - e(G) + 1$, *where* $K[t; n]$ *is the complete* t*-partite graph of order* n *with the maximum number of edges and* G *is the graph* $K_{p_1,p_2,\cdots,p_t}$;

4. $p_t \leq p_1 + 2\sqrt{(p_1 - 1)/t}$.

11.4.5 The χ-unique generalized theta-graphs

The *theta-graph* is the multigraph consisting two vertices joined by three parallel edges, and the *generalized theta-graph* $\theta(a_1, a_2, \cdots, a_k)$ is the graph consisting two vertices joined by k internally disjoint paths of lengths $a_1, a_2, \cdots, a_k$ respectively, as shown in Figure 11.11. Such a graph is also called a *k-theta graph* or a *k-bridge graph.*

Chao and Whitehead [288, 289] initiated the study of chromaticity of $\theta(a_1, a_2, \cdots, a_k)$ and showed that $\theta(1, a_2, a_3)$ is χ-unique. For some cases, $\theta(a_1, a_2, \cdots, a_k)$ has been confirmed to be χ-unique.

Theorem 11.61. *For integers $k \geq 3$ and $2 \leq a_1 \leq a_2 \leq \cdots \leq a_k$, $\theta(a_1, a_2, \cdots, a_k)$ is χ-unique for the following cases:*

1. $k = 3$;

2. $k = 4$ *and* $(a_1, a_2, a_3, a_4) \neq (2, c, c + 1, c + 2)$ *for all* $c \geq 2$;

3. $k = 5$, $a_3 = 2$ *and* $(a_4, a_5) \neq (3, 4)$;

4. $k \geq 4$ *and* $a_1 \geq \min\{k - 1, a_k + 1 - a_2\}$;

5. $k \geq 4$, $a_2 - 1 = a_1 = h$, $a_j = h + s$ *for* $j = 3, \cdots, k-1$, $a_k \geq h+s$ *and* $a_k \notin \{2h, 2h+s, 2h+s-1\}$, *where* h, s *are positive integers with* $h \geq s+1 \geq 2$ *or* $h = s-1$.

The first item of Theorem 11.61 is from [785], the second from [73], the third from [769], the fourth from [409, 1178], and the last from [914].

11.4.6 Chromaticity of q-trees

For any positive integer q, the set of *q-trees* is defined recursively as follows: every q-tree has an order at least q, K_q is the only q-tree of order q, and any q-tree of order $n+1$ is a graph G with a simplicial vertex x of degree q such that $G \backslash x$ is a q-tree of order n, where $n \geq q$. Thus every tree is a 1-tree.

Three χ-equivalence classes consisting of q-trees or some graphs obtained from q-trees are determined.

Theorem 11.62. *Let* q *and* n *be integers with* $n \geq q \geq 1$ *and* G *be a graph of order* n. *Then*

1. G *is a* q*-tree if and only if*

$$\chi(G;\lambda) = \lambda(\lambda-1)\cdots(\lambda-q+1)(\lambda-q)^{n-q};$$

2. *for* $q \geq 2$, G *is a* K_{q-1}*-gluing of two* q*-trees if and only if*

$$\chi(G;\lambda) = \lambda(\lambda-1)\cdots(\lambda-q+1)^2(\lambda-q)^{n-q-1};$$

3. *for* $n \geq q+1 \geq i+3 \geq 4$, G *contains a simplicial vertex* v *with* $\deg(v) = i$ *such that* $G - v$ *is a* q*-tree if and only if*

$$\chi(G;\lambda) = \lambda(\lambda-1)\cdots(\lambda-q+1)(\lambda-q)^{n-q-1}(\lambda-i).$$

The first item of Theorem 11.62 is from [286, 386, 599, 1109], the second from [1129], and the third from [180, 599].

11.4.7 Chromaticity of wheels

By Corollary 11.41 and Example 11.12, we have

$$\chi(W_n;\lambda) = \lambda\left((\lambda-2)^n + (-1)^n(\lambda-2)\right),$$

where $W_n = C_n + K_1$ is the wheel of order $n+1$. Whether W_n is χ-unique has been confirmed for all even number $n \geq 4$ and all odd $n \leq 9$. But it is still unknown if W_n is χ-unique for all odd $n \geq 11$. In the following, the first item is from [391, 1177], the second from [289, 1177], and the third from [948].

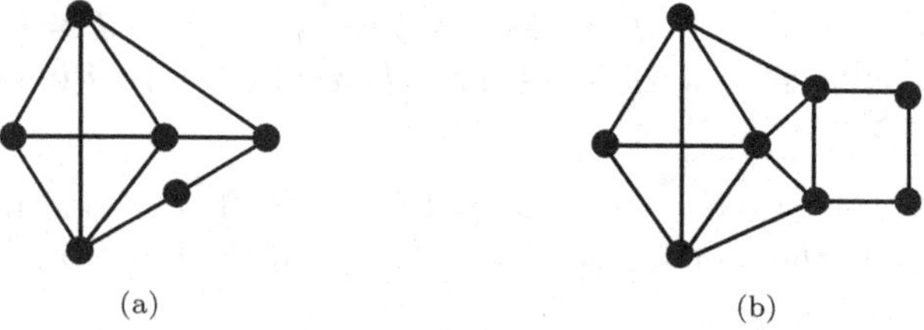

FIGURE 11.12: Graphs χ-equivalent to W_5 and W_7 respectively.

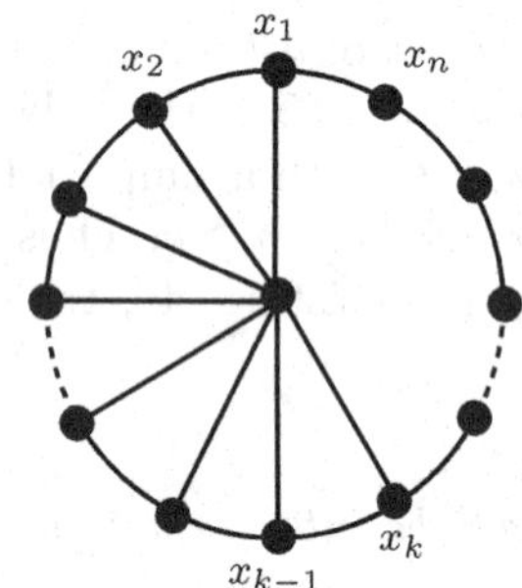

FIGURE 11.13: A broken wheel $W(n, k)$.

Theorem 11.63.

1. *The graph $C_n + K_r$ is χ-unique for all $r \geq 1$ and even $n \geq 4$;*
2. *W_5 and W_7 are not χ-unique, as they are χ-equivalent to the graphs in Figure 11.12(a) and 11.12(b) respectively;*
3. *W_9 is χ-unique.*

For $0 \leq k \leq n-1$, let $W(n, k)$ denote the graph (called a *broken wheel*) obtained from W_n by deleting all but k consecutive spokes, as shown in Figure 11.13. Whether $W(n, k)$ is χ-unique has been confirmed for $k \in \{3, 4, n-1, n-2\}$.

Proposition 11.64.

1. *$W(n, k)$ is χ-unique for $n \geq k+1$ and $k \in \{3, 4\}$;*
2. *$W(6, 5)$ is not χ-unique, as it is χ-equivalent to the graph in Figure 11.14;*
3. *$W(n, n-1)$ is χ-unique for all $n \geq 3$, except for $n = 6$;*
4. *$W(n, n-2)$ is χ-unique for all $n \geq 5$.*

The first two items of the above proposition are from [289], the third from [289, 299, 300, 396, 407], and the fourth from [406].

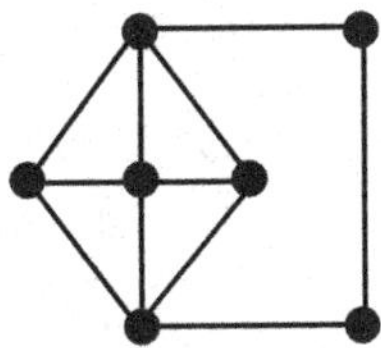

FIGURE 11.14: A graph χ-equivalent to $W(6,5)$.

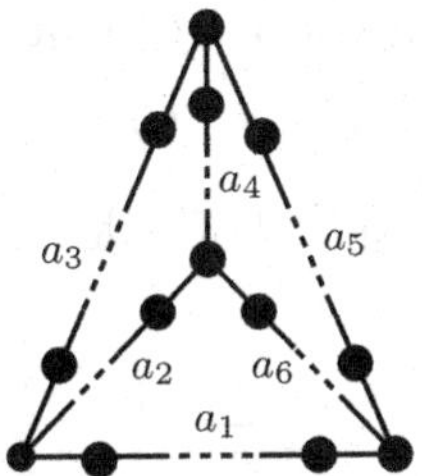

FIGURE 11.15: A K_4-homeomorph $K_4(a_1, a_2, \cdots, a_6)$.

11.4.8 Chromaticity of K_4-homeomorphs

A K_4-*homeomorph* is a graph obtained from K_4 by subdividing its edges (i.e., adding vertices of degree two along its edges). We denote by $K_4(a_1, a_2, \cdots, a_6)$ the K_4-homeomorph obtained through a subdivision that replaces the six edges of K_4 by six paths of length $a_1, a_2, \cdots, a_6$, as shown in Figure 11.15.

An important result for the study of chromaticity of K_4-homeomorphs was established by Chao and Zhao [290] who showed that any graph χ-equivalent to a K_4-homeomorph is also a K_4-homeomorph. Li [766] showed that almost all K_4-homeomorph are χ-unique. These results are stated formally in the following theorem.

Theorem 11.65.

1. *Any graph χ-equivalent to a K_4-homeomorph must itself be a K_4-homeomorph.*

2. $\lim_{n\to\infty} g(n) = 1$, *where $g(n)$ is the proportion of χ-unique K_4-homeomorphs of order n among all K_4-homeomorphs of order n.*

Li [766], Whitehead and Zhao [1149], Li [767], Ren and Zhang [960], Zhang [1186] and Guo and Whitehead [592] have confirmed some χ-unique K_4-homeomorphs.

Theorem 11.66. *The graph $K_4(a_1, a_2, \cdots, a_6)$ is χ-unique in the following cases:*

1. *the set $\{a_1, a_2, \cdots, a_6\}$ has at most two distinct numbers;*

2. $\min\{a_1, a_2, \cdots, a_6\} = s$ *and* $|\{1 \le i \le 6 : a_i = s\}| \ge \max\{3, 5-s\}$*;*

3. *if it contains a triangle;*

4. *three of* $a_1, a_2, \cdots, a_6$ *are* 1, *except for the two cases* $(a_1, a_2, \cdots, a_6) = (1, 1, 1, s+1, s, 3)$ *or* $(a_1, a_2, \cdots, a_6) = (1, 1, s, 2, 1, s+2)$ *for some* $s \geq 2$.

In the above theorem, the first item is from [1149], the second from [766, 960, 1186], the third from [766], and the fourth from [592].

11.4.9 Chromaticity studied by adjoint polynomials

The adjoint polynomial was introduced by Liu in [777].

Definition 11.67. For any graph G of order n, the *adjoint polynomial* of G is

$$h(G; x) = \sum_{k=1}^{n} \alpha(\overline{G}, k)\, x^k,$$

where $\alpha(G, k)$ is the number of ways of partitioning $V(G)$ into k non-empty independent sets of G.

If $h(G; x) = h(H; x)$, then G is said to be *adjointly equivalent* to H, written as $G \sim_h H$. A graph G is said to be *adjointly unique* if $H \cong G$ whenever $H \sim_h G$. Evidently, we have the following result.

Lemma 11.68. *Let* G *and* H *be graphs. Then* $G \sim H$ *if and only if* $\overline{G} \sim_h \overline{H}$. *In particular,* G *is* χ*-unique if and only if* $\overline{G}$ *is adjointly unique.*

By Lemma 11.68, the χ-equivalent class of a graph G is the set of graphs whose complements are adjointly equivalent to $\overline{G}$.

Theorem 11.69.

1. *The graphs* $P_4, C_4, P_{2n+1}, P_n \sqcup C_{n+1}$ *and* $K_1 \sqcup C_n$ *are not adjointly unique, where* $n \geq 3$.

2. P_5 *is adjointly unique but* $\overline{K}_r \sqcup P_5$ *is not for any* $r \geq 1$ *as it is adjointly equivalent to* $\overline{K}_{r-1} \sqcup K_{1,3} \sqcup P_2$.

3. *For* $r \geq 0$, $\overline{K}_r \bigsqcup_{1 \leq i \leq k} P_{n_i}$ *is adjointly unique if* n_i (≥ 6) *is even for all* $1 \leq i \leq k$.

4. $\bigsqcup_{1 \leq i \leq k} C_{n_i}$ *is adjointly unique if* $n_i = 3$ *or* $n_i \geq 5$ *for all* $1 \leq i \leq k$.

5. *For* $s \geq 1$, $t \geq 1$, $\bigsqcup_{1 \leq i \leq s} P_{u_i} \bigsqcup_{1 \leq j \leq t} C_{v_j}$ *is adjointly unique if* u_i (≥ 6) *is even,* $u_i \not\equiv 4 \mod 5$, $u_i + 1 \neq v_j$ *and* $v_j \geq 5$ *for all* $i = 1, 2, \cdots, s$ *and* $j = 1, 2, \cdots, t$.

The first and third items of this theorem are from [410], the second [1174], the fourth [420], and the fifth from [410, 1188] .

11.5 Roots of chromatic polynomials

For a graph G, a root of $\chi(G;\lambda)$ is called a *chromatic root* of G. In this section, we consider chromatic roots. See also Chapter 25 for roots of the Tutte polynomial, and Chapter 20 for the role of the roots of chromatic polynomials in statistical mechanics.

11.5.1 Complex roots of chromatic polynomials

For any integer $k \geq 1$, let Ω_k be the set of roots of monic polynomials of degree k whose coefficients are integers. Two conjectures related to Ω_k, called the $\alpha + n$ conjecture and the αn conjecture, were proposed during the program on "Combinatorics and Statistical Mechanics" at the Newton Institute of Cambridge in 2008, [150, 268].

Conjecture 11.70 (The $\alpha + n$ Conjecture)**.** For every $\alpha \in \Omega_k$, where $k \geq 1$, there exists an integer n such that $\alpha + n$ is a chromatic root.

Conjecture 11.71 (The αn Conjecture)**.** Let α be a chromatic root. Then $n\alpha$ is also a chromatic root for every positive integer n.

If the $\alpha + n$ conjecture is true, we can ask, for a given α, what is the smallest n for which $\alpha+n$ is a chromatic root? The $\alpha+n$ conjecture is trivial for $k = 1$, and has been proved for $k = 2$ and $k = 3$ by Bohn [150]. He also showed that the set of chromatic roots satisfying the αn conjecture is dense in the complex plane.

So far the most notable breakthrough in the study of complex roots of chromatic polynomials is due to Sokal [1021]. In 2001, he showed that there is a constant C with $C < 8$ such that $|z-1| \leq C\Delta(G)$ holds for all chromatic roots z of an arbitrary graph G. This result, stated as Theorem 11.72 below, proved a conjecture proposed by Biggs, Damerell and Sands [114] in 1972 and a conjecture proposed by Brenti, Royle and Wagner [199] in 1994. Also see Chapter 25.

Theorem 11.72. *Let C be the smallest number for which*

$$\inf_{\alpha>0} \alpha^{-1} \sum_{n=2}^{\infty} \exp(\alpha n) C^{-(n-1)} \frac{n^{n-1}}{n!} \leq 1. \tag{11.9}$$

If G is a graph and $\chi(G;z) = 0$, then $|z| < C\Delta(G)$.

It can be shown that in Theorem 11.72, $C \leq 7.963907$. This theorem was improved by Fernandez and Procacci [486] who showed that if $\chi(G;z) = 0$, then $|z| < K\Delta(G)$, where $K \leq 6.907\cdots$.

Sokal [1021] further showed that the "maximum degree" in Theorem 11.72 can be replaced by the "second-largest degree":

Theorem 11.73. *Let C be the smallest number satisfying Inequality (11.9). If G is a graph in which all but one of the vertices have degrees not exceeding k, and $\chi(G;z) = 0$, then either $|z| < Ck$ or $|z-1| < Ck$ (and in particular $|z| < Ck+1$).*

In 2004, Sokal [1022] proved that chromatic roots are dense in the entire complex plane:

Theorem 11.74. *Let z_0 be any complex number with $|z_0-1| \geq 1$. For any real $\epsilon > 0$, there exist constants $f_0(z_0)$ and $k_0(f_0, z_0)$ such that for all integers $f > f_0(z_0)$ and $k > k_0(f_0, z_0)$, it follows that $\chi(\theta_k(f);\lambda)$ has a root in $|z - z_0| < \epsilon$, where $\theta_k(f)$ denotes the k-theta graph obtained from two vertices by joining them with k internally disjoint paths of length f.*

Brown [225] and Brown and Hickman [236] showed that sufficiently subdividing the edges of a graph ensures that its zeros will lie in a disk with radius arbitrarily close to 1:

Theorem 11.75. *For any graph G and any $\epsilon > 0$, there is a constant $L = L(G,\epsilon)$ such that, if we divide each edge of G into a path of length at least L, then all chromatic roots of the resulting graph G' lie in the disk $|z-1| < 1+\epsilon$.*

11.5.2 Real roots of chromatic polynomials

As all real chromatic roots of any graph are nonnegative (by Theorem 11.76 Item 1), Theorem 11.72 implies that all real chromatic roots of a graph G are within $[0, \Delta(G)C)$, where $C = 7.963907$. Due to Dong and Koh [399], this result holds when $C = 5.664$. Sokal conjectured an improvement of this result (see Conjecture 11.99).

An interval is called a *zero-free interval* for chromatic roots of graphs in a set $\mathcal{S}$ if no graph in $\mathcal{S}$ has a chromatic root in this interval. It is known that when $\mathcal{S}$ is the set of all graphs, $(-\infty, 0)$, $(0,1)$ and $(1, 32/27]$ are the only maximal zero-free intervals for chromatic roots of graphs in $\mathcal{S}$.

Theorem 11.76. *Let G be a graph of order n. Then*

1. *$(-1)^n\chi(G;\lambda) > 0$ for all real $\lambda < 0$;*
2. *$(-1)^{n-k(G)}\chi(G;\lambda) > 0$ for all real $\lambda \in (0,1)$, where $k(G)$ is the number of components of G;*
3. *if G is connected with b blocks, then $(-1)^{n+b+1}\chi(G;\lambda) > 0$ for all real $\lambda \in (1, 32/27]$;*
4. *for any real numbers λ_1 and λ_2 with $32/27 \leq \lambda_1 < \lambda_2$, there exists a graph G such that $\chi(G;\lambda) = 0$ for some $\lambda \in (\lambda_1, \lambda_2)$.*

In this theorem, the first two items can be verified easily (see [404, 950]), the third is from [637], and the last from [1064].

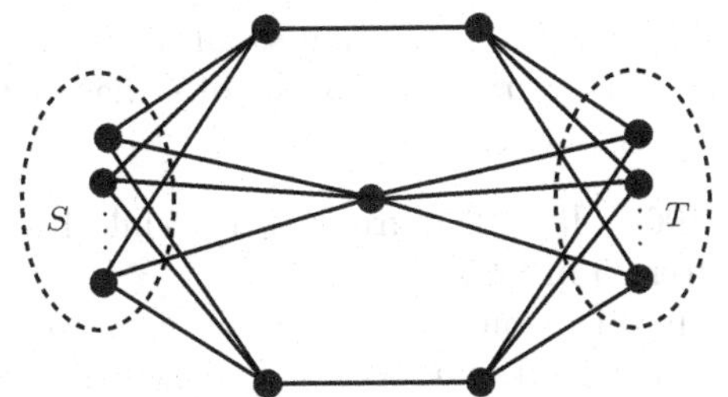

FIGURE 11.16: The graph $X(s,t)$, where s and t are the sizes of independent sets S and T respectively.

Jackson [637] conjectured that $(1,2)$ is a zero-free interval for chromatic roots of all 3-connected non-bipartite graphs. This conjecture was disproved by Royle [974] who found that the graph $X(s,t)$, shown in Figure 11.16, where S and T are independent sets of this graph with $|S| = s$ and $|T| = t$, is a counter-example whenever $s, t \geq 3$ are both odd.

A connected graph G is said to be *α-tough* if $k(G \backslash S) \leq |S|$ holds for any non-empty independent set S of G. (Note that "α" is not a number in this definition.) The following theorem, due to Dong and Koh [398], verified some cases of Conjecture 11.103 that α-tough graphs have no chromatic roots in $(1,2)$.

Theorem 11.77. *If G is an α-tough graph with n vertices and is a member of one of the following families of graphs, then $(-1)^n \chi(G;\lambda) > 0$ for all real λ with $1 < \lambda < 2$.*

1. *Graphs containing 2-trees as spanning subgraphs;*

2. *complete t-partite graphs for any $t \geq 3$;*

3. *graphs with a Hamiltonian path $v_1 v_2 \cdots v_n$ such that v_i has at least two neighbors in the set $\{v_1, v_2, \cdots, v_{i-1}\}$ for all $i \geq 3$; and*

4. *$(n - \Delta(G) + 1)$-connected graphs G of order n.*

For a set $\mathcal{S}$ of graphs, let $\eta(\mathcal{S})$ be the largest possible real number in $(1,2]$ such that $(1, \eta(\mathcal{S}))$ is a zero-free interval for chromatic roots of graphs in $\mathcal{S}$. The number $\eta(\mathcal{S})$ has been determined for some sets $\mathcal{S}$ of graphs:

1. $\eta(\mathcal{S}) = 1.29559\cdots$ if $\mathcal{S}$ is the set of graphs with Hamiltonian paths;

2. $\eta(\mathcal{S}) = 32/27$ if $\mathcal{S}$ is the set of bipartite graphs;

3. $\eta(\mathcal{S}) = 1.31767\cdots$ if $\mathcal{S}$ is the set of graphs with domination numbers at most 2;

4. $\eta(\mathcal{S}) = 1.430159\cdots$ if $\mathcal{S}$ is the set of graphs $G = (V,E)$ with $\Delta(G) \geq |V| - 2$; and

5. $\eta(\mathcal{S}) = 1.2040\cdots$ if $\mathcal{S}$ is the set of non-separable plane graphs G with an outer face C such that every 2-vertex-cut of G has at least one vertex in C.

Item 1 is from [1065], Item 2 is from [401], while Items 3 and 4 can be found in [400], and Item 5 in [395].

Another famous problem on real chromatic roots of planar graphs was proposed by Beraha in [92]. Define $B_n = 2 + 2\cos(2\pi/n)$ for any positive integer n. The number B_n is called the n-th *Beraha number*. Beraha asked if it is true that for every $\epsilon > 0$, there exists a plane triangulation G such that $\chi(G;\lambda)$ has a root in $(B_n - \epsilon, B_n + \epsilon)$.

When $n = 1$, Beraha's question was answered by Royle [972] who constructed a sequence of plane triangulations $X_1, X_2, X_3, \cdots$ such that for any $\epsilon > 0$, the polynomial $\chi(X_N;\lambda)$ has roots in $(4-\epsilon, 4)$ for some N.

It is not difficult to answer the question for $n = 2,3,4,6$, as $B_2 = 0$, $B_3 = 1$, $B_4 = 2$ and $B_6 = 3$ (see [93]). The existence of such plane triangulations was shown for $n = 7$ and $n = 10$ by Beraha, Kahane and Reid [977]. For $n = 5$ ($B_5 = \tau^2$), where $\tau = (1+\sqrt{5})/2$, Beraha, Kahane and Weiss [93] found an infinite family of plane triangulations with chromatic roots tending to τ^2. Due to Salas and Sokal [980], no non-integer Beraha number is the chromatic root of any graph, with the possible exception of B_{10}. Harvey and Royle [601] showed that $B_{10} = \tau + 2$ is indeed a chromatic root, and thus B_1, B_2, B_3, B_4, B_6 and B_{10} are the only Beraha numbers which are chromatic roots.

However, the question remains unanswered for $n = 8$, 9 and $n \geq 11$.

Thomassen [1064] conjectured that real chromatic roots of planar graphs are dense in the interval $(32/27, 4)$. Woodall [1170, 1171] proved that $(1,2)$ and $(2, 2.5466\cdots)$ are zero-free intervals for chromatic roots of near-triangulations and plane triangulations respectively. The plane triangulation $\overline{K}_2 + C_5$ has a chromatic root $2.6778\cdots$. Woodall [1171] conjectured that $(2.6778\cdots, 3)$ is a zero-free interval for all plane triangulations.

11.5.3 Integral roots of chromatic polynomials

A polynomial is called an *integral-root polynomial* if all its roots are integers. Observe that any chordal graph G has an integral-root chromatic polynomial, as Corollary 11.6 implies that

$$\chi(G;\lambda) = \lambda^{r_0}(\lambda-1)^{r_1}\cdots(\lambda-k)^{r_k}, \tag{11.10}$$

for some positive integers $r_0, r_1, \cdots, r_k$, where $k = \chi(G)-1 \geq 0$. Furthermore, for any positive integers $r_0, r_1, \cdots, r_k$, there exists chordal graphs G such that Equation (11.10) holds.

Braun, Kretz, Walter and Walter [196] and Vaderlind [1109] conjectured that only chordal graphs have integral-root chromatic polynomials. This conjecture was disproved by Read [944] (and also independently by Dmitriev [386]), who discovered the non-chordal graph in Figure 11.17 whose

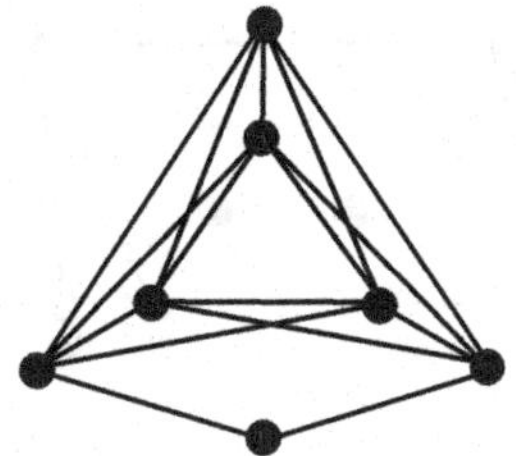

FIGURE 11.17: The smallest non-chordal graph whose chromatic polynomial has integral roots only.

chromatic polynomial is given by

$$\lambda(\lambda-1)(\lambda-2)(\lambda-3)^3(\lambda-4).$$

For positive integers $r_0, r_1, \cdots, r_k$, where $k \geq 0$, let $\mathcal{I}(r_0, r_1, \cdots, r_k)$ denote the family of graphs with chromatic polynomials equal to $\lambda^{r_0}(\lambda-1)^{r_1}\cdots(\lambda-k)^{r_k}$. The study of integral chromatic roots now concerns the following question.

Open Question 11.78. Given $k, r_0, r_1, \cdots, r_k \in \mathbb{N}$, characterize the graphs in the family $\mathcal{I}(r_0, r_1, \cdots, r_k)$. In particular, determine if the family $\mathcal{I}(r_0, r_1, \cdots, r_k)$ consists of chordal graphs only.

Some answers to Open Question 11.78 follow from the results in Section 11.4.6. If $r_i = 1$ for all $i = 0, 1, \cdots, k$, then $\mathcal{I}(r_0, r_1, \cdots, r_k) = \{K_{k+1}\}$. It is also clear that if $k = 1$, then $\mathcal{I}(r_0, r_1)$ is the family of forests of order $r_0 + r_1$ with r_0 components. When $k = 2$, Theorems 11.46 and 11.62 imply that $\mathcal{I}(r_0, r_1, r_2)$ contains chordal graphs only. When $k = 3$, Xu [1176] conjectured that for all positive integers r_0, r_1, r_2, r_3, the set $\mathcal{I}(r_0, r_1, r_2, r_3)$ contains chordal graphs only. Xu's conjecture has been proved to be true by Dong and Koh [397] for planar graphs. Kung and Royle [737] generalized this result to flow polynomials.

Proposition 11.79. *If G is a bridgeless graph, then the roots of its flow polynomial are integral if and only if G is the dual of a planar chordal graph.*

For the case that $k \geq 4$, Open Question 11.78 has been partially answered. Some known results are listed below.

Proposition 11.80. *Let $k \geq 4$ and $r_0, r_1, \cdots, r_k$ be positive integers.*

1. *If $r_i \geq 3$ for some i with $3 \leq i \leq k-1$, then $\mathcal{I}(r_0, r_1, \cdots, r_k)$ contains non-chordal graphs.*

2. *If $r_0 = r_1 = \cdots = r_{k-1} = 1$, then $\mathcal{I}(r_0, r_1, \cdots, r_k)$ contains chordal graphs only.*

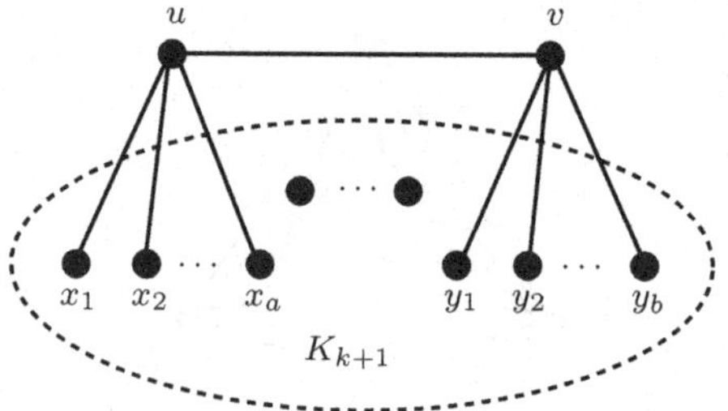

FIGURE 11.18: The graph $R(a, b, k)$.

3. *If* $r_2 + \cdots + r_{k-1} = k - 1$, *then* $\mathcal{I}(r_0, r_1, \cdots, r_k)$ *contains chordal graphs only.*

4. *Assume that* $r_0 = r_1 = r_k = 1$ *and* $r_2 + \cdots + r_{k-1} = k$. *Then all graphs in* $\mathcal{I}(r_0, r_1, \cdots, r_k)$ *are chordal if and only if* $r_i = r_j = 2$ *for some* $2 \leq i < j \leq k - 1$ *such that either* $j \geq 2i - 2$ *or* $2j \geq k + i$.

The first item in the proposition can be verified easily by applying Proposition 11.5 and the fact that the graph in Figure 11.17 belongs to the set $\mathcal{I}(1, 1, 1, 3, 1)$, the second is from [286, 387, 599, 1109], the third follows from Theorem 11.62, and the fourth is in [1176].

For any positive integers a, b, k with $a + b \leq k + 1$, let $R(a, b, k)$ be a graph obtained from the complete graph K_{k+1} by adding two adjacent vertices u and v and adding edges joining u to vertices $x_1, \cdots, x_a$ in K_{k+1} and v to vertices $y_1, \cdots, y_b$ in K_{k+1}, where $\{x_1, \cdots, x_a\} \cap \{y_1, \cdots, y_b\} = \emptyset$, as shown in Figure 11.18. Note that every graph $R(a, b, k)$ is non-chordal. Dmitriev [386] observed that $\chi(R(a, b, k); \lambda)$ is integral-root if and only if $\{a, b\} = \{st, (s + 1)(t + 1)\}$ for positive integers s and t. Notice that $R(1, 4, 5)$ is the graph in Figure 11.17.

A *chordless cycle* in a graph is a cycle of length at least 4 which is an induced subgraph. A chordal graph is simply a graph without chordless cycles. All chordless cycles in $R(a, b, k)$ are of order 4. Let $g^*(G)$ denote the order of the smallest chordless cycle in a non-chordal graph G. Dmitriev [386] asked the following: given an integer $p \geq 5$, does there exist a graph G with $g^*(G) = p$ such that $\chi(G; \lambda)$ is integral-root?

Read [946] initiated the study of Dmitriev's question by considering a family of graphs $H_{k_1,k_2,\cdots,k_n}$ which are actually the graphs $R(a, b, k)$ when $n = 3$ and $k_1 = 1$. For any positive integers $k_1, k_2, \cdots, k_n$, where $n \geq 2$, let $H_{k_1,k_2,\cdots,k_n}$ denote the graph obtained from n vertex-disjoint complete graphs $K_{k_1}, K_{k_2}, \cdots, K_{k_n}$ and a vertex w, not in any one of these complete graphs, by adding edges joining each vertex in K_{k_i} to each vertex in $K_{k_{i+1}}$ for $i = 1, 2, \cdots, n - 1$, and edges joining w to each vertex in K_{k_1} and K_{k_n}. When $n = 2$, $H_{k_1,k_2,\cdots,k_n}$ is a chordal graph and when $n \geq 3$, all chordless cycles in this graph are of order $n + 1$. The following lemma is from [408, 946].

Lemma 11.81. *For any positive integers $n, k_1, k_2, \cdots, k_n$, where $n \geq 2$,*

$$\chi(H_{k_1,k_2,\cdots,k_n};\lambda) = \frac{\prod_{i=1}^{n-1}(\lambda)_{k_i+k_{i+1}}}{\lambda\prod_{i=2}^{n-1}(\lambda)_{k_i+1}} \cdot f_{k_1,k_2,\cdots,k_n}(\lambda)$$

where

$$f_{k_1,k_2,\cdots,k_n}(x) = \left(\prod_{i=1}^{n}(x-k_i) + (-1)^{n+1}\prod_{i=1}^{n}k_i\right).$$

Corollary 11.82. *$\chi(H_{k_1,k_2,\cdots,k_n};\lambda)$ is an integral-root polynomial if and only if $f_{k_1,k_2,\cdots,k_n}(x)$ is an integral-root polynomial.*

Dmitriev's question for $p = 5, 6, 7$ was answered affirmatively by Dong and Koh [397] and Dong, Teo, Koh and Hendy [408].

Hernández and Luca [611] showed that for any integer $n \geq 3$, there exist positive integers $k_1, k_2, \cdots, k_n$ such that $f_{k_1,k_2,\cdots,k_n}(x)$ is an integral-root polynomial if and only if the Prouhet–Tarry–Escott problem has an ideal solution of degree $n-1$, i.e., there exist two distinct sets of integers $\{\alpha_1, \alpha_2, \cdots, \alpha_n\}$ and $\{\beta_1, \beta_2, \cdots, \beta_n\}$ such that

$$\sum_{i=1}^{n}\alpha_i^j = \sum_{i=1}^{n}\beta_i^j$$

holds for all $j = 1, 2, \cdots, n-1$. It can be found in Borwein [181] that the Prouhet–Tarry–Escott problem admits ideal solutions for all n with $2 \leq n \leq 12$ and $n \neq 11$. Thus, by Corollary 11.82, Open Question 11.107 has an affirmative answer for $3 \leq p \leq 13$ and $p \neq 12$.

For $p = 12$ or $p \geq 14$, the question remains unsolved. See Open Question 11.107 below.

11.6 Open problems

The area of chromatic polynomials abounds with open questions and long-standing conjectures. We collect a few of them here, focussing on those closely related to this chapter.

Open Question 11.83. Let $B_{m,n} = P_m \times P_n$. Can $\chi(B_{m,n};\lambda)$ be determined by the method of Section 11.2.3? By any method, find a general formula for $\chi(B_{m,n};\lambda)$. See [658, 950] for further details.

Tomescu [1070] conjectured the following about the maximum chromatic polynomial. Although there has been progress toward its proof, e.g. Theorems 11.50 and 11.51, the full statement remains open.

Conjecture 11.84. Let G be a connected graph of order n and chromatic number k, where $k \geq 4$. Then

$$\chi(G;\lambda) \leq (\lambda)_k(\lambda-1)^{n-k}$$

holds for all integers $\lambda \geq 0$, where equality holds for $\lambda \geq k$ if and only if G is a connected graph of order n with exactly one block isomorphic to K_k and $n-k$ blocks isomorphic to K_2.

Dong, Koh, and Teo [404] made the following conjecture about the behavior of the chromatic polynomial that is stronger than Theorem 11.53.

Conjecture 11.85. For any graph G of order n and any real number $\lambda \geq n$, the inequality $\chi(G;\lambda+1)\chi(G;\lambda-1) \leq \chi(G;\lambda)^2$ holds.

In the same paper, [82], in which Bartels and Welsh introduced the shameful conjecture, they also proposed the following conjecture which is stronger than the shameful conjecture and still remains open.

Conjecture 11.86. For any non-empty graph G of order n, there exists an edge e such that

$$\frac{\chi(G;n-1)}{\chi(G;n)} \leq \frac{\chi(G\backslash e;n-1)}{\chi(G\backslash e;n)}.$$

In 1968, Read [945] posed the following problem on determining the set of chromatic polynomials:

Open Question 11.87. What is a necessary and sufficient condition for a polynomial to be the chromatic polynomial of a graph?

It is known that even a polynomial having all the basic properties for chromatic polynomials in Theorem 11.36 and Corollaries 11.27 and 11.34 may be not a chromatic polynomial of any graph, [950], Open Question 11.87 remains an important open question in the field.

Read [945] also asked the following question.

Open Question 11.88. What is a necessary and sufficient condition for two graphs G and H to have the same chromatic polynomial?

While much progress has been made in this area (see Section 11.4), much still remains to be done.

Recall from Equation (11.8) that the χ-invariant $\epsilon(G)$ in [798] gives the mean size of broken-cycle-free spanning subgraphs of G. In the same paper, Lundow and Markström proposed a number of open problems about the bounds of $\epsilon(G)$:

Conjecture 11.89. For any connected graph G of order n that is nor complete, $\epsilon(G) < \epsilon(K_n)$.

Open Question 11.90. Given positive integers n and k with $n-1 < k < \binom{n}{2}$, what are the maximum and minimum of $\epsilon(G)$ among all connected graphs with n vertices and k edges?

Conjecture 11.91. For any 2-connected graph G, there is an edge e in G such that $\epsilon(G\backslash e) < \epsilon(G)$.

While considerable progress has been made in determining the chromaticity of complete multipartite graphs, the following problem posed by Chia and Ho [304] remains unsolved for $k \geq 3$.

Open Question 11.92. For $k \geq 3$ and $n \geq 2$, does it hold that

$$\mathcal{G}[K_{1,n,n+k}] = \{\bar{K}_n + T : T \in \mathcal{T}_{n+k+1}\} \cup \{\bar{K}_{n+k} + T' : T' \in \mathcal{T}_{n+1}\},$$

where $\mathcal{T}_n$ is the set of trees of order n?

Similarly, the chromaticity of wheels is not yet completely determined. The following question remains open.

Open Question 11.93. Determine the set $\mathcal{G}[W_n]$ for all odd integers n, $n \geq 11$.

Item 2 of Theorem 11.65 is a special case of the following conjecture proposed by Bollobás, Pebody and Riordan [156] (see also [886]).

Conjecture 11.94. $F_\chi(n) \to 1$ as $n \to \infty$, where $F_\chi(n)$ is the proportion of χ-unique graphs of order n among all graphs of order n.

We give next a number of conjectures on chromatic roots either proposed in [1023] or during the program on "Combinatorics and Statistical Mechanics" in Cambridge in 2008. Recall that Ω_k is the set of roots of monic polynomials of degree k whose coefficients are integers.

Conjecture 11.95 (The $\alpha + n$ conjecture). For every $\alpha \in \Omega_k$, where $k \geq 1$, there exists an integer n such that $\alpha + n$ is a chromatic root.

Conjecture 11.96 (The αn conjecture). Let α be a chromatic root. Then $n\alpha$ is also a chromatic root for every positive integer n.

Conjecture 11.97.

1. There is no chromatic root z with $Re(z) = 0$ and $Im(z) \neq 0$.
2. Among all graphs with n vertices, the graph with the largest real part of a chromatic root is K_n.
3. Among all graphs of maximum degree Δ with $\Delta \geq 4$, the graph with the largest real part of a chromatic root is $K_{\Delta+1}$.
4. Among all graphs of maximum degree Δ with $\Delta \geq 4$, the graph with the largest modulus of a chromatic root is $K_{\Delta,\Delta}$.

5. Among all 3-regular graphs, the graph with the largest modulus of a chromatic root is K_4 and the graph with the second largest modulus of a chromatic root is $K_{3,3}$.

In the above list, the first conjecture is due to Fengming Dong, the second to Jason Brown, the third to Alan Sokal, and the fourth and fifth to Gordon Royle.

The following two conjectures would strengthen known results on chromatic zeros. For the first, Sokal [1020] conjectured that Theorem 11.72 can be further improved to 2Δ:

Conjecture 11.98. For any graph G with maximum degree Δ, if $\chi(G; z) = 0$, then $|z| \leq 2\Delta$.

Also, Brown, Hickman, Sokal and Wagner [235] extended Theorem 11.72 in another direction by replacing "second-largest degree" in Theorem 11.73 by a weaker criterion, the "maximaxflow" denoted by $\Lambda(G)$ and defined by

$$\Lambda(G) := \max_{\substack{u \neq v \\ u,v \in V(G)}} \lambda(u, v),$$

where $\lambda(u, v)$ is the maximum number of edge-disjoint u, v-paths, which is equal to the minimum number of edges separating u from v. Since $\lambda(u, v) \leq \min\{\deg(u), \deg(v)\}$, it follows that $\Lambda(G)$ cannot exceed the second-largest degree of G. The following conjecture is from [235, 1020].

Conjecture 11.99. For any positive integer k, there exists a constant $C(k)$ such that if G is a graph with $\Lambda(G) = k$ and $\chi(G; z) = 0$, then $|z - 1| \leq C(k)$.

Furthermore, Sokal [1023] conjectured the following.

Conjecture 11.100. For any graph G, $\chi(G; \lambda) > 0$ for all $\lambda > \Lambda(G)$.

Jackson [637] proposed the following conjecture on zero-free intervals for chromatic polynomials of 3-connected graphs.

Conjecture 11.101. For any 3-connected graph G of order n, $(-1)^n\chi(G; \lambda) > 0$ for all real numbers λ in the interval $(1, 1.781\cdots)$, where $1.781\cdots$ is a root of $\chi(K_{3,4}; \lambda)$.

Thomassen [1063] conjectured that Hamiltonian graphs have no chromatic roots in $(1, 2)$:

Conjecture 11.102. If G is a Hamiltonian graph with n vertices, then $(-1)^n\chi(G; \lambda) > 0$ for all real λ with $1 < \lambda < 2$.

Conjecture 11.103, from [398], has been verified for some families of graphs (see Theorem 11.77) but remains open in general.

Conjecture 11.103. If G is a α-tough graph with n vertices, then $(-1)^n\chi(G; \lambda) > 0$ for all real λ with $1 < \lambda < 2$.

Birkhoff and Lewis [124] conjectured in 1946 that $\chi(G;\lambda) > 0$ holds for all planar graphs G and all real numbers $\lambda \geq 4$ after they proved this inequality for the case that $\lambda \geq 5$.

Conjecture 11.104. For any planar graph G, we have $\chi(G;\lambda) > 0$ for all real $\lambda \in (4,5)$.

So far it is even unknown if there is any real number λ in $(4,5)$ and any planar graph G such that $\chi(G;\lambda) \leq 0$.

The following question of Beraha [92] about real chromatic roots of planar graphs remains open except for very small n.

Open Question 11.105. Let $B_n = 2 + 2\cos(2\pi/n)$ for any positive integer n. Is it true that for every $\epsilon > 0$, there exists a plane triangulation G such that $\chi(G;\lambda)$ has a root in $(B_n - \epsilon, B_n + \epsilon)$? This question remains open for $n = 8, 9$ and $n \geq 11$.

The question of determining graphs yielding chromatic polynomials with only integral roots has been only partially addressed. In particular, the following remains open for k=3 in the case of nonplanar graphs, and $k \geq 4$ beyond some partial results given in Proposition 11.80.

Open Question 11.106. Given k and $r_0, r_1, \cdots, r_k \in \mathbb{N}$, characterize the graphs in the family $\mathcal{I}(r_0, r_1, \cdots, r_k)$. In particular determine if the family $\mathcal{I}(r_0, r_1, \cdots, r_k)$ consists of chordal graphs only.

We also recall a question of Dmitriev [386] about integral roots, recalling that $g^*(G)$ is the order of the smallest chordless cycle in a non-chordal graph G. It remains open in general, with some partial results given in Subsection 11.5.3.

Open Question 11.107. For any integer $p \geq 5$, does there exist a non-chordal graph G with $g^*(G) = p$ such that $\chi(G;\lambda)$ is integral-root?

We close by noting that there are many topics and further questions beyond those covered in this chapter, for example this one originally posed independently by Wilf and Linial, and recently surveyed by Lazebnik [759].

Open Question 11.108. What is the maximum number of proper vertex colorings in (at most) λ colors a graph with n vertices and m edges can have? On which graphs is this maximum attained? In other words, what is the maximum value of $\chi(G;\lambda)$ over all graphs with n vertices and m edges?

12

Flows and colorings

Delia Garijo • Andrew Goodall • Jaroslav Nešetřil

Synopsis

Tutte first introduced the dichromate of a graph in large part motivated by the fact that it contained the flow polynomial and chromatic polynomial as univariate specializations. The latter receives attention in Chapter 11. In this chapter we consider flows.

- Flows of graphs taking values in a finite abelian group, colorings and tensions (dual to flows).
- The flow polynomial, chromatic polynomial and dichromatic polynomial as specializations of the Tutte polynomial.
- Coloring-flow convolution formulas for the Tutte polynomial.
- "A-bicycles" (called bicycles when $\mathsf{A} = \mathbb{Z}_2$), the Tutte polynomial evaluated at $(-1, -1)$, $(e^{2\pi i/3}, e^{4\pi i/3})$ and $(i, -i)$.
- Tutte's flow conjectures.

12.1 Introduction

W. Tutte [1107] describes how he first "became acquainted with the Tutte polynomial" by looking for graph invariants with deletion–contraction recurrences such as the number of spanning trees (of a connected graph), the chromatic polynomial, and the flow polynomial. In this chapter we focus on evaluations of the Tutte polynomial $T(G; x, y)$ of a graph G that involve $\mathbb{Z}_k$-flows of G, for which there are often dual results in terms of vertex k-colorings of

DOI: 10.1201/9780429161612-12

G, or, equivalently, $\mathbb{Z}_k$-tensions of G. These evaluations are all at points on the hyperbola $(x-1)(y-1)=k$ with k a positive integer.

The duality between colorings and flows of a graph G consists in the orthogonality of A-tensions and A-flows of G for a finite abelian group A of order k. (See [107, 546] for the interaction of graph theory and linear algebra more generally.) While colorings and flows are dual notions for planar graphs—the tensions of a plane graph correspond to the flows of the dual plane graph—for graphs generally, duality resides at the level of the cycle and cocycle matroids of the graph. The set of tensions of a non-planar graph do not correspond to the set of flows of a graph (a result that goes back to H. Whitney in the 1930s). There is a consequent asymmetry between properties of colorings and flows for graphs (for example, loopless graphs may have arbitrarily high chromatic number, but bridgeless graphs have bounded flow number). Any information about flows that can be extracted from the Tutte polynomial extends our still incomplete picture of them.

After giving the fundamental specializations of the Tutte polynomial to the flow polynomial and chromatic polynomial, we describe more elaborate decompositions of the Tutte polynomial that for graphs can be expressed in terms of Hamming weights of flows and tensions, beginning with the classical expression of the Tutte polynomial as the dichromatic polynomial (effectively a Hamming weight enumerator for tensions) and its dual expansion as a weight enumerator for flows. (For a development of the relationship between the Tutte polynomial and weight enumerators of codes see Chapter 16.) We then consider the conjunction of flows and tensions in what are known as bicycles (for $\mathbb{Z}_2$-flows and $\mathbb{Z}_2$-tensions) and "A-bicycles" more generally (the intersection of the set of A-tensions and the set of A-flows). Evaluations of the Tutte polynomial at the points $(-1,-1)$, $(i,-i)$ and $(e^{2\pi i/3}, e^{4\pi i/3})$ yield the dimension of the A-bicycle space for $\mathsf{A}=\mathbb{Z}_2$ and $\mathsf{A}=\mathbb{Z}_3$, and some further information about the weight distribution of A-bicycles for evaluations at the last two points.

In our concluding section reporting open problems, we give an overview of Tutte's flow conjectures, which have motivated intensive research in the topic of flows on graphs ever since their formulation.

Complementary surveys of the Tutte polynomial that treat combinatorial interpretations of its evaluations involving flows include [247, 454, 1133, 1139].

12.2 Flows and the flow polynomial

An *orientation* ω of a graph $G=(V,E)$ assigns a direction to each edge $uv\in E$, either $u\xrightarrow{\omega}v$ or $u\xleftarrow{\omega}v$. For $U\subset V$, define $\omega^+(U)=\{uv\in E: u\in U, v\in V\setminus U,\ u\xrightarrow{\omega}v\}$ and $\omega^-(U)=\omega^+(V\setminus U)$. In particular, for a vertex $v \in V$ the set $\omega^+(v)$ consists of those edges directed out of v by the orientation ω and $\omega^-(v)$ is the set of edges directed into v.

For an additive abelian group A, scalar multiples of a $\{0, \pm 1\}$-vector by an element of A are defined by using the identities $0a = 0, 1a = a$ and $(-1)a = -a$ for each $a \in \mathsf{A}$. The abelian group A is a $\mathbb{Z}$-module. The set of mappings $\phi : E \to \mathsf{A}$ is denoted by A^E, and likewise A^V is the set of mappings $\kappa : V \to \mathsf{A}$.

Definition 12.1. Let G^ω be the graph $G = (V, E)$ with a fixed orientation ω of its edges. The *incidence matrix* of G^ω is the matrix $D = (D_{v,e}) \in \{0, \pm 1\}^{V \times E}$ whose (v, e)-entry is defined by

$$D_{v,e} = \begin{cases} +1 & \text{if } e \text{ is directed out of } v \text{ by } \omega, \\ -1 & \text{if } e \text{ is directed into } v \text{ by } \omega, \\ 0 & \text{if } e \text{ is not incident with } v, \text{ or } e \text{ is a loop on } v. \end{cases}$$

Definition 12.2. The incidence matrix of G^ω defines a linear transformation $D : \mathsf{A}^E \to \mathsf{A}^V$ between modules, called the *boundary mapping,* which for $\phi : E \to \mathsf{A}$ and $v \in V$ is given by

$$D\phi(v) = \sum_{\substack{e=uv \\ u \xleftarrow{\omega} v}} \phi(e) - \sum_{\substack{e=uv \\ u \xrightarrow{\omega} v}} \phi(e).$$

The transpose of the incidence matrix defines the *coboundary* map $D^T : \mathsf{A}^V \to \mathsf{A}^E$, which for $\kappa : V \to \mathsf{A}$ and edge $e = uv$ is given by

$$D^T\kappa(e) = \begin{cases} \kappa(v) - \kappa(u) & \text{if } u \xleftarrow{\omega} v, \\ \kappa(u) - \kappa(v) & \text{if } u \xrightarrow{\omega} v. \end{cases}$$

In other words, the boundary assigns the net value at each vertex for a given mapping $\phi : E \to \mathsf{A}$, and the coboundary assigns to each edge the difference (taken according to the edge orientation) of its endpoints in a given mapping $\kappa : V \to \mathsf{A}$.

Definition 12.3. Let G^ω be the graph G with a fixed orientation ω, let A be an additive abelian group, and let D the incidence matrix of G^ω. The set of A-*flows* of G^ω is the kernel of D and the set of A-*tensions* of G^ω is the image of D^T.

An A-flow of G^ω is thus a mapping $\phi : E \to \mathsf{A}$ such that Kirchhoff's law is satisfied at each vertex of G:

$$\sum_{e \in \omega^+(v)} \phi(e) - \sum_{e \in \omega^-(v)} \phi(e) = 0, \qquad \text{for each } v \in V.$$

Likewise, an A-tension of G^ω is a mapping $\theta : E \to \mathsf{A}$ such that for each closed walk on G the sum of values on forward edges is equal to the sum of values on backward edges, forward and backward being relative to the direction of

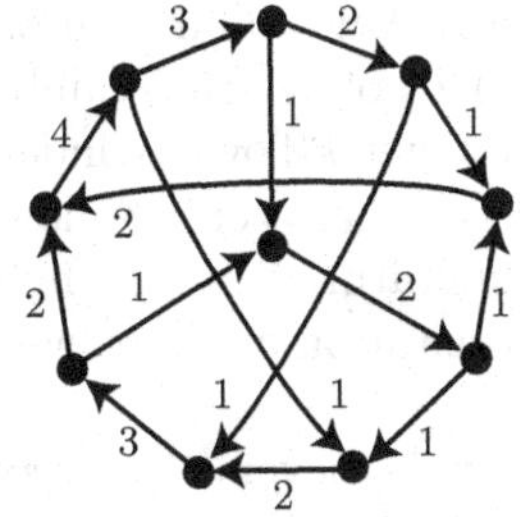

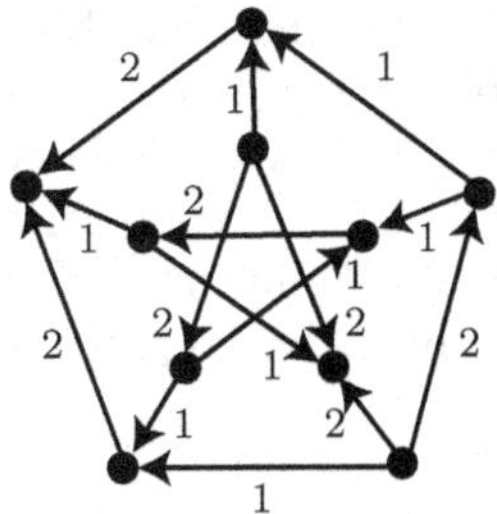

FIGURE 12.1: The Petersen graph, with on the left a nowhere-zero 5-flow and on the right a nowhere-zero $\mathbb{Z}_5$-flow.

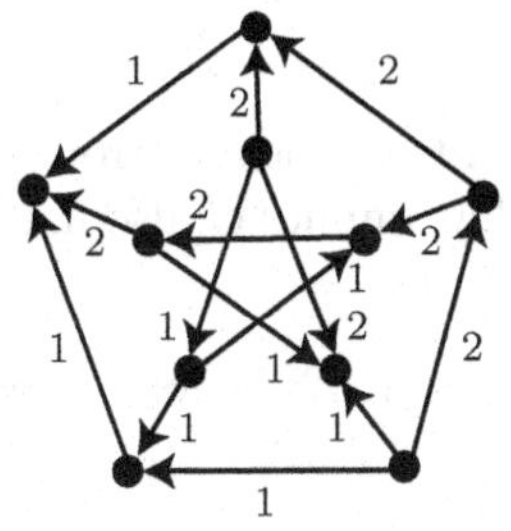

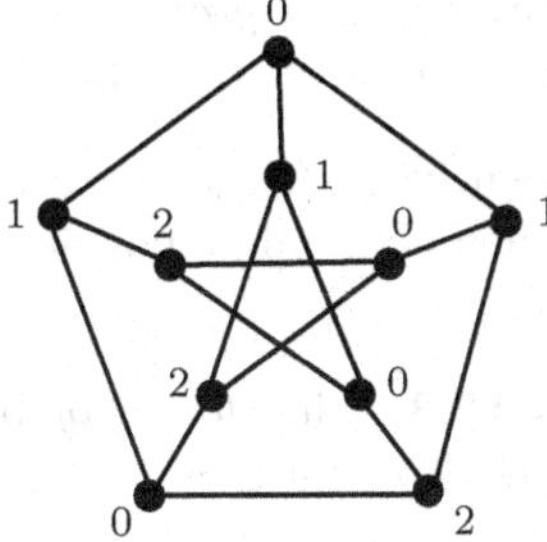

FIGURE 12.2: The Petersen graph, with on the left a nowhere-zero $\mathbb{Z}_3$-tension and a corresponding proper 3-coloring on the right.

traversal of the walk. To each A-tension θ there corresponds a set of $|\mathsf{A}|^{k(G)}$ vertex colorings $\kappa : V \to \mathsf{A}$ such that $D^T\kappa = \theta$.

The *support* of a mapping $\phi : E \to \mathsf{A}$ is defined by $\text{supp}(\phi) = \{e \in E : \phi(e) \neq 0\}$.

Definition 12.4. A *nowhere-zero* A-flow of G^ω is an A-flow $\phi : E \to \mathsf{A}$ with the additional property that $\phi(e) \neq 0$ for every $e \in E$, that is, $\text{supp}(\phi) = E$.

An *integer k-flow* of G^ω is a $\mathbb{Z}$-flow $\phi : E \to \mathbb{Z}$ such that $-k < \phi(e) < k$ for every $e \in E$. A *nowhere-zero* integer k-flow is an integer k-flow with the additional property that $\phi(e) \neq 0$ for every $e \in E$.

Remark 12.5. In the literature, the term "k-flow" may mean either an integer k-flow or a $\mathbb{Z}_k$-flow, with the choice apparent from context. Since we discuss both integer k-flows and $\mathbb{Z}_k$-flows here, we make a clear distinction.

Example 12.6. Figure 12.1 illustrates a nowhere-zero integer k-flow with $k = 5$ and a nowhere-zero A-flow with $\mathsf{A} = \mathbb{Z}_5$ for the Petersen graph with orientation of edges as indicated. Figure 12.2 illustrates a nowhere-zero $\mathbb{Z}_3$-tension and a corresponding proper 3-coloring.

Although the definition of a (nowhere-zero) A-flow and a (nowhere-zero) integer k-flow of G^ω requires an orientation ω of G, the number of such (nowhere-zero) A-flows and (nowhere-zero) integer k-flows is independent of the orientation ω. Therefore, the number of (nowhere-zero) A-flows of G^ω is an invariant of G, and in the context of enumerating flows we may speak of flows of a graph G without specifying the orientation of the edges of G that is used.

The enumerative theory of integer k-flows is beyond the scope of this chapter, which focuses solely on enumerative results concerning nowhere-zero A-flows for a finite abelian group A. However, it is worth noting the following from [1092].

Proposition 12.7. *A graph G has a nowhere-zero $\mathbb{Z}_k$-flow if and only if G has a nowhere-zero integer k-flow.*

Consequently, if A and A′ are finite additive abelian groups such that $|\mathsf{A}| \leq |\mathsf{A}'|$, then G having a nowhere-zero A-flow implies G also has a nowhere-zero A′-flow.

Definition 12.8. The *flow polynomial* of $G = (V, E)$ is defined by its evaluations at a positive integer k by

$$F(G;k) = \#\{\phi : E \to \mathbb{Z}_k : \phi \text{ a nowhere-zero } \mathbb{Z}_k\text{-flow of } G\}.$$

That the number of nowhere-zero $\mathbb{Z}_k$-flows is indeed a polynomial in k is a consequence of the following recurrence:

Proposition 12.9. *The flow polynomial of a graph satisfies*

$$F(G;k) = \begin{cases} F(G/e;k) - F(G\backslash e;k) & \text{if } e \text{ is ordinary,} \\ 0 & \text{if } e \text{ is a bridge,} \\ (k-1)F(G\backslash e;k) & \text{if } e \text{ is a loop,} \end{cases}$$

and multiplicativity over disjoint unions,

$$F(G_1 \sqcup G_2;k) = F(G_1;k)F(G_2;k).$$

The recurrence in Proposition 12.9 holds not only for the number of nowhere-zero $\mathbb{Z}_k$-flows of G but also for the number of nowhere-zero A-flows of G for any choice of finite additive abelian group A of order k, hence the following.

Corollary 12.10. *The number of nowhere-zero A-flows of a graph G depends only on $|\mathsf{A}|$ and not the structure of A as a group.*

Another consequence of Proposition 12.9 is that homeomorphic graphs (i.e., graphs that are isomorphic after suppression of all vertices of degree two) have the same flow polynomial.

Proposition 12.11. *Let $G = G_1 * G_2$ be a one-point join of graphs. Then*

$$F(G;k) = F(G_1;k)F(G_2;k).$$

Thus the flow polynomial of a graph G is the product of its values on the blocks (2-connected components) of G. As shown in [997], the flow polynomial of a graph may be further decomposed over vertex-cuts of size 2 or 3. The decompositions for vertex-cuts of sizes 1 or 2 yield by the deletion–contraction formula of Proposition 12.9 a similar decomposition of the flow polynomial of G when there is an edge e such that $G\backslash e$ is separable (see e.g. [638]).

An inductive deletion–contraction argument also establishes decompositions of the flow polynomial over small edge-cuts (from [997]):

Theorem 12.12. *Let G be a connected graph with an edge cut A of size $t \in \{2,3\}$. Let E_1 and E_2 be the edge sets of two components of $G\backslash A$, and $G_i = G/E_i$, for $= 1,2$. Then*

$$F(G;k) = \frac{F(G_1;k)F(G_2;k)}{F(\theta_t;k)},$$

where θ_t is the t-theta graph consisting of two vertices joined by t parallel edges.

Proposition 12.9 and the universality property of the Tutte polynomial (Theorem 2.21) imply the following result.

Theorem 12.13. *The flow polynomial of a graph is a specialization of the Tutte polynomial:*

$$F(G;k) = (-1)^{n(G)}T(G;0,1-k).$$

Remark 12.14. The analogously defined *integer flow polynomial*

$$F_{\mathbb{Z}}(G;k) = \#\{\phi : E \to \mathbb{Z} : \phi \text{ nowhere-zero integer } k\text{-flow of } G\}, \tag{12.1}$$

was only much later proved actually to be a polynomial by M. Kochol [699]. It is not a specialization of the Tutte polynomial. Furthermore, when not restricting to nowhere-zero flows, while the number of A-flows for $|\mathsf{A}| = k$ is equal to $k^{n(G)}$ (as can easily verified by from the deletion–contraction relation satisfied by the number of A-flows), the number of integer k-flows is a polynomial $F_{\mathbb{Z}}^{0}(G;k)$ which is as difficult to compute as $F_{\mathbb{Z}}(G;k)$.

12.3 Coloring-flow convolution formulas

Dual to the deletion–contraction recurrence for flows (Proposition 12.9) and the specialization of the Tutte polynomial to the flow polynomial (Theorem 12.13) are the following recurrence and specialization for colorings (for which see Chapter 11).

Proposition 12.15. *The chromatic polynomial of a graph G satisfies the recurrence relation*

$$\chi(G;k) = \chi(G\backslash e;k) - \chi(G/e;k), \tag{12.2}$$

for any edge e.

Theorem 12.16. *The chromatic polynomial of a graph is a specialization of the Tutte polynomial:*

$$\chi(G;k) = (-1)^{r(G)} k^{k(G)}\, T(G;1-k,0).$$

The dichromatic polynomial of Definition 2.1 can be expressed as a generating function for k-colorings of G (not necessarily proper) counted according to the number of monochromatic edges, that is, edges receiving the same color on their endpoints. (Edges are taken with their multiplicity when counting the number of monochromatic edges in the exponent of y.)

Proposition 12.17. *The dichromatic polynomial has the following expansion for positive integer k:*

$$Z(G;k,y-1) = \sum_{\kappa:V(G)\to[k]} y^{\#\{uv\in E(G):\kappa(u)=\kappa(v)\}}. \tag{12.3}$$

The right-hand side of Equation (12.3) is also known under the name of *bad coloring polynomial* [454], *monochrome polynomial* [1139], and *monochromial* [11]. When suitably parameterized $Z(G;k,y-1)$ gives the partition function of the k-state Potts model (see Chapter 20). Furthermore, by Theorem 2.6 we have the following identity.

Theorem 12.18. *For a graph $G=(V,E)$ and positive integer k,*

$$\sum_{\kappa:V\to[k]} y^{\#\{uv\in E:\kappa(u)=\kappa(v)\}} = k^{k(G)}(y-1)^{r(G)}\, T(G;\tfrac{k+y-1}{y-1},y).$$

For flows the dual result is as follows.

Theorem 12.19. *Let A be a finite additive abelian group of size k, and $G=(V,E)$ be a graph. Then*

$$\sum_{\substack{\phi:E\to\mathsf{A}\\ \phi \text{ an } \mathsf{A}\text{-flow}}} x^{|E|-|\mathrm{supp}(\phi)|} = (x-1)^{n(G)} T(G;x,\tfrac{k+x-1}{x-1}).$$

Remark 12.20. The polynomial $k^{-k(G)}Z(G;k,y-1)$ is an invariant of the cycle matroid of G and is a generating function for $\mathbb{Z}_k$-tensions of G counted according to the size of their support. Tensions (see Definition 12.3) rather than vertex colorings are the proper object of study when considering the structure of the cycle matroid of a graph, and make coloring-flow duality for graphs more transparent. Tension-flow duality has been thoroughly explored in, for example, [376, 877].

Definition 12.21. Let $G = (V, E)$ be a graph. For $A \subseteq E$, let the *bond closure* of A, denoted $\mathrm{cl}(A)$, be the set of all edges both of whose endpoints belong to one and the same block of the spanning subgraph (V, A). The lattice of subsets A with $\mathrm{cl}(A) = A$ is called the *bond lattice* of the graph G, denoted by $\mathcal{L}(G)$.

The edges assigned the value zero in a tension (flow) once contracted (deleted) leave a nowhere-zero tension (flow) on the resulting graph whose edges are the support of the tension (flow). The support of a tension is a flat in the cycle matroid of G, that is, an element of the bond lattice of G. The support A of a tension has the property that if all but one element of a circuit C belongs to A, then all of the edges of C do. The support A of a flow has the dual property that if all but one edge of a bond B belong to A, then all of the edges of B do. Equivalently, $n(A \cup e) = n(A) + 1$ for all $e \in E \setminus A$. The support of a flow is a flat in the *bond matroid* of G (the dual of the cycle matroid of G).

Theorems 12.18 and 12.19 can be re-expressed as follows.

Theorem 12.22. *Let $G = (V, E)$ be a graph with bond lattice $\mathcal{L}(G)$ and circuit lattice $\mathcal{L}^*(G)$ (bond lattice of the dual of the cycle matroid of G). Then*

$$\sum_{A \in \mathcal{L}(G)} y^{|A|} \chi(G/A; k) = k^{k(G)} (y-1)^{r(G)} T(G; \tfrac{y-1+k}{y-1}, y).$$

Dually,

$$\sum_{A \in \mathcal{L}^*(G)} x^{|E|-|A|} F(G|A; k) = (x-1)^{n(G)} T(G; x, \tfrac{x-1+k}{x-1}),$$

where $G|A = G \backslash (E \setminus A)$ is the spanning subgraph on A.

Theorem 12.22 has a generalization to matroids (in which the characteristic polynomial replaces the chromatic polynomial—see Definition 4.136) due to H. Crapo [349] and Tutte [1095].

For a fixed graph G, let G_p denote the random subgraph of G obtained by deleting each edge of G independently with probability $1 - p$. If f is a graph invariant, then the expected value of f on a random subgraph G_p of G is by definition equal to $\sum_{A \subseteq E} p^{|A|}(1-p)^{|E|-|A|} f(G|A)$. As an application of the expansions of Theorem 12.22 we have [1133]:

Theorem 12.23. *For a graph G and $0 < p \leq 1$, the random subgraph G_p has chromatic polynomial whose expectation is given by*

$$\mathbb{E}\, \chi(G_p; k) = (-p)^{r(G)} k^{k(G)} T(G; 1 - \tfrac{k}{p}, 1 - p).$$

Dually, for a graph G and $p \in (0, \frac{1}{2}) \cup (\frac{1}{2}, 1)$, the random subgraph G_p has flow polynomial whose expectation is given by

$$\mathbb{E}\, F(G_p; k) = p^{r(G)} (1 - 2p)^{n(G)} T(G; \tfrac{1-p}{p}, 1 + \tfrac{kp}{1-2p}).$$

When $p = \frac{1}{2}$,

$$\mathbb{E}\, F(G_p; k) = k^{n(G)} 2^{-|E|}.$$

The convolution formula for Tutte polynomial of a matroid given independently in [704] and (implicitly) in [469] can be written for graphs in the form of a tension-flow convolution [959]:

Theorem 12.24. *For a graph* $G = (V, E)$,

$$T(G; x, y) = \sum_{A \subseteq E} T(G/A; x, 0)\, T(G|A; 0, y).$$

Equivalently, for positive integers k, k' *and* A *and* A' *additive abelian groups of orders* k *and* k' *respectively,*

$$T(G; 1-k, 1-k') = (-1)^{r(G)} \sum_{\substack{\theta\ \mathsf{A}\text{-tension} \\ \phi\ \mathsf{A}'\text{-flow} \\ \mathrm{supp}(\theta) \sqcup \mathrm{supp}(\phi) = E}} (-1)^{|\mathrm{supp}(\phi)|}.$$

A generalization of Theorem 12.24 was shown by J. Kung [735], which we state for graphs, but which holds, mutatis mutandis, for matroids more widely and for the multivariate Tutte polynomial in place of the Tutte polynomial.

Theorem 12.25. *For a graph* $G = (V, E)$ *and indeterminates* a, b, c, d,

$$\begin{aligned}\sum_{A \subseteq E} a^{r(E)-r(A)}(-d)^{|A|-r(A)} T(G|A; 1-a, 1-c) T(G/A; 1+b, 1+d) \\ = T(G; 1+ab, 1+cd).\end{aligned} \tag{12.4}$$

A pair of special cases of identity (12.4) follow upon taking $ac = 1$ or $bd = 1$, the first of which, given as identity (12.5) below, is used in [81] to prove a result about win vectors (score vectors of partial orientations).

Theorem 12.26. *For a graph* $G = (V, E)$,

$$\begin{aligned}\sum_{A \subseteq E} (z-1)^{|A|} (x-1)^{r(E)-r(A)} T(G|A; x, y) \\ = (z-1)^{r(G)} z^{n(G)} T\left(G; (x-1)\tfrac{z}{z-1} + 1, (y-1)\tfrac{z-1}{z} + 1\right).\end{aligned} \tag{12.5}$$

Dual to identity (12.5) *is*

$$\begin{aligned}\sum_{A \subseteq E} (z-1)^{|E \setminus A|} (y-1)^{|A|-r(A)} T(G/A; x, y) \\ = z^{r(G)} (z-1)^{n(G)} T\left(G; (x-1)\tfrac{z-1}{z} + 1, (y-1)\tfrac{z}{z-1} + 1\right).\end{aligned} \tag{12.6}$$

F. Jaeger [649] obtained for matroids representable over a finite field a version of the following convolution formula for $k = k'$ a prime power. V. Reiner [959] then extended it to prime powers $k \neq k'$, and for graphic matroids deduced its validity for arbitrary positive integers k and k':

Theorem 12.27. *Let k and k' be positive integers, A and A' additive abelian groups of orders k and k' respectively, and z an indeterminate. Then*

$$(1-z)^{r(G)} z^{n(G)} T\left(G; \tfrac{1+(k-1)z}{1-z}, \tfrac{1+(k'-1)(1-z)}{z}\right) = \sum_{\substack{\theta \ \mathsf{A}\text{-tension} \\ \phi \ \mathsf{A}'\text{-flow} \\ \mathrm{supp}(\theta)\cap\mathrm{supp}(\phi)=\emptyset}} z^{|\mathrm{supp}(\theta)|}(1-z)^{|\mathrm{supp}(\phi)|}.$$

Theorem 12.24 is a special case of Theorem 12.27 on taking the limit $z \to \infty$. F. Breuer and R. Sanyal [200] and B. Chen [296] derived related expansions of the Tutte polynomial at integer points.

12.4 A-bicycles

For a graph G with orientation ω and an abelian group A, recall from Definitions 12.1 and 12.2 that if $D : \mathsf{A}^E \to \mathsf{A}^V$ is the incidence mapping, then the set $\ker(D)$ consists of A-flows of G and the set $\mathrm{im}(D^T)$ consists of A-tensions of G (with a given orientation ω). The intersection $\ker(D)\cap\mathrm{im}(D^T)$ of the submodule of A-tensions with the submodule of A-flows is trivial if A is torsion-free, but reveals an interesting structure for finite abelian groups A. For $\mathsf{A} = \mathbb{Z}_2$ and $\mathsf{A} = \mathbb{Z}_3$ the dimension of this subspace is given in terms of evaluations of the Tutte polynomial at points $(-1,-1)$, $(i,-i)$ and $(e^{2\pi i/3}, e^{4\pi i/3})$, which are some of the special points at which the Tutte polynomial is not $\sharp P$-hard to compute [653].

We begin with $\mathsf{A} = \mathbb{Z}_2$, in which a $\mathbb{Z}_2$-flow (cycle) that is also a $\mathbb{Z}_2$-tension (cutset) is known as a *bicycle.* For more about bicycles see Sections 14.15–16 and 15.7 in [546], and for the usefulness of bicycles in relation to knots see Chapter 17 of the same reference. The following describes the *principal tripartition* [969] of edges of a graph.

Theorem 12.28. *Let e be an edge of a graph G. Then precisely one of the following holds:*

1. *e belongs to a bicycle,*
2. *e belongs to a cutset B such that $B \setminus \{e\}$ is a cycle,*
3. *e belongs to a cycle C such that $C \setminus \{e\}$ is a cutset.*

An edge e of G is of *bicycle-type*, *cut-type* or *flow-type* according as Items 1–3 holds in the statement of Theorem 12.28, respectively. A bridge is an edge of cut-type (take cut $B = \{e\}$ in Item 2) and a loop is an edge of flow-type (take cycle $\{e\}$ in Item 3). If G is planar, then edges of bicycle-type in G remain of bicycle-type in G^*. By tension-flow duality, edges of cut-type in G are edges of flow-type in G^*, and similarly edges of flow-type in G^* are edges of cut-type in G^*.

Lemma 12.29. *Let $G = (V, E)$ be a graph with bicycle space of dimension $b(G)$, and let e be an edge of G. Then the graph invariant*

$$f(G) = (-1)^{|E|}(-2)^{b(G)}$$

satisfies

$$f(G) = \begin{cases} (-1)f(G/e) & e \text{ a bridge,} \\ (-1)f(G\backslash e) & e \text{ a loop,} \\ f(G/e) + f(G\backslash e) & e \text{ ordinary.} \end{cases}$$

Universality of the Tutte polynomial for deletion–contraction invariants (Theorem 2.21) yields the following polynomial-time computable evaluation of the Tutte polynomial [969]:

Theorem 12.30. *Let $G = (V, E)$ be a graph and let $b(G)$ denote the dimension of its bicycle space. Then*

$$(-1)^{|E|}(-2)^{b(G)} = T(G; -1, -1).$$

The point $(-1, -1)$ lies on the hyperbola $(x - 1)(y - 1) = 4$, so that by Theorem 12.19

$$T(G; -1, -1) = (-2)^{-n(G)} \sum_{\mathbb{Z}_2 \oplus \mathbb{Z}_2\text{-flows } \phi} (-1)^{|E|-|\mathrm{supp}(\phi)|}.$$

A $\mathbb{Z}_2 \oplus \mathbb{Z}_2$-bicycle decomposes by projection into a pair of $\mathbb{Z}_2$-bicycles, and conversely such a pair of $\mathbb{Z}_2$-bicycles can be pieced together to make a $\mathbb{Z}_2 \oplus \mathbb{Z}_2$-bicycle. Hence there are precisely $(2^{b(G)})^2$ vectors that are $\mathbb{Z}_2 \oplus \mathbb{Z}_2$-bicycles, that is, they comprise a space of dimension $b(G)$ over $\mathbb{F}_4$. Hence Theorem 12.30 could also have stated that $T(G; -1, -1) = (-1)^{|E|}(-2)^{b(G)}$, where $b(G)$ is now defined as the dimension of the space of $\mathbb{Z}_2 \oplus \mathbb{Z}_2$-bicycles.

We now take $\mathsf{A} = \mathbb{Z}_3$ and $D : \mathbb{Z}_3^E \to \mathbb{Z}_3^V$ as the boundary mapping. An element of $\ker(D) \cap \mathrm{im}(D^T)$ is a $\mathbb{Z}_3$-*bicycle*. In other words, a $\mathbb{Z}_3$-bicycle is both a $\mathbb{Z}_3$-tension and a $\mathbb{Z}_3$-flow, and is self-orthogonal in $\mathbb{Z}_3^E$. Let $i = \sqrt{-1}$ and $j = e^{2\pi i/3}$. In [650] Jaeger proved by a deletion–contraction argument that $T(G; j, j^2) = \pm j^{|E|+\dim(\ker(D))}(i\sqrt{3})^{\dim(\ker(D)\cap \mathrm{im}(D^T))}$, using the *principal quadripartition* of the edges of a graph (a generalization to flows and tensions over finite fields of characteristic $\neq 2$ of the principal tripartition of Theorem 12.28). E. Gioan and M. Las Vergnas [533] provide a linear algebra proof that has the benefit of determining the sign.

Theorem 12.31. *Let $G = (V, E)$ be a graph and $j = e^{2\pi i/3}$. We have*

$$T(G; j, j^2) = (-1)^{q(G)} j^{|E|+n(G)} (i\sqrt{3})^{b_3(G)},$$

where $b_3(G)$ is the dimension of the space of $\mathbb{Z}_3$-bicycles of G, and $q(G)$ is the number of vectors with support size congruent to 2 modulo 3 in any orthogonal basis for the space of $\mathbb{Z}_3$-flows of G.

D. Vertigan [1116] proved that the Tutte polynomial evaluated at the point $(i, -i)$ on the hyperbola $(x-1)(y-1) = 2$ has the following interpretation:

Theorem 12.32. *Let G be a graph with bicycle dimension $b(G)$. Then*

$$|T(G; i, -i)| = \begin{cases} \sqrt{2}^{b(G)} & \text{if every bicycle has size a multiple of 4,} \\ 0 & \text{otherwise.} \end{cases}$$

As every bicycle has even size, the bicycles of size a multiple of 4 either comprise all bicycles, or exactly half of them. R. Pendavingh [912] determines the argument of the complex number $T(G; i, -i)$ in terms of a certain $\mathbb{Z}_4$-valued quadratic form canonically associated with G.

While the number of (nowhere-zero) A-flows and (nowhere-zero) A-tensions are independent of the structure of A, the same is not true in general of the number of A-bicycles, as shown in [98].

Definition 12.33. Let $D : \mathbb{Z}^E \to \mathbb{Z}^V$ be the integer boundary mapping and $D^T : \mathbb{Z}^V \to \mathbb{Z}^E$ the integer coboundary mapping. The *critical group* $K(G)$ of G (with arbitrary orientation ω) is defined as the quotient of integer lattices

$$K(G) = \mathbb{Z}^E / (\ker(D) \oplus \operatorname{im}(D^T)).$$

The invariant factors $n_1, \ldots, n_r$ of the critical group $K(G) \cong \mathbb{Z}_{n_1} \oplus \mathbb{Z}_{n_2} \oplus \cdots \oplus \mathbb{Z}_{n_r}$ satisfy $n_1 \mid n_2 \mid \cdots \mid n_r$ and, by Kirchhoff's matrix-tree theorem, $n_1 n_2 \cdots n_r = T(G; 1, 1)$ (see for example [113]). For finite additive abelian group A and $n \in \mathbb{Z}$, let $\mathsf{A}^{(n)} = \{a \in \mathsf{A} : na = 0\}$.

Theorem 12.34. *Let G be a graph with a fixed arbitrary orientation and $n_1, \ldots, n_r$ be the invariant factors of the critical group of G. Then the module of A-bicycles of G has the following direct sum decomposition:*

$$\ker(D) \cap \operatorname{im}(D^T) \cong \mathsf{A}^{(n_1)} \oplus \mathsf{A}^{(n_2)} \oplus \cdots \oplus \mathsf{A}^{(n_r)}.$$

12.5 Open problems

As well as establishing an algebraic and homological approach to colorings and flows that is still being elaborated to this day, Tutte formulated a series of conjectures concerning flows, each of which reflects miraculous intuition and each of which still resists solution. Most of these ask about the existence of flows. As noted in Proposition 12.7 a graph has a nowhere-zero $\mathbb{Z}_k$-flow if and only if G has a nowhere-zero integer k-flow. Thus, these questions are typically stated just in terms of k-flows, without the need to specify which kind, and we follow that convention here.

Grötzsch's theorem [588] states that every triangle-free loopless planar graph is 3-colorable. By coloring-flow duality, this is equivalent to the statement that every 4-edge-connected planar graph has a nowhere-zero 3-flow. Tutte's 3-flow conjecture (first popularized in the 1970s [162] as formulated by Tutte in 1972) asserts that this is still true without the assumption of planarity.

Conjecture 12.35 (Tutte's 3-flow conjecture)**.** Every 4-edge-connected graph has a nowhere-zero 3-flow.

Kochol [698] showed that it suffices to prove Tutte's 3-flow conjecture for 5-edge-connected graphs.

To help find an oblique path up the steep hill to the summit of resolving Tutte's 3-flow conjecture, Jaeger [645] asked first whether there indeed exists some fixed integer k such that every k-edge-connected graph has a nowhere-zero 3-flow (Jaeger's weak 3-flow conjecture). Spectacular progress has recently been made by C. Thomassen [1066], who verified Jaeger's weak 3-flow conjecture by proving that every 8-edge-connected graph admits a nowhere-zero 3-flow. This was then improved by L. M. Lovász et al. [797], who proved that every 6-edge-connected graph has a nowhere-zero 3-flow.

Other references concerning the 3-flow conjecture include [645, 647, 652, 698, 1066, 1067]. For the 4-flow and 5-flow conjectures, to which we now turn, the reader may consult [645, 690, 966, 999, 1092, 1094].

Conjecture 12.36 (Tutte's 4-flow conjecture)**.** Every bridgeless graph without a Petersen minor has a nowhere-zero 4-flow.

Jaeger [645] used the fact, due to Tutte and to C. Nash-Williams, that a 4-edge-connected graph has two edge-disjoint spanning trees in order to prove that every 4-edge-connected graph has a nowhere-zero 4-flow. As nowhere-zero 4-flows of a cubic graph correspond to proper edge 3-colorings, the 4-flow conjecture implies that every bridgeless cubic graph without a Petersen minor is 3-edge-colorable. The latter was another conjecture of Tutte. A proof was announced by N. Robertson, P. Seymour, and R. Thomas in [966], thereby giving a strengthening of the four color theorem, which is equivalent to the assertion that planar cubic graphs without a bridge are 3-edge-colorable.

Tutte conjectured that there is no obstruction to a nowhere-zero A-flow when $|\mathsf{A}| \geq 5$ other than a bridge (unlike cliques of order $k+1$ being obstructions to having a nowhere-zero k-tension).

Conjecture 12.37 (Tutte's 5-flow conjecture)**.** Every bridgeless graph has a nowhere-zero 5-flow.

Every bridgeless planar graph has a nowhere-zero 5-flow by coloring-flow duality and Heawood's theorem that every loopless planar graph is 5-colorable [604]. It is not difficult to show that a minimal counterexample to the 5-flow conjecture must be a *snark*, that is, a cyclically 4-edge-connected cubic graph without a proper edge 3-coloring and with girth at least 5; see [647]. (A graph is cyclically k-edge-connected if removing at most $k-1$ edges from G leaves at most one component containing a cycle.) U. Celmins [272] proved that a minimal counterexample must be a cyclically 5-edge-connected snark with girth at least 7. Kochol showed further that such a minimal counterexample snark must be cyclically 6-edge-connected in [700] and in [701, 702] that it must have girth at least 11.

A parameter measuring how far a bridgeless cubic graph is from being 3-edge-colorable is its *oddness*, defined as the minimum number of odd circuits in a 2-factor of G. Since G has an even number of vertices, its oddness is necessarily even. Furthermore, its oddness is 0 if and only if G is 3-edge-colorable. Jaeger [647] showed that cubic graphs with oddness at most 2 have a nowhere-zero 5-flow. Furthermore, a consequence of the main result in [1044] is that cyclically 7-edge-connected cubic graphs with oddness at most 4 have a nowhere-zero 5-flow. In [831], it is shown that a cyclically 6-edge-connected cubic graph G of oddness at most 4 has a nowhere-zero 5-flow.

Tutte made the 5-flow conjecture at a time when it was not even clear that there was a k such that every graph has a nowhere-zero k-flow. (The dual statement for the chromatic polynomial is patently false.) Jaeger [645] and P. Kilpatrick [690] independently proved using $\mathbb{Z}_2 \oplus \mathbb{Z}_2 \oplus \mathbb{Z}_2$-flows that every bridgeless graph has a nowhere-zero 8-flow. Later, Seymour [999] proved using $\mathbb{Z}_2 \oplus \mathbb{Z}_3$-flows that every bridgeless graph has a nowhere-zero 6-flow.

Open Question 12.38. Are the real roots of the flow polynomial bounded above?

Tutte's flow conjectures ask for a characterization of those graphs for which the flow polynomial has respectively 3, 4 or 5 as a root. A conjecture of G. Birkhoff and D. Lewis [124], which is still open, states that the real roots of the chromatic polynomial of a planar graph do not exceed 4 (Birkhoff and Lewis did prove that 5 is an upper bound). The dual statement for the flow polynomial is that its real roots for a planar graph do not exceed 4. D. Welsh (as reported in [638]) conjectured that the latter statement is true without the assumption of planarity. G. Haggard et al. [596] produced counterexamples to Welsh's conjecture, including a generalized Petersen graph, and modified the conjecture to state that the real roots of the flow polynomial are at most 5. J. Jacobsen and J. Salas [643] thereupon produced, after extensive computations, another generalized Petersen graph whose roots exceed 5, and proposed that the real roots of the flow polynomial do not exceed 6. It is still not known, however, whether the real roots of the flow polynomial are bounded above at all. For further results on flow roots see the survey [638].

13

Skein polynomials and the Tutte polynomial when $x = y$

Joanna A. Ellis-Monaghan • Iain Moffatt

Synopsis

This chapter surveys some graph polynomials that are based on medial graph constructions. While none of these polynomials are specializations of the Tutte polynomial, all of them coincide with the Tutte polynomial for special classes of graphs or along special curves. We give these relations to the Tutte polynomial, as well as a number of combinatorial interpretations that derive from them.

- A brief review of vertex and graph states, and skein relations.
- Some graph and link polynomials arising from skein relations, including: the Martin, or circuit partition, polynomial; the Penrose polynomial; the Kauffman bracket; and transition polynomials.
- Evaluations of the Tutte polynomial when $x = y$ that come from medial graph and skein polynomial connections.

13.1 Introduction

We consider graph polynomials that can be defined through *skein relations*, which in this chapter are recursive reductions that involve splitting a graph at a vertex. Jaeger's transition polynomial and the generalized transition polynomial unify skein polynomials such as the Martin or circuit partition polynomials, the Penrose polynomial, and the Kauffman bracket, as well as their "topological" extensions to graphs in surfaces. Moreover, medial graph

DOI: 10.1201/9780429161612-13

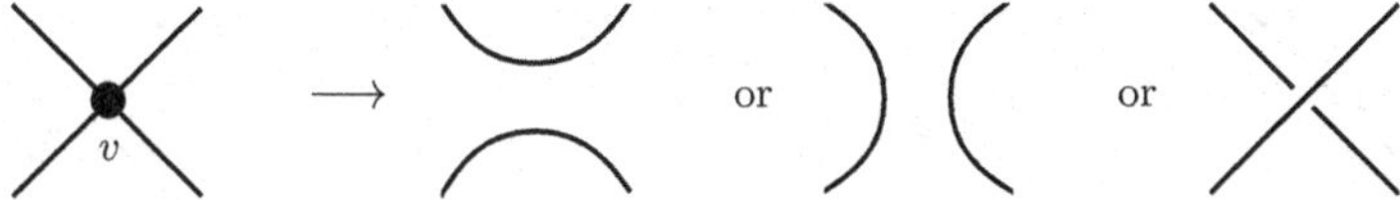

FIGURE 13.1: The three vertex states of a degree-four vertex v of an abstract graph (shown after splicing). In all four diagrams, the rest of the graph is the same outside the configurations shown.

constructions for both plane and embedded graphs relate the Tutte polynomial and its generalizations to various transition polynomials, albeit along restricted curves or surfaces. This connection between deletion–contraction invariants and skein invariants has proved highly fruitful, as new understandings in one setting transfer to the other, with the result that identities and combinatorial interpretations propagate through the different types of polynomials.

We give an overview here of the operations of skein reductions, as well as a selection of skein and transition polynomials with their properties and evaluations. We focus on the connections of these ideas to the Tutte polynomial and the new information that results.

13.2 Vertex states, graph states, and skein relations

Vertex states and graph states lead to state model formulations for the skein polynomials covered in this chapter. A *vertex state* at a degree-four vertex v in an abstract graph G is a partition, into pairs, of the edges incident with v. If edges (a, v) and (b, v) are paired, and (c, v) and (d, v) are paired, then these edges are removed from the graph and replaced with edges (a, b) and (c, d) as in Figure 13.1. We refer to this replacement as *splicing*. In the case of a loop, say $a = b$ in the above, then the added edge (a, b) is a *free loop* which is an edge with no vertex, and is considered as a closed curve.

A *graph state* s of a 4-regular graph results from choosing a vertex state at each of its vertices. Through splicing, a graph state s gives rise to a set of disjoint closed curves called the *components* of the state.

We begin by defining vertex states for vertices of degree-four and and graphs states for 4-regular graphs since these are involved in many applications. However, vertex states, graph states, and splicing can be defined for Eulerian graphs in general, as follows.

Definition 13.1. A *vertex state* at a vertex v is a partition, into pairs, of the edges incident with v. A *graph state* is a choice of vertex state at each of its vertices. Through splicing, a graph state s gives rise to a set of disjoint

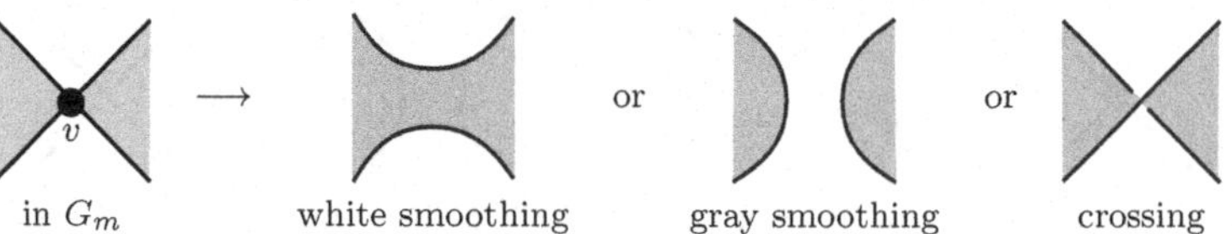

FIGURE 13.2: The three vertex states (shown after splicing) of a vertex v of a checkerboard colored embedded graph.

closed curves, called the *components* of the state. We denote the number of components of s by $c(s)$.

A graph state with exactly k components may also be called an *Eulerian k-partition* of the graph (for example in [748, 824]), and vertex and graph states also correspond to the *transitions* and *transition systems* of Fleischner from [488, 489].

Computing the skein polynomials given later in this chapter often necessitates distinguishing among the possible vertex states and there are some common ways to do this. Recall from Section 1.2.17 the construction from an embedded graph G of its canonically checkerboard colored medial graph G_m, here using the convention that the medial graph of an isolated vertex is a free loop. We may use the canonical checkerboard coloring of G_m (or any checkerboard colored 4-regular graph) to distinguish among the three vertex states at a degree-four vertex v, calling them a *white smoothing*, a *gray smoothing*, and a *crossing* as indicated in Figure 13.2.

Edge orientations of 4-regular digraphs also distinguish one vertex state that is different from the other two. At each vertex of a 4-regular digraph, there is one state that pairs the two incoming edges and also pairs the two outgoing edges. This vertex state is said to be *inconsistent* with the underlying orientation of the digraph, while the remaining two, which pair incoming with outgoing edges, are called *consistent*.

Each graph state consists of a collection of closed curves. Various graph polynomials arise by assigning a weight to each graph state of a graph, then summing these weights over the set of all possible graph states of the graph. (The weights may include indeterminates, and here depend upon the number of components and the vertex states of the graph states.) Alternatively, various graph polynomials arise by reducing a graph via a weighted sum of vertex states. This gives a linear recursion relation for a graph polynomial, called a *skein relation*, a term appropriated from knot theory.

In this chapter a *skein relation* for a graph polynomial (or graph invariant) is a linear relation that expresses the polynomial of a graph in terms of the polynomials of graphs that result from replacing some vertex with the results of splicing its possible vertex states. For example, the top line of Figure 13.3 shows a skein relation for the Martin polynomial $M(G;x)$. In the figure, the images , , , and represent four graphs that are identical except in the local region shown, where they differ as illustrated. Thus shows a graph G locally at a vertex (in this particular example it must not be a cut-vertex), , , represent the three graphs G_1, G_2 and G_3 that arise by replacing that vertex with its three vertex states. The skein relation says that the value of the polynomial on the graph G is equal to the sum of its values on the graphs G_1, G_2 and G_3.

Repeated application of a skein relation to a graph results in an expression where the graphs are all closed curves or, more generally, are graphs that are "simpler" than the original in some sense. In such cases a polynomial or an element of a ring can be obtained by providing rules for sending a the closed curves (or the "simpler" graphs) to a ring element. If the value of such a calculation is independent of any choices made in the computation, then we obtain a graph invariant called a *skein invariant*, or a *skein polynomial* if it is polynomial-valued.

For example, repeated application of the skein relation for the Martin polynomial given in the top line of Figure 13.3 to non-cut-vertices results in an expression involving graphs whose vertices are all cut-vertices. Repeated application of the relation in the middle line of the figure will reduce this expression to one involving polynomials of closed loops. The third relation in Figure 13.3 then states that the polynomial of a closed loop is 1. Taken together these three relations provide a recursive way to calculate the Martin polynomial for any connected 4-regular graph. Such a recursive definition involving a skein relation is called a *skein definition* of a polynomial or invariant.

13.3 Skein polynomials

Graph states and skein relations give rise to a family of skein polynomials. We present several of its important members here. Note that as graph states are also called transition systems, skein polynomials are also called *transition polynomials*, for example in the seminal work by Jaeger [651]. The are also a motivating example in Yetter's discussion of graph invariants given by linear recursion relations in [1179].

13.3.1 The Martin and circuit partition polynomials

The Martin polynomials, from Martin's 1977 dissertation [824], were the first skein graph polynomials. The Martin polynomials were originally formulated for 4-regular graphs and digraphs, but Las Vergnas subsequently showed that they extend naturally to all Eulerian graphs and digraphs and have a generating function formulation [744, 748, 751]. Recall that an *Eulerian digraph* has, at each vertex, the same number of incoming as outgoing edges, and an

Eulerian orientation of an Eulerian graph is an assignment of directions to the edges that results in an Eulerian digraph. The Martin polynomials, and the circuit partition polynomials which are reformulations of the Martin polynomials, encode information about cycles in Eulerian graphs and digraphs.

The term *Eulerian graph* is used somewhat loosely in this section, as the graphs need not be connected. We only require that all vertices are of even degree. We follow the analogous convention for Eulerian digraphs. Recall that an Eulerian k-partition is a graph state with exactly k components.

Definition 13.2. The *Martin polynomial,* $M(G;x)$, of a connected Eulerian graph G is

$$M(G;x) = \sum_{k\geq 0} f_{k+1}(G)(x-2)^k,$$

and the *Martin polynomial,* $m(\vec{G};x)$, of a connected Eulerian digraph $\vec{G}$ is

$$m(\vec{G};x) = \sum_{k\geq 0} f_{k+1}(\vec{G})(x-1)^k,$$

where $f_k(G)$ and $f_k(\vec{G})$ are the numbers of Eulerian k-partitions of G and $\vec{G}$, respectively.

Example 13.3. If G is the connected 4-regular graph with two loops and two parallel edges, then $M(G;x) = 4(x-2)^0 + 4(x-2)^1 + (x-2)^2$.

The circuit partition polynomials, $J(G;x)$ and $j(\vec{G};x)$, defined in [448] and so-named in [153], have the following generating function formulations.

Definition 13.4. The *circuit partition polynomial,* $J(G;x)$, of an Eulerian graph G is

$$J(G;x) = \sum_{k\geq 1} f_k(G)\, x^k,$$

and the *circuit partition polynomial,* $j(\vec{G};x)$, of an Eulerian digraph $\vec{G}$ is

$$j(\vec{G};x) = \sum_{k\geq 1} f_k(\vec{G})\, x^k,$$

where $f_k(G)$ and $f_k(\vec{G})$ are the numbers of Eulerian k-partitions of G and $\vec{G}$, respectively.

Note that vertices of degree zero or two do not have any effect on either the Martin or circuit partition polynomials, so may be ignored.

For a connected graph G and connected digraph $\vec{G}$ the Martin polynomials and circuit partition polynomials are related as follows:

$$J(G;x) = x\, M(G;x+2), \qquad \text{and} \qquad j(\vec{G};x) = x\, m(\vec{G};x+1).$$

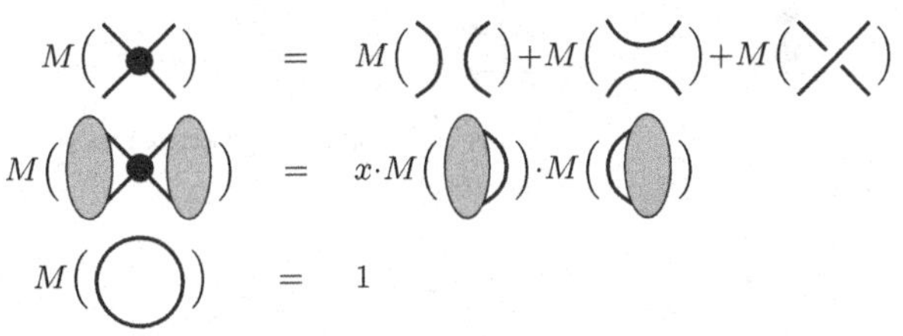

FIGURE 13.3: Skein definition for the Martin polynomial of 4-regular Eulerian graphs. The first equation is for non-cut-vertices, the second is for cut-vertices, and the third is for a free loop.

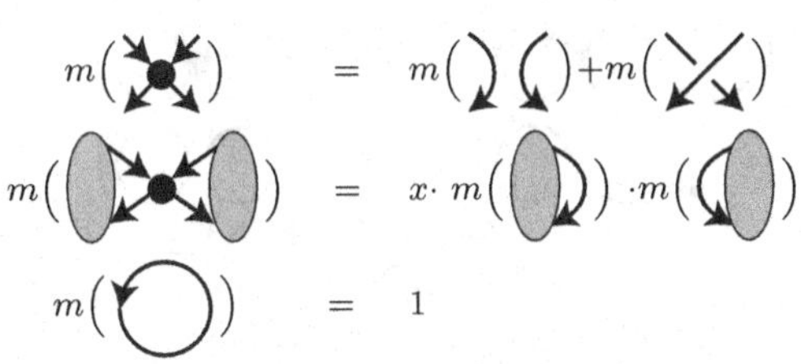

FIGURE 13.4: Skein definition for the Martin polynomial of 4-regular Eulerian digraphs. The first equation is for non-cut-vertices, the second is for cut-vertices, and the third is for a free loop.

This identity can be used to extend the Martin polynomial to non-connected graphs (see also Section 14.4 for the extension). Furthermore, for any Eulerian graph G,

$$J(G;x) = \sum j\left(\vec{G};\frac{x}{2}\right),$$

where the sum is over all Eulerian orientations $\vec{G}$ of G (see [748]).

The Martin and circuit partition polynomials satisfy skein relations. Figures 13.3 and 13.4 show the skein relations for degree-four vertices for the Martin polynomial for Eulerian graphs, and for Eulerian digraphs, respectively. For higher degree vertices, it is generally more convenient to use the generating function model from Definition 13.2, or to work with the circuit partition polynomial. Figure 13.5 shows a skein definition for the circuit partition polynomial for general Eulerian digraphs. A skein definition for Eulerian graphs is given by summing over all vertex states, not just consistent ones.

Example 13.5. Figure 13.6 illustrates the computation of $j(\vec{G};x) = x^3 + 2x^2 + x$ via the skein relation of Figure 13.5, where $\vec{G}$ is the Eulerian digraph consisting of a digon with a loop at each vertex.

The circuit partition polynomials are multiplicative on disjoint unions. They also have splitting formulas analogous to Tutte's identity for the chromatic polynomial shown in Example 11.4, Item 7.

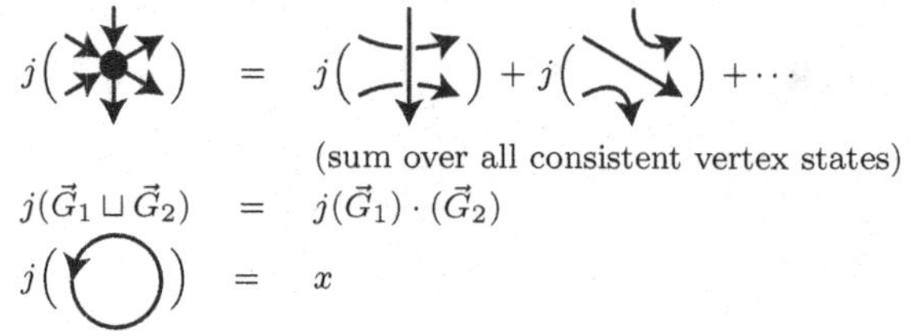

(sum over all consistent vertex states)

$$j(\vec{G}_1 \sqcup \vec{G}_2) = j(\vec{G}_1) \cdot (\vec{G}_2)$$

$$= x$$

FIGURE 13.5: Circuit partition polynomial skein definition for Eulerian digraphs.

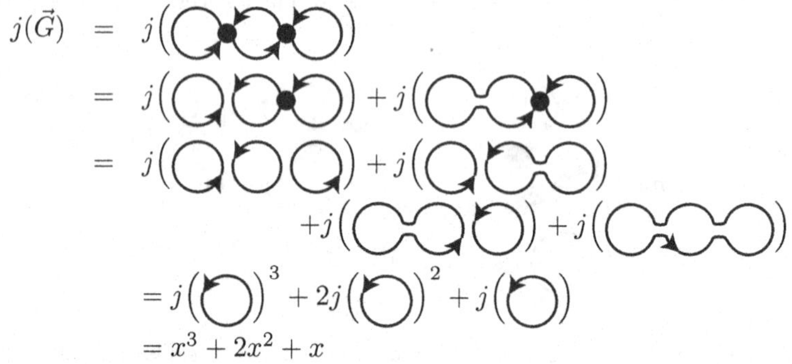

$$j(\vec{G}) = \ldots = x^3 + 2x^2 + x$$

FIGURE 13.6: Computing the circuit partition polynomial of an Eulerian digraph.

Theorem 13.6. *Using the notation $J(A;x)$ for $J(G[A];x)$ and A^c for $E \setminus A$,*

$$J(G; x+y) = \sum J(A;x)\ J(A^c;y),$$

where the sum is over all subsets A of $E(G)$ such that G restricted to both A and A^c is Eulerian. Also,

$$j(\vec{G}; x+y) = \sum j(\vec{A};x)\, j(\vec{A}^c;y),$$

where the sum is over all subsets A of $E(\vec{G})$ such that $\vec{G}$ restricted to both A and A^c is an Eulerian digraph.

Theorem 13.6 follows from the fact that $J(G;x)$ is a Hopf algebra map from a Hopf algebra of graphs to the binomial bialgebra, and hence $J(G; 1 \otimes x + x \otimes 1) = \sum J(A;x) \otimes J(A^c;x)$ (see [448]). Substituting x for $1 \otimes x$ and y for $x \otimes 1$ gives the result. Alternative simple combinatorial proofs as well as generalizations may be found in [153, 451].

Martin [824] found a connection between the Martin, or circuit partition, polynomial of certain digraphs and the Tutte polynomial of a plane graph G using the *directed medial graph* $\vec{G}_m$ (see Section 1.2.17 for a definition of $\vec{G}_m$).

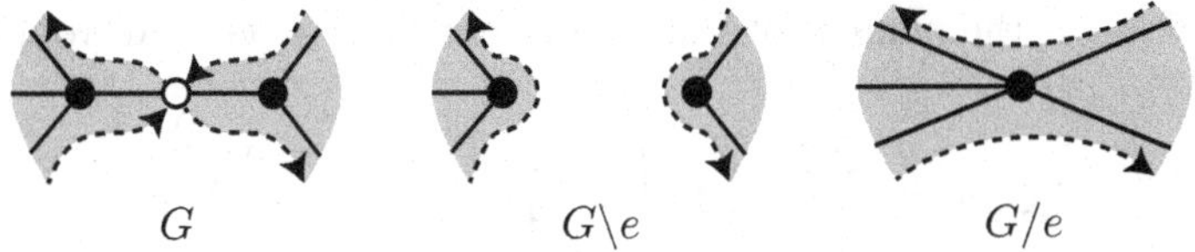

FIGURE 13.7: Relating vertex states of $\vec{G}_m$ to deletion and contraction in G.

Theorem 13.7. *Let G be a connected plane graph, and let $\vec{G}_m$ be its directed medial graph. Then*

$$T(G;x,x) = m(\vec{G}_m;x) = \frac{1}{x-1}j(\vec{G}_m;x-1). \tag{13.1}$$

This connection to Tutte polynomial follows from a fundamental observation relating deletion and contraction of an edge e in G with the appropriate vertex states at the vertex corresponding to e in $\vec{G}_m$, as shown in Figure 13.7. In the figure, G has black vertices and solid edges, and $\vec{G}_m$ has a white vertex and dashed edges, and we see that contraction in G corresponds to a white smoothing in $\vec{G}_m$ while deletion in G corresponds to a gray smoothing in $\vec{G}_m$.

Example 13.8. The digraph in Example 13.6 is the directed medial graph of the P_2, the path with two edges. Since $T(P_2;x,x) = x^2$, then indeed $xT(P_2;x+1,x+1) = x^3 + 2x^2 + x = j((\vec{P}_2)_m, x)$.

Both Martin [825] and Las Vergnas [751] found combinatorial interpretations for some small integer evaluations of the Martin polynomials, while combinatorial interpretations for all integer values as well as some combinatorial evaluations of some derivatives for the Martin and circuit partition polynomials were given in [153, 449, 450, 451]. The derivative formulas give insights into the derivatives of the Tutte polynomial in Section 13.4.1. These evaluations, combined with induction and the splitting formulas in Theorem 13.6 yield a number of combinatorial interpretations. We give two representative examples below, and more may be found in Martin [824, 825], Las Vergnas [744, 748, 751], Bollobás [153], and also [448, 449, 450, 451].

Write $Eul(G)$ for the number of Eulerian orientations of an Eulerian graph G, i.e. the number of ways to direct the edges so there are the same number of incoming as outgoing edges at each vertex. Let $A_n(G)$ denote the set $\{(A_1,\ldots,A_n)\}$ of ordered partitions of $E(G)$ into n subsets such that $G[A_i]$ is Eulerian for all i, recalling the convention that these graphs need not be connected. For digraphs, $\vec{A}_n(\vec{G})$ is defined similarly. From [153, 450, 451],

$$J(G;2n) = \sum_{A_n(G)} \prod_{i=1}^{n} \left(\mathrm{Eul}(A_i) \prod_{v\in V(A_i)} \left(\frac{\deg_{A_i}(v)}{2}\right)! \right),$$

where Eul(G) is the number of Eulerian orientations of G; and from [450],

$$j(\vec{G};n) = \sum_{\vec{A}_n(\vec{G})} \prod_{i=1}^{n} \left(\prod_{v \in V(\vec{A}_i)} \left(\frac{\deg_{\vec{A}_i}(v)}{2} \right)! \right).$$

These and similar identities and evaluations of the Martin and partition polynomials combined with the relationship of Theorem 13.7 gives the foundation for many of the combinatorial interpretation of the Tutte polynomial along the line $y = x$ described in Section 13.4.

Regts and Sevenster [956] use such identities and Theorem 13.6 to express the Martin polynomial at even negative integers as a skew-partition function of skew-symmetric tensors. Furthermore, the theory of isotropic systems, which unifies essential properties of 4-regular graphs and pairs of dual binary matroids, provides a framework to considerably extend the relation between the Tutte and Martin polynomials. Details of this can be found in Chapter 14.

13.3.2 The Penrose polynomial

The Penrose polynomial $P(G;\lambda)$ of a plane graph G appeared implicitly in work of Penrose [916] in the context of tensor diagrams in physics. Aigner developed it in a purely graph theoretical context in [10], and the following definition reveals that it can be viewed as skein polynomial of medial graphs (see also Definition 13.15, Item 3).

Definition 13.9. Let G be a plane graph and G_m be its canonically checkerboard colored medial graph. Then the *Penrose polynomial* is

$$P(G;\lambda) = \sum_{s} (-1)^{\mathrm{cr}(s)} \lambda^{c(s)}, \tag{13.2}$$

where the sum is over the graph states s of G_m that consist entirely of white smoothing and crossing vertex states (see Figure 13.2). Here $\mathrm{cr}(s)$ is the number of vertices with crossing states in s, and $c(s)$ its number of components.

Example 13.10. If G is the plane theta-graph, then $P(G;\lambda) = \lambda^3 - 3\lambda^2 + 2\lambda$, as in Figure 13.8.

The Penrose polynomial has some surprising properties, particularly with respect to graph coloring. The four color theorem is equivalent to showing that every planar, cubic, connected graph can be properly edge-colored with three colors. The Penrose polynomial encodes exactly this information (see [916]).

Theorem 13.11. *Let G be a plane, cubic, connected graph. Then*

$$P(G;3) = \left(\frac{-1}{4}\right)^{\frac{|V|}{2}} P(G;-2) = \text{the number of edge-3-colorings of } G.$$

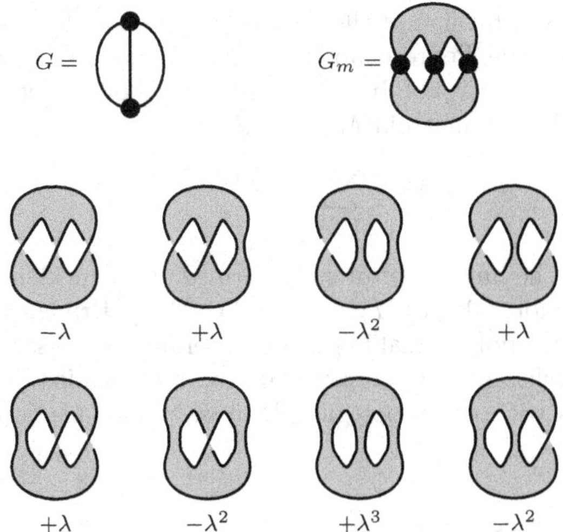

FIGURE 13.8: Computing the Penrose polynomial, $P(G;\lambda) = \lambda^3 - 3\lambda^2 + 2\lambda$.

A good resource for other properties and evaluations of the Penrose polynomial is [10].

Since the medial graph of an embedded graph may also be checkerboard colored, the definition of the Penrose polynomial given in (13.2) extends immediately to graphs embedded in surfaces. The resulting polynomial, often called the *topological Penrose polynomial*, was first considered in [458]. The main advantage of considering the Penrose polynomial of non-plane graphs is that in this class it satisfies a recursive deletion–contraction-type definition (see Theorem 5.2 of [458]):

$$P(G;\lambda) = P(G/e,\lambda) - P(G^{\tau(e)}/e;\lambda),$$

or equivalently

$$P(G^{\delta\tau(e)};\lambda) = P(G\backslash e;\lambda) - P(G/e;\lambda).$$

Here δ and τ are the partial dual and twist of an edge defined in Section 1.2.16. Furthermore, the class of non-plane graphs allows identities that cannot be realized for plane graphs, such as the following connection with the chromatic polynomial $\chi(G;\lambda)$ from [456].

Theorem 13.12. *Let G be a ribbon graph. Then*

$$P(G;\lambda) = \sum_{A\subseteq E(G)} (-1)^{|A|}\chi((G^{\tau(A)})^*;\lambda).$$

Here $G^{\tau(A)}$ is the result of taking the partial Petrial with respect to A, i.e. giving a half-twist to each of edges in A, again as in Subsection 1.2.16.

An alternative extension of the Penrose polynomial for embedded graphs appears in [453]. The Penrose polynomial has also been extended to both matroids and delta-matroids. In [12]. Aigner and Mielke defined the Penrose polynomial of a binary matroid $M = (E, \mathcal{F})$ as

$$P(M;\lambda) = \sum_{X\subseteq E} (-1)^{|X|}\lambda^{\dim(B_M(X))}, \tag{13.3}$$

where $B_M(X)$ is the binary vector space formed of the incidence vectors of the sets in the collection $\{A \in \mathcal{C}(M) : A \cap X \in \mathcal{C}^*(M)\}$. Brijder and Hoogeboom defined the Penrose polynomial in greater generality for vf-safe delta-matroids in [206]. If the delta-matroid corresponds to an embedded graph, then their definition agrees with the topological Penrose polynomial up to a factor of $\lambda^{k(G)}$ (see [318]).

13.3.3 The Tutte polynomial

The Tutte polynomial of a plane graph may be expressed, along a restricted curve, in terms of the states of its medial graph. The connection was first observed by Martin in his 1977 thesis [824] and in [825]. We give a more general version here.

Theorem 13.13. *Let G be a plane graph and G_m be its canonically checkerboard colored medial graph. Then*

$$\sum_{s} a^{\mathrm{gr}(s)} b^{\mathrm{wh}(s)} t^{c(s)} = t^{k(G)} a^{n(G)} b^{r(G)} T\Big(G; \frac{at}{b}+1, \frac{bt}{a}+1\Big), \tag{13.4}$$

where the sum is over all graph states s of G_m that have no crossing states, where $\mathrm{gr}(s)$ *is the number of gray states in s and* $\mathrm{ws}(s)$ *is the number of white states in s, and where $c(s)$ its number of components.*

The identity in Theorem 13.13 does not hold for non-plane graphs in general. However, an analogous formula holds for graphs on surfaces by using the checkerboard colored medial graph of an embedded graph and the Bollobás–Riordan polynomial $R(G)$ (see Chapter 27) instead of the Tutte polynomial.

Theorem 13.14. *Let G be a graph cellularly embedded in any surface and G_m be its canonically checkerboard colored medial graph. Then*

$$\sum_{s} a^{\mathrm{gr}(s)} b^{\mathrm{wh}(s)} t^{c(s)} = t^{k(G)} a^{n(G)} b^{r(G)} R\Big(G; \frac{at}{b}+1, \frac{bt}{a}, \frac{1}{t}, 1\Big),$$

where the sum is over all graph states s of G_m that have no crossing states, where $\mathrm{gr}(s)$ *is the number of gray states in s and* $\mathrm{ws}(s)$ *is the number of white states in s, and where $c(s)$ its number of components.*

Theorem 13.14 may be found in [462], and Theorem 13.13 is an immediate consequence of it. See [457] for an overview of these and similar identities.

13.3.4 The Kauffman bracket

The Kauffman bracket is a polynomial valued function on knot and link diagrams. See Chapter 18 for notation and background on it. Consider a link diagram as a 4-regular plane graph in which the crossings are vertices with additional information to record the crossing. The A- and B-splicings of Figure 18.6 can then be regarded as vertex states. Moreover, the crossing type of the diagram at that vertex distinguishes between the vertex states, again as in Figure 18.4. With this

$$\sum_s A^{\alpha(s)} B^{\beta(s)} d^{c(s)} = d[D](A, B, d), \tag{13.5}$$

and

$$\sum_s A^{\alpha(s)} A^{-\beta(s)} (-A^2 - A^{-2})^{c(s)} = (-A^2 - A^{-2}) \langle D \rangle (A) \tag{13.6}$$

where the sum is over all graph states s, $\alpha(s)$ is the number of A-splicings, and $\beta(s)$ is the number of B-splicings.

The Kauffman bracket of a virtual link digram (again see Chapter 18) can be obtained in a similar way. Similarly, consider the virtual link diagram as a plane graph. As the graph is plane, there is a cyclic order of the half-edges at each vertex. Writing the cyclic order of half-edges as $(abcd)$, we call a vertex state *crossing* if pairs ac and bd . With this, Equations (13.5) and (13.6) hold when D is a virtual link diagram, provided that the sum is restricted to graph states in which each virtual crossing has a crossing vertex state.

13.3.5 The Transition polynomial

All the above skein polynomials are weighted sums over graph states and have a very similar form. Jaeger in [651] defined a polynomial of 4-regular graphs, called the *transition polynomial*, which contains each of these polynomials as specializations. Subsequently, more general forms of the transition polynomial, for higher degree graphs, graphs with weight systems, embedded graphs, and even non-Eulerian graphs emerged, all also loosely referred to as transition polynomials, sometimes with descriptors such as *generalized transition polynomial* or *topological transition polynomials* (see [448, 456, 457, 458, 460]). These assimilate the Martin and circuit partition polynomials for arbitrary graphs as well as the various generalizations of the Penrose polynomial and Kauffman bracket to surface embeddings.

We begin with Jaeger's polynomial in the 4-regular setting. Let G be a 4-regular graph. As in Figure 13.1, there are three types of vertex state at each vertex. Suppose there is a way to distinguish among these vertex states (for example, by labelling half-edges, by using an underlying orientation, by a checkerboard coloring, etc.). Then we may use such distinctions to define weight systems and the transition polynomial, as follows.

A *weight system* W for G is function that assigns an element of a unitary ring $\mathfrak{R}$ to each vertex state of G. If s is a graph state of G and v a vertex, let $\omega(v,s)$ denote the element W assigns to the vertex state at v in s, and let $\omega(s) := \prod_{v\in V(F)} \omega(v,s)$ be the *state weight* of s.

Definition 13.15. The *transition polynomial* of G with weight system W is then

$$q(G;W,t) = \sum_{s} \omega(s)\, t^{c(s)}, \tag{13.7}$$

where the sum is over all graph states s of G.

Note that in some references, such as Jaeger's original paper [651], the exponent $c(s)-1$ is used in place of $c(s)$.

Comparing definitions immediately gives the following identities.

1. Let G be a 4-regular graph. Then

$$q(G;W_J,\lambda) = J(G;\lambda),$$

where the weight system W_J, takes the value 1 for all vertex states.

2. Let $\vec{G}$ be a 4-regular digraph, with underlying graph G. Then

$$q(G;W_j,\lambda) = j(\vec{G};\lambda),$$

where the weight system W_j, takes the value 1 when the vertex state corresponds to pairing incoming with outgoing edges in the underlying orientation of $\vec{G}$ and 0 otherwise.

3. Let G be an embedded graph. Then

$$q(G_m;W_p,\lambda) = P(G;\lambda),$$

where the weight system W_p, with respect to the canonical checkerboard coloring of G_m and conventions in Figure 13.2, takes the value 1 when the vertex state is a white smoothing, 0 when it is a gray smoothing, and -1 when it is a crossing.

4. When G is a plane graph,

$$q(G_m;W_t,t) = t^{k(G)}a^{n(G)}b^{r(G)}T\Big(G;\frac{at}{b}+1,\frac{bt}{a}+1\Big),$$

and for any embedded graph

$$q(G_m;W_t,t) = t^{k(G)}a^{n(G)}b^{r(G)}R\Big(G;\frac{at}{b}+1,\frac{bt}{a},\frac{1}{t},1\Big),$$

where W_t is the weight system of a canonically checkerboard colored medial graph that takes the value b when the vertex state is a white smoothing, a when it is a gray smoothing, and 0 when it is a crossing.

5. If D is a classical link diagram

$$q(G_m; W_b, d) = d[D](A, B, d),$$

where G_m is the underlying graph of D, W_b is the weight system that takes the value A on an A-splicing, B on an B-splicing, and 0 otherwise.

6. If D is a virtual link diagram

$$q(G_m; W_{vb}, d) = d[D](A, B, d),$$

where G_m is the underlying graph of D, W_{vb} is the weight system that on a vertex corresponding to a classical crossing takes the value A on an A-splicing, B on an B-splicing, and 0 otherwise; and on a vertex corresponding to a virtual crossing takes the value 1 on a crossing state and 0 otherwise.

The requirement that G is 4-regular is not essential to the definition of the transition polynomial; in fact G need only be Eulerian. For higher valencies, a vertex state is just a partition into pairs of its incident edges, and then the definitions of graph states and weight systems extend in the obvious way. With this, Equation (13.7) defines the *generalized transition polynomial*, $q(G; W, t)$, introduced in [460]. Using the generalized transition polynomial, Items 1 and 2 above extend to the full circuit partition polynomials for arbitrary Eulerian graphs and digraphs.

A special weight system, called the *medial weight system*, can be used to encode topological information about embedded graph. If G is an embedded graph, then its canonically checkerboard colored medial graph G_m is 4-regular and the checkerboard coloring distinguishes the states at a vertex as either a white smoothing, gray smoothing or crossing as in Figure 13.2. This weight system gives rise to the *topological transition polynomial* introduced in [456].

Definition 13.16. If G is an embedded graph with canonically checkerboard colored medial graph G_m, then each vertex may be assigned state weights via an ordered triple $(\alpha_v, \beta_v, \gamma_v)$, that indicates the weights of the white smoothing, gray smoothing, and crossing state, in that order. The resulting *medial weight system*, $W_m(G_m)$, is denoted by the resulting set of ordered triples, $(\boldsymbol{\alpha}, \boldsymbol{\beta}, \boldsymbol{\gamma})$, indexed equivalently either by the vertices of G_m or by the edges of G. The *topological transition polynomial* of G is then:

$$Q(G; (\boldsymbol{\alpha}, \boldsymbol{\beta}, \boldsymbol{\gamma}), t) := q(G_m; W_m, t).$$

Example 13.17. If $G =$ 

, then $G_m =$ 

and so

$Q(G; (\boldsymbol{\alpha}, \boldsymbol{\beta}, \boldsymbol{\gamma}), t) = \alpha_u \alpha_v \, Q\big($  $\big) + \alpha_u \beta_v \, Q\big($ 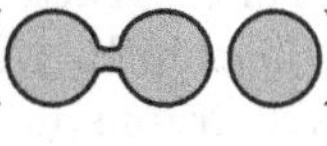 $\big)$

$+ \alpha_u \gamma_v \, Q\big($ 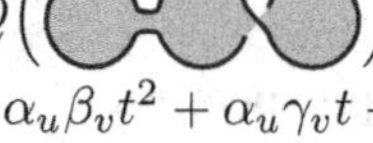 $\big) + \cdots$

$= \alpha_u \alpha_v t + \alpha_u \beta_v t^2 + \alpha_u \gamma_v t + \cdots$

The definition of the topological transition polynomial requires passing through medial graphs. This can be, and in the literature often is, avoided by describing embedded graphs as ribbon graphs (see Chapter 27 for a definition of ribbon graphs). For an edge e of a ribbon graph G, if $G\backslash e$, G/e, and $G^{\tau(e)}$ denote ribbon graph deletion, contraction, and partial Petriality (see Subsection 1.2.16, and full definitions for them may be found in, for example, [456, 457]), then

$$Q(G;(\boldsymbol{\alpha},\boldsymbol{\beta},\boldsymbol{\gamma}),t) = \sum_{(A,B,C)} \Big(\prod_{e\in A} \alpha_e\Big)\Big(\prod_{e\in B} \beta_e\Big)\Big(\prod_{e\in C} \gamma_e\Big) t^{\mathrm{bc}(G^{\tau(C)}\backslash B)},$$

where the sum is over ordered partitions (A,B,C) of $E(G)$, and where $\mathrm{bc}(G)$ denotes the number of boundary components of G (see Chapter 27). Furthermore, $Q(G)$ can be defined by the recursion relation

$$Q(G) = \alpha_e Q(G/e) + \beta_e Q(G\backslash e) + \gamma_e Q(G^{\tau(e)}/e)$$

together with its value of $t^{\mathrm{bc}(G)}$ on edgeless ribbon graphs. Duality and twisted duality relations allow permutations of the vertex state weights in the weight system so that, for example, from [462] we have $Q(G;(\boldsymbol{\alpha},\boldsymbol{\beta},\boldsymbol{\gamma}),t) = Q(G^*;(\boldsymbol{\beta},\boldsymbol{\alpha},\boldsymbol{\gamma}),t)$, and from [456] we have $Q(G;(\boldsymbol{\alpha},\boldsymbol{\beta},\boldsymbol{\gamma}),t) = Q(G^{\Gamma},(\boldsymbol{\alpha},\boldsymbol{\beta},\boldsymbol{\gamma})^{\Gamma},t)$, where G^{Γ} is a twisted dual (see Subsection 1.2.16).

Brijder and Hoogeboom defined a transition polynomial for vf-safe delta-matroids in [206]. If the delta-matroid corresponds to an embedded graph, then this definition agrees with the topological transition polynomial up to a factor of $t^{k(G)}$ (see [318]).

We also note that the space of Eulerian graphs has a Hopf algebra structure with multiplication given by disjoint union and and comultiplication given by summing over ordered partitions into edge disjoint Eulerian subgraphs. With this, the generalized transition polynomial of is a Hopf algebra map from the Hopf algebra of Eulerian graphs to the binomial bialgebra. Full details may be found in [460], including extensions to non-Eulerian graphs.

13.4 Evaluations of the Tutte polynomial along $x = y$

In addition to the elementary interpretations of the Tutte polynomial along $x = y$ given in Chapter 3 (e.g. at $(1,1)$ and $(2,2)$), several other interpretations of the Tutte polynomial at $(-1,-1)$ and $(3,3)$ are known. Moreover, the relationship between the Martin and Tutte polynomials for plane graphs via medial graphs in Theorem 13.7 leads to combinatorial interpretations of the Tutte polynomial along the line $x = y$. See also Section 21.3.2 for some of these results from a Hollant perspective.

1. Let G be a connected planar graph and G_m be the checkerboard colored medial graph of any plane embedding of G. The *all-crossing state* of G_m is the state that results from choosing the crossing vertex state at each vertex (see Figure 13.2). The components of an all-crossing state are called the *anti-circuits* or *crossing circuits* of G_m (or equivalently of $\vec{G}_m$, where they result from following edges in alternating direction, forward-backward-forward-etc., to form circuits). Martin [824, 825] found that

$$T(G;-1,-1) = (-1)^{v(G)}(-2)^{\mu(G_m)-1},$$

where $\mu(G_m)$ is the number *anti-circuits* of G_m.

Note that this interpretation of $T(G;-1,-1)$ can also be deducted through knot theory. The Jones polynomial (see Chapter 18) evaluated at $t = 1$ is known to equal $(-2)^{c(L)-1}$ where $c(L)$ is the number of components of a link. An application of Theorem 18.18 then gives the result.

2. If M is a binary matroid, then $T(M;-1,-1) = (-1)^{r(M)}(-2)^d$, where $d = \dim V_2(M) \cap V_2^{\perp}(M)$ and $V_2(M)$ is the vector space over $GF(2)$ generated by the circuits of M. See Section 12.4 for details. This is due to Rosenstiehl and Read [969].

3. Let G be a connected planar graph and G_m be the medial graph of any plane embedding of G. Say that v is a *saddle vertex* if the directions on its incident edges alternate in direction as in-out-in-out in their cyclic ordering about v. Then

$$T(G;3,3) = \sum_{k \geq 0} 2^{k-1} e_k(G_m),$$

where $e_k(G_m)$ is the number of Eulerian orientations of G_m with exactly k saddle vertices. This interpretation is due to Las Vergnas, [751].

4. A *T-tetromino* is an arrangement of four unit squares into a "T" shape. Korn and Pak [706] showed that if G is an $m \times n$ grid, then $T(G;3,3)$ is 1/2 the number of ways to tile a $4m \times 4n$ rectangle with T-tetrominoes. (In fact this was shown for a larger class of graphs.)

5. A *claw covering* of a graph G is a spanning subgraph of G where every component isomorphic to $K_{1,3}$. Korn and Pak [705] showed that if G, is a connected plane graph, then $T(G;3,3)$ equals 1/2 the number of claw coverings of a graph associated with G.

6. Let G be a planar graph and $\vec{G}_m$ be the directed medial graph of any plane embedding of G. Then Ellis-Monaghan [451] gave the following interpretation for a positive integer n.

$$T(G;1+n,1+n) = \left(\frac{1}{n}\right)^{k(G)} \sum_{\phi} 2^{\mu(\phi)},$$

where the sum is over all edge colorings ϕ of $\vec{G}_m$ with n colors so that each (possibly empty) set of monochromatic edges forms an Eulerian digraph, and where $\mu(\phi)$ is the number of monochromatic vertices (those whose incident edges are all the same color) in the coloring ϕ. See also Theorems 21.9 and 21.10.

7. Let G be a planar graph and $\vec{G}_m$ be the directed medial graph of any plane embedding of G. Let n be a positive integer. Then

$$T(G;1-n,1-n) = \left(\frac{-1}{n}\right)^{k(G)} \sum_{D_n(\vec{G}_m)} (-1)^{\sum_{i=1}^{n} k(D_i)}.$$

where $D_n(\vec{G}_m)$ is the set of all ordered partitions $(D_1, \ldots, D_n)$ of $E(\vec{G}_m)$ such that $\vec{G}_m$ restricted to each D_i is 2-regular and consistently oriented (i.e., each vertex has in-degree one and out-degree one). This interpretation is due to Ellis-Monaghan [451].

13.4.1 Derivatives of the Tutte polynomial

Results from [450] give interpretations for derivatives of the Tutte polynomial along the line $x = y$. See Chapter 28 for more on the derivatives of the Tutte polynomial.

For the following, let $P_n(\vec{G})$ be the set of ordered n-tuples $\overline{p} := (p_1, \ldots, p_n)$ where the p_i are consistently oriented edge-disjoint closed trails in $\vec{G}$. We denote by $\overline{p}^c$ the edges of $\vec{G}$ that are not in any trail of $\overline{p}$. Furthermore we write $m(\overline{p})$ for the number of vertices of $\vec{G}$ not belonging to any of the trails of $\overline{p}$. If G is a connected plane graph, then for all nonnegative integers n,

$$\frac{d^n}{dx^n} T(G;x,x)\bigg|_{x=2} = \sum_{k=0}^{n} (-1)^{n-k} \frac{n!}{k!} \sum_{\overline{p} \in P_k(\vec{G}_m)} 2^{m(\overline{p})}.$$

Recall that if G has more than one edge, then the coefficients of x and of y in the Tutte polynomial are equal, and this joint value is the β-invariant, $\beta(G)$. Noting that for a plane graph G the derivative $\frac{d}{dx} T(G;x,x)\Big|_{x=0} = 2\beta = -j'(G;-1)$ yields the following interpretation of the β-invariant:

$$\beta = \frac{1}{2} \sum (-1)^{k(P^c)+1}.$$

Here the sum is over all closed trails P in $\vec{G}_m$ which visit all vertices at least once, and P^c is the set of edges not in P.

13.5 Open problems

In [751], Las Vergnas conjectured that if M is a binary matroid, then $T^{(n)}(G;-1,-1) = K2^{d-n}$ where K is an integer, and $n \in \{0,\ldots d\}$. This long-standing question is still open. See also Conjectures 28.40 and 28.41.

The results given here for the Tutte polynomial along $x = y$ are all for integer values, and in fact nearly all that is known about combinatorial interpretations of the Tutte polynomial involve integer evaluations. This begs the question of what might be encoded at rational, irrational, or even complex values. The line $x = y$, like the hyperbola family $(x-1)(y-1) = q$ that plays such an important role in the computational complexity of the Tutte polynomial, may offer a somewhat more accessible setting in which to begin address this question than just generic points (x, y).

The Tutte polynomial has been generalized or adapted to a wide range of combinatorial objects including matroids, matroid perspectives, embedded graphs, delta-matroids, etc. This expansion has only just begun for the generalized transition polynomial for example with adaptations for embedded graphs (see [448, 456, 457, 458, 460]) and vf-safe delta-matroids in [206]. Given how such extensions have deepened the theory of the Tutte polynomial, it is likely they would be similarly fruitful for generalized transition polynomials.

Little is known about zeros of the circuit partition polynomials, and hence about zeros of the Tutte polynomial along the line $y = x$. Zeros on other curves have been heavily studied, e.g. $y = 0$ in the context of the chromatic polynomial, $x = 0$ for tensions and flows, and the hyperbolas $(x-1)(y-1) = q$ particularly for the Potts model (see Chapter 25). However, behavior along the line $y = x$ is ripe for exploration.

14

The interlace polynomial and the Tutte–Martin polynomial

Robert Brijder • Hendrik Jan Hoogeboom

Synopsis

This chapter introduces the interlace polynomial and the Tutte–Martin polynomials. It describes their main properties and their relationships with the Tutte polynomial and other graph polynomials.

- Interlace polynomial and variants.
- Relationships with the Martin polynomial, Tutte polynomial, and circle graphs.
- Recursive relations and other properties of the interlace polynomial.
- Isotropic systems and the Tutte–Martin polynomial.

14.1 Introduction

The interlace polynomial is a graph polynomial discovered by Arratia, Bollobás, and Sorkin [48, 49] that arose from their study of DNA sequencing methods. The interlace polynomial can be characterized in terms of recursive relations involving the graph operations of local complementation and pivoting, and it satisfies a number of invariance properties and evaluations (see Section 14.3). The definition of the interlace polynomial can be traced from 4-regular graphs (or 2-in 2-out digraphs), to circle graphs and finally to arbitrary looped simple graphs (so multiple edges are not allowed).

Various other polynomials are closely related to interlace polynomial, including the global interlace polynomial of Aigner and van der Holst [13] (see

DOI: 10.1201/9780429161612-14

Section 14.5) and the two-variable extension of the interlace polynomial [50] (see Section 14.6). The interlace polynomial is also closely related to the Tutte polynomial of binary matroids [13, 191] (see Section 14.8).

Isotropic systems were introduced by Bouchet [184] to unify various properties of circuit partitions in 4-regular graphs and properties of pairs of duals of binary matroids. The Tutte–Martin polynomials of isotropic systems [189] were defined about a decade before the introduction of the interlace polynomial, and, once the preliminary conference paper of the interlace polynomial appeared, various authors quickly noticed that the interlace polynomial can be seen as a special case of the restricted Tutte–Martin polynomial [13, 191]. In Section 14.9 we discuss the Tutte–Martin polynomials and their relationship with interlace polynomials.

Section 14.10 discusses other generalizations of the interlace polynomial and open problems are given in Section 14.12. Note that this chapter does not follow the chronological development of these polynomials. It also does not discuss the complexity of computing evaluations of the interlace polynomials at specific points, details about which can be found in [142, 338].

14.2 Notation

Symmetric difference of sets is denoted by Δ. In this chapter we generally do not allow graphs to have multiple edges, but we do allow them to have loops. The exceptions for this are 2-in 2-out digraphs and 4-regular graphs, where we allow both multiple edges and loops. Graphs without multiple edges, but possibly with loops are called *looped simple graphs*.

For finite sets V and W, a $V \times W$ *matrix* is a matrix of size $|V| \times |W|$ whose rows are indexed by elements of V and columns are indexed by elements of W. If A is a $V \times W$ matrix, and $X \subseteq V$ and $Y \subseteq W$, we denote the $X \times Y$-submatrix of A by $A[X, Y]$. We use $A[X]$ to denote $A[X, X]$. The rank and nullity of a matrix A are denoted by $r(A)$ and $n(A)$, respectively.

Let $G = (V, E)$ be a looped simple graph. The *adjacency matrix*, $A(G)$, of G, is a $V \times V$ matrix over $\mathbb{F}_2$ where $A(G)_{u,v} = 1$ if and only if $(u, v) \in E$. In particular, $A(G)_{v,v} = 1$ if and only if there is a loop of G at v. (Our convention for $V \times V$ matrices is that the rows and columns are labelled in the same order by the elements of V.)

For $X \subseteq V$, we write $G+X$ for the graph that results from G by "toggling" the existence of loops at the vertices of X (i.e., for each vertex v in X, if there is a loop at v in G, remove it, and if there is not a loop at v add one). Thus, the adjacency matrices of G and $G+X$ are related by $A(G+X) = A(G)+I_X$, where I_X is the diagonal $V \times V$ matrix over $\mathbb{F}_2$ where, for $i \in V$, the (i, i)-entry is 1 if and only if $i \in X$. For $X \subseteq V$, $G[X]$ denotes the induced subgraph on X, which has adjacency matrix $A(G)[X]$, and $G \backslash X$ denotes $G[V \setminus X]$, the

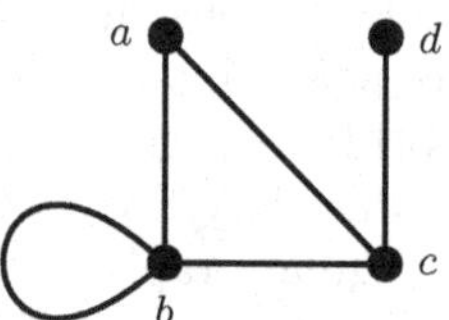

FIGURE 14.1: A looped simple graph. (It has no multiple edges, but loops are allowed.)

graph obtained by deleting the vertices in X. In case $X = \{v\}$ is a singleton, we also write $G\backslash v$ to denote $G\backslash X$.

14.3 Interlace polynomial

We start with the (single-variable) interlace polynomial. It is sometimes called the *vertex-nullity interlace polynomial* to distinguish it from its two-variable cousin (see Section 14.6) and to stress that the summation is over sets of vertices of the graph (rather than edges, as for the Tutte polynomial). The interlace polynomial was originally defined recursively, as in Theorem 14.6 below. Definition 14.1 is from [13, 50], see also [11, Chapter 9].

Definition 14.1. Let $G = (V, E)$ be a looped simple graph. Then the (single-variable) *interlace polynomial* of G is

$$q(G; y) := \sum_{X \subseteq V} (y-1)^{n(A(G)[X])}.$$

Example 14.2. Consider the looped simple graph G in Figure 14.1. For this graph $q(G; y) = (y-1)^2 + 6(y-1) + 9(y-1)^0 = y^2 + 4y + 4$. As an example, the subset $\{a, d\}$ induces the 2×2 zero-matrix, which has nullity 2 and so contributes $(y-1)^2$ to $q(G; y)$.

Explicit formulas have been obtained for the interlace polynomials of complete graphs, star graphs, complete bipartite graphs, and cycles [49]. See also Proposition 14.23 for several of these formulas in terms of the two-variable interlace polynomial.

Remark 14.3. The interlace polynomial corresponds to the (restricted) Tutte–Martin polynomial of isotropic systems (see Section 14.9). Many results regarding the interlace polynomial can be found in the context of the Tutte–Martin polynomial in [189], published almost a decade before the introduction of the interlace polynomial. However, since the settings and conventions of interlace polynomial and the Tutte–Martin polynomial are sufficiently different, we here attribute results of the interlace polynomial to both [189] and the papers that prove the results in the context of interlace polynomials.

14.3.1 Recursive relations

The interlace polynomial has a recursive definition. In fact, this is how it was originally defined [48]. For this definition we need some additional notation.

Let v be a vertex of a looped simple graph $G = (V, E)$. The *neighborhood* of v in G, denoted by $N_G(v)$, is the set $\{w \in V : (v, w) \in E, w \neq v\}$. The *complement* G' of G is the graph obtained by complementing the edge relation, i.e., for every $e = (v, w)$ (note $v = w$ is allowed), e is an edge of G if and only if e is not an edge of G'. We emphasize that loops at the vertices in G are toggled in forming G'. The *closed neighborhood* of a vertex v in G, denoted by $\bar{N}_G(v)$, is the set $N_G(v) \cup \{v\}$.

Definition 14.4. Let G be a looped simple graph and u be a looped vertex of G. Then the *local complement* of G at u, denoted by $G{*}u$, is the graph obtained from G by complementing the subgraph induced by the neighborhood of u.

We note that although we can easily define local complement for arbitrary vertices, it is convenient to restrict applicability of this operation to looped vertices. Similarly, we define the operation of pivot below only for edges having unlooped vertices.

Definition 14.5. Let $G = (V, E)$ be a looped simple graph, and let $e = (v, w)$ be an edge for which both v and w are unlooped vertices. Consider the partition of the closed neighborhoods $\bar{N}_G(v) \cup \bar{N}_G(w)$, of v and w, into the sets $V_1 = \bar{N}_G(v) \backslash \bar{N}_G(w)$, $V_2 = \bar{N}_G(w) \backslash \bar{N}_G(v)$, and $V_3 = \bar{N}_G(v) \cap \bar{N}_G(w)$. The *pivot* of G at e, denoted by $G * e$, is the graph obtained from G by "complementing" the edges between distinct V_i's. Thus, for every $e' = (x, y)$ with $x \in V_i$, $y \in V_j$, and $i \neq j$, we have that e' is an edge of G if and only if e' is not an edge of $G * e$.

A pivot is illustrated in Figure 14.2. Note that in the figure v and w are both adjacent to all vertices in V_3—these edges are omitted in the diagram. The operation does not affect edges adjacent to vertices outside the sets V_1, V_2, V_3, nor does it change any of the loops. Dashed connections may be edges, or not.

The pivot operation is also called *edge local complementation* and may be denoted by $G^{(vw)}$. We note that the definition of a pivot here differs from that given in [49] and that the definitions become equivalent after swapping the names of the vertices of the edge e after pivoting. (The convention used here follows Bouchet [185].)

A vertex v of a graph is *isolated* if it is not incident to any edge (in particular, v is unlooped). The following result is from [49, 189].

Theorem 14.6. *Let $G = (V, E)$ be a looped simple graph.*

1. *If $V = \emptyset$, then $q(G; y) = 1$.*

2. *If v is an isolated vertex of G, then*

$$q(G; y) = y\, q(G \backslash v; y).$$

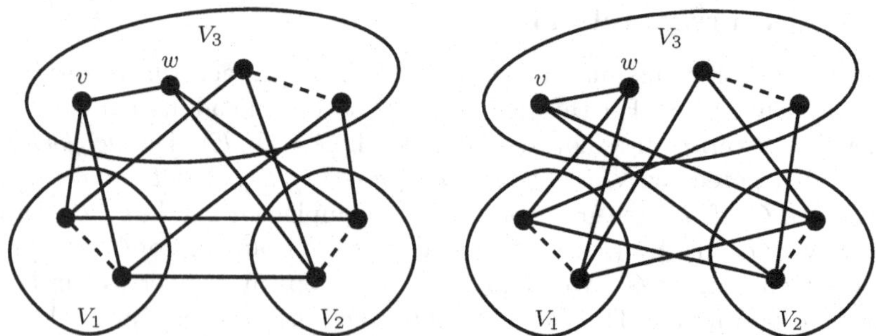

FIGURE 14.2: Pivot on an edge $e = (v, w)$ in a graph.

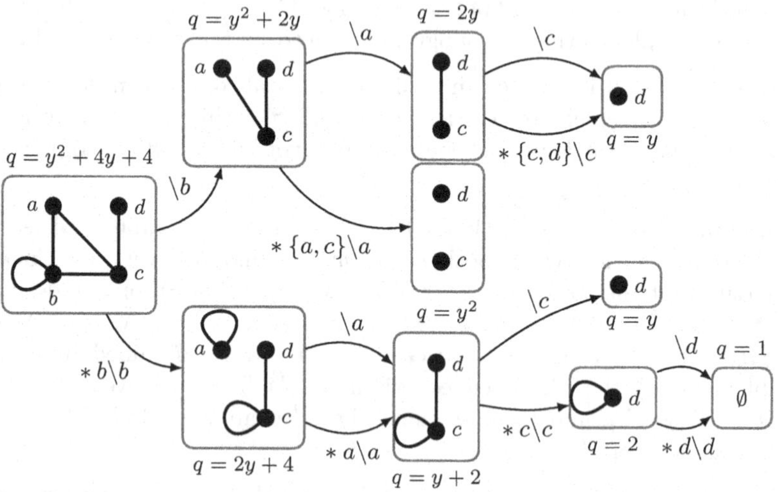

FIGURE 14.3: Recursive computation of the interlace polynomial $q(G; y)$. With each graph F we give the polynomial $q := q(F; y)$.

3. If v is a looped vertex of G, then

$$q(G; y) = q(G\backslash v; y) + q((G * v)\backslash v; y).$$

4. If $e = (v, w) \in E$ with v and w unlooped vertices, then

$$q(G; y) = q(G\backslash v; y) + q((G * e)\backslash v; y).$$

Note that Theorem 14.6 implies that the coefficients of $q(G; y)$ are always nonnegative integers, which is not immediately clear from the form of the interlace polynomial given in Definition 14.1.

Example 14.7. Figure 14.3 illustrates the recursive computation of the interlace polynomial for the graph G from Figure 14.1. (See also Example 14.2.)

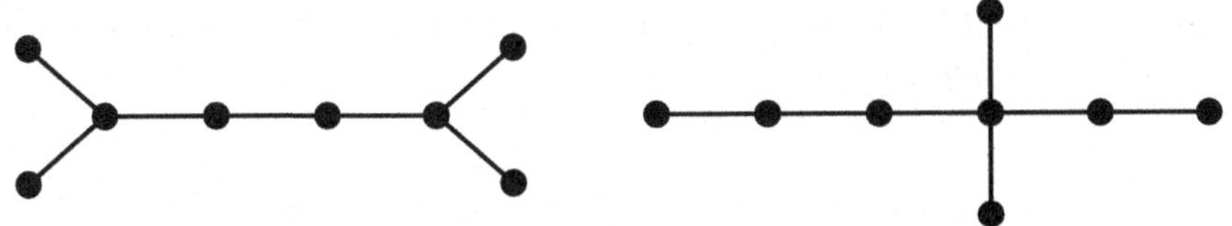

FIGURE 14.4: Two trees with the same interlace polynomial.

We obtain (again) $q(G; y) = y^2 + 4y + 4$. For simplicity, Figure 14.3 does not include the computations on graphs with only isolated vertices.

Remark 14.8. In this chapter we restrict to looped simple graphs. In the case of simple graphs, the third case of Theorem 14.6 disappears. Also, using the natural counterpart of local complement $*u$ for simple graphs G, we have $G * (u, v) = G * u * v * u$. However, with the definition of local complement given for looped simple graphs in this chapter, $*(u, v)$ and $*u * v * u$ are different. See [203, Section 8] for more details.

14.3.2 Invariance and evaluations

An important property of the interlace polynomial is that it is invariant under both local complementation and pivoting, as shown in [49, 50].

Theorem 14.9. *Let $G = (V, E)$ be a looped simple graph.*

1. *If v is a looped vertex of G, then $q(G; y) = q(G * v; y)$.*
2. *If $e = (v, w) \in E$ with v and w unlooped vertices, then $q(G; y) = q(G*e; y)$.*

A consequence of Theorem 14.9 is that graphs are in general not characterized by their interlace polynomials. In fact, the trees of Figure 14.4 have the same interlace polynomial, but cannot be obtained from each other by pivots (see [13] and also [187]).

We now consider a number of evaluations of the interlace polynomial. A *perfect matching* a graph G is a set of edges P (loops are allowed) of G such that every vertex of G is incident to exactly one edge of P. A graph G has an odd number of perfect matchings if and only if the determinant $\det(A(G))$, computed over $\mathbb{F}_2$, is nonzero [775]. The next theorem is shown in [13] for the case of simple graphs, but it is easy to see that the evaluations of $q(G; 1)$ and $q(G; 2)$ carry over trivially to looped simple graphs and, using [189], it is observed in [206] that the evaluations of $q(G; -1)$ and $q(G; 3)$ carry over to looped simple graphs as well.

Theorem 14.10. *Let $G = (V, E)$ be a looped simple graph. Then*

1. $q(G; -1) = (-1)^{|V|}(-2)^{n(A(G)+I)}$, *where I is the $V \times V$ identity matrix;*
2. $q(G; 0) = 0$ *if both $|V| > 0$ and G has no loops;*

3. *$q(G;1)$ is equal to the number of induced subgraphs of G with an odd number of perfect matchings;*

4. *$q(G;2) = 2^{|V|}$; and*

5. *$q(G;3) = k\,|q(G;-1)|$, for some odd integer k.*

Evaluations and invariance properties of the interlace polynomial have also been investigated for particular subclasses of graphs. In particular, bipartite distance hereditary graphs have been characterized by the coefficient of y^1 in their interlace polynomials [461].

14.3.3 Square matrices

The interlace polynomial can be considered more generally for arbitrary square matrices A, not just the adjacency matrices $A(G)$ of graphs, see Definition 14.1. It was observed by Geelen [517] that the two operations of local complementation and edge local complementation can be seen as special cases of the *principal pivot transform* [1086].

Definition 14.11. Let A be a $V \times V$-matrix and $X \subseteq V$ with $A[X]$ nonsingular. If we write

$$A = \begin{array}{c} \\ X \\ V\backslash X \end{array} \!\!\begin{array}{c} \begin{array}{cc} X & V\backslash X \end{array} \\ \begin{pmatrix} P & Q \\ R & S \end{pmatrix} \end{array},$$

then the principal pivot transform of A by X equals

$$A * X := \begin{array}{c} \\ X \\ V\backslash X \end{array} \!\!\begin{array}{c} \begin{array}{cc} X & \qquad V\backslash X \end{array} \\ \begin{pmatrix} P^{-1} & -P^{-1}Q \\ RP^{-1} & S - RP^{-1}Q \end{pmatrix} \end{array}.$$

The matrix $A*X[V\setminus X] = S - RP^{-1}Q$ is well known and called the *Schur complement* of $A[X]$ in A [1185].

The invariance result (Theorem 14.9) and the recursive relation (Theorem 14.6) of the interlace polynomial can be explained in this general setting using the principal pivot transform [204]. For $v \in V$, we write $A\backslash v$ to denote $A[V \setminus \{v\}]$.

Theorem 14.12. *Let A be a $V \times V$-matrix and $X \subseteq V$ with $A[X]$ nonsingular. Then $q(A * X; y) = q(A; y)$ and $q(A; y) = q(A\backslash v; y) + q(A * X\backslash v; y)$ for all $v \in X$.*

14.4 Martin polynomial

There is a direct connection between the Martin polynomial $m(\vec{G};y)$ of 2-in 2-out digraphs and the interlace polynomial of circle graphs. See Section 13.3.1 for background on the Martin polynomial and graph states (which are more commonly called *transition systems* in the context of the interlace polynomial). A graph state s in a 2-in 2-out digraph $\vec{G}$ uniquely determines a *circuit partition*, namely the set of components of the graph state viewed as subgraphs of $\vec{G}$. The number of these components is denoted $c(s)$. Here we use the following generalization of the Martin polynomial to digraphs that are not necessarily connected: $m(\vec{G};y) = \sum_s (y-1)^{c(s)-k(\vec{G})}$, where the summation is over the graph states s of $\vec{G}$.

An *Eulerian circuit* C in a connected 4-regular graph G visits each vertex exactly twice. This defines the *interlace graph* $I(C)$ of C on $V(G)$: two distinct vertices u and v are adjacent if they are *interlaced* in C, i.e., vertices occurring in the order $\cdots u \cdots v \cdots u \cdots v \cdots$ in C. Interlace graphs are extended to (possibly disconnected) 4-regular graphs by considering *Eulerian systems*, which are sets containing an Eulerian circuit for each connected component by applying the above to each component.

The local complement and pivot operations have a natural interpretation in the context of interlace graphs. The operations describe how an interlace graph $I(C)$ is changed when local changes are made to the Eulerian circuit C (reversing a segment between two occurrences of a vertex p, or swapping the two segments between the occurrences of interlaced vertices u and v) [188, 715]. The following theorem is from [49].

Theorem 14.13. *Let $\vec{G}$ be a 2-in 2-out digraph, and let C be an Eulerian system for $\vec{G}$. Then $m(\vec{G};y) = q(I(C);y)$.*

Note that this result connects combinatorial concepts (counting circuit partitions) to algebraic concepts (nullity of principal submatrices), as we now make precise. Relative to a fixed Eulerian circuit C in a connected 4-regular graph, the vertex states at a vertex can be unambiguously described. Fix an orientation of C. Then each vertex state either *follows* C (it equals the one chosen by C), or it differs, and is either *orientation consistent* or *inconsistent* (relative to the one chosen by C), as in Figure 14.5.

The next result, which is from [321, 1077], is a consequence of a more general equality in terms of matroids [646, 1079]. For the theorem, recall that $G+X$ is the graph that results from $G=(V,E)$ by "toggling" the existence of loops at the vertices of $X \subseteq V$.

Theorem 14.14. *Let $G=(V,E)$ be a 4-regular graph and C be an Eulerian system for G. Let P be a circuit partition of E, where D_1, D_2, D_3 are the sets of vertices that follow C, are orientation consistent, or are orientation inconsistent (respectively). Then $|P| - k(G) = n((I(C)+D_3)\backslash D_1)$.*

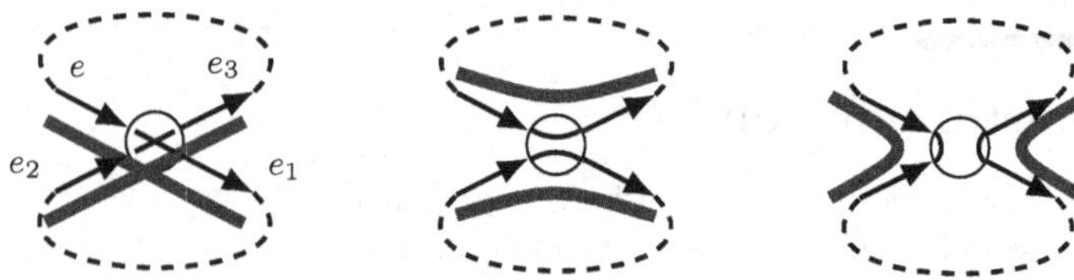

FIGURE 14.5: Three ways to connect pairs of edges in a 4-regular graph relative to an (oriented) Eulerian circuit: following the circuit, in an orientation-consistent way, and in an orientation-inconsistent way.

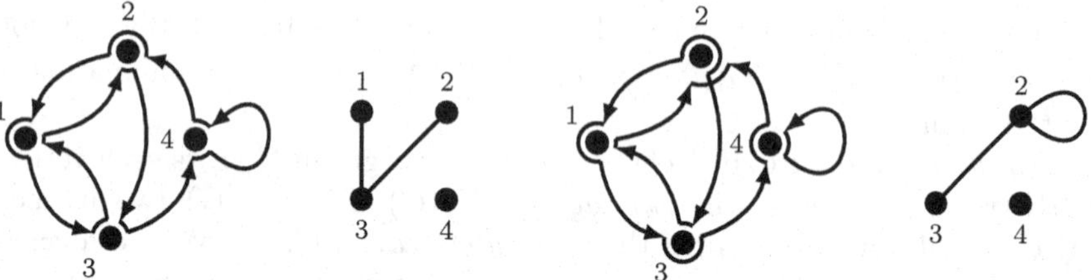

(a) An Eulerian circuit with its interlace graph.

(b) Circuit partition and an application of Theorem 14.14.

FIGURE 14.6: Examples of interlace graphs and circuit partitions.

Example 14.15. Figure 14.6a depicts (with edge orientations added for clarity) an Eulerian circuit C of a 4-regular graph and its interlace graph $I(C)$. Note that C traverses along the sequence 2 1 3 1 2 3 4 4 of vertices.

Now consider the circuit partition P which is described using the notation of Theorem 14.14 by $D_1 = \{1\}$, $D_2 = \{3,4\}$, and $D_3 = \{2\}$ relative to C. It is depicted in Figure 14.6b together with the graph $I(C) + D_3 \backslash D_1$. Its nullity equals 1, hence P contains two circuits. Part of the largest circuit runs against the original orientation of C, due to the orientation-inconsistent vertex state at vertex 2.

14.5 Global interlace polynomial

The global interlace polynomial was defined by Aigner and van der Holst in [13]. (Recall the notation in Section 14.2.)

Definition 14.16. Let $G = (V, E)$ be a looped simple graph. Then the *global interlace polynomial* of G is

$$Q(G; y) = \sum_{X \subseteq V} \sum_{Y \subseteq X} (y-2)^{n((A(G+Y))[X])}.$$

This formulation of the global interlace polynomial is motivated by Theorem 14.14. Whereas the interlace polynomial $q(G;y)$ for an interlace graph $G = I(C)$ only considers vertex states that either follow C or are orientation consistent, this global variant allows all three possibilities. Thus, where $q(G;y)$ generalizes the Martin polynomial for 2-in 2-out digraphs, $Q(G;y)$ generalizes the Martin polynomial $M(H;y) = \sum_s (y-2)^{c(s)-k(H)}$ for 4-regular graphs H (see Section 13.3.1).

Theorem 14.17. *Let H be a 4-regular graph, and let C be an Eulerian system for H. Then $M(H;y) = Q(I(C);y)$.*

The next result, from [13, 189], presents important invariance results.

Theorem 14.18. *Let $G = (V,E)$ be a looped simple graph.*

1. *If v is a vertex of G, then $Q(G;y) = Q(G+v;y)$.*
2. *If v is a looped vertex of G, then $Q(G;y) = Q(G*v;y)$.*
3. *If $e = (v,w)$ with v and w unlooped vertices, then $Q(G;y) = Q(G*e;y)$.*

The next theorem, from [13, 189], shows that there are recursive relations that define $Q(G;y)$.

Theorem 14.19. *Let $G = (V,E)$ be a looped simple graph.*

1. *If $V = \emptyset$, then $Q(G;y) = 1$.*
2. *If v is an isolated vertex of G, then*

$$Q(G;y) = y\,Q(G\backslash v;y).$$

3. *If $e = (v,w) \in E$ with v and w unlooped vertices, then*

$$Q(G;y) = Q(G\backslash v;y) + Q((G*e)\backslash v;y) + Q(((G+v)*v)\backslash v;y).$$

Since $Q(G;y)$ is invariant under adding or removing loops, we may assume without loss of generality that G is a simple graph in the following result of [13, 189].

Theorem 14.20. *Let G be a simple graph on n vertices.*

1. *$Q(G;0) = 0$ if $n > 0$;*
2. *$Q(G;2)$ is the number of graphs G' (including the empty graph) that have an odd number of perfect matchings and are such that removing all loops from G' obtains an induced subgraph of G;*
3. *$Q(G;3) = 3^n$;*
4. *$Q(G;4) = 2^n e_G$, where e_G is the number of Eulerian induced subgraphs of G.*

14.6 Two-variable interlace polynomial

The two-variable interlace polynomial was introduced by Arratia, Bollobás, and Sorkin in [50].

Definition 14.21. Let $G = (V, E)$ be a looped simple graph. The *two-variable interlace polynomial* is

$$q(G; x, y) = \sum_{X \subseteq V} (x-1)^{r(A(G)[X])}(y-1)^{n(A(G)[X])}.$$

Note that the single-variable interlace polynomial (or vertex-nullity polynomial) of Definition 14.1 is a specialization of the two-variable interlace polynomial, since $q(G; y) = q(G; 2, y)$.

The two-variable interlace polynomial can be defined recursively as in the following theorem from [50].

Theorem 14.22. *Let $G = (V, E)$ be a looped simple graph.*

1. *If $V = \emptyset$, then $q(G; x, y) = 1$.*
2. *If v is a looped vertex of G, then*
$$q(G; x, y) = q(G \backslash v; x, y) + (x-1)q(G * v \backslash v; x, y).$$
3. *If $e = (v, w) \in E$ with v and w unlooped vertices, then*
$$q(G; x, y) = q(G \backslash v; x, y) + q(G * e \backslash v; x, y) + ((x-1)^2 - 1)\, q(G * e \backslash \{v, w\}; x, y).$$

In contrast to $q(G; y)$, the two-variable interlace polynomial $q(G; x, y)$ is not, in general, invariant under pivots.

Explicit formulas for the two-variable interlace polynomials of common families of graphs can be found in [50].

Proposition 14.23. *The two-variable interlace polynomials $q(G; x, y)$ of complete graphs, complete bipartite graphs, and paths are as follows.*

1. $q(\overline{K}_n; x, y) = y^n$,
2. $q(K_n; x, y) = \frac{1}{2}(x^n + (2-x)^n) + \frac{1}{2}\left(\frac{y-1}{x-1}\right)(x^n - (2-x)^n)$,
3. $q(K_{m,n}; x, y) = \frac{(x-1)^2}{(y-1)^2}((y^m - 1)(y^n - 1)) + y^m + y^n - 1$,
4. $$q(P_n; x, y) = \frac{1}{2}\left(y + \frac{3y+2x(x-2)}{\sqrt{1+4(y+x(x-2))}}\right)\left(\frac{1+\sqrt{1+4(y+x(x-2))}}{2}\right)^n + \frac{1}{2}\left(y - \frac{3y+2x(x-2)}{\sqrt{1+4(y+x(x-2))}}\right)\left(\frac{1-\sqrt{1+4(y+x(x-2))}}{2}\right)^n.$$

14.7 Weighted interlace polynomial

For a finite set V, we define $\mathcal{P}_3(V)$ to be the set of triples (V_1, V_2, V_3) where the V_i's are pairwise disjoint and $V_1 \cup V_2 \cup V_3 = V$. Hence, (V_1, V_2, V_3) is an ordered partition of V (where V_i's are allowed to be empty).

As in [338, 1078], we may define a common generalization of the interlace polynomial and the global interlace polynomial.

Definition 14.24. Let $G = (V, E)$ be a looped simple graph and $W = (\mathbf{a}, \mathbf{b}, \mathbf{c})$ with $\mathbf{a}$, $\mathbf{b}$, and $\mathbf{c}$ vectors indexed by a superset of V. Then the *weighted interlace polynomial* (or *multivariate interlace polynomial*) of G with respect to W is

$$Q(G; W, y) = \sum_{(X_1,X_2,X_3)\in\mathcal{P}_3(V)} a_{X_1} b_{X_2} c_{X_3}\, y^{n(A(G[X_2\cup X_3]+X_3))},$$

where $\mathbf{a}$ ($\mathbf{b}$, $\mathbf{c}$, respectively) has entries a_v (b_v, c_v, respectively) for all $v \in V$, $a_{X_1} := \prod_{v\in X_1} a_v$, and similarly for b_{X_2} and c_{X_3}.

Again as in [50], this polynomial can be characterized recursively.

Theorem 14.25. *Let $G = (V, E)$ be a looped simple graph.*

1. *If $V = \emptyset$, then $Q(G; W, y) = 1$.*
2. *If v is a vertex of G, then*
$$Q(G; W, y) = Q(G + v; W', y),$$
where W' is obtained from W by swapping b_v and c_v.
3. *If v is an isolated vertex of G, then*
$$Q(G; W, y) = (a_v + b_v y + c_v)\, Q(G\backslash v; W, y).$$
4. *If $e = (v, w) \in E$ with v and w unlooped vertices, then*
$$Q(G; W, y) = a_v\, Q(G\backslash v; W, y) + b_v\, Q((G * e)\backslash v; W, y) \\ + c_v\, Q(((G + v) * v)\backslash v; W, y).$$

All of the interlace polynomials defined in this chapter are specializations of $Q(G; W, y)$. Indeed, $q(G; y)$ corresponds to $Q(G; W, y-1)$ with $a_v = b_v = 1$ and $c_v = 0$ for all $v \in V(G)$, and $Q(G; y)$ corresponds to $Q(G; W, y-2)$ with $a_v = b_v = c_v = 1$ for all $v \in V(G)$. The two-variable interlace polynomial $q(G; x, y)$ is obtained from $Q(G; W, (y-1)/(x-1))$ by setting $a_v = 1, b_v = x - 1, c_v = 0$ for all $v \in V(G)$, see [206].

As before, via Theorem 14.14, circle graphs form the link between the weighted transition polynomial of 4-regular graphs [651] and the weighted interlace polynomial. The ordered partition (X_1, X_2, X_3) serves as the description of a graph state (relative to an Eulerian system), see Theorem 14.14.

14.8 Connection with the Tutte polynomial

A (U, W)-*bipartite graph* is a bipartite graph G with U and W as parts (i.e., $V(G)$ is the disjoint union of U and W and every edge in G has an end in U and an end in W).

We turn to binary matroids. Let B be a basis of a binary matroid M. Then the *fundamental graph* G of M with respect to B is the $(B, E(M) \setminus B)$-bipartite graph with $(v, w) \in E(G)$ if and only if $v \in B$ and $w \in E(M) \setminus B$ and $(B \setminus \{v\}) \cup \{w\}$ is a basis of M. The following result is from [13]. See also [49, 461].

Theorem 14.26. *Let M be a binary matroid and G be the fundamental graph of M with respect to some basis B. Then $T(M; y, y) = q(G; y)$.*

In view of Theorem 14.26, Theorem 14.10, which holds for arbitrary graphs (not only bipartite graphs), provides a generalization of the evaluations of $T(M; y, y)$ to a binary matroid M and $y \in \{-1, \dots, 3\}$. Indeed, it can be shown that the dimension of the bicycle space of M is equal to $n(A(G) + I)$ where G is an arbitrary fundamental graph of M—this recovers the evaluation of $T(M; -1, -1)$ from [969]. (See Chapter 13 for evaluations of the Tutte polynomial along the line $y = x$.)

As observed by Bouchet [191], it is possible to extend Theorem 14.26 to the full two-variable Tutte polynomial $T(M; x, y)$ for binary matroids if one carefully distinguishes the two parts $(B, E(M) \setminus B)$ of the (bipartite) fundamental graph G of M with respect to B. For this we first define a graph polynomial much like the interlace polynomial, but it is only defined for (U, W)-bipartite graphs.

Theorem 14.27. *Let G be a (U, W)-bipartite simple graph. There is a graph polynomial $q'(G; x, y)$ defined by the following relations.*

1. *If $V(G) = \emptyset$, then $q'(G; x, y) = 1$.*
2. *If $v \in U$ is isolated in G, then $q'(G; x, y) = x\, q'(G \setminus v; x, y)$.*
3. *If $v \in W$ is isolated in G, then $q'(G; x, y) = y\, q'(G \setminus v; x, y)$.*
4. *If v is the end of an edge $e \in E(G)$, then*

$$q'(G; x, y) = q'(G \setminus v; x, y) + q'((G * e) \setminus v; x, y).$$

Here, $G * e$ is to be regarded as an $(U \triangle e, W \triangle e)$-bipartite graph.

The next result, from [191], generalizes Theorem 14.26.

Theorem 14.28. *Let M be a binary matroid. Then $T(M; x, y) = q'(G; x, y)$, where G is the $(B, E(M) \setminus B)$-bipartite fundamental graph of M with respect to some basis B.*

14.9 Isotropic systems and the Tutte–Martin polynomial

Isotropic systems were defined by Bouchet in [184] to unify various properties of graph states of 4-regular graphs and binary matroids. In [189] two polynomials for isotropic systems were studied: the restricted Tutte–Martin polynomial and the global Tutte–Martin polynomial. It is observed in [191] that the interlace polynomial can be formulated as a specialization of the Tutte–Martin polynomial. Many results of the previous sections of this chapter can be obtained, in a straightforward way, from the results of [189] once the interlace polynomial is formulated as a specialization of the Tutte–Martin polynomial.

To make the connection with interlace polynomials more transparent, we define isotropic systems in a slightly nonstandard way, using terminology from multimatroids [190], similarly as in [1080]. Let U be a finite set and Ω be a partition of U such that $|\omega| = 3$ for all $\omega \in \Omega$. A *subtransversal*, respectively *transversal*, of Ω is a subset $S \subseteq U$ such that $|S \cap \omega| \leq 1$, respectively $|S \cap \omega| = 1$, for all $\omega \in \Omega$. Let $\mathcal{S}(\Omega)$ and $\mathcal{T}(\Omega)$ be the sets of all subtransversals and transversals, respectively. We regard $\mathcal{S}(\Omega)$ as a vector space over $\mathbb{F}_2$ isomorphic to $(\mathbb{F}_2^2)^\Omega$: every singleton $\{x\} \subseteq \omega \in \Omega$ is assigned to a unique $x' \in (\mathbb{F}_2^2)^\Omega$ with entries $x'_{\omega'} = (0,0)$ if $\omega' \neq \omega$ and entry $x'_\omega \neq (0,0)$ otherwise. The elements of U generate in this way the whole of $\mathcal{S}(\Omega)$. We equip $\mathcal{S}(\Omega)$ with a bilinear form $B : \mathcal{S}(\Omega) \times \mathcal{S}(\Omega) \to \mathbb{F}_2$. For $S_1, S_2 \in \mathcal{S}(\Omega)$ we set $B(S_1, S_2) = 1$ if and only if there are an odd number of $\omega \in \Omega$ with $|\omega \cap (S_1 \cup S_2)| = 2$.

A subspace $\mathcal{L}$ of $\mathcal{S}(\Omega)$ is called *totally isotropic* if all vectors of $\mathcal{L}$ are mutually orthogonal, i.e., $B(S_1, S_2) = 0$ for all $S_1, S_2 \in \mathcal{L}$.

Definition 14.29. Let Ω be as above. Then $\mathcal{I} = (\Omega, \mathcal{L})$ is an *isotropic system* if $\mathcal{L}$ is totally isotropic subspace of $\mathcal{S}(\Omega)$ of dimension $|\Omega|$.

For an isotropic system $\mathcal{I} = (\Omega, \mathcal{L})$ and $T \in \mathcal{T}(\Omega)$, we define the *nullity* of T in $\mathcal{I}$, denoted by $n_\mathcal{I}(T)$, as $\dim(\{X \in \mathcal{L} : X \subseteq T\})$.

Definition 14.30. Let $\mathcal{I} = (\Omega, \mathcal{L})$ be an isotropic system.

1. The *restricted Tutte–Martin polynomial* of $\mathcal{I}$ with respect to $T \in \mathcal{T}(\Omega)$ is

$$m(\mathcal{I}, T; y) = \sum_{\substack{X \in \mathcal{T}(\Omega_V) \\ X \cap T = \emptyset}} (y-1)^{n_\mathcal{I}(X)}.$$

2. The *global Tutte–Martin polynomial* of $\mathcal{I}$ is

$$M(\mathcal{I}; y) = \sum_{X \in \mathcal{T}(\Omega)} (y-2)^{n_\mathcal{I}(X)}.$$

Definitions 14.29 and 14.30 are both due to Bouchet, from [184] and [189] respectively.

We now associate an isotropic system to a graph. Identify the elements of $\ker(E)$ for an $X \times Y$ matrix E over $\mathbb{F}_2$ with subsets of Y in the usual way. (So we identify $\ker(E)$ with the cycle space of the vector matroid of E.) For $Y' \subseteq Y$, we denote $n(E[X, Y'])$ by $n_E(Y')$. The following theorem is from [1080].

Theorem 14.31. *Let G be a graph, let Ω be a partition of a set with $|\omega| = 3$ for all $\omega \in \Omega$ and $|\Omega| = |V(G)|$, and let $V_1, V_2, V_3 \in \mathcal{T}(\Omega)$ be mutually disjoint. Consider the matrix whose columns are indexed by V_1, V_2 and V_3,*

$$E = \begin{pmatrix} \overset{V_1}{I} & \overset{V_2}{A(G)} & \overset{V_3}{A(G)+I} \end{pmatrix}.$$

Then $\mathcal{I}_G = (\Omega, \mathcal{L}_G)$ with $\mathcal{L}_G = \mathcal{S}(\Omega) \cap \ker(E)$ is an isotropic system.
Moreover, for all $X \in \mathcal{T}(\Omega)$,

$$n_{\mathcal{I}_G}(X) = n_E(X) = n(A(G + X_3[X_2 \cup X_3])),$$

where $X_i = X \cap V_i$ for all $i \in \{1, 2, 3\}$.

The graph G is called the *fundamental graph* or *graphic presentation* of $\mathcal{I}_G$ with respect to (V_1, V_2, V_3) (see [185]). It turns out that every isotropic system has a fundamental graph. In fact, isotropic systems can essentially be viewed as an alternative formulation of the null spaces $\ker(E)$ (or, equivalently, the vector matroids) of the matrices E as in Theorem 14.31 (see [1080]).

As a consequence of Theorem 14.31 we have the following corollary which establishes the close relationship between the Tutte–Martin polynomials and the (global) interlace polynomial.

Corollary 14.32. *Let G be a graph and $\mathcal{I}_G$ the isotropic system from Theorem 14.31. Then we have the following:*

1. *$m(\mathcal{I}_G, V_3; y) = q(G; y)$, and*
2. *$M(\mathcal{I}_G; y) = Q(G; y)$.*

We note that isotropic systems have three kinds of elementary minors which correspond to the terms in the recursive relation of the global interlace polynomial of Theorem 14.19.

14.10 Interlace polynomials for delta-matroids

We next define interlace polynomials for set systems, which include delta-matroids and in particular matroids (here defined by their bases). In turn, interlace polynomials for delta-matroids generalize interlace polynomials for graphs. One of the motivations for considering this general setting is that it allows one to study interlace polynomials for embedded graphs (see [318, 856]).

14.10.1 Delta-matroids and their representations

A *set system* M (with ground set V) is a pair $M := (V, D)$ with $D \subseteq 2^V$ a family of subsets of V. For simplicity we write $X \in M$ to denote $X \in D$. M is *proper* if $D \neq \emptyset$. We let $d_M := \min_{Y \in M}(|Y|)$ denote the cardinality of the smallest set in M. A *delta-matroid* [186] is a proper set system that satisfies the following symmetric-difference variant of the basis exchange axiom for matroids:

Definition 14.33. A proper set system M is a *delta-matroid* if and only if, for each $X, Y \in M$ and $u \in X \triangle Y$, there is an element $v \in X \triangle Y$ (we allow $u = v$) such that $X \triangle \{u, v\} \in M$.

A set system $M = (V, D)$ is a matroid described by its bases (i.e., D is the family of bases of M) if and only if it is an *equicardinal* delta-matroid, i.e., if $|X| = |Y|$ for all $X, Y \in D$. In that case d_M is equal to the rank $r(M)$ of M.

For a $V \times V$ matrix A (over a field $\mathbb{F}$) define the set system $\mathcal{M}_A = (V, D_A)$ with $D_A = \{X \subseteq V : A[X] \text{ is nonsingular}\}$. By convention, the empty matrix $A[\emptyset]$ is nonsingular. The matrix is called *skew-symmetric* if $A^T = -A$ (here A^T denotes the transpose of A) and its diagonal is zero (the latter condition is automatic in case $\mathbb{F}$ has characteristic different from 2). If A is symmetric or skew-symmetric, then the set system $\mathcal{M}_A$ is a delta-matroid [186].

Let M be a set system with ground set V, and let $X \subseteq V$. The *twist* of M on X, denoted $M * X$, is the set system $(V, \{Y \triangle X : Y \in M\})$. If M is a delta-matroid, then so is $M * X$. In the case that M is a matroid, $M * V$ is equal to the dual matroid M^* of M.

A delta-matroid M over V is called *representable over* $\mathbb{F}$ (or $\mathbb{F}$*-representable*) if $M = \mathcal{M}_A * X$ for a $V \times V$-symmetric or skew-symmetric matrix A over $\mathbb{F}$ and for some $X \subseteq V$. A matroid turns out to be $\mathbb{F}$-representable in this delta-matroid sense if and only if it is $\mathbb{F}$-representable in the usual matroid sense (see [186]). A (delta-)matroid is *binary* if it is representable over $\mathbb{F}_2$. Since an adjacency matrix $A(G)$ is a symmetric matrix over $\mathbb{F}_2$, it follows that $\mathcal{M}_{A(G)}$ is a binary delta-matroid. We write $\mathcal{M}_G$ to denote $\mathcal{M}_{A(G)}$.

Example 14.34. Let G be the graph in Figure 14.1. The delta-matroid $\mathcal{M}_G$ is equal to

$$(\{a,b,c,d\}, \{\emptyset, \{b\}, \{a,b\}, \{a,c\}, \{b,c\}, \{c,d\}, \{a,b,c\}, \{b,c,d\}, \{a,b,c,d\}\}).$$

The following theorem relates graphs and principal pivot transforms (as discussed in Section 14.3.3) to delta-matroids and twist. Note that $A(G)*X$ is defined if and only if $A(G)[X]$ is nonsingular. The following theorem combines results from [186, 192, 205].

Theorem 14.35. *Let $G = (V, E)$ be a looped simple graph and $X \subseteq V$. Then the binary delta-matroid $\mathcal{M}_G = \mathcal{M}_{A(G)}$ uniquely determines G (and vice versa). Moreover, $\mathcal{M}_{A(G)*X} = \mathcal{M}_{A(G)} * X$ (if the left-hand side is defined), and $d_{\mathcal{M}_G * X} = n(G[X])$.*

14.10.2 Interlace polynomial and Tutte connection

We now define the interlace polynomial for set systems [206].

Definition 14.36. Let M be a set system over V. The *interlace polynomial* for M is defined as

$$q(M;y) = \sum_{X \subseteq V} (y-1)^{d_{M*X}}.$$

By Theorem 14.35 we have $q(G;y) = q(\mathcal{M}_G;y)$, so the interlace polynomial for set systems generalizes the interlace polynomial for graphs. Moreover, it is easy to see that $q(M*X;y) = q(M;y)$ for all $X \subseteq V$.

Originally obtained for 4-regular graphs [651] and for binary matroids [12], we now state the connection, from [206], of a weighted interlace polynomial to the Tutte polynomial for matroids in general.

Theorem 14.37. *Let M be a matroid with ground set V described by its bases. Let a and b be variables. Then*

$$\sum_{X \subseteq V} a^{|V \setminus X|} b^{|X|} y^{d_{M*X}} = a^{n(M)} b^{r(M)} T(M; 1 + \tfrac{a}{b}y, 1 + \tfrac{b}{a}y).$$

In particular, $q(M;y) = T(M;y,y)$.

Let M be a set system with ground set V, and let $X \subseteq V$. The *deletion* of M by X, denoted $M \backslash X$, equals $(V \setminus X, D')$ where $D' = \{Y \in D : Y \cap X = \emptyset\}$. If $v \in V$, then we also write $M \backslash v$ for $M \backslash \{v\}$. If M is a delta-matroid, then so is $M \backslash X$, provided that it is proper. An element $v \in V$ is a *loop* of M if $M * v \backslash v$ is not proper and a *coloop* of M if $M \backslash v$ is not proper. This generalizes the corresponding notions for matroids. Moreover, we say that $v \in V$ is *singular* in M if v is a loop or a coloop of M.

The next result, from [206], characterizes the interlace polynomial for delta-matroids.

Theorem 14.38. *Let M be a delta-matroid over V.*

1. *If $V = \emptyset$, then $q(M;y) = 1$.*
2. *If $v \in V$ is singular in M, then*
$$q(M;y) = y\, q(M \backslash v; y).$$
3. *If $v \in V$ is not singular in M, then*
$$q(M;y) = q(M \backslash v; y) + q(M * v \backslash v; y).$$

Example 14.39. Consider the delta-matroid $M = \mathcal{M}_G$ from Example 14.34 for the graph of Figure 14.1. Figure 14.7 shows a recursive computation the interlace polynomial $q(M;y) = y^2 + 4y + 4$. This figure abbreviates sets so that, e.g., *cd* denotes $\{c, d\}$. Since $q(G;y) = q(\mathcal{M}_G;y)$, the result of the computation of Figure 14.7 is equal to that of Figure 14.3.

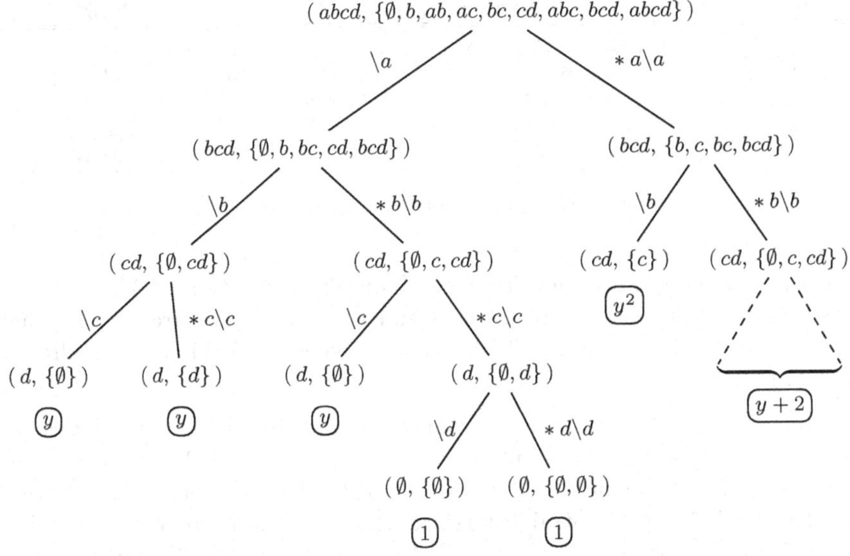

FIGURE 14.7: Recursive computation of $q(M; y)$. The tree with root (cd, $\{\emptyset, c, cd\}$) occurs twice as a subtree.

14.10.3 Evaluations

Using Theorem 14.37, some of the known evaluations of $T(M; y, y)$ can be generalized to set systems. We have trivially $q(M; 2) = 2^{|V|}$. Also, as $X \in M$ if and only if $d_{M*X} = 0$, we have that $q(M; 1)$ is equal to the number of sets $X \in M$.

A set system is called *even* if the cardinalities of all its sets have the same parity. Obviously, a matroid (described by its bases) is even. Also, the delta-matroid $\mathcal{M}_G$ of a graph G without loops is even. We state a result on so-called *vf-safe delta-matroids*, a family a matroids with specific closure properties (we omit the definition here, see [205]). Given a set system $M = (V, D)$, the set system $M \bar{*} V$ is defined as follows: $X \subseteq V$ is in $M \bar{*} V$ if and only if X is a subset of an odd number of sets of D. If M is a vf-safe delta-matroid, then so is $M \bar{*} V$. Quaternary delta-matroids (i.e., representable over $\mathbb{F}_4$) form a subclass of the class of vf-safe delta-matroids. The following result was shown in [189, 206].

Theorem 14.40. *Let M be a delta-matroid.*

1. *If M is even and $|V| > 0$, then $q(M; 0) = 0$.*
2. *If M is vf-safe, then $q(M; -1) = (-1)^{|V|}(-2)^{d_{M \bar{*} V}}$.*
3. *If M is binary, then $q(M; 3) = k\,|q(M; -1)|$, for some odd integer k.*

For quaternary matroids M on ground set E, then $d_{M \bar{*} V}$ is equal to the

Section	Structure \ # of elem. minor types	2	3	3 Weighted
14.4	2-in 2-out or 4-regular graph $\vec{G}$, G	$m(\vec{G};y)$	$M(G;y)$	$M(G,W;y)\star$
14.3, 14.5	graph G	$q(G;y)$	$Q(G;y)$	$Q(G,W;y)$
14.8, 14.10	matroid M	$T(M;y,y)$	-	-
14.9	isotropic system $\mathcal{I}$	$m(\mathcal{I},T;y)$	$M(\mathcal{I};y)$	$M(\mathcal{I},W;y)\star$
14.10	(vf-safe) delta-matroid M	$q(M;y)$	$Q(M;y)\star$	$Q(M,W;y)\star$

TABLE 14.1: Summary of the main polynomials considered in this chapter.

dimension $\dim(B)$ of the bicycle space of any representation of M [202], and thus we retrieve, using Theorems 14.37 and 14.40, the result from [1116] that $T(M;-1,-1) = (-1)^{|E|}(-2)^{\dim(B)}$. The case where M is binary was already shown in [969].

Example 14.41. The uniform matroid $U_{2,5}$ is not binary, but it is quaternary and therefore vf-safe. For a subset X of the ground set V, we have that $d_{U_{2,5}*X} = ||X|-2|$. Hence $q(U_{2,5};y) = (y-1)^3+(5+1)(y-1)^2+(10+5)(y-1)+10 = y^3+3y^2+6y$. Straightforward combinatorial arguments show that $U_{2,5} \bar{*} V = U_{2,5}$. Indeed $q(U_{2,5};-1) = -4 = (-1)^5(-2)^2$, see Theorem 14.40.

It is possible to formulate a *global interlace polynomial* for set systems, similar to those for graphs and isotropic systems. One needs to restrict to vf-safe delta-matroids in order to formulate a recursive formulation for that polynomial. The vf-safe delta-matroids are essentially equivalent to the tight 3-matroids which form a particular class of multimatroids [190, 206]. In turn, the class of isotropic systems (see Section 14.9) can be viewed as a subclass of the class of tight 3-matroids. The vf-safe delta-matroids play a significant role for the topological transition polynomial of ribbon graphs (see Section 13.3.5 and [318]).

14.11 Summary

Table 14.1 lists the main classes of polynomials considered in this chapter. The three right-most columns arrange the polynomials according to the number of types of elementary minors in their recursive formulas. Except for the row concerning the Tutte polynomial, the rows are ordered in increasing level of generality. Moreover, the last column, concerning three elementary minor types with weights, generalizes the other two columns. Polynomials marked with $\star$ are not explicitly defined here, but can be found in [651] for $M(G,W;y)$ and [206] for $Q(M;y)$ and $Q(M,W;y)$. The weighted variant $M(\mathcal{I},W;y)$ of $M(\mathcal{I};y)$ can be defined in the natural and obvious way.

We mention that multimatroids, introduced by Bouchet [190], generalize delta-matroids by allowing an arbitrary number of elementary minor types (instead of two, and, in the case of vf-safe delta-matroids, three). It turns out that the weighted interlace polynomial $Q(M, W; y)$ (and thus also its specializations) can be defined for arbitrary multimatroids. Moreover, many properties of $Q(M, W; y)$, such as its recursive formulation and some evaluations, carry over to this multimatroid polynomial [88, 206].

14.12 Open problems

Several open problems on interlace polynomials and research directions remain. In [153, 448] the Martin polynomial $m(G; y)$ is considered for arbitrary Eulerian (di)graphs G, with [448] extending the Martin polynomial to arbitrary graphs as a generating function for cycles and paths. A natural question is to generalize the interlace polynomial in a way that extends the link with the Martin polynomial for Eulerian (di)graphs in general. A difficulty here is that a vertex reduction (skein relation) may split a connected graph in more than two connected components. This increases the "nullity" by more than one, which is impossible for the standard elementary minors of contraction and deletion in (delta-)matroids. Perhaps the generalization of delta-matroids called *parity systems* defined in [193] provides a clue to extend the link between the Martin polynomial and the interlace polynomial to graphs more general than 2-in 2-out digraphs and 4-regular graphs.

Also, it is an open problem to characterize which evaluations of the interlace polynomial can be computed efficiently. Moreover, the relationships with other polynomials, such as the Penrose polynomial [916], has not been fully clarified (some initial results can be found in [202]).

Another direction for further research is to generalize the classical two-variable Tutte polynomial to polynomials defined on more general combinatorial structures than matroids. While part of the (x, y)-plane is generalized through the interlace polynomial of delta-matroids through Theorem 14.37, it is an open question to generalize the whole (x, y)-plane (i.e., the whole two-variable Tutte polynomial). Perhaps Theorem 14.27 provides a starting point to generalize the whole Tutte polynomial to more general structures than matroids.

Part IV

Applications

15

Network reliability

Jason I. Brown • Charles J. Colbourn

Synopsis

Network reliability measures the likelihood of overall network functionality (generally various kinds of connectivity) given the probabilities of individual edges functioning. This chapter considers the ramifications of the connection between Tutte polynomials and network reliability.

- General model of reliability of graphs and a compact representation in terms of reliability polynomials.
- Network operations for connectedness of target vertex sets; all-terminal reliability is an evaluation of the Tutte polynomial.
- Bounding the value of the reliability polynomial by computing and bounding its coefficients, primarily using the combinatorial structure of the operating or failed subgraphs.
- Inequalities for the coefficients in various forms of the all-terminal reliability polynomial.
- Analytic properties of reliability polynomials (location of roots, shape, and points of inflection).
- The structure of most reliable graphs and average all-terminal reliability.
- Extensions to directed graphs.

DOI: 10.1201/9780429161612-15

15.1 Introduction

Let $G = (V, E)$ be a graph on n vertices and m edges; loops and multiple edges are permitted. Associate with each edge of E a probability p that the edge is *operating*; it is *failed* with the complementary probability $q = 1-p$. Our model assumes independence among the edge probabilities. A *structure function* is an indicator function $\phi : 2^E \to \{0, 1\}$ on the subsets of E. A set $S \subseteq E$ is a *state* of the network. A state S occurs when all edges of S are operating, and all edges of $E \setminus S$ are failed; it is *operating* when $\phi(S) = 1$, and *failed* when $\phi(S) = 0$. The basic question is: If every edge operates independently, what is the probability $\mathrm{Rel}_\phi(G; p)$ that the graph is in an operating state (under ϕ)? This can be viewed as the *reliability* of the graph at performing the "operation" specified by the structure function ϕ (and is equal to the expected value of ϕ). A structure function is *coherent* if, whenever S is an operational state and $S \subseteq T$, then T is also an operational state.

Consider a fixed state S containing i edges. As each edge operates independently with probability p, the probability of being in state S is exactly $\Pr[S] = p^i(1-p)^{m-i}$ (each of i edges operates, the remaining edges fail, and all edge events are independent). Because the 2^m states are disjoint events covering all possibilities, we arrive at the following central definition.

Definition 15.1. The reliability with respect to an arbitrary structure function ϕ is

$$\mathrm{Rel}_\phi(G; p) := \sum_{S \subseteq E} \Pr[S]\phi(S).$$

This expresses the reliability with respect to ϕ as a polynomial of degree at most m in one variable, the edge operation probability p. It is a *reliability polynomial* for ϕ.

Some pertinent examples follow. Let $K \subseteq V$ be a set of *target* vertices. For each state S, define $\phi_K(S) = 1$ if and only if the subgraph (V, S) has all target vertices in the same connected component. Then $\mathrm{Rel}_{\phi_K}(G; p)$ is the *k-terminal reliability* of G for target set K and operation probability p; we typically abbreviate this as $\mathrm{Rel}_K(G; p)$. When $K = \{s, t\}$, then $\mathrm{Rel}_K(G; p)$ is the *two-terminal reliability*, the probability that at least one s,t-path is operating. When $K = V$, then $\mathrm{Rel}_K(G; p)$ is the *all-terminal reliability* or *connectedness probability* of the graph, written $\mathrm{Rel}_V(G; p)$ or simply $\mathrm{Rel}(G; p)$. A multivariate version of the reliability polynomial is also considered in Chapter 24.

The connection between Tutte polynomials and all-terminal reliability comes from the following observation from [851, 866] about the probabilities, which leads to a deletion–contraction reduction for $\mathrm{Rel}(G; p)$. Consider an edge e of G. Deleting e gives a new graph $G \backslash e$. Contracting e by identifying its endpoints, removing e, but saving all multiple edges and other loops that arise, gives a graph G/e. Bayes' theorem from probability now implies Theorem 15.2:

Theorem 15.2 (Factoring Theorem). *For a connected graph G and an edge e,*

$$\text{Rel}(G; p) = p \cdot \text{Rel}(G/e; p) + (1 - p) \cdot \text{Rel}(G \backslash e; p). \tag{15.1}$$

The connection to the Tutte polynomial in Theorem 15.3 follows by induction using the form of the Tutte polynomial given in Definition 2.7 upon observing that if e is an ordinary edge then both $G\backslash e$ and G/e are connected and that the reliability polynomial of a connected graph consisting of b bridges and l loops is p^b.

Theorem 15.3. *For a connected graph $G = (V, E)$,*

$$\text{Rel}(G; p) = p^{|V|-1}(1 - p)^{|E|-|V|+1}T(G; 1, 1/(1 - p)).$$

In the context of reliability polynomials, deletion–contraction is often referred to as *factoring* or *pivotal decomposition*, and Theorem 15.2 is known as the *factoring theorem*. Figure 15.1 shows the calculation of the reliability polynomial of K_4 via factoring.

While much of the subsequent discussion holds in the general case, in light of our interest in the connection to Tutte polynomials, we restrict our attention to all-terminal reliability until Section 15.7. In particular, a reliability polynomial is taken to mean an all-terminal reliability polynomial, unless otherwise stated.

Substantial research has been invested in studying reliability polynomials and their close relatives. Moore and Shannon [860] used reliability polynomials to verify their construction of two-terminal networks of arbitrarily better reliability for both making and breaking inter-terminal connection than that of a single edge, as long as $p > 1/2$. Since that time, generalizations of their model have been used extensively as simple models for network design and analysis. Reliability polynomials tell only part of the story of a network subject to random failures, but can be useful guides to reliable network design, see [69].

A full treatment of the subject of reliability polynomials would run to great length. We refer the reader to [69, 323, 1001] for background on network reliability, [294] for an older survey on reliability polynomials, [1138] for related material on matroids, [107, 131] for information on Tutte polynomials, and [117, 1034] for background on complexes.

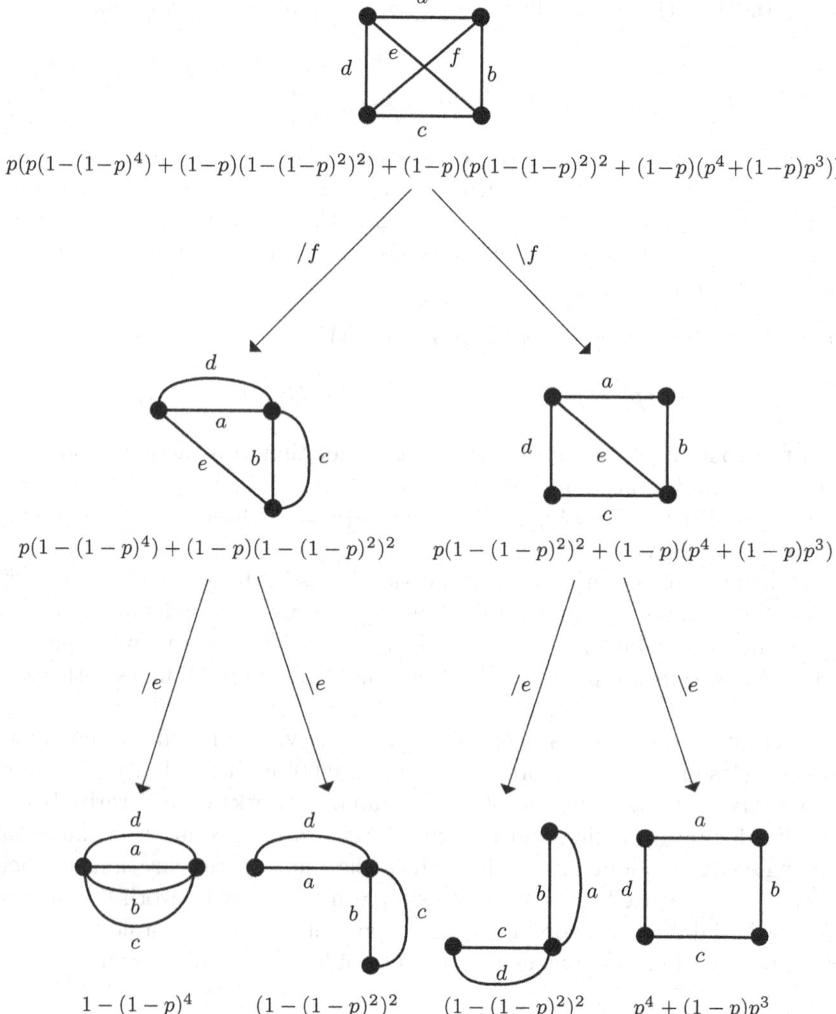

FIGURE 15.1: Computing the reliability polynomial of the graph K_4.

15.2 Forms of the reliability polynomial

For the all-terminal reliability model, let N_i be the number of operating all-terminal states (that is, spanning connected subgraphs) of cardinality i. Then

$$\mathrm{Rel}(G;p) = \sum_{i=0}^{m} N_i\, p^i\, (1-p)^{m-i}.$$

In network design, the value of this polynomial when evaluated at a point is of principal concern. However, suppose that we find lower and upper bounds, $\underline{N}_i \leq N_i \leq \overline{N}_i$, on each *coefficient*. It is then easy to see that for $p \in [0,1]$,

$$\sum_{i=0}^{m} \underline{N}_i \, p^i (1-p)^{m-i} \leq \mathrm{Rel}(G;p) \leq \sum_{i=0}^{m} \overline{N}_i \, p^i (1-p)^{m-i}.$$

This motivates the problem of understanding the coefficients of reliability polynomials.

Of course, $0 \leq N_i \leq \binom{m}{i}$, but we shall see that that the underlying graph theoretic structure imposes further constraints on the coefficients. Often these constraints are best found by writing the reliability polynomials in different forms. The structure function for all-terminal reliability is coherent (i.e. if $S \subseteq T$, then S operational implies T is operational). Define a family $\mathcal{F}$ of sets by

$$\mathcal{F} = \{S : S \subseteq E, \; E \setminus S \text{ is operational}\}.$$

The coherence of all-terminal reliability ensures that $\mathcal{F}$ is an hereditary family of sets, or *(simplicial) complex*, that is, $\sigma \in \mathcal{F}$ and $\alpha \subseteq \sigma$ implies $\alpha \in \mathcal{F}$. Letting $F_i = |\{F \in \mathcal{F} : |F| = i\}|$, we call $(F_0, \ldots, F_m)$ the *F-vector* and obtain the *F-form* of the reliability polynomial:

$$\mathrm{Rel}(G;p) = \sum_{i=0}^{m} F_i \, (1-p)^i p^{m-i}.$$

Evidently, $F_i = N_{m-i}$.

We can similarly consider the reliability as the probability that G is *not* in a failed state and obtain the following M- and C-forms. If M_i is the number of failed states on i edges, we have

$$\mathrm{Rel}(G;p) = 1 - \sum_{i=0}^{m} M_i \, p^i (1-p)^{m-i},$$

and if C_i is the number of sets of i edges whose removal renders the graph failed (*cutsets*, although not necessarily minimal), we have

$$\mathrm{Rel}(G;p) = 1 - \sum_{i=0}^{m} C_i \, (1-p)^i p^{m-i}.$$

These are all simple translations, using the fact that $N_i + M_i = \binom{m}{i}$, $F_i + C_i = \binom{m}{i}$, and $N_i = F_{m-i}$.

A further form simply expands the reliability polynomial in powers of p, the *P-form*:

$$\mathrm{Rel}(G;p) = \sum_{i=0}^{m} P_i \, p^i.$$

One form of particular interest here is the *H-form*. Let ℓ be the smallest number of edges whose operation suffices to make the graph operational (in the all-terminal case, $\ell = n-1$). Then we can factor p^ℓ from the reliability polynomial, and collect the remaining terms in powers of $1-p$ to obtain the *H-form*:

$$\mathrm{Rel}(G;p) = p^\ell \sum_{i=0}^{m-\ell} H_i\,(1-p)^i$$

The *H-vector* is the sequence $(H_0,\ldots,H_{m-\ell})$.

Example 15.4. Consider the all-terminal reliability of the complete graph K_4. Every connected spanning subgraph has at least 3 edges, with spanning subgraphs with 3 edges being precisely the spanning trees of K_4, of which there are 16. Thus $N_3 = 16$. Moreover, any spanning subgraph with at least 4 edges is connected, so that $N_i = \binom{6}{i}$ for $i = 4,5,6$. Thus the N-form of the all-terminal reliability of K_4 is given by

$$\mathrm{Rel}(K_4;p) \;=\; 16p^3(1-p)^3 + 15p^4(1-p)^2 + 6p^5(1-p) + p^6.$$

The F-, M-, C-, P-, and H- forms are found to be, respectively,

$$\begin{aligned}\mathrm{Rel}(K_4;p) &= p^6 + 6(1-p)p^5 + 15(1-p)^2p^4 + 16(1-p)^3p^3\\ &= 1-\left((1-p)^6 + 6p(1-p)^5 + 15p^2(1-p)^4 + 4p^3(1-p)^3\right)\\ &= 1-\left(4(1-p)^3p^3 + 15(1-p)^4p^2 + 6(1-p)^5p + (1-p)^6\right)\\ &= 16p^3 - 33p^4 + 24p^5 - 6p^6\\ &= p^3\left(1 + 3(1-p) + 6(1-p)^2 + 6(1-p)^3\right)\end{aligned}$$

so that

- $N_0 = N_1 = N_2 = 0$, $N_3 = 16$, $N_4 = 15$, $N_5 = 6$, $N_6 = 1$;
- $F_0 = 1$, $F_1 = 6$, $F_2 = 15$, $F_3 = 16$, $F_4 = F_5 = F_6 = 0$;
- $M_0 = 1$, $M_1 = 6$, $M_2 = 15$, $M_3 = 4$, $M_4 = M_5 = M_6 = 0$;
- $C_0 = C_1 = C_2 = 0$, $C_3 = 4$, $C_4 = 15$, $C_5 = 6$, $C_6 = 1$;
- $P_0 = P_1P_2 = 0$, $P_3 = 16$, $P_4 = -33$, $P_5 = 24$, $P_6 = -6$; and
- $H_0 = 1$, $H_1 = 3$, $H_2 = 6$, $H_3 = 6$, $H_4 = H_5 = H_6 = 0$.

A calculation of the reliability of K_4 is shown in Figure 15.1.

From a mathematical viewpoint, the various forms of all-terminal reliability arise merely as expansions of the polynomial $\mathrm{Rel}(G;p)$ (or, in some cases, $1-\mathrm{Rel}(G;p)$) in terms of different bases for the vector space over $\mathbb{Q}$ of all polynomials in p of degree m. Each of the various forms is useful for reliability polynomials, precisely because each sequence of coefficients has its own combinatorial interpretation. Interpretations of N_i, F_i, C_i and M_i are all immediate. The remaining forms require some explanation.

An easy but crucial observation is that factoring using Equation (15.1) underlies simple recursive formulas for the coefficients (conceptually simple, not necessarily computationally efficient). For example,

$$\begin{aligned} F_i(G) &= F_i(G/e) + F_{i-1}(G\backslash e) \\ P_i(G) &= P_{i-1}(G/e) + P_i(G\backslash e) - P_{i-1}(G\backslash e) \\ H_i(G) &= H_i(G/e) + H_{i-1}(G\backslash e) \end{aligned}$$

Now consider the P-form. Let $A_1, \ldots, A_s$ be the set of all minimal operating states. Reliability can be written as

$$\sum_{I \subseteq \{1,\ldots,s\}} (-1)^{|I|-1} \Pr\left(\bigcup_{i \in I} A_i\right).$$

A state S occurs in the sum only when S is the union of minimal operating states. The number of times it contributes to this sum depends upon the difference between the number of times it arises as the union of an odd number of minimal operating states and the number of times it arises as the union of an even number. Summing these differences over all i-edge states gives P_i [985].

The H-form is perhaps the most important for the all-terminal reliability polynomial because of its relation to the theory of shellable complexes (see Chapter 5). The H-vector $(H_0, H_1, \ldots, H_{m-n+1})$ is always nonnegative for the all-terminal reliability polynomial. While this is not obvious from the definition, it can be shown inductively by using the recursion implied by Theorem 15.2. A remarkable interpretation for these coefficients results from a classical theorem of Tutte [1090] (see Theorem 4.135). For the all-terminal problem on a connected graph, every maximal element of $\mathcal{F}$ is the complement of a spanning tree and hence $\mathcal{F}$ is the cographic matroid of G. Under an arbitrary order $\prec$ on the set of edges of G, an element e of a spanning tree B is *internally inactive* if it is *not* the first element under $\prec$ of the fundamental cocircuit of e with respect to B. Then it follows from Tutte's theorem that H_i is the number of spanning trees with precisely i internally inactive edges.

This thumbnail sketch of forms and their interpretation suggests the main direction taken in studying reliability polynomials, namely to interpret the coefficients in some form of the polynomial combinatorially, and then to employ the combinatorial structure to characterize the possible sequences of coefficients in that or another form of the polynomial.

We give the actual coefficients in a number of forms for another all-terminal problem. The graph is depicted in Figure 15.2, and coefficients for four forms of its reliability polynomial are given in Table 15.1.

i	F_i	C_i	H_i	P_i
0	1	0	1	0
1	20	0	9	0
2	190	0	45	0
3	1136	4	161	0
4	4771	74	453	0
5	14861	643	1050	0
6	35285	3475	2056	0
7	64480	13040	3435	0
8	90052	35918	4856	0
9	93177	74783	5588	24704
10	65793	118963	4764	-205951
11	24704	143256	2286	793967
12	0	125970	0	-1864016
13	0	77520	0	2955596
14	0	38760	0	-3318205
15	0	15504	0	2688033
16	0	4845	0	-1569511
17	0	1140	0	646718
18	0	190	0	-178958
19	0	20	0	29910
20	0	1	0	-2286

TABLE 15.1: Coefficients for four forms of its reliability polynomial for the graph in Figure 15.2.

15.3 Calculating coefficients

The goal of understanding reliability via the coefficients can be approached in a number of ways. One method is to determine relationships among the coefficients. A second is to determine inequalities that all graphs with specified parameters (number of vertices, number of edges, edge connectivity, and so on) must satisfy. And a third is to determine, given a graph, which coefficients we can calculate efficiently. We consider the computational question in this section.

Provan and Ball [934] proved that calculating all-terminal reliability, even when all edges have probability $p = \frac{1}{2}$ is ♯P-complete (see also [653]). Provan [933] established similar results for planar graphs. In fact, the results in [653] on the computational complexity of evaluations of the Tutte polynomial show that computing $\mathrm{Rel}(G;p)$ is ♯P-hard for any real number p, except when $p \in \{0,1\}$; analogous ♯P-hardness results for planar graphs are given in [1118]. See Chapter 9 for further details. Karger [674, 675] established that, despite these very negative worst case complexity results, there does exist a fully polynomial approximation scheme.

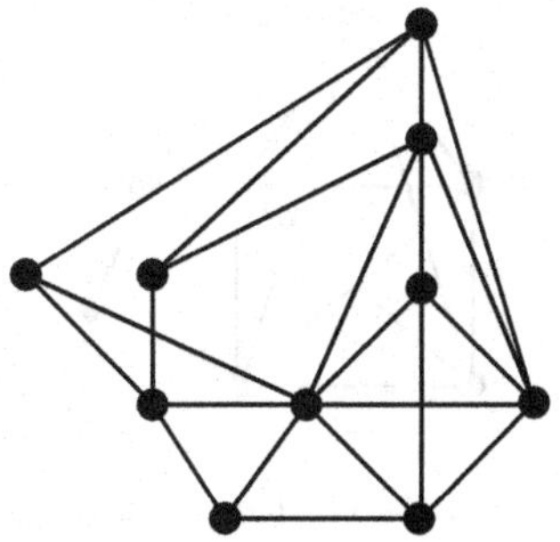

FIGURE 15.2: A (relatively) small graph.

These complexity results establish that exact calculation of all of the coefficients (in any form) is ♯P-hard. However, the calculation of *some* of the coefficients is feasible. Let us consider the F-form. Evidently if fewer than $n-1$ edges operate, the graph is disconnected. Thus $F_i = 0$ for $i > m-n+1$. If the smallest edge cutset has size c (the edge connectivity is c), there is no way to remove fewer than c edges and disconnect the graph. Thus $F_i = \binom{m}{i}$ for $i < c$. For any k, Ramanathan and Colbourn [943] give an algorithm for determining F_{c+k} whose running time is polynomial in the size of the graph but exponential in k (and so, for fixed k, is polynomial). The coefficient F_{m-n+1} is exactly the number of spanning trees of the graph, which by the Kirchoff matrix-tree theorem [691] can be calculated efficiently (although the number itself can be exponential in the size of the graph).

Turning to the H-form, assume that the graph is loopless and let $d = m-n+1$. Then $F_k = \sum_{j=0}^{k} \binom{d-j}{d-k} H_j$ and $H_k = \sum_{j=0}^{k} (-1)^{k-j} \binom{d-j}{d-k} F_j$. Hence we can calculate $H_0, \ldots, H_{c+k}$ for fixed k efficiently, and we can also calculate $\sum_{j=0}^{d} H_j$, which is again the number of spanning trees. One coefficient, H_d, sometimes called the *reliability domination*, has been extensively studied [147, 148, 347, 634, 901] because of its use in determining the number of steps employed in computing reliability via Theorem 15.2. It follows from a remarkable general theorem of Vertigan [1118] on complexity of evaluations of the Tutte polynomials, that computation of H_d is ♯P-hard even for planar graphs. However, some bounds for H_d are known, as we shall see later.

15.4 Transformations

A further strategy is to find transformations that have a predictable effect on each coefficient (for example that never increase a coefficient). A basic device for studying transformations is Theorem 15.2 and the recursion that it

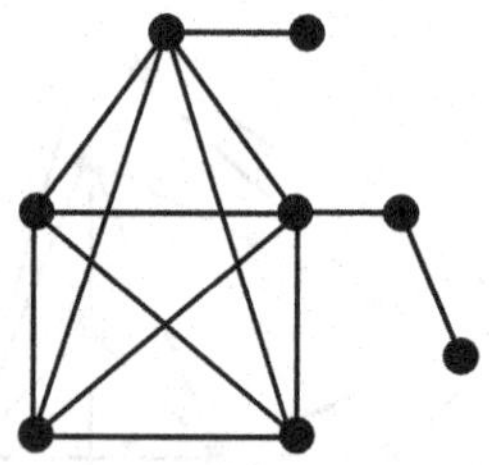

FIGURE 15.3: The graph $L_{8,13}$.

implies for the various coefficients. A transformation, roughly speaking, is the replacement of one subgraph by another. For example, the operation of adding an edge in series, that is, deleting an edge (x, y) and adding a pair of edges (x, z) and (y, z), is a *Lomonosov join* on the graph. It never decreases the reliability, and never decreases any coefficient in the F-vector or the H-vector of the graph [228].

One main problem on which progress has been made by transformation methods is determining the least reliable all-terminal network having no loops or multiple edges. A *least reliable graph* for n vertices and m edges is a simple graph whose all-terminal reliability is the lowest among all such graphs for every edge operation probability p with $0 < p < 1$. Given a number of edges m and a number of vertices n, let s be the largest integer satisfying $\binom{s}{2} + n - s - 1 \leq m$. Form a complete graph on s vertices. Add one more vertex adjacent to $m - \binom{s}{2} - n + s + 1$ vertices in this complete graph. Finally add the $n - s - 1$ remaining vertices, each connected by a bridge to the graph constructed. Call this graph $L_{n,m}$ (an example is presented in Figure 15.3). It has been conjectured that $L_{n,m}$ is the least reliable *simple* graph on n vertices and m edges [147], that is, for any other graph H with the same number of vertices and edges, $\text{Rel}(L_{n,m}; p) \leq \text{Rel}(H; p)$ for *all* $p \in [0, 1]$. (We remark that the least reliable n-vertex m-edge graph, allowing multiple edges, contains $n - 2$ bridges and one bundle of $m - n + 2$ multiple edges.)

15.5 Coefficient inequalities

In this section, we develop some relations among the coefficients that all stem from the fact that $\mathcal{F}$ is a complex. Indeed $\mathcal{F}$ has more structure than just being a complex; it is a subclass of the *shellable complexes*. For a set α, let $\overline{\alpha}$ denote the power set of α, that is, the set of all subsets of α; it is always a complex in its own right. For sets σ and α of a complex with $\sigma \subseteq \alpha$, the interval $[\sigma, \alpha]$ is the collection $\{\beta : \sigma \subseteq \beta \subseteq \alpha\}$. We shall need to define some important subclasses of complexes. We begin with two.

Definition 15.5. A complex Σ is *purely d-dimensional* if it has dimension d and every maximal set (i.e., *facet*) has cardinality d.

Definition 15.6. A complex Σ is *shellable* if, for positive integer d, Σ is purely d-dimensional and these maximal sets (or *facets*) can be ordered as $\sigma_1, \sigma_2, \ldots, \sigma_t$ such that $\overline{\sigma_i} \cap \left(\bigcup_{j=1}^{i-1} \overline{\sigma_j}\right)$ is a purely $d-1$ dimensional complex, for all $i = 2, \ldots, t$. Such an ordering, if it exists, is called a *shelling*.

Many pure complexes (including matroids) are shellable, and shellable complexes have certain nice homological properties (see Chapter 5, or, for example, [131]).

We define one more important class of complexes.

Definition 15.7. A purely d-dimensional complex Σ is *partitionable* if it has an interval partition, that is, its sets can be partitioned into a sequence of intervals $[L_i, U_i]$, for $i = 1, \ldots, t$, where each U_i is a facet of Σ.

The key observation is that any shellable complex is partitionable, and the number of intervals whose lower set has size i turns out to be exactly H_i [1034]. We shall make much use of these properties of complexes.

15.5.1 Bounds via combinatorial properties of the complex $\mathcal{F}$

Sperner [1025] established a famous lemma for F-vectors of complexes that we express as

$$iF_i \leq (m - i + 1)F_{i-1}.$$

These inequalities have direct implications for the F-form of reliability. Sperner's lemma is not at all surprising, because it is equivalent to

$$(m - i)N_i \leq (i + 1)N_{i+1},$$

which states, in terms of reliability, that the fraction of i-edge spanning subgraphs that are connected is a nondecreasing function of i. However, the use of inequalities such as this *is* surprising. Given F_i, we can now easily obtain a *lower* bound on F_{i-1} and an *upper* bound on F_{i+1}. Together with exact calculations, we can then produce lower and upper bounds on every coefficient in the F-form.

Kruskal [726] and Katona [679] independently improved on Sperner's result, and obtained the best possible inequality for complexes in general. The application of the Kruskal–Katona theorem to reliability was first studied by Van Slyke and Frank [1114], and leads naturally to the question are the bounds of the Kruskal–Katona theorem still tight?

In the all-terminal case, every maximal set in $\mathcal{F}$ has the same cardinality (the complex is *pure*). F-vectors and H-vectors of pure complexes are not yet characterized [151, 325]. However, we know more: $\mathcal{F}$ is the cographic matroid, so it is reasonable to expect that it exhibits much more structure than complexes in general, pure or not. We saw earlier that all the H-coefficients for this case are positive and they count the spanning trees of a certain type.

It is known that any matroid complex is a shellable complex [935], and admits an interval partition [70] as implied by shellability. In fact, a partition into intervals for $\mathcal{F}$ can be obtained by considering all bases of the cographic matroid (complements of spanning trees), and for each taking the lower set of the interval to be the internally inactive elements of this basis.

Example 15.8. For the labelled graph K_4 in Figure 15.1, we find that (writing sets multiplicatively)

$$abc, abd, acd, bcd, abe, ace, bde, cde, acf, adf, bcf, bdf, aef, bef, cef, def$$

is a shelling of the cographic matroid $\mathcal{F}$ of K_4, and an interval partition is

$$[\emptyset, abc], [d, abd], [e, abe], [f, acf], [bf, bcf], [cd, acd], [ce, ace], [de, bde], [df, adf], [ef, aef], [cde, cde], [bcd, bcd], [bdf, bdf], [bef, bef], [cef, cef], [def, def].$$

The notion of active and inactive elements due to Tutte was extended to matroids by Crapo [347], and the relation to interval partitioning has been employed extensively [131, 326, 362, 573, 1073]. The importance here is that while interval partitions arise from every shellable complex, partitions that arise from Tutte's notion of inactive elements offer more structure.

15.5.2 Connections to commutative algebra

The fact that matroids are shellable enables employment of Stanley's characterization [1028, 1029, 1034] of the H-vectors of shellable complexes. The application to reliability has been studied in [71, 72, 228]. Stanley's characterization rests on the construction of an order ideal of monomials (that is, a set of monomials closed under divisibility) given a shelling of $\mathcal{F}$ such that the H-vector of $\mathcal{F}$ lists the number of monomials of each degree in the order ideal. For example, an order ideal of monomials for the cographic matroid of K_4 is

$$\{1, x, y, z, xy, xz, yz, x^2, y^2, z^2, xy^2, xz^2, x^2y, yz^2, x^2z, y^2z\}.$$

Note that the numbers of monomials of degree $i = 0, 1, 2, 3$ are $1, 3, 6, 6$, respectively, and give the values of the H_i for the H-vector.

From an enumerative point of view, each interval in the interval partition corresponds to a monomial in the order ideal of monomials. The techniques used to produce the order ideal of monomials are primarily from commutative algebra; see Billera [117] for a clear summary. We have remarked that Tutte's notion of active and inactive elements allows us to obtain a combinatorial bijection from spanning trees to intervals. Brown, Colbourn and Wagner [230] describe a process, via Gröbner bases of algebraic ideals, to generate an order ideal of monomials from the edge cuts and a single spanning tree of a graph. Furthermore, Merino [838] shows how certain "critical" configurations in a chip firing game on a graph could yield an order ideal of monomials as well (this fact was utilized in [229] to produce better bounds for reliability).

15.5.3 Unimodality and log concavity

Some well-known conjectures about the numerical properties of the F-vectors and H-vectors of matroids remained open until only recently. A sequence of nonnegative integers $x_0, \ldots, x_d$ is *unimodal* if there is an index j for which $x_0 \leq x_1 \leq \cdots \leq x_j \geq x_{j+1} \geq \cdots \geq x_d$. It is *log-concave* if $x_i^2 \geq x_{i-1}x_{i+1}$ for all i. Log concavity implies unimodality when the sequences contain only positive integers as is the case with F-vectors and H-vectors. Mason [826] and Welsh [1137, 1138] conjectured unimodality and log concavity for F-vectors of matroids in general. Dawson [362] conjectured log concavity for H-vectors of matroids in general. Is the sequence $(F_0, \ldots, F_d)$ or the sequence $(H_0, \ldots, H_d)$ for the all-terminal reliability polynomial unimodal or log-concave?

Dawson [362] showed that log concavity of the sequence $(H_0, \ldots, H_d)$ implies the log concavity of the sequence $(F_0, \ldots, F_d)$. However, as shown in [1031], the unimodality of $(H_0, \ldots, H_d)$ does not imply the unimodality of $(F_0, \ldots, F_d)$. Recently results of Lenz [761] and Huh [630], respectively, imply that the F-vectors and H-vectors of cographic matroids are in fact log-concave (and hence unimodal).

15.5.4 Linear inequalities

Linear inequalities for the H-vector have been established by Brown and Colbourn [226, 227] and Chari [292, 293]. Many of these follow from a certain expansion of the all-terminal reliability polynomial given by Chari [292, 293]. For a 2-edge connected n-vertex m-edge graph G, let $q = 1-p$, $d = m-n+1$ and define $Q(G) = \mathrm{Rel}(G;p)/p^{n-1} = \sum_{i=0}^{d} H_i\, q^i$.

Chari [292, 293] showed that there exist positive integers ρ_i and s_i, and positive integers $r_{i,1}, \ldots, r_{i,s_i}$, for $i = 1, 2, \ldots, d$, satisfying $\rho_i + r_{i,1} + \cdots + r_{i,s_i} = d$ for all i (with $\rho_1 = 0$) such that

$$Q(G) = \sum_{i=1}^{d} \left[q^{\rho_i} \prod_{j=1}^{s_i} (1 + q^1 + \cdots + q^{r_{i,j}}) \right]. \tag{15.2}$$

This result extends to a broad subclass of shellable complexes, including all matroid complexes [293]. We outline a proof for all-terminal reliability, using a generalization of the factoring theorem. For a graph G, let X be a maximal set of multiple edges, and let $G \backslash X$ be the graph obtained by deleting all the edges in X. Denote by $G \wr X$ the graph obtained by identifying the two end vertices of the edges in X and deleting all the loops resulting from X. Then

$$\mathrm{Rel}(G;p) = (1 - q^{|X|}) \cdot \mathrm{Rel}(G \wr X;p) + q^{|X|} \cdot \mathrm{Rel}(G \backslash X;p).$$

In particular if X is not a cutset of G,

$$Q(G) = (1 + q^1 + \cdots + q^{|X|-1}) \cdot Q(G \wr X) + q^{|X|} \cdot Q(G \backslash X).$$

With a careful inductive argument one can use this to derive Equation (15.2). As a consequence of (15.2), we get certain linear inequalities for the H-vectors of 2-edge connected graphs [293]:

$$\begin{aligned} & H_0 \leq H_1 \leq \cdots \leq H_{\lfloor d/2 \rfloor}, \\ & H_i \leq H_{d-i}, \text{ for } i \leq \lfloor d/2 \rfloor, \\ & H_{d-1} \leq dH_d, and \\ & \sum_{i=0}^{j} (-b)^i H_{j-i} \geq 0, \text{for all } 0 \leq j \leq d, \end{aligned} \tag{15.3}$$

for $b \geq 1$ with strict inequality except, possibly, for $b = 1$.

The first two inequalities are also known to hold for the degree sequence (that is, the sequence of the number of monomials of each degree) of a *pure* order ideal of monomials [613]. Hence the proof of these inequalities represents at least partial progress towards Stanley's conjecture from [1028] that a matroidal H-vector is the degree sequence of a pure order ideal of monomials. The last two inequalities were proved earlier using different methods by Brown and Colbourn [226, 227]. It can be shown that whenever the H-vector of any complex is nonnegative, the F-vector satisfies all of these inequalities. In particular, they are true for the F-coefficients of the reliability polynomial [362, 826].

Brown and Colbourn [227] also derive many other inequalities that hold for all-terminal reliability of graphs which incorporate efficiently computable graph parameters like the number of spanning trees, girth etc., or properties such as being simple or connected. For instance if a matroid is connected, then $H_d \geq H_1$, see [131, 227]. However, further research is needed into determining stronger conditions for H-vectors of connected matroids and for binary, regular and cographic matroids. It would also be of significant practical interest to determine if these new inequalities can be incorporated into the Ball–Provan shellability bounds [71, 72] for all-terminal reliability. For some results of this nature, see [227, 228].

15.6 Analytic properties of reliability

Reliability polynomials are, at their very essence, functions. So it is natural to ask what can be said analytically about such functions, especially over the interval $p \in [0,1]$. Figure 15.4 plots the reliability of the graph in Figure 15.2. In general, $\text{Rel}(G;0) = 0$, and $\text{Rel}(G;1) = 1$, provided that G is connected (otherwise, $\text{Rel}(G;p) = 0$ for all p). A simple calculation from the F-form of the reliability polynomial,

$$\text{Rel}(G;p) = \sum_{i=0}^{m} F_i\,(1-p)^i p^{m-i},$$

shows that, for the derivative,

$$\text{Rel}'(G;p) = \sum_{i=0}^{m-1} ((m-i)F_i - (i+1)F_{i+1})\,(1-p)^i p^{m-1-i} \geq 0$$

by Sperner's lemma. Hence all-terminal reliability (and indeed, the reliability of any coherent system) is a nondecreasing function on $[0,1]$ (provided that G is connected and of order at least 2, the function is *strictly* increasing on the interval). Moreover, the curve is S-shaped [125] in the range $p \in [0,1]$. Moore and Shannon [860] proved the elegant theorem that the reliability polynomial of any coherent structure (and so for all-terminal of any connected graph of order at least 3) crosses the line $f(p) = p$ at most once in the range $p \in (0,1)$. In fact, the closure of such fixed points of reliability polynomials is the entire interval $[0,1]$, see [222]. Margulis [816] proved that provided the edge connectivity is large enough, the all-terminal reliability function jumps from close to 0 to close 1 over a very small interval.

15.6.1 Inflection points of reliability polynomials

The S-shaped property of the curve suggests that there exists at least one point of inflection in $(0,1)$ for reliability polynomials. There are some exceptions, including the singleton graph K_1 and those consisting of two vertices with a bundle of parallel edges between them. It was shown in [237] that every connected graph is an induced subgraph of a graph without inflection points in its reliability polynomial. However, such graphs are rare. Of the 11117 connected simple graphs of order 8, only 540 have no inflection points. In fact, in [237] it was also shown that the reliability polynomial of every connected graph of order at least 3 with no bridges has a point of inflection in $(0,1)$, and hence the reliability of almost every simple graph of order n has a point of inflection in the interval.

The inflection points of a reliability polynomial relate to the rate of improvement in network reliability as edges probabilities increase. By the Moore–Shannon theorem [860] as p ranges from 0 to 1, we typically see three phases.

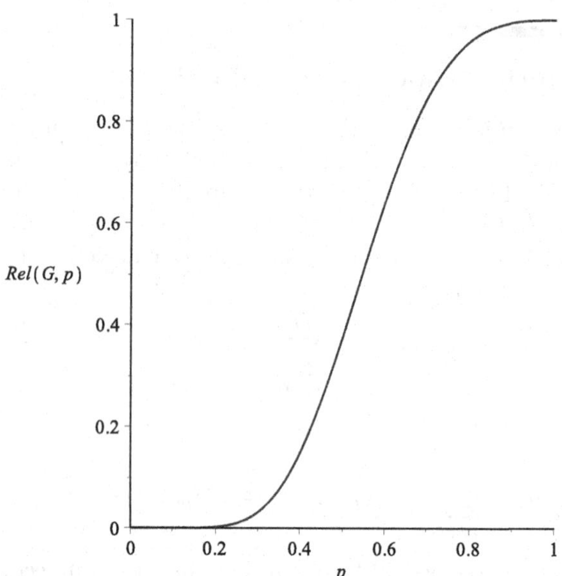

FIGURE 15.4: The reliability polynomial of the graph in Figure 15.2.

Consider a small increase Δ in the value of p. The corresponding increase in the network reliability typically behaves as follows: at first, it is smaller than Δ, then it overtakes Δ, and near $p = 1$ is again less than Δ. The single crossing of p permitted by the Moore–Shannon theorem occurs in the middle portion when present. This suggests that an (all-terminal or two-terminal) reliability polynomial has at most one point of inflection in the range $p \in (0, 1)$ [324], which is equivalent to the first derivative of the reliability polynomial having a unique maximum. Does its second derivative have at most one zero? Recently, in [237] an infinite family of graphs was produced, each member of which has two inflection points in the interval, and the existence of a finite number of simple graphs with the property was established as well. Moreover, Graves and Milan [577] provide a recursive construction of graphs with an arbitrary number of inflection points in $(0, 1)$.

Finally, what about the distribution of inflection points in $(0, 1)$ of all-terminal reliability polynomials? Are they concentrated in one area of the interval? Consider the all-terminal reliabilities of *cycle bundles*, $C_{m^k}^{kj}$, formed from a cycle on m^k vertices by replacing each edge by a bundle of kj parallel edges, Using all choices of $m \geq 2$, $k \geq 1$, and $j \geq 1$, these have points of inflection that are dense in $(0, 1)$ [222]. Indeed the fixed points of the all-terminal reliability polynomials of these graphs, as described in the Moore–Shannon result, are dense in $[0, 1]$ as well.

15.6.2 Average reliability

Among all graphs with a fixed number of vertices and edges, is there an "optimal" graph with respect to reliability? Of course, for a given choice of edge probability p, there is always an optimal graph, but can such a graph be chosen, independent of what the edge probability p is? The answer, sadly, is no—there are values of n and m for which there are no optimal simple graphs [688, 873]; allowing multiple edges does not improve matters [232].

So what can be done if there is uncertainty about the value of p? As in [233], it is possible to optimize the *average reliability* of the graph, that is, the average value of the all-terminal reliability function over the interval $[0, 1]$:

$$\text{average reliability}(G) = \int_0^1 \text{Rel}(G; p)\ dp.$$

This seems a natural way to associate a single summary value to the all-terminal reliability of a graph. For example, the average reliability of a tree of order n is $1/n$, which is strictly less than $2/(n + 1)$, the average reliability of a cycle of order n, and corresponds nicely to the fact that a cycle is more reliable than a tree (of course, properties of definite integrals implies that if the reliability of G is always at least as large as the reliability of H, then the average reliability of G is always at least as large as that of H).

The average reliability is a rational number between 0 and 1, because $0 \leq \text{Rel}(G; p) \leq 1$ for $p \in [0, 1]$ and $\text{Rel}(G; p)$ is a polynomial with integer coefficients. In [233] the distribution of average reliabilities is shown to be dense in $[0, 1]$. Although there is a polynomial time algorithm to calculate the average reliability for some graph families, the general problem is ♯P-hard [233].

Returning to the problem of finding an optimal graph for reliability, it is possible to propose choosing, among all graphs in a class, one having maximum average reliability—such a graph is *guaranteed* to exist! It seems difficult to determine the structure of such graphs, even for the class of all graphs with n vertices and m edges, so in [233] the authors define, for any given positive number $C \in (0, 1]$, a graph G to be "C-good" if its average reliability is at least C times the maximum average reliability of any graph with the same number of vertices and edges as G, and conjecture that cycle bundles in which the bundles of parallel edges are of equal or nearly equal sizes are 0.75-good graphs.

15.6.3 Roots of reliability polynomials

All-terminal reliabilities are polynomials, and, as such, it is natural to explore their roots. The alternating sum inequalities (15.3), with strict inequality for $b > 1$ [226, Theorem 3.1] imply that all real roots of the all-terminal reliability polynomial are at $p = 0$ and in the interval $(1, 2]$. This result and some empirical evidence for small graphs (see Figure 15.5) supported the conjecture

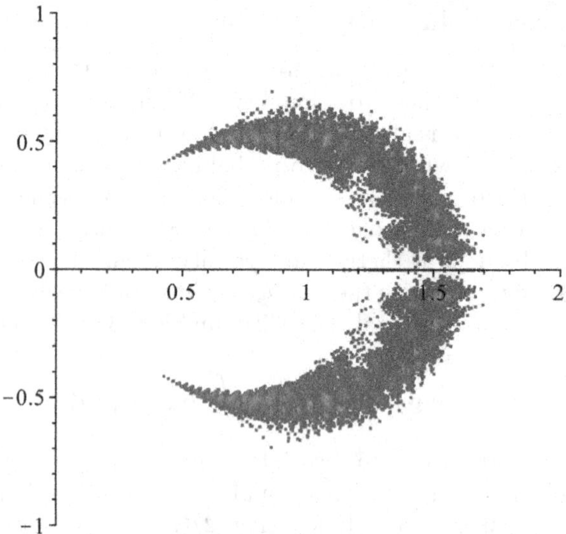

FIGURE 15.5: The roots of reliability polynomials of all simple graphs of order 8.

from [226] that all complex roots of all-terminal reliability polynomials lie in the unit circle in the complex plane centered at $z = 1$.

Characterizing where such roots can lie has the potential to give important information about the coefficients of reliability polynomials [1031]. In [226], it is shown that the closure of the roots contain this disc. Wagner [1126] establishes an affirmative answer for series-parallel graphs. Sokal [1021], by considering the multivariate version of all-terminal reliability where each edge has its own probability, provided a beautifully succinct proof of Wagner's result. Finally, Royle and Sokal [973] proved the conjecture to be false, but only by a slim margin—the furthest they found a root of an all-terminal reliability polynomial from $z = 1$ was about 1.04!

15.7 Open problems

In spite of all that is already known about reliability, there are still many open problems. Some are purely combinatorial, while others undoubtedly will need tools from other areas of mathematics to solve. We mention a few.

While several of the initial coefficients of the F-form of the all-terminal reliability polynomial can be found quickly, from the top down, only the very last coefficient F_{m-n+1} (the number of spanning trees) is known to be calculable in polynomial time.

Open Question 15.9. Can F_{m-n}, the number of spanning connected unicyclic subgraphs of a graph, be computed efficiently?

Liu and Chow [776] show that an efficient algorithm does exist for planar graphs, but their method relies heavily on planar duality.

Brown, Colbourn and Devitt [228] establish that a number of transformations have predictable effect on coefficients of all-terminal reliability polynomials. Perhaps transformations can be useful in the following:

Open Question 15.10. For all-terminal reliability, if G has a cycle basis containing cycles of lengths $c_1, \ldots, c_{m-n+1}$, then is the following true?

$$H_i \leq \sum_{I \subseteq \{1,\ldots,m-n+1\}, |I|=i} \prod_{j \in I} (c_j - 1).$$

As noted earlier, the graph $L_{n,m}$ is conjectured to be the least reliable *simple* graph on n vertices and m edges. Some progress in proving this has been made using transformations; see [228, 688]. The more general problem remains open, however:

Open Question 15.11. Is $L_{n,m}$ the least reliable connected simple graph on n vertices and m edges for all edge operation probabilities? Is the H-vector of any such graph bounded below by that of $L_{n,m}$ coefficientwise?

We have outlined how commutative algebra can be used to derive a surprising connection between H-vectors of cographic matroids and order ideals of monomials.

Open Question 15.12. Is there a (conceptually) simple combinatorial construction of an associated order ideal of monomials whose number of monomials of each degree yields the coefficients in the H-form of the reliability. In particular, is there a natural bijection from spanning trees to monomials in the order ideal?

Stanley's results give a complete characterization of H-vectors (and hence F-vectors) of shellable complexes. Because for all-terminal reliability, $\mathcal{F}$ is a matroid, we might hope for tighter inequalities yet:

Open Question 15.13. Characterize H-vectors or F-vectors of matroids (or binary matroids, regular matroids, cographic matroids) of dimension d with ground set of size m.

Such a characterization seems out of reach at this time. However, there are other more modest but still difficult open problems in this area. One of them is due to Stanley [1028, 1034], who conjectured that for any matroid complex, the H-vector is the degree sequence of a *pure* order ideal of monomials, that is, an order ideal in which all maximal monomials are of the same degree. Merino [838] proved this conjecture for cographic matroids via chip firings.

The examples of graph with multiple inflection points all contain multiple edges. For graphs without multiple edges, only a few graphs are known with more than one inflection point.

Open Question 15.14. For every $k \geq 1$, are there are simple graphs with at least k inflection points?

Despite recent work finding roots of all-terminal reliability polynomials somewhat further away from 1 (see [224]), we do not know if the set of all such roots is bounded.

Open Question 15.15. Is there a small disc, centered at $z = 1$, that contains all of the roots of all-terminal reliability polynomials?

Until this point, we have focused on the all-terminal reliability polynomial. Generalizations to two- and k-terminal reliability abound; see [69] for a somewhat dated discussion, and [291, 326] for generalizations to matroid ports (a *matroid port* $\{C \setminus e : C$ is a matroid circuit and $e \in C\}$ generalizes the clutter of simple paths in a graph joining two terminals). What about extensions to directed graphs? A common generalization is *reachability* (see, for example, [323]) where there is a distinguished vertex, called the *root*, and, under the assumption of perfectly reliable vertices but edges independently operational with probability p, one asks for the probability that all vertices are reachable from the root. Many of the approaches and results from all-terminal reliability extend in a fairly straightforward way to reachability.

However, it seems that this extension of all-terminal reliability is not quite the natural one. What seems like the closest generalization of all-terminal reliability to directed graphs is to ask, under the same assumptions, the probability that all vertices in a digraph D are reachable from one another, that is, that the spanning subdigraph of operational edges is *strongly connected*; this is called the *strongly connected reliability* of D, $\mathrm{screl}(D, p)$. Strongly connected reliability arises as the probability that a random $\{0, 1\}$ $n \times n$ matrix (each of whose entries are 1 independently with probability p) is irreducible [603].

While the system is still coherent, strongly connected reliability suffers from a number of problems that do not arise in the undirected case. For example, while strongly connected reliability is also a polynomial in p, the degree of the smallest nonzero term varies, and indeed is intractable on its own (as such a degree is, for a digraph of order n, at least n, and equal to n if and only if the digraph has a Hamiltonian directed cycle). As well, the associated complex (whose faces are sets of edges whose removal leaves the digraph strongly connected) is no longer a matroid, nor, indeed, even pure-dimensional. The upshot is that many of the bounding techniques of all-terminal reliability do not extend to strongly connected reliability. Finally (and disappointingly!) there is also no "nice" recursive factoring theorem to rely on for calculations.

Open Question 15.16. Develop methods for bounding the strongly connected reliability of a digraph.

On the other hand, strongly connected reliability behaves rather differently than all-terminal reliability is some pleasantly surprising ways. For all $n \geq 1$

and $m \geq n$ there exists an optimal digraph with n vertices and m edges, irrespective of the value of p [238]. Moreover, the roots of strongly connected reliability are dense in the complex plane [231, 234].

We have only touched on a few aspects of the fascinating field of network reliability. Reliability, in all its variants and forms, continues to open up more avenues for exploration—applied, theoretical, and computational.

16

Codes

Thomas Britz • Peter J. Cameron

Synopsis

This chapter introduces linear codes and some of their properties, surveys how the Tutte polynomial determines these properties, and provides an overview of more general results as well as applications to coding theory.

- Linear code, codeword weight, support and weight enumerator, dual code.
- The MacWilliams identity.
- The coboundary polynomial, Greene's theorem and the critical theorem.
- Ordered tuples and subcodes.
- Equivalence between the Tutte polynomial and generalized weights.
- Orbital polynomials.
- Generalizations and applications of Greene's theorem.
- Wei's duality theorem.

16.1 Introduction

In his paper [579] from 1976, Curtis Greene showed how the weight enumerator of a linear code could be expressed as an evaluation of the Tutte polynomial. This result has since been celebrated as *Greene's theorem*, not to be confused with the other, unrelated result [578] celebrated by the same name and also published by Curtis Greene in 1976.

DOI: 10.1201/9780429161612-16

One cause for celebrating Greene's theorem is that it created a link between coding theory and the Tutte polynomial and to matroid theory more generally. That matroid theory was connected to coding theory was well-known to researchers of matroid theory but was a surprise to most coding-theoreticians, to whom matroid theory was generally unfamiliar. In his paper, Curtis Greene applied Greene's theorem to provide a very simple proof of the MacWilliams identity [807], a fundamental result in coding theory, and this lent credence to the premise that matroid theory might be useful to coding theory. To researchers of graph and matroid theory, Greene's theorem demonstrated a new and interesting instance of the increasingly ubiquitous Tutte polynomial.

This chapter introduces linear codes and the associated Hamming codeword weight enumerator. According to the MacWilliams identity, the Hamming codeword weight enumerator of a linear code determines that of its dual code. This chapter defines the Tutte polynomial of a linear code and presents Greene's theorem which shows how the codeword weight enumerator of a linear code can be expressed as an evaluation of the Tutte polynomial of that code. This is applied to give a very brief proof of the MacWilliams identity.

By extending the definition of Hamming weight from codewords to subcodes, Greene's theorem and the MacWilliams identity may generalized. This and some of the many other ways to generalize Greene's theorem will be described in this chapter, along with applications of these results to coding theory. It turns out that the Tutte polynomial not only determines a vast array of generalized weight enumerators but is itself determined by certain sets of these enumerators.

The chapter concludes with the elegant Tutte polynomial proof by Duursma [426] of Wei's duality theorem [1131] which, like the MacWilliams identity, is a celebrated duality theorem from coding theory.

16.2 Linear codes and the Tutte polynomial

Let $\mathbb{F}_q$ be the finite field with q elements where q is a prime power, and set $E := \{1, \ldots, n\}$ for some positive integer n. Let $\mathbb{F}_q^n := \mathbb{F}_q^E$ be the vector space of ordered n-tuples of elements from $\mathbb{F}_q$ indexed by E.

Definition 16.1. A *linear code* C over $\mathbb{F}_q$ is a vector space over $\mathbb{F}_q$. In particular, an $[n, k]$ linear code C over $\mathbb{F}_q$ is a k-dimensional subspace of the vector space $\mathbb{F}_q^n$. Each vector in C is called a *codeword.*

Definition 16.2. The *Hamming support* and *Hamming weight*, or simply the *support* and *weight*, of each codeword $\mathbf{x} = (x_1, \ldots, x_n) \in \mathbb{F}_q^n$ is given by

$$\begin{aligned} \text{supp}(\mathbf{x}) &:= \{ i \ : \ x_i \neq 0 \} ; \\ \text{wt}(\mathbf{x}) &:= |\text{supp}(\mathbf{x})| . \end{aligned}$$

The *(Hamming) support* and *(Hamming) weight* of each subset $B \subseteq \mathbb{F}_q^n$ are defined similarly:

$$\mathrm{Supp}(B) := \bigcup_{\mathbf{x}\in B} \mathrm{supp}(\mathbf{x})\,;$$

$$\mathrm{wt}(B) := |\,\mathrm{Supp}(B)|\,.$$

The *homogeneous Hamming weight enumerator*, or simply the *weight enumerator*, $W_C(x, y)$ of an $[n, k]$ linear code C over $\mathbb{F}_q$ enumerates the number of codewords of each given weight, as follows:

$$W_C(x,y) := \sum_{i=0}^{n} A_i x^{n-i} y^i$$

where $A_i = |\{\mathbf{v} \in C \, : \, \mathrm{wt}(\mathbf{v}) = i\}|$.

Definition 16.3. The *dual code*, $C^\perp$, of an $[n, k]$ linear code C over $\mathbb{F}_q^n$ is the orthogonal subspace of C:

$$C^\perp := \{\mathbf{v} \in \mathbb{F}_q^n \, : \, \mathbf{u}\cdot\mathbf{v} = 0 \text{ for all } \mathbf{u} \in C\}\,.$$

Example 16.4. Consider the vector space $\mathbb{F}_q^n$ and the $[n, k]$ linear code C when $q = 2$, $n = 5$, and $k = 3$. The vector space $\mathbb{F}_2^5$ is the set of all binary codewords of length 5 and C is the $[5, 3]$ linear code over $\mathbb{F}_2$ spanned by the rows of the matrix

$$G = \begin{bmatrix} 1 & 0 & 1 & 0 & 0 \\ 0 & 1 & 1 & 0 & 0 \\ 0 & 0 & 0 & 1 & 1 \end{bmatrix}$$

The matrix G is a *generator matrix* for C. The linear code C contains eight codewords, two of which are

$$\mathbf{u} = (1,0,1,0,0) \qquad \text{and} \qquad \mathbf{v} = (1,1,0,1,1)\,.$$

The weights and supports of these two particular codewords are

$$\begin{array}{rclcrcl} \mathrm{wt}(\mathbf{u}) & = & 2, & & \mathrm{supp}(\mathbf{u}) & = & \{1,3\}, \\ \mathrm{wt}(\mathbf{v}) & = & 4, & & \mathrm{supp}(\mathbf{v}) & = & \{1,2,4,5\}\,. \end{array}$$

The linear code C has $A_0 = 1$ codeword of weight 0 (the zero codeword), $A_2 = 4$ codewords of weight 2, and $A_4 = 3$ codewords of weight 4. The weight enumerator of C is therefore

$$W_C(x,y) = \sum_{i=0}^{n} A_i x^{n-i} y^i = x^5 + 4x^3y^2 + 3xy^4\,.$$

16.2.1 The MacWilliams identity

In her Ph.D. thesis and a subsequent paper [807] from 1963, Jessie MacWilliams proved that the weight enumerator of a linear code over $\mathbb{F}_q$ and that of its dual code determine each other by the following simple identity which is now known as the MacWilliams identity.

Theorem 16.5 (MacWilliams Identity). *If C is an $[n, k]$ linear code over $\mathbb{F}_q$, then*

$$W_{C^\perp}(x, y) = \frac{1}{q^k} W_C(x + (q-1)y, x - y).$$

Example 16.6. To find the weight enumerator $W_{C^\perp}(x, y)$ of the dual code $C^\perp$ of the linear code C from Example 16.4, we could first find the dual code $C^\perp$; it is given by the generator matrix

$$\begin{bmatrix} 1 & 1 & 1 & 0 & 0 \\ 0 & 0 & 0 & 1 & 1 \end{bmatrix}.$$

By inspecting the four codewords of $C^\perp$ and their weights, we find that

$$W_{C^\perp}(x, y) = x^5 + x^3y^2 + x^2y^3 + y^5.$$

A more direct way to find $W_{C^\perp}(x, y)$ is to apply the MacWilliams identity (Theorem 16.5) to the weight enumerator $W_C(x, y) = 1 + x^3y^2 + xy^4$ found in Example 16.4:

$$\begin{aligned} W_{C^\perp}(x, y) &= \frac{1}{q^k} W_C(x + (q-1)y, x - y) \\ &= \frac{1}{2^3} \left(1 + 4(x + y)^3(x - y)^2 + 4(x + y)(x - y)^4\right) \\ &= x^5 + x^3y^2 + x^2y^3 + y^5. \end{aligned}$$

16.2.2 Greene's theorem

The *vector matroid*, M_C, of an $[n, k]$ linear code C with coordinates E is the matroid $M_C := (E, r_{M_C})$ with rank function $r_{M_C}(X) := \dim C|X$ for each subset $X \subseteq E$, where $C|X$ is the linear code obtained by removing the coordinates $E \setminus X$ from each codeword in C. (A review of matroids can be found in Chapter 4.)

Definition 16.7. The *(homogenous) coboundary polynomial* $W(M; \lambda, x, y)$ of a matroid $M = (E, r_M)$ is the enumerator

$$W(M; \lambda, x, y) = \sum_{X \subseteq E} \lambda^{r_M(E) - r_M(X)} y^{|E \setminus X|} (x - y)^{|X|}.$$

Note that $W(M; \lambda, x, y)$ is a homogenized form of the coboundary polynomial $\overline{\chi}(M; \lambda, x) := W(M; \lambda, x, 1)$ as originally defined by Crapo [348, 349] (See Definition 4.145). It is equivalent to the Tutte polynomial $T(M; x, y)$ via the following identity (cf. [242, 247, 349, 670, 671, 867, 1146]).

Proposition 16.8. *For a matroid M,*

$$(y-1)^{r(M)}T(M;x,y) = W(M;(x-1)(y-1),y,1).$$

In 1976, Curtis Greene [579] proved that the weight enumerator $W_C(x,y)$ of a linear code C is determined by the Tutte polynomial of the vector matroid M_C, or equivalently by the rank generating function $W(M_C;x,y)$.

Theorem 16.9 (Greene's Theorem). *If C is an $[n,k]$ linear code over $\mathbb{F}_q$, then*

$$W_C(x,y) = W(M_C;q,x,y)\,.$$

The original expression of the Greene's theorem did not feature the coboundary polynomial but instead directly involved the Tutte polynomial:

$$W_C(z,1) = (1-z)^k z^{n-k} T\Big(M_C; \frac{1+(q-1)z}{1-z}, \frac{1}{z}\Big)\,.$$

Example 16.10. Consider once more the $[5,3]$ linear code C over $\mathbb{F}_2$ from Examples 16.4 and 16.6. The associated coboundary polynomial is

$$W(M_C;\lambda,x,y) = \\ x^5 + (\lambda-1)\big(4x^3y^2 + (\lambda-2)x^2y^3 + 3(\lambda-1)xy^4 + (\lambda-1)(\lambda-2)y^5\big)\,.$$

By Greene's theorem, the weight enumerator of C is

$$W_C(x,y) = W(M_C;2,x,y) = x^5 + 4x^3y^2 + 3xy^4\,,$$

just as previously calculated in Example 16.4.

Greene's theorem was initially celebrated for showing what was in 1976 a rare and unexpected connection between coding theory and matroid theory. Also, it demonstrated the perhaps first notable application of matroid theory to coding theory, by providing a very simple proof of the MacWilliams identity (Theorem 16.5), using the basic identity $T(M^*;x,y) = T(M;y,x)$ for Tutte polynomials (see Theorem 4.125). For coboundary polynomials, this translates to the following identity.

Theorem 16.11. *For a matroid M,*

$$\lambda^{r(M)}W(M^*;\lambda,x,y) = W(M;\lambda,x+(\lambda-1)y,x-y)$$

Since $M_{C^\perp} = (M_C)^*$, the MacWilliams identity can then be proved as follows:

$$\begin{aligned} W_{C^\perp}(x,y) &= W(M_{C^\perp};q,x,y) \\ &= W((M_C)^*;q,x,y) \\ &= q^{-r(M)}W(M_C;q,x+(q-1)y,x-y) \\ &= \frac{1}{q^k}W_C(x+(q-1)y,x-y)\,. \end{aligned}$$

16.2.3 The critical theorem

As Greene pointed out, Greene's theorem can be derived from a stronger result, the critical theorem by Crapo and Rota [350] (see also [239]). Where Greene's theorem describes how the codeword weights of a code are determined by the Tutte polynomial of the vector matroid of that code, the critical theorem describes how the codeword supports of the code are determined by the vector matroid itself. Recall from Chapter 4 that the *characteristic polynomial* of a matroid $M = (E, r_M)$ is given by

$$\chi(M;x) := \sum_{X \subseteq E} (-1)^{|X|} x^{r_M(E)-r_M(X)} = (-1)^{r_M(E)} W(M;-x,-1)\,.$$

From Definition 4.74, for any subset $X \subseteq E$, the *contraction* of M to the set X is denoted by $M.X$ (i.e., $M.X := M/(E \setminus X)$).

Theorem 16.12 (The Critical Theorem). *If C is an $[n,k]$ linear code over $\mathbb{F}_q$, then for any $X \subseteq E$ and any $m \in \mathbb{Z}^+$, the number of ordered m-tuples $(\mathbf{v}_1, \ldots, \mathbf{v}_m) \subseteq C^m$ of codewords in C with* $\operatorname{supp}(\mathbf{v}_1) \cup \cdots \cup \operatorname{supp}(\mathbf{v}_m) = X$ *is $\chi(M_C.X; q^m)$.*

Example 16.13. Consider again the $[5,3]$ linear code C over $\mathbb{F}_2$ from the previous examples. The characteristic polynomial of M_C is

$$\chi(M_C;x) = (-1)^{r(M)} W(M;-x,-1) = (x-2)(x-1)^2\,.$$

The code C does not contain the codeword $(1,\ldots,1)$, so the number of codewords $\mathbf{v}$ in C with $\operatorname{supp}(\mathbf{v}) = E = \{1,\ldots,5\}$ is 0. This number is also obtained from the critical theorem by setting $X = E$, $q = 2$, and $m = 1$:

$$\chi(M_C.X;q^m) = \chi(M_C.E;2^1) = \chi(M_C;2) = (2-2)(2-1)^2 = 0\,.$$

In contrast, the number of ordered pairs $(\mathbf{v}_1, \mathbf{v}_2)$ of codewords in C with $\operatorname{supp}(\mathbf{v}_1) \cup \operatorname{supp}(\mathbf{v}_2) = E$ is, by the critical theorem with $m = 2$,

$$\chi(M_C.X;q^m) = \chi(M_C;2^2) = (4-2)(4-1)^2 = 18\,.$$

For instance, one such pair is $(\mathbf{v}_1, \mathbf{v}_2)$ where

$$\mathbf{v}_1 = (1,0,1,0,0) \qquad \text{and} \qquad \mathbf{v}_2 = (1,1,0,1,1)\,.$$

Indeed, $\operatorname{supp}(\mathbf{v}_1) \cup \operatorname{supp}(\mathbf{v}_2) = \{1,3\} \cup \{1,2,4,5\} = \{1,\ldots,5\} = E$.

To derive Greene's theorem from the critical theorem, let x_e and y_e be independent variables for each element $e \in E$ and set $\mathbf{x} = (x_1, \ldots, x_n)$ and $\mathbf{y} = (y_1, \ldots, y_n)$. For any sequence $\mathbf{z} = (z_1, \ldots, z_n)$ and subset $X \subseteq E$, define

$$\mathbf{z}^X := \prod_{e \in X} z_e\,.$$

The *m-tuple support enumerator* is then defined as follows:

$$W_C^{[m]}(\mathbf{x},\mathbf{y}) := \sum_{X\subseteq E} A_X^{[m]}\mathbf{x}^{E\backslash X}\mathbf{y}^X,$$

where

$$A_X^{[m]} := |\{(\mathbf{v}_1,\ldots,\mathbf{v}_m) \in C^m : \operatorname{supp}(\mathbf{v}_1)\cup\cdots\cup\operatorname{supp}(\mathbf{v}_m) = X\}|.$$

Further, the *generalized coboundary polynomial* of a matroid $M = (E, r_M)$ is the polynomial

$$W(M;\lambda,\mathbf{x},\mathbf{y}) := \sum_{X\subseteq E} p(M.X;\lambda)\mathbf{x}^{E\backslash X}\mathbf{y}^X.$$

The critical theorem may now be re-expressed ([350]) as follows:

Theorem 16.14. *If C is an $[n,k]$ linear code over $\mathbb{F}_q$ and $m \in \mathbb{Z}^+$, then*

$$W_C^{[m]}(\mathbf{x},\mathbf{y}) = W(M_C;q^m,\mathbf{x},\mathbf{y}).$$

The coboundary polynomial of M may be re-expressed (cf. [218]) simply as

$$W(M;\lambda,x,y) = \sum_{X\subseteq E} p(M.X;\lambda)x^{|E\backslash X|}y^{|X|}.$$

Therefore, Greene's theorem is obtained from the critical theorem by setting $m = 1$, $\mathbf{x} = (x,\ldots,x)$ and $\mathbf{y} = (y,\ldots,y)$.

16.3 Ordered tuples and subcodes

The critical theorem addresses the support of ordered m-tuples of codewords, so it is natural to generalize Greene's theorem with respect to these tuples.

Definition 16.15. The *m-tuple weight enumerator* is defined as follows:

$$W_C^{[m]}(x,y) := \sum_{X\subseteq E} A_X^{[m]}x^{|E\backslash X|}y^{|X|}.$$

Setting $\mathbf{x} = (x,\ldots,x)$ and $\mathbf{y} = (y,\ldots,y)$ in the critical theorem (Theorem 16.14) yields the following m-tuple generalization of Greene's theorem, from [210].

Theorem 16.16. *If C is an $[n,k]$ linear code over $\mathbb{F}_q$ and $m \in \mathbb{Z}^+$, then*

$$W_C^{[m]}(x,y) = W(M_C;q^m,x,y).$$

Example 16.17. Consider the $[5, 3]$ linear code C over $\mathbb{F}_2$ from the previous examples. The associated coboundary polynomial was found to be

$$W(M_C; \lambda, x, y) = \\ x^5 + (\lambda - 1)\big(4x^3y^2 + (\lambda - 2)x^2y^3 + 3(\lambda - 1)xy^4 + (\lambda - 1)(\lambda - 2)y^5\big).$$

By Theorem 16.16, the 2-tuple weight enumerator of C is

$$W_C^{[2]}(x, y) = W(M_C; 2^2, x, y) = x^5 + 12x^3y^2 + 6x^2y^3 + 27xy^4 + 18y^5.$$

Thus for instance, there are 18 ordered codeword pairs $(\mathbf{v}_1, \mathbf{v}_2)$ with $|\mathrm{supp}(\mathbf{v}_1) \cup \mathrm{supp}(\mathbf{v}_2)| = 5$, or equivalently, $\mathrm{supp}(\mathbf{v}_1) \cup \mathrm{supp}(\mathbf{v}_2) = \{1, \ldots, 5\}$, as previously found in Example 16.13.

The codeword m-tuple generalization of the MacWilliams identity, first proven by Shiromoto [1002], follows from Theorem 16.16 in exactly the same way that the MacWilliams identity follows from Greene's theorem (see [210, 1002, 1003]):

Theorem 16.18. *If C is an $[n, k]$ linear code over $\mathbb{F}_q$, then*

$$W_{C^\perp}^{[m]}(x, y) = \frac{1}{q^{km}} W_C^{[m]}(x + (q^m - 1)y, x - y).$$

Example 16.19. The $[5, 3]$ linear code C over $\mathbb{F}_2$ from the previous examples was found in Example 16.17 to have 2-tuple weight enumerator

$$W_C^{[2]}(x, y) = x^5 + 12x^3y^2 + 6x^2y^3 + 27xy^4 + 18y^5.$$

By Theorem 16.18, the 2-tuple weight enumerator of the dual code $C^\perp$ is then

$$W_{C^\perp}^{[2]}(x, y) = \frac{1}{2^{3\times 2}} W_C^{[2]}(x + (2^2 - 1)y, x - y) = x^5 + 3x^3y^2 + 3x^2y^3 + 9y^5.$$

Thus for instance, the code $C^\perp$ contains no codeword pair $(\mathbf{v}_1, \mathbf{v}_2)$ with weight $|\mathrm{supp}(\mathbf{v}_1) \cup \mathrm{supp}(\mathbf{v}_2)|$ equal to 1 or 4. This is easily seen to be true since the codewords of $C^\perp$ are

$$(0, 0, 0, 0, 0), \quad (1, 1, 1, 0, 0), \quad (0, 0, 0, 1, 1) \quad \text{and} \quad (1, 1, 1, 1, 1),$$

completing the example.

Codeword m-tuples of a code are closely related to the subcodes of that code, and the latter also provide natural generalizations of Greene's theorem.

Definition 16.20. The *r-th higher weight enumerator* of an $[n, k]$ code C is

$$W_C^{(r)}(x, y) := \sum_{i=0} A_i^{(r)} x^{n-i} y^i,$$

where $A_i^{(r)}$ denotes the number of subcodes $D \subseteq C$ with weight $\mathrm{wt}(D) = i$.

The *Gaussian q-nomial coefficient* is

$$\begin{bmatrix} k \\ r \end{bmatrix} := \frac{[k]_r}{[r]_r}, \qquad \text{where} \qquad [a]_b := \prod_{i=0}^{b-1} (q^a - q^i).$$

Barg [76, 77] generalized Greene's theorem with respect to subcodes as follows.

Theorem 16.21. *If C is an $[n,k]$ linear code over $\mathbb{F}_q$, then for $r = 1, \ldots, k$,*

$$\sum_{i=0}^{r} [r]_i W_C^{(i)}(x,y) = W(M_C; q^r, x, y).$$

Applying Möbius inversion to Theorem 16.21 yields the following direct expressions for the r-th higher weight enumerators (see [209, 210, 211, 212, 213, 217]).

Theorem 16.22. *If C is an $[n,k]$ linear code over $\mathbb{F}_q$, then for $r = 1, \ldots, k$,*

$$W_C^{(r)}(x,y) = \sum_{i=0}^{r} \frac{(-1)^{r-i}}{[r]_r} q^{\binom{r-i}{2}} \begin{bmatrix} r \\ i \end{bmatrix} W(M_C; q^i, x, y).$$

Example 16.23. The $[5,3]$ linear code C over $\mathbb{F}_2$ from the previous examples has seven subcodes of dimension 2, given by the following generator matrices:

$$\begin{bmatrix} 1 & 0 & 1 & 0 & 0 \\ 0 & 1 & 1 & 0 & 0 \end{bmatrix}, \begin{bmatrix} 0 & 1 & 1 & 0 & 0 \\ 0 & 0 & 0 & 1 & 1 \end{bmatrix}, \begin{bmatrix} 1 & 0 & 1 & 0 & 0 \\ 0 & 0 & 0 & 1 & 1 \end{bmatrix}, \begin{bmatrix} 1 & 1 & 0 & 0 & 0 \\ 0 & 0 & 0 & 1 & 1 \end{bmatrix},$$

$$\begin{bmatrix} 1 & 0 & 1 & 1 & 1 \\ 0 & 1 & 1 & 0 & 0 \end{bmatrix}, \begin{bmatrix} 1 & 0 & 1 & 0 & 0 \\ 0 & 1 & 1 & 1 & 1 \end{bmatrix}, \begin{bmatrix} 1 & 0 & 1 & 1 & 1 \\ 0 & 1 & 1 & 1 & 1 \end{bmatrix}.$$

Thus, C has one 2-dimensional subcode of weight 3; three of weight 4; and three of weight 5. The second higher weight enumerator for C is then

$$W_C^{(2)}(x,y) = x^2y^3 + 3xy^4 + 3y^5.$$

According to Theorem 16.22, this enumerator may also be obtained from the polynomial $W(M_C; \lambda, x, y)$ found in Example 16.10:

$$W_C^{(2)}(x,y) = \sum_{i=0}^{2} \frac{(-1)^{2-i}}{[2]_2} 2^{\binom{2-i}{2}} \begin{bmatrix} 2 \\ i \end{bmatrix} W(M_C; 2^i, x, y) = x^2y^3 + 3xy^4 + 3y^5,$$

completing the example.

Combining Theorems 16.11 and 16.21 yields the following subcode generalization of the MacWilliams identity, first proved by Kløve [696]; see also the closely related and equivalent identity by Simonis [1010].

Theorem 16.24. *If C is an $[n,k]$ linear code over $\mathbb{F}_q$, then for $r = 1, \ldots, k$,*

$$\sum_{i=0}^{r} [r]_i W_{C^\perp}^{(i)}(x,y) = \frac{1}{q^{kr}} \sum_{i=0}^{r} [r]_i W_C^{(i)}(x + (q^r - 1)y, x - y).$$

Like Greene's theorem, the critical theorem may also be generalized with respect to subcodes; see [211, 212, 413, 731] for instance.

16.4 Equivalence of the Tutte polynomial and weights

Greene's theorem describes how the Tutte polynomial of a linear code C determines the weight enumerator of that code. Similarly, Theorems 16.16 and 16.22 imply that the Tutte polynomial of C determines all of the m-tuple weight enumerators $W_C^{[m]}(x, y)$ and all of the r-th higher weight enumerators $W_C^{(r)}(x, y)$. A natural question is therefore whether the Tutte polynomial contains more information than these polynomials or, conversely, whether the Tutte polynomial may be determined from these polynomials.

Not surprisingly, no single one of these weight enumerators can alone determine the Tutte polynomial. However, it turns out, as shown in [213], that together they can.

Theorem 16.25. *If C is an $[n, k]$ linear code over $\mathbb{F}_q$, then the Tutte polynomial $T(M_C; x, y)$ determines the r-th higher weight enumerators $W_C^{(r)}(x, y)$ and is itself determined by the set $\{W_C^{(r)}(x, y) : r = 1, \ldots, k\}$ of these.*

Similarly, $T(M_C; x, y)$ determines and is itself determined by the set $\{W_C^{[m]}(x, y) : m = 1, \ldots, k\}$ of m-tuple weight enumerators.

The Tutte polynomial may be expressed explicitly as a function of the enumerators $W_C^{(r)}(x, y)$ (respectively $W_C^{[m]}(x, y)$); see [213] for the (technical) details.

Turning from weights to supports, the critical theorem implies that the vector matroid M_C of a linear code C over a field determines the codeword supports of this code. The supports of subcodes and many other code substructures are also determined by the vector matroid; see [211, 212, 413, 731]. Conversely, the minimal non-empty codeword supports determine the vector matroid; indeed these are the cocircuits of M_C; see [894, 1138, 1159]. Thus, the vector matroid M_C is equivalent to the set of supports of C, just as the Tutte polynomial is equivalent to the weights of C. However, not all properties of linear codes are given by their vector matroids. Skorobogatov [1014] noted that two linear codes may share a common vector matroid but have distinct covering radii; see [216] for similar observations.

16.5 Orbital polynomials

We give a brief account of some extensions of the Tutte polynomial which count orbits of a group Γ of automorphisms of a graph, code or matroid. We begin with a simple result as motivation. For any automorphism γ of a graph G, we denote by G/γ the graph obtained by shrinking each vertex cycle of γ to a single vertex: that is, its vertices are the cycles of γ; there is a loop at the cycle C if the set of vertices of C contains an edge of G, and an edge between cycles C_1 and C_2 if there is an edge between a vertex of C_1 and a vertex of C_2 in G.

Theorem 16.26. *Let Γ be a group of automorphisms of a graph G. Then the number of orbits of Γ on proper q-colorings of Γ is*

$$\frac{1}{|\Gamma|}\sum_{\gamma\in\Gamma}\chi(G/\gamma),$$

where $\chi(G)$ denotes the chromatic polynomial of G.

The proof follows immediately from the *orbit-counting lemma* [265, Theorem 2.2], on observing that the number of q-colorings of G fixed by γ is equal to the number of q-colorings of G/γ (any fixed coloring is constant on the cycles of γ, and any coloring of G/γ uniquely lifts to a coloring of G fixed by γ).

This result suggests that the expression should be an evaluation of an "orbital Tutte polynomial" in the same way that the chromatic polynomial is an evaluation of the usual Tutte polynomial. This is indeed the case, but the construction (given by Cameron, Jackson and Rudd [264]) is rather complicated. Indeed, their polynomial involves two potentially infinite families of variables $(x_i, x_i^* : i \in \{0,1,2,\ldots\})$. We give a very brief account, referring to the cited paper for further details.

The orbital polynomial is associated with a *dual pair* (M, M^*) of matrices over a principal ideal domain $\mathfrak{R}$. A dual pair of matrices M and M^* have the same numbers n of columns, and the row space of M is equal to the column space of M^* and vice versa. The group Γ is a group of automorphisms of the free module $\mathfrak{R}^n$ preserving the row spaces of M and M^*. The two families of indeterminates are indexed by the associate classes in $\mathfrak{R}$ (so that, if $\mathfrak{R} = \mathbb{Z}$, the index set is the nonnegative integers, while if $\mathfrak{R}$ is a field, the index set is $\{0,1\}$). We associate a monomial $x(N)$ with a matrix N as follows: the exponent of x_i is equal to the number of invariant factors in the Smith normal form of N which are in the i-th associate class. (So, if $\mathfrak{R}$ is a field, the monomial is $x_n^{n-r}x_1^r$, where n and r are the number of columns and the rank of N). Now we enlarge M and M^* to matrices $M(\gamma)$ and $M^*(\gamma)$, for each $\gamma \in \Gamma$, by appending the rows of $\gamma - I_n$. If $N[S]$ denotes the matrix formed by the columns of N lying in $S \subseteq \{1,\ldots,n\}$, then we define the orbital Tutte polynomial by

$$OT(M,\Gamma) = \frac{1}{|\Gamma|}\sum_{\gamma\in\Gamma}\sum_{S\subseteq E} x(M(\gamma)[S])x^*(M^*(\gamma)[E\setminus S]).$$

Here the outer sum invokes the orbit-counting lemma, and if we substitute -1 for one family of variables, then the inner sum performs inclusion-exclusion to count the nowhere-zero elements in the row space of M or M^*.

If Γ is the trivial group, then only the variables x_0, x_1, x_0^*, x_1^* occur in the polynomial.

In the case of a graph G, the matrices M and M^* are the signed vertex-edge and cycle-edge incidence matrices of G. If Γ is the trivial group, we obtain the chromatic polynomial by substituting $x_i = -1$, $x_0^* = q$, and $x_1^* = 1$. A "dual" substitution gives rise to the flow polynomial.

Tutte observed that the number of nowhere-zero k-flows of a graph with values in an abelian group A depends only on the order of A, not on its structure. But for the number of Γ-orbits on flows, the structure of A is involved; the appropriate substitution puts x_i equal to the number of solutions of $ix = 0$ in the abelian group, and $x_i^* = -1$.

If $\mathfrak{R}$ is a finite field, the orbital Tutte polynomial specializes to an "orbital weight enumerator" for a group of automorphisms of a linear code, counting the number of orbits of the group on codewords of each given weight. However, the substitution does not reduce to Greene's theorem in the case of the trivial group. The situation is not completely understood.

16.6 Other generalizations and applications

Greene's theorem (Theorem 16.9) has been generalized in many ways, as indicated in Section 16.3 and hinted at in Section 16.5. One typical way is to replace weight enumerators by support enumerators as in the critical theorem (Theorem 16.14), i.e., by assigning each coordinate its own independent variable; see [210, 211, 213]. This type of approach was independently initiated by Fortuin and Kasteleyn [498] in 1972 for the Ising model for physical phase transitions (see Chapter 20), MacWilliams and Sloane [806] in 1977 for codeword enumerators, and by Traldi [1072] in 1989 for the dichromatic polynomial (see Chapter 24). This approach was later promoted by Alan Sokal (eg. [1021, 1023]) in relation to the Potts model partition function and has been used extensively by other these and other authors since (a small but representative sample includes Bollobás and Riordan [158], Britz [210], Courcelle [338], Ellis-Monaghan and Traldi [463], Farr [479], Gordon and Traldi [573], Kaplan [673], Kung [734], Makowsky [811], Noble and Welsh [885], Welsh and Merino [1145], White [1148] and Zaslavsky [1183]).

A second way in which to generalize Greene's theorem is to consider the weights and supports of code substructures more general than single codewords. As seen in Section 16.3 these include codeword m-tuples, as in the critical theorem (Theorem 16.12), and subcode generalizations such as Theorem 16.22 and more general ones for instance by Britz [212], Brylawski [242, 244], Dowling [413] and Zaslavsky [1182]. Many other substructures lead to generalizations as well; see Britz [211, 213].

A third way to generalize Greene's theorem and the critical theorem is to replace matroids by more general structures. Helgason and Whittle proved a polymatroid generalization of the critical theorem (see [731, 1160, 1162]); Ardila [32] provided hyperplane arrangement generalizations; Farr [474, 475, 476] and Britz and Shiromoto [218] proved purely set-theoretical generalizations of Greene's theorem; and similar earlier work is described in the excellent exposition by Kung [731]. As described in Section 16.5, Cameron [262] found a polynomial that simultaneously generalizes the Tutte polynomial and the cycle index of a certain group; see also [263, 264, 266, 267, 975].

Similarly, single linear codes may be replaced by more general codes structures. Britz [212], Kaplan [673] and Kung [731] provide generalizations of Greene's theorem and the critical theorem with respect to multiples of linear codes each over its own field. Chan et al. [278] prove generalizations of these theorems with respect to quasi-uniform codes, and generalizations of Greene's theorem with respect to linear codes over certain class of rings are given by Britz, Shiromoto and Westerbäck [219], Etienne and Las Vergnas [470], and Vertigan [1117]; see also [922, 968]. Simonis and Ashikhmin [1011] generalize Greene's theorem with respect to almost affine codes, whereas Barg and Park [78] generalize this theorem with respect to a particular class of poset codes, namely the codes whose partial order is a disjoint union of equal length chains. Shiromoto [1004] generalizes Greene's theorem with respect to rank-metric codes; see also [672].

Many of the generalizations mentioned above induce generalizations of the MacWilliams identity (eg., [1168, 1169]) which in turn yields generalizations of Delsarte's bound [371, 372] and Duursma's bound [425]. It has also been possible to use these more general MacWilliams-type identities to generalize the Assmus–Mattson theorem [53] with respect to constructing t-designs from subcode supports and from matroids; see [215, 217]. The subcode generalizations of Greene's theorem in [209] were also applied in [209] to solve a computational problem on doubly-even, self-dual codes. Greene's theorem has furthermore been applied to linear codes over finite fields to show that the dimension of the bicycle space of a code is determined by an evaluation of the associated Tutte polynomial; see Gioan and Las Vergnas [533], Jaeger [649, 650], Jaeger, Vertigan and Welsh [653] Pendavingh [912], Pendavingh and van Zwam [913], Rosenstiehl and Read [969], and Vertigan [1116, 1118].

For more information on Greene's theorem, the critical theorem, their generalizations, and applications thereof, see [101, 102, 247, 730, 731, 1138, 1145].

16.7 Wei's duality theorem

The application of the Tutte polynomial by Greene [579] to prove the MacWilliams identity quickly and simply is well-known and celebrated. Less well-known but equally elegant and deserving of recognition is the Tutte polynomial proof by Duursma [426] of another famous duality theorem from coding theory, Wei's duality theorem [1131] below. This final section presents Duursma's proof.

For any $[n,k]$ linear code C over some field $\mathbb{F}$, the i-th (respectively j-th) *generalized Hamming weight* of C (respectively $C^\perp$) is, for $i = 0, \dots, k$ and $j = 0, \dots, n-k$,

$$\begin{aligned} d_i &:= d_i(C) &&= \min\{\mathrm{wt}(D) \,:\, D \text{ is a linear } [n,i] \text{ subcode of } C\}, \\ d_j^\perp &:= d_j(C^\perp) &&= \min\{\mathrm{wt}(D) \,:\, D \text{ is a linear } [n,j] \text{ subcode of } C^\perp\}. \end{aligned}$$

Generalized weights, also called *higher support weights*, were first introduced by Helleseth, Kløve, et al. [607, 608, 609]. Set

$$\begin{aligned} U_C &:= \{d_1, \dots, d_k\}, \\ V_C &:= \{n+1-d_{n-k}^\perp, \dots, n+1-d_1^\perp\}. \end{aligned}$$

Theorem 16.27 (Wei's Duality Theorem). *If C is an $[n,k]$ linear code over a field $\mathbb{F}$, then*

$$U_C \cup V_C = \{1, \dots, n\} \quad \textit{and} \quad U_C \cap V_C = \emptyset\,.$$

Example 16.28. The $[5,3]$ linear code C over $\mathbb{F}_2$ from the previous examples has generalized weights $d_1 = 2$, $d_2 = 3$, and $d_3 = 5$. Similarly, the dual code $C^\perp$ has generalized weights $d_1^\perp = 2$ and $d_2 = 5$. Therefore,

$$\begin{aligned} U_C &:= \{d_1, d_2, d_3\} = \{2, 3, 5\}, \\ V_C &:= \{5+1-d_2^\perp, \dots, 5+1-d_1^\perp\} = \{6-5, 6-2\} = \{1, 4\}\,. \end{aligned}$$

Therefore,

$$U_C \cup V_C = \{1, \dots, 5\} \quad \text{and} \quad U_C \cap V_C = \emptyset\,,$$

as asserted by Theorem 16.27.

To prove Wei's duality theorem using the Tutte polynomial, or equivalently as in Duursma [426], by the *Whitney rank generating function* $R(M_C; x, y) = T(M_C; x+1, y+1)$ (see Definition 2.3), we let r_{ij} denote the coefficient of $x^i y^j$ in

$$R(M_C; x, y) = \sum_{X \subseteq E} x^{k - r_{M_C}(X)} y^{|X| - r_{M_C}(X)} = \sum_{i=0}^{k} \sum_{j=0}^{n-k} r_{ij} x^i y^j\,.$$

In other words,

$$r_{ij} = |\{X \subseteq E \,:\, k - r_{M_C}(X) = i \text{ and } |X| - r_{M_C}(X) = j\}|\,.$$

Define $r_{ij}^\perp$ similarly for $R(M_{C^\perp}; x, y)$ and note that $r_{ij}^\perp = r_{ji}$ since $R(M_{C^\perp}; x, y) = R(M_C; y, x)$. Finally, define R to be the $(k+1) \times (n-k+1)$ matrix whose (i,j)-entry is r_{ij} for all integers $i = 0, \dots, k$ and $j = 0, \dots, n-k$. The transpose $R^\perp = R^T$ has entries $r_{ij}^\perp$.

Example 16.29. The $[5,3]$ linear code C over $\mathbb{F}_2$ from the previous examples has rank generating function

$$R(M_C; x, y) = 6 + 5y + y^2 + 9x + 4xy + 5x^2 + x^2y + x^3\,,$$

so the coefficient matrices R and $R^\perp$ are the following.

$$R = \begin{array}{|ccc|} \hline 6 & 5 & 1 \\ 9 & 4 & 0 \\ 5 & 1 & 0 \\ 1 & 0 & 0 \\ \hline \end{array} \qquad R^\perp = \begin{array}{|cccc|} \hline 6 & 9 & 5 & 1 \\ 5 & 4 & 1 & 0 \\ 1 & 0 & 0 & 0 \\ \hline \end{array}$$

Note that the non-zero entries in the matrices R and $R^\perp$ each form a Ferrers shape, namely and . This is true for any linear code C. In particular, the nonzero coefficients r_{ij} and $r_{ij}^\perp$ for C and for $C^\perp$ form two dual Ferrers shapes since, by definition of r_{ij} and some basic matroid considerations, if $r_{ij} > 0$, then $r_{i-1,j} > 0$ when $i > 0$, and $r_{i,j-1} > 0$ when $j > 0$.

As Wei [1131] and others (eg., [212, 214, 1084]) have proved,

$$d_r = n - \max_{X \subseteq E}\{|X| \,:\, r_{M_C}(X) = k - r\}\,.$$

Rephrasing this in terms of the matrix entries r_{ij} of R gives the identity

$$\begin{aligned} d_r - r &= n - k - \max_{X \subseteq E}\{|X| - r_{M_C}(X) \,:\, k - r_{M_C}(X) = r\} \\ &= n - k - \max\{j \,:\, r_{rj} > 0\}\,. \end{aligned}$$

In other words, $d_r - r$ is the number of zero entries in row r of R.

Example 16.30. For the $[5,3]$ linear code C from the previous examples, the matrix R in Example 16.29 yields the numbers $(d_1-1, d_2-2, d_3-3) = (1,1,2)$. Therefore, $(d_1, d_2, d_3) = (2,3,5)$ as seen previously in Example 16.28.

Similarly, $d_s^\perp - s$ is the number of zero entries in row s of $R^\perp$ or, equivalently since $R^\perp = R^T$, column s of R. This is illustrated in Figure 16.1 by two pictures, one in which the zero subrow is disjoint from the zero subcolumn, and one in which they intersect.

In the first case, the zero subrow (shaded) lies to the right of columns $0, 1, \ldots, s$, so it contains at most $n - k + 1 - (s+1) = n - k - s$ entries since each row of R contains $n-k+1$ entries in total. Similarly, the zero subcolumn contains at most $k + 1 - (r+1) = k - r$ entries. Therefore,

$$\begin{aligned} d_r + d_s^\perp &= \big((d_r - r) + (d_s^\perp - s)\big) + (r + s) \\ &\le \big((n - k - s) + (k - r)\big) + (r + s) \\ &= n\,. \end{aligned}$$

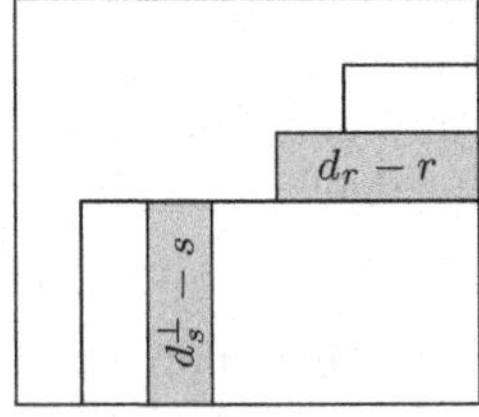

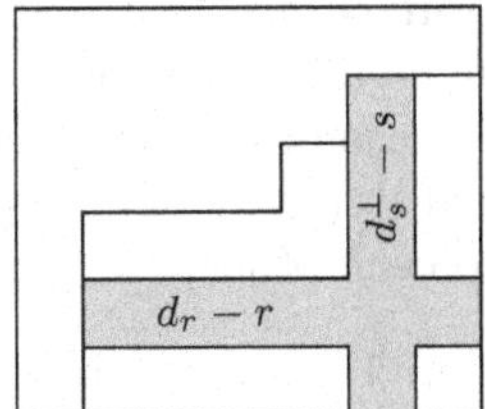

FIGURE 16.1: Zero entries in column s of R.

In the second case, the zero subrow contains at least $n - k + 1 - s$ entries and the zero subcolumn contains at least $k + 1 - s$ entries, so

$$\begin{aligned} d_r + d_s^\perp &= \big((d_r - r) + (d_s^\perp - s)\big) + (r + s) \\ &\geq \big((n - k + 1 - s) + (k + 1 - r) + (r + s)\big) \\ &= n + 2\,. \end{aligned}$$

In either case, $d_r + d_s^\perp \neq n + 1$. Therefore, $d_r \neq n + 1 - d_s^\perp$. The numbers $d_1, \ldots, d_k$ are strictly increasing and are thus distinct, as are the numbers $n+1-d_{n-k}^\perp, \ldots, n+1-d_1^\perp$. Therefore, $d_1, \ldots, d_k, n+1-d_{n-k}^\perp, \ldots, n+1-d_1^\perp$ are n distinct integers that lie between 1 and n; they must therefore be the numbers $1, \ldots, n$. This proves Wei's theorem.

16.8 Open problems

Open Question 16.31. As seen in this chapter, the supports and weights of many code substructures are uniquely determined by vector matroids and their Tutte polynomial, respectively. Are all classes of supports and weights thus determined? More precisely, if C is an linear code over a finite field, then does the Tutte polynomial $T(M_C; x, y)$ (respectively, the vector matroid M_C) determine each property of C that can be expressed as a set of codeword weights (respectively, supports)?

Open Question 16.32. A wide range of generalizations of Greene's theorem and of the critical theorem were presented in Section 16.6. How far it is possible to extend these generalizations, and which generalizations might be combined to yield even stronger generalizations? By pursuing the answers to such questions, it is possible that new directions of generalization might arise, and that existing theorems might be considerably strengthened, such as has already happened to the MacWilliams identity and Wei's duality theorem, as seen in this chapter.

Open Question 16.33. Is it possible to further generalize or modify the orbital weight enumerator from Section 16.5 in such a way that it yields Greene's theorem in the case of the trivial group? More generally, how do the combinatorial matroid structure of a graph and the algebraic orbit structure of that graph relate to each other?

Open Question 16.34. Specializations of the Tutte polynomial count many kinds of objects in a matroid, graph, or code. As well as colorings and flows, we mention independent sets, spanning sets and bases in a matroid, and acyclic orientations of a graph. In these cases, ad hoc orbital polynomials have been constructed to count orbits of an automorphism group Γ on these objects [262, 267]. Is it possible to construct a more general "orbital polynomial" which specializes to give all of these counts?

17

The chip-firing game and the sandpile model

Criel Merino

Synopsis

This chapter covers a general process which can be seen as the exchange of discrete tokens between different sites according to simple rules. Instances of this process include the chip-firing game and the abelian sandpile model. We concentrate on connections with the Tutte polynomial.

- The chip-firing game and the sandpile model.
- The relationship between the chip-firing game with the Tutte polynomial.
- The critical group and parking functions.

17.1 Introduction

We begin with the *dollar game* of N. Biggs, that although different from the original chip-firing game of [135], under current usage is also called the *chip-firing game*. Let G be a connected loopless graph on vertices $v_0, v_1, \ldots, v_n$, possibly with multiple edges. We choose the vertex v_0 as a special vertex, naming it q. The game consists of the exchange of identical chips between the vertices. During the game a vertex whose number of chips exceeds a given threshold will exchange chips with the other vertices according to a specified rule of the game.

The *Laplacian* of G is the $(n+1) \times (n+1)$ matrix $\Delta(G) = \Delta$ given by

$$\Delta_{i,j} = \begin{cases} -\nu(v_i, v_j), & \text{if } i \neq j, \\ \deg(v_i), & \text{if } i = j, \end{cases} \tag{17.1}$$

DOI: 10.1201/9780429161612-17

where $\nu(v_i, v_j)$ denotes the number of edges joining v_i and v_j, and $\deg(v_i)$ the number of edges incident to v. The *reduced Laplacian* of G, denoted $\Delta^q(G)$, is obtained from $\Delta(G)$ by deleting the row and column corresponding to q.

At each moment a vertex $v_i \neq q$ has a nonnegative integer value h_i associated to it. The vector $\theta{=}(h_1, \ldots, h_n)$ is called a *configuration* of the game. For every vertex $v_i \neq q$, we have a nonnegative value $H_i = \deg(v_i)$ that can be viewed as a threshold for the amount of chips vertex v_i can store. Configurations where $h_i < H_i$, for all $i \geq 1$ are called *stable*.

The game behaves as follows. In a non-stable configuration θ there is a vertex v_i with $h_i \geq H_i$. The game will update the configuration θ according to the rule:

$$h'_j = h_j - \Delta^q_{ij}, \quad \text{for all } j \geq 1. \tag{17.2}$$

The new configuration is $\theta'{=}(h'_1, \ldots, h'_n)$ and the game continues with this new configuration. Equation (17.2) is called the *firing* or *toppling* rule and when it is applied we say that v_i was *fired*. We note that the game has a *confluence* or abelian property in that if at some stage two or more vertices can fire, the configuration that results after all of the firings is independent of the order in which the vertices were fired.

If the game arrives at a stable configuration θ, it has a special rule called the *loading of the system*. The game has a nonnegative vector $\mathbf{l}_q$ that is the sum of the rows of Δ^q. The loading of the system consists of adding the vector $\mathbf{l}_q$ to θ. Loading is applied until a non-stable configuration results, and then the firing continues.

Any non-stable configuration will eventually become stable under this process since the game is sending chips to q every time one of its neighbors fires and since the graph is connected. At this point, the loading of the system will resume until we get to an unstable configuration, and then the firing will continue. Because the number of stable configurations is finite, some of these configurations will be recurrent. We call a stable recurrent configuration *critical*. Thus, by the confluence property, for any configuration θ, the game eventually arrives at a unique critical configuration θ_c, see [113, 377].

Critical configurations are a very rich mathematical structure. For example, they have a group structure and are related to parking functions and to the Tutte polynomial, as will be explained in the following sections.

17.2 The Tutte polynomial and the chip-firing game

In order to describe the relation between critical configurations and the Tutte polynomial, we need to define the chip-firing game for graphs with loops. The natural extension is that a vertex v with l loops, when it is fired, sends a chip through each of the ends of every loop at v; so, $2l$ chips stay on v after the firing. To model this behavior, the firing rule in Equation (17.2) uses $\Delta^q(G_0)$,

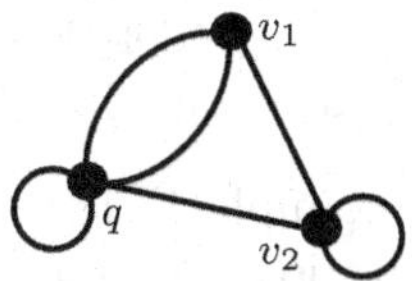

FIGURE 17.1: A graph G with loops.

where G_0 is obtained from G by deleting all the loops. However, the threshold of vertex v is still its degree, which in this case equals the number of edges incident to v plus twice the number of loops. Finally, the vector $\mathbf{l}_q$ is the sum of the rows of $\Delta^q(G_0)$.

Example 17.1. For the graph G in Figure 17.1, the corresponding matrix $\Delta^q(G_0)$ is

$$\begin{bmatrix} 3 & -1 \\ -1 & 2 \end{bmatrix}.$$

The configuration $(5,3)$ is unstable, because $5 \geq \deg(v_1) = 3$, and v_1 will be fired to get the configuration (2,4) that is also unstable, because $4 \geq \deg(v_2) = 4$. After firing v_2 we get $(3,2)$, and then v_1 will be fired to get $(0,3)$. This is a stable configuration. The loading of the system will change this configuration to $(0,3)+(2,1) = (2,4)$. Now, the game will cycle through the configurations $(2,4)$, $(3,2)$ and $(0,3)$ forever, and so $(0,3)$ is a critical configuration of the game. The critical configurations in this game are $(1,2)$, $(0,3)$, $(2,2)$, $(1,3)$ and (2,3).

It was shown in [116, 508] that the number of critical configurations equals the number of spanning trees of G. Furthermore, the enumerator of critical configurations by number of chips is essentially the Tutte polynomial of G along the line $x = 1$. To make this precise, define the *weight* of a configuration θ, denoted $w(\theta)$, to be the sum of its entries. If θ_c is a critical configuration, we define its *level* as $\text{level}(\theta_c) := w(\theta_c) - |E(G)| + \deg(q)$. In [113] it was shown that $0 \leq \text{level}(\theta_c) \leq |E(G)| - |V(G)| + 1$.

Definition 17.2. Let $G = (V,E)$ be a connected graph with $q \in V$ and, for $i \geq 0$, let $c_i(q)$ be the number of critical configurations with level i in the chip-firing game. Then the *critical configuration polynomial* is

$$P_c(G,q;y) = \sum_{i=0}^{|E|-|V|+1} c_i(q)\, y^i.$$

Theorem 17.3. *For a connected graph G and any vertex q we have that*

$$P_c(G,q;y) = T(G;1,y),$$

and thus $P_c(G,q;y)$ and $c_i(q)$ are independent of the choice of q.

A proof of this theorem based upon deletion and contraction can be found in [845], and a bijective proof can be found in [332].

Example 17.4. Continuing Example 17.1, the critical configurations $(1,2)$ and $(0,3)$ have level 2; configurations $(2,2)$ and $(1,3)$ have level 3; and (2,3) has level 4. Thus, the critical configuration polynomial is $y^4 + 2y^3 + 2y^2$. Observe that the Tutte polynomial of G is $x^2y^2 + xy^2 + xy^3 + y^3 + y^4$ and when evaluated at $x = 1$ we obtain the critical configuration polynomial.

The set of critical configurations of a connected graph G has the structure of a pure order ideal of monomials whose degree sequence is the H-vector of the matroid complex of $M^*(G)$. This was originally observed by N. Biggs, and an explanation of this relationship can be found in [838] and Section 15.5. In 1977, R. Stanley conjectured that the H-vector of a matroid complex is the degree sequence of a pure order ideal of monomials, see [1028, 1034]. Thus, the conjecture is true in the case of cographic matroids. The H-vector of a cographic matroid appears in Chapter 15 as the H-form of the reliability polynomial.

17.3 Sandpile models

The chip-firing game is a special case of a general discrete process on a set of $n+1$ sites labeled $v_0, v_1, \ldots, v_n$, where v_0 is a special site usually denoted by q. Fix a real $(n \times n)$-matrix M that has the property that each off-diagonal entry is nonpositive, but the sum of the columns is a nonnegative vector. Take an arbitrary nonnegative n-element vector $\mathbf{l}_q$, and form a vector $\mathbf{d}_q$ that is equal to the sum of the columns in M. These vectors are called the *loading* and *dissipation* of the system, respectively. Finally, choose a set of n thresholds $H_i > 0$, for $i \geq 1$, which we can write as $M_{ii} + k_i$, for some positive constants k_i.

The notions of *configuration* and *stable configuration* are defined as for the chip-firing game above. The firing rule is

$$h'_j = h_j - M_{ij}, \quad \text{for all } j \geq 1,$$

which replaces the matrix $\Delta^q(G)$ with M in Equation (17.2). Note that after v_i is fired, any j-coordinate of the configuration, where $j \neq i$, will increase or stay the same. A sequence of firings is called an *avalanche* and it satisfies the same abelian property as the chip-firing game. However, the notion of critical configuration does not have a standard definition.

In this general model the loading of the system could be the zero vector $\mathbf{0}$ and a stable configuration is the end of the process. The system is called *weakly dissipative* if M does not have a principal submatrix with zero dissipation.

When $\mathbf{l}_q = \mathbf{0}$ and $\mathbf{d}_q \neq \mathbf{0}$ we say that the system has q as a *sink*. In a weakly dissipative system with $\mathbf{l}_q = \mathbf{0}$ we called q a *global sink*.

The choice of $M, \mathbf{l}_q$, and $\mathbf{d}_q$ gives different common models in the literature. The *classical chip-firing game* in [23, 135] has $\mathbf{l}_q = \mathbf{d}_q = \mathbf{0}$ and uses as M the Laplacian matrix of an undirected connected graph G. The *abelian sandpile model* of D. Dhar [377] uses an integer matrix, with positive entries in the diagonal, and at each loading, the vector $\mathbf{l}_q$ is chosen at random from the standard basis of $\mathbb{R}^n$. The *abelian avalanche model* in [507] is a weakly dissipative system on a real matrix M with non-zero loading vector $\mathbf{l}_q$. The chip-firing game described in Section 17.2 is a weakly dissipative system with $l_q = d_q$ and uses as M the reduced Laplacian of a connected graph.

The Laplacian matrix $\Delta(\vec{G})$ of a digraph $\vec{G}$ is defined as in Equation (17.1) but with $\nu(v_i, v_j)$ equal to the number of edges from v_i to v_j and with $\deg(v_i)$ equal to the number of edges leaving v_i, for $i, j \geq 1$. The *reduced Laplacian* $\Delta^q(\vec{G})$ is define as above. The *chip-chip-firing game on directed graphs* in [134] has $l_q = d_q = 0$ and uses as M the Laplacian matrix of a directed graph. The model for directed graphs defined in [622] has a global sink and uses the reduced Laplacian of a directed graph. Recently, K. Perrot et al. [917] studied a model with $M = \Delta^q(\vec{G})$, a global sink q, and $\mathbf{l}_q = \mathbf{0}$. They showed that the critical configuration polynomial is independent of the choice of q. S. Chang [277] proved that this result is valid in the more general case of connected Eulerian digraphs. Chang also proved that if $\vec{G}$ is a connected Eulerian digraph, then the critical configuration polynomial is equal to the greedoid Tutte polynomial $T(\vec{G}; 1, y)$ for the greedoid associated with the digraph $\vec{G}$ rooted at q. (See Section 33.2 for the greedoid Tutte polynomial.)

17.4 The critical group and parking functions

There is a natural binary operation on the set of critical configurations. If θ_1 and θ_2 are critical configurations, define $\theta_1 \odot \theta_2$ to be the unique critical configuration associated with the configuration $\theta_1 + \theta_2$. The set of critical configurations together with this operation form an abelian group called the *critical group* or *sandpile group* (see [113, 377]). This group is isomorphic to the group $\mathbb{Z}^n / \mathbb{Z}^n \Delta^q(G)$, where $\mathbb{Z}^n \Delta^q(G)$ is the integer row-span of the reduced Laplacian of G (see [113]).

Because the critical group is a finite abelian group, it has an invariant factor decomposition of the form $\mathbb{Z}/n_1\mathbb{Z} \oplus \ldots \oplus \mathbb{Z}/n_r\mathbb{Z}$, where the positive integers n_i satisfy $n_i | n_{i+1}$ for $1 \leq i \leq r-1$. For example, the critical group of the complete graph K_n is isomorphic to the direct sum of $n-2$ copies of $\mathbb{Z}/n\mathbb{Z}$ and the group of the n-cycle C_n is isomorphic to $\mathbb{Z}/n\mathbb{Z}$.

The combinatorial structure of the critical configurations of the complete graph was studied, under a different name, before the chip-firing game or the sandpile model were introduced. A *parking function* is a sequence $(a_1, \ldots, a_n)$ of nonnegative integers such that, for each $1 \leq r \leq n$, the number of terms in the sequence that are strictly less than r is at least r. For example, $(0,4,1,2,6,4,2)$ is a parking function for $n = 7$. Also, $(0,0,0)$, $(1,0,0)$, $(0,1,0)$, $(0,0,1)$, $(2,0,0)$, $(0,2,0)$, $(0,0,2)$, $(1,1,0)$, $(1,0,1)$, $(0,1,1)$, $(0,1,2)$, $(0,2,1)$, $(1,0,2)$, $(1,2,0)$, $(2,0,1)$ and $(2,1,0)$ are the 16 parking functions for $n = 3$. Parking functions are classical objects in combinatorics and have many interesting properties. For example, combining an insightful result of G. Kreweras [723] and a bijection in [87] gives that

$$y^{\binom{n}{2}} T(K_{n+1}; 1, 1/y) = \sum y^{a_1 + \ldots + a_n},$$

where the sum is over all parking functions $(a_1, \ldots, a_n)$.

A. Postnikov et al., in [924], extended this idea by considering a connected graph $G = (V, E)$ on vertices v_0, v_1, ..., v_n with v_0 labeled q as a special vertex. For a subset F of vertices in $V \setminus \{q\}$, and a vertex $v \in F$, let $d_F(v)$ be the number of edges from the vertex v to the vertices in $V \setminus F$. Then, a sequence $(a_1, \ldots, a_n)$ of nonnegative integers is a *G-parking function* if, for any nonempty subset F, there exists a vertex $v_i \in F$ with $a_i < d_F(v_i)$. The K_{n+1}-parking functions are the usual parking functions.

Given a G-parking function $(a_1, \ldots, a_n)$, the sequence $(h_1, \ldots, h_n)$ defined by $h_i = \deg(v_i) - 1 - a_i$ is a critical configuration of G with q as special vertex. That the correspondence is a bijection follows from a characterization of critical configurations conjectured in [377] and later proved in [507]. This bijection and Theorem 17.3 gives

$$y^{|E|-|V|+1} T(G; 1, 1/y) = \sum y^{a_1 + \ldots + a_n},$$

where the sum is over all G-parking functions $(a_1, \ldots, a_n)$.

Remark 17.5. The origin of the chip-firing game in mathematics can be traced back to 1991, with the work in A. Björner et al. [135], but there are indications of early instances of the game in [23, 1024]. Around that time, D. Lorenzini studied a group associated with the Laplacian matrix of a graph, isomorphic to the critical group, see [786, 787]. In 1996, Biggs slightly modified the chip-firing game, to what he called the dollar game, see [112, 113]. In physics, in 1988, P. Bak et al. [64] introduced self-organized criticality and a model of sandpiles in lattices with boundary conditions. This motivated D. Dhar [377] to define the abelian sandpile model, and in turn, inspired A. Gabrielov [507] to generalize it to the abelian avalanche model. For further background on the chip-firing game, see [622, 695, 763, 839], and for the sandpile model, see [336, 378, 764, 809, 953]. The model described in Section 17.3 [507] is very general but there are recent models more general than that, see [67, 427, 594, 595].

The critical group has been discovered in several different contexts and been given many names: the sandpile group for graphs [377]; the critical group [113]; the group of bicycles [97]; the group of components [786]; the Jacobian of the graph [66]; and, the Picard group [112]. For further information on the critical group, see [333, 427, 543, 872]. The interest in this group has increased because of the analogy with algebraic geometry through the Riemann–Roch theorem for graphs due to M. Baker et al. [66], see [335, 542, 543]. The critical group has also been extended to critical ideals, see [334].

Parking functions were introduced by A. G. Konheim et al. [703]. For more about parking functions see [91, 962, 1035, 1037], and for their relation with the Tutte polynomial see [279, 709].

17.5 Open problems

A natural and interesting unresolved problem is to generalize the definition of the chip-firing game to include other combinatorial structures. This has been done for directed graphs [134, 622, 917], simplicial complexes [427], and more general matrices [594]. It would be especially interesting to have a chip-firing game for matroids whose critical configurations correspond to an evaluation of the Tutte polynomial as in Theorem 17.3. A tantalizing problem comes from trying to prove the conjecture of Stanley mentioned at the end of Section 17.2 using a chip-firing like process. So far, cographic matroids are the only example for which there is such a proof [838]. However, the conjecture has been proven, using different methods, for several families of matroids [369, 842, 887, 993]. Intriguingly, no proof of Stanley's conjecture for graphic or regular matroids is known.

18

The Tutte polynomial and knot theory

Stephen Huggett

Synopsis

This chapter gives an overview of connections between knot theory and the Tutte polynomial. It covers relationships between knot and graph polynomials, applications of graph polynomials to knot theory, and categorification.

- Knots, links, and their graphs.
- The Jones and HOMFLYPT polynomials and the Tutte polynomial.
- The Bollobás–Riordan polynomial and virtual knot invariants.
- Categorification of graph polynomials.

18.1 Introduction

When, in 1877, W. Tait [1053] drew up his remarkable table of knots up to and including seven crossings he also conjectured, among other things, that any reduced diagram of an alternating link has the fewest possible crossings. This conjecture was to remain open until, in 1985, V. Jones [668] discovered a new polynomial invariant of knots and links. Coming out of the blue, this new invariant attracted a great deal of interest, and a two-variable generalization of it, the HOMFLYPT (or HOMFLYPT) polynomial, was quickly discovered [503, 939].

In 1987 a relationship was found by M. Thistlethwaite [1061] between the Jones polynomial of a link and the Tutte polynomial of the "Tait" graph of that link, and this relationship was used to deduce Tait's conjecture from properties of the Tutte polynomial. F. Jaeger [648] in 1998 discovered a new

DOI: 10.1201/9780429161612-18

relation, between the Tutte polynomial of a graph (this time not a Tait graph) and the HOMFLYPT polynomial of a link determined by that graph. Then in 2001 B. Bollobás and O. Riordan generalized the Tutte polynomial to graphs embedded on surfaces: ribbon graphs (see Chapter 27). These are a natural half-way house between combinatorics and low-dimensional topology, and knot theory has motivated a lot of the recent developments in these topological Tutte polynomials.

Another very active area of research is "categorification", which first associated abstract cohomology groups to a link and now does so for a graph, yielding some of the polynomial invariants as characteristic classes of the cohomology groups. We survey these various connections between knot and graph polynomials.

18.2 Graphs and links

Standard texts on knot theory include [2, 352, 772], and, for historical accounts, [466, 938]. See also the excellent survey [771].

Definition 18.1. A *link* is a set of n disjoint piecewise linear simple closed curves in $\mathbb{R}^3$.

Definition 18.2. A *knot* is a link with $n = 1$.

The piecewise linear condition prevents links from being "wild"—infinitely knotted—but we normally imagine that there are so many pieces that the link looks like a smooth submanifold of $\mathbb{R}^3$. (Indeed, defining it as a smooth submanifold would be an equivalent way of excluding wildness.)

We regard two links as being the same if they are ambient isotopic.

Definition 18.3. Two links L_1 and L_2 are *ambient isotopic* if for all $t \in [0,1]$ there is a family of piecewise linear homeomorphisms $f_t : \mathbb{R}^3 \to \mathbb{R}^3$ which are continuous in t and such that f_0 is the identity and f_1 maps L_1 to L_2.

Links are quite hard to work with directly, but Reidemeister's theorem [957] allows us instead to work with diagrams in the plane, called link diagrams.

Definition 18.4. A *link diagram* is the image of a link under a projection map $p : \mathbb{R}^3 \to \mathbb{R}^2$ in which the only singular points are double points at which the arcs intersect transversally, and with each such double point drawn so as to indicate the over-crossing and under-crossing arcs.

Conversely, given a link diagram we can use the over-crossing and under-crossing information at each double point to reconstruct its link. Perhaps surprisingly, it is also of interest to consider link diagrams in which at *some*

FIGURE 18.1: Reidemeister moves R1, R2, and R3.

FIGURE 18.2: Replacing an edge of a graph by a crossing.

FIGURE 18.3: Replacing a vertex of a medial graph by a crossing.

FIGURE 18.4: Signed edges, and crossings.

of the double points there is no over-crossing or under-crossing information. These are diagrams of what are called *virtual links* [682], to which we shall return in Section 18.7.

Theorem 18.5 (Reidemeister's Theorem). *Two link diagrams D and D' represent the same link if and only if D and D' are related by a sequence of planar isotopies or the moves R1, R2, and R3 shown in Figure 18.1.*

A basic construction relating graphs and knots is as follows. Given a plane graph G we define a link diagram $D(G)$ by replacing each edge of G by a crossing, as in Figure 18.2, and then joining the crossings up by following the face boundaries of G. This link diagram $D(G)$ is necessarily alternating.

Definition 18.6. A link diagram is *alternating* if when tracing around each of its simple closed curves one alternates between over-crossings and under-crossings.

The construction of $D(G)$ can usefully be thought of in terms of the medial graph $M(G)$ of G: we replace each vertex of $M(G)$ by a crossing, as in Figure 18.3. (See Section 1.2.17.)

Conversely, given any alternating link diagram, we can invert this process to obtain the *Tait graph*. However, most links do not have alternating diagrams. So we need a way of recording which way each crossing goes. This is done by putting signs on the edges of the graph, as in Figure 18.4.

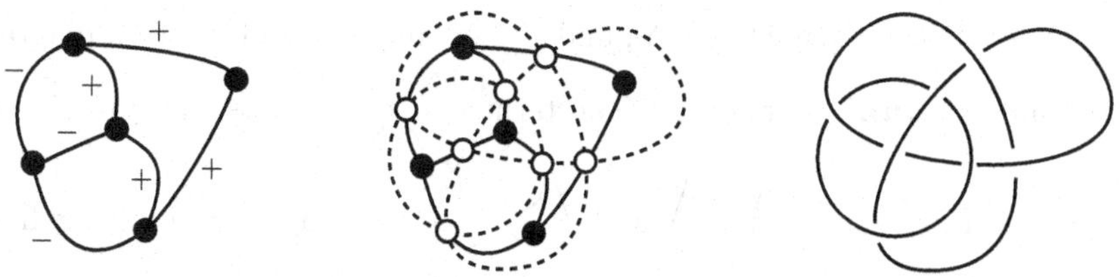

FIGURE 18.5: Moving between a link diagram and its signed Tait graph.

FIGURE 18.6: A-splicing and B-splicing, respectively.

This establishes a correspondence $G \leftrightarrow D(G)$ between signed plane graphs and link diagrams, illustrated in Figure 18.5.

18.3 Polynomial invariants of links

The first paper on the Jones polynomial is [668], but we will start by considering the Kauffman bracket polynomial [681], partly because it turns out to be the easiest way to define the Jones polynomial and partly because it is of independent interest.

These polynomials can be defined by a recursion process very similar to that giving the Tutte polynomial, but instead of deleting or contracting an edge of a graph we "resolve" or "splice" a crossing in a link diagram in one of two ways. In the underlying projection a crossing is a double point, which divides its neighborhood into four regions. In the link diagram itself, we label two of these regions A and the other two B: the A regions are swept out if the over-crossing arc is rotated anticlockwise until it lies on top of the under-crossing arc. (This labelling can be extended to a 2-coloring of the faces of the underlying projection if and only if the link diagram is alternating.) Now we can splice this crossing by joining the A regions or by joining the B regions, as in Figure 18.6.

Suppose that we require a function $\langle D \rangle$ of link diagrams D to satisfy the following conditions:

1. $\langle \times \rangle = A \langle)(\rangle + A^{-1} \langle \asymp \rangle$,
2. $\langle D \sqcup O \rangle = (-A^2 - A^{-2})\langle D \rangle$,
3. $\langle O \rangle = 1$.

Here $\sqcup$ denotes the disjoint union, and O is a simple closed curve, a diagram of the unknot with no crossings. The term $\langle \times \rangle$ refers to a crossing in the diagram D, and $\langle)(\rangle$ and $\langle \asymp \rangle$ the diagrams resulting from A-splicing and B-splicing that crossing of D, respectively. We use the first condition to remove all the crossings in D one by one until we are left with disjoint closed curves, and then use the second and third conditions to obtain $\langle D \rangle$ as a Laurent polynomial in A. Furthermore, it is an easy exercise to see that $\langle D \rangle$ is invariant under R2 and R3, but not under R1.

Definition 18.7. The *Kauffman bracket* is the function $\langle D \rangle$ satisfying the conditions 1–3 above.

The Kauffman bracket can also be defined as a "state sum", as follows. A state S of a link diagram D is the result of a choice of resolution (an A-splicing or a B-splicing) for each crossing of the diagram. Denote by $\mathcal{S}(D)$ the set of the states of D. Clearly, a diagram D with n crossings has $|\mathcal{S}(D)| = 2^n$ different states. Denote by $\alpha(S)$ and $\beta(S)$ the number of A-splicings and B-splicings in a state S, respectively. Also, denote by $\delta(S)$ the number of components of the curve obtained from the link diagram D in the state $S \in \mathcal{S}(D)$.

Definition 18.8. For a link diagram (or virtual link diagram) D, define

$$[D](A, B, d) = \sum_{S \in \mathcal{S}(D)} A^{\alpha(S)} B^{\beta(S)} d^{\delta(S)-1}.$$

The Kauffman bracket can then be recovered from this state sum as in the following theorem from [681]:

Theorem 18.9. *Let D be a link diagram (or virtual link diagram). Then* $\langle D \rangle = [D](A, A^{-1}, -A^2 - A^{-2})$.

Definition 18.10. An *oriented link* is a link in which each component has been given an orientation.

In order to use the Kauffman bracket to define an invariant of oriented links, we first choose an orientation for D. (In general, both the equivalence class of oriented links and the invariant will depend upon this choice of orientation.) Then we can give each crossing i in D a sign $\epsilon_i = \pm 1$, as in Figure 18.7. (Note that these signs are not related to those in Figure 18.4.)

Definition 18.11. The *writhe* of an oriented link diagram D is $w(D) = \sum_i \epsilon_i$ where the sum is taken over all crossings in the diagram.

Theorem 18.12. *Let L be an oriented link with diagram D. The polynomial $X(L)$ given by $X(L) = (-A)^{-3w(D)} \langle D \rangle$ is invariant under R1, R2, and R3, and is therefore an invariant of oriented links.*

FIGURE 18.7: Signs of oriented crossings, and the only possible splicing consistent with the orientation.

This theorem is from [681]. The Jones polynomial $V(L;t)$ of the oriented link L is obtained by substituting $A = t^{-1/4}$ in $X(L)$. It can also be defined directly, as follows. Given a diagram D of an oriented link L, choose a crossing, which will have a sign (see Figure 18.7). Draw two more link diagrams, identical to D except at this crossing, one of them with the crossing (and hence sign) changed and the other with the crossing spliced, as in Figure 18.7. We now have three oriented links L_+, L_-, and L_0 (one of L_+ and L_- being the original L). Also, let U be any diagram of the unknot. Note that the first equations in each of the next two theorems are known as *skein relations*. The first theorem is from [668].

Theorem 18.13. *The Jones polynomial $V(L;t)$ of the oriented link L is the unique polynomial invariant defined by the relations*

1. $t^{-1}V(L_+) - tV(L_-) = (t^{1/2} - t^{-1/2})V(L_0)$,
2. $V(U) = 1$.

Almost immediately after the discovery of the Jones polynomial, several mathematicians [503, 939] discovered a generalization now known as the HOMFLYPT polynomial.

Theorem 18.14. *There is a unique function P from oriented links to homogeneous Laurent polynomials of degree zero in x, y, z such that*

1. $xP(L_+;x,y,z) + yP(L_-;x,y,z) + zP(L_0;x,y,z) = 0$,
2. $P(U;x,y,z) = 1$.

Theorem 18.15. *For an oriented link L,*

$$V(L;t) = P(L;t, -t^{-1}, t^{1/2} - t^{-1/2}).$$

18.4 The Tutte and Jones polynomials

As we noted above, Thistlethwaite's approach [1061] to the first Tait conjecture used the Tutte polynomial. We start by describing this conjecture.

Given a link diagram we may consider its underlying projection, in which all the over-crossing and under-crossing information in the link diagram has been lost. Then we place a vertex at each double point and obtain a plane 4-regular graph, called the *universe* of the link diagram.

Definition 18.16. A link diagram is *reduced* if its universe has no separating vertex.

Tait conjectured that any reduced diagram of an alternating link has the fewest possible crossings among all diagrams of that link.

The *breadth* of a Laurent polynomial in the variable t is the difference between the highest and lowest powers of t occurring in the polynomial. Thistlethwaite showed that if a link L admits an alternating irreducible diagram D with m crossings, then the breadth of $V(L;t)$ is m. To do this he studied the Tutte polynomial of the Tait graph of D, and used its spanning tree expansion to calculate its breadth. He then, in [1061], deduced the following.

Theorem 18.17. *If a link L admits an alternating irreducible diagram with m crossings, then L cannot have a diagram with fewer than m crossings.*

Almost as an aside, Thistlethwaite showed that $V(D(G);t)$ is, up to a multiplication by a power of t, $\pm T(G;-t,-1/t)$. The full relation from [152] is as follows.

Theorem 18.18. *If G is a plane graph and $D(G)$ is the corresponding (necessarily alternating) link diagram, then*

$$V(D(G);t) = (-1)^w t^{(b-a+3w)/4} T(G;-t,-1/t)$$

where a is the number of A regions, b is the number of B regions, and w is the writhe.

However, most links do not have alternating diagrams, and so we need to consider signed graphs. In 1989, a Tutte polynomial for signed graphs was defined by Kauffman [680].

Definition 18.19. Let G be a signed graph. The polynomial $Q(G;A,B,d)$ in the variables A,B, and d is defined as follows.

1. If the edge e of G is neither a bridge nor a loop, then

$$Q(G) = \begin{cases} A\,Q(G\backslash e) + B\,Q(G/e) & \text{if } \operatorname{sign}(e) < 0 \\ B\,Q(G\backslash e) + A\,Q(G/e) & \text{if } \operatorname{sign}(e) > 0. \end{cases}$$

2. If every edge of G is either a bridge or a loop and G is connected, then

$$Q(G) = (A+Bd)^{b_+ + l_-}(Ad+B)^{b_- + l_+},$$

where $b_{+/-}$ and $l_{+/-}$ are the numbers of positive/negative bridges and loops in G.

FIGURE 18.8: Replacing an edge by a clasp.

3. If G is the disjoint union of graphs G_1 and G_2, then $Q(G) = d\,Q(G_1)\,Q(G_2)$.

Theorem 18.20. *Let G be a signed plane graph. Then*

$$Q(G; A, A^{-1}, -A^2 - A^{-2}) = \langle D(G) \rangle.$$

This polynomial Q is an evaluation of the multivariate Tutte polynomial Z (defined in Section 25.2):

$$Q(G; A, B, d) = d^{-(v(G)+1)} A^{e_-(G)} B^{e_+(G)} Z(G; d^2, \mathbf{w}),$$

where $e_+(G)$ and $e_+(G)$ are the numbers of positive and negative edges in G respectively, and the weight vector $\mathbf{w}$ is given by

$$w_e = \begin{cases} Ad/B & \text{if } e \text{ is positive} \\ Bd/A & \text{if } e \text{ is negative.} \end{cases}$$

This is equivalent to the Tutte polynomial if all the edges of G have the same sign. (See Section 2.2.1 for the relationship between Z and T.)

18.5 The Tutte and HOMFLYPT polynomials

Jaeger's 1988 relationship between the Tutte and HOMFLYPT polynomials [648] comes from a different way of associating a link diagram with a plane graph. Let G be a plane graph. Then we define an oriented link diagram $\mathcal{D}(G)$ by replacing each edge by a clasp, as in Figure 18.8, and then joining up the clasps by following the face boundaries of G.

Theorem 18.21. *If G is connected, then*

$$P(\mathcal{D}(G); x, y, z) = (y/z)^{|V(G)|-1}(-z/x)^{|E(G)|} T(G; -x/y, 1 - (xy + y^2)/z^2)).$$

Traldi [1072] generalized Jaeger's result above to take care of all four ways in which each edge of the graph G can be replaced by a clasp, as in Figure 18.9.

This raises the question of when a link L has a *matched diagram*, which is a diagram whose crossings are split into pairs of the types given in Figure 18.9. In 1987 Przytycki [937] conjectured that not every link has a matched diagram. This turned out to be a difficult problem, but links without matched diagrams have recently been discovered [428].

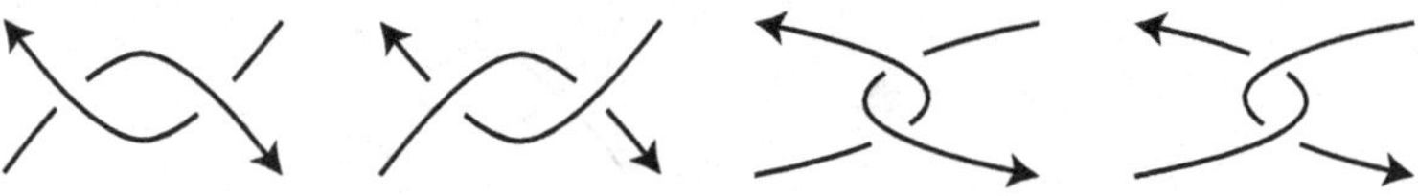

FIGURE 18.9: The four possible clasps.

Given a plane graph G, we now have *two* relationships between the Tutte polynomial of G and the Jones polynomial of a link diagram obtained from G, given by Theorems 18.18 and 18.21. In each case we considered a plane graph G, but different link diagrams $D(G)$ and $\mathcal{D}(G)$. To change from $D(G)$ to $\mathcal{D}(G)$ one replaces each crossing by a clasp. Instead, this could be achieved by replacing G by the tensor product $G \otimes K_3$, and then taking $D(G \otimes K_3)$.

$$\begin{array}{ccc} G & \to & G \otimes K_3 \\ \downarrow & & \downarrow \\ \mathcal{D}(G) & = & D(G \otimes K_3) \end{array} \tag{18.1}$$

Using Brylawski's theorem [245, 653] (here Theorem 6.43) on the behavior of the Tutte polynomial under tensor products, it was shown in [627] that these two approaches are equivalent: using Thistlethwaite's theorem 18.18 on $D(G \otimes K_3)$ yields the same result as using Jaeger's theorem 18.21 on $\mathcal{D}(G)$.

One may ask what happens if each crossing in $D(G)$ is replaced by a more general two-tangle: this was addressed in [627, 665]. (We define two-tangles in the next section.)

18.6 Applications in knot theory

By using tangles (for which, see Conway's remarkable paper [328]) we will discover how to construct pairs of links with the same Jones polynomial.

Definition 18.22. A *two-tangle* is a part of a link diagram obtained by intersecting the diagram with a closed disc in such a way that the boundary of the disc intersects the tangle in exactly four points, which can be regarded as equally spaced around the boundary circle.

Let T be a two-tangle and S its complement in the link diagram $D(L)$. View the plane in which D is drawn as sitting in $\mathbb{R}^3$. Then there are three rotations ρ of T, each of which defines a *mutant* $S \cup \rho(T)$ of the link L, as in Figure 18.10.

Theorem 18.23. *Links which are mutants of each other have the same Jones polynomial.*

Note that the rotation $t \mapsto \text{ɟ}$ in Figure 18.10 corresponds to a "Whitney flip" in the Tait graph.

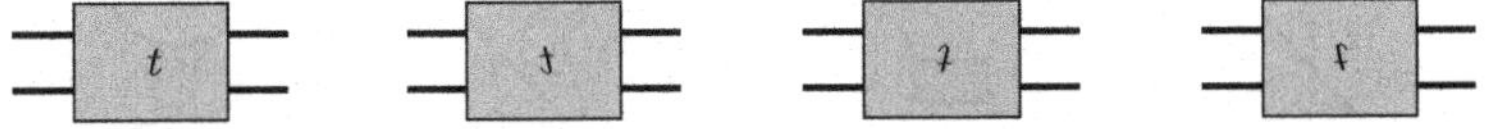

FIGURE 18.10: A two-tangle and its three rotations.

Inspired by Tutte's work on rotors in graphs [1103], in which families of graphs with the same Tutte polynomial are constructed, the concept of mutation was extended in 1989 in [29] as follows. (See also [864, 936], and Chapter 6.)

Definition 18.24. An *n-tangle* is a part of a link diagram obtained by intersecting the diagram with a closed disc in such a way that the boundary of the disc intersects the tangle in exactly $2n$ points.

Make the disc into a regular n-gon, with two these $2n$ points on each edge, arranged so that they have the full dihedral symmetry. The tangle itself will not have dihedral symmetry, of course. Now use the indirect symmetries in the dihedral group to generate generalized mutations.

Theorem 18.25. *If $n \leq 5$, links which are generalized mutations arising from an n-tangle have the same Jones polynomial.*

The complexity theory of the Jones polynomial was investigated in [653]. As described in Chapter 9, evaluating $T(G; x, y)$ for $(x, y) \in \mathbb{C}^2$ is a ♯P-hard problem except when $(x-1)(y-1) = 1$ or when (x, y) is one of $(1, 1)$, $(-1, -1)$, $(0, -1)$, $(-1, 0)$, $(i, -i)$, $(-i, i)$, (j, j^2), or (j^2, j), where $j = e^{2\pi i/3}$. From Theorem 18.18, it was then deduced that determining the Jones polynomial of a link is in general a ♯P-hard problem. This strengthened earlier results of Thistlethwaite and Jaeger.

In 2001, the Jones polynomials of several families of alternating knots and links were obtained by calculating the Tutte polynomials of the corresponding families of graphs [283]. These families were C_n with one edge doubled, wheels W_n, wheels with a vertex of degree two inserted on each spoke, and hammock graphs $H_{3,n}$ (which are constructed by inserting a vertex of degree two on each edge of the graph on two vertices joined by n edges).

In 2002, the evaluation

$$T(G; -1, -1) = (-1)^{|E(G)|}(-2)^{c-1}$$

was used to find the component numbers c of links whose graphs were fans or wheels [868]. It was also shown how the component numbers behaved under the operation of taking 2-sums of graphs.

FIGURE 18.11: A-splicing and B-splicing, with ribbons.

18.7 The Bollobás–Riordan polynomial

In 2001 and 2002 Bollobás and Riordan introduced the ribbon graph polynomial [160, 161]. Let G be a ribbon graph, and denote by $r(G), n(G), k(G)$, and $\mathrm{bc}(G)$, its rank, nullity, number of components, and number of boundary components, respectively. Also, define $t(G)$ to be 0 if G is orientable and 1 otherwise.

Definition 18.26. For a ribbon graph G, the *Bollobás–Riordan polynomial* is

$$R(G;x,y,z,w) := \sum_{F\subseteq G} (x-1)^{r(G)-r(F)} y^{n(F)} z^{k(F)-\mathrm{bc}(F)+n(F)} w^{t(F)},$$

where the sum is over all spanning ribbon subgraphs F of G.

This is a generalization to ribbon graphs of the Tutte polynomial. (See also Chapter 27, and [457].)

Thistlethwaite's theorem 18.18 relating the Jones and Tutte polynomials was extended to a relationship between the Jones and Bollobás–Riordan polynomials in [360]. Given a state S of a connected link diagram, we define a ribbon graph, as follows. Attach ribbons across each splicing in S as shown in Figure 18.11, yielding a set of closed curves joined by ribbons. Lift this off the plane and make each such closed curve the boundary of a disc. These discs are then the vertices of the ribbon graph $\mathbb{D}(S)$. There are two choices of S of particular interest: the all-A splicing and the all-B splicing.

Theorem 18.27. *Let D be a connected link diagram or virtual link diagram and $\mathbb{D}$ be the ribbon graph of D associated with the all-A splicing, with $e(\mathbb{D})$ edges and $v(\mathbb{D})$ vertices. Then*

$$\langle D\rangle(A) = A^{2-2v(\mathbb{D})+e(\mathbb{D})} R(\mathbb{D}; -A^4, -1-A^{-4}, (-A^2-A^{-2})^{-1}, 1).$$

Thus the Jones polynomial of D is obtained as an evaluation of the Bollobás–Riordan polynomial of $\mathbb{D}$. It is important to note that this generalization of Thistlethwaite's theorem 18.18 applies to all links, not just alternating ones, without resorting to the machinery of signed graphs. The crossing information has been stored in the topology of $\mathbb{D}$.

Virtual link diagrams can be thought of as link diagrams drawn on orientable surfaces other than the plane: the virtual crossings are only there because we are using a plane representation, just as when we draw a non-planar graph in the plane. (Here we are regarding these surfaces up to stable equivalence, which means adding or removing redundant handles. See [269, 682].) The Kauffman bracket in Definition 18.7 can easily be extended to virtual links. Given an oriented virtual link diagram, there is a unique state S in which at each classical crossing the splicing respects the orientation (and the virtual crossings are left unspliced). As before, attach ribbons across each splicing in S. This state is a set of oriented circles, known as *Seifert circles*, some of which have virtual crossings (because they have been drawn in the plane with double points). Untwist all these virtual crossings (thus twisting the attached ribbons), move the circles up off the plane as before, and let them bound discs. Finally, transfer the signs from the classical crossings (see Figure 18.7) to the ribbons. This constructs a signed ribbon graph, and there is a signed Bollobás–Riordan polynomial R_s (see [310]). Just as with the signed Tutte polynomial Q, this R_s is an evaluation of the multivariate Bollobás–Riordan polynomial [1121]. The Kauffman bracket of a virtual link can then be obtained from R_s, as shown in [310]:

Theorem 18.28. *Let D be a virtual link diagram and $\mathbb{D}$ be the corresponding signed ribbon graph. Then*

$$\begin{aligned}\langle D\rangle(A) = {} & A^{n(\mathbb{D})-r(\mathbb{D})}(-A-A^{-1})^{k(\mathbb{D})-1}\\ & \cdot R_s\left(\mathbb{D}; -A^3-A+1, -A^{-1}-A^{-3}, (-A-A^{-1})^{-1}, 1\right).\end{aligned}$$

Theorems 18.27 and 18.28 are related using Chmutov's idea of partial duality [306].

For any state S of a connected link diagram, the ribbon graph $\mathbb{D}(S)$ can be embedded in a surface $\mathcal{G}(S)$ called the *Turaev surface* [1087]. We note in passing that Turaev surfaces are the subject of the fascinating survey [275].

Given a graph G embedded in an orientable surface one might choose instead to use Jaeger's technique to construct a corresponding link $\mathcal{D}(G)$ in the thickened surface. This was explored by Moffatt in [854], who used Lieberum's [774] extension of the HOMFLYPT polynomial to links in thickened surfaces to give the following generalization of Theorem 18.21 to ribbon graphs. Here the non-homogeneous variables (l, m) are related to the homogeneous (x, y, z) of Theorem 18.14 by $l = x/\sqrt{-xy}$ and $m = -z/\sqrt{-xy}$, and the skein relation is $lL_+ - l^{-1}L_- - mL_0 = 0$.

Theorem 18.29.

$$\begin{aligned}P(\mathcal{D}(G); l, m) &= \left(\frac{1}{lm}\right)^{v(G)-1}\left(\frac{m}{l}\right)^{e(G)}(l^2-1)^{k(G)-1}\\ &\quad \cdot R\left(G; l^2, \frac{l-l^{-1}}{lm^2}, \frac{m}{l-l^{-1}}, 1\right).\end{aligned}$$

Then [854] similarly generalized Traldi's results [1072] to ribbon graphs, as well as the relationship (18.1) above.

It is clear from Theorem 18.21 that $T(G)$ and $P(\mathcal{D}(G))$ determine each other. It was noted in both [309] and [854] that this is no longer true for ribbon graphs: it can be seen from Theorem 18.29 that the Bollobás–Riordan polynomial is only determined when its parameters x, y, and z satisfy $xyz^2 = 1$.

As was pointed out in [854], this has an immediate application in knot theory, which we describe next. A link $L \subset \Sigma \times I$ is *essential* in the surface Σ if there is no embedded surface Σ' with smaller genus such that $L \subset \Sigma' \times I$. A link $L \subset \Sigma \times I$ is *split* if it is isotopic to a link L' with the property that

$$L' \cap (\Sigma \times [0, 1/2]) \neq \emptyset, \quad L' \cap (\Sigma \times [1/2, 1]) \neq \emptyset, \quad L' \cap (\Sigma \times \{1/2\}) = \emptyset.$$

Theorem 18.30. *Let $L \subset \Sigma \times I$ be a non-split alternating link essential in Σ. If L has a projection of the form $\mathcal{D}(G)$ for some graph $G \subset \Sigma$, then the genus of Σ can be recovered from the* HOMFLYPT *polynomial of L.*

We have noted that the classical Theorem 18.18 relating the Jones polynomial of an alternating link to the Tutte polynomial of its Tait graph is generalized to virtual links in Theorem 18.28. There is a second generalization, based on a *relative Tutte polynomial* defined by Diao and Hetyei in 2009 [379] (see Section 24.6). They showed that if L is a virtual link diagram and G is its Tait graph, having subgraph H corresponding to the virtual crossings in L, then the relative Tutte polynomial $T_H(G)$ is the same as the Kauffman bracket of L. A direct relationship between the Bollobás–Riordan and relative Tutte polynomials was given by Butler and Chmutov in [256], showing how these two generalizations of Theorem 18.18 are connected. Then in 2012, Diao and Hetyei [381] extended their work to the calculation of relative Tutte polynomials of tensor products of colored graphs.

18.8 Categorification

A significant development in knot theory has been Khovanov's "categorification" [689], in which abstract cohomology groups are constructed whose Euler characteristics yield polynomial invariants of links. This idea has stimulated a similar construction [305, 610] in graph theory, which we briefly describe here.

Given a graph G with ordered edges, a *state* is a spanning subgraph of G. The number of edges in a state is called its *dimension*. An *enhanced state* is a state whose connected components are colored either x or 1. The number of connected components colored x is called the *degree* of the enhanced state. The cochain group $C^{i,j}$ is defined to be the real vector space spanned by all enhanced states of dimension i and degree j. A cochain complex is constructed by defining a differential

$$d : C^{i,j} \to C^{i+1,j},$$

which depends upon choosing the algebra corresponding to a single vertex without edges: a simple such choice is $\mathbb{R}[x]/(x^2)$. (We omit the details, which are in [305], but note that other choices are possible, and interesting.) Then the *chromatic cohomology* of the graph G is defined in the usual way by

$$H^{i,j}(G) = \frac{\ker(d : C^{i,j} \to C^{i+1,j})}{\operatorname{im}(d : C^{i-1,j} \to C^{i,j})},$$

and we have the following results.

Theorem 18.31. *Let the graph G have n vertices and k connected components.*

1. *Its chromatic polynomial is the following Euler characteristic:*

$$\chi(G; \lambda) = \sum_{i,j} (-1)^i (\lambda - 1)^j \dim(H^{i,j}(G)).$$

2. *By deleting or contracting the edge $e \in E(G)$ we obtain the long exact sequence:*

$$0 \to H^{0,j}(G) \to H^{0,j}(G\backslash e) \to H^{0,j}(G/e) \to H^{1,j}(G) \to \dots$$

3. *The non-trivial cohomology is concentrated on $k+1$ diagonals:*

$$H^{i,j}(G) = 0 \quad \textit{unless} \quad n - k \leq i + j \leq n.$$

This idea, of using characteristic classes to somehow generate polynomial invariants, has seen many developments. Stošić [1047] defines a chain complex of doubly-graded modules whose doubly-graded Euler characteristic is equal to the whole two-variable dichromatic polynomial. Jasso-Hernandez and Rong [656] define bigraded homology groups whose Euler characteristic is a variant of the Tutte polynomial.

Loebl and Moffatt [783] construct a bigraded chain complex using the set of spanning subgraphs of a ribbon graph. They show that the graded Euler characteristic of the homology of this complex is the chromatic polynomial. The homology groups themselves (which are not Tutte–Grothendieck invariants) are strictly stronger than the chromatic polynomial, and will distinguish graphs which differ only by multiple edges or loops. The construction can be extended to give a homology theory from which the Bollobás–Riordan polynomial can be recovered as the graded Euler characteristic.

In Luse and Rong [801], for each positive integer n a homology theory is constructed whose Euler characteristic is the Penrose polynomial evaluated at n. (For $n = 2$ the homology groups are shown to be stronger than the Penrose polynomial.) The homology groups satisfy a long exact sequence corresponding to the skein relation of the Penrose polynomial, and there is an isomorphism with the chromatic homology in [610].

Champanerkar and Kofman [274] conjecture that Khovanov homology is invariant under mutation of knots, and reformulate this conjecture using a matroid obtained from the spanning trees of the Tait graph of a knot diagram. (It is known, see [1130], that this is not true for links.)

Dancso and Licata [358] define several homology theories for central hyperplane arrangements, deriving polynomial invariants including the characteristic polynomial, the Poincaré polynomial, and the Tutte polynomial. (The characteristic polynomial of hyperplane arrangements specializes to the chromatic polynomial of graphs when restricted to graphical arrangements, which are essentially the same as graph configuration spaces.)

As is pointed out in [610], chromatic cohomology does not come with a geometrical interpretation. There is another construction [433] of the chromatic polynomial as the Euler characteristic of a family of manifolds, which does have a clear geometrical interpretation. A relationship between these two approaches is found in Baranovsky and Sazdanovic [74], who construct a spectral sequence with E_1 term isomorphic to a graph cohomology complex, and which converges to the relative cohomology of the graph configuration space.

18.9 Open problems

From the perspective of knot theory, graph polynomials can be thought of as a tool. From this point of view, one asks how the expression of a topological problem in terms of a combinatorial one can help in its solution. For example, can insights from the theory of graph polynomials assist in resolving the question of whether the Jones polynomial detects the unknot? Or can Tutte uniqueness provide further insights into the strengths and weaknesses of graph polynomials as knot invariants?

On the other hand, knot theory has led to a number of unexpected advances and research directions in the theory of graph polynomials. For example, the discovery the Bollobás–Riordan polynomial as well as Noble and Welsh's U-polynomial (see Chapter 26) were inspired by the theory of Vassiliev and quantum knot invariants, but their scope reaches far beyond their origins. Many combinatorial constructions in knot theory are reminiscent of constructions in graph polynomials, but remain to be understood in this context. For example, the sl_2-weight system from the theory of Vassiliev knot invariants defines a 1-variable graph polynomial for circle graphs and for 1-vertex ribbon graphs (see [314]). Lando has asked if this polynomial extends to either all simple graphs, or to all ribbon graphs.

More generally, we are interested in graph polynomials because they encode combinatorial information, while we are interested in knot invariants as they encode topological information. In which ways does the information in each of these settings inform the other? For example, does knot theory provide insights into what combinatorial information about a ribbon graph is contained in the Bollobás–Riordan polynomial? How can categorified graph polynomials be used to obtain results about graph structure?

19

Quantum field theory connections

Adrian Tanasa

Synopsis

This chapter covers the relation between the Tutte polynomial and the Symanzik polynomials of the parametric representation of the Φ^4 model in quantum field theory (QFT).

- The Φ^4 model and the parametric representation (Symanzik polynomials) of its Feynman integrals.
- The renormalization group equation in QFT.
- The relation between the multivariate Tutte polynomial and the Symanzik polynomials.
- Graph Hopf algebra characters.
- A differential equation obeyed by the Tutte polynomial.

19.1 Introduction

Recently, problems in combinatorics have been successfully tackled using methods inspired by quantum field theory (QFT). Conversely, open problems in physics have been successfully approached through various combinatorial methods. This represents the new field of "combinatorial physics". In this chapter we focus on a particular topic of combinatorial physics, namely the relation between the Tutte polynomial (and its variations) and QFT.

We begin the chapter by defining the Symanzik polynomial, and showing it is an evaluation of the multivariate Tutte polynomial. We then introduce some concepts from QFT in general, and for the Φ^4 QFT model in particular. In

DOI: 10.1201/9780429161612-19

QFT, a crucial role is played by a certain type of graphs called Feynman graphs to which are associated, in a canonical way, the so-called Feynman integrals. We then focus on the parametric representation of the Feynman integrals appearing in the Φ^4 QFT model. This parametric representation leads to the Symanzik polynomial mentioned above. For a general QFT model, we also exhibit the renormalization group equation.

In the second part of this chapter, we introduce some appropriate graph Hopf algebra characters. We use these maps to obtain a certain Tutte polynomial convolution formula and also to show that the Tutte polynomial satisfies a certain differential equation which is of the same type as the renormalization group equation mentioned above.

19.2 The Symanzik polynomials as evaluations of the multivariate Tutte polynomial

In QFT is is common to order and label the edges with integers, and then identify the edges with their labels. We follow this convention in this chapter.

Definition 19.1. Let G be a connected graph with edge weights a_e, for $e = 1, \ldots, |E|$. The *first Symanzik polynomial* is defined as

$$U(G; \mathbf{a}) := \sum_{\mathcal{T}} \prod_{e \notin \mathcal{T}} a_e,$$

where $\mathcal{T}$ is a spanning tree of G and $\mathbf{a} := (a_1, \ldots, a_{|E|})$. The polynomial $U(G; \mathbf{a})$ is extended multiplicatively to non-connected graphs.

A *second Symanzik polynomial* will be defined in the next section. We now give the following deletion–contraction relation for the first Symanzik polynomial.

Theorem 19.2. *For any edge e which is not a loop*

$$U(G; \mathbf{a}) = a_e\, U(G \backslash e; \mathbf{a}) + U(G/e; \mathbf{a}).$$

Moreover, the terminal form evaluation is

$$U(G; \mathbf{a}) = \prod_{e \in E(G)} a_e,$$

for G consisting only of loops attached to isolated vertices.

A proof of this theorem can be obtained by following the proof of Theorem 3.3 of [720] where a Grassmann representation of the polynomial U is given (since U can be obtained as the determinant of some matrix associated to the graph).

The polynomial $U(G;\mathbf{a})$ can be obtained as a limit of the multivariate Tutte polynomial $Z(G;q,\mathbf{w})$, defined in Equation (25.1), as follows. Consider $q^{-k(G)}Z(G;q,\mathbf{w})$. By taking the limit $q \to 0$ we obtain a sum over maximally spanning subgraphs A, that is, subgraphs with $k(A) = k(G)$:

$$\lim_{q\to 0} q^{-k(G)} Z(G;q,\mathbf{w}) = \sum_{\substack{A \text{ maximally} \\ \text{spanning } E}} \prod_{e\in A} w_e .$$

If only the terms of lowest degree of homogeneity in the variables w are retained, we can obtain a sum over maximally spanning graphs with lowest number of edges, i.e., maximally spanning acyclic graphs (or spanning forests) of G. This lowest number of edges is the rank of G, namely $r(G)$. We have

$$F(G;\mathbf{w}) := \sum_{\substack{\mathcal{F} \text{ max. spanning} \\ \text{forest of } G}} \prod_{e\in E(\mathcal{F})} w_e .$$

Finally, dividing $F(G;\mathbf{w})$ by $\prod_{e\in E} w_e$ and making the change of variables $a_e = w_e^{-1}$ leads to the polynomial $U(G;\mathbf{a})$. All this is summarized by the formula

$$U(G;\mathbf{a}) = \left[\left(\prod_{e\in E} w_e \right) \lim_{q'\to 0} \frac{1}{(q')^{r(G)}} \lim_{q\to 0} q^{-k(G)} Z(G;q,q'\mathbf{w}) \right]_{w_e^{-1}=a_e} .$$

Let us end this section by mentioning that the relation between the first Symanzik polynomial and the multivariate Tutte polynomial can also be deduced from the universality property of the Tutte polynomial and from Theorem 19.2.

19.3 The Symanzik polynomials in quantum field theory

Quantum field theory (QFT) is a general mathematical framework lying at the very heart of fundamental physics. It gives a quantum description of particles and interactions that is naturally compatible with Einstein's theory of special relativity. QFT led to the standard model of elementary particle physics, which is one of the best experimentally tested physical theories. QFT's mathematical formalism (which combines in an elegant way analysis, algebra, combinatorics and so on) also successfully applies to other branches of theoretical physics, such as condensed matter or statistical physics. The interested reader may consult any of the very good textbooks on QFT, such as [636, 918, 1132].

19.3.1 The Φ^4 QFT model

A particularly simple field theoretical model is the so-called Φ^4 *model.* It consists of a single type of *field* Φ, which is here a function

$$\Phi : \mathbb{R}^4 \to \mathfrak{R},$$

where the target space $\mathfrak{R}$ is either $\mathbb{R}$ (Φ is then called a *real field*) or $\mathbb{C}$ (Φ is then called a *complex field*). (We emphasize that the term field used here for Φ has nothing to do with the usual algebraic definition of a field from mathematics.) If we deal with the so-called zero-dimensional QFT, the field Φ is not a function of $\mathbb{R}^4$ anymore, but a simple real or complex variable. If only one type of field is involved, the respective QFT model is called *scalar.* Another example of a scalar QFT model is the Φ^3 model, which is presented, for example, in the first chapter of [327].

A field theoretical model further requires the definition of an action, that is, a functional of the field. For the real Φ^4 model, the action is

$$S[\Phi] = \int_{\mathbb{R}^4} d^4x \left[\frac{1}{2} \sum_{\mu=1}^{4} \partial_\mu \Phi(x) \partial_\mu \Phi(x) + \frac{1}{2} m^2 \Phi(x)^2 + V_{\text{int}}[\Phi(x)] \right]. \quad (19.1)$$

Here, $\partial_\mu = \frac{\partial}{\partial x_\mu}$ and $V_{\text{int}}[\Phi(x)]$ is the *interacting potential,* $\frac{\lambda}{4!}\Phi(x)^4$ (hence the name of Φ^4 model). Furthermore, m and λ are real parameters, respectively the mass and the coupling constant.

Equation (19.1) is written in *configuration space* (or *direct space*). A Fourier transform, which we indicate by a tilde, converts Equation (19.1) to the formula of the action in *momentum space.* The latter space is where the elementary particle physics computations are usually performed. The transformed action in momentum space then is

$$\tilde{S}[\tilde{\Phi}] = \int_{\mathbb{R}^4} d^4p \left[\frac{1}{2} \sum_{\mu=1}^{4} p_\mu p_\mu \tilde{\Phi}^2 + \frac{1}{2} m^2 \tilde{\Phi}^2 + \tilde{V}_{\text{int}}[\tilde{\Phi}] \right]. \quad (19.2)$$

This action has a quadratic part in Φ (or $\tilde{\Phi}$) and a non-quadratic part. The quadratic part corresponds to the *propagation,* while the non-quadratic part corresponds to the *interaction.*

The physical information of a theory is encoded in N*-point functions* which are defined by

$$S^{(N)}(x_1, x_2, \ldots, x_N) := \frac{1}{Z} \int \mathcal{D}\Phi\, \Phi(x_1)\Phi(x_2) \ldots \Phi(x_N) e^{-S[\Phi]}. \quad (19.3)$$

Here $\mathcal{D}\Phi$ is the functional integration (for details, see for example [963]), and Z is the *partition function*

$$Z := \int \mathcal{D}\Phi e^{-S[\Phi(x)]}.$$

The partition function above can be seen, from a combinatorial point of view, as the generating function of Feynman graphs, and therefore is closely related to map enumeration problems. The interpretation of the partition function Z as a generating function is straightforward in the case of the zero-dimensional QFT because in this case the propagation part of the action is trivial.

In the particle physics setting, N-point functions are related to the probability of a particle interaction involving N fields. In general, it is not possible to derive explicit analytic expressions for the N-point functions. The tool used then in physics is the perturbative expansion, i.e., an expansion of the exponential above in powers of λ. The coefficients of such an expansion are sums of multiple integrals (see Equations (19.1) and (19.3)); the number of these integrals grows rapidly with increasing order in perturbation theory (i.e., power of the coupling constant λ).

19.3.1.1 Feynman integrals and Feynman rules

To the above multiple integrals, called *Feynman integrals*, one associates *Feynman graphs*, which are very useful in organizing the expansion coefficients. For the quartic model exhibited here, the graphs have valence four at each vertex. Moreover, two types of edges, *internal* and *external* edges, exist (see, for example, Figure 19.1).

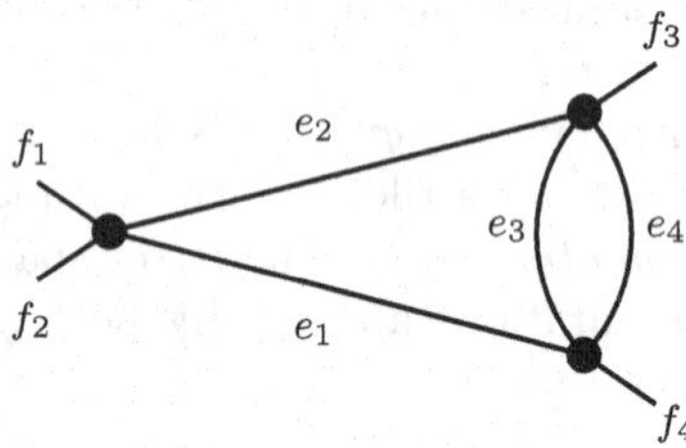

FIGURE 19.1: A Feynman graph, with four internal edges e_1, e_2, e_3, e_4, and four external edges f_1, f_2, f_3, f_4. It is a graph at the third order in perturbation theory (it has three vertices).

Note that in momentum space representations, the degree-one vertices at the ends of external edges are generally not depicted. Note that each of these Feynman graphs comes with its combinatorial weight.

In momentum space, one orients arbitrarily each edge e and associates to it some momentum p_e (living in the Fourier transformed space $\mathbb{R}^4$ as above). The contribution from any edge comes from the quadratic part of the action (19.2) and reads

$$C(p_e) = \frac{1}{p_e^2 + m^2}. \tag{19.4}$$

This contribution is called the *propagator*.

The contribution of some vertex is given by the coupling constant λ times a δ-function of conservation of the incoming/outgoing momenta at the respective vertex. If the momenta of the external edges are fixed (they correspond to physical states), the momenta of the internal edges are free and one has to integrate over them, leading to the respective Feynman integral. These are the *Feynman rules*, which allow us to associate, in a canonical way, a Feynman integral to a given Feynman graph. For more details on this, we refer the interested reader to any QFT textbook, such as [636], or to the articles [149, 719].

For the example of Figure 19.1, these rules lead to the following Feynman integral:

$$\lambda^3 \int \left(\prod_{i=1}^{4} d^4 p_{e_i}\right)\left(\prod_{i=1}^{4} \frac{1}{p_{e_i}^2 + m^2}\right)$$
$$\cdot \delta(p_{f_1} + p_{f_2} - p_{e_1} - p_{e_2})\delta(p_{e_1} + p_{e_3} - p_{e_4} - p_{f_4})\delta(p_{e_2} - p_{e_3} + p_{e_4} - p_{f_3}). \tag{19.5}$$

We leave it as an exercise to determine the orientation of the edges necessary for such a Feynman integral to occur.

Because of the presence of the three δ-functions in (19.5), the number of remaining integrals is equal to two, which is actually the number of independent cycles of the Feynman graph (this being a general result in QFT). Thus, one can easily check that the Feynman integral (19.5) can be rewritten as

$$\lambda^3 \int d^4 p_{e_1} d^4 p_{e_3} \left(\frac{1}{p_{e_1}^2 + m^2} \frac{1}{p_{e_3}^2 + m^2} \frac{1}{(P - p_{e_1})^2 + m^2} \frac{1}{(p_{e_3} + p_{e_1} - p_{f_4})^2 + m^2}\right), \tag{19.6}$$

where we have denoted $P = p_{f_1} + p_{f_2}$ and we have omitted $\delta(p_{f_1} + p_{f_2} - p_{f_3} - p_{f_4})$ (which corresponds to the so-called external momentum conservation).

19.3.1.2 Parametric representation of Feynman integrals

Let us now proceed further and introduce the *parametric representation* of a Feynman integral. The main idea is to write each of the internal propagators (19.4) of the integral as an integral over some real positive parameter a. Thus,

$$C(p_e) = \frac{1}{p_e^2 + m^2} = \int_0^\infty da_e \exp(-a_e(p_e^2 + m^2)), \tag{19.7}$$

where $e = 1, 2, \ldots, |E|$ and $|E|$ is the number of internal edges of the graph (recall the QFT convention that edges are labelled with and identified with integers). Inserting these formulas in the general expression of a Feynman integral allows us to integrate (through Gaussian integrations) the internal momenta p_e. The Feynman integral associated to a graph G then becomes

$$\int_0^\infty \frac{\exp(-V(G; p_{\text{ext}}, \mathbf{a})/U(G; \mathbf{a}))}{U(G; \mathbf{a})^2} \prod_{e=1}^{|E|} (\exp(-m^2 a_e) da_e), \tag{19.8}$$

where $U(G;\mathbf{a})$ is the first Symanzik polynomial (see Definition 19.1) and $V(G;p_{\text{ext}},\mathbf{a})$ is a polynomial in the set of external momenta p_{ext} and the set of parameters $\mathbf{a}$. See [636] or [963] for a proof. We call this polynomial the *second Symanzik polynomial.* (Note that, in the QFT literature, these polynomials are also called the Kirchhoff polynomials, or the Kirchhoff–Symanzik polynomials). Specifically, the second Symanzik polynomial is

$$V(G;p_{\text{ext}},\mathbf{a}) := \sum_{\mathcal{T}_2} \prod_{e \notin \mathcal{T}_2} a_e \left(\sum_{i \in \mathcal{E}(\mathcal{T}_2)} p_i \right)^2,$$

where we have denoted by $\mathcal{T}_2$ a two-tree (which is a spanning forest with two connected components). We have denoted by $\mathcal{E}(\mathcal{T}_2)$ one of the connected components thus obtained. By momentum conservation, the total external momenta of one of these connected components (for example $\mathcal{E}(\mathcal{T}_2)$) is equal to the one of the other connected component (so the choice of $\mathcal{E}(\mathcal{T}_2)$ is arbitrary).

Let us now illustrate the implementation of the parametric representation for the Feynman graph of the example of Figure 19.1. Inserting the expression (19.7) in the expression (19.6) of the Feynman integral leads to the integral

$$\begin{aligned} \lambda^3 \int_0^\infty \prod_{i=1}^4 da_{e_i} \exp(-m^2 a_{e_i}) \int d^4p_{e_1} d^4p_{e_3} \exp(-a_{e_1}(p_{e_1}^2 + m^2)) \\ \exp(-a_{e_3}(p_{e_3}^2 + m^2)) \exp(-a_{e_2}((P - p_{e_1})^2 + m^2)) \\ \exp(-a_{e_4}((p_{e_3} + p_{e_1} - p_{f_4})^2 + m^2)). \end{aligned}$$

As above, the integrals are over the internal momenta, here p_{e_1} and p_{e_3} become Gaussian, and they can be explicitly performed. This finally leads to the expression (19.8) of the Feynman integral, where the Symanzik polynomials are:

$$\begin{aligned} U(G;\mathbf{a}) &= \alpha_3\alpha_4 + \alpha_2\alpha_4 + \alpha_2\alpha_3 + \alpha_1\alpha_3 + \alpha_1\alpha_4, \\ V(G;p_{\text{ext}},\mathbf{a}) &= P^2\alpha_1\alpha_2(\alpha_3+\alpha_4) + p_{f_4}^2\alpha_1\alpha_3\alpha_4 + p_{f_3}^2\alpha_2\alpha_3\alpha_4. \end{aligned}$$

The parametric representation is useful in QFT for the following reasons. The form (19.8) of the Feynman integral gives a very condensed expression of this integral. Through a rescaling of the parameters a_e the so called power counting theorem can be easily obtained; this is a very important result for QFT renormalization. Moreover, the formulas used here are explicitly positive and this is a particularly convenient property when doing various mathematical manipulations of these integrals (such as the study of analyticity properties of the N-point functions). The number of space-time dimensions (here 4) is just a parameter and this means that the parametric representation can also be used as starting point for dimensional regularization, a very convenient way for carrying out the renormalization program. Finally, the parametric representation exhibits interesting analogies with the theory of electric circuits.

For the sake of completeness, let us also mention that the parametric interpretation can be implemented for any scalar model involving self-interaction of the field Φ (the interaction doesn't need to be quartic)—see, for example, the book [636]. Moreover, the Bollobás–Riordan polynomial (see Chapter 27) is related in the same way to the Symanzik polynomial of scalar QFT on the non-commutative Moyal space (where Feynman graphs become Feynman ribbon graphs). This relation was discovered in [720]. Let us end this subsection by mentioning that the deletion–contraction properties of QFT Symanzik polynomials are used in F. Brown's work on number theoretical aspects of QFT (see, for example, [221]).

19.3.2 Renormalization group equation

The Feynman integrals computed in QFT are generally divergent. One way to address this is to consider a real positive *cut-off* Λ, called the flowing parameter. This leads to a family of cut-off dependent actions, a family denoted by S_Λ. The derivation $\Lambda\frac{\partial S_\Lambda}{\partial \Lambda}$ gives the *renormalization group equation.*

This renormalization group equation (also called the Polchinski flow equation [923] in the QFT literature) is

$$\Lambda\frac{\partial S_\Lambda}{\partial \Lambda} = \int_{\mathbb{R}^8} \frac{1}{2} d^4p d^4q \Lambda \frac{\partial C_{\Lambda,\Lambda_0}}{\partial \Lambda}\left(\frac{\delta^2 S}{\delta\tilde{\Phi}(p)\delta\tilde{\Phi}(q)} - \frac{\delta S}{\delta\tilde{\Phi}(p)}\frac{\delta S}{\delta\tilde{\Phi}(q)}\right). \tag{19.9}$$

Here Λ_0 is an ultra-violet cutoff and Λ is a floating infra-red cut-off (the ultra-violet corresponding in QFT to high energies and the infra-red corresponding to low energies). The first term in the right-hand side of the equation above corresponds to the derivation of a propagator associated to a bridge in the respective Feynman graph. The second term corresponds to an edge which is not a bridge and is part of some circuit in the graph. We will see below that the Tutte polynomials satisfies a differential equation of this type (see Proposition 19.8).

19.4 Hopf algebras and the Tutte polynomial

19.4.1 A Hopf algebra on matroids

In this section we recall a Hopf algebras of matroids. (See also Section 30.3 where a different Hopf algebra of matroids is considered.) We begin with the unital associative algebra over $\mathbb{Q}$ freely generated by matroids. where the product is direct sum. Consider the coproduct

$$\mathbf{\Delta}(M) = \sum_{A\subseteq E} M|A \otimes M/A,$$

where M is a matroid on E. The comultiplication is coassociative and it can be shown that it gives rise to a Hopf algebra.

19.4.2 Characters and the Tutte polynomial

Definition 19.3. Let $\mathcal{H}$ be a Hopf algebra and $f, g \in \mathrm{Hom}(\mathcal{H}, \mathcal{H})$. The *convolution product*, $f * g$, of f and g is defined by

$$f * g = m \circ (f \otimes g) \circ \mathbf{\Delta},$$

where m is the product in the Hopf algebra and $\mathbf{\Delta}$ the coproduct.

Definition 19.4. An *infinitesimal character* is a linear morphism g from a Hopf algebra $\mathcal{H}$ to a commutative ring $\mathfrak{R}$ such that

$$g(m(G_1, G_2)) = g(G_1)\epsilon(G_2) + \epsilon(G_1)g(G_2),$$

for all $G_1, G_2, \in \mathcal{H}$, and where ϵ is the counit of the Hopf algebra.

Definition 19.5. A Hopf algebra *character* f is an algebra morphism from the Hopf algebra to commutative ring $\mathfrak{R}$. This means that the following holds

$$f(m(G_1, G_2)) = f(G_1)f(G_2), \qquad f(1) = 1_{\mathfrak{R}}.$$

for all G_1 and G_2 in the Hopf algebra, and where $1_{\mathfrak{R}}$ is the unit element of $\mathfrak{R}$.

We work in a Hopf algebra where the non-trivial part of the coproduct is nilpotent, so we can define an exponential map by the following expression

$$\exp_*(\delta) = \epsilon + \delta + \frac{1}{2}\delta * \delta + \dots,$$

where δ is an infinitesimal character.

Following [717], define

$$\delta_{\text{loop}}(M) = \begin{cases} 1_{\mathfrak{R}} \text{ if } M = U_{0,1}, \\ 0_{\mathfrak{R}} \text{ otherwise.} \end{cases}$$

$$\delta_{\text{coloop}}(M) = \begin{cases} 1_{\mathfrak{R}} \text{ if } M = U_{1,1}, \\ 0_{\mathfrak{R}} \text{ otherwise.} \end{cases}$$

These maps are infinitesimal characters of the Hopf algebras. They can be exponentiated, giving

$$\alpha(M; x, y, s) = \exp_* s(\delta_{\text{coloop}} + (y-1)\delta_{\text{loop}}) * \exp_* s((x-1)\delta_{\text{coloop}} + \delta_{\text{loop}})(M). \tag{19.10}$$

By a direct check, it can be proven that the maps α are Hopf algebra characters. Moreover, the following holds.

$$\exp_*(u\,\delta_{\text{coloop}} + v\,\delta_{\text{loop}})(M) = u^{r(M)}v^{n(M)}.$$

With this result, we can obtain the Tutte polynomial from the infinitesimal characters (see [423]), as follows.

Theorem 19.6. *The character α defined by formula* (19.10) *and the Tutte polynomial are related by:*

$$\alpha(M; x, y, s) = s^{|E|} T(M; x, y). \tag{19.11}$$

The identity (19.11) gives a new proof (see again [423] and also Theorem 4.150) of a *convolution formula* for the Tutte polynomial:

Theorem 19.7.

$$T(M; x, y) = \sum_{A \subseteq E} T(M|A; 0, y)\, T(M/A; x, 0).$$

Using the definition (19.10) of the Hopf algebra character α, one can directly prove the following result:

Proposition 19.8. *The character α is the solution of the differential equation*

$$\frac{d\alpha}{ds} = x\alpha * \delta_{\text{coloop}} + y\delta_{\text{loop}} * \alpha + [\delta_{\text{coloop}}, \alpha]_* - [\delta_{\text{loop}}, \alpha]_* \,. \tag{19.12}$$

This differential equation obeyed by the character α (and hence by the Tutte polynomial) is of the same type as the renormalization group equation (19.9) used in QFT. This comes from the fact that, when considering a given edge, this edge can be a bridge or not. Thus, the terms in the right-hand side of the differential equation (19.12) can be regrouped in two types of terms with respect to this classification—it is exactly these two types of terms that appear in the QFT renormalization group equation (19.9).

Numerous other graph polynomials from the literature arise canonically from this type of framework. These polynomials include Las Vergnas' Tutte polynomial of the morphism of matroids and his Tutte polynomial for embedded graphs, Bollobás and Riordan's ribbon graph polynomial, the Krushkal polynomial, and the Penrose polynomial (see [718]). Moreover, it forms an algebraic foundation for the approach to topological Tutte polynomials through deletion–contraction relations discussed in Section 27.6.

19.5 Open problems

The technique exposed in the previous section allows us to use appropriate Hopf algebra characters to obtain combinatorial invariants. An interesting perspective for future work appears to be the investigation of relations between the approach presented here and different other approaches used to obtain combinatorial invariants from Hopf algebraic manipulations, such as the approaches given in [4, 5, 89].

20

The Potts and random-cluster models

Geoffrey Grimmett

Synopsis

The relationship between the Tutte polynomial and the random-cluster, Ising, and Potts models of statistical physics is summarized. Certain fundamental properties of these models are described, particularly those that may be expressed neatly in terms of Tutte polynomials.

- Ising, Potts, and random-cluster models; physical origins; couplings; partition functions.
- Basic properties of random-cluster measures, stochastic ordering, comparison inequalities, positive association.
- Limit as $q \downarrow 0$, uniform spanning tree, uniform spanning forest, uniform connected subgraph, negative association.
- Flow polynomial, Potts two-point correlation, Simon inequality.
- Zero-temperature limit and the chromatic polynomial.
- Asymptotics of the Tutte polynomial on the complete graph.

20.1 Introduction

The four principal elements in this chapter are the Ising model, the Tutte polynomial, the Potts model, and the random-cluster model. The *Ising model* [635] is the fundamental model for the ferromagnet, and it has generated enormous interest and activity in mathematics and physics. The *Potts model* [928] extends the number of local states of the Ising model from 2 to a general

DOI: 10.1201/9780429161612-20

number q. The *random-cluster model* of Fortuin and Kasteleyn [498] provides an overarching framework for the Ising and Potts models that incorporates percolation and electrical networks, together with certain other processes. The common aspect of importance for these three systems is the singularity that occurs at points of phase transition.

It turns out that the Tutte polynomial is equal (subject to a change of variables) to the partition function of the random-cluster model, and therefore to that of the Potts model. This connection is not a coincidence since both the Tutte and random-cluster functions arose in independent analyses of local graph operations. The Tutte polynomial originated in Tutte's exploration of deletion and contraction on a finite graph. The random-cluster model originated similarly in Kasteleyn's observation that the Ising, Potts, and percolation models, and also electrical networks, have a property of invariance under series and parallel operations on edges.

Combinatorial theory and statistical mechanics are areas of science which have much in common, while retaining their distinctive characteristics. Statistical mechanics is mostly concerned with the structure of phases and of singularities, and has developed appropriate methodology and language. Although, in principle, the properties of a physical model are encoded entirely within its partition function, the extraction of such properties is often challenging and hinges frequently on other factors such as the nature of the underlying graph.

The connection between the Tutte polynomial and statistical mechanics is summarized in this chapter, as follows. We aim: (i) to give a clear formulation of the relevant models, (ii) to explain the connection between their partition functions and the Tutte polynomial, (iii) to present some of the basic properties of the random-cluster model that are contingent on the partition function, and (iv) to present a selection of open problems concerning Potts and random-cluster models that may be related to the Tutte polynomial. Further references containing material relevant to this chapter include [83, 152, 476, 585, 1023, 1139, 1143, 1145, 1173].

20.2 Probabilistic models from physics

20.2.1 The Ising and Potts ferromagnets

The Ising model for ferromagnetism was analyzed in one dimension in Ising's thesis and 1925 paper [635]. It modeled the following physical experiment. A piece of iron is placed in a magnetic field, with an intensity that is increased from zero to a maximum, and then reduced to zero. The iron retains some residual magnetization if and only if the temperature is sufficiently low, and the critical temperature for this phenomenon is called the *Curie point.*

The Ising model may be summarized as follows. Suppose that particles are placed at the vertices of a graph embedded in a Euclidean space. Each particle may be in either of two states: spin "up" or spin "down". Spin-values are chosen at random according to a certain probability measure governed by interactions between neighboring particles. This measure is described as follows.

Let $G = (V, E)$ be a finite, simple graph. Each vertex $v \in V$ is occupied by a particle with a random spin. Since spins are assumed to come in two basic types, we take as sample space the set $\Sigma = \{-1, +1\}^V$ of vectors $\sigma = (\sigma_x : x \in V)$ with entries ± 1, called *configurations* (or, sometimes, *states*) of the system.

Definition 20.1. Let $\beta \in (0, \infty)$ and $h \in \mathbb{R}$. The Ising *Hamiltonian* H_{I} and *partition function* Z_{I} are given by

$$H_{\mathrm{I}}(\sigma) = -\sum_{e=(x,y)\in E} \sigma_x \sigma_y - h \sum_{x\in V} \sigma_x, \tag{20.1}$$

and

$$Z_{\mathrm{I}}(G; \beta, h) = \sum_{\sigma\in\Sigma} e^{-\beta H_{\mathrm{I}}(\sigma)}.$$

The *Ising probability measure* $\lambda_{\beta,h}$ on G, which gives the probability of a configuration $\sigma \in \Sigma$, is defined by

$$\lambda_{\beta,h}(\sigma) = \frac{1}{Z_{\mathrm{I}}} e^{-\beta H_{\mathrm{I}}(\sigma)}.$$

The parameter β represents the reciprocal $1/T$ of temperature, and h is the *external field.* The second summation of Equation (20.1) may be subsumed into the first by adding a new "ghost" vertex to the graph, and connecting it to each member of V. Such an augmentation is a classical device in the study of the Ising model. For reasons of simplicity, we shall assume generally here that $h = 0$, and we write λ_β for $\lambda_{\beta,0}$. It is usual to include also an edge-interaction J, which we have chosen to absorb into the parameter β. The above model is called *ferromagnetic* (in that $\beta > 0$) in contrast to the antiferromagnetic model of Section 20.2.5.

The Ising model has two admissible spin-values, and a very rich theory. In his 1952 paper [928], Potts developed an extension of the Ising model to a general number of spin-values.

Let q be an integer satisfying $q \geq 2$, and consider the sample space $\Sigma = \{1, 2, \dots, q\}^V$. Each vertex of G may now be in any of q states.

Definition 20.2. Let $\beta \in (0, \infty)$ and $q \in \{2, 3, \dots\}$. The *Potts Hamiltonian* H_{P} and *partition function* Z_{P} are given by

$$H_{\mathrm{P}}(\sigma) = -\sum_{e=(x,y)} \delta_{\sigma_x,\sigma_y},$$

and

$$Z_P(G;\beta,q) = \sum_{\sigma\in\Sigma} e^{-\beta H_P(\sigma)},$$

where $\delta_{u,v}$ is the Kronecker delta. The *Potts probability measure* is defined by

$$\pi_{\beta,q}(\sigma) = \frac{1}{Z_P} e^{-\beta H_P(\sigma)}.$$

When $q = 2$, we have that

$$\delta_{\sigma_x,\sigma_y} = \tfrac{1}{2}(1 + \sigma_x\sigma_y),$$

from which it follows that the $q = 2$ Potts model is simply the Ising model with β replaced by $\frac{1}{2}\beta$.

More general settings may include a non-zero external field h and a vector $\mathbf{J}$ of edge-parameters (see Section 20.2.5).

20.2.2 The random-cluster model

The random-cluster model was formulated in a series of papers [496, 497, 498] by Fortuin and Kasteleyn. It is described next, and its relationship to the Potts model is explained in Section 20.2.3.

Let $G = (V, E)$ be a finite, simple graph. The relevant state space is the set $\Omega = \{0,1\}^E$ of vectors $\omega = (\omega(e) : e \in E)$ with entries 0 or 1. An edge e as said to be *open* in $\omega \in \Omega$ if $\omega(e) = 1$, and *closed* if $\omega(e) = 0$. For $\omega \in \Omega$, let $\eta(\omega) = \{e \in E : \omega(e) = 1\}$ denote the set of open edges, and let $k(\omega)$ be the number of connected components (or "open clusters") of the graph $(V, \eta(\omega))$; the count $k(\omega)$ includes the number of isolated vertices.

Definition 20.3. Let $p \in (0,1)$ and $q \in (0,\infty)$. The *random-cluster measure* $\phi_{p,q}$ on G is given by

$$\phi_{p,q}(\omega) = \frac{1}{Z_{RC}}\left(\prod_{e\in E} p^{\omega(e)}(1-p)^{1-\omega(e)}\right) q^{k(\omega)}, \qquad \text{for } \omega \in \Omega,$$

and the *partition function* Z_{RC} is given by

$$Z_{RC}(G;p,q) = \sum_{\omega\in\Omega}\left(\prod_{e\in E} p^{\omega(e)}(1-p)^{1-\omega(e)}\right) q^{k(\omega)}. \tag{20.2}$$

The most important values of q are arguably the positive integers. When $q = 1$, we have that $\phi_p := \phi_{p,1}$ is a product measure, and the words "percolation" and "random graph" are often used in this context, see [582, 655]. The random-cluster model with $q \in \{2, 3, \dots\}$ corresponds, as sketched in the next section, to the Potts model with q local states. See [583] for the general theory of the random-cluster model.

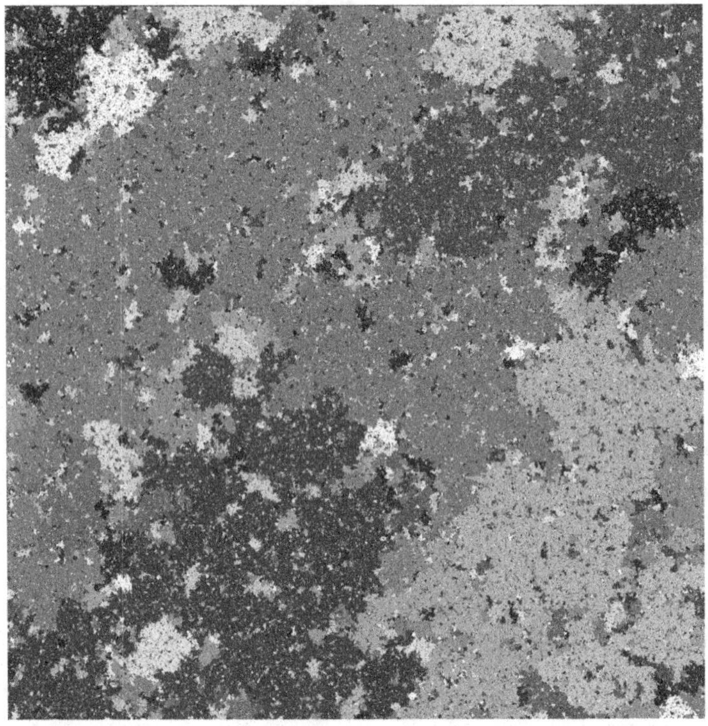

FIGURE 20.1: A sample from the random-cluster model on a 2048×2048 box of the square lattice, with $p = 0.585816$ and $q = 2$. It was obtained by simulating the Ising model and applying the coupling of Section 20.2.3. (Figure by courtesy of Raphaël Cerf.)

20.2.3 Coupling of the Potts and random-cluster measures

Let $q \in \{2, 3, \dots\}$, let $p \in (0, 1)$, and let $G = (V, E)$ be a finite, simple graph. We consider the product sample space $\Sigma \times \Omega$ where $\Sigma = \{1, 2, \dots, q\}^V$ and $\Omega = \{0, 1\}^E$ as above. Let μ be the probability measure on $\Sigma \times \Omega$ given by

$$\mu(\sigma, \omega) \propto \psi(\sigma)\phi_p(\omega)1_F(\sigma, \omega), \qquad \text{for } (\sigma, \omega) \in \Sigma \times \Omega,$$

where ψ is uniform measure on Σ, where $\phi_p := \phi_{p,1}$ is product measure on Ω with density p, and where 1_F is the indicator function of the event that σ is constant on each open cluster of ω, that is,

$$F = \big\{(\sigma, \omega) \in \Sigma \times \Omega : \sigma_x = \sigma_y \text{ for every } e = (x, y) \text{ satisfying } \omega(e) = 1\big\}.$$

The measure μ may be viewed as the product measure $\psi \times \phi_p$ conditioned on the event F. The *marginal* of a measure on a product space is its projection onto a component.

Theorem 20.4. *Let $q \in \{2, 3, \dots\}$ and $p \in (0, 1)$.*

1. *The first marginal of μ (on Σ) is the Potts measure $\pi_{\beta,q}$ where $p = 1 - e^{-\beta}$.*

2. *The second marginal of μ (on Ω) is the random-cluster measure $\phi_{p,q}$.*

3. *The following properties of conditional measures hold.*

   (a) *Given $\omega \in \Omega$, the conditional measure on Σ is obtained by putting (uniformly) random spins on entire clusters of ω. These spins are constant on given clusters, and are independent between clusters.*

   (b) *Given $\sigma \in \Sigma$, the conditional measure on Ω is obtained as follows. For $e = (x, y) \in E$, we set $\omega(e) = 0$ if $\sigma_x \neq \sigma_y$, and otherwise $\omega(e) = 1$ with probability p (independently of other edges).*

4. *We have that*

$$Z_{\mathrm{RC}}(G; p, q) = e^{-\beta|E|} Z_{\mathrm{P}}(G; \beta, q). \tag{20.3}$$

This coupling may be used to show that correlations in Potts models correspond to connection probabilities in random-cluster models (see, for example, [583, Thm. 1.16]). In this way, the methods of stochastic geometry may be harnessed to understand the correlation structure of the Potts system. The basic theorem of this type is Theorem 20.5 below.

The "two-point correlation function" of the Potts measure $\pi_{\beta,q}$ on the finite graph $G = (V, E)$ is the function

$$\tau_{\beta,q}(x, y) := \pi_{\beta,q}(\sigma_x = \sigma_y) - \frac{1}{q}, \qquad \text{for } x, y \in V. \tag{20.4}$$

Let $\{x \leftrightarrow y\}$ be the event of Ω on which there exists an open path joining vertex x to vertex y. The "two-point connectivity function" of the random-cluster measure $\phi_{p,q}$ is the function $\phi_{p,q}(x \leftrightarrow y)$ for $x, y \in V$, that is, the probability that x and y are joined by a path of open edges. It turns out that these two-point functions are the same up to a constant factor.

Theorem 20.5 (Correlation/Connection Theorem). *If $q \in \{2, 3, \ldots\}$ and $p = 1 - e^{-\beta} \in (0, 1)$, then*

$$\tau_{\beta,q}(x, y) = (1 - q^{-1})\phi_{p,q}(x \leftrightarrow y), \qquad \textit{for } x, y \in V. \tag{20.5}$$

The Potts models considered above have zero external field. Some complications arise when an external field is added; see Section 20.2.5.

20.2.4 Partition functions and the Tutte polynomial

The Potts and random-cluster partition functions may be viewed as evaluations of the dichromatic and Tutte polynomials, as follows.

Theorem 20.6. *Let $G = (V, E)$ be a graph, let $p \in (0, 1)$ and $q \in (0, \infty)$, and let*

$$u - 1 = \frac{q(1-p)}{p}, \qquad v - 1 = \frac{p}{1-p}.$$

1. *The partition function* $Z_{\mathrm{RC}}(G)$ *of the random-cluster measure on* G *with parameters* p, q *satisfies*

$$\begin{aligned} Z_{\mathrm{RC}}(G) &= (1-p)^{|E|}\, Z\big(G; \tfrac{p}{1-p}, q\big) \\ &= \tfrac{(u-1)^{k(G)}(v-1)^{|V|}}{v^{|E|}}\, T(G;u,v). \end{aligned}$$

2. *If* $q \in \{2, 3, \dots\}$ *and* $p = 1 - e^{-\beta}$, *the partition function of the* q*-state Potts model on* G *satisfies*

$$\begin{aligned} Z_{\mathrm{P}}(G;\beta,q) &= Z\big(G; \tfrac{p}{1-p}, q\big) \\ &= (u-1)^{k(G)}(v-1)^{|V|}\, T(G;u,v). \end{aligned}$$

20.2.5 Potts extensions

There are three senses in which the Potts model of Definition 20.2 may be said to be in its simplest form: (i) each edge plays an equal (deterministic) role, (ii) the external field satisfies $h = 0$, and (iii) the model is *ferromagnetic*. More generally, we have the partition function $Z_{\mathrm{P}}(G;\beta,\mathbf{J},h) = \sum_{\sigma\in\Sigma} e^{-\beta H_{\mathrm{P}}(\sigma)}$ where the Hamiltonian is given by

$$H_{\mathrm{P}}(\sigma) = -\sum_{e=(x,y)} J_e \delta_{\sigma_x,\sigma_y} - \sum_{j=1}^{q}\sum_{x\in V} h_j \delta_{\sigma_x, j}. \tag{20.6}$$

Here, $\mathbf{J} = (J_e : e \in E)$ is a family of edge-parameters assumed to satisfy $J_e \neq 0$, and $\mathbf{h} = (h_j : j = 1, 2, \dots, q)$ is a vector of external fields. The model is termed *ferromagnetic* if $J_e > 0$ for $e \in E$, and *purely antiferromagnetic* if $J_e < 0$ for all $e \in E$. In the ferromagnetic case, the measure has a property of positive association (as in Section 20.4.2) which is absent in the non-ferromagnetic case. The general Potts partition function of (20.6) poses some new difficulties.

Assume first that $\mathbf{h} = \mathbf{0}$. The associated random-cluster formula yields a function $\phi_{\mathbf{p},q}$ where $p_e = 1 - e^{-\beta J_e}$. If $J_e < 0$ for some e, this does not define a probability measure. In addition, the Potts model does not satisfy the range of correlation inequalities that hold in the ferromagnetic case. On the other hand, Item 2 of Theorem 20.6 is easily extended for general $\mathbf{J}$ to a multivariate Tutte polynomial on $G = (V, E)$ which may be written in the form

$$Z(G;q,\mathbf{v}) = \sum_{A\subseteq E} q^{k(A)} \prod_{e\in A} v_e, \tag{20.7}$$

where q and $\mathbf{v} = (v_e : e \in E)$ are viewed as parameters. See [455, 1023] for recent accounts.

When $\mathbf{h} \neq \mathbf{0}$, a form of the Tutte–Potts correspondence may be found in [126], where positive association and infinite-volume limits are explored, and also in a slightly more general setting in [455] (see also [885]). It turns out that the Potts partition function Z_{P} arising from (20.6) equals an evaluation of the $\mathbf{V}$-polynomial, namely $\mathbf{V}(G, \mathbf{h}; s, \mathbf{p})$ where $s = \sum_j e^{\beta h_j}$, and $\mathbf{p} = (e^{\beta J_e} - 1 : e \in E)$ (see Theorem 26.33).

The J_e may themselves be random, in which case the model is termed *quenched*, in contrast to the *annealed* case which initially averages over the J_e. If the probability distribution of the J_e allocates strictly positive probability to both positive and negative values, the system is a *spin glass*. See [878].

20.3 Phase transition

Statistical mechanics focuses on the study of *phase transition.* Suppose for simplicity that a given physical system has a single parameter denoted T and called "temperature". In many cases in nature, there exists a "critical temperature", denoted T_{c}, such that the macroscopic behavior of the system depends on whether $T < T_{\mathrm{c}}$ or $T > T_{\mathrm{c}}$. For example, it was observed by Pouillet [669, 929] and, later, Curie [353] that there exists a threshold temperature T_{c} for the retention of magnetization by an iron body. This discovery motivated Lenz's proposal of the Ising model of Definition 20.1, restricted initially to the case of one dimension and extended subsequently to higher dimensions by Peierls [910] and others.

Within the context of such a mathematical model, a singularity can occur only in an infinite system. The procedure is as follows. The configuration space in question is determined inside a space of size n, say. To each configuration σ is allocated an energy, or "Hamiltonian", $H(\sigma)$, leading to the "weight" $w(\sigma) := e^{-\beta H(\sigma)}$, where $\beta = 1/T$ as before. The "partition function" $Z_n := \sum_\sigma w(\sigma)$ is a smooth function of T and of any other parameters. The "infinite-volume partition function" is given in the so-called "thermodynamic limit" by

$$\log Z := \lim_{n \to \infty} \left(\frac{1}{n} \log Z_n \right).$$

Now, Z is not generally a smooth function, and it is through studying the singularities of Z and its partial derivatives that one obtains a picture of any phase transition (see [976]). An explicit example of the thermodynamic limit and the infinite-volume partition function is exhibited in Theorem 20.19.

20.4 Basic properties of random-cluster measures

This section includes some of the basic properties of a random-cluster measure on the finite graph $G = (V, E)$. Each may be expressed in terms of the Tutte polynomial.

20.4.1 Stochastic ordering

The state space $\Omega = \{0,1\}^E$ is a partially ordered set with partial order given by $\omega_1 \le \omega_2$ if $\omega_1(e) \le \omega_2(e)$ for all $e \in E$. This partial order is extremely useful in the analysis of Potts and random-cluster models, and it induces partial orderings on the spaces of associated functions and measures.

Definition 20.7.

1. A random variable $f : \Omega \to \mathbb{R}$ is called *increasing* if $f(\omega_1) \le f(\omega_2)$ whenever $\omega_1 \le \omega_2$.
2. An event $A \subseteq \Omega$ is called *increasing* if its indicator function 1_A is increasing.
3. Given two probability measures μ_1, μ_2 on Ω, we write $\mu_1 \le_{\text{st}} \mu_2$, and say that μ_1 is stochastically smaller than μ_2, if $\mu_1(f) \le \mu_2(f)$ for all increasing random variables f on Ω.

Arguably the most useful approach to stochastic ordering is due to Holley. We obtain the following comparison inequalities as corollaries of Holley's inequality (see [621] and [583, Thm 2.1]).

Theorem 20.8 (Comparison inequalities). *It is the case that*

$$\phi_{p',q'} \le_{\text{st}} \phi_{p,q} \qquad \text{if} \quad q' \ge q, \ q' \ge 1, \ \text{and } p' \le p; \ \text{and that}$$

$$\phi_{p',q'} \ge_{\text{st}} \phi_{p,q} \qquad \text{if} \quad q' \ge q, \ q' \ge 1, \ \text{and } \frac{p'}{q'(1-p')} \ge \frac{p}{q(1-p)}.$$

20.4.2 Positive association

Holley's inequality admits a neat proof of the FKG inequality of [499]. This amounts to the following in the case of random-cluster measures.

Theorem 20.9 (Positive association). *Let $p \in (0,1)$ and $q \in [1,\infty)$. If f and g are increasing functions on Ω, then*

$$\phi_{p,q}(fg) \ge \phi_{p,q}(f)\phi_{p,q}(g).$$

FIGURE 20.2: Two edges in series and in parallel.

Specializing to indicator functions $f = 1_A$, $g = 1_B$, we obtain that

$$\phi_{p,q}(A \cap B) \geq \phi_{p,q}(A)\phi_{p,q}(B)$$

for increasing events A, B whenever $q \geq 1$. Positive association is generally false when $0 < q < 1$.

20.5 The Limit as $q \downarrow 0$

20.5.1 UST, USF, and UCS

Some interesting limits with combinatorial flavors arise from consideration of $\phi_{p,q}$ as $q \downarrow 0$. Write Ω_{for}, Ω_{st}, Ω_{cs} for the subsets of Ω containing all forests, spanning trees, and connected subgraphs, respectively, and write USF, UST, and UCS for the uniform probability measures on the respective sets Ω_{for}, Ω_{st}, Ω_{cs}. An account of the following limits and their history may be found at [583, Thm 1.2].

Theorem 20.10. *We have in the limit as $q \downarrow 0$ that:*

$$\phi_{p,q} \implies \begin{cases} \text{UCS} & \text{if } p = \frac{1}{2}, \\ \text{UST} & \text{if } p \to 0 \text{ and } q/p \to 0, \\ \text{USF} & \text{if } p = q. \end{cases}$$

The spanning tree limit UST is especially interesting for historical and mathematical reasons. The random-cluster model originated in a systematic study by Fortuin and Kasteleyn of systems of a certain type which satisfy certain parallel and series laws. Electrical networks are the best known such systems: two resistors of resistances r_1 and r_2 in parallel (respectively, in series) may be replaced by a single resistor with resistance $(r_1^{-1} + r_2^{-1})^{-1}$ (respectively, $r_1 + r_2$); see Figure 20.2. Fortuin and Kasteleyn realized that the electrical-network theory of a graph G is related to the limit as $q \downarrow 0$ of the random-cluster model on G, where p is given by $p = \sqrt{q}/(1 + \sqrt{q})$. It has been known since Kirchhoff's theorem [691] that the electrical currents which flow in a network may be expressed in terms of counts of spanning trees.

The theory of the uniform spanning tree measure UST is beautiful in its own right. In partnership with the so-called "loop erased random walk",

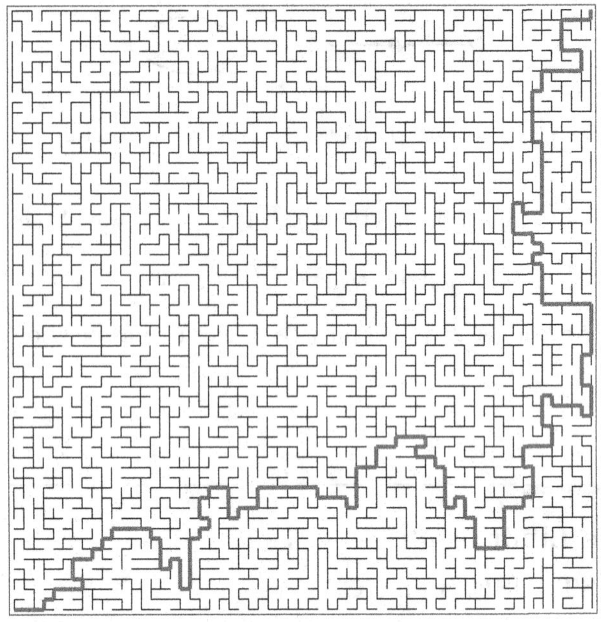

FIGURE 20.3: A uniform spanning tree (UST) on a large box of the square lattice. It contains a unique path between any two vertices, taken here as opposite corners of the box. (Figure by courtesy of Oded Schramm.)

it is linked in an important way to the emerging field of stochastic growth processes of "stochastic Löwner evolution" (SLE) type. See Figure 20.3, and the references in [585, Chap. 2].

20.5.2 Negative association

Let E be a finite set, and let μ be a probability measure on the space $\Omega = \{0,1\}^E$. There are several concepts of negative association, of which we present three here.

For $\omega \in \Omega$ and $F \subseteq E$, the event $\Omega_{F,\omega}$ generated by ω on F is given by

$$\Omega_{F,\omega} = \{\omega' \in \Omega : \omega'(e) = \omega(e) \text{ for } e \in F\}.$$

For $E' \subseteq E$ and an event $A \subseteq \Omega$, we say that A *is defined on* E' if, for all $\omega \in \Omega$, we have that $\omega \in A$ if and only if $\Omega_{E',\omega} \subseteq A$. Let A and B be events in Ω. We define $A \square B$ to be the set of all vectors $\omega \in \Omega$ for which there exists $F \subseteq E$ such that $\Omega_{F,\omega} \subseteq A$ and $\Omega_{F^c,\omega} \subseteq B$, where $F^c = E \setminus F$. Note that the choice of F is allowed to depend on the vector ω. The operator $\square$ originated in the work of van den Berg and Kesten [94] on the well known BK inequality.

Definition 20.11. 1. The measure μ is *edge negatively associated* if

$$\mu(J_e \cap J_f) \le \mu(J_e)\mu(J_f), \qquad \text{for all } e, f \in E,\ e \ne f,$$

where J_e is the event that e is open.

2. We call μ *negatively associated* if

$$\mu(A \cap B) \le \mu(A)\mu(B)$$

for all pairs (A, B) of increasing events with the property that there exists $E' \subseteq E$ such that A is defined on E' and B is defined on its complement $E \setminus E'$.

3. We say that μ has the *disjoint occurrence property* if

$$\mu(A \square B) \le \mu(A)\mu(B), \qquad \text{for } A, B \subseteq \Omega.$$

Proposition 20.12. *We have that*

$$\begin{aligned} &\mu \textit{ has the disjoint occurrence property} \\ &\qquad \Longrightarrow \mu \textit{ is negatively associated} \\ &\qquad \Longrightarrow \mu \textit{ is edge negatively associated.} \end{aligned}$$

The proof of the proposition follows from the definitions. Neither of the two implications of the proposition can be reversed: see [818] for the first, and the second is more elementary.

It was proved by Reimer [958] that the product measures $\phi_{p,1}$ have the disjoint occurrence property. The random-cluster measure $\phi_{p,q}$ cannot (generally) be edge negatively associated when $q > 1$. It may be conjectured that $\phi_{p,q}$ satisfies some form of negative association when $q < 1$. Such a property would be very useful in studying random-cluster measures when $q < 1$.

In the absence of a satisfactory approach to the general case of random-cluster measures with $q < 1$, we turn next to the issue of negative association of $\phi_{p,q}$ in the limit as $q \downarrow 0$.

Conjecture 20.13. For any finite graph $G = (V, E)$, the uniform spanning forest measure USF and the uniform connected subgraph measure UCS are edge negatively associated.

A stronger version of this conjecture is that USF and UCS are negatively associated in one or both of the further senses described above. Numerical evidence for the conjecture is found in [586]. (The problem is simpler in the symmetric context of USF on the complete graph, see [1043].)

The UST measure is, in contrast, much better understood, owing to the theory of electrical networks and, more particularly, Kirchhoff's matrix–tree theorem, [691], and its ramifications. The following was proved by Feder and Mihail, [484].

Theorem 20.14. *The uniform spanning tree measure* UST *is negatively associated.*

In addition, UST has the stronger property of being "strong Rayleigh", see [174]. The material in this section may be found in expanded form in [583, 585].

20.6 Flow polynomial

20.6.1 Potts correlations and flow counts

The Potts correlation functions (20.4) may be expressed in terms of flow polynomials (see Chapter 12) associated with a certain Poissonian random graph derived from $G = (V, E)$ by replacing each edge by a random number of copies. For a vector $\mathbf{m} = (m_e : e \in E)$ of nonnegative integers, let $G_{\mathbf{m}} = (V, E_{\mathbf{m}})$ be the graph with vertex set V and, for each $e \in E$, with exactly m_e edges in parallel joining the end-vertices of the edge e; the original edge e is removed.

Let $\beta \geq 0$, and let $\mathbf{P} = (P_e : e \in E)$ be a family of independent random variables such that P_e has the Poisson distribution with parameter β. The random graph $G_{\mathbf{P}} = (V, E_{\mathbf{P}})$ is called a *Poisson graph with intensity* β. Let $\Pr_\beta$ and $\mathbb{E}_\beta$ denote the corresponding probability measure and expectation.

For $x, y \in V$, let $G_{\mathbf{P}}^{x,y}$ denote the graph obtained from $G_{\mathbf{P}}$ by adding an edge with ends x, y. If x and y are adjacent in the original graph $G_{\mathbf{P}}$, we add a further edge between them. Potts correlations are related to flow counts as follows.

Theorem 20.15. *Let $q \in \{2, 3, \dots\}$ and $\beta \geq 0$. Then*

$$q\tau_{\beta,q}(x, y) = \frac{\mathbb{E}_\beta(F(G_{\mathbf{P}}^{x,y}; q))}{\mathbb{E}_\beta(F(G_{\mathbf{P}}; q))}, \qquad \text{for } x, y \in V. \tag{20.8}$$

This formula is particularly striking when $q = 2$, since non-zero $\mathbb{Z}_2$-flows take only the value 1. A finite graph $H = (W, F)$ is said to be *even* if every vertex has even degree. Evidently $F(H; 2) = 1$ if H is even, and $F(H; 2) = 0$ otherwise. By (20.8), for any graph G,

$$q\tau_{\beta,q}(x, y) = \frac{\Pr_\beta(G_{\mathbf{P}}^{x,y} \text{ is even})}{\Pr_\beta(G_{\mathbf{P}} \text{ is even})}, \tag{20.9}$$

when $q = 2$. This observation is central to the so called "random-current expansion" of the Ising model, which has proved very powerful in the study of both classical and quantum Ising models. See [14, 15, 16, 17, 128].

Theorem 20.15 may be extended via (20.5) to the random-cluster model. Theorem 20.16 follows by expressing the flow polynomial in terms of the Tutte polynomial T, and allowing q to vary continuously.

Theorem 20.16. *Let $p \in (0, 1)$, $q \in (0, \infty)$, and let β satisfy $p = 1 - e^{-\beta q}$.*

1. *For $x, y \in V$,*

$$(q-1)\phi_{G,p,q}(x \leftrightarrow y) = \frac{\mathbb{E}_\beta((-1)^{1+|E_{\mathbf{P}}|}T(G_{\mathbf{P}}^{x,y}; 0, 1-q))}{\mathbb{E}_\beta((-1)^{|E_{\mathbf{P}}|}T(G_{\mathbf{P}}; 0, 1-q))}. \tag{20.10}$$

2. *For* $q \in \{2, 3, \dots\}$,

$$Z_{\mathrm{RC}}(G; p, q) = \phi_{G,p}(q^{k(\omega)}) = (1-p)^{|E|(q-2)/q} q^{|V|} \mathbb{E}_\beta(F(G_{\mathbf{P}}; q)). \quad (20.11)$$

Further details may be found in [584].

20.6.2 The random-current expansion when $q = 2$

Unlike the case for the Potts model, there is a fairly complete analysis of the Ising model. A principal part in this analysis is played by Theorem 20.15 with $q = 2$ under the heading "random-current expansion". This has permitted proofs amongst other things of the exponential decay of correlations in the low-β regime on the cubic lattice $\mathbb{L}^d$ with $d \geq 2$. It has not so far been possible to extend this work to general Potts models, but Theorem 20.15 could play a part in such an extension.

Let $G = (V, E)$ be a finite graph and set $q = 2$. By Theorem 20.15,

$$2\tau_{\beta,2}(x, y) = \frac{\Pr_\beta(G_{\mathbf{P}}^{x,y} \text{ is even})}{\Pr_\beta(G_{\mathbf{P}} \text{ is even})}, \qquad \text{for } 0 \leq \beta < \infty. \quad (20.12)$$

There is an important correlation inequality known as Simon's inequality, [1009]. Let $x, z \in V$ be distinct vertices. A set W of vertices is said to *separate* x and z if $x, z \notin W$ and every path from x to z contains some vertex of W.

Theorem 20.17. *Let* $x, z \in V$ *be distinct vertices, and let* W *separate* x *and* z*. Then* $\kappa_{\beta,2}(x, y) := 2\tau_{\beta,2}(x, y)$ *satisfies*

$$\kappa_{\beta,2}(x, z) \leq \sum_{y \in W} \kappa_{\beta,2}(x, y)\kappa_{\beta,2}(y, z).$$

The Ising model corresponds to a random-cluster measure $\phi_{p,q}$ with $q = 2$. By (20.5),

$$\kappa_{\beta,q}(x, y) = \phi_{p,q}(x \leftrightarrow y),$$

where $p = 1 - e^{-\beta q}$ and $q = 2$. The Simon inequality may be written in the form

$$\phi_{p,q}(x \leftrightarrow z) \leq \sum_{y \in W} \phi_{p,q}(x \leftrightarrow y)\phi_{p,q}(y \leftrightarrow z) \quad (20.13)$$

whenever W separates x and z. It is well-known that this inequality is valid also when $q = 1$, see [582, Chap. 6]. It is not known, but is reasonable to ask, if it also holds for any $q \in [1, 2]$.

20.7 The limit of zero temperature

The physical interpretation of the constant β is as $\beta = 1/(kT)$ where k is Boltzmann's constant and T denotes (absolute) temperature. The limit $T \downarrow 0$

corresponds to the limit $\beta \to \infty$. The ferromagnetic Potts measure $\pi_{\beta,q}$ on a finite graph $G = (V, E)$ converges weakly to the probability measure that allocates a uniform random spin to each connected component of G, this being constant on each component and independent between components. Such a limit measure is called a "ground state" of the system.

The situation is more interesting in the presence of a vector $\mathbf{J}$ of edge-parameters, some of which are negative. The ground states in this case are colorings κ of V with the color palette $\{1, 2, \ldots, q\}$ and the property that, for each edge $e = (x, y)$,

$$\kappa(x) \begin{cases} = \kappa(y) & \text{if } J_e > 0, \\ \neq \kappa(y) & \text{if } J_e < 0. \end{cases} \tag{20.14}$$

In the purely antiferromagnetic case, such a coloring κ is has the property that any two neighbors have different colors.

Theorem 20.18. *For the purely antiferromagnetic Potts model,*

$$Z_{\mathrm{P}}(G; \beta, q) \to \chi(G; q) \qquad \text{as } \beta \to \infty,$$

where χ denotes the chromatic polynomial.

This is easily seen to hold, since, as $\beta J_e \to -\infty$, there is limiting weight of zero on any pair of *equal* adjacent spins. This observation may be extended naturally in the presence of negative external fields (as in (20.6)) to counts of *list colorings* of G in which the available colors at any given $v \in V$ is restricted to a given list. See [459].

For given q, there exist graphs G for which (20.14) has no solution, and such graphs are called *frustrated.*

20.8 The random-cluster model on the complete graph

When the underlying graph is the complete graph K_n, the asymptotic behavior of the corresponding random-cluster partition function $Z_{\mathrm{RC}}(n, p, q)$ may be studied using a mixture of combinatorics and probability, within the regime $q \geq 1$, and $p = \lambda/n$. Here is some notation and explanation, in preparation for the main theorem.

Let $q \geq 1$ and $p = \lambda/n$. It turns out that there is a critical value of λ that marks the arrival of a giant cluster in the random-cluster model on K_n, and this value is given in [155] by

$$\lambda_c(q) = \begin{cases} q & \text{if } q \in (0, 2], \\ 2\left(\dfrac{q-1}{q-2}\right)\log(q-1) & \text{if } q \in (2, \infty), \end{cases}$$

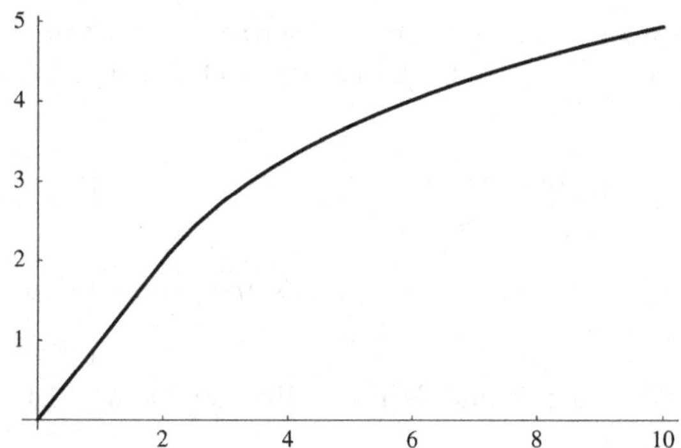

FIGURE 20.4: The critical value $\lambda_c(q)$ plotted against $q > 0$. There is a discontinuity in the second derivative at the value $q = 2$.

and plotted in Figure 20.4. As λ increases through the value λ_c, a giant cluster of size approximately $\theta(\lambda, q)n$ is created, where

$$\theta(\lambda, q) = \begin{cases} 0 & \text{if } \lambda < \lambda_c(q), \\ \theta_{\max} & \text{if } \lambda \geq \lambda_c(q), \end{cases}$$

and $\theta_{\max}$ is the largest root of the equation

$$e^{-\lambda\theta} = \frac{1-\theta}{1+(q-1)\theta}.$$

Theorem 20.19. *Let $q \in (0, \infty)$ and $\lambda \in (0, \infty)$. We have that*

$$\frac{1}{n} \log Z_{\mathrm{RC}}(n, \lambda/n, q) \to \eta(\lambda) \qquad \text{as } n \to \infty,$$

where

$$\eta(\lambda) = \frac{g(\theta(\lambda))}{2q} - \frac{q-1}{2q}\lambda + \log q, \quad \text{and}$$
$$g(\theta) = -(q-1)(2-\theta)\log(1-\theta) - \big[2+(q-1)\theta\big]\log\big[1+(q-1)\theta\big].$$

By Theorem 20.6, this provides an asymptotic evaluation of the Tutte polynomial $T(K_n; u, v)$ within the quadrant $u, v \in (1, \infty)$. See [155] and [583, Chap. 10] for further details.

20.9 Open problems

There is an enormous range of open problems associated with Ising, Potts, and random-cluster models. The following list provides a brief selection of problems, some of which may be allied in part to the Tutte polynomial.

1. Prove or disprove some version of negative association for the uniform forest measure USF or the uniform connected subgraph measure UCS. See Conjecture 20.13.

2. Prove or disprove some version of negative association for the random-cluster measure $\phi_{p,q}$ with $0 < q < 1$.

3. Prove or disprove a version of Simon's inequality for a random-cluster measure $\phi_{p,q}$ with $q \in [1, 2]$, as in (20.13).

4. Establish a version of Simon's inequality, Theorem 20.17, for the Potts model with $q \geq 3$.

5. More generally, find an application of $\mathbb{Z}_q$-flows to the q-state Potts model with $q \geq 3$, as in Theorem 20.15. Develop such an application for real $q > 0$, as in Theorem 20.16.

6. There is a universe of problems associated with classical and quantum phase transitions, of which we mention the determination of the natures of the phase transitions of the Potts model with $q \geq 2$ on lattices in dimensions $d \geq 2$.

7. A problem of great current interest to probabilists and mathematical physicists is to understand the geometry of interfaces in the 3-state Potts model on the square lattice $\mathbb{Z}^2$, and more generally the random-cluster model on $\mathbb{Z}^2$ with $1 < q < 4$. See [1015], or [86] for a recent reference.

21

Where Tutte and Holant meet: a view from counting complexity

Jin-Yi Cai • Tyson Williams

Synopsis

The evaluation of the Tutte polynomial of a graph at certain points can be expressed in a sum-of-products form over edge-assignments of the graph. Thus, the problem of evaluating at these points is based on local rather than global structure. We describe the Holant framework, which gives a general and powerful way to express problems of this form.

- Holant graph invariants.
- Partition functions.
- Holant expressions that are specializations of the Tutte polynomial.

21.1 Introduction

The term Holant was first introduced by L. Valiant [1112] to denote a sum expressible by graph perfect matchings, called matchgates. It was formally defined in [260] in order to study the computational complexity of counting problems. The Holant is a graph invariant that maps a graph to a sum of products. There is one term in the sum for each edge assignment to some finite domain set. Furthermore, each term is a product of factors, with one factor for each vertex of the graph. The Holant sum is essentially the partition function of the edge coloring model (see Section 22.3), where the phrase "edge coloring" refers to all possible edge assignments without restricting to proper edge colorings.

DOI: 10.1201/9780429161612-21

There are some specializations of the Tutte polynomial that coincide with some Holant sums. We provide a brief survey of these known specializations with references to the literature for further reading. The known specializations occur in three settings. First is the positive integer hyperbolas of the Tutte polynomial, i.e. the set of points (x, y) such that $(x-1)(y-1)$ is a positive integer. Second is integral points in the first quadrant, i.e. the set of points $(x, y) \in \mathbb{Z}^2$ such that $x, y \geq 1$, provided the graph is planar. Third is the point $(0, -2)$ provided the graph is 4-regular.

21.2 Definition of Holant and related concepts

The number of matchings in a graph is an important quantity of a graph; it is a graph invariant in the sense that it is invariant under a graph isomorphism, and it can be defined in a local way: at each vertex, at most one incident edge can be in a matching. A set of edges is called a *matching* if and only if it satisfies this condition at every vertex. Based on this, we can express the number of matchings as follows. First, we sum over all subsets of edges. To determine if a given subset of edges forms a matching, we take a product over all vertices in which a vertex contributes 1 if it is incident to at most one edge in the subset and 0 otherwise.

Therefore, counting matchings corresponds to each vertex applying the AT-MOST-ONE function to its set of incident edges, where every edge is treated as a 0-1 valued variable. If instead we use the EXACTLY-ONE function at each vertex, then the sum counts perfect matchings, another graph invariant. A further example is that the sum-of-products using the EXACTLY-TWO function corresponds to counting cycle covers.

Now suppose that instead of summing over subsets of edges, which corresponds to assigning one of two possible states to each edge, we sum over all possible ways to assign a value from the set $\{1, \ldots, q\}$ to each edge, for some positive integer q. This allows us to express more graph invariants. For example, if the function at each vertex is now the ALL-DISTINCT function, then proper edge colorings using at most q colors are counted.

In each of these four examples, we use essentially the same function at every vertex, except the function has an arity equal to the degree of the vertex. In general, we allow different functions to be attached at different vertices.

Fix an integer $q \geq 1$ and a commutative semiring $\mathfrak{R}$ (such as the complex numbers $\mathbb{C}$). Let $\mathcal{F}$ be a set of functions, called *signatures* or *constraint functions*, where each $f \in \mathcal{F}$ has some arity $\operatorname{ar}(f)$ and maps from $\{1, \ldots, q\}^{\operatorname{ar}(f)}$ (ordered tuples of size $\operatorname{ar}(f)$) to $\mathfrak{R}$. A *signature grid* $\Omega = (G, \mathcal{F}, \pi)$ is a tuple, where $G = (V, E)$ is an undirected graph, and π maps each $v \in V(G)$ to a function $f_v \in \mathcal{F}$ of arity $\deg(v)$, with input variables associated with the incident edges at v. Each edge is viewed as a variable taking values from the finite domain set $\{1, \ldots, q\}$.

Definition 21.1. For a signature grid $\Omega = (G, \mathcal{F}, \pi)$, the *Holant* of Ω, or more simply of the graph G, is

$$\text{Holant}_q(G; \mathcal{F}) = \sum_{\sigma: E \to \{1,\ldots,q\}} \prod_{v \in V} f_v(\sigma|_{E(v)}), \tag{21.1}$$

where $E(v)$ is the set of edges incident to v and $\sigma|_{E(v)}$ is the restriction of σ to $E(v)$.

We can treat $\text{Holant}_q(G; \mathcal{F})$ as a multivariable graph invariant if we treat each entry of the function value f_v as an indeterminate. Also, see Section 22.3 for the definition of the partition function of the edge coloring model, which is similar.

If G is an r-regular graph, then the only functions in $\mathcal{F}$ that matter are those of arity r. We also consider a generalization of Definition 21.1 to bipartite graphs, and define $\text{Holant}_q(H; \mathcal{F} \mid \mathcal{G})$, where every vertex of the left side of H is assigned a function from $\mathcal{F}$ and every vertex of the right side of H is assigned a function from $\mathcal{G}$. The *edge-vertex incidence graph* of a graph $G = (V, E)$ is a bipartite graph on the disjoint vertex sets E (on the left side) union V (on the right side) such that there is an edge between $e \in E$ and $v \in V$ if and only if e is incident to v in G. When $\mathcal{F} = \{f\}$ is a singleton set, we simply write f in place of $\mathcal{F}$. For example, given a bipartite graph $H = (U, V, E)$ in which all vertices in U have degree two, we write the expression corresponding to (21.1) as $\text{Holant}_q(G; b \mid \mathcal{F})$ when every vertex in U is assigned the binary function b. If b is the binary equality function, i.e.

$$b(x, y) = \begin{cases} 1 & x = y, \\ 0 & \text{otherwise}; \end{cases}$$

then $\text{Holant}_q(G; \mathcal{F})$ is equivalent to $\text{Holant}_q(G'; b \mid \mathcal{F})$ when G' is the edge-vertex incidence graph of G.

Consider a bipartite signature grid $\Omega = (H, \mathcal{F} \mid \mathcal{G}, \pi)$ in which every signature has codomain $\mathbb{C}$. Let M be an invertible $q \times q$ matrix over $\mathbb{C}$, that is, $M \in \text{GL}_q(\mathbb{C})$. For a signature $f \in \mathcal{F}$ of arity n, we can write it as a row vector in $\mathbb{C}^{q^n}$ by listing its values in lexicographical order as in a truth table. We write $fM := fM^{\otimes n}$ as the transformed signature. We define $\mathcal{F}M := \{fM : f \in \mathcal{F}\}$. For a signature $g \in \mathcal{G}$ of arity m written as a column vector, we similarly define $Mg := M^{\otimes m} g$, and $M\mathcal{G}$.

The holographic transformation defined by M is the following operation. Given a bipartite signature grid $\Omega = (H, \mathcal{F} \mid \mathcal{G}, \pi)$, for the same bipartite graph H, we get a new signature grid $\Omega' = (H, \mathcal{F}M \mid M^{-1}\mathcal{G}, \pi')$ by replacing each signature in $\mathcal{F}$ or $\mathcal{G}$ with the corresponding signature in $\mathcal{F}M$ or $M^{-1}\mathcal{G}$. Signatures in $\mathcal{F}$ are called *convariant* and those in $\mathcal{G}$ *contravariant*. Valiant's Holant theorem, Theorem 21.2 (which first appeared in [1112]), states that the Holant value is unchanged under a holographic transformation.

Theorem 21.2. *If $M \in \mathrm{GL}_q(\mathbb{C})$, then*

$$\mathrm{Holant}_q(\Omega; \mathcal{F} \mid \mathcal{G}) = \mathrm{Holant}_q(\Omega'; \mathcal{F}M \mid M^{-1}\mathcal{G}).$$

Example 21.3. For a function f of arity 3 on the Boolean domain ($q = 2$), we can write $f = (f_{0,0,0}, f_{0,0,1}, f_{0,1,0}, f_{0,1,1}, f_{1,0,0}, f_{1,0,1}, f_{1,1,0}, f_{1,1,1})^T$ as a column vector. The Holant of a 3-regular graph can be viewed as a polynomial in these $f_{i,j,k}$ if we treat these values as indeterminates. The equality function of arity 3 (on the Boolean domain) is $f = (1,0,0,0,0,0,0,1)^T$. We can perform a holographic transformation by the Hadamard matrix $H = \left[\begin{smallmatrix} 1 & 1 \\ 1 & -1 \end{smallmatrix}\right]$. Then f is transformed to $H^{\otimes 3} f = (2,0,0,2,0,2,2,0)^T$. This can be seen as

$$H^{\otimes 3} f = \left[\begin{smallmatrix} 1 & 1 \\ 1 & -1 \end{smallmatrix}\right]^{\otimes 3} \{\left[\begin{smallmatrix} 1 \\ 0 \end{smallmatrix}\right]^{\otimes 3} + \left[\begin{smallmatrix} 0 \\ 1 \end{smallmatrix}\right]^{\otimes 3}\} = \left[\begin{smallmatrix} 1 \\ 1 \end{smallmatrix}\right]^{\otimes 3} + \left[\begin{smallmatrix} 1 \\ -1 \end{smallmatrix}\right]^{\otimes 3},$$

completing the example.

A binary function f on domain $\{1, \dots, q\}$ can also be represented by a matrix B_f in $\mathbb{C}^{q \times q}$, where the (i,j)-entry of B_f is $f(i,j)$. A holographic transformation $f \mapsto fM$ can be computed as $M^T B_f M$. For example, the binary equality function on domain $\{1, \dots, q\}$ is represented by the identity matrix I_q. It follows that the binary equality function is unchanged under a holographic transformation by any orthogonal matrix.

21.3 Specializations of the Tutte polynomial with local expressions

21.3.1 Positive integer hyperbolas

Up to a simple prefactor, the specialization of $T(G; x, y)$ to the hyperbola H_q for any positive integer $q = (x-1)(y-1)$ closely corresponds to the q-state Potts model. The precise relation is given in Theorem 2.6. Recall (or see Chapter 20) that the partition function $Z(G; q, v)$ of the q-state Potts model with parameter v on a graph G is a sum, over all configurations of vertices taking one of q values, of v raised to the number of adjacent vertices assigned the same value. Let $\mathcal{EQ} = \{\textsc{Equal}_k : k \geq 1\}$, where $\textsc{Equal}_k$ is the equality function of arity k. For example,

$$\textsc{Equal}_3(x_1, x_2, x_3) = \begin{cases} 1 & x_1 = x_2 = x_3, \\ 0 & \text{otherwise.} \end{cases}$$

With this definition, the Potts model can be expressed in terms of the bipartite Holant notation. This result is straight forward. See Section 2 of [257] for a detailed discussion of the relevant definitions.

Lemma 21.4. *Let G be a graph and H be its edge-vertex incidence graph. Then for any integer $q \geq 1$,*

$$Z(G; q, v) = \text{Holant}_q(H; b \mid \mathcal{EQ}),$$

where b is the binary function

$$b(x, y) = \begin{cases} v & x = y, \\ 1 & \textit{otherwise.} \end{cases}$$

After a change of basis, this quantity can also be expressed as the Holant of G instead of H. There is a family of possible transformations due to the invariance of the edge coloring model under an action of the orthogonal group (a nice explanation of this appears in [954, Section 3.5]). We state one transformation here explicitly after giving two examples: one for even q and one for odd q. We use F_n to denote the $n \times n$ *Fourier matrix*, which has entry $\frac{1}{\sqrt{n}} e^{2\pi i jk/n}$ at row j and column k for $0 \leq j, k \leq n-1$, where $i = \sqrt{-1}$. For example, if $\omega = e^{2\pi i/3}$, then

$$F_3 = \frac{1}{\sqrt{3}} \begin{bmatrix} 1 & 1 & 1 \\ 1 & \omega & \omega^2 \\ 1 & \omega^2 & \omega \end{bmatrix}, \qquad \text{and} \qquad F_4 = \frac{1}{2} \begin{bmatrix} 1 & 1 & 1 & 1 \\ 1 & i & -1 & -i \\ 1 & -1 & 1 & -1 \\ 1 & -i & -1 & i \end{bmatrix}.$$

Example 21.5. Let G be a graph and H its edge-vertex incidence graph. If $\lambda \neq 1$ and $\lambda \neq -2$, then

$$\text{Holant}_3(H; b \mid \mathcal{EQ}) = \text{Holant}_3(G; M\mathcal{EQ}),$$

where

$$M = \begin{bmatrix} \sqrt{\lambda + 2} & 0 & 0 \\ 0 & \sqrt{\lambda - 1} & 0 \\ 0 & 0 & \sqrt{\lambda - 1} \end{bmatrix} \begin{bmatrix} 1 & 0 & 0 \\ 0 & \frac{1}{\sqrt{2}} & \frac{1}{\sqrt{2}} \\ 0 & \frac{i}{\sqrt{2}} & \frac{-i}{\sqrt{2}} \end{bmatrix} F_3.$$

One can verify this by checking that $M^T M$ is the matrix that defines the Potts model with parameter λ.

Example 21.6. Let G be a graph and H its edge-vertex incidence graph. If $\lambda \neq 1$ and $\lambda \neq -3$, then

$$\text{Holant}_4(H; b \mid \mathcal{EQ}) = \text{Holant}_4(G; M\mathcal{EQ}),$$

where

$$M = \begin{bmatrix} \sqrt{\lambda + 3} & 0 & 0 & 0 \\ 0 & \sqrt{\lambda - 1} & 0 & 0 \\ 0 & 0 & \sqrt{\lambda - 1} & 0 \\ 0 & 0 & 0 & \sqrt{\lambda - 1} \end{bmatrix} \begin{bmatrix} 1 & 0 & 0 & 0 \\ 0 & \frac{1}{\sqrt{2}} & 0 & \frac{1}{\sqrt{2}} \\ 0 & 0 & 1 & 0 \\ 0 & \frac{i}{\sqrt{2}} & 0 & \frac{-i}{\sqrt{2}} \end{bmatrix} F_4.$$

Example 21.7 is the general result for positive integers q, and follows from standard techniques.

Example 21.7. Let G be a graph and H its edge-vertex incidence graph. If $\lambda \neq 1$ and $\lambda \neq 1-q$, then for any integer $q \geq 2$,

$$\text{Holant}_q(H; b \mid \mathcal{EQ}) = \text{Holant}_q(G; M\mathcal{EQ}),$$

where $M = DZF_q$ with

$$D_{j,k} = \begin{cases} \sqrt{\lambda+q-1} & j=k=0, \\ \sqrt{\lambda-1} & j=k\neq 0, \\ 0 & \text{otherwise;} \end{cases} \quad \text{and} \quad Z_{j,k} = \begin{cases} 1 & j=k=0, \\ \frac{1}{\sqrt{2}} & 1 \leq j=k<\frac{q}{2}, \\ 1 & j=k=\frac{q}{2} \text{ (for even } q), \\ \frac{-i}{\sqrt{2}} & j=k>\frac{q}{2}, \\ \frac{1}{\sqrt{2}} & j+k=q \text{ and } j<\frac{q}{2}, \\ \frac{i}{\sqrt{2}} & j+k=q \text{ and } j>\frac{q}{2}, \\ 0 & \text{otherwise;} \end{cases}$$

for row index j and column index k with $0 \leq j, k \leq q-1$.

We have excluded the cases $q=1$, $\lambda=1$, and $\lambda=1-q$, each of which corresponds in some way to a degenerate case.

- If $q=1$, then $\text{Holant}_q(H; b \mid \mathcal{EQ}) = \lambda^u$, where u is number of vertices assigned the signature b.
- If $\lambda=1$, then $\text{Holant}_q(H; b \mid \mathcal{EQ}) = q^v$, where v is the number of vertices not assigned the signature b.
- If $\lambda = 1-q$, then $\text{Holant}_q(H; b \mid \mathcal{EQ})$ is degenerate in the sense that it has an equivalent Holant formulation with domain size $q-1$.

(For a deeper understanding of the last item, see the proof of [259, Lemma 8.3], which analyzes the same degeneracy in a similar setting.)

Combining Theorem 2.6, Lemma 21.4, and Example 21.7 yields local expressions for the positive integer hyperbolas on the Tutte polynomial.

21.3.2 Integral points in the first quadrant

Let G be a connected plane graph, and let G_m be its medial graph (see Section 1.2.17). Las Vergnas proved (see Item 3 in Section 13.4) that $T(G;3,3)$ can be expressed, up to a factor of 2, as a sum over weighted Eulerian orientations of G_m. Recall that an Eulerian orientation of an Eulerian graph directs the edges so that the in-degree and the out-degree are equal at each vertex. The weight of an Eulerian orientation O is $2^{\beta(O)}$, where $\beta(O)$ is the number of vertices in O whose incident edges are cyclically oriented as "in, out, in, out".

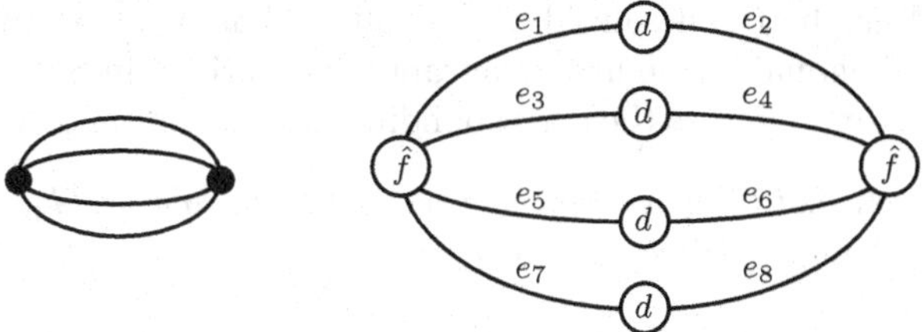

FIGURE 21.1: A graph G (with two black vertices), and its edge-vertex incidence graph H (with six white vertices) as part of the signature grid $\Omega = (H, d \mid \hat{f}, \pi)$.

Using Holant, this sum of weighted Eulerian orientations in G is naturally expressed as $\text{Holant}_2(H; d \mid \hat{f})$, where H is the edge-vertex incidence graph of G, and

$$d(x,y) = \begin{cases} 1 & x \neq y, \\ 0 & \text{otherwise}; \end{cases} \quad \text{and} \quad \hat{f}(w,x,y,z) = \begin{cases} 1 & wxyz = aabb, \\ 1 & wxyz = abba, \\ 2 & wxyz = abab, \\ 0 & \text{otherwise}; \end{cases}$$

for all distinct $a, b \in \{1, 2\}$. The order of the inputs for each d and $\hat{f}$ is obtained from a cyclic order of incident edges at the corresponding vertex of the plane graph. See Figure 21.1 for an example. In this example, a general form of a summand in $\text{Holant}_2(\Omega)$ is

$$\hat{f}(e_1, e_3, e_5, e_7) d(e_1, e_2) d(e_3, e_4) d(e_5, e_6) d(e_7, e_8) \hat{f}(e_8, e_6, e_4, e_2).$$

Typically the direction of rotation for this cyclic order matters, but it does not matter in this case because both d and $\hat{f}$ are invariant under the permutation of their inputs induced by this choice.

A change of basis by $\left[\begin{smallmatrix} 1 & 1 \\ i & -i \end{smallmatrix}\right]$, as described in [591, Section 3], gives the following result, whose proof is implicit in the proof of Theorem 8 in [591].

Theorem 21.8. *Let G be a connected plane graph, and let G_m be its medial graph. Then*

$$2\,T(G; 3, 3) = \text{Holant}_2(G_m; f),$$

where f is the quaternary function

$$f(w,x,y,z) = \begin{cases} 2 & wxyz = aaaa, \\ 1 & wxyz = abba, \\ 1 & wxyz = aabb, \\ 0 & \textit{otherwise}; \end{cases}$$

for all distinct $a, b \in \{1, 2\}$, with input order obtained from a cyclic order of incident edges at the corresponding vertex.

J. Ellis-Monaghan generalized the result of Las Vergnas to all integral points on the diagonal in the first quadrant, as described in Section 13.4. The next theorem, from [259], is the corresponding generalization of Theorem 21.8.

Theorem 21.9. *Let G be a connected plane graph, and let G_m be its medial graph. Then for any positive integer n,*

$$n\,T(G; n+1, n+1) = \text{Holant}_n(G_m; f),$$

where f is the quaternary function

$$f(w,x,y,z) = \begin{cases} 2 & wxyz = aaaa, \\ 1 & wxyz = abba, \\ 1 & wxyz = aabb, \\ 0 & \text{otherwise}; \end{cases}$$

for all distinct $a, b \in \{1, \ldots, n\}$, with input order obtained from a cyclic order of incident edges at the corresponding vertex.

M. Korn and I. Pak generalized the result of Ellis-Monaghan to all integral points in the first quadrant (see [705, Theorem 3] and also Item 5 in Section 13.4). Theorem 21.10 is the corresponding generalization of Theorem 21.9.

Theorem 21.10. *Let G be a connected plane graph, and let G_m be its medial graph. Let p and q be positive integers with $\text{lcm}(p, q) = n$. Then*

$$n\,T(G; p+1, q+1) = \text{Holant}_n(G_m; f),$$

where f is the quaternary function

$$f(w,x,y,z) = \begin{cases} 2 & wxyz = aaaa, \\ 1 & wxyz = abba \text{ and } a \equiv b \pmod{\frac{n}{p}}, \\ 1 & wxyz = aabb \text{ and } a \equiv b \pmod{\frac{n}{q}}, \\ 0 & \text{otherwise}; \end{cases}$$

for all distinct $a, b \in \{1, \ldots, n\}$, with input order obtained from a cyclic order of incident edges at the corresponding vertex. Furthermore, the edge corresponding to the first input w must have a white face on its left and a gray face on its right.

21.3.3 The point $(0, -2)$

In Section 21.3.1, we presented a Holant expression of $T(0, -2)$ that follows from the point $(0, -2)$ lying on the hyperbola H_3. Recall that the domain size in this Holant expression is also 3. In this section, to illustrate techniques of this chapter, we present an example giving an alternative Holant expression of this point, and this one uses a domain of size 2.

Example 21.11 below builds on an observation stated by D. Welsh. Let G be a 4-regular graph, and recall the specialization of the Tutte polynomial to the flow polynomial $F(G;x)$, which is given in Equation (3.5). Then Welsh, in (3.6.2) of [1139], equates $F(G;3)$ with the number of Eulerian orientations in G.

The number of Eulerian orientations in G is naturally expressed as $\mathrm{Holant}_2(H; d \mid \hat{g})$, where H is the edge-vertex incidence graph of G,

$$d(x,y) = \begin{cases} 1 & x \neq y, \\ 0 & \text{otherwise;} \end{cases} \quad \text{and} \quad \hat{g}(w,x,y,z) = \begin{cases} 1 & \text{exactly two inputs are 1,} \\ 0 & \text{otherwise.} \end{cases}$$

A change of basis by $\left[\begin{smallmatrix} 1 & 1 \\ i & -i \end{smallmatrix}\right]$, as described in [258, Section 1], gives the following result.

Example 21.11. Let G be a 4-regular graph. Then the number of Eulerian orientations in G is

$$(-1)^{|V|+|E|+k(G)}\, T(G;0,-2) = \mathrm{Holant}_2(G;g),$$

where $k(G)$ is the number of connected components in G and g is the quaternary function

$$g(w,x,y,z) = \begin{cases} \frac{3}{2} & w = x = y = z, \\ \frac{1}{2} & \text{exactly two inputs are 1,} \\ 0 & \text{otherwise.} \end{cases}$$

21.4 Open problems

The Holant framework is a natural and powerful setting to express a broad class of counting problems. In the study of counting complexity, a primary goal is to achieve complexity classifications. Such a classification often takes the form of a dichotomy theorem, which says that for any set of local constraint functions that defines a counting problem, the problem is either solvable in polynomial time or is ♯P-hard. The complexity of Holant problems are a generalization of counting constraint satisfaction problems, graph homomorphisms, and partition functions of spin systems. It is essentially equivalent to computing the contraction of a tensor network.

As described in Chapter 9, there are three dichotomy theorems for evaluations of the Tutte polynomial. These connections between the Tutte polynomial and Holant serve as starting points in complexity reductions that ultimately lead to complexity dichotomy theorems. For example, Theorem 21.8 is crucial for obtaining the dichotomy theorems for sets of symmetric signatures in [626] (for real valued signatures) and [591] (for complex valued signatures). These theorems are valid for the Boolean domain. The full classification for Holant problems on the Boolean domain without the restriction to symmetric signatures is open. Currently very little is known about complexity classifications for Holant problems on any domain with $q > 2$. One such dichotomy theorem for a restricted class of Holant problems on arbitrary domain size q has been proved [259]. It includes as a special case the problem of counting the number of proper edge colorings. Theorem 21.9 plays a crucial role in obtaining that dichotomy theorem.

Open Question 21.12. What are all the polynomial-time computable Holant problems on the Boolean domain (without restricting to symmetric signatures)? Is there a complexity dichotomy for this class of problems? If so, prove such a dichotomy.

Open Question 21.13. The same question can be asked for general domain size q. This is open even for very restricted classes of signature sets.

Open Question 21.14. Are there any other ways to express specializations of the Tutte polynomial as a Holant sum?

22

Polynomials and graph homomorphisms

Delia Garijo • Andrew Goodall • Jaroslav Nešetřil • Guus Regts

Synopsis

We develop in the language of graph homomorphisms the connection between the Tutte polynomial and the state models of statistical physics.

- The Tutte polynomial and homomorphism numbers.
- Spin models and edge coloring models.
- Connection matrices and the characterization of graph invariants arising from spin models.
- Homomorphism numbers and invariants of the cycle matroid of a graph.
- Graph homomorphism numbers as evaluations of graph polynomials.
- Other graph polynomials from counting graph homomorphisms such as the independence polynomial, the Averbouch–Godlin–Makowsky polynomial, and the Tittmann–Averbouch–Makowsky polynomial.

22.1 Introduction

Hyperbolas of the form $(x-1)(y-1) = q$ play a special role in the theory of the Tutte polynomial, especially when q is an integer. For a positive integer q, the Tutte polynomial $T(G; x, y)$ of a graph G along the hyperbola $(x-1)(y-1) = q$ is equivalent to the partition function of the q-state Potts model on G see Chapter 20). We consider more general state models in the setting of graph homomorphisms, namely in the form of homomorphisms from a graph G to a specified weighted graph H on q vertices. We situate the Tutte polynomial

DOI: 10.1201/9780429161612-22

among other polynomial graph invariants derived from such "H-colorings", such as the independence polynomial and polynomials recently introduced by I. Averbouch et al. [55], and by P. Tittmann et al. [1068].

A *graph homomorphism profile* of a graph G collects together as a single invariant of G the number of the H-colorings of G for a family of weighted graphs H. Graph homomorphism profiles play an important role in the recently developed theory of graph limits (see [795]). When restricted to certain families of graphs, such as complete graphs, graph homomorphism profiles may coincide with known polynomial graph invariants, such as the chromatic polynomial. This allows a unifying formulation of seemingly diverse questions such as whether a graph is chromatically unique, Tutte unique or spectrally unique. (See Chapter 6 and Chapter 11 for the topics of graphs uniquely determined, respectively, by their Tutte polynomial and by their chromatic polynomial.) Recently finite model theory has been used for a general construction of polynomial graph invariants from graph homomorphism profiles [561].

The number of H-colorings of a graph G may be equivalently formulated as the partition function of a spin model and we sketch the analogous theory for edge coloring models. Graph invariants sharing with the Tutte polynomial the property of being expressible as the partition function of a spin model can be elegantly characterized by means of connection matrices [502, 990].

The ubiquity of the Tutte polynomial in combinatorics is in large part due to it being not only a graph invariant but also an invariant of the graph's underlying cycle matroid, allowing it to be defined for matroids more generally. Graph invariants obtained from homomorphism numbers that share this property of being a matroid invariant have been characterized [562]. Although polynomial graph invariants different from the Tutte polynomial are known that share the property of being a cycle matroid invariant, it is not yet known how they might be defined on more general classes of matroid than graphic matroids [367, 562].

Another important property of the Tutte polynomial of a graph G is that it satisfies a deletion–contraction recurrence, in which the terms of the recurrence involve smaller graphs than G. We give some examples of other polynomial graph invariants obtained from graph homomorphism profiles which have similar size-reducing recurrences. However, a general method of constructing by graph homomorphism profiles a polynomial graph invariant with a size-reducing recurrence formula is not known.

Homomorphisms are closely related to universal algebra and category theory. Likewise, the theory of invariants and abstract algebra provide a proper setting for many questions related to the Tutte polynomial. We do not pursue these connections here but instead refer to [308, 606, 716, 740, 795].

22.2 Homomorphism profiles

Definition 22.1. A *homomorphism* from a simple graph G to a simple graph H is a mapping $f : V(G) \to V(H)$ such that $f(u)f(v) \in E(H)$ whenever $uv \in E(G)$. The *Hom complex* $\mathrm{Hom}(G, H)$ is the set of homomorphisms from G to H. The number of homomorphisms from G to H is denoted by $\mathrm{hom}(G, H)$.

Homomorphisms are adjacency-preserving mappings but no other condition is imposed: $f(u)f(v) \in E(H)$ need not imply $uv \in E(G)$, and f need not be one-to-one.

Example 22.2. Every homomorphism from a graph G to the complete graph K_q corresponds to a proper vertex q-coloring of G. Thus, $\mathrm{hom}(G, K_q)$ is the chromatic polynomial $\chi(G; x)$ of G evaluated at q. (See Chapter 11 for the chromatic polynomial.)

When G and H are multigraphs (graphs which may have loops or multiple edges), a homomorphism f from G to H, denoted $f : G \to H$, is a pair $f = (f_V, f_E)$ of functions $f_V : V(G) \to V(H)$ and $f_E : E(G) \to E(H)$ with the property that if $e \in E(G)$ has endpoints u and v, then $f_E(e)$ has endpoints $f_V(u)$ and $f_V(v)$. Thus, the function f_E maps parallel edges in G to parallel edges (possibly the same edge) in H, and loops in G to loops in H. The notation $\mathrm{hom}(G, H)$ extends from simple graphs to multigraphs G and H, defined as the number of homomorphisms from G to H. The two definitions of homomorphisms (and thus of $\mathrm{hom}(G, H)$) are equivalent when G and H are simple loopless graphs.

The parameter $\mathrm{hom}(G, H)$ can be further extended to weighted multigraphs H, in which weights on vertices and edges of H belong to a commutative ring (usually a field such as $\mathbb{R}$).

Definition 22.3. Let H be a multigraph with a weight $\alpha(u)$ associated with each vertex $u \in V(H)$ and a weight $\beta(e)$ associated with each edge $e \in E(H)$. The parameter $\mathrm{hom}(G, H)$ defined on multigraphs G is defined by

$$\mathrm{hom}(G, H) = \sum_{\substack{f:G\to H \\ \text{homomorphism}}} \prod_{v\in V(G)} \alpha(f_V(v)) \prod_{e\in E(G)} \beta(f_E(e)), \qquad (22.1)$$

where the sum is over all homomorphisms $f = (f_V, f_E)$ from G to H.

If f is a homomorphism mapping a given edge e of G to an edge e' of H, then so is the mapping that is equal to f with the sole exception that it maps e to an edge parallel to e'. Consequently, in Definition 22.3 the value of $\mathrm{hom}(G, H)$ is unchanged when parallel edges $e_1, e_2, \ldots, e_\ell$ in H of weights $\beta(e_1), \beta(e_2), \cdots, \beta(e_\ell)$ are replaced by a single edge of weight $\beta(e_1) + \beta(e_2) +$

$\cdots + \beta(e_\ell)$. Thus in studying parameters of the form $\hom(G,H)$ it suffices to consider H to be a simple graph possibly with a loop on some vertices. Furthermore, as an edge of weight 0 in H plays the same role as a non-edge, we can take H to be a complete graph with a loop on each vertex. In this case the sum in Equation (22.1) need not be restricted to homomorphisms and we have, for G a multigraph and H a complete graph with a single loop on each vertex,

$$\hom(G,H) = \sum_{f:V(G)\to V(H)} \prod_{v\in V(G)} \alpha(f(v)) \prod_{uv\in E(G)} \beta(f(u)f(v)). \qquad (22.2)$$

Here, edges of G are taken with their multiplicity in the product (parallel edges in G are sent to the same edge in H). When in H all the vertex weights are 1 and the edge weights are nonnegative integers, Equation (22.2) gives a definition of $\hom(G,H)$ equivalent to that given previously for unweighted multigraphs G and H (as the number of homomorphisms from G to H) by taking the edge weights of H to represent edge multiplicities. Equation (22.2) is therefore often taken as the starting definition of $\hom(G,H)$.

In statistical physics, the quantity $\hom(G,H)$ is known as the *partition function of a spin model* on G. Here the vertices of H are states of the model, with vertex weight the probability of the state, and edges joining states have weight equal to the interaction energy between the two states, see [368]. (In combinatorics spin models are sometimes called vertex coloring models, for example [502], but we shall use the term spin model, in particular since what are termed edge coloring models in the next section are known as vertex models in statistical physics.)

Graph invariants defined by the partition function of a spin model include many graph polynomial invariants such as the Tutte polynomial. In order to make this statement precise we specify how invariants may be compared (so that one invariant may "include" or "specialize to" another).

A *graph invariant* is a function on graphs invariant under isomorphism. A graph invariant f defines a partition on graphs in which G and G' are equivalent when $f(G) = f(G')$, in which case we write $G \sim_f G'$. The partition of graphs by f is a coarsening of the partition of graphs by isomorphism. A graph invariant determines G up to isomorphism precisely when the equivalence class containing G under f is the isomorphism class of G, that is, $G \sim_f G'$ implies $G \cong G'$. The graph G is then said to be *f-unique.* The subject of Chapter 6 is Tutte uniqueness, that is, the question of which graphs are uniquely determined up to isomorphism by their Tutte polynomials.

A graph invariant g is said to be an *f-invariant* when $G \sim_f G'$ implies $G \sim_g G'$ for all graphs G, G'. For example, the rank $r(G)$ of G is a Tutte polynomial invariant (since $r(G)$ is the degree of $T(G;x,y)$ as a polynomial in x). If g is an f-invariant and f is a g-invariant, then f and g have the same distinguishing power as each other, and in this case we say that f is *equivalent* to g (as graph invariants) and write $f \simeq g$. The comparative study of graph invariants may be viewed as the study of the partial order on graph invariants defined by the relation "g is an f-invariant" (see Chapter 9).

Definition 22.4. Let $\mathcal{G}$ be a family of graphs and $\mathcal{H}$ a family of edge-$\mathbb{C}$-weighted graphs. The *right $\mathcal{H}$-profile* of $G \in \mathcal{G}$ is the vector $(\hom(G, H) : H \in \mathcal{H})$. The *left $\mathcal{G}$-profile* of $H \in \mathcal{H}$ is the vector $(\hom(G, H) : G \in \mathcal{G})$.

When $\mathcal{H}$ is a countable family of edge-weighted graphs enumerated as a multisequence (for a d-dimensional sequence the terms are indexed by $\mathbb{N}^d$), the profile of G by the $(q_1, \ldots, q_d)$-th term of $\mathcal{H}$ defines a function $g(G; q_1, \ldots, q_d)$. For example, when $\mathcal{H}$ is the family of complete graphs written as the sequence $(K_q : q \in \mathbb{N})$, the right profile of G is the sequence of evaluations of the chromatic polynomial of G at nonnegative integers, $(\chi(G; q) : q \in \mathbb{N})$.

When $\mathcal{G}$ is the family of all simple graphs the left $\mathcal{G}$-profile of a graph H is also known as the *Lovász vector* of H, due to its role in the dramatically simple proof of the cancellation law for the product of relational structures (such as graphs) [789, 790]. The Lovász vector of H determines H up to isomorphism [789].

For a given graph invariant f, it is not clear whether there is a family of graphs $\mathcal{H}$ such that $f(G) = f(G')$ if and only if G and G' have the same right $\mathcal{H}$-profile. (A similar question can be asked for left profiles.) We have just seen that for the trivial graph invariant f defined by $f(G) = f(G')$ if and only if $G \cong G'$ we can take $\mathcal{G}$ to be all simple graphs and the left $\mathcal{G}$-profile is equivalent to f. (L. Lovász subsequently showed [793] that when $\mathcal{G}$ is the set of all simple graphs the right $\mathcal{G}$-profile is equivalent to f as well.) The graph parameter $|V(G)|$ is equivalent to the left $\{K_1\}$-profile, as $\hom(K_1, G) = |V(G)|$ and is equivalent, for example, to the right $\{\overline{K}_2\}$-profile as $\hom(G, \overline{K}_2) = 2^{|V(G)|}$.

When f is a graph polynomial whose evaluations at positive integers q count homomorphisms to graphs in a sequence $\mathcal{H} = (H_q : q \in \mathbb{N})$ we can by interpolation determine the polynomial f from the right $\mathcal{H}$-profile. Therefore the right $\mathcal{H}$-profile in such a case is equivalent to the polynomial invariant f, as illustrated in the following example.

Example 22.5. If $\mathcal{H} = \{K_q : q \in \mathbb{N}\}$, then G and G' have the same right $\mathcal{H}$-profile if and only if G and G' are chromatically equivalent (i.e., have the same chromatic polynomial; see Chapter 11).

If $\mathcal{G} = \{K_1\} \cup \{C_q : q \in \mathbb{N}\}$, then the left $\mathcal{G}$-profiles of H and H' are the same if and only if H and H' are cospectral (i.e., have the same characteristic polynomial) [511].

Definition 22.6. A *q-state Potts model graph*, denoted by $K_q^{x,y}$, is an edge-weighted graph on q vertices in which each pair of distinct vertices are joined by an edge of the same weight x and each vertex has a loop attached of weight y. (Figure 22.1 shows the 4-state Potts model graph $K_4^{x,y}$.)

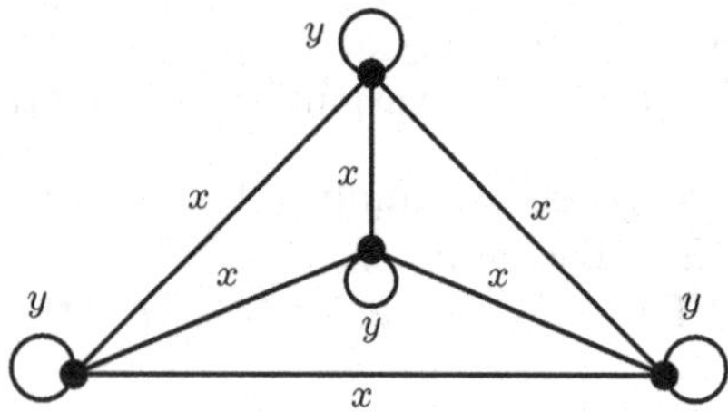

FIGURE 22.1: The 4-state Potts model graph $K_4^{x,y}$.

In the notation of Definition 22.6, $K_q \cong K_q^{1,0}$. The following result was shown in [511].

Theorem 22.7. *The right* $\{K_q^{1,y} : q \geq 1, y \in \mathbb{N}\}$*-profile of* G *is equivalent as a graph invariant to the Tutte polynomial of* G*, with*

$$\hom(G, K_q^{1,y}) = q^{k(G)}(y-1)^{r(G)}T\big(G; \tfrac{q-1+y}{y-1}, y\big).$$

In particular, a graph G *is Tutte-unique if and only if* G *is determined by its right* $\{K_q^{1,y} : q \geq 1, y \in \mathbb{N}\}$*-profile.*

Whether a graph is determined by its right $\{K_q^{1,y} : q \in \mathbb{N}\}$-profile includes the question of chromatic uniqueness by setting $y = 0$, and flow uniqueness by taking $y = 1 - q$. On the other hand, the case where y is variable and q fixed gives a right profile determining $Z(G; q, y-1)$ as a polynomial in y, that is, the q-state Potts partition function of G. This leads to the notion of *q-state Potts uniqueness* [511], for which see Section 6.5.3.

22.3 Edge coloring models

Roughly speaking, by interchanging the roles of edges and vertices in (22.2), one obtains the partition function of an edge coloring model. Edge coloring models also originate from statistical physics models, where they are called *vertex models*, and were introduced to the graph theory community by P. de la Harpe and V. Jones [368]. They form a special case of Holants (see Chapter 21) and they can also be seen as contractions of tensor networks, which originated in the work of R. Penrose [916] and have found applications in areas such as knot theory [683], the theory of Vassiliev invariants [308], quantum computing [817], and even the theory of neural networks [911].

Definition 22.8. Let $q \in \mathbb{N}$. A *q-color edge coloring model* h is a map $h : \mathbb{N}^q \to \mathbb{C}$. The *partition function* of h is defined as the graph parameter that maps a graph $G = (V, E)$ to the number

$$p_h(G) := \sum_{\phi: E \to [q]} \prod_{v \in V} h(\phi(\delta(v))), \tag{22.3}$$

where $\delta(v)$ is the multiset of edges incident with v and we identify the multiset of colors $\phi(\delta(v))$ with its incidence vector in $\mathbb{N}^q$. The field $\mathbb{C}$ in this definition can be replaced by an arbitrary commutative ring.

Example 22.9. Let $q = 2$ and let $h : \mathbb{N}^2 \to \mathbb{C}$ be defined for $x = (x_1, x_2) \in \mathbb{N}^2$ by $h(x) = 1$ if $x_1 \leq 1$ and 0 otherwise. Then $p_h(G)$ is equal to the number of matchings of G. Indeed, for every assignment of colors to the edges that gives a nonzero contribution to the sum in (22.3), the edges that receive color 1 together form a matching.

Denoting the line graph of a graph G by $L(G)$, it is not difficult to see that for a fixed graph H the parameter $G \mapsto \hom(L(G), L(H))$ can be expressed as the partition function of an edge coloring model [368]. More surprisingly, it is also possible to express $G \mapsto \hom(G, H)$ as the partition function of an edge coloring model for any weighted graph H, as shown by B. Szegedy [1051]. In particular, taking the q-state Potts model graph $H = K_q^{x,y}$, the Tutte polynomial along the hyperbola $(x - 1)(y - 1) = q$ can be written as the partition function of an edge coloring model, as we now describe.

Example 22.10. Let us fix $q \in \mathbb{N}$ and define an edge coloring model $h : \mathbb{N}^{q+1} \to \mathbb{C}$ by for $a = (a_1, \ldots, a_{q+1}) \in \mathbb{N}^{q+1}$,

$$h(a) = \begin{cases} (y - x)^{a_i/2} \cdot x^{a_{q+1}/2} & \text{if } a_j = 0 \text{ for each } j \notin \{i, q+1\}, \\ q x^{a_{q+1}/2} & \text{otherwise.} \end{cases}$$

Then one can verify that for any graph G, $p_h(G) = \hom(G, K_q^{x,y})$. Alternatively, this follows from Lemma 22.11 below. The construction of Szegedy, given below, shows that one can in fact find an edge coloring model $h' : \mathbb{N}^q \to \mathbb{C}$ such that $p_{h'}(G) = \hom(G, K_q^{x,y})$ for all graphs G.

Note that if for example $y = 0$ and $x = 1$, in which case $\hom(G, K_q^{1,0})$ is the number of proper q-colorings of G, the edge coloring model h is not real-valued. There are in fact no real-valued edge coloring models h for which $p_h(G) = \hom(G, K_q^{1,0})$ for all graphs G when $q \geq 3$, as is shown in [955].

We now describe Szegedy's construction in general. Let H be an n-vertex weighted graph with vertex weights given by $\alpha \in \mathbb{C}^n$ and edge weights given by a symmetric matrix $\beta \in \mathbb{C}^{n \times n}$. As β is symmetric we can write $\beta = U^T U$ for some $q \times n$ (complex) matrix U for some q. (One can take $q = \mathrm{rk}(\beta)$ unless β is the all-zero matrix, see [563, Lemma 5.2.4]; when β is real this follows for example from the spectral theorem.) Let $u_1, \ldots, u_n \in \mathbb{C}^q$ be the columns of U. Define the edge coloring model $h = h_{\alpha,\beta}$ for $a \in \mathbb{N}^q$ by

$$h(a) := \sum_{i=1}^{n} \alpha_i \prod_{j=1}^{q} u_i(j)^{a_j}, \tag{22.4}$$

where α_i is the i-th entry of α and $u_i(j)$ is the j-th entry of the column vector u_i. In Example 22.10 we have q vectors $u_1, \ldots, u_q \in \mathbb{C}^{q+1}$, where for $i = 1, \ldots, q$ the top q-part of u_i is equal to $\sqrt{y-x}$ times the i-th unit vector and where $u_i(q+1)$ is equal to $\sqrt{x}$. With this, we have the following from [1051].

Lemma 22.11. *Let H be an n-vertex weighted graph with vertex weights given by $\alpha \in \mathbb{C}^n$ and edge weights given by a symmetric matrix $\beta \in \mathbb{C}^{n\times n}$, and let h be the edge coloring model in (22.4). Then $hom(G, H) = p_h(G)$ for every graph G.*

The weighted graphs H for which the corresponding edge coloring model h can be taken to be real-valued are characterized in [955]. For example, if the matrix β is real and positive semidefinite, and the vector α is real, then obviously h can be taken to be real-valued.

The *complex orthogonal group,* O_q, is the group of $q \times q$ complex-valued matrices A such that $A^T A$ is equal to the identity matrix. There is a natural action of O_q on edge coloring models that leaves the partition function invariant. We give an example before embarking on the general case. For $A \in O_q$, let h^A be the edge coloring model as defined in (22.4) with each u_i replaced by Au_i. Then, as $(AU)^T AU = \beta$, Lemma 22.11 implies that for any graph G we have $p_h(G) = p_{h^A}(G)$.

In the general case, we can view any edge coloring model h as a linear map $\mathbb{C}[x_1, \ldots, x_q] \to \mathbb{C}$ by identifying the monomials in $\mathbb{C}[x_1, \ldots, x_q]$ with their coefficient vectors in $\mathbb{N}^q$. As O_q has a natural action on this polynomial ring, it also acts naturally on edge coloring models. Thus, for any $A \in O_q$ and $h : \mathbb{N}^q \to \mathbb{C}$ we can associate another edge coloring model h^A and just as above we have that $p_h(G) = p_{h^A}(G)$ for all graphs G. A proof of this fact can be found in for example [419, 1051]. This orthogonal group invariance can lead to interesting representations of certain counting problems, as in Chapter 21.

Partition functions of edge coloring models may also be defined for directed graphs [419], in which case they are invariant under an action of the general linear group. See [956] for partition functions related to invariant theory of the symplectic group.

22.4 Connection matrices

For a positive integer ℓ, an *ℓ-labeled graph* is a graph G together with an injective map $\lambda : [\ell] \to V(G)$; a 0-labelled graph is unlabelled. Two ℓ-labeled graphs are isomorphic if there is a label-preserving isomorphism between them.

Definition 22.12. The *gluing product* $G_1 \sqcup_\ell G_2$ of two ℓ-labeled graphs G_1 and G_2 is the ℓ-labeled graph obtained by taking the disjoint union of G_1 and G_2 and then identifying vertices with the same label. The product $G_1 \sqcup_0 G_2$ is disjoint union.

Let f be a graph invariant, which we extend to ℓ-labeled graphs by forgetting the labels.

Definition 22.13. For nonnegative integer ℓ, the *ℓ-th connection matrix* of f, denoted by $M(f,\ell)$, has rows and columns indexed by ℓ-labeled graphs. The entry of $M(f,\ell)$ in the intersection of the row corresponding to G_1 and the column corresponding to G_2 is $f(G_1 \sqcup_\ell G_2)$.

A graph invariant f is *multiplicative* (over disjoint unions) if $f(G_1 \sqcup G_2) = f(G_1)f(G_2)$ for all graphs G_1, G_2 and $f(\emptyset) = 1$, where $\emptyset$ denotes the empty graph (no vertices or edges). The graph invariant defined by $G \mapsto \hom(G,H)$ for a fixed weighted graph H is multiplicative. The following result is a variation on [502, Proposition 2.1].

Proposition 22.14. *Let f be a graph invariant not identically zero. Then f is multiplicative if and only if $f(\emptyset) = 1$ and $M(f,0)$ is of rank* 1.

M. Freedman et al. [502] proved that the connection matrices $M(f,\ell)$ when $f(G) = \hom(G,H)$ for a fixed graph H have special properties, and in fact give the following characterization.

Theorem 22.15. *Let f be a real-valued graph invariant defined on loopless graphs. Then there is a finite weighted graph H with real edge weights and positive real vertex weights such that $f(G)$ is equal to* $\hom(G,H)$ *for all G if and only if for each $\ell \in \mathbb{N}$ the connection matrix $M(f,\ell)$ is positive semidefinite and there exists $q > 0$ such that* $\operatorname{rk}(M(f,\ell)) \leq q^\ell$ *for each $\ell \in \mathbb{N}$.*

T. Kotek and J. Makowsky [712, Theorem 9] give a characterization in terms of monadic second order logic for polynomial graph invariants f all of whose evaluations have connection matrices of finite rank.

In [502, 794, 795] explicit formulas for the rank of connection matrices of various graph invariants are given, from which we highlight the following (from [502, 795]):

Example 22.16. Let f be the graph invariant defined by $f(G) = T(G;x,y)$ for a fixed point (x,y) on the hyperbola $(x-1)(y-1) = q$, and let $\ell \in \mathbb{N}$. Then

$$\operatorname{rk}(M(f,\ell)) = \begin{cases} \sum_{i=1}^{q} \left\{ {\ell \atop i} \right\} & q \in \mathbb{N}, \\ B_\ell & \text{otherwise}, \end{cases}$$

where $\left\{ {\ell \atop i} \right\}$ is the number of partitions of $[\ell]$ into i subsets and $B_\ell = \sum_{i=1}^{\ell} \left\{ {\ell \atop i} \right\}$. Thus the rank is the exponentially bounded number of partitions of $[\ell]$ into at most q subsets when $q \in \mathbb{N}$, and is equal to the superexponential Bell number B_ℓ when $q \notin \mathbb{N}$.

Example 22.16 includes as a special case $f(G) = \chi(G;q)$, the evaluation of the chromatic polynomial at q.

22.5 Matroid invariants

A graph invariant that has the further property of depending only on the cycle matroid of the graph may, like the Tutte polynomial, be extendable to a larger class of matroids.

P. De la Harpe and F. Jaeger [367] were first to consider the question of when $|V(H)|^{-k(G)}\hom(G,H)$ depends only on the cycle matroid of G, as is the case for $H = K_q^{1,y}$, where

$$q^{-k(G)}\hom(G, K_q^{1,y}) = (y-1)^{r(G)}T\big(G; \tfrac{q-1+y}{y-1}, y\big).$$

The automorphism group of a (weighted) graph H is *transitive* if for any pair of vertices $u, v \in V(H)$ there is an automorphism of H sending u to v. Further, that group is *generously transitive* if for any pair of vertices $u, v \in V(H)$ there is an automorphism of H swapping u and v. De la Harpe and Jaeger showed the following [367].

Proposition 22.17. *If H has a transitive automorphism group, then the graph invariant f defined by $f : G \mapsto |V(H)|^{-k(G)}\hom(G,H)$ is multiplicative over 1-gluing products, that is, for 1-labeled graphs G_1, G_2,*

$$f(G_1 \sqcup_1 G_2) = f(G_1)f(G_2).$$

The 1-gluing product has the property that $G_1 \sqcup_1 G_2$ has the same cycle matroid as $G_1 \sqcup G_2$. The inverse operation of separating $G_1 \sqcup_1 G_2$ into a disjoint union of G_1 and G_2 also preserves the cycle matroid. For a 2-labeled graph G, define G^T to be the same graph with the labels swapped. For 2-labeled graphs G_1, G_2, the graphs $G_1 \sqcup_2 G_2$ and $G_1 \sqcup_2 G_2^T$ are related by a Whitney flip (see Definition 4.101) about the 2-cut defined by the two labeled vertices. A Whitney flip of a graph about a 2-cut also preserves the cycle matroid.

H. Whitney characterized graphs with the same cycle matroid as precisely those related by a sequence of 1-gluings, 1-separations, and Whitney flips (see Theorem 4.102). De la Harpe and Jaeger [367] showed that if a graph H has a generously transitive automorphism group, then the graph invariant $f : G \mapsto |V(H)|^{-k(G)}\hom(G,H)$ depends only on the cycle matroid of G. The converse statement can be proved using a slight strengthening of a characterization by Lovász [793] of the column space of the connection matrix of f. We therefore have the following from [367, 562]:

Theorem 22.18. *Let H be an edge-weighted graph, with weights in a field of characteristic zero. The graph invariant*

$$G \mapsto \frac{\hom(G,H)}{|V(H)|^{k(G)}}$$

is an invariant of the cycle matroid of G if and only if H has a generously transitive automorphism group.

Example 22.19. The automorphism group of the Cayley graph $\Gamma(\mathsf{A}, B)$ on an additive abelian group A and edges defined by $B \subseteq \mathsf{A}$ is generously transitive, since u and v can be swapped by the automorphism $x \mapsto u + v - x$. So for $B = -B$ the invariant

$$f : G \mapsto |\mathsf{A}|^{-k(G)}\hom(G, \Gamma(\mathsf{A}, B))$$

depends only on the cycle matroid of G. This can be confirmed independently of Theorem 22.18 by using the fact that $f(G)$ is the number of A-tensions of G taking values only in B. These tensions are defined solely in terms of the oriented cycle matroid of G, and the number of tensions is independent of the choice of orientation (since $B = -B$).

Definition 22.20. Let $q, r \in \mathbb{N}$ with $r \leq q$, and let $D \subseteq \{0, 1, \ldots, r\}$. The *generalized Johnson graph* $J_{q,r,D}$ is the graph on vertex set $\binom{[q]}{r}$ and edges joining u and v precisely when $|u \cap v| \in D$.

Example 22.21. The complete graphs are generalized Johnson graphs, with $K_q \cong J_{q,1,\{0\}}$. The *Kneser graph* is $KG_{q,r} \cong J_{q,r,\{0\}}$ and the *Johnson graph* is $J_{q,r} \cong J_{q,r,\{r-1\}}$. The complement of $J_{q,r,D}$ is $J_{q,r,\{0,1,\ldots,r-1\}\setminus D}$.

De la Harpe and Jaeger [367] give generalized Johnson graphs (and Grassmann graphs) as examples of graphs that are not in general equal to a Cayley graph on an abelian group, yet have a generously transitive automorphism group. By Theorem 22.18 the graph invariant $\binom{q}{r}^{-k(G)}\hom(G, J_{q,r,D})$ depends only on the cycle matroid of G. They leave open how to interpret this parameter in terms of the cycle matroid of G alone, from which it would be possible to see how the graph invariant might be extended to an invariant of a larger class of matroids than graphic matroids (see Problem 22.32).

22.6 Graph polynomials from homomorphism profiles

From Example 22.5, the right $\{K_q : q \in \mathbb{N}\}$-profile of a graph is equivalent to the chromatic polynomial, and, more generally, the right $\{K_q^{1,y} : q \geq 1, y \in \mathbb{N}\}$-profile is equivalent to the Tutte polynomial (Theorem 22.7). Instead of asking for a given graph polynomial f whether there is a family of graphs $\mathcal{H}$ for which the right $\mathcal{H}$-profile is equivalent to f, in this section we consider the reverse question: for which sequences of graphs (H_q) is the invariant $G \mapsto \hom(G, H_q)$ the evaluation of a fixed polynomial in q?

The generalized Johnson graphs $J_{q,r,D}$ (see Definition 22.20) provide us with our first example (from [367, 561]):

Theorem 22.22. *For any fixed $r \geq 1$ and $D \subseteq \{0, 1, \ldots, r\}$, the graph invariant $\binom{q}{r}^{-k(G)}\hom(G, J_{q,r,D})$ defines a graph polynomial in q.*

For complete graphs ($D = \{0, 1, \ldots, r-1\}$) we have

$$\hom\left(G, K_{\binom{q}{r}}\right) = \chi\left(G; \binom{q}{r}\right),$$

and when $D = \{0, 1, \ldots, r\}$ we have $\hom(G, K^{1,1}_{\binom{q}{r}}) = \binom{q}{r}^{|V(G)|}$. For $D = \{0\}$ we have the Kneser graph $KG_{q,r}$, and here

$$\hom(G, KG_{q,r}) = (r!)^{-|V(G)|}\chi(G[K_r]; q),$$

where $G[K_r]$ is the lexicographic product of G with K_r, replacing each vertex of G with a clique of r vertices and edges of G by complete joins between the r-cliques. Generalized Johnson graphs in general though do not give specializations of the Tutte polynomial.

The *fractional chromatic number* of a graph G is defined by

$$\chi_f(G) = \inf\left\{\frac{q}{r} : q, r \in \mathbb{N},\ \hom(G, KG_{q,r}) > 0\right\}. \tag{22.5}$$

Lovász [791] established that the chromatic number of $KG_{q,r}$ is equal to $q - 2r + 1$ when $q \geq 2r$, while for the fractional chromatic number, it is known [987] that $\chi_f(KG_{q,r}) = \frac{q}{r}$ when $q \geq 2r$.

As homomorphisms to complete graphs are to the chromatic number, so homomorphisms to Kneser graphs are to the fractional chromatic number. The *circular chromatic number* $\chi_c(G)$ of a graph G is the minimum ratio r/d such that G is homomorphic to $\Gamma(\mathbb{Z}_r, \{d, d+1, \ldots, r-d\})$. The following was shown in [561].

Proposition 22.23. *For every $r \geq 2d$ the graph invariant*

$$(qr)^{-k(G)}\hom(G, \Gamma(\mathbb{Z}_{qr}, \{qd, qd+1, \ldots, q(r-d)\})$$

defines a graph polynomial in q.

The examples of $(K_q) = (\Gamma(\mathbb{Z}_q, \mathbb{Z}_q\backslash\{0\}))$ and Proposition 22.23 prompt the question of when it is the case that a sequence of Cayley graphs defines a graph polynomial. In this direction, de la Harpe and Jaeger [367] showed that if $B_q \subseteq \mathbb{Z}_q$ is the set of residue classes modulo q of a fixed subset of integers B, then the invariant $\hom(G, \Gamma(\mathbb{Z}_q, B_q))$ is not a fixed polynomial in q; however, when B is finite or cofinite then for each G there is a polynomial $p(G)$ for which $\hom(G, \Gamma(\mathbb{Z}_q, B_q)) = p(G; q)$ for q larger than the girth of G.

In [561] a wide-ranging construction using finite model theory is given for sequences (H_q) with the property that $\hom(G, H_q) = p(G; q)$ for some polynomial $p(G)$. However, it is open as to whether this method might furnish a characterization of all such sequences (see Problem 22.33). It is further shown [561, Theorem 2.5] that if (H_q) has this property, then for any quantifier-free formula ϕ about graphs with n free variables, it is also true that the satisfaction set

$$\phi(H_q) = \#\{(v_1, \ldots, v_n) \in V(H_q)^n : \phi(v_1, \ldots, v_n) \text{ true in } G\}$$

has size polynomial in q. (The first-order language associated with graphs has a binary relation, interpreted as vertex adjacency, in addition to the usual equality relation and primitive logical symbols. A quantifier-free formula about graphs is a formula in this language that uses no quantifiers.) A special choice of ϕ yields the following (from [561]):

Proposition 22.24. *Let (H_q) be a sequence of (not necessarily simple) graphs with the property that, for each graph G, $\hom(G, H_q) = p(G; q)$ for some polynomial $p(G)$. Further let $\mathcal{H}$ be a class of (isomorphism types of) graphs and $r \in \mathbb{N}$. Then the number of homomorphisms $f : G \to H_q$ with the property that for each $U \in \binom{V(H_q)}{r}$ the inverse image $f^{-1}(U)$ induces a subgraph of G which is isomorphic to a graph in $\mathcal{H}$ is a fixed polynomial function of q.*

As an application of Proposition 22.24, take $(H_q) = (K_q)$ and $\mathcal{H}$ the set of all forests in order to deduce that the number of *acyclic q-colorings* (no two color classes induce a subgraph containing a cycle) is a polynomial in q. If instead we take $\mathcal{H}$ to be the set of forests whose connected components are stars, then we have the number of *star q-colorings* of G and this is a polynomial in q. (These types of colorings and the associated acyclic chromatic number and star chromatic number were first introduced by B. Grünbaum [589].) Kotek et al. [714] use infinite model theory to derive similar polynomial graph invariants whose evaluations count "generalized colorings" that are preserved under permutations of colors.

22.7 Graph polynomials by recurrence formulas

The deletion–contraction relation expresses the Tutte polynomial of a graph G in terms of the Tutte polynomials of graphs with fewer edges. Likewise, the *independence polynomial* [593], given by

$$I(G; x) = \sum_{i=0}^{|V(G)|} s_i x^i,$$

where s_i is the number of i-subsets of $V(G)$ that induce a subgraph with no edges, has recurrence formula

$$I(G; x) = I(G \backslash v; x) + xI(G \backslash \bar{N}(v); x),$$

which expresses the independence polynomial of G in terms of the independence polynomials of graphs with fewer vertices (and edges). Here $\bar{N}(v)$ denotes the closed neighborhood of v, which contains v and all vertices adjacent to it.

When a graph polynomial has a size-reducing recurrence formula, such as the Tutte polynomial with its deletion–contraction recurrence, it can be straightforward to prove combinatorial interpretations of its evaluations, or proofs of its algebraic properties. For example, that $T(G;1,1)$ is the number of spanning trees of G is easily established by deletion–contraction and induction, and the recurrence formula for the independence polynomial is used in an essential way by M. Chudnovsky and P. Seymour [317] to show that it has real roots for claw-free graphs.

A natural question then arises: when does a sequence (H_q) define a graph polynomial $p(G;q) = \hom(G, H_q)$ that satisfies a size-reducing recurrence formula?

Example 22.25. The sequence (H_q) defines a graph polynomial $p(G;q) = \hom(G,H_q)$ that satisfies a size-reducing recurrence formula when $H_q = K_q$ (chromatic polynomial), $H_q = K_q^{1,1-q}$ (flow polynomial), $H_q = K_q^{1,y}$ (Tutte polynomial along $(x-1)(y-1) = q$), and $H_q = K_1^{1,1} + \overline{K}_q$ (independence polynomial).

We do not know any examples of graph polynomials defined by homomorphism profiles that probably do *not* have a size-reducing recurrence formula; on the other hand, beyond the examples given in Example 22.25 and the rest of this section, we do not know of others that *do* (see Problem 22.35).

Lovász and Szegedy [796] show that a graph parameter $\hom(G,H)$ has a *contractor* (see [796] for a definition). The existence of a contractor implies the existence of a recurrence formula for the parameter $\hom(G,H)$. However, this recurrence formula is not size-reducing in general, although it is when $H = K_q$, the contractor being in this case equivalent to the deletion–contraction recurrence for the chromatic polynomial evaluated at q.

We give two examples of graph polynomials with a size-reducing recurrence formula which, like the Tutte polynomial, are equivalent to a homomorphism profile.

Averbouch et al. [55] define a simultaneous trivariate generalization of the Tutte polynomial and the bivariate matching polynomial:

Definition 22.26. The *edge-elimination polynomial* $\xi(G;x,y,z)$ is defined by the following recurrence relation, for edge $e = uv \in E(G)$,

$$\xi(G;x,y,z) = \xi(G\backslash e;x,y,z) + y\,\xi(G/e;x,y,z) + z\,\xi(G\backslash\{u,v\};x,y,z).$$

Together with

$$\xi(G_1 \sqcup G_2;x,y,z) = \xi(G_1;x,y,z)\xi(G_2;x,y,z)$$

and

$$\xi(K_1;x,y,z) = x$$

this uniquely defines the polynomial $\xi(G;x,y,z)$.

The edge elimination polynomial includes the dichromatic polynomial $Z(G;q,y-1)$ and the bivariate chromatic polynomial of K. Dohmen et al. [388]. Moreover, Averbouch et al. showed that $\xi(G;x,y,z)$ is universal with respect to the above recurrence relation [55, Theorem 2]. The following result is from the same reference.

Theorem 22.27. *Let $K_{q-\ell}^{1,1} + K_{\ell}^{1,y}$ be the complete join of the $(q-\ell)$-state and ℓ-state Potts model graphs $K_{q-\ell}^{1,1}$ and $K_{\ell}^{1,y}$. For a graph G,*

$$\hom(G, K_{q-\ell}^{1,1} + K_{\ell}^{1,y}) = \xi(G; q, y-1, (\ell-q)(y-1)).$$

The following was shown in [511].

Theorem 22.28. *Let H be an edge-$\mathbb{C}$-weighted graph on q vertices and $x, y, z \in \mathbb{C}$. If $\hom(G,H) = \xi(G;x,y,z)$ for all $G \in \{K_1^{n,0}, K_2^{0,n} : n \in \mathbb{N}\}$, then there is an integer ℓ, $0 \le \ell \le q$, such that*

1. $x = q \in \mathbb{N}$, $z = (\ell - q)y$,
2. $H \cong K_{q-\ell}^{1,1} + K_{\ell}^{1,1+y}$.

Motivated in part by the study of properties of community structure in social networks, P. Tittmann et al. [1068] introduced a graph polynomial that satisfies a size-reducing recurrence akin to that satisfied by the independence polynomial:

Definition 22.29. The *subgraph component polynomial* $Q(G;x,y)$ of a graph $G = (V,E)$ is defined by

$$Q(G;x,y) = \sum_{U \subseteq V} x^{|U|} y^{k(G[U])},$$

where $G[U]$ is the induced subgraph on $U \subseteq V$ and $k(G[U])$ is the number of its connected components.

The subgraph component polynomial is related to the edge elimination polynomial via the identity $Q(L(G);x,y) = \xi(G;1,x,x(y-1))$, where $L(G)$ is the line graph of G. The coefficient of x^i in the independence polynomial $I(G;x)$ appears as the coefficient of $x^i y^i$ in $Q(G;x,y)$.

Theorem 22.30. *Let G be a graph and $v \in V(G)$. Then the subgraph component polynomial satisfies the following recurrence formula:*

$$Q(G;x,y) = Q(G \backslash v;x,y) + x(y-1)Q(G \backslash \bar{N}(v);x,y) + xQ(G/v;x,y),$$

where $G \backslash v$ is the graph obtained from G by deleting v, where $G \backslash \bar{N}(v)$ is obtained by deleting the closed neighborhood of v, and G/v by deleting v and adding edges between all pairs of non-adjacent vertices in the open neighborhood $N(v)$.

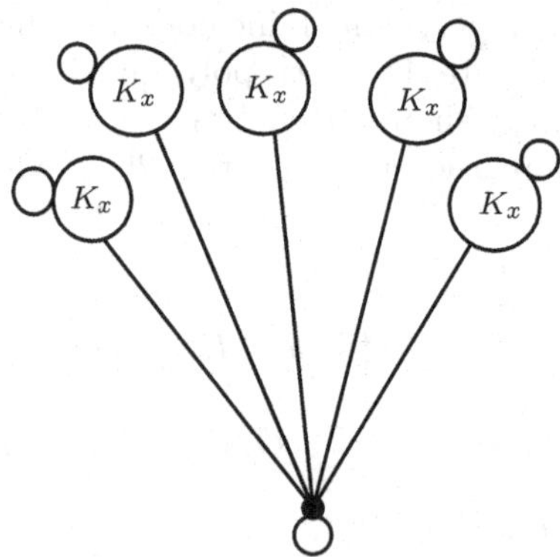

FIGURE 22.2: The graph $H_{x,5}$. The loop on each K_x indicates that there is a loop on each of its x vertices (making the graph $K_x^{1,1}$).

Graph polynomial	Graph sequence
chromatic	K_q
flow	$K_q^{1,1-q}$
Tutte	$K_q^{1,\ell}$
independence	$K_1^{1,1} + \overline{K}_q$
Averbouch–Godlin–Makowsky	$K_{q-\ell}^{1,1} + K_\ell^{m,1}$
Tittmann–Averbouch–Makowsky	$K_1^{1,1} + \overline{K}_q[K_\ell]$

TABLE 22.1: Table of graph sequences (H_q) for which $\hom(G, H_q) = p(G; q)$ for a polynomial $p(G)$ satisfying a size-reducing recurrence formula. (Some are double or triple sequences indexed by q, ℓ and m.)

Tittmann et al. in [1068], where the above recurrence formula was also shown, proved that the polynomial $Q(G; x, y)$ is universal with respect to its defining recurrence relation. Further they showed that the subgraph component polynomial $Q(G; x, y)$ for $x \in \mathbb{R}$ and $y \in \mathbb{N}$ can be viewed as a partition function by counting homomorphisms to a vertex- and edge-weighted graph, relating it thereby to the Widom–Rowlinson model of statistical physics [1163] (see also, for example, [432]).

Let $H_{x,y}$ with $x, y \in \mathbb{N}$ be the graph $K_1^{1,1} + \overline{K}_y[K_x^{1,1}]$, comprising the star $K_{1,y}$ with a loop on its central vertex and each leaf replaced by a copy of $K_x^{1,1}$, each completely joined to the central vertex (see Figure 22.2). Then, it is not difficult to see [1068] that $\hom(G, H_{x,y}) = Q(G; x, y)$, and a converse result holds [511]:

Theorem 22.31. *Let H be an edge-$\mathbb{C}$-weighted graph and $x, y \in \mathbb{C}$. Then $\hom(G, H) = Q(G; x, y)$ for all $G \in \{K_1^{n,0}, K_2^{0,n}, K_{1,n}, C_n : n \in \mathbb{N}\}$ if and only if $x, y \in \mathbb{N}$ and $H \cong H_{x,y}$.*

In Table 22.1 we summarize the graph polynomials with size-reducing recurrence formulas that are determined by counting homomorphisms to graphs;

for example, the fourth line means that the independence polynomial of G is determined by the homomorphism profiles of the graphs $K_1^{1,1} + \overline{K}_q$, as witnessed by the equation

$$\hom(G, K_1^{1,1} + \overline{K}_q) = I(G; q).$$

Some other graph polynomials that have a size-reducing recurrence formula and that are missing from Table 22.1 include the interlace polynomial [48] and the domination polynomial [46]. However, we do not know if either of these polynomials is equivalent to the right homomorphism profile by some sequence (H_q).

In [512] a broad family of graph sequences is produced using the encoding of graphs by rooted tree models (an example of a rooted tree model for graphs is the cotree encoding of cographs). This family of graph sequences produces by graph homomorphism profiles all the graph polynomials described in this section (see [512, Figure 5] for cotree representations of the graphs in Table 22.1 above). This family of graph sequences made by using rooted tree models is, in turn, a special case of the general finite model theory construction of [561] described briefly in Section 22.6 above.

22.8 Open problems

Theorem 22.18 characterizes graph invariants obtained from homomorphism numbers that depend on the underlying cycle matroid alone; such graph invariants may be extended to a wider class of matroids in a similar way to the Tutte polynomial, and are therefore of great potential interest. Such invariants include $G \mapsto |\mathsf{A}|^{-k(G)}\hom(G, \Gamma(\mathsf{A}, B))$, defined by homomorphisms to a Cayley graph on an abelian group A. As explained in Example 22.19, the graph invariant in this case can be interpreted in terms of the cycle matroid alone via tensions. In other cases it seems challenging to discover such a formulation [367]:

Problem 22.32. For H a generalized Johnson graph or Grassmann graph, give an interpretation to the parameter $G \mapsto |V(H)|^{-k(G)}\hom(G, H)$ in terms of the cycle matroid of G alone.

In Problem 22.32, it may be that some additional structure on the cycle matroid of G is needed (like the orientation of edges that is required for defining tensions and flows). This additional structure must be defined with reference to the cycle matroid alone and the value of the graph invariant should ultimately not depend on this additional structure (as for tensions and flows, where the choice of orientation does not matter). For orientable matroids with more than one orientation class, the number of nowhere-zero tensions or flows is not properly a matroid invariant but an invariant of the oriented matroid with specified orientation class [544].

The next problem (formulated in [561, Problem 8.1]) is perhaps the main open problem related to polynomials defined by homomorphism numbers. See the end of Section 22.6 for a discussion of graph interpretations.

Problem 22.33. Suppose (H_q) is a sequence of simple graphs with the property that for each graph G there is a polynomial $p(G)$ such that $\hom(G, H_q) = p(G; q)$ for all $q \in \mathbb{N}$. Can (H_q) be obtained from disjoint unions of transitive tournaments plus unary relations by a quantifier-free interpretation scheme?

Roughly speaking, the interpretation scheme mentioned in Problem 22.33 converts a relational structure Σ (such as a disjoint union of transitive tournaments with further unary relations) to a graph H by taking vertices of H to be tuples of elements in the domain of Σ and using the relations of Σ to define the edges of H; that it is a quantifier-free interpretation scheme means that edges of H are defined only by a quantifier-free formula in the language of Σ.

The chromatic polynomial, Tutte polynomial, independence polynomial and other polynomials that we have seen to be determined by right homomorphism profiles are all definable by subgraph expansions in which the range of summation is defined by a predicate expressible in the language of graphs using monadic second-order logic (MSOL). See for example [712] for MSOL-definable graph polynomials. While the interlace polynomial of a graph [48] is MSOL-definable [712], it is not known whether it can be determined by right homomorphism profiles in this way. This prompts the following question:

Problem 22.34. Is every MSOL-definable univariate graph polynomial equivalent to some right (H_q)-profile for some sequence of graphs (H_q)?

From Example 22.5, the left profile by cycles (plus K_1) is equivalent to the characteristic polynomial; Problem 22.34 includes the question whether the characteristic polynomial is equivalent to some right profile as well.

Finally, the deletion–contraction relation for the Tutte polynomial is an example of a size-reducing recurrence formula: the Tutte polynomial of a graph G is expressed in terms of smaller graphs obtained by modifying G locally (remove an edge, contract an edge). We have also seen that the Tutte polynomial of a graph G is determined by the right profile of G by the sequence of q-state Potts model graphs $(K_q^{1,\ell})_{q,\ell \in \mathbb{N}}$. In Section 22.7 we give some other examples of graph polynomials determined in a similar way by right profiles and which have a size-reducing recurrence formula. However, a general construction is lacking:

Problem 22.35. For which weighted graph sequences (H_q) defining a graph polynomial $p(G; q) = \hom(G, H_q)$ is there a size-reducing recurrence formula, that is, a recurrence formula for $p(G)$ expressing it in terms of its values on a finite number of graphs smaller than G?

Part V

Extensions

23

Digraph analogues of the Tutte polynomial

Timothy Y. Chow

Synopsis

This chapter considers analogues of the Tutte polynomial for directed graphs. Although no fully satisfactory analogue of the Tutte polynomial exists for directed graphs, this chapter discusses several candidates that have been proposed.

- The cover polynomial and its multivariate generalizations—the cycle-path indicator polynomial and the path-cycle symmetric function.
- Tutte invariants of alternating dimaps.
- Various digraph polynomials of Gordon and Traldi.
- The B-polynomial of Awan and Bernardi.

23.1 Introduction

A *directed graph* or *digraph* is a graph equipped with an orientation on each edge. It is natural to ask if there is a digraph analogue of the Tutte polynomial. Several proposals have been made, as discussed in this chapter. However, none of them seems to lay claim to being *the* Tutte polynomial of a digraph, so perhaps the correct analogue remains to be discovered, or does not exist.

One key difficulty is that it is not obvious how to define the *contraction* D/e of a digraph D by an edge e. Chung and Graham [320] define contraction in a way that prevents the creation of directed paths and cycles in D/e that do not arise from directed paths and cycles in D. This allows them to define a polynomial called the *cover polynomial* that obeys a deletion–contraction relation and has interesting connections to the chromatic polynomial and classical

DOI: 10.1201/9780429161612-23

rook theory. There are also multivariate generalizations of the cover polynomial due to D'Antona and Munarini [359] and Chow [316].

For *alternating dimaps*, which are special kinds of digraphs embedded in a surface, Farr [480] has defined three minor operations and a notion of an "extended Tutte invariant" that generalizes the Tutte polynomial of a planar graph.

Instead of focusing on the definition of contraction, one can instead seek analogues of the corank–nullity definition of the Tutte polynomial. In this direction Gordon and Traldi [574, 575] have defined several possible digraph analogues of the Tutte polynomial. More about some of these polynomials may be found in Chapter 33.

Very recently, Awan and Bernardi [57] have defined the *B-polynomial* of a digraph. If G is a graph and D is the digraph obtained by replacing each undirected edge by a bidirected edge, then the B-polynomial of D is equivalent to the Tutte polynomial of G. However, the deletion–contraction recurrence for the B-polynomial does not express it in terms of B-polynomials of digraphs with fewer edges, so there is no universality property.

Throughout this chapter, we write $V(D)$ and $E(D)$ for the vertex and edge sets of a digraph D. If the end vertices of an edge are u and v, then we write $u \to v$ and think of the edge as being directed from u (the *tail*) to v (the *head*). The reader is cautioned that in different sections, digraphs may have additional restrictions placed on them.

23.2 The cover polynomial

In this section, unless otherwise stated, digraphs do not have multiple directed edges, but may have loops and may have oppositely directed edges $u \to v$ and $v \to u$.

Chung and Graham [320] defined a contraction operation on digraphs that leads, via a deletion–contraction relation, to a polynomial called the *cover polynomial.*

Definition 23.1. If e is an edge of D, then the *deletion* $D\backslash e$ is the digraph obtained by deleting e from D.

Definition 23.2. If $e = u \to v$ is an edge of D from a vertex u to a vertex v, and $u \neq v$, then the *contraction* D/e is the digraph obtained by deleting all edges leaving u (including e) and all edges entering v, and then merging u and v into a single vertex. See Figure 23.1. If $u = v$, i.e., if e is a loop, then the contraction D/e is the digraph obtained by deleting v and all its incident edges.

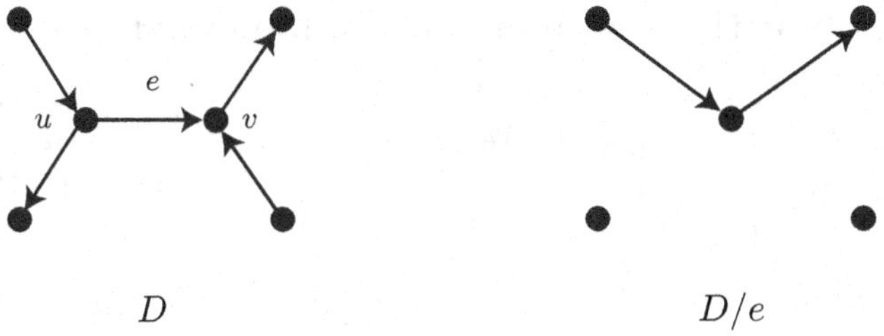

FIGURE 23.1: Chung–Graham contraction by a non-loop e.

Definition 23.3. The *cover polynomial* $C(D; x, y)$ of a digraph D is defined recursively as follows.

1. If I_n is the digraph with $n > 0$ independent vertices and no edges, then
$$C(I_n) = x^{\underline{n}} := x(x-1)\cdots(x-n+1). \tag{23.1}$$
For $n = 0$, define $C(I_0) = 1$.

2. If e is an edge of D that is not a loop, then
$$C(D) = C(D\backslash e) + C(D/e). \tag{23.2}$$

3. If e is a loop, then
$$C(D) = C(D\backslash e) + yC(D/e). \tag{23.3}$$

Chung and Graham show that $C(D; x, y)$ is well-defined, i.e., independent of the order in which edges are chosen in the above recursive procedure.

Example 23.4. Let D be a digraph on two vertices u and v, with a loop on u and two edges $u \to v$ and $v \to u$. Then $C(D; x, y) = x^2 + xy + x + y$.

Definition 23.5. A *path-cycle cover* of a digraph D is a subgraph consisting of a disjoint union of directed paths and directed cycles such that every vertex of D belongs to exactly one directed path or cycle. A path may have zero edges and a loop counts as a cycle. We write $c_D(i, j)$ for the number of path-cycle covers of D with i directed paths and j directed cycles.

The cover polynomial derives its name from the following basic result from [320].

Theorem 23.6. *For a digraph D,*
$$C(D; x, y) = \sum_{i,j} c_D(i, j) x^{\underline{i}} y^j.$$

Corollary 23.7. *If a digraph D is formed by joining the disjoint digraphs D_1 and D_2 with all possible edges $v_1 \to v_2$ with $v_1 \in V(D_1)$ and $v_2 \in V(D_2)$, then $C(D) = C(D_1)C(D_2)$.*

23.2.1 Connections with classical polynomials

If $D = (V, E)$, we may visualize $V \times V$ as a square chessboard, and $E \subseteq V \times V$ as a subset of that chessboard. We obtain a connection between the cover polynomial and *rook theory*, which is the theory of enumerating the number of ways of placing non-attacking rooks on E.

Theorem 23.8. *If $D = (V, E)$ has n vertices, then the number of $(n-i)$-element subsets of E with no two elements in the same row or column is $\sum_j c_D(i,j)$.*

The theorem above is from [320]. In the context of rook theory, Gessel [519] had earlier defined a two-variable polynomial that is very similar to the cover polynomial.

Definition 23.9. If D is a digraph with n vertices, then the *cycle rook polynomial* $r(D; x, y)$ is defined by

$$r(D; x, y) := \sum_{i,j} (-1)^{n-i} c_D(i,j) x^i y^j. \tag{23.4}$$

The single-variable polynomial from rook theory most closely related to the cover polynomial is the *n-factorial (rook) polynomial* of Goldman, Joichi, and White [554, 555], which in our notation coincides with $C(D; x, 1)$. Among other things, Goldman, Joichi, and White showed that the n-factorial polynomial factors into linear factors for *increasing Ferrers boards*, (i.e., a subset of the chessboard such that all squares below or to the right of an included square are also included), and *decreasing Ferrers boards* (i.e., a subset of the chessboard such that all squares below or to the left of an included square are also included). Dworkin [429] generalized this result to the cover polynomial (the result is stated in the following theorem). For increasing Ferrers boards, the result was proved independently by Haglund (unpublished).

Theorem 23.10. *The cover polynomials of increasing Ferrers boards and decreasing Ferrers boards factor completely into linear factors.*

The n-factorial polynomial of a board determines the n-factorial polynomial of the complementary board. This fact extends to the cover polynomial, as proved independently by Gessel (unpublished) and Chow [316].

The *complement* $\overline{D}$ of a digraph D is the digraph with the same vertex set as D and with an edge $u \to v$ (or loop $u \to u$) precisely when there is *not* an edge $u \to v$ (or loop $u \to u$) in D.

Theorem 23.11. *Let D be a digraph with n vertices and $\overline{D}$ be its complement. Then $C(\overline{D}; x, y) = (-1)^n C(D; -x-y, y)$.*

As Chung and Graham noted, the cover polynomial is also related to the chromatic polynomial.

Theorem 23.12. *Let $P = (V, \prec)$ be a partially ordered set. Let $D(P)$ be the digraph with vertex set V that has an edge $u \to v$ precisely when $u \prec v$ in P. Let $G(P)$ be the* incomparability graph *of P, i.e., the undirected graph with vertex set V in which u and v are adjacent precisely when u and v are incomparable in P. Then*

$$C(D(P); x, 1) = \chi(G(P); x).$$

23.2.2 The cycle-path indicator polynomial

D'Antona and Munarini [359] have considered some digraph polynomials that are closely related to the cover polynomial.

Definition 23.13. Let D be a digraph. The *geometric cover polynomial* $\hat{C}(D; x, y)$ is defined by

$$\hat{C}(D; x, y) := \sum_{i,j} c_D(i, j) x^i y^j. \tag{23.5}$$

Warning: D'Antona and Munarini write $C(D; x, y)$ for what in our notation is $\hat{C}(D; y, x)$ and write $\tilde{C}(D; x, y)$ for what in our notation is $C(D; y, x)$.

The polynomial $\hat{C}(D)$ satisfies the deletion–contraction relations (23.2) and (23.3) (with $\hat{C}(D)$ replacing $C(D)$ in the relations). However, $\hat{C}(D)$ satisfies a different base case than $C(D)$: in place of (23.1) we have $\hat{C}(I_n) = x^n$. Also, $\hat{C}(D; x, y) = (-1)^n r(D; -x, y)$, where $r(D)$ is as defined in (23.4).

Example 23.14. For the digraph D in Example 23.4, $\hat{C}(D; x, y) = x^2 + xy + 2x + y$.

More generally, D'Antona and Munarini define a multivariate *cycle-path indicator polynomial* as follows.

Definition 23.15. A *vertex-weighted* digraph is a digraph, possibly with multiple edges, that has a nonnegative integer weight $w(v)$ associated with each vertex v.

Definition 23.16. Let D be a vertex-weighted digraph. Let $x_1, y_1, x_2, y_2, \ldots$ be independent indeterminates. If β is a directed path or a directed cycle with k vertices $v_1, \ldots, v_k$, then define

$$\mathrm{Ind}(\beta) := \begin{cases} x_{k+w(v_1)+\cdots+w(v_k)}, & \text{if } \beta \text{ is a path;} \\ y_{k+w(v_1)+\cdots+w(v_k)}, & \text{if } \beta \text{ is a cycle.} \end{cases}$$

If $\mathscr{C}$ is a path-cycle cover of D, then define

$$\mathrm{Ind}(\mathscr{C}) := \prod_{\beta \in \mathscr{C}} \mathrm{Ind}(\beta),$$

where the product is over all directed paths and cycles β in $\mathscr{C}$. Finally, define the *cycle-path indicator polynomial* of D by

$$\mathrm{Ind}(D) := \sum_{\mathscr{C}} \mathrm{Ind}(\mathscr{C}),$$

where the sum is over all path-cycle covers of D.

Example 23.17. For the D of Example 23.4, and assuming all the vertex weights are zero, $\mathrm{Ind}(D) = x_1^2 + x_1y_1 + 2x_2 + y_2$.

It is easy to see that if all the vertex weights are zero and we set $x_i = x$ and $y_i = y$ for all i, then the cycle-path indicator polynomial coincides with the geometric cover polynomial.

The main result of [359] is a deletion–contraction relation for $\mathrm{Ind}(D)$.

Definition 23.18. Let D be a vertex-weighted digraph with an edge e. The vertex-weighted digraphs $D\backslash e$ and D/e are formed by following Definitions 23.1 and 23.2 and weighting the resulting digraphs as follows. The vertex weights of $D\backslash e$ are the same as the vertex weights of D. If u and v are the original end vertices of e, then the weight of the vertex that they merge into in forming D/e is defined to be $w(u) + w(v) + 1$.

Thus the vertex weights keep a record of edges that have been contracted.

Theorem 23.19. *Let D be a vertex-weighted digraph with an edge e. If e is not a loop, then*

$$\mathrm{Ind}(D) = \mathrm{Ind}(D\backslash e) + \mathrm{Ind}(D/e).$$

If e is a loop on the vertex v, then

$$\mathrm{Ind}(D) = \mathrm{Ind}(D\backslash e) + y_{w(v)+1}\,\mathrm{Ind}(D/e).$$

23.2.3 Computational complexity

Nederlof [874, Section 5] has given a polynomial-space algorithm for computing the cover polynomial. In the other direction, Bläser et al. [140] have shown that computing the cover polynomial is ♯P-hard. (For definitions of standard complexity classes, see for example Papadimitriou [906].) More precisely, we have the following results from [140].

Theorem 23.20. *Computing $C(D;0,0)$, $\hat{C}(D;0,0)$, $C(D;0,-1)$, $\hat{C}(D;0,-1)$, and $C(D;1,-1)$ can be done in polynomial time. For any other fixed rational numbers x and y, computing $C(D;x,y)$ is ♯P-hard with respect to polynomial-time Turing reductions, as is computing $\hat{C}(D;x,y)$.*

For the following theorem, we say that $(x,y) \in \mathbb{Q}^2$ *has a root* if there exists a digraph D such that $\hat{C}(D;x,y) = 0$.

Theorem 23.21. *Let $(x,y) \in \mathbb{Q}^2 \setminus \{(0,0),(0,-1)\}$.*

1. *If $x \geq 0$ and $y = 1$, then there exists a fully polynomial randomized approximation scheme for computing $\hat{C}(D; x, y)$.*

2. *If $1 \neq y > 0$ and (x, y) has a root, then $\hat{C}(D; x, y)$ cannot be approximated within any polynomial factor unless* RP = NP.

3. *If $y \leq 0$ and (x, y) has a root, then $\hat{C}(D; x, y)$ cannot be approximated within any polynomial factor unless* RFP = NP.

Here RFP is the class of all functions computable by a BPP-machine. One might expect that Theorem 23.21 also holds for the cover polynomial $C(D; x, y)$. However, except in certain special cases, this remains an open problem.

23.2.4 The path-cycle symmetric function

Chow [316] defined a symmetric function generalization of the cover polynomial called the *path-cycle symmetric function* that is analogous to the symmetric function generalization of the chromatic polynomial defined and studied by Stanley [1033] (see Section 26.3). To define it, we need some preliminaries on symmetric functions.

A finite sequence $\lambda = (\lambda_1, \ldots, \lambda_\ell)$ of positive integers is said to be an *integer partition of* n if $\lambda_1 \geq \cdots \geq \lambda_\ell$ and the sum of the λ_i's equals n. The λ_i's are called the *parts* of λ.

Definition 23.22. Let $\mathbf{x} = (x_1, x_2, \ldots)$ be a countably infinite sequence of commuting independent indeterminates, and let n be a positive integer. The *power sum symmetric function* p_n is the formal power series

$$p_n(\mathbf{x}) := x_1^n + x_2^n + x_3^n + \cdots . \tag{23.6}$$

Let $\lambda = (\lambda_1, \ldots, \lambda_\ell)$ be an integer partition of n. The *power sum symmetric function* p_λ is the formal power series

$$p_\lambda(\mathbf{x}) := \prod_{i=1}^{\ell} p_{\lambda_i}(\mathbf{x}). \tag{23.7}$$

The *augmented monomial symmetric function* $\tilde{m}_\lambda$ is the formal power series

$$\tilde{m}_\lambda(\mathbf{x}) := \sum_{(i_1,\ldots,i_\ell)} x_{i_1}^{\lambda_1} \cdots x_{i_\ell}^{\lambda_\ell}, \tag{23.8}$$

where the sum is over all length-ℓ sequences $(i_1, \ldots, i_\ell)$ of *distinct* positive integers. If $\ell = 0$, then we set $p_\lambda(\mathbf{x}) = \tilde{m}_\lambda(\mathbf{x}) = 1$.

Definition 23.23. Let $\mathbf{x} = (x_1, x_2, \ldots)$ and $\mathbf{y} = (y_1, y_2, \ldots)$ be two distinct countably infinite sequences of mutually commuting independent indeterminates. Let D be a digraph. If $\mathscr{C}$ is a path-cycle cover of D, let $\lambda(\mathscr{C})$ denote the integer partition whose parts are the lengths of the paths in $\mathscr{C}$ arranged in decreasing order, and let $\mu(\mathscr{C})$ denote the integer partition whose parts are the lengths of the cycles in $\mathscr{C}$ arranged in decreasing order. The *path-cycle symmetric function* $\Xi(D; \mathbf{x}, \mathbf{y})$ is the power series in $\mathbf{x}$ and $\mathbf{y}$ defined by

$$\Xi(D; \mathbf{x}, \mathbf{y}) := \sum_{\mathscr{C}} \tilde{m}_{\lambda(\mathscr{C})}(\mathbf{x}) p_{\mu(\mathscr{C})}(\mathbf{y}), \tag{23.9}$$

where the sum is over all path-cycle covers $\mathscr{C}$ of D.

Example 23.24. For the digraph D in Example 23.4, $\Xi(D; \mathbf{x}, \mathbf{y}) = \tilde{m}_{1,1}(\mathbf{x}) + \tilde{m}_1(\mathbf{x})p_1(\mathbf{y}) + 2\tilde{m}_2(\mathbf{x}) + p_2(\mathbf{y})$.

Note that if D is acyclic, then $\Xi(D)$ is a power series in the $\mathbf{x}$ variables only.

The connection with the cover polynomial is given by the following proposition (from [316]), whose proof is just a definition chase.

Proposition 23.25. *If i of the $\mathbf{x}$ variables and j of the $\mathbf{y}$ variables are set to 1 and the rest are set to zero, then $\Xi(D)$ becomes a finite sum that evaluates to $C(D; i, j)$.*

Unfortunately, the path-cycle symmetric function does not seem to satisfy a deletion–contraction relation. However, several facts about the cover polynomial generalize readily. The results stated in the remainder of this section are from [316].

Proposition 23.26. *If D, D_1 and D_2 are as in Corollary 23.7, then $\Xi(D; \mathbf{x}, \mathbf{y}) = \Xi(D_1; \mathbf{x}, \mathbf{y})\Xi(D_2; \mathbf{x}, \mathbf{y})$.*

Proposition 23.27. *Let P be a finite partially ordered set. In the notation of Theorem 23.12, $\Xi(D(P); \mathbf{x}, \mathbf{y}) = X(G(P); \mathbf{x})$, where X denotes Stanley's chromatic symmetric function (see Definition 26.45).*

There is also a generalization of Theorem 23.11, but its statement and proof are more involved. We need some more definitions. It is well known [1038, Corollary 7.7.2] that the power-sum symmetric functions are algebraically independent and generate the ring of symmetric functions as a $\mathbb{Q}$-algebra.

Definition 23.28. Define an endomorphism ω on the ring of symmetric functions by setting $\omega p_n(\mathbf{x}) = -p_n(-\mathbf{x})$, where the notation "$-\mathbf{x}$" indicates that each variable x_i should be replaced by $-x_i$. Let $f_\lambda(\mathbf{x}) := \omega\tilde{m}_\lambda(\mathbf{x})$ for any integer partition λ.

Theorem 23.29. *If $\bar{D}$ is the complement of the digraph D, then*

$$\Xi(\bar{D}; \mathbf{x}, \mathbf{y}) = \sum_{\mathscr{C}} f_{\lambda(\mathscr{C})}(\mathbf{x} \cup \mathbf{y}) p_{\mu(\mathscr{C})}(-\mathbf{y}), \tag{23.10}$$

where the sum is over all path-cycle covers $\mathscr{C}$ of D, and $\mathbf{x} \cup \mathbf{y}$ denotes the union of all the $\mathbf{x}$ and the $\mathbf{y}$ variables.

We remark in passing that Theorems 23.11 and 23.29 are examples of *combinatorial reciprocity theorems* (theorems that give combinatorial interpretations of combinatorially defined polynomials evaluated at negative integers) and have been generalized further by Haglund [597] and Lass [758] (see also [315] and [553]).

Chow [316] proves several results about expanding $\Xi(D)$ in terms of various symmetric function bases for special digraphs. We mention just one such result.

Theorem 23.30. *If D is an acyclic digraph, then $\omega\Xi(D)$ is a nonnegative linear combination of power-sum symmetric functions.*

Note that $\omega X(G)$ is also a nonnegative linear combination of power-sum symmetric functions [1033, Corollary 2.7].

Chung and Graham [319] have generalized the cover polynomial to the *matrix cover polynomial*, which is an invariant of a matrix with elements taken from an arbitrary commutative ring with identity. The matrix cover polynomial has a symmetric function generalization that is similar in spirit to the path-cycle symmetric function.

23.3 Tutte invariants for alternating dimaps

In [480], Farr defined Tutte invariants and extended Tutte invariants for alternating dimaps, which may be thought of as a special class of digraphs equipped with additional structure.

Definition 23.31. An *alternating dimap* is a digraph with no isolated vertices, cellularly embedded in a disjoint union of oriented surfaces, where each vertex has even degree and, for each vertex v, the edges incident with v are directed alternately into and out of v, when considered in the order in which they appear around v in the embedding. An alternating dimap may have loops and/or multiple edges, and may be empty (with no vertices, edges, or faces).

Definition 23.32. In an alternating dimap, the edges around a face are consistently directed. A face is called a *c-face* or an *a-face* according to whether this direction is clockwise or anticlockwise (i.e., with the orientation or against it). Every edge lies on a c-face and also on an a-face; its *right successor* (respectively, *left successor*) is the next edge along the c-face (respectively, the a-face).

Definition 23.33. Let $\omega := \exp(2\pi i/3)$ (not to be confused with the map ω of Definition 23.28). A *1-loop* is an edge whose head has degree two. An *ω-loop* is an edge forming a single-edge a-face. An *ω^2-loop* is an edge forming a single-edge c-face. A *triloop* is an edge that is a 1-loop, an ω-loop, or an ω^2-loop. An *ultraloop* is a triloop which, together with its vertex, constitutes a connected component of the graph. (Note that an ultraloop is simultaneously a 1-loop, an ω-loop, and an ω^2-loop.) A triloop is *proper* if it is not also an ultraloop.

Next, we define the minor operations. For alternating dimaps, deleting an edge e does not usually produce an alternating dimap, but if e is an ω-loop, an ω^2-loop, or an ultraloop, then its deletion is straightforward, and is denoted by $G\backslash e$.

Definition 23.34. If G is an alternating dimap and e is an edge of G, then the *1-reduction* or *contraction* $G[1]e$ is defined as follows.

1. If the endpoints of e do not coincide, then $G[1]e$ is formed by deleting the edge e and identifying its endpoints, while preserving the order of the edges and faces around vertices.
2. If e is an ω-loop or an ω^2-loop, then $G[1]e$ is formed just by deleting e.
3. Otherwise, let v be the vertex of e, and let the edges incident with v, in cyclic order around v starting with e directed into v, be $e, a_1, b_1, \ldots, a_k, b_k, e, c_1, d_1, \ldots, c_l, d_l$. So the a_i and d_i are directed out of v while the b_i and c_i are directed into v. Replace v by two new vertices, v_1 and v_2, and reconnect the edges a_i, b_i, c_i, d_i as follows. The tail of each a_i and the head of each b_i become v_1 instead of v, while the head of each c_i and the tail of each d_i become v_2 instead of v. The edge e is deleted. The cyclic orderings of edges around v_1 and v_2 are those induced by the ordering around v.

As compensation for the absence of a true deletion operation, we have two other minor operations, ω-reduction and ω^2-reduction.

Definition 23.35. If G is an alternating dimap and e is an edge of G, then the *ω-reduction* $G[\omega]e$ (respectively, the *ω^2-reduction* $G[\omega^2]e$) is formed as follows. Let f be the left (respectively, right) successor of e, with tail v and head w. Delete e and f, and, if $e \neq f$, replace them with a new edge g from the tail of e to w. If the degree of v is two, then v is deleted.

23.3.1 Simple Tutte invariants for alternating dimaps

The reduction operations give rise the following definition of a Tutte invariant for alternating dimaps.

Definition 23.36. A *simple Tutte invariant* for alternating dimaps is a function F defined on every alternating dimap such that F is invariant under isomorphism, $F(\varnothing) = 1$, and there exist w, x, y, z such that, for any alternating dimap G,

1. for any ultraloop e of G, $F(G) = wF(G\backslash e)$;
2. for any proper 1-loop e of G, $F(G) = xF(G[1]e)$;

3. for any proper ω-loop e of G, $F(G) = yF(G[\omega]e)$;

4. for any proper ω^2-loop e of G, $F(G) = zF(G[\omega^2]e)$;

5. for any edge e of G that is not an ultraloop or a triloop,

$$F(G) = F(G[1]e) + F(G[\omega]e) + F(G[\omega^2]e).$$

However, it turns out that there are not many simple Tutte invariants, as shown in [480].

Theorem 23.37. *The only simple Tutte invariants of alternating dimaps are:*

1. $F(G) = 0$ *for nonempty* G, *with* $w = 0$;

2. $F(G) = 3^{|E(G)|}$, *with* $w = x = y = z = 3$;

3. $F(G) = (-1)^{|V(G)|}$, *with* $y = z = 1$ *and* $x = w = -1$;

4. $F(G) = (-1)^{c(G)}$, *with* $x = z = 1$ *and* $y = w = -1$ *(where* $c(G)$ *is the number of c-faces of* G*);*

5. $F(G) = (-1)^{a(G)}$, *with* $x = y = 1$ *and* $z = w = -1$ *(where* $a(G)$ *is the number of a-faces of* G*).*

23.3.2 Extended Tutte invariants for alternating dimaps

Other definitions of Tutte invariants for alternating dimaps are possible.

Definition 23.38. Let G be an alternating dimap. A *1-semiloop* is an edge which is a loop in the underlying undirected graph of G. If e is an edge and f is its right (respectively, left) successor, then e is an *ω-semiloop* (respectively, *ω^2-semiloop*) if

1. $e = f$, or

2. $e \neq f$ and $\{e, f\}$ is a cutset of G, or

3. $e \neq f$ and deleting both e and f increases the genus of the underlying undirected graph of G.

A 1-semiloop, ω-semiloop, or ω^2-semiloop is *proper* if it is not a triloop.

Definition 23.39. An *extended Tutte invariant* for alternating dimaps is a function F defined on every alternating dimap such that F is invariant under isomorphism, $F(\varnothing) = 1$, and there exist w, x, y, z, a, b, c, d, e, f, g, h, i, j, k, l, such that, for any alternating dimap G,

1. for any ultraloop e of G, $F(G) = wF(G\backslash e)$;

2. for any proper 1-loop e of G, $F(G) = xF(G[1]e)$;

3. for any proper ω-loop e of G, $F(G) = yF(G[\omega]e)$;
4. for any proper ω^2-loop e of G, $F(G) = zF(G[\omega^2]e)$;
5. for any proper 1-semiloop e of G,
$$F(G) = aF(G[1]e) + bF(G[\omega]e) + cF(G[\omega^2]e);$$
6. for any proper ω-semiloop e of G,
$$F(G) = dF(G[1]e) + eF(G[\omega]e) + fF(G[\omega^2]e);$$
7. for any proper ω^2-semiloop e of G,
$$F(G) = gF(G[1]e) + hF(G[\omega]e) + iF(G[\omega^2]e);$$
8. for any edge e of G that is not an ultraloop or a triloop or a semiloop,
$$F(G) = jF(G[1]e) + kF(G[\omega]e) + lF(G[\omega^2]e).$$

Farr shows that the Tutte polynomial of a planar graph may be viewed as an extended Tutte invariant, as follows.

Definition 23.40. To any undirected graph G cellularly embedded in an oriented surface we can associate two alternating dimaps $\mathrm{alt}_c(G)$ and $\mathrm{alt}_a(G)$ as follows. For $\mathrm{alt}_c(G)$ (respectively, $\mathrm{alt}_a(G)$), replace each edge $e = (u, v)$ by a pair of oppositely directed edges $u \to v$ and $v \to u$, forming a clockwise (respectively, anticlockwise) face of size two. The faces of G now all correspond to anticlockwise (respectively, clockwise) faces in $\mathrm{alt}_c(G)$ (respectively, $\mathrm{alt}_a(G)$).

Theorem 23.41. *The Tutte polynomial of a plane graph G is an extended Tutte invariant of* $\mathrm{alt}_c(G)$ *and of* $\mathrm{alt}_a(G)$.

23.4 Gordon–Traldi polynomials

Gordon and Traldi [574, 575] introduced eight different polynomials $f_1, \ldots, f_8$ for directed graphs, each of which is some kind of analogue of the Tutte polynomial. Their general approach is to define a function r_i, for $1 \leq i \leq 8$, called a *rank function*, on the edges of a digraph D, and then define a corresponding Tutte-like polynomial f_i by the corank–nullity formula

$$f_i(D; t, z) := \sum_{A \subseteq E(D)} t^{r_i(E(D)) - r_i(A)} z^{|A| - r_i(A)}. \qquad (23.11)$$

They also define functions c_i, and define a Tutte-like polynomial by the formula

$$f_i(D;t,z) := \sum_{A \subseteq E(D)} t^{c_i(A)-c_i(E(D))} z^{|A|+c_i(A)-|V(D)|}. \tag{23.12}$$

By setting $r_i(A) = |V(D)| - c_i(A)$, we see that Equations (23.11) and (23.12) give equivalent polynomials. For an undirected graph, we recover the Tutte polynomial by setting $r_i(A)$ to be the cardinality of the largest acyclic subset of A or $c_i(A)$ to be the number of connected components of D if the edge set is restricted to A.

Not much is known about most of the polynomials $f_1, \ldots, f_8$, other than that various specializations of them count analogues of bases, spanning sets, and independent sets. Therefore, in most cases, we limit ourselves to providing just the definitions, referring the reader to [574, 575] for further details.

23.4.1 Polynomials for rooted digraphs

Gordon and Traldi first consider *rooted digraphs*, i.e., digraphs D with a distinguished vertex $*$ called the *root.*

Definition 23.42. A subgraph T of a rooted digraph is a *$*$-rooted arborescence* if for every vertex v of T, there is a unique directed path in T from $*$ to v. A *$*$-rooted forest of arborescences* is a vertex-disjoint union of arborescences rooted at $*, v_1, v_2, \ldots$ for some vertices $v_1, v_2, \ldots$.

Definition 23.43. Let f_1, f_2 and f_3 be defined via Equation (23.11) using the following rank functions respectively:

1. $r_1(A) := \max\{|T| : T \subseteq A \text{ is a } *\text{-rooted arborescence}\}$,
2. $r_2(A) := \max\{|T \cap A| : T \subseteq E(D) \text{ is a } *\text{-rooted arborescence}\}$,
3. $r_3(A) := \max\{|F| : F \subseteq A \text{ is a } *\text{-rooted forest of arborescences}\}$.

The polynomials f_1 and f_2 can be defined for any greedoid (see Chapter 33), and f_1 in particular has been studied by Gordon and McMahon [570] and McMahon [834]. We mention here one of the main theorems of the latter.

Definition 23.44. In a rooted digraph, an edge e from u to v is a *greedoid loop* if v lies on every directed path from the root to u.

Theorem 23.45. *Let D be a rooted digraph with no greedoid loops. Then D has a directed cycle if and only if $z+1$ divides $f_1(D;t,z)$.*

23.4.2 Polynomials for unrooted digraphs

Definition 23.46. Let D a be an unrooted digraph. A set $F \subseteq E(D)$ is a *forest of rooted arborescences* if it is a vertex-disjoint union of arborescences rooted at $v_1, v_2, \ldots$ for some vertices $v_1, v_2, \ldots$. Let $r_4(A)$ be the maximum size of a forest of rooted arborescences contained in A, and let f_4 be the polynomial defined by setting $r = r_4$ in Equation (23.11).

Definition 23.47. A digraph D is *strongly connected* if, for every pair of vertices u and v, there exists a directed path from u to v. Let f_5 and f_6 be defined via Equation (23.12) using the following functions respectively:

1. $c_5(A) :=$ the number of strongly connected components of D if the edge set is restricted to A;

2. $c_6(A) := 1 +$ the smallest cardinality of a set R of (directed) edges such that $R \cup A$ strongly connects all the vertices of D. (The edges of R may or may not be edges of D.)

23.4.3 Order-dependent polynomials

Finally, we describe f_7 and f_8. In the case of f_7, we follow Gordon and Traldi by giving the recursive definition directly instead of describing r_7.

Definition 23.48. Let D be a rooted digraph, with root $*$. An edge in D is a *2-isthmus* if it is in every maximal $*$-rooted arborescence, and is a *2-loop* if it is in no maximal $*$-rooted arborescence. A 2-loop whose initial and terminal vertices coincide is called an *ordinary loop*; otherwise, a 2-loop is called a *reversed loop*.

Definition 23.49. Let D be a rooted digraph whose underlying undirected graph is connected, and let D be equipped with a total ordering O of its edges. We define a polynomial $f_7(D, O; x, y, z)$ as follows.

1. If $D = \{*\}$, then $f_7(D) = 1$.

2. Let e be the first edge in the ordering O which emanates from $*$.

   (a) $f_7(D) = xf_7(D/e)$ if e is a 2-isthmus, where D/e is the digraph obtained from D by identifying the end vertices of e and deleting e.

   (b) $f_7(D) = yf_7(D\backslash e)$ if e is an ordinary loop.

   (c) $f_7(D) = f_7(D\backslash e) + f_7(D/e)$ otherwise.

3. If no edge emanates from $*$, then let e be the first edge directed into $*$, so that e is a reversed loop. Then $f_7(D) = zf_7(D/e)$.

Definition 23.50. Let D be a rooted digraph equipped with a total ordering of its vertices. Let $r_8(A)$ be the maximum size of a subset of A that is a $*$-rooted forest of arborescences in which each arborescence is rooted at its least vertex, and let f_8 be the polynomial defined by setting $r = r_8$ in Equation (23.11).

23.5 The B-polynomial

Recently [57], Awan and Bernardi have defined a three-variable digraph polynomial that they call the *B-polynomial.* Given a function $f : V(D) \to \{1, 2, \ldots, q\}$, let $\gamma(f)$ be the number of edges $u \to v$ in $E(D)$ such that $f(v) > f(u)$, and let $\lambda(f)$ be the number of edges $u \to v$ in $E(D)$ such that $f(v) < f(u)$. Then

$$B(D; q, y, z) := \sum_{f:V(D)\to\{1,2,\ldots,q\}} y^{\gamma(f)} z^{\lambda(f)}.$$

They show that if G is an undirected graph, and D is the directed graph obtained by replacing each edge (u, v) of G with the pair of directed edges $u \to v$ and $v \to u$, then $B(D; q, y, z)$ is equivalent, up to change of variables, to the Tutte polynomial of G. In this sense, the B-polynomial generalizes the Tutte polynomial.

The B-polynomial does satisfy a certain kind of deletion–contraction recurrence, but the recurrence does not express $B(D; q, y, z)$ in terms of the B-polynomials of digraphs with fewer edges, and so it does not yield any "universality" property. However, the B-polynomial does detect several important properties of a digraph, such as acyclicity, the length of the longest directed path, and the number of strongly connected components. It also satisfies a partial planar duality relation.

The B-polynomial has a generalization to a quasisymmetric function in two sets of variables, which yields a digraph generalization of Stanley's symmetric function generalization of the Tutte polynomial. It also generalizes Ellzey's chromatic quasisymmetric function (in one set of variables) for digraphs [464], which in turn is a generalization of Shareshian and Wachs's chromatic quasisymmetric function of a graph [1000].

23.6 Open problems

1. Does there exist another definition of a digraph Tutte polynomial that is more satisfying than the ones given in this chapter?

2. Does Theorem 23.21, or something like it, hold for the cover polynomial?

3. If P is a unit interval order, is $\Xi(D(P))$ a nonnegative combination of elementary symmetric functions? This is an old conjecture of Stanley and Stembridge [1042].

4. Characterize all extended Tutte invariants for alternating dimaps. See [1180] for some recent progress on this question.

24

Multivariable, parameterized, and colored extensions of the Tutte polynomial

Lorenzo Traldi

Synopsis

Multivariable and paramaterized versions of various forms of the Tutte polynomial arise from assigning parameters to the edges of a graph or to the elements of the ground set of a matroid. These parameters can capture combinatorial properties such as polynomial computation trees and series-parallel reductions, as well as information relevant to applications such as statistical mechanics models and resistances of circuit elements. This chapter covers several forms of these generalizations.

- A general form for doubly parameterized Tutte polynomials.
- Formulas for duality, direct sums, and various types of edges (loops, bridges, series-parallel, etc.).
- Equivalence with the cycle matroid.
- Deletion–contraction form and an activities expansion.
- The colored Tutte polynomial and strong Tutte functions.
- Ported polynomials, the relative Tutte polynomial, knot invariants, and variations.

24.1 Introduction

Multivariable versions of the Tutte and dichromatic polynomials arise in many applications of graph theory that involve assigning parameters (or variables

DOI: 10.1201/9780429161612-24

or weights) to the edges of a graph. In various contexts, these generalizations of the Tutte polynomial may be referred to as multivariable, multivariate, parameterized, colored, or weighted Tutted polynomials. Examples of such applications include single, double and triple bonds in molecules; variable currents, resistances, and voltages in electrical circuits; positive and negative crossings in knot diagrams; probabilities of edge or bond failure in unreliable networks and random graphs; and interaction energies in thermodynamic systems. In addition, parameterized Tutte polynomials are of conceptual and computational value as they can be used to highlight and use various properties of Tutte polynomials to construct efficient algorithms. We focus here on definitions and theoretical properties of multivariable Tutte polynomials, but note that detailed discussions of several of these applications and related topics appear elsewhere in this handbook, for example in Chapters 9, 20, 25, and 26. We also note that in the literature (and elsewhere in this handbook), it is common to blur the distinction between parameterized and multivariate versions of graph polynomials; however in this chapter we make a clear distinction between the two.

24.2 Parameterized and multivariate Tutte polynomials

In this chapter we generally use bold letters to denote functions mapping E into some commutative ring with unity $\mathfrak{R}$, and subscripts to denote the value of these functions on an element e of E. For example, if $\mathbf{a} : E \to \mathfrak{R}$, and $e \in E$ then we write a_e for $\mathbf{a}(e)$. The function $\mathbf{a}$ induces functions on $E(G\backslash e)$, $E(G/e)$, and $E(G^*)$ in the natural way, and, by a standard abuse of notation, we also denote these functions by $\mathbf{a}$.

Definition 24.1. Suppose $G = (V, E)$ is a graph and $\mathbf{a}, \mathbf{b} : E \to \mathfrak{R}$ for some commutative ring with unity $\mathfrak{R}$. Then the *standard doubly parameterized Tutte polynomial* of G with parameters $\mathbf{a}$ and $\mathbf{b}$ is the polynomial $T_{\text{par}}(G; x, y, \mathbf{a}, \mathbf{b})$ in $\mathfrak{R}[x, y]$ given by:

$$T_{\text{par}}(G; x, y, \mathbf{a}, \mathbf{b}) = \sum_{A \subseteq E} \left(\prod_{e \in A} a_e \right) \left(\prod_{e \notin A} b_e \right) (x-1)^{r(E)-r(A)} (y-1)^{n(A)},$$

recalling that we write a_e for $\mathbf{a}(e)$, and b_e for $\mathbf{b}(e)$.

Notice that the formula of Definition 24.1 is obtained by inserting a_e and b_e values into the state sum definition of the Tutte polynomial (Definition 2.3).

The most general version of $T_{\text{par}}(G; x, y, \mathbf{a}, \mathbf{b})$ is obtained by using independent indeterminates for the various parameters a_e and b_e, with $\mathfrak{R}$ the polynomial ring $\mathbb{Z}[a_e, b_e : e \in E]$. We denote this version of the polynomial by $T_{\text{gen}}(G; x, y, \mathbf{a}, \mathbf{b})$.

Definition 24.2. Suppose $G = (V, E)$ is a graph and $\mathbf{a} = \{a_e : e \in E\}$ and $\mathbf{b} = \{b_e : e \in E\}$ are two sets of indeterminates. Then the *doubly-multivariate Tutte polynomial* is

$$T_{\text{gen}}(G; x, y, \mathbf{a}, \mathbf{b}) = \sum_{A \subseteq E} \left(\prod_{e \in A} a_e \right) \left(\prod_{e \notin A} b_e \right) (x-1)^{r(E)-r(A)} (y-1)^{n(A)}.$$

It is a polynomial in the ring $\mathbb{Z}[x, y, a_e, b_e : e \in E]$.

All other standard parameterized Tutte polynomials are obtained from $T_{\text{gen}}(G; x, y, \mathbf{a}, \mathbf{b})$ by evaluating the a_e and b_e at the parameters. In particular this means that many results about T_{gen}, such as its deletion–contraction relations, induce results for standard doubly parameterized Tutte polynomials.

Definitions 24.1 and 24.2 immediately generalize to an arbitrary matroid M whose ground set is equipped with parameter functions $\mathbf{a}$ and $\mathbf{b}$. We denote the resulting polynomials $T_{\text{par}}(M; x, y, \mathbf{a}, \mathbf{b})$ and $T_{\text{gen}}(M; x, y, \mathbf{a}, \mathbf{b})$.

As detailed in Chapter 2, the classical Tutte polynomial is determined by deletion–contraction relations. The multivariate polynomial $T_{\text{gen}}(G; x, y, \mathbf{a}, \mathbf{b})$ also has a deletion–contraction definition. (The analogous definition applies to matroids as well.)

Definition 24.3. The polynomial $T_{\text{gen}}(G; x, y, \mathbf{a}, \mathbf{b})$ may be computed by repeated application of the following identities.

1. If e is a bridge of G, then

$$T_{\text{gen}}(G; x, y, \mathbf{a}, \mathbf{b}) = (a_e + b_e(x-1)) \cdot T_{\text{gen}}(G/e; x, y, \mathbf{a}, \mathbf{b}).$$

2. If e is a loop of G, then

$$T_{\text{gen}}(G; x, y, \mathbf{a}, \mathbf{b}) = (a_e(y-1) + b_e) \cdot T_{\text{gen}}(G \backslash e; x, y, \mathbf{a}, \mathbf{b}).$$

3. If e is an ordinary edge of G, then

$$T_{\text{gen}}(G; x, y, \mathbf{a}, \mathbf{b}) = a_e \cdot T_{\text{gen}}(G/e; x, y, \mathbf{a}, \mathbf{b}) + b_e \cdot T_{\text{gen}}(G \backslash e; x, y, \mathbf{a}, \mathbf{b}).$$

4. If G has no edges, then $T_{\text{gen}}(G; x, y, \mathbf{a}, \mathbf{b}) = 1$.

As discussed in Chapter 2, the recursive calculation of $T(G; x, y)$ yields an activities formula that is a sum indexed by maximal spanning forests of G. The same reasoning implies that the deletion–contraction relations for $T_{\text{gen}}(G; x, y, \mathbf{a}, \mathbf{b})$ yield the following parameterized and multivariate forms of the activities expansion.

Definition 24.4. For a graph G with a fixed linear order of its edges,

$$T_{\text{gen}}(G; x, y, \mathbf{a}, \mathbf{b}) = \sum_{\substack{F \text{ a max.} \\ \text{span. forest}}} \Big(\prod_{e \in \text{IA}(F)} (a_e + b_e(x-1)) \Big) \Big(\prod_{e \in F \setminus \text{IA}(F)} a_e \Big) \Big(\prod_{e \in \text{EA}(F)} (a_e(y-1) + b_e) \Big) \Big(\prod_{e \notin F \cup \text{EA}(F)} b_e \Big). \tag{24.1}$$

Here the sum is over the maximal spanning forests F of G, and for each F, we let $\text{IA}(F)$ and $\text{EA}(F)$ denote the sets of edges of G that are internally and externally active with respect to F, respectively.

If M is a matroid, then the same reasoning yields an expansion of $T_{\text{gen}}(M; x, y, \mathbf{a}, \mathbf{b})$ indexed by the bases of M. Notice that if we take $a_e = 1$ and $b_e = 1$ in (24.1), then we obtain the activities expansion of $T(G; x, y)$ given in Equation (2.8).

We do not know the earliest instance of a formula like (24.1). It may be that activities with parameters first appeared in the mathematical literature as part of Thistlethwaite's 1987 analysis of the Jones polynomial [1061]. Kirchhoff used trees and cycles to analyze electrical circuits 140 years earlier [691], though, and it may well be that an equivalent formulation has appeared in the literature of electrical theory.

24.2.1 The multivariable dichromatic polynomial

In the classical setting, the Tutte polynomial $T(G; x, y)$ and the dichromatic polynomial $Z(G; x, y)$ are related by a simple transform (see Theorem 2.6), and one or the other form is better suited for particular applications. Here too there are parameterized and multivariable generalizations of the dichromatic polynomial, which in turn are translates of $T_{\text{par}}(G; x, y, \mathbf{a}, \mathbf{b})$ and $T_{\text{gen}}(G; x, y, \mathbf{a}, \mathbf{b})$.

Definition 24.5. Given a graph $G = (V, E)$ and a weight function $\mathbf{v} : E \to \mathfrak{R}$, the *parameterized dichromatic polynomial* is

$$Z_{\text{par}}(G; q, \mathbf{v}) = \sum_{A \subseteq E} q^{k(A)} \prod_{e \in A} v_e.$$

As with $T_{\text{par}}(G; x, y, \mathbf{a}, \mathbf{b})$, a most general multivariate polynomial is obtained by using independent indeterminates for the parameters v_e, with $\mathfrak{R}$ the polynomial ring $\mathbb{Z}[v_e : e \in E]$.

Definition 24.6. Suppose $G = (V, E)$ is a graph and $\mathbf{v} = \{v_e : e \in E\}$ is a set of indeterminates. Then the *multivariate dichromatic polynomial* is

$$Z_{\text{gen}}(G; q, \mathbf{v}) = \sum_{A \subseteq E} q^{k(A)} \prod_{e \in A} v_e.$$

It is a polynomial in the ring $\mathbb{Z}[q, v_e : e \in E]$.

Notation 24.7. In the literature, it is common to blur the distinction between parameterized and multivariate versions of the dichromatic polynomial, with the term *multivariate Tutte polynomial* or *dichromatic polynomial of weighted graphs* commonly used to refer to either depending upon context. This convention appears elsewhere in this handbook, where in both cases the polynomial is referred to as the *multivariate Tutte polynomial* and denoted simply by $Z(G; q, \mathbf{v})$. However, because of the particular focus of this chapter, we distinguish the two functions precisely here with the notation Z_{gen} and Z_{par}.

That a parameterized dichromatic polynomial is a multiple of an evaluation of a standard doubly parameterized Tutte polynomial follows immediately from Definition 24.1 with

$$Z_{\text{par}}(G; q, \mathbf{v}) = q^{k(G)} \cdot T_{\text{par}}(G; q+1, 2, \mathbf{v}, \mathbf{1}), \tag{24.2}$$

where $\mathbf{1}$ is the function that maps each edge to the unit of the ring. Furthermore, the multivariable dichromatic polynomial and the doubly-multivariate Tutte polynomial can be related through the following.

Proposition 24.8. *Suppose $G = (V, E)$ is a graph and $\mathbf{a}, \mathbf{b}, \mathbf{v}$ are functions from E to some commutative ring with unity $\mathfrak{R}$. Suppose also that for each e we have $b_e v_e = (y-1)a_e$. Then*

$$\left(\prod_{e \in E} b_e\right) Z_{\text{par}}(G; (x-1)(y-1), \mathbf{v}) \\ = (x-1)^{k(G)}(y-1)^{v(G)} T_{\text{par}}(G; x, y, \mathbf{a}, \mathbf{b}). \tag{24.3}$$

Versions of $Z_{\text{gen}}(G; q, \mathbf{v})$ and $Z_{\text{par}}(G; q, \mathbf{v})$ are useful in knot theory (where the edge parameters indicate crossing types) and in statistical mechanics (where the edge parameters give interaction energies). See Chapters 3, 18, and 20 for discussions of these applications. The multivariable dichromatic polynomial also appears in the study of the zeros of the Tutte polynomial (see Chapter 25).

The multivariable dichromatic polynomial $Z_{\text{gen}}(G; q, \mathbf{v})$ has a recursive definition analogous to that for the doubly-multivariate Tutte polynomial in Definition 24.3. This can be obtained from Definition 24.3 and Equation (24.2).

Proposition 24.9. *The polynomial $Z_{\text{gen}}(G; q, \mathbf{v})$ may be computed by repeated application of the following identities.*

1. *If e is a bridge of G, then $Z_{\text{gen}}(G; q, \mathbf{v}) = (v_e + q) \cdot Z_{\text{gen}}(G/e; q, \mathbf{v})$.*
2. *If e is a loop of G, then $Z_{\text{gen}}(G; q, \mathbf{v}) = (v_e + 1) \cdot Z_{\text{gen}}(G \backslash e; q, \mathbf{v})$.*
3. *If e is an ordinary edge of G, then $Z_{\text{gen}}(G; q, \mathbf{v}) = v_e \cdot Z_{\text{gen}}(G/e; q, \mathbf{v}) + Z_{\text{gen}}(G \backslash e; q, \mathbf{v})$.*

4. If G has no edges and n vertices, then $Z_{\text{gen}}(G; q, \mathbf{v}) = q^n$.

As in Lemma 25.4, these properties can be used to write Items 1–3 of Proposition 24.9 as a single identity by stating that for *any* edge e,

$$Z_{\text{gen}}(G; q, \mathbf{v}) = v_e \cdot Z_{\text{gen}}(G/e; q, \mathbf{v}) + Z_{\text{gen}}(G\backslash e; q, \mathbf{v}). \qquad (24.4)$$

24.2.2 Equivalence with cycle matroids

The classical Tutte polynomial $T(G; x, y)$ is an (unweighted) double generating function where the coefficient of $(x-1)^{\alpha}(y-1)^{\beta}$ is the number of subsets $A \subseteq E$ with corank $r(E) - r(A) = \alpha$ and nullity $n(A) = \beta$. One might be tempted to think that parameterized Tutte polynomials are parameterized forms of this double generating function, but in general, this view is unnecessarily complicated. In fact, $T_{\text{gen}}(G; x, y, \mathbf{a}, \mathbf{b})$ is simply a way to present the cycle matroid of G. This can be seen by noting that for each $A \subseteq E$, the term corresponding to A in $T_{\text{gen}}(G; x, y, \mathbf{a}, \mathbf{b})$,

$$\left(\prod_{e \in A} a_e\right)\left(\prod_{e \notin A} b_e\right)(x-1)^{r(E)-r(A)}(y-1)^{n(A)},$$

determines both A and the nullity $n(A)$. For ease of reference we state this interpretation formally as follows, noting that it also applies to matroids to show that if M is any matroid, then $T_{\text{gen}}(M; x, y, \mathbf{a}, \mathbf{b})$ is equivalent to M itself.

Proposition 24.10. *Let G and G' be graphs, and let $\beta : E(G) \to E(G')$ be a bijection. Then β defines an isomorphism between the cycle matroids $M(G)$ and $M(G')$ if and only if $T_{\text{gen}}(G'; x, y, \mathbf{a}, \mathbf{b})$ is the polynomial obtained from $T_{\text{gen}}(G; x, y, \mathbf{a}, \mathbf{b})$ using the substitutions $a_e \mapsto a_{\beta(e)}$ and $b_e \mapsto b_{\beta(e)}$ for every $e \in E(G)$.*

We often summarize Proposition 24.10 by saying that $T_{\text{gen}}(G; x, y, \mathbf{a}, \mathbf{b})$ and $M(G)$ are equivalent as invariants of G, or that either one determines the other. The proposition is a folklore result and the analogous results for other polynomials mentioned in this chapter follow by similar arguments.

24.3 Identities for parameterized Tutte polynomials

24.3.1 Elementary formulas

We have just seen that $T_{\text{gen}}(M; x, y, \mathbf{a}, \mathbf{b})$ provides a complete description of a matroid M. This description in turn provides algebraic descriptions of many graph and matroid operations. For instance, the formulas given below follow immediately from elementary definitions of graph theory and matroid theory.

Proposition 24.11.

1. *If M^* and M are dual matroids, then*

$$T_{\text{gen}}(M^*; x, y, \mathbf{a}, \mathbf{b}) = T_{\text{gen}}(M; y, x, \mathbf{b}, \mathbf{a}).$$

2. *If M_1 and M_2 are disjoint matroids, then*

$$T_{\text{gen}}(M_1 \oplus M_2; x, y, \mathbf{a}, \mathbf{b}) = T_{\text{gen}}(M_1; x, y, \mathbf{a}, \mathbf{b}) \cdot T_{\text{gen}}(M_2; x, y, \mathbf{a}, \mathbf{b}).$$

3. *Suppose G is either the disjoint union or the one-point join of graphs G_1 and G_2. Then*

$$T_{\text{gen}}(G; x, y, \mathbf{a}, \mathbf{b}) = T_{\text{gen}}(G_1; x, y, \mathbf{a}, \mathbf{b}) \cdot T_{\text{gen}}(G_2; x, y, \mathbf{a}, \mathbf{b}).$$

4. *If e is not a coloop of M (or not a bridge in the case of graphs), then*

$$b_e \cdot T_{\text{gen}}(M \backslash e; x, y, \mathbf{a}, \mathbf{b}) = T_{\text{gen}}(M; x, y, \mathbf{a}, \mathbf{b})\Big|_{a_e=0}.$$

5. *If e is not a loop of M, then*

$$a_e \cdot T_{\text{gen}}(M/e; x, y, \mathbf{a}, \mathbf{b}) = T_{\text{gen}}(M; x, y, \mathbf{a}, \mathbf{b})\Big|_{b_e=0}.$$

6. *Suppose e_1 and e_2 are parallel non-loops of M. Let M' be obtained from $M \backslash e_2$ by using $a'_e = a_e$ and $b'_e = b_e$ for $e \neq e_1$, with $a'_{e_1} = a_{e_1} a_{e_2}(y-1) + a_{e_1} b_{e_2} + b_{e_1} a_{e_2}$ and $b'_{e_1} = b_{e_1} b_{e_2}$. Then $T_{\text{gen}}(M; x, y, \mathbf{a}, \mathbf{b}) = T_{\text{gen}}(M'; x, y, \mathbf{a}', \mathbf{b}')$.*

7. *Suppose e_1 and e_2 are non-coloops in series in M (or non-bridges in the case of graphs). Let M' be obtained from M/e_2 by using $a'_e = a_e$ and $b'_e = b_e$ for $e \neq e_1$, with $a'_{e_1} = a_{e_1} a_{e_2}$ and $b'_{e_1} = a_{e_1} b_{e_2} + b_{e_1} a_{e_2} + b_{e_1} b_{e_2}(x-1)$. Then $T_{\text{gen}}(M; x, y, \mathbf{a}, \mathbf{b}) = T_{\text{gen}}(M'; x, y, \mathbf{a}', \mathbf{b}')$.*

The last two items of Proposition 24.11 are called the *parallel reduction* and *series reduction* relations, respectively.

24.3.2 Recursions involving reduction formulas

In addition to Definition 24.3, there is another recursion strategy for $T_{\text{gen}}(G; x, y; a, b)$, which utilizes the parallel reduction and series reductions given in Items 6 and 7 of Proposition 24.11, along with deletion–contraction, bridge/coloop contraction and loop deletion. Unlike Definition 24.3, the steps of this recursion have overlapping edge selection criteria: if a step involves one of a pair of parallel non-loop edges, for instance, then either deletion–contraction or a parallel reduction may be applied. Moreover an implementation might not follow a predetermined order of the edges, as different branches of a calculation may create different sets of edges that are parallel or in series. Consequently, it is not easy to write down a closed form for the result of a calculation, like (24.1). However, an implementation of this recursion may involve significantly fewer steps than an implementation of the deletion–contraction recursion.

Example 24.12. For instance, let G be the graph considered in Example 2.9, a 3-cycle with one edge doubled: $V = \{1, 2, 3\}$ and $E = \{e_1, e_2, e_3, e_4\}$ with $e_4 = (2, 3)$, $e_3 = (1, 3)$, $e_2 = (1, 2)$ and $e_1 = (1, 2)$. As indicated in Figures 2.3 and 24.1, a deletion–contraction calculation of $T_{\text{gen}}(G; x, y, \mathbf{a}, \mathbf{b})$ involves 11 recursive steps. The result is the following activities expansion.

$$\begin{aligned} T_{\text{gen}}(G; x, y, \mathbf{a}, \mathbf{b}) &= [(a_{e_1}(y-1) + b_{e_1})(a_{e_2}(y-1) + b_{e_2})a_{e_3}a_{e_4}] \\ &+ [(a_{e_1}(y-1) + b_{e_1})a_{e_2}b_{e_3}a_{e_4}] \\ &+ [(a_{e_1} + b_{e_1}(x-1))b_{e_2}b_{e_3}a_{e_4}] \\ &+ [(a_{e_1}(y-1) + b_{e_1})a_{e_2}(a_{e_3} + b_{e_3}(x-1))b_{e_4}] \\ &+ [(a_{e_1} + b_{e_1}(x-1))b_{e_2}(a_{e_3} + b_{e_3}(x-1))b_{e_4}]. \end{aligned}$$

There are five summands (in brackets) because G has five spanning trees.

In contrast, a calculation involving series and parallel reductions requires only four steps. First, contract e_4, change a_{e_3} to $a'_{e_3} = a_{e_3}a_{e_4}$, and change b_{e_3} to $b'_{e_3} = a_{e_3}b_{e_4} + b_{e_3}a_{e_4} + b_{e_3}b_{e_4}(x-1)$. Then delete e_3, change a_{e_2} to $a'_{e_2} = a_{e_2}a'_{e_3}(y-1) + a_{e_2}b'_{e_3} + b_{e_2}a'_{e_3}$, and change b_{e_2} to $b'_{e_2} = b_{e_2}b'_{e_3}$. Next, delete e_2, change a_{e_1} to $a'_{e_1} = a_{e_1}a'_{e_2}(y-1) + a_{e_1}b'_{e_2} + b_{e_1}a'_{e_2}$, and change b_{e_1} to $b'_{e_1} = b_{e_1}b'_{e_2}$. The only remaining edge, e_1, is now a bridge, so we conclude that $T_{\text{gen}}(G; x, y, \mathbf{a}, \mathbf{b}) = a'_{e_1} + b'_{e_1}(x-1)$.

Notice that if M is a matroid with elements e_1 and e_2 that are parallel or in series, then the matroid $N' = M\backslash e_2$ has the property that every matroid obtained from M by contracting or deleting both elements of $N = \{e_1, e_2\}$ equals either $N'\backslash e_1$ or N'/e_1. There are several other instances of graph and matroid reductions in which such N and N' appear. For instance, if G is the union of subgraphs G_1 and G_2 with $|E(G_1) \cap E(G_2)| \leq 1$ and $|V(G_1) \cap V(G_2)| \leq 2$, then either $M(G_1)$ or $M(G_2)$ can play the role of N'; examples include disjoint unions, one-point unions, parallel connections and series connections. In matroid theory direct sums, parallel connections, series connections and 2-sums are all similar reductions.

When such N and N' exist, they provide reduction formulas for parameterized and multivariate Tutte polynomials. The basic idea is simple: a deletion–contraction calculation of $T_{\text{gen}}(M; x, y, \mathbf{a}, \mathbf{b})$ is performed, with all elements of N contracted or deleted before any element of $M\backslash N$ is contracted or deleted. At the moment when all elements of N have been removed, every term in the computation is associated with either N'/e' or $N'\backslash e'$. Collecting terms yields parameter functions a', b' for N' such that

$$T_{\text{gen}}(M; x, y, \mathbf{a}, \mathbf{b}) = T_{\text{gen}}(N'; x, y, \mathbf{a}', \mathbf{b}'). \tag{24.5}$$

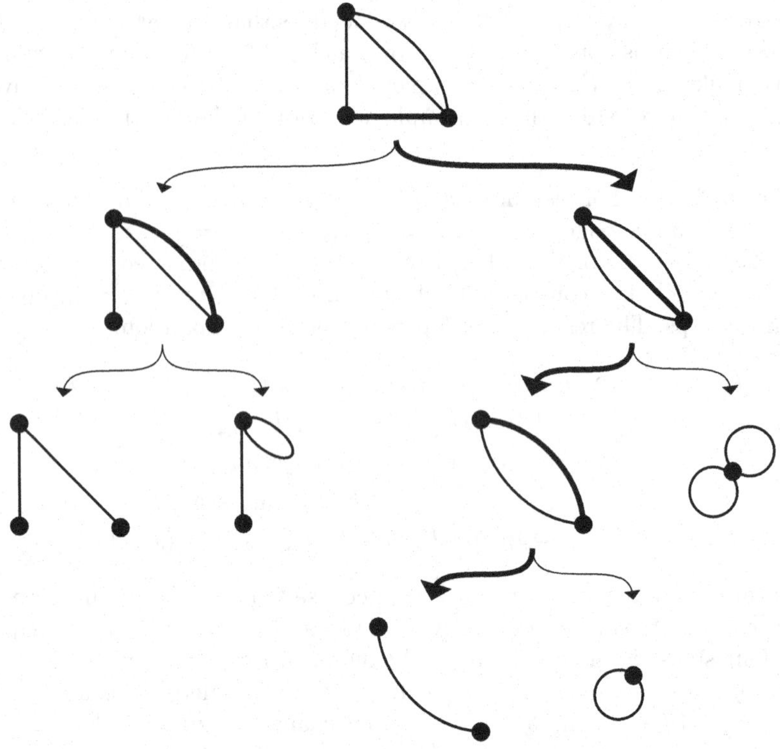

FIGURE 24.1: The deletion–contraction computation tree for Example 24.12. A series-parallel computation requires only the reductions indicated by the bold arrows.

To appreciate the potential algorithmic value of such results note that in general the number of steps required to implement the deletion–contraction algorithm is exponential in the size of the matroid in question. It follows that if $N \subseteq M$ then the number of steps required to calculate $T_{\text{gen}}(M; x, y, \mathbf{a}, \mathbf{b})$ is roughly the product of the numbers of steps required to calculate $T_{\text{gen}}(N; x, y, \mathbf{a}, \mathbf{b})$ and $T_{\text{gen}}(M \backslash N; x, y, \mathbf{a}, \mathbf{b})$ separately. However if M reduces to N' and $M \backslash N$ in the manner just discussed, then the number of steps required to provide a formula for $T_{\text{gen}}(M; x, y, , \mathbf{a}, \mathbf{b})$ is roughly the sum of the numbers of steps in separate computations of $T_{\text{gen}}(N'; x, y, \mathbf{a}, \mathbf{b})$ and $T_{\text{gen}}(M \backslash N; x, y, \mathbf{a}, \mathbf{b})$. The value of such reductions lies in the fact that the sum is typically much smaller than the product.

These rough counts do not provide a precise analysis of computational complexity; we have ignored important issues like the computational cost of algebraic operations and the difficulty of determining whether each minor of M corresponds to N'/e' or $N' \backslash e'$. Nevertheless the rough counts do provide a conceptual explanation for the relative tractability of Tutte polynomial calculations for a graph or matroid of bounded width, so long as the width bound guarantees that the graph or matroid can be described by iterated reductions to graphs or matroids of bounded size. For precise thorough discussions of complexity see Chapters 7–10 of this handbook.

Series–parallel reductions were originally introduced in the theory of electrical circuits. Their significance in the theory of the classical Tutte polynomial was noticed by T. Brylawski [250], and then J. Oxley and D. Welsh [905]. Since then, several authors have studied these reductions for multivariable Tutte polynomials [158, 992, 1075, 1172]. Other useful reduction strategies include S. Negami's splitting formula from Theorem 24.17, and M. Las Vergnas' idea of "normal" subsets of matroids [752].

24.4 Related parameterized polynomials

The literature includes several types of parameterized graph polynomials that are closely related to the standard doubly parameterized Tutte polynomial. The polynomials we discuss here share two important properties.

1. Each of the polynomials satisfies a version of Proposition 24.10 in that it is equivalent to either $M(G)$ or the combination of $M(G)$ and $k(G)$.
2. Each of the polynomials is distinctive for the fact that its particular version of one of the elementary formulas of Proposition 24.11 is significantly simplified.

Because of the first property, these five polynomials are all nearly equivalent to each other, and to $M(G)$ and $T_{\mathrm{gen}}(G'; x, y, \mathbf{a}, \mathbf{b})$. In particular, if $k(G) = k(G')$ then G and G' share one of the polynomials if and only if G and G' share all of the polynomials, and also have isomorphic cycle matroids. Because of the second property, each of these polynomials is especially well-suited to one of the elementary formulas of Subsection 24.3.1. And because of the first property, results about any one of the five polynomials apply indirectly to all of them, and to T_{gen}. It follows that all of these polynomials are useful in the development of the theory of multivariable Tutte polynomials. Each of the five provides its own special insight, and all of these insights apply to all of the polynomials.

24.4.1 The chain and sheaf polynomials

R. Read [949], and R. Read and E. Whitehead [951] introduced the chain and sheaf polynomials, which are distinctive because they satisfy very simple parallel and series reduction formulas. (The polynomials presented here slightly extend the domain of those in [949, 951].)

A *chain* in a graph G is a path in G in which all vertices (as vertices of G), except possibly the end-vertices, have degree two. The *length* of a chain is the number of edges in it.

Definition 24.13. Let $G = (V, E)$ be a graph, $\mathbf{w} = \{w_e : e \in E\}$ a set of indeterminates, and $F(G, \omega)$ be the flow polynomial. Then the *chain polynomial*, which is a polynomial in the variables $\omega \cup \{w_e :\in E\}$, is

$$\mathrm{Ch}(G; \omega, \mathbf{w}) = \sum_{A \subseteq E} F(A^c, \omega) \prod_{e \in A} w_e.$$

An identity from [1074] reveals the connection between the chain polynomial and the standard doubly parameterized Tutte polynomial.

Proposition 24.14. *The chain polynomial is given by the following formula.*

$$\mathrm{Ch}(G; \omega, \mathbf{w}) = T_{\mathrm{par}}(G; 2, 2 - \omega, 1, \mathbf{w} - \mathbf{1}) = \sum_{A \subseteq E} \Big(\prod_{e \notin A} (w_e - 1) \Big) (1 - \omega)^{n(A)},$$

where $\mathbf{w} - \mathbf{1} := \{w_e - 1 : e \in E(G)\}$.

Notably, the formulas for bridge contraction and series reduction are the same for the chain polynomial. If e is a bridge of G, then the bridge contraction formula and Proposition 24.14 together imply that

$$\mathrm{Ch}(G/e; \omega, \mathbf{w}) = \mathrm{Ch}(G; \omega, \mathbf{w}) \Big|_{w_e = 1}. \tag{24.6}$$

Also, if e and e' are two non-bridge edges incident on a degree-two vertex of G, then the same formula (24.6) follows from the series reduction formula of Item 7 of Proposition 24.11 with $e = e_2$ and $e' = e_1$.

Read and Whitehead used (24.6) to describe the effect on the Tutte polynomial of inserting or removing vertices of degree two. One of their results is that the classical Tutte polynomial of G determines the number of chains in G [952], which plays a role in Tutte uniqueness (see Chapter 6). Chains can be introduced in a graph by repeatedly subdividing edges, that is, replacing an edge by a path of length two. When, to a graph G, the reverse of subdividing edges is applied to eliminate all vertices of degree two, we obtain the *reduced graph* of G. The *chain generating function* is $\mathrm{Chg}(G; \omega) := \sum_{k \geq 1} c_k x^k$, where c_k is the number of chains in G of length k.

Theorem 24.15. *Let G be a graph with n vertices and m edges, and whose reduced graph is 3-edge-connected. Let C be the coefficient of y^{m-n} in $T(G; x, y)$ (so C is a polynomial in x). Then the chain generating function of G can be expressed as*

$$\mathrm{Chg}(G; x) = (x - 1)[C + (m - n + 1)] + m.$$

A *sheaf* is a set of multiple edges between two vertices. Duality properties yield results for sheafs analogous to those for chains. In particular, the sheaf polynomial is a "dual polynomial" for the chain polynomial. Following [1074], it may be defined as

$$\mathrm{Sh}(G;\omega,\mathbf{w}) = (-1)^{v(G)-1}\sum_{A\subseteq E}\left(\prod_{e\in A}(w_e-1)\right)(1-\omega)^{r(E)-r(A)}.$$

It can be recovered from the standard doubly parameterized Tutte polynomial:

$$\mathrm{Sh}(G;\omega,\mathbf{w}) = (-1)^{v(G)-1}\cdot T_{\mathrm{par}}(G;2-\omega,2,\mathbf{w}-\mathbf{1},1).$$

Just as the chain polynomial's bridge contraction and series reduction formulas are both given by (24.6), the sheaf polynomial has a single, simple formula for both loop deletion and parallel reduction following from Proposition 24.11: if e is a loop or e and e' are parallel edges then

$$\mathrm{Sh}(G\backslash e;\omega,\mathbf{w}) = \mathrm{Sh}(G;\omega,\mathbf{w})\Big|_{w_e=1}.$$

Arguing as in Section 24.2.2 shows the cycle matroid of a graph G is determined by either the chain polynomial or the sheaf polynomial. The chain and sheaf polynomials generalize immediately to arbitrary matroids (suppressing the factor $(-1)^{v(G)-1}$ in the case of the sheaf polynomial) and similar arguments imply that these polynomials of a matroid are equivalent to the matroid itself.

24.4.2 Edge failures and reliability

Suppose the edges of a graph $G=(V,E)$ are independently subject to failure: each edge $e\in E$ has an independent probability p_e of operating successfully to connect its end-vertices. Let $\mathbf{p}=\{p_e : e\in E\}$. Then the probability generating function for the number of connected components in the operational subgraph (i.e. the subgraph of functioning edges) of G is given by the formula

$$\mathrm{Prob}_{\mathrm{k}}(G;u,\mathbf{p}) = \sum_{A\subseteq E}\Big(\prod_{e\in A}p_e\Big)\Big(\prod_{e\notin A}(1-p_e)\Big)u^{k(A)}$$

from which one can show that

$$\mathrm{Prob}_{\mathrm{k}}(G;u,\mathbf{p}) = u^{k(G)}\cdot T_{\mathrm{par}}(G;u+1,2,\mathbf{p},\mathbf{1}-\mathbf{p}),$$

where $\mathbf{1}-\mathbf{p} := \{1-p_e : e\in E\}$. For a connected graph and taking each $p_e=p$, the all-terminal reliability polynomial (see Chapter 15) is the evaluation

$$\mathrm{Rel}(G;p) = \left(\frac{\mathrm{Prob}_{\mathrm{k}}(G;u,\mathbf{p})}{u}\right)\Bigg|_{u=0}.$$

This multivariable polynomial was central to locating the zeros of the reliability and other polynomials in the work of A. Sokal [1021].

A distinctive feature of $\mathrm{Prob_k}(G; u, \mathbf{p})$ is that a single deletion–contraction formula holds for all edges: if e is any edge of G then

$$\mathrm{Prob_k}(G; u, \mathbf{p}) = p_e \cdot \mathrm{Prob_k}(G/e; u, \mathbf{p}) + (1 - p_e) \cdot \mathrm{Prob_k}(G\backslash e; u, \mathbf{p}).$$

The polynomial $\mathrm{Prob_k}(G; u, \mathbf{p})$ satisfies an extended version of Proposition 24.10. If the various probabilities p_e are replaced with independent indeterminates, then $\mathrm{Prob_k}(G; u, \mathbf{p})$ is equivalent to the combination of $M(G)$ and $k(G)$.

24.4.3 Negami's polynomials

S. Negami [876] introduced two 3-variable graph polynomials, $f(G; t, x, y)$ and $f^*(G; t, x, y)$. These can be defined recursively by

$$f(G; t, x, y) = xf(G/e; t, x, y) + yf(G\backslash e; t, x, y),$$
$$f^*(G; t, x, y) = yf^*(G/e; t, x, y) + xf^*(G\backslash e; t, x, y),$$

together with the values $f(G; t, x, y) = f^*(G; t, x, y) = t^n$ when G consists of n isolated vertices. J. Oxley [903] showed that each of $f(G; t, x, y)$ and $f^*(G; t, x, y)$ can be recovered from $k(G)$ and the Tutte polynomial of G.

In [1076], a parameterized version of Negami's polynomials was introduced.

Definition 24.16. Let $G = (V, E)$ be a graph, $\mathfrak{R}$ be a commutative ring with unity, and $\mathbf{a}, \mathbf{b} : E \to \mathfrak{R}$. Then a *parameterized Negami polynomial* is

$$N(G; u, \mathbf{a}, \mathbf{b}) = \sum_{A \subseteq E} \left(\prod_{e \in A} a_e \right) \left(\prod_{e \notin A} b_e \right) u^{k(A)}.$$

It can be obtained from the standard doubly parameterized Tutte polynomial:

$$N(G; u, \mathbf{a}, \mathbf{b}) = u^{k(G)} \cdot T_{\mathrm{par}}(u + 1, 2, \mathbf{a}, \mathbf{b}).$$

Parameterized Negami polynomials have a single recursive formula that applies to all edges: if e is any edge of G then

$$N(G; u, \mathbf{a}, \mathbf{b}) = a_e \cdot N(G/e; u, \mathbf{a}, \mathbf{b}) + b_e \cdot N(G\backslash e; u, \mathbf{a}, \mathbf{b}).$$

From this it is easily seen that $f(G; t, x, y)$ and $f^*(G; t, x, y)$, respectively, can be recovered from $N(G; u, \mathbf{a}, \mathbf{b})$ by setting $u = t$ and each $a_e = x$, $b_e = y$, respectively $a_e = y$, $b_e = x$.

The recursive formula gives $N(G; u, \mathbf{a}, \mathbf{b})$ a distinctive property: *Negami's splitting formula* from [876] for the 3-variable polynomial and [1076] for the parameterized polynomials:

Theorem 24.17. *Let w be a positive integer. Suppose G has subgraphs H and K with $G = H \cup K$, $E(K) \cap E(H) = \emptyset$ and $|V(K) \cap V(H)| = w$. Then there is a formula that depends only on $k(G)$, $v(G)$ and w, and gives $N(G; u, \mathbf{a}, \mathbf{b})$ as a function of the parameterized Negami polynomials polynomials and connected component numbers of contractions of H and K.*

A. Andrzejak [25] showed that Theorem 24.17 (in the 3-variable case) leads to the result that computing the Tutte polynomials of graphs is fixed parameter tractable, with treewidth as the parameter (see also [882]).

If the parameters a_e, b_e associated to the edges of G are independent indeterminates, then $N(G; u, \mathbf{a}, \mathbf{b})$ is equivalent to the combination of $M(G)$ and $k(G)$.

24.4.4 Parameterized dichromatic polynomials

Recall from Subsection 24.2.1 that the multivariable dichromatic polynomial is $Z_{\mathrm{gen}}(G; q, \mathbf{v}) = \sum_{A \subseteq E} \big(\prod_{e \in A} v_e \big) q^{k(A)}$. It also has a single recursive formula that applies to all edges, as given in (24.4). Like the polynomials discussed above, $Z_{\mathrm{gen}}(G; q, \mathbf{v})$ inherits versions of the elementary formulas of Section 24.3.1 directly from T_{gen}, using (24.2), in addition to the identities given in Proposition 24.9. Of particular significance are the identities involving series-parallel reductions, which are stated explicitly for this polynomial in Lemma 25.6. These properties play an important role in the study of zeros of the Tutte polynomial, as described in Chapter 25. In analogy with Proposition 24.10, the polynomial polynomial $Z_{\mathrm{gen}}(G; q, \mathbf{v})$ is equivalent to the combination of $M(G)$ and $k(G)$.

24.5 Four parameters: colored and strong extensions

T. Zaslavsky [1183] introduced *strong Tutte functions* as well as the idea of modifying the deletion–contraction relations and activities expansions for parameterized Tutte polynomials by incorporating independent parameters to serve as multiplying factors that occur when bridges (or coloops) are deleted and loops are contracted. Thus, each edge of a graph (or element of a matroid) has four parameters associated to it: one for each of deletion, contraction, coloop deletion, and loop contraction. Zaslavsky focused on functions with values in fields, and he proved that all such functions that are not standard parameterized Tutte polynomials are degenerate in some way. (For instance it might be that only one or two particular elements have nonzero parameters.) See [1183] for details.

B. Bollobás and O. Riordan [158] took a different approach to associating four parameters to edges with their *colored Tutte polynomial*, emphasizing the underlying algebraic framework and using parameters dependent on a coloring of the edges and ring-valued functions invariant under color-preserving isomorphism. The centerpiece of their theory is a most general function with a deletion–contraction recursion and an activities expansion. This function is "most general" in the sense that it yields all other such functions through evaluation of the parameters. The absence of such an example in Zaslavsky's context is due to the fact that this most general function has values in a ring with zero divisors, not a field.

In this section we summarize a unifying generalization of the ideas of these two papers, following J. Ellis-Monaghan and L. Traldi [463]. First, we define what we mean when we say that a function satisfies a deletion–contraction recursion with independent parameters for coloop contraction and loop deletion.

Definition 24.18. Let $\mathcal{M}$ be a class of graphs or matroids which is closed under contraction of arbitrary edges (or elements) and deletion of non-coloops. Let $\mathfrak{R}$ be a commutative ring, and suppose every edge (or element) e of each $M \in \mathcal{M}$ has four associated parameter values, $a_e, b_e, X_e, Y_e \in \mathfrak{R}$. Then a *quadruply parameterized Tutte polynomial* on $\mathcal{M}$ is a function $f : \mathcal{M} \to \mathfrak{R}$ that satisfies the following properties.

1. If e is neither a coloop nor a loop of $M \in \mathcal{M}$, then
$$f(M) = a_e\, f(M/e) + b_e\, f(M\backslash e).$$
2. If e is a coloop of $M \in \mathcal{M}$, then
$$f(M) = X_e\, f(M/e).$$
3. If e is a loop of $M \in \mathcal{M}$, then
$$f(M) = Y_e\, f(M\backslash e).$$

Notice that despite the name "quadruply parameterized Tutte polynomial", a function that satisfies Definition 24.18 need not actually be a polynomial. The central result of the theory of such functions combines ideas from Bollobás and Riordan [158], and Zaslavsky [1183]. We call it the *generalized Zaslavsky–Bollobás–Riordan theorem for matroids*:

Theorem 24.19. *Let $\mathcal{M}$ be a class of matroids as in Definition 24.18, and $\mathfrak{R}$ a commutative ring with an element α. Let $E(\mathcal{M})$ be the set of elements of $\mathcal{M}$, and suppose that $\mathbf{a}, \mathbf{b}, \mathbf{X}, \mathbf{Y} : E(\mathcal{M}) \to \mathfrak{R}$ are functions. Then there is a quadruply parameterized Tutte polynomial f on $\mathcal{M}$ with parameter functions $\mathbf{a}, \mathbf{b}, \mathbf{X}, \mathbf{Y}$ and $f(\emptyset) = \alpha$ if and only if the following identities hold.*

1. *Suppose e_1 and e_2 are two elements of $E(\mathcal{M})$, and $\mathcal{M}$ includes the matroid on $\{e_1, e_2\}$ in which e_1 and e_2 are parallel non-loops. Then*
$$\alpha \cdot (a_{e_1} Y_{e_2} + b_{e_1} X_{e_2}) - \alpha \cdot (a_{e_2} Y_{e_1} + b_{e_2} X_{e_1}) = 0.$$

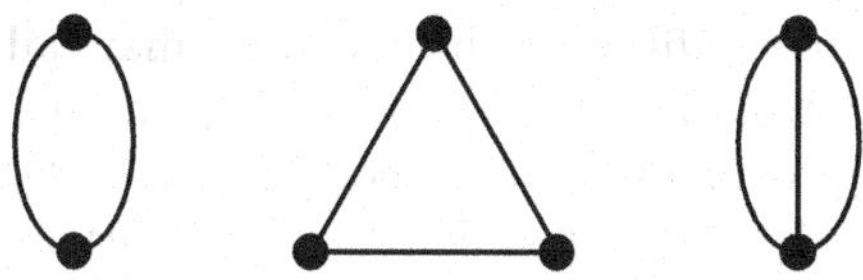

FIGURE 24.2: The three graphs arising from Theorem 24.19.

2. *Suppose* e_1, e_2, e_3 *are three distinct elements of* $E(\mathcal{M})$, *and* $\mathcal{M}$ *includes the matroid on* $\{e_1, e_2, e_3\}$ *in which* $\{e_1, e_2, e_3\}$ *is a circuit. Then*

$$\alpha \cdot X_{e_3} \cdot (a_{e_1} Y_{e_2} + b_{e_1} a_{e_2}) - \alpha \cdot X_{e_3} \cdot (Y_{e_1} a_{e_2} + a_{e_1} b_{e_2}) = 0.$$

3. *Suppose* e_1, e_2, e_3 *are three distinct elements of* $E(\mathcal{M})$, *and* $\mathcal{M}$ *includes the matroid on* $\{e_1, e_2, e_3\}$ *in which* e_1, e_2 *and* e_3 *are all parallel non-loops. Then*

$$\alpha \cdot Y_{e_3} \cdot (a_{e_1} Y_{e_2} + b_{e_1} a_{e_2}) - \alpha \cdot Y_{e_3} \cdot (Y_{e_1} a_{e_2} + a_{e_1} b_{e_2}) = 0.$$

Of course Definition 24.18 implies that f is unique, if it exists. Notably, Theorem 24.19 shows that the existence of a well-defined quadruply parameterized Tutte polynomial depends only a very few small objects, which, in the case of graphs, are the three in Figure 24.2. The identities in Theorem 24.19 guarantee that computation of f by deletion–contraction is independent of order on any of these small objects in $\mathcal{M}$, and this in turn suffices for f to be well-defined on all of $\mathcal{M}$.

Example 24.20. If $\mathcal{M}$ is the set of matroids, $\alpha = 1$, $X_e = a_e + b_e(x-1)$, and $Y_e = a_e(y-1) + b_e$ for some a_e, and b_e in a ring with unity $\mathfrak{R}$, then the conditions of Theorem 24.19 are satisfied. It follows that $T_{\text{par}}(G; x, y, \mathbf{a}, \mathbf{b})$ is an example of a quadruply parameterized Tutte polynomial and its deletion–contraction relations result in a well-defined graph polynomial.

As in the other settings of the Tutte polynomial, a quadruply parameterized Tutte polynomial also has an activities expansion.

Proposition 24.21. *Let* f *be a quadruply parametrized Tutte polynomial on a minor-closed class of matroids* $\mathcal{M}$ *with parameter functions* $\mathbf{a}, \mathbf{b}, \mathbf{X}, \mathbf{Y}$ *and* $f(\emptyset) = \alpha$, *and let* $M \in \mathcal{M}$ *with a given linear order on the elements of* M, *and* $\mathcal{B}(M)$ *be the set of bases of* M. *Then*

$$f(M) = \alpha \cdot \sum_{B \in \mathcal{B}(M)} \Big(\prod_{e \in \mathrm{IA}(B)} X_e \Big) \Big(\prod_{e \in B \setminus \mathrm{IA}(B)} a_e \Big) \Big(\prod_{e \in \mathrm{EA}(B)} Y_e \Big) \Big(\prod_{e \notin B \cup \mathrm{EA}(B)} b_e \Big).$$

In contrast, the possibility of zero divisors in a ring confounds a natural extension of the rank-nullity (state sum) expression of the Tutte polynomial.

However, under certain conditions (which are automatically met in most common settings, e.g., when $\mathfrak{R}$ is field and $\mathcal{M}$ includes all matroids), there is the analogous rank-nullity expression in Proposition 24.22. For details of the several degenerate forms that result when these conditions fail, see [463] and [1183].

Proposition 24.22. *Let f be a quadruply parametrized Tutte polynomial on a minor-closed class of matroids $\mathcal{M}$ with parameter functions $\mathbf{a}, \mathbf{b}, \mathbf{X}, \mathbf{Y}$ and $f(\emptyset) = \alpha$. Suppose p is an element in the union of the ground sets of matroids in $\mathcal{M}$ and that both a_p and b_p have multiplicative inverses in $\mathfrak{R}$. Let $u = (X_p - a_p)b_p^{-1}$ and $v = (Y_p - b_p)a_p^{-1}$. Suppose $M \in \mathcal{M}$ has the property that for every e in its ground set there is an element q such that either $\{e, p, q\}$ is a circuit in M and X_q is not a zero divisor in $\mathfrak{R}$; or e, p and q are parallel non-loops in M, and Y_q is not a zero divisor in $\mathfrak{R}$. Then*

$$f(M) = \alpha \cdot \sum_{S \subseteq E} \Big(\prod_{e \in S} a_e\Big)\Big(\prod_{e \in E(M)\setminus S} b_e\Big)\, u^{r(M)-r(S)}\, v^{|S|-r(S)}.$$

As mentioned above, Theorem 24.19 unifies the approaches of Bollobás and Riordan [158], and Zaslavsky [1183] which differed from each other in several regards. In addition to the fact that Zaslavsky focused on functions with values in fields, he restricted his attention to *strong functions*, i.e., functions with the property that $f(M \oplus N) = f(M) \cdot f(N)$. Bollobás and Riordan did not observe this restriction, but their discussion focused on functions with very large domains (e.g., the set of all connected graphs).

Both Bollobás and Riordan [158], and Zaslavsky [1183] highlight the special case of *colored* graphs and matroids. Rather than starting with functions $\mathbf{a}, \mathbf{b}, \mathbf{X}, \mathbf{Y} : E(\mathcal{M}) \to \mathfrak{R}$, one starts with a single function $\mathbf{c} : E(\mathcal{M}) \to \Lambda$, where Λ is the set of colors, and functions $\mathbf{a}, \mathbf{b}, \mathbf{X}, \mathbf{Y} : \Lambda \to \mathfrak{R}$. The equations of Theorem 24.19 are then interpreted as constraining the choices of parameters associated to the colors.

If $\mathcal{M}$ is a minor-closed class of matroids, then let $\mathcal{M}_{(3)}$ be the subclass consisting of elements of $\mathcal{M}$ of size at most 3. (Note that every nontrivial element of the set $\mathcal{M}_{(3)}$ is isomorphic to the cycle matroid of one of the small graphs in Figure 24.2.) Theorem 24.19 immediately implies the following theorem, whose restriction to field-valued functions is due to Zaslavsky [1183].

Theorem 24.23. *A function $f : \mathcal{M} \to \mathfrak{R}$ is a quadruply parameterized Tutte polynomial if and only if the restriction $f|_{\mathcal{M}_{(3)}}$ is a quadruply parameterized Tutte polynomial. Moreover, $f|_{\mathcal{M}_{(3)}}$ completely determines f.*

Another consequence of Theorem 24.19 is the following version of a central result of Bollobás and Riordan [158].

Theorem 24.24. *For every minor-closed class of matroids $\mathcal{M}$, there is a most general example of a quadruply parameterized Tutte polynomial on $\mathcal{M}$. This most general polynomial maps $\mathcal{M}$ (where each element has four associated* *parameter values) to the quotient ring* $\mathfrak{R}_{\mathcal{M}} = \mathbb{Z}[\mathcal{S}_{\mathcal{M}}]/\mathcal{I}_{\mathcal{M}}$*, where* $\mathcal{S}_{\mathcal{M}} = \{\alpha\} \cup \{a_e, b_e, X_e, Y_e : e \in E(\mathcal{M})\}$*,* $\mathbb{Z}[\mathcal{S}_{\mathcal{M}}]$ *is the polynomial ring, and* $\mathcal{I}_{\mathcal{M}}$ *is the ideal of* $\mathbb{Z}[\mathcal{S}_{\mathcal{M}}]$ *generated by the left-hand sides of the equalities required by Theorem 24.19.*

Example 24.25. Suppose f is the most general quadruply parameterized Tutte polynomial on a minor-closed class of matroids $\mathcal{M}$ containing the matroid M of two elements e_1 and e_2 in parallel. Then

$$f(M) = \alpha \cdot (a_{e_1} Y_{e_2} + b_{e_1} X_{e_2}) + \mathcal{I}_{\mathcal{M}} = \alpha \cdot (a_{e_2} Y_{e_1} + b_{e_2} X_{e_1}) + \mathcal{I}_{\mathcal{M}},$$

since f takes its values in the quotient ring $\mathbb{Z}[\mathcal{S}_{\mathcal{M}}]/\mathcal{I}_{\mathcal{M}}$.

This most general function is often referred to as "the Bollobás–Riordan colored Tutte polynomial" in the literature, even in contexts where the parameters a_e, b_e, X_e, Y_e are defined without reference to a color set Λ.

Definition 24.26. The *(Bollobás–Riordan) colored Tutte polynomial* for a minor-closed class of matroids $\mathcal{M}$ is the most general example of a quadruply parameterized Tutte polynomial ensured by Theorem 24.24. The *colored Tutte polynomial* of a graph is the most general example of a quadruply parameterized Tutte polynomial for the class of graphic matroids (where the value of the polynomial of a graph is computed through its cycle matroid).

Here, "most general" means that every quadruply parameterized Tutte polynomial on $\mathcal{M}$ comes from this one through composition with a ring homomorphism $\mathbb{Z}[\mathcal{S}_{\mathcal{M}}]/\mathcal{I}_{\mathcal{M}} \to \mathfrak{R}$. However, the most general quadruply parameterized Tutte polynomial on $\mathcal{M}$ is not strictly stronger than the matroid invariants given earlier in this chapter, such as $T_{\text{gen}}(M; x, y, \mathbf{a}, \mathbf{b})$. According to Proposition 24.10 a strictly stronger invariant cannot exist, as $T_{\text{gen}}(M; x, y, \mathbf{a}, \mathbf{b})$ is equivalent to M itself. Nevertheless, like all of the polynomials in this chapter, quadruply parameterized Tutte polynomials contribute new perspectives and theoretical understandings through their particular forms.

24.5.1 The special case of graphs

Suppose $\mathcal{M}$ is a class of graphs that is closed under contraction of arbitrary edges and deletion of non-bridges. For each integer $k \geq 1$ let $\mathcal{M}_k$ be the subclass of $\mathcal{M}$ consisting of graphs with exactly k connected components. Then each individual subclass $\mathcal{M}_k$ is also closed under contraction of arbitrary edges and deletion of non-bridges. For $k \neq k'$ none of the three requirements of Definition 24.18 relates values of f on $\mathcal{M}_k$ to values of f on $\mathcal{M}_{k'}$. Moreover, a quadruply parameterized Tutte polynomial f must have $f(G) = f(H)$ whenever $k(G) = k(H)$ and $M(G) = M(H)$. It follows that a quadruply parameterized graph Tutte polynomial for $\mathcal{M}$ is simply a family of unrelated quadruply parameterized matroid Tutte polynomials, one for each nonempty

$\mathcal{M}_k$. Consequently, in order to understand the quadruply parameterized Tutte polynomials on these classes of graphs, it is enough to understand the quadruply parameterized Tutte polynomials on minor-closed classes of matroids. In particular, there is a most general quadruply parameterized Tutte polynomial on $\mathcal{M}$, with values defined in the ring

$$\bigoplus_{\mathcal{M}_k \neq \emptyset} \mathbb{Z}[\mathcal{S}_{\mathcal{M}_k}]/\mathcal{I}_{\mathcal{M}_k}.$$

This most general polynomial is obtained by applying Theorem 24.24 to each nonempty $\mathcal{M}_k$ separately.

24.6 Ported polynomials and others

So far in this chapter we have focused on polynomials obtained by modifying the classical Tutte polynomial to incorporate parameters defined on the elements of a matroid or the edges of a graph. There are other ways to modify the classical Tutte polynomial that incorporate other kinds of parameters. One of these other modifications was introduced by S. Chaiken and M. Las Vergnas (see [273, 752] and the references given there). Suppose we have a class $\mathcal{M}$ of graphs or matroids closed under edge contractions and non-bridge deletions as in the preceding section, and a subset $P \subseteq E(\mathcal{M})$ has been distinguished. The elements of P are called "ports" and they are not subject to contraction or deletion. (The terminology is motivated by the theory of electrical networks: a port is a location where two networks may be attached.) This leads to *P-ported* or *P-pointed* Tutte polynomials as follows.

Definition 24.27. Let $\{M_i : i \in I\}$ be the set of matroids included in $\mathcal{M}$ whose ground sets are contained in P, and let $\{[M_i] : i \in I\}$ be a set of independent indeterminates. Then a *P-ported* or *P-pointed Tutte polynomial* $T(M, P)$ is defined on $\mathcal{M}$ by using coloop contraction, loop deletion and deletion–contraction only for elements $e \in E(\mathcal{M}) \setminus P$. When a branch of such a computation reaches an M_i, the end state is recorded using $[M_i]$.

A natural way to interpret a ported Tutte polynomial is that $T(M, P)$ represents a moment in a recursive calculation of $T(M)$, when every element of $M \backslash P$ has been removed through contraction or deletion, and no element of P has yet been removed. At this moment we have completed part of a computation tree, and the end states are graphs (or matroids) whose edges (or elements) all lie in P. Consequently each end state corresponds to one of the M_i, and for each i, $[M_i]$ serves as a place-holder for the result of the remaining part of that branch of the computation.

A formula for $T(M, P)$ follows naturally from the idea of the preceding paragraph. Suppose $G = (V, E)$ is a graph and $\mathbf{a}, \mathbf{b}$ are functions mapping E

into some commutative ring with unity $\mathfrak{R}$. Then the P-ported version of the standard doubly parameterized Tutte polynomial is given by the state sum:

$$\sum_{A \subseteq E \setminus P} \Big(\prod_{e \in A} a_e\Big)\Big(\prod_{e \in E \setminus (A \cup P)} b_e\Big)(x-1)^{r(E)-r(A\cup P)}(y-1)^{n(A)}[(G/A)|P].$$

An interesting application of ported Tutte polynomials to virtual knot theory has been presented by Y. Diao and G. Hetyei [379]. (They use the name *relative Tutte polynomial* rather than ported Tutte polynomial.) They showed that the Jones polynomial of a virtual link diagram may be obtained from a ported Tutte polynomial of the "face graph", with port edges corresponding to virtual crossings. Another topological application of ported Tutte polynomials is discussed in Chapter 27.

A different way to modify deletion–contraction invariants is to specify that certain "atomic" graphs or matroids are immune to deletion and contraction. Then these atomic structures are the end states of a deletion–contraction recursion. For example, the parameterized graph invariants studied by Bollobás, Pebody, and Riordan [156] have atomic graphs whose edges are all loops. The contribution of an end state includes information about the sets of loops incident on individual vertices. A knot-theoretic strategy of this general type has been used by H. Dye and L. Kauffman [430], and Y. Miyazawa [852].

The classical Tutte polynomial may also be modified to incorporate parameters assigned to non-edge features of a graph in Definition 24.1. For instance, the U, V and W polynomials discussed in Chapter 26 incorporate parameters assigned to vertices and connected components of a graph.

24.7 Open problems

According to Proposition 24.10, fully parameterized Tutte polynomials of matroids are essentially equivalent to matroids themselves. Therefore, questions about parameterized Tutte polynomials are equivalent to questions about matroids. Such questions include the following.

- What properties of combinatorial structures are described by (parameterized Tutte polynomials of) matroids?
- Is there an effective, general technique for answering an instance of the first question?

We can answer a special case of the second question using Whitney's 2-isomorphism relation [1156]: a graph property is determined by the cycle matroid if and only if the property is invariant under 2-isomorphism. But this answer is particular to the cycle matroid; there are other matroids associated with graphs, which are not invariant under 2-isomorphism. For instance, it turns out that the isotropic matroid of a graph [1080, 1081] determines several polynomials discussed in other chapters of this handbook, including interlace polynomials of graphs and binary delta-matroids, knot polynomials including the Jones, HOMFLYPT and Kauffman polynomials, and topological Tutte polynomials like that of Bollobás and Riordan.

25

Zeros of the Tutte polynomial

Bill Jackson

Synopsis

This chapter considers the distribution of the zeros of the Tutte polynomial and the dichromatic polynomial in $\mathbb{R}^2$ and $\mathbb{C}^2$.

- The multivariate Tutte polynomial and its elementary properties.
- Zero-free regions for the Tutte polynomial.
- Regions where the zeros of the Tutte polynomial are dense.
- Points in $\mathbb{R}^2$ at which we can determine the sign of the Tutte polynomial for all graphs.

25.1 Introduction

The study of the distributions of the zeros of three specializations of the Tutte polynomial, namely the chromatic, flow, and reliability polynomials, has a long history in the literature. This study can be unified by considering the distribution of the zeros of the Tutte polynomial in $\mathbb{R}^2$ or $\mathbb{C}^2$. Recent results have come close to determining the regions of $\mathbb{R}^2$ which are free of zeros of the Tutte polynomial, the regions for which the real zeros are dense, and the points at which we can determine the sign of the Tutte polynomial efficiently. We also know that the complex zeros are dense in a large region in $\mathbb{C}^2$. The aim of this chapter is to describe these results and the intriguing open problems which still remain unsolved. We refer the reader to Chapters 11, 12 and 15 for results and problems which are specific to chromatic, flow and reliability polynomials.

DOI: 10.1201/9780429161612-25

25.2 The multivariate Tutte polynomial

The study of the zero distribution of the Tutte polynomial naturally leads us to consider its multivariate generalization.

Definition 25.1. Given a graph $G = (V, E)$ and a weight function $\mathbf{v} : E \to \mathbb{R}$, the *multivariate Tutte polynomial* of the weighted graph $(G, \mathbf{v})$ is

$$Z(G; q, \mathbf{v}) = \sum_{A \subseteq E} q^{k(A)} v_A, \tag{25.1}$$

where $k(A)$ is the number of connected components in the spanning subgraph (V, A), and where $v_A = \prod_{e \in A} v_e$ with $v_e = \mathbf{v}(e)$.

Example 25.2. If K_3 is a complete graph on three vertices with edge weights v_1, v_2, v_3, then

$$Z(G; q, \mathbf{v}) = q^3 + q^2(v_1 + v_2 + v_3) + q(v_1v_2 + v_2v_3 + v_1v_3 + v_1v_2v_3).$$

The multivariate Tutte polynomial is a close relative of the Potts model partition function described in Chapter 20. When all edges have the same weight v, it reduces to the dichromatic polynomial $Z(G; q, v)$ of G, which can be transformed to the Tutte polynomial $T(G; x, y)$ by the simple change of variables $q = (x - 1)(y - 1)$ and $v = y - 1$, and multiplication by a suitable pre-factor, as in Equation (2.4). Thus, $Z(G; q, \mathbf{v})$ might more correctly be called the *multivariate dichromatic polynomial.* However, common usage, particularly in the physics literature, often refers to $Z(G; q, \mathbf{v})$ as the multivariate Tutte polynomial, and we follow that convention here. Note, however, that the variables in these polynomials play different roles. In the multivariate Tutte polynomial, q acts as a global variable (representing the number of possible colors or "states" for each vertex) and v represents a weight on each edge. The variables x, y in the Tutte polynomial act as a pair of "dual" global variables. This duality is best illustrated by the extension of the Tutte polynomial to a matroid M and its dual M^* where we obtain $T(M^*; x, y) = T(M; y, x)$. In particular we have $T(G; x, y) = T(G^*; y, x)$ for any connected planar graph G with planar dual G^* (see Sections 2.5 and 4.4). The corresponding identity for the dichromatic polynomial $Z(G^*, q, v) = q^{1-|V|}\, v^{|E|}\, Z(G; q, q/v)$ is not so pretty but is still useful. It extends to the multivariate Tutte polynomial as $Z(G^*; q, \mathbf{v}) = q^{1-|V|}\, v_E\, Z(G; q, \mathbf{v}^*)$, where we take $v_e^* = q/v_e$ for all $e \in E$.

C. M. Fortuin and P. W. Kasteleyn [498] give the following expression for the multivariate Tutte polynomial when q is a positive integer, which illustrates why we should think of the variable q as representing the number of colors in a vertex coloring of G.

Proposition 25.3. *Let $G = (V, E)$ be a graph, and q a positive integer. Then*

$$Z(G; q, \mathbf{v}) = \sum_{\sigma: V \to [q]} \prod_{e=ab \in E} (1 + v_e \delta_{\sigma_a, \sigma_b}) \tag{25.2}$$

where $[q] = \{1, 2, \ldots, q\}$ and δ is the Kronecker delta function.

This second expression for $Z(G; q, \mathbf{v})$ is similar to the definition of the partition function given in the original paper of R. B. Potts [928]. Equation (25.2) expresses $Z(G; q, \mathbf{v})$ as a weighted sum over all q-colorings of the vertices of G, where a particular q-coloring σ has weight $\prod(1 + v_e)$, and this product ranges over all edges e that are not properly colored by σ. This coloring interpretation enables us to obtain two reduction formulas for the multivariate Tutte polynomial which extend the analogous results for Tutte polynomials.

Lemma 25.4. *Let $(G, \mathbf{v})$ be a weighted graph and e be an edge of G. Then*

$$Z(G; q, \mathbf{v}) = Z(G \backslash e, q, \mathbf{v}|_{E \backslash \{e\}}) + v_e\, Z(G/e, q, \mathbf{v}|_{E \backslash \{e\}}).$$

Lemma 25.5. *Let $(G, \mathbf{v})$ be a weighted graph. Suppose $G = G_1 \cup G_2$ for two edge-disjoint subgraphs $G_1 = (V_1, E_1)$ and $G_2 = (V_2, E_2)$ of G with $|V_1 \cap V_2| = r$ for some $r \in \{0, 1\}$. Then*

$$Z(G; q, \mathbf{v}) = q^{-r} Z(G_1; q, \mathbf{v}|_{E_1})\, Z(G_2; q, \mathbf{v}|_{E_2})\,.$$

(The hypothesis of Lemma 25.5 that the subgraphs are edge-disjoint may appear redundant, but it is needed in the case when $r = 1$ and their common vertex is incident to one or more loops.)

A big advantage of introducing different edge-weights into the dichromatic polynomial is that it gives rise to new reduction operations. The *parallel reduction operation* replaces two parallel edges by a single edge with the same endpoints. The *series reduction operation* replaces a vertex b of degree two and its incident edges by a single edge joining the neighbors of b (see also Chapter 24).

Lemma 25.6. *Let $(G, \mathbf{v})$ be a weighted graph and let e_1, e_2 be edges of G.*

1. *If H is obtained by applying the parallel reduction operation to e_1, e_2, then*

$$Z(G; q, \mathbf{v}) = Z(H; q, \mathbf{v}|_{E \backslash \{e_1, e_2\}}, v_{e_1} \parallel v_{e_2})\,,$$

where $v_{e_1} \parallel v_{e_2} = v_{e_1} + v_{e_2} + v_{e_1} v_{e_2}$ is the weight of the edge that replaces e_1, e_2.

2. *If H is obtained by applying the series reduction operation to e_1, e_2, then*

$$Z(G; q, \mathbf{v}) = (q + v_{e_1} + v_{e_2}) Z(H; q, \mathbf{v}|_{E \backslash \{e_1, e_2\}}, v_{e_1} \bowtie_q v_{e_2})\,,$$

where $v_{e_1} \bowtie_q v_{e_2} = v_{e_1} v_{e_2} (q + v_{e_1} + v_{e_2})^{-1}$ is the weight of the edge that replaces e_1, e_2.

We refer the reader to [1023] for proofs of these lemmas. This reference also gives several impressive examples of how the freedom to change the edge-weights in the multivariate Tutte polynomial can give simple proofs for results on the classical Tutte polynomial which would be much more difficult to prove directly. The results in the next section will provide another such example.

25.3 Zero-free regions in $\mathbb{R}^2$

A combination of results from [637] and [1092] show that the sub-intervals $(-\infty, 0)$, $(0, 1)$ and $(1, 32/27]$ of the real line contain no zeros of chromatic polynomials of loopless graphs.

Theorem 25.7. *Suppose G is a loopless graph with n vertices, k components and b non-trivial blocks (i.e., blocks with at least one edge). Let $\chi(G;q)$ be the chromatic polynomial of G.*

1. *If $q \in (-\infty, 0)$, then $(-1)^n \chi(G;q) > 0$.*
2. *If $q \in (0, 1)$, then $(-1)^{n+k} \chi(G;q) > 0$.*
3. *If $q \in (1, 32/27]$, then $(-1)^{n+k+b} \chi(G;q) > 0$.*

Since $\chi(G;q) = Z(G;q,-1)$ (through Equation (3.4)), Theorem 25.7 tells us that $Z(G;q,-1)$ is nonzero for all $q \in (-\infty,0) \cup (0,1) \cup (1,32/27]$. The proof for the interval $(1, 32/27)$ in [637] is a complicated inductive argument based on the chromatic polynomial versions of Theorems 25.4 and 25.5. A simpler proof technique which extends all three intervals to zero-free regions of $\mathbb{R}^2$ for $Z(G;q,\mathbf{v})$ was developed in [642].

The main idea behind this proof technique is to find a "suitable" interval $I(q) \subseteq \mathbb{R}$, for each $q \in (-\infty,0) \cup (0,1) \cup (1,32/27]$, such that $-1 \in I(q)$ and such that $I(q)$ is closed under the parallel and series transformations, i.e., for all $v_1, v_2 \in I(q)$ we have $v_1 \parallel v_2 \in I(q)$ and $v_1 \bowtie_q v_2 \in I(q)$. We can then prove inductively that, for any graph G and suitably chosen α, $(-1)^{n+\alpha} Z(G;q,\mathbf{v}) > 0$ for all $\mathbf{v} : E \to I(q)$. We can use Lemma 25.5 to reduce to the case when G is 2-connected and loopless, and Lemma 25.6 to reduce to the case when G has minimum degree three and no multiple edges. We can then use a result of J. Oxley [902] to deduce that G has an edge e for which $G\backslash e$ and G/e are both 2-connected. The proof is completed by applying Lemma 25.4 using this edge e.

This proof also explains the significance of the number $\frac{32}{27}$; it is the largest value of q for which there exists an interval $I(q) \subseteq (-\infty, 0)$ such that $-1 \in I(q)$ and $I(q)$ is closed under the parallel and series transformations. (We need $I(q) \subseteq (-\infty, 0)$ to have any chance that $Z(G;q,\mathbf{v})$ will have the correct sign for all $\mathbf{v} : E \to I(q)$.)

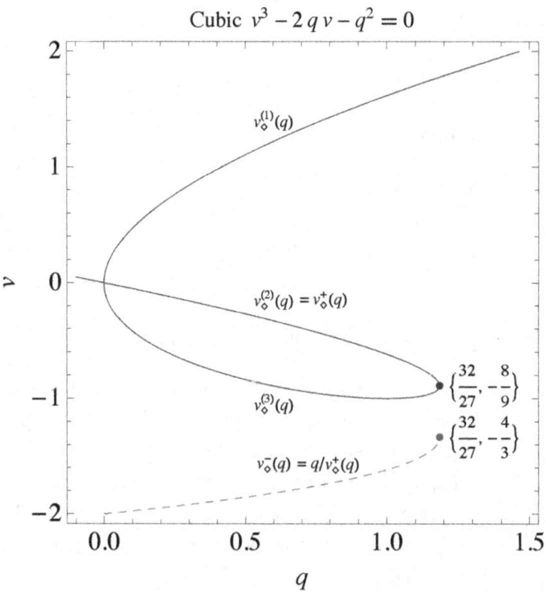

FIGURE 25.1: The solid curve shows the solutions of $v^3 - 2qv - q^2 = 0$. The dashed curve shows the "dual" to the middle branch.

A necessary condition that $I(q)$ is closed under the parallel and series transformations is that $I(q)$ should be closed under both the *diamond transformation* $v \mapsto (v \bowtie_q v) \parallel (v \bowtie_q v)$, and its dual transformation $v \mapsto (v \parallel v) \bowtie_q (v \parallel v)$. We can investigate the diamond transformation by considering its fixed points. This gives rise to the curve $v^3 - 2qv - q^2 = 0$ which is illustrated in Figure 25.1. The cubic equation has three roots for $0 \leq q \leq 32/27$. The root corresponding to the middle branch of the curve $v_\diamond^+(q)$ is a repulsive fixed point of the diamond transformation so is a good candidate for the upper end of our interval $I(q)$. We can ensure that $I(q)$ is closed under the dual transformation whenever it is closed under the diamond transformation by making it *self-dual*, i.e., closed under the duality transformation $v \mapsto q/v$. This indicates that we should take the lower end of our interval to be $v_\diamond^-(q) = q/v_\diamond^+(q)$. It turns out that the choice $I(q) = I_\diamond(q) := [v_\diamond^-(q), v_\diamond^+(q)]$ is indeed closed under the parallel and series operations. We refer the reader to [642] for the details.

We can now summarize the main results of [642].

Theorem 25.8. *Let G be a 2-connected graph with n vertices.*

1. *If $q \in (-\infty, 0)$, then $(-1)^n Z(G; q, \mathbf{v}) > 0$ whenever $v_e \in [-2, 0]$ for all $e \in E$, and $Z(G; q, \mathbf{v}) < 0$ whenever $v_e \in [-q/2, \infty)$ for all $e \in E$.*

2. *If* $q \in (0,1)$ *and*

$$I(q) = (-1 - \sqrt{1-q}, -1 + \sqrt{1-q})$$

then $(-1)^{n+1} Z(G; q, \mathbf{v}) > 0$ *whenever* $v_e \in I(q)$ *for all* $e \in E$.

3. *If* $q \in (1, 32/27]$ *and*

$$I(q) = \begin{cases} (-q - \sqrt{q^2 - q}, \ -q + \sqrt{q^2 - q}) & \text{for } 1 < q \leq 9/8, \\ I_\diamond(q) & \text{for } 9/8 < q \leq 32/27; \end{cases}$$

then $(-1)^n Z(G; q, \mathbf{v}) > 0$ *whenever* $v_e \in I(q)$ *for all* $e \in E$.

The proof in [642] is matroidal and gives the same result for matroids. More precisely, we can define the *multivariate Tutte polynomial of a weighted matroid* $(M, \mathbf{v})$ with ground set E and rank function r by $Z(M; q, \mathbf{v}) = \sum_{A \subseteq E} q^{r(E)-r(A)} v_A$. Then Theorem 25.8 extends to a result on $Z(M; q, \mathbf{v})$ if we replace $Z(G; q, \mathbf{v})$ by $qZ(M; q, \mathbf{v})$, n by $r(E) + 1$, and the hypothesis that G is 2-connected by the hypotheses that M is connected and $|E| \geq 2$. More details are given in [642].

The second part of Item 1 of Theorem 25.8, that $Z(G; q, \mathbf{v}) < 0$ whenever $v_e \in [-q/2, \infty)$ for all $e \in E$, was first obtained by Feng-Ming Dong at the program "Combinatorics and Statistical Mechanics" held at the Newton Institute in 2008. His proof is similar to that given for the first part of that item in [642]. Alternatively, Dong's result can be deduced from the matroid version of the first part by using matroid duality: we have $Z(M^*; q, \mathbf{v}) = q^{-r(E)} v_E Z(M; q, \mathbf{v}^*)$ where $v_e^* = q/v_e$ for all $e \in E$, and the duality transformation $v \mapsto q/v$ maps the region $q < 0$, $-2 \leq v_e \leq 0$, to the region $q < 0$, $0 \leq v_e \leq q/2$.

We can identify another (trivial) zero-free region for the multivariate Tutte polynomial which is not covered by Theorem 25.8: it follows immediately from Equation (25.1) that $Z(G; q, \mathbf{v}) > 0$ for all all $q > 0$ and all weighted graphs $(G, \mathbf{v})$ with $v_e \geq 0$ for all $e \in E$. We can also identify an additional family of line segments which contain no zeros of the multivariate Tutte polynomial. The expression for the multivariate Tutte polynomial given in Equation (25.2) implies that $Z(G; q, \mathbf{v}) > 0$ whenever q is a positive integer and $v_e > -1$ for all $e \in E$. Since this remains true for the multivariate Tutte polynomial of a matroid, we can use matroid duality to deduce that we also have $(-1)^{|E|} Z(G; q, \mathbf{v}) > 0$ whenever q is a positive integer and $v_e < -q$ for all $e \in E$. The zero-free regions for the dichromatic polynomial of a 2-connected graph given by Theorem 25.8 and these observations are illustrated in Figure 25.2.

Theorem 25.8 can be extended to all graphs by using the fact that the multivariate Tutte polynomial is multiplicative over blocks by Lemma 25.5 and that its values for the three blocks which are not 2-connected are given by $Z(K_1; q, v) = q$, $Z(K_2; q, v) = q(q + v)$ and $Z(C_1; q, v) = q(1 + v)$, where C_1 is the graph with one vertex and one loop.

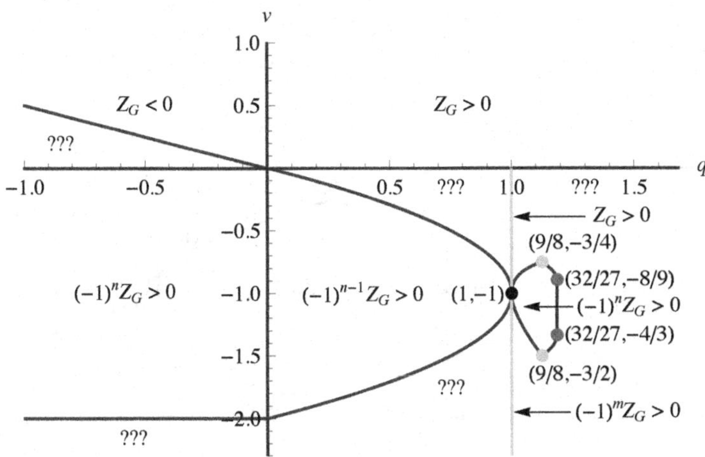

FIGURE 25.2: Zero-free regions for the dichromatic polynomial $Z_G := Z(G; q, v)$ of a 2-connected graph G with n vertices and m edges.

The reader may wonder why the choice of $I(q)$ in Items 2 and 3 of Theorem 25.8 is not simply $I(q) = I_\diamond(q)$. This is because the induction hypothesis that $(-1)^{n+\alpha}Z(G; q, \mathbf{v}) > 0$ does not hold for all weighted graphs $(G, \mathbf{v})$ with $\mathbf{v} : E \to I_\diamond(q)$ when $q \in (0,1) \cup (1, 9/8)$. The graph C_2 consisting of two vertices joined by two parallel edges has $Z(C_2; q, \mathbf{v}) = q(q + v_1 + v_2 + v_1v_2)$ so $v_1 = v_2 = -q \pm \sqrt{1-q}$ are both zeros of $Z(C_2; q, \mathbf{v})$ when $q < 1$, and $v_1 = -q - \sqrt{q^2 - q}, v_2 = -q + \sqrt{q^2 - q}$ is a zero when $q > 1$. This shows that the zero-free regions given in Items 2 and 3 of Theorem 25.8 are in a very limited sense best possible. We can show in [642], however, that these zero-free regions can be expanded if we add a lower bound on the number of edges of G. More specifically we can obtain larger zero-free regions in both Items 2 and 3 when $|E| \geq 3$, and a still larger zero-free region in Item 2 when $|E| \geq 4$. It is conjectured in [642] that this trend will continue and that the limiting zero-free region will be determined by the intervals $I_\diamond(q)$ for all $q \in (0,1) \cup (1, 32/27)$.

Conjecture 25.9. Suppose $q \in (0, 1) \cup (1, 32/27)$. Let $\alpha = 1$ if $q \in (0, 1)$ and $\alpha = 0$ if $q \in (1, 32/27)$. Then there exists an increasing sequence of self-dual intervals $I_m(q)$, $m \geq 2$, such that $\lim_{m\to\infty} I_m(q) = I_\diamond(q)$, and $(-1)^{n+\alpha} Z(G; q, \mathbf{v}) > 0$ for all 2-connected weighted graphs $(G, \mathbf{v})$ with n vertices, at least m edges, and $v_e \in I_m(q)$ for all $e \in E$.

The regions $R_1 = \{(q, v) : q \leq 0,\ v \in [-2, 0] \cup [-q/2, \infty)\}$, $R_2 = \{(q, v) : q \geq 0,\ v \in [0, \infty)\}$ and $R_\diamond = \{(q, v) : q \in (0, 1) \cup (1, 32/27) \text{ and } v \in I_\diamond(q)\}$ covered by Theorem 25.8 and Conjecture 25.9 are illustrated in Figures 25.3 and 25.4. Theorem 25.8 determines the sign of $Z(G; q, v)$ when (q, v) is in R_1 or in a large subset of $R_\diamond$. The definition of $Z(G; q, v)$ implies that $Z(G; q, v) > 0$

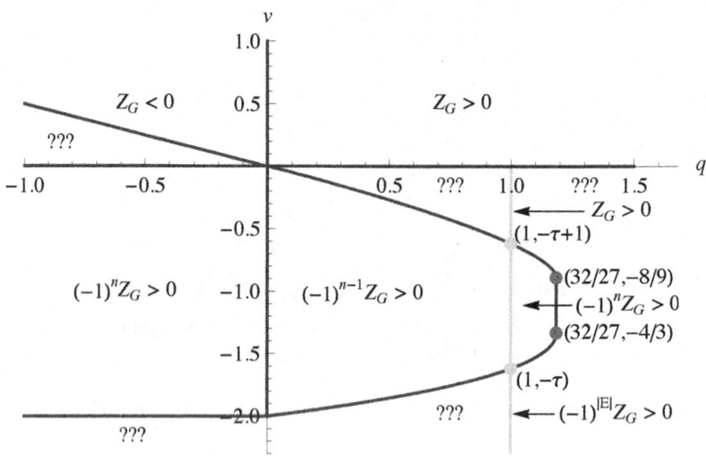

FIGURE 25.3: Conjectured limiting regions as $m \to \infty$ in which the sign of $Z_G := Z(G; q, v)$ can be controlled for loopless 2-connected graphs G with at least m edges. Here $\tau = (1 + \sqrt{5})/2$ is the golden ratio.

when (q, v) is in R_2. Conjecture 25.9 would imply that the sign of $Z(G; q, v)$ is determined by $|V|$ for all 2-connected graphs G with sufficiently many edges, for any fixed (q, v) in the interior of $R_\diamond$.

25.4 Density of real zeros

C. Thomassen [1064] obtained a complementary result to Theorem 25.7 which implies that the intervals $(-\infty, 0)$, $(0, 1)$ and $(1, 32/27]$ are the only intervals on the real line which contain no zeros of chromatic polynomials of loopless graphs.

Theorem 25.10. *The real zeros of chromatic polynomials are dense in* $[32/27, \infty)$.

We will consider the problem of identifying the regions of $\mathbb{R}^2$ for which the real zeros of the dichromatic polynomial $Z(G; q, v)$ are dense. Let R_1, R_2 and $R_\diamond$ be as defined at the end of Section 25.3. Theorem 25.8 implies that R_1, R_2 and a large subset of $R_\diamond$ contain no zeros of the dichromatic polynomial of a 2-connected graph and Conjecture 25.9 would imply that the closure of the set of zeros of dichromatic polynomials of graphs has measure zero inside $R_1 \cup R_2 \cup R_\diamond$. This region is illustrated by the uncolored area in the top half of Figure 25.4.

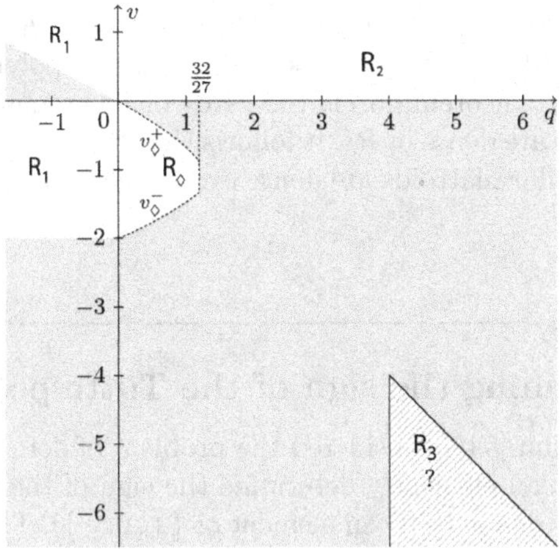

FIGURE 25.4: Regions for which the zeros of the dichromatic polynomial are known to be dense.

It was also conjectured in [642] that the zeros of dichromatic polynomials of graphs are dense outside of $R_1 \cup R_2 \cup R_\diamond$. A recent result of S. Ok and T. Perret [888] comes close to verifying this conjecture by showing that the zeros of dichromatic polynomials are dense throughout the colored area of Figure 25.4.

Theorem 25.11. *The real zeros of the dichromatic polynomial of graphs are dense in* $\mathbb{R}^2 \setminus (R_1 \cup R_2 \cup R_\diamond \cup R_3)$ *where* $R_3 = \{(q,v) : q \geq 4,\ v \in (-\infty, -q)\}$.

The proof of Ok and Perret uses techniques developed by C. Thomassen [1064] to prove Theorem 25.7, and subsequently extended by L. Goldberg and M. Jerrum [548] to determine the points in $\mathbb{R}^2$ for which it is NP-hard to calculate the sign of the Tutte polynomial (see Section 25.5).

The exceptional region R_3 of Theorem 25.11 is shown in Figure 25.4 by the shaded area with a question mark. As noted above, we have $(-1)^{|E|}Z(G;q,v) > 0$ for all bridgeless graphs G whenever q is a positive integer and $v < -q$. This gives an infinite family of zero-free line segments in R_3. It is not known if any of these line segments extend to zero-free regions of positive measure in R_3.

The region R_3 is bounded on one side by the line $v = -q$ along which the dichromatic polynomial reduces to the flow polynomial $F(G;q)$ in that $Z(G;q,-q) = q^{|V|}(-1)^{|E|}F(G;q)$. In this context, points on the line $v = -q$ have received some attention in the literature. D. Welsh conjectured that $Z(q,-q) > 0$ for all bridgeless graphs G and $q \in (4,\infty)$, see [638]. Bridgeless graphs G with $Z(q,-q) = 0$ for some values of q with $4 < q < 6$ were constructed by G. Haggard et al. [596] and J. Jacobsen and J. Salas [643]. A proof that $Z(G;q,-q) > 0$ for all bridgeless graphs G on n vertices and all $q \in (2\log_2 n, \infty)$ is sketched in [638].

Note that the dual of the region R_3 is the region $R_3^* = \{(q,v) : q \geq 4 \text{ and } -1 < v < 0\}$. Since the zeros of dichromatic polynomials of graphs are dense in R_3^* by Theorem 25.11, the zeros of dichromatic polynomials of cographic matroids are dense in R_3. It follows that the zeros of the dichromatic polynomials of regular matroids are dense everywhere outside of $R_1 \cup R_2 \cup R_\diamond$.

25.5 Determining the sign of the Tutte polynomial

Goldberg and Jerrum [548] considered the problem of determining the points in $\mathbb{R}^2$ at which we can efficiently determine the sign of the dichromatic polynomial (which they define to be an element of $\{+, 0, -\}$). Theorem 25.8 shows that this problem is easy for the points in R_1, R_2 and a large subset of $R_\diamond$. Conjecture 25.9 would imply that the problem remains tractable for all points in the interior of $R_\diamond$ since it could be reduced to that of determining the sign for finitely many graphs. The results of [548] show that it is $\sharp$P-hard to determine the sign of the dichromatic polynomial at most points outside the region $R_1 \cup R_2 \cup R_\diamond \cup R_3$ described in the previous section (and deduce that it is $\sharp$P-hard to approximate the dichromatic polynomial at most points outside this region). The complexity of determining the sign of the dichromatic polynomial at most points in R_3 remains open.

A summary of the above results, translated to the Tutte plane, is given in Figure 25.5. In the figure, computing the sign of the Tutte polynomial $T(G, x, y)$ is polynomial at black points and is NP-hard at gray points. The complexity at white points is not known. The black region corresponds to $R_1 \cup R_2$ together with the points in $R_\diamond$ covered by Theorem 25.8. The upper white regions correspond to the points in $R_\diamond$ not covered by Theorem 25.8, together with two rather strange white line segments on the lines $x = -1$ and $y = -1$ which are not covered by the results of [548]. The lower white region corresponds to R_3. The black hyperbolas $(x-1)(y-1) = k$ correspond to the lines $q = k$ for $k \in \mathbb{N}$. The white point $(0, -4)$ and black point $(0, -5)$ correspond respectively to Tutte's 5-flow conjecture and Seymour's 6-flow theorem. We refer the reader to [548] for the precise details, and to Chapter 9 for more information on the algorithmic complexity of the Tutte polynomial.

25.6 Complex zeros

We now consider the distribution of the complex zeros of dichromatic polynomials. In this context, it is natural to allow both the global variable q and the edge-weight v to take complex values and to consider the distribution of

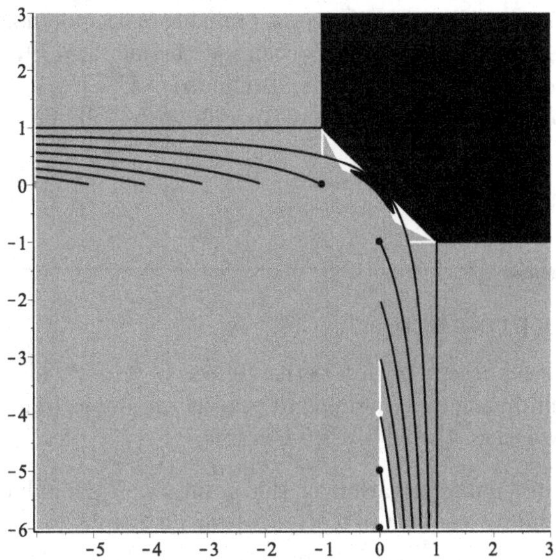

FIGURE 25.5: The complexity of determining the sign of the Tutte polynomial.

zeros in $\mathbb{C}^2$. We first state a result of Sokal [1022] which implies that there are no zero-free regions for chromatic polynomials in $\mathbb{C}$.

Theorem 25.12. *The complex zeros of chromatic polynomials of loopless graphs are a dense subset of* $\mathbb{C}$.

Sokal obtains this result by showing that the zeros of dichromatic polynomials of a special family of graphs are dense in a large region in $\mathbb{C}^2$. The graphs he uses are the *generalized theta graphs* $\Theta_{s,p}$ consisting of two vertices joined by s internally disjoint paths, each of length p.

Theorem 25.13. *Suppose* $(q_0, v_0) \in \mathbb{C}^2$ *and* $|v_0| \leq |q_0 + v_0|$. *Then for all* $\epsilon > 0$ *there exists* $s, p \in \mathbb{N}$ *such that*

1. $Z(\Theta_{s,p}; q_0, v) = 0$ *for some* $v \in \mathbb{C}$ *with* $|v - v_0| < \epsilon$, *and*

2. *if* $v_0 \neq 0$, *then* $Z(\Theta_{s,p}; q, v_0) = 0$ *for some* $q \in \mathbb{C}$ *with* $|q - q_0| < \epsilon$.

The special case of Item 2 of Theorem 25.13 when $v_0 = -1$ tells us that the zeros of chromatic polynomials are dense in $\mathbb{C} \setminus \{q \in \mathbb{C} : |q + 1| < 1\}$. Sokal uses the chromatic roots of the graph $\Theta_{s,p} + K_2$ (obtained from the disjoint union of $\Theta_{s,p}$ and K_2 by adding all edges between them) and the chromatic polynomial identity $\chi(\Theta_{s,p} + K_2; q) = q(q-1)\chi(\Theta_{s,p}; q-2)$, to show that the roots of chromatic polynomials are also dense in the region $\{q \in \mathbb{C} : |q - 1| < 1\}$, and hence verify Theorem 25.12.

Theorem 25.13 implies that the zeros of dichromatic polynomials are dense in $\mathbb{C}^2 \setminus \{(q,v) \in \mathbb{C} : |q+v| < |v|\}$. We can use the fact that $\Theta_{s,p}$ is planar and duality to deduce that these zeros are also dense in $\mathbb{C}^2 \setminus \{(q,v) \in \mathbb{C} : |v| < 1\}$, see [1022, Theorem 7.2]. It seems likely that the zeros will be dense throughout $\mathbb{C}^2$.

25.7 Open problems

1. Complete the characterization of the regions in $\mathbb{R}^2$ for which the set of real zeros of dichromatic polynomials of graphs are dense (and, in particular, determine whether Conjecture 25.9 is true).

2. Complete the characterization of the points in $\mathbb{R}^2$ at which we can efficiently determine the sign of $Z(G,q,v)$ for all graphs.

3. Determine if there exists a constant c such that $Z(G;q,-q) > 0$ for all bridgeless graphs G and all $q \in (c,\infty)$.

4. Obtain similar characterizations for special families of graphs. It is known, for example, that the distributions of the real zeros of chromatic polynomials of graphs of bounded degree, planar graphs, and Hamiltonian graphs differ significantly from that of all graphs. It is conceivable that the same will be true for the real zeros of their dichromatic polynomials. Some first steps in this direction are that $Z(G;q,v) > 0$ whenever: G is series parallel, $q > 2$ and $v \geq -1$, see [1022]; or G has bounded degree and $|q|$ is sufficiently large compared to $|v|$, see [1021, 641].

5. Determine whether the complex zeros of dichromatic polynomials are dense in $\mathbb{C}^2$. Note that Theorem 25.13 implies that the complex zeros of $\Theta_{s,p}$ and its planar dual are dense in $\mathbb{C}^2 \setminus \{(q,v) \in \mathbb{C}^2 : |q+v| \geq |v| > 1\}$.

6. Determine regions of $\mathbb{C}^2$ where the complex zeros of dichromatic polynomials of special families of graphs are dense. Some results for graphs of bounded degree and 3-connected planar graphs are given in [641, 1021] and [981], respectively.

26

The U, $\mathbf{V}$, and W polynomials

Steven Noble

Synopsis

This chapter covers the U-, W-, $\mathbf{V}$- and strong U-polynomials. This family of polynomials includes extensions of the Tutte polynomial to vertex-weighted graphs. These polynomials possess connections with a surprisingly wide range of graph invariants.

- The definitions of the U-, W-, $\mathbf{V}$- and strong U-polynomials.
- Basic theory including expansions of the polynomials and evaluations.
- The equivalence of U with Stanley's symmetric Tutte function and Brylawski's polychromate.
- The problem of deciding whether non-isomorphic trees can have the same U-polynomial.
- Complexity issues.
- Other related polynomials.

26.1 Introduction

In this chapter we discuss a family of four graph polynomials that generalize the Tutte polynomial. The U- and strong U-polynomials are polynomials of (unweighted) graphs, with the strong U-polynomial generalizing the U-polynomial. The W-polynomial extends the U-polynomial to vertex-weighted graphs. The $\mathbf{V}$-polynomial generalizes W to graphs with weights on both the vertices and edges. The original motivation behind them comes from

DOI: 10.1201/9780429161612-26

a variety of sources, from knot theory to statistical physics, but they have turned out to possess connections with a surprisingly wide range of graph invariants. Each of the polynomials potentially has infinitely many variables. For any given graph, however, only finitely many variables appear in any monomial with a non-zero coefficient, so they are genuinely polynomials. In all cases the variables commute and we shall assume this without mentioning it again.

The first of these polynomials to be considered in depth were the U- and W-polynomials, which were introduced by Noble and Welsh in [885]. The W-polynomial was motivated by a series of papers by Chmutov, Duzhin and Lando [311, 312, 313] on Vassiliev invariants of knots. It is a polynomial of graphs having strictly positive integer vertex weights. The U-polynomial is the special case of the W-polynomial in which all the vertex weights are equal to one. Thus the U-polynomial is really a polynomial of (unweighted) graphs rather than vertex-weighted graphs. Its deletion–contraction relation, however, involves vertex-weighted graphs. One surprising result concerning U is that it is equivalent, in a sense to be made precise later, to Stanley's symmetric function generalization of the Tutte polynomial [1036] and Brylawski's polychromate [244], which although introduced several years earlier, did not receive the attention it deserved until much later. We shall discuss Stanley's symmetric function generalization of the Tutte polynomial briefly in Section 26.3, but our main focus will be on the polynomials.

As we shall see, in an informal sense, each of these polynomials replaces the x variable of the Tutte polynomial with many variables, but leaves the y variable unaltered. Bollobás and Riordan [159] refined the polychromate to take care of this situation, by introducing the strong polychromate. Interested in whether the equivalence between U and the polychromate could be extended, Welsh [1144] introduced the strong U-polynomial, which, like the U-polynomial, is a polynomial of unweighted graphs, and satisfies a very similar deletion–contraction relation to that satisfied by the U-polynomial. We discuss the link between these polynomials in Section 26.3.

More recently, Ellis-Monaghan and Moffatt sought to refine the W-polynomial to include the partition function of the Potts model with an external field as a specialization. Allowing the vertex weights to belong to any commutative semigroup and attaching edge weights in an analogous way to which they are used in the multivariate Tutte polynomial, they defined the **V**-polynomial.

Throughout, we will use the graph in Figure 26.1 as a running example to illustrate the definitions.

It is easy to see that, for a graph $G = (V, E)$,

$$x^{k(G)}T(G; x+1, y) = \sum_{A \subseteq E} x^{k(A)}(y-1)^{|A|-r(A)}. \qquad (26.1)$$

Example 26.1. For the graph G in Figure 26.1, disregarding all the labels for the moment, we have $x^{k(G)}T(G; x+1, y) = x^3 + 3x^2 + 2x + 2xy + x^2y + xy^2$.

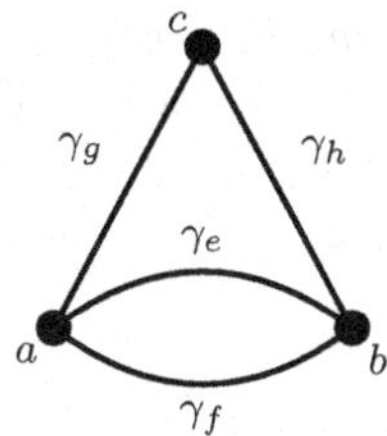

FIGURE 26.1: Graph G used as an example throughout the chapter.

When expanded in powers of x and $y-1$, (26.1) is a two-variable generating function, enumerating subsets A of edges, according to the number of connected components and nullity of the subgraph (V, A). We first consider ways in which we might generalize the x variable. Suppose that instead of counting merely the number of connected components, we record the number of vertices in each of the connected components. With this in mind, we define the *U-polynomial*, introduced by Noble and Welsh in [885].

Definition 26.2. The *U-polynomial* $U(G; \mathbf{x}, y) \in \mathbb{Z}[\{x_i\}_{i\in\mathbb{N}}, y]$ is

$$U(G; \mathbf{x}, y) = \sum_{A \subseteq E(G)} x_{n_1} \cdots x_{n_{k(A)}} (y-1)^{|A|-r(A)},$$

where $n_1, \ldots, n_{k(A)}$ are the numbers of vertices in the $k(A)$ connected components of the subgraph $(V(G), A)$.

Example 26.3. For the graph G in Figure 26.1, again disregarding all the labels,

$$U(G; \mathbf{x}, y) = x_1^3 + 3x_1x_2 + 2x_3 + 2x_3y + x_1x_2y + x_3y^2.$$

The relationship between T and U, illustrated by Examples 26.1 and 26.3, holds in general.

Theorem 26.4. *For any graph G*

$$U(G; \{x_i = x\}_{i\in\mathbb{N}}, y) = x^{k(G)} T(G; x+1, y),$$

where the notation on the left-hand side means that $x_i = x$ for all $i \in \mathbb{N}$.

It is straightforward to generalize U to vertex-weighted graphs.

Definition 26.5. A *vertex-weighted graph* comprises a graph $G = (V, E)$, together with a weight function ω mapping V to a commutative semigroup Ω.

We call $\omega(v)$ the *weight* of vertex v. More generally, if $X \subseteq V(G)$, we define the *weight* of X, $\omega(X)$, to be $\sum_{v\in X} \omega(v)$. We let $\mathbf{1}$ denote the weight function taking the value 1 on every vertex.

Definition 26.6. The *W-polynomial* is a polynomial of vertex-weighted graphs. We take $\Omega = \mathbb{N}$ with the operation of addition. (For our purposes $0 \notin \mathbb{N}$.) Now $W(G, \omega; \mathbf{x}, y) \in \mathbb{Z}[\{x_i\}_{i \in \mathbb{N}}, y]$ with

$$W(G, \omega; \mathbf{x}, y) = \sum_{A \subseteq E(G)} x_{\omega_1} \cdots x_{\omega_{k(A)}} (y-1)^{|A|-r(A)},$$

where $\omega_1, \ldots, \omega_{k(A)}$ are the sums of the weights of the vertices in the $k(A)$ connected components of the subgraph $(V(G), A)$.

Example 26.7. We continue our running example, but now treat G as a vertex-weighted graph, with weights a, b and c as shown in Figure 26.1. We have

$$W(G, \omega; \mathbf{x}, y) = x_a x_b x_c + x_{a+b} x_c + x_{a+c} x_b + x_{b+c} x_a + x_{a+b} x_c y + x_{a+b+c}(y^2 + 2y + 2).$$

Proposition 26.8. *For any graph G, $W(G, \mathbf{1}; \mathbf{x}, y) = U(G; \mathbf{x}, y)$.*

When we work with W, we will generally not mention that $\Omega = \mathbb{N}$. Indeed many of the results below still hold if Ω is allowed to be an arbitrary commutative semigroup. The reason for specifying that $\Omega = \mathbb{N}$ is purely historical and to maintain consistency with [885].

We now generalize W to give the **V**-polynomial, which was introduced by Ellis-Monaghan and Moffatt in order to include the partition function of the Potts model with an external field as a specialization. This polynomial is defined on graphs which have weights attached to both the vertices and the edges, although the two sets of weights work in fundamentally different ways.

Definition 26.9. A *doubly-weighted graph* comprises a vertex-weighted graph (G, ω) equipped with a function γ from $E(G)$ to a set $\mathbf{\Gamma}$. We let $\gamma_e = \gamma(e)$.

Note that γ_e will play two roles: it is both a weight on an edge and will also be a variable in the polynomial. We now define the **V**-polynomial.

Definition 26.10. Let Ω be any commutative semigroup. Then $\mathbf{V}(G, \omega) = \mathbf{V}(G, \omega; \mathbf{x}, \gamma) \in \mathbb{Z}[\{x_i\}_{i \in \Omega}, \{\gamma_e\}_{e \in E(G)}]$ is given by

$$\mathbf{V}(G, \omega; \mathbf{x}, \gamma) = \sum_{A \subseteq E(G)} x_{\omega_1} \cdots x_{\omega_{k(A)}} \prod_{e \in A} \gamma_e,$$

where $\omega_1, \ldots, \omega_{k(A)}$ are the sums of the weights of the vertices in the $k(A)$ connected components of the subgraph $(V(G), A)$.

Example 26.11. Returning to the example once more, but now treating G as a doubly-weighted graph, using both the vertex and edge labels, we see that

$$\mathbf{V}(G, \omega; \mathbf{x}, \gamma) = x_a x_b x_c + (\gamma_e + \gamma_f + \gamma_e \gamma_f) x_{a+b} x_c + \gamma_g x_{a+c} + \gamma_h x_{b+c} + ((\gamma_e + \gamma_f + \gamma_e \gamma_f)(\gamma_g + \gamma_h + \gamma_g \gamma_h) + \gamma_g \gamma_h) x_{a+b+c}.$$

The $\mathbf{V}$-polynomial specializes to the W-polynomial and consequently the U-polynomial. It also contains the multivariate Tutte polynomial $Z(G;q,\{\gamma_e\})$ (see Equation (20.7)).

Theorem 26.12.

1. *Let (G,ω) be a vertex-weighted graph and let $\gamma_e = y-1$ for all $e \in E(G)$. Then*

$$\mathbf{V}(G,\omega;\mathbf{x},\{\gamma_e = y-1\}_{e\in E}) = (y-1)^{|V(G)|}W(G,\omega;\{x_i/(y-1)\}_{i\in\Omega},y).$$

2. *Let G be a graph and let $\gamma_e = y-1$ for all $e \in E(G)$. Then*

$$\mathbf{V}(G,1;\mathbf{x},\{\gamma_e = y-1\}_{e\in E}) = (y-1)^{|V(G)|}U(G;\{x_i/(y-1)\}_{i\in\mathbb{N}},y).$$

3. *Let (G,ω,γ) be a doubly-weighted graph, then independently of the vertex weights ω,*

$$\mathbf{V}(G,\omega;\{x_i = q\}_{i\in\Omega},\{\gamma_e\}_{e\in E(G)}) = Z(G;q,\{\gamma_e\}_{e\in E(G)}).$$

Consideration of the definition of $U(G)$ reveals a lack of symmetry: we count subgraphs $(V(G),A)$ according to the number of vertices in each connected component, but only the total number of edges. This leads us to a different way of generalizing U.

Definition 26.13. The *strong U-polynomial*, denoted by $\overline{U}$, is defined as follows. We have $\overline{U}(G;\mathbf{z}) \in \mathbb{Z}[\{z_{i,j}\}_{i\in\mathbb{N},j\in\mathbb{Z}^+}]$ and

$$\overline{U}(G;\mathbf{z}) = \sum_{A\subseteq E(G)} z_{n_1,e_1-n_1+1}\cdots z_{n_{k(A)},e_{k(A)}-n_{k(A)}+1},$$

where n_i and e_i are the number of vertices and edges respectively in the i-th connected component of the subgraph $(V(G),A)$.

Example 26.14. Taking G as in Figure 26.1, once again, we see that

$$\overline{U}(G;\mathbf{z}) = z_{1,0}^3 + 4z_{2,0}z_{1,0} + z_{2,1}z_{1,0} + 5z_{3,0} + 4z_{3,1} + z_{3,2}.$$

Proposition 26.15. *For any graph G*

$$U(G;\mathbf{x},y) = \overline{U}(G;\{z_{i,j} = x_i(y-1)^j\}_{i,j}).$$

26.2 Properties of the polynomials

In this section we describe properties of each of the polynomials, beginning with the $\mathbf{V}$- and W-polynomials. Many of the results for the $\mathbf{V}$-polynomial immediately imply corresponding results for the W-polynomial, although sometimes in a slightly different form.

One of the many attractive features of the Tutte polynomial is that it satisfies a simple deletion–contraction relation. Perhaps surprisingly, the deletion–contraction relation satisfied by the **V**-polynomial is arguably simpler than that satisfied by the Tutte polynomial.

To describe the identity we need first to define deletion and contraction in a vertex-weighted graph (G, ω). Deletion is easy: when an edge is deleted, the vertex weights remain unchanged. Contraction is slightly more complex: when a non-loop edge e with end-vertices u and v is contracted, the vertex weight on the new vertex resulting from the contraction is given weight $\omega(u) + \omega(v)$; the other vertex weights are unchanged. We will write ω/e to denote the weight function that results from contracting the edge e. If e is a loop, then $G/e = G\backslash e$. Deletion and contraction are similar in a doubly-weighted graph: when an edge is deleted or contracted, the other edge weights are unchanged.

The deletion–contraction relations are easily proved by splitting the sum in the definition into two according to whether or not the set A contains the edge e.

Theorem 26.16. *Let e be an edge of a doubly-weighted graph (G, ω, γ). Then*

$$\mathbf{V}(G, \omega; \mathbf{x}, \gamma) = \mathbf{V}(G\backslash e, \omega; \mathbf{x}, \gamma) + \gamma_e \mathbf{V}(G/e, \omega/e; \mathbf{x}, \gamma).$$

Note that we have defined the contraction of a loop to be the same as its deletion. This ensures that the deletion–contraction relation relation comprises only one case. Furthermore, note that in order to keep the notation manageable, we have abused notation slightly in the previous theorem. On the left-hand side γ denotes the weight function on $E(G)$, whereas on the right-hand side, it denotes its restriction to $E(G) \setminus e$.

The identity for the W-polynomial is slightly different.

Theorem 26.17. *Let e be an edge of a vertex-weighted graph (G, ω).*

1. *If e is a loop, then*

$$W(G, \omega; \mathbf{x}, y) = yW(G\backslash e, \omega; \mathbf{x}, y).$$

2. *Otherwise,*

$$W(G, \omega; \mathbf{x}, y) = W(G\backslash e, \omega; \mathbf{x}, y) + W(G/e, \omega/e; \mathbf{x}, y).$$

This result also gives a deletion–contraction relation for the U-polynomial, but clearly contracting a non-loop edge takes us away from the setting in which every vertex has weight equal to one. Therefore, a deletion–contraction relation for U must be stated in terms of W rather than U alone.

Taken together with a boundary condition defining **V** or W on graphs with no edges, the deletion–contraction relation relations are sufficient to define **V** or W uniquely. In fact we have slightly stronger results demonstrating the universality of **V** and W.

Theorem 26.18. *Let f be a function on vertex-weighted graphs defined recursively by the following conditions, where each $\alpha_e \neq 0$.*

1. *For any edge e,*
$$f(G,\omega) = \alpha_e f(G\backslash e,\omega) + \beta_e f(G/e,\omega/e).$$

2. *If (G,ω) consists of isolated vertices with weights $\omega_1,\dots\omega_n$, then $f(G,\omega) = x_{\omega_1}\cdots x_{\omega_n}$.*

Then
$$f(G,\omega) = \Big(\prod_{e\in E(G)} \alpha_e\Big)\mathbf{V}\big(G,\omega;\mathbf{x},\{\beta_e/\alpha_e\}_{e\in E(G)}\big).$$

Theorem 26.19. *Let f be a function on vertex-weighted graphs defined recursively by the following conditions, where a and b are non-zero constants.*

1. *For any non-loop edge e,*
$$f(G,\omega) = af(G\backslash e,\omega) + bf(G/e,\omega/e).$$

2. *If e is a loop, then $f(G,\omega) = yf(G\backslash e,\omega)$.*

3. *If (G,ω) consists of isolated vertices with weights $\omega_1,\dots\omega_n$, then $f(G,\omega) = x_{\omega_1}\cdots x_{\omega_n}$.*

Then
$$f(G,\omega) = a^{|E(G)|-|V(G)|}b^{|V(G)|}W(G,\omega;\{ax_i/b\}_{i\in\Omega},y/a).$$

We now move on to describe a variety of results concerning $\mathbf{V}$ and W that can be easily derived from the definitions or from the deletion–contraction relation. The following properties of $\mathbf{V}$ are easily verified.

Proposition 26.20.

1. *If G consists of connected components $G_1,\dots,G_k$, then $\mathbf{V}(G) = \prod_{i=1}^k \mathbf{V}(G_i)$.*

2. *G is a forest if and only if each monomial in $\mathbf{V}(G)$ is the product of the same number of terms.*

Similar properties can be established for W. In particular, G is a forest if and only if $W(G,\omega;\mathbf{x},y)$ has no term in y. At this point, it is convenient to introduce several definitions and pieces of notation involving partitions.

Definition 26.21. Let π be a partition. Then $\#\pi$ denotes the number of blocks of π.

Definition 26.22. Let $\pi = \{V_1,\dots,V_{\#\pi}\}$ be a partition of the vertex set V of a graph G into non-empty blocks. It determines subgraphs $G_1,\dots,G_{\#\pi}$ where G_i denotes the subgraph of G induced by V_i. We say that a partition π of V is *connected* if each of $G_1,\dots,G_{\#\pi}$ is connected.

Definition 26.23. Let Ω be a commutative semigroup. Then an *Ω-partition* ω of c is a multiset $\{\omega_1, \dots, \omega_k\}$ of elements of Ω such that $\sum_{i=1}^k \omega_i = c$. If ω is an Ω-partition of c, then we write $\omega \vdash c$.

Note that we allow elements of an Ω-partition to be equal to the identity and that the standard notion of an integer partition is an $\mathbb{N}$-partition.

Definition 26.24. Given a partition π of the vertex set of a vertex-weighted graph G, we let $\omega(\pi)$ denote the Ω-partition whose parts are the sums of the weights of the vertices in each block of π. We denote the special case where $\Omega = \mathbb{N}$ and all weights are equal to one by $\tau(\pi)$.

In other words $\tau(\pi)$ denotes the integer partition of $|V(G)|$ for which the parts are the sizes of the blocks of π.

Definition 26.25. Let $\omega = \{\omega_1, \dots, \omega_k\}$ be an Ω-partition. Then we define $\mathbf{x}(\omega)$ to be the monomial $x_{\omega_1} \dots x_{\omega_k}$.

Definition 26.26. Let A be a subset of the edges of a graph G. Then the partition $\pi_G(A)$ is the partition of $V(G)$ in which each block is the vertex set of a connected component of the subgraph $(V(G), A)$. When the context is clear, we write $\pi(A)$ rather than $\pi_G(A)$.

Using the notation established above, we have

$$W(G, \omega; \mathbf{x}, y) = \sum_{A \subseteq E(G)} \mathbf{x}(\omega(\pi(A)))(y-1)^{|A|-r(A)}.$$

Replacing ω by τ gives an analogous expression for U.

Our next results follow from the definitions and give an interpretation of $\mathbf{V}$ and W in terms of the multivariate Tutte polynomial and Tutte polynomial respectively. For a connected graph G, we let $C(G;\gamma)$ denote the coefficient of q in $Z(G;q,\gamma)$. Thus the only terms contributing to $C(G;\gamma)$ are the edge sets of connected spanning subgraphs of G. The first part of the theorem below is a slight reformulation of a result due to McDonald and Moffatt [832]; the second part is from [885].

Theorem 26.27.

1. *Let (G, ω, γ) be a doubly-weighted graph. Then*

$$\mathbf{V}(G, \omega; \mathbf{x}, \gamma) = \sum_{\pi} \mathbf{x}(\omega(\pi)) C(G_1;\gamma) \cdots C(G_{\#\pi};\gamma).$$

2. *Let (G, ω) be a vertex-weighted graph. Then*

$$W(G, \omega; \mathbf{x}, y) = \sum_{\pi} \mathbf{x}(\omega(\pi)) T(G_1; 1, y) \cdots T(G_{\#\pi}; 1, y).$$

In both cases the sum is over all connected partitions of $V(G)$ *and* $G_1, \ldots, G_{\#\pi}$ *are the subgraphs determined by the blocks of* π.

We immediately obtain the following corollary. Let $\mathcal{F}(G)$ denote the set of all spanning forests of G. As usual, in the result below and later on, we shall think of F as both a spanning forest of G and as the edge-set of a spanning subgraph of G that is a forest.

Corollary 26.28. *The polynomial* $W(G, \omega; \mathbf{x}, 1)$ *is a homogeneous polynomial in* $\mathbf{x}$ *and is a weighted sum over all spanning forests of* G, *namely*

$$W(G, \omega; \mathbf{x}, 1) = \sum_{F \in \mathcal{F}(G)} \mathbf{x}(\omega(\pi(F))).$$

Tutte's notion of internal and external activities is defined for spanning trees but the notion of external activity can easily be extended to spanning forests.

Definition 26.29. Given an ordering on the edges of a graph G, an edge e is *externally active* with respect to a spanning forest F if both its end-vertices are in the same component T of F and e is externally active with respect to the spanning tree T of the subgraph of G induced by the vertices of T. Let $\mathrm{EA}(F)$ denote the set of edges that are externally active with respect to F.

Thus we obtain another corollary of Theorem 26.27 from [832].

Corollary 26.30.

1. *If* (G, ω, γ) *is a doubly-weighted graph, then*

$$\mathbf{V}(G, \omega; \mathbf{x}, \gamma) = \sum_{F \in \mathcal{F}(G)} \mathbf{x}(\omega(\pi(F))) \prod_{e \in E(F)} \gamma_e \prod_{e \in \mathrm{EA}(F)} (1 + \gamma_e).$$

2. *If* (G, ω) *is a vertex-weighted graph, then*

$$W(G, \omega; \mathbf{x}, y) = \sum_{F \in \mathcal{F}(G)} \mathbf{x}(\omega(\pi(F))) y^{|\mathrm{EA}(F)|}.$$

Assuming some fixed ordering of the edges of a graph, let $\mathrm{II}(T)$ denote the set of edges that are internally inactive with respect to a maximal spanning forest T. Let $\mathcal{T}(G)$ denote the set of maximal spanning forests of a graph G. The following result from [832] shows that it is possible to replace the sum over spanning forests by a sum over maximal spanning forests.

Theorem 26.31.

1. *If* (G, ω, γ) *is a doubly-weighted graph, then*

$$\mathbf{V}(G, \omega; \mathbf{x}, \gamma) = \sum_{T \in \mathcal{T}(G)} \Big(\prod_{e \in \mathrm{II}(T)} \gamma_e \Big) \Big(\prod_{e \in \mathrm{EA}(T)} (\gamma_e + 1) \Big) \mathbf{V}(T/\mathrm{II}(T)).$$

2. If (G,ω) is a vertex-weighted graph, then

$$W(G,\omega;\mathbf{x},y)=\sum_{T\in\mathcal{T}(G)} y^{|\mathrm{EA}(T)|}W(T/\Pi(T)).$$

By either exploiting the previous theorem or a careful use of the deletion–contraction relation, it is possible to prove the following generalization of the convolution formula for the Tutte polynomial from [704]. The result is due, independently, to the author and Kayebi [686].

Theorem 26.32. *Let (G,ω) be a vertex-weighted graph. Then*

$$W(G,\omega;\mathbf{x},y)=\sum_{A\subseteq E(G)} T(G\backslash(E(G)\setminus A);0,y)\cdot W(G/A,\omega/A;\mathbf{x},0).$$

Because many of the terms are zero, it is possible to restrict the sum to those sets A that are flats such that $(V(G),A)$ is bridgeless.

In [311, 312, 313] Chmutov, Duzhin and Lando define weighted chromatic invariants. A weighted chromatic invariant is a function on graphs satisfying a recurrence related to that satisfied by Vassiliev invariants on singular knots. Weighted chromatic invariants correspond exactly to the W-polynomial with $y=0$. The definition of the W-polynomial came from a remark that the weighted chromatic invariants did not specialize to the Tutte polynomial.

In contrast, the origins of the $\mathbf{V}$-polynomial lie in statistical physics, more precisely in the Potts model with variable edge interaction energy and variable magnetic field. The main result of [455] by Ellis-Monaghan and Moffatt shows that the partition function Z_P of the Potts model with variable interaction energies and an external magnetic field is a specialization of $\mathbf{V}(G)$. See Chapter 20, in particular Equation (20.6), or [781] for a description of the Potts model. We generalize the setting there a little more by allowing the magnetic field to depend on both the spin j and the vertex x. So the magnetic field vector $\mathbf{h}$ is replaced by $(h_{x,j}:x\in V,\ 1\le j\le q)$. Thus the Hamiltonian is

$$H_P(\sigma)=-\sum_{e=(x,y)} J_e\delta_{\sigma_x,\sigma_y}-\sum_{j=1}^{q}\sum_{x\in V} h_{x,j}\delta_{\sigma_x,j}.$$

Theorem 26.33.

$$Z_P(G;\beta,\mathbf{J},h)=\mathbf{V}(G,\omega;\{x_{\mathbf{h}}\}_{\mathbf{h}\in\mathbb{C}^q},\{e^{\beta J_e}-1\}_{e\in E(G)}),$$

where the vertex weights are given by $\omega(v)=(h_{v,1},\dots,h_{v,q})$ and for any $\mathbf{h}=(h_1,\dots,h_q)\in\mathbb{C}^q$, we have $x_{\mathbf{h}}=\sum_{i=1}^{q}e^{\beta h_i}$.

Specific instances of this version of the Potts model include those where the magnetic field acts to favor one spin or a subset of spins over the others.

Explicit expressions for the Potts model partition function in terms of the **V**-polynomial are given in [455] for these cases.

As a corollary of Theorem 26.33, a Fortuin–Kasteleyn type representation for the Potts model with variable edge interaction and external magnetic field may be obtained [455].

Corollary 26.34.

$$Z(G) = \sum_{A \subseteq E(G)} \mathbf{x}(\omega(\pi(A))) \prod_{e \in A} (e^{\beta J_e} - 1),$$

where the vertex weights are given by $\omega(v) = (h_{v,1}, \ldots, h_{v,q})$ *and for any* $\mathbf{h} = (h_1, \ldots, h_q) \in \mathcal{C}^q$*, we have* $x_{\mathbf{h}} = \sum_{i=1}^{q} e^{\beta h_i}$.

By exploiting the fact that U and W are special cases of $\mathbf{V}$, their links with the Potts model may be derived from Theorem 26.33 [455]. The partition function may be expressed in terms of W when there exist $b_1, \ldots, b_q \in \mathbb{C}$ and $k_1, \ldots, k_{|V|} \in \mathbb{Z}$ so that $h_{v,i} = k_v b_i$, that is magnetic field vectors at each vertex are integer multiples of one another; it may be expressed in terms of U when the interaction energy J_e is constant and the magnetic field h does not depend on v (but may still depend on q).

Combining Theorem 26.33 with each of Theorem 26.27, Corollary 26.30 and Theorem 26.31 yields expressions for $Z(G)$ due to McDonald and Moffatt [832]. Further results on the Potts model and the **V**-polynomial are contained in [459], where a connection is drawn between $Z(G)$ and a list-coloring polynomial.

We now shift focus to U. First we summarize its basic properties that follow from what we know about **V** and W.

Proposition 26.35. *Let* $G = (V, E)$ *be a graph. Then*

1. $U(G)$ *is independent of* y *if and only if* G *is a forest; and*

2. *each monomial in* $U(G)$ *is of the form* $x_{a_1} \cdots x_{a_k} y^t$ *where* $\sum_{i=1}^{k} a_i = |V|$.

We now turn to evaluations of U in terms of known combinatorial polynomials.

Theorem 26.36. *For any graph* $G = (V, E)$,

$$T(G; x, y) = (x-1)^{-k(G)} U(G; x_i = x-1, y).$$

In particular, the chromatic polynomial $\chi(G; \lambda)$ *is given by*

$$\chi(G; \lambda) = (-1)^{|V|} U(G; x_i = -\lambda, y = 0).$$

However, U is a much stronger invariant than the Tutte polynomial. For example

$$U(P_4) = x_1^4 + 3x_1^2 x_2 + 2x_1 x_3 + x_2^2 + x_4,$$
$$U(St_4) = x_1^4 + 3x_1^2 x_2 + 3x_1 x_3 + x_4,$$

where P_4 and St_4 are respectively the path and star with four vertices. But both P_4 and St_4 are trees with four vertices and consequently share the same Tutte polynomial, namely, x^3.

Another specialization of U gives the *stability polynomial*, $A(G;p)$, of a loopless graph. This was introduced by Farr in [478] and is the one-variable polynomial given by

$$A(G;p) = \sum_{\substack{X \subseteq V(G) \\ X \in \mathcal{S}(G)}} p^{|X|}(1-p)^{|V(G)\setminus X|},$$

where $\mathcal{S}(G)$ is the set of all stable sets of G.

Theorem 26.37. *If G is loopless, then $A(G;p)$ is given by*

$$A(G;p) = U(G; x_1 = 1, x_j = -(-p)^j \text{ if } j \geq 2, y = 0).$$

We now see that the number of cliques of given size appear as coefficients of monomials in U.

Proposition 26.38. *If $G = (V,E)$ is simple, then the coefficient of $x_k x_1^{|V|-k} y^{\binom{k}{2}-k+1}$ in $U(G;\mathbf{x},y)$ equals the number of cliques of size k in G.*

The *matching polynomial* of G is the polynomial $m(G;t)$ given by

$$m(G;t) = \sum_{k \geq 0} m_k t^k,$$

where m_k is the number of matchings of G with k edges.

Proposition 26.39. *The matching polynomial is given by*

$$m(G;t) = U(G; x_1 = 1, x_2 = t, x_j = 0 \text{ for } j > 2, y = 1).$$

The *most general edge elimination polynomial* $\xi(G;x,y,z) \in \mathbb{Z}[x,y,z]$ of a graph was introduced by Averbouch, Godlin and Makowsky [55] and is defined recursively as follows.

1. If G consists of k isolated vertices, then $\xi(G;x,y,z) = x^k$;
2. If e is an edge of G, then

   $$\xi(G) = \xi(G\backslash e) + y\xi(G/e) + z\xi(G \sim e),$$

   where $G \sim e$ is the graph formed by deleting the end-vertices of e and any edges incident with them.

Note that by deleting the end-vertices of an edge, we may end up with a completely empty graph having no vertices. This possibility is dealt with by taking $k = 0$, in the first case.

Example 26.40. Returning to our example from Figure 26.1, we have

$$\xi(G;x,y,z) = 4xz + 5yz + x^3 + 4x^2y + 5xy^2 + 4y^2z + xyz + 4xy^3 + x^2y^2 + y^3z + xy^4.$$

Trinks [1082] showed that for any graph G,

$$\xi(G;x,y,z) = \sum_{A \subseteq E(G)} x^{k(A)} y^{|A|} \left(\frac{z}{xy} + 1\right)^{\bar{k}(A)},$$

where $\bar{k}(A)$ is the number of components of the subgraph $(V(G), A)$ that are not isolated vertices. Answering a question from [55], he then established the following [1083].

Theorem 26.41. *If G is loopless, then*

$$\xi(G;x,y,z) = U(G;x_1 = x, x_i = xy^{i-1}(z/(xy)+1) \text{ for } i \geq 2, y+1).$$

We can remove the restriction on loops and obtain $\xi(G)$ as a specialization of $\overline{U}$.

Theorem 26.42. *For any graph G,*

$$\xi(G;x,y,z) = \overline{U}(G; z_{1,0} = x, z_{i,j} = xy^{i+j-1}(z/(xy)+1) \text{ otherwise}).$$

In [899], Oxley and Whittle studied a polynomial S that in its general setting is the analogue of the Tutte polynomial for 2-polymatroids, but we only consider its specialization to graphs here. For a graph G, define $f: 2^{E(G)} \to \mathbb{Z}^+$ by setting $f(A)$ to be the number of vertices of G incident with an edge in A. Note that $0 \leq f(A) \leq 2|A|$. They define $S(G;u,v) \in \mathbb{Z}[u,v]$ so that

$$S(G;u,v) = \sum_{A \subseteq E(G)} u^{f(E)-f(A)} v^{2|A|-f(A)}.$$

For graphs without isolated vertices, we have

$$S(G;u,v) = v^{-|V(G)|} \sum_{A \subseteq E(G)} (uv)^{k(A)} v^{2|A|} (uv)^{-\bar{k}(A)}.$$

Thus we have the following.

Theorem 26.43.

1. *If G is loopless and has no isolated vertices, then*

$$S(G;u,v) = U(G; x_1 = u, x_i = v^{i-2} \text{ otherwise}, v^2+1).$$

2. *If G has no isolated vertices, then*

$$S(G;u,v) = \overline{U}(G; z_{1,0} = u, z_{i,j} = v^{i+2j-2} \text{ otherwise}).$$

26.3 Equivalent polynomials and symmetric functions

In this section we consider graph polynomials and a symmetric function that are equivalent to U. We refer the reader to [804] for definitions concerning symmetric functions. First we make the notion of equivalence precise.

Definition 26.44. We say that two graph polynomials are *equivalent* if the coefficients of each one are functions of the coefficients of the other and the number of vertices of the graph.

We extend this idea, by allowing the coefficients to be those of a symmetric function with respect to some basis. Our notion of equivalence turns out to coincide with notions studied by Makowsky [813]. In this section we will assume that G denotes a graph having n vertices $v_1, \ldots, v_n$.

A less apparent specialization of $U(G)$ is that it gives the symmetric function generalization $X(G)$ of the chromatic polynomial developed by Stanley in [1033].

Definition 26.45. For any graph $G = (V, E)$, the *chromatic symmetric function*, $X(G; \mathbf{x})$, is a homogeneous symmetric function in $\mathbf{x} = (x_1, x_2, \ldots)$ of total degree $n = |V|$ defined by

$$X(G; \mathbf{x}) = X(G; x_1, x_2, \ldots) = \sum_{\kappa} x_{\kappa(v_1)} x_{\kappa(v_2)} \cdots x_{\kappa(v_n)},$$

where the sum ranges over all proper colorings $\kappa : V(G) \to \mathbb{N}$.

Note that the sum is infinite since we allow any proper coloring using positive integers. In [1033] Stanley develops $X(G)$ and its properties in terms of the standard "natural" bases for the space of symmetric functions.

Definition 26.46. For $r \geq 1$, let $p_r = p_r(\mathbf{x}) = p_r(x_1, x_2, \ldots) = \sum_i x_i^r$. If $\tau = \{n_1, \ldots, n_{\#\tau}\}$ is an integer partition of n, then $p_\tau = p_{n_1} \cdots p_{n_{\#\tau}}$. The collection $\{p_\tau\}_{\tau \vdash n}$ forms the *power sum basis* for the homogeneous symmetric functions of degree n. Let

$$m_\tau = m_\tau(\mathbf{x}) = \sum_{i_1, \ldots, i_{\#\tau}} \prod_{j=1}^{\#\tau} x_{i_j}^{n_j},$$

where the sum is over all $\#\tau$-tuples of pairwise distinct positive integers. The collection $\{m_\tau\}_{\tau \vdash n}$ forms the *monomial basis* for the homogeneous symmetric functions of degree n.

Stanley [1033] showed the following.

Theorem 26.47. *For any graph G,*

$$X(G; \mathbf{x}) = \sum_{A \subseteq E(G)} (-1)^{|A|} p_{\tau(\pi(A))}.$$

It follows from this result and the definition of U that X is a specialization of the U-polynomial.

Theorem 26.48. *For any graph G on n vertices,*

$$X(G;\mathbf{x}) = (-1)^n U(G; x_j = -p_j, y = 0).$$

Many of the results in [1033] now follow from our earlier interpretations of U. In a later paper [1036], Stanley introduces the following symmetric function generalization of the Tutte polynomial of a graph, which we call the *Tutte symmetric function* for brevity. His starting point is the "bad coloring" expansion of the Tutte polynomial from [247], which gives T as a generating function for the number of "bad" or monochromatic edges over all colorings of the vertices.

Definition 26.49. Let $\mathbf{x} = (x_1, x_2, \ldots)$ and t be indeterminates and define

$$Y(G;\mathbf{x},t) = \sum_{\kappa: V(G)\to\mathbb{N}} (1+t)^{b(\kappa)} x_{\kappa(v_1)} x_{\kappa(v_2)} \cdots x_{\kappa(v_n)},$$

where $b(\kappa)$ denotes the number of monochromatic edges in the coloring κ and the sum is over *all* colorings.

If $\mathbf{x} = 1^n$ denotes the substitution $x_1 = x_2 = \cdots = x_n = 1$, and $x_{n+1} = x_{n+2} = \cdots = 0$, then

$$Y(G;1^n,t) = n^{k(G)} t^{r(G)} T\big(G; \frac{t+n}{t}, t+1\big)$$

and

$$Y(G;\mathbf{x},-1) = X(G;\mathbf{x}). \tag{26.2}$$

In [1036], Stanley attributes the following extension of Theorem 26.47 to Chow.

Theorem 26.50. *For any graph G,*

$$Y(G;\mathbf{x},t) = \sum_{A\subseteq E(G)} t^{|A|} p_{\tau(\pi(A))}.$$

It now follows that $U(G;\mathbf{x},y)$ and $Y(G;\mathbf{x},t)$ determine each other.

Theorem 26.51. *For any graph G on n vertices we have*

$$Y(G;\mathbf{x},t) = t^n U(G; \{x_j = p_j/t\}_{j\in\mathbb{N}}, y = t+1).$$

Another way of describing the substitution into U that gives Y is to say that for each τ such that $\tau \vdash n$ and each i, the monomial $\mathbf{x}(\tau)y^i$ in $U(G)$ is replaced by $p_\tau(\mathbf{x})t^{n-\#\tau}(t+1)^i$. Thus U and the symmetric Tutte function are equivalent.

Crew and Spirkl [351] have generalized the chromatic and Tutte symmetric functions to weighted graphs, in such a way that the Tutte symmetric function of a weighted graph contains exactly the same information as W. These generalizations admit a deletion–contraction recurrence.

We shall now explore further equivalences involving U and $\overline{U}$. In the introduction we mentioned the polychromate, introduced by Brylawski [244].

Definition 26.52. The *polychromate* $\nu(G;\mathbf{x},y)$ is defined by

$$\nu(G;\mathbf{x},y) = \sum_{\pi} y^{e(\pi)}\mathbf{x}(\tau(\pi)),$$

where the summation is over all partitions of $V(G)$ and $e(\pi)$ is the number of edges with both endpoints in the same block of the partition π.

Example 26.53. For the graph G from Figure 26.1,

$$\nu(G;\mathbf{x},y) = x_1^3 + (y^2+2y)x_1x_2 + y^4x_3.$$

The following result is due to Sarmiento [984].

Theorem 26.54. *The U-polynomial and the polychromate are equivalent.*

The proof is complicated and uses ideas from intersection theory described in [244]. An alternative way to understand this result is to observe that Y and ν are equivalent. It follows immediately from Definition 26.49 that

$$Y(G;\mathbf{x},t) = \sum_{\pi} m_{\tau(\pi)}(1+t)^{e(\pi)},$$

where the summation is over all partitions of $V(G)$. Furthermore as $\{p_\tau\}_{\tau\vdash n}$ and $\{m_\tau\}_{\tau\vdash n}$ are both bases for the homogeneous symmetric functions of degree n, there are constants $a_{\tau,\tau'}$ such that $p_\tau(\mathbf{x}) = \sum_{\tau'} a_{\tau,\tau'}m_{\tau'}(\mathbf{x})$. It is not difficult to compute $a_{\tau,\tau'}$. Given a partition π, we say that the partition π' is a coarsening of π if every block of π' is a union of blocks of π. Let π be a partition of $\{1,\dots,n\}$ of type τ. Then $a_{\tau,\tau'}$ is the number of coarsenings of π of type τ'. Thus, as observed in [841], we can give a more succinct explanation of the substitution required to obtain the polychromate from the U-polynomial than is given in [984].

Theorem 26.55. *The polychromate may be obtained from the U-polynomial by, for each τ such that $\tau \vdash n$, replacing the monomial $\mathbf{x}(\tau)y^j$ with $\sum_{\tau'\vdash n} a_{\tau,\tau'}\mathbf{x}(\tau')y^j(y-1)^{n-\#\tau}$.*

The equivalence of U and ν leads to the following observation originally due to R. Hall (unpublished), as the corresponding results for ν are straightforward.

Proposition 26.56.

1. *If G is loopless, then $U(G)$ determines the degree sequence of G.*
2. *If G is simple, then $U(G)$ determines $U(\overline{G})$, where $\overline{G}$ is the complement of G.*

Brylawski [243] showed that ν and hence U and Y are reconstructible. See also the results of Kotek [710].

Loebl has introduced q-analogues of the chromatic polynomial and the dichromate [780]. The latter is effectively a q-analogue of the Tutte polynomial.

Definition 26.57. For real numbers y and $q \neq 1$, the *quantum number* $(y)_q = (q^y - 1)/(q-1)$. We have $(y)_1 = y$.

Definition 26.58. The *q-dichromate* is the function $B_q(G) = B_q(G; x, y)$, where x and y are real numbers, and

$$B_q(G; x, y) = \sum_{A \subseteq E(G)} y^{|A|} (x)_{q^{n_1}} \cdots (x)_{q^{n_k}},$$

where $\{n_1, \ldots, n_k\} = \tau(\pi(A))$.

Note that we have interchanged the variables x and y compared with the definition in [780]. Loebl shows the following relationship with the U-polynomial.

Theorem 26.59. *For any graph G,*

$$B_q(G; s, t) = t^{|V(G)|} U(G; \{x_i = (s)_{q^i}/t\}_{i \in \mathbb{N}}, t + 1).$$

Loebl [780] conjectures that the q-dichromate and the U-polynomial are equivalent, but as far as we know this is still an open problem. In a later paper [693], Klazar, Loebl and Moffatt introduce a refinement of the q-dichromate and show it to be equivalent to U.

Definition 26.60. The *(q,r)-dichromate* is the function $B_{q,r}(G; x, y)$, where x is a positive integer, y is a real number and

$$B_{q,r}(G; x, y) = \sum_{A \subseteq E(G)} y^{|A|} \prod_{i=1}^{k} \sum_{j=0}^{x-1} q^{n_i r^j}, \tag{26.3}$$

where $\{n_1, \ldots, n_k\} = \tau(\pi(A))$.

We obtain B_q from $B_{q,r}$ by replacing each exponent of q of the form $n_j r^i$ with $i n_j$.

In [693], it is shown that $B_{q,r}$ may be written as a sum over colorings, a result analogous to Theorem 26.47.

Theorem 26.61. *For any graph G,*

$$B_{q,r}(G;x,y) = \sum_{\kappa:V(G)\to\{0,\dots,x-1\}} (1+y)^{b(\kappa)} q^{\sum_{v\in V(G)} r^{\kappa(v)}},$$

where $b(\kappa)$ denotes the number of monochromatic edges in the coloring κ and the sum is over all *colorings.*

Theorem 26.62. *For any fixed $q > 1$, $B_{q,r}$ and U are equivalent. Moreover for any graph G,*

$$B_{q,r}(G;s,t) = t^{|V(G)|}U(G;\{x_i = \sum_{j=0}^{s-1} q^{ir^j}/t\}_{i\in\mathbb{N}}, t+1).$$

A version of $B_{q,r}$ for vertex-weighted graphs is defined in [693] by replacing n_i by the total weight of the corresponding connected component. This is analogous to the way in which U may be generalized to W and permits a similar deletion–contraction relation.

A strong version of the polychromate was defined by Bollobás and Riordan [159]. The strong U-polynomial was originally introduced by Welsh, who asked whether the equivalence between the U-polynomial and the polychromate carried over to their strong versions [1144]. This turns out to be true and the link between the two is best seen by introducing a strong version of the Tutte symmetric function [841].

We close this section with a discussion of Tutte's universal V-function. (There is an unfortunate clash of notation with the $\mathbf{V}$-polynomial.) The *universal V-function* is a polynomial in $\mathbf{y} = (y_0,\dots,y_{|E(G)|})$ and is defined recursively as follows. If the only edges of G are loops and the number of loops on the vertices are $e_1,\dots,e_n$ then

$$V(G;\mathbf{y}) = \prod_{i=1}^{n} y_{e_i}.$$

Otherwise for any edge e that is not a loop

$$V(G;\mathbf{y}) = V(G\backslash e;\mathbf{y}) + V(G/e;\mathbf{y}). \tag{26.4}$$

It is relatively simple to prove by induction that the definition is independent of the choice of edge in (26.4). It turns out that V is a specialization of $\overline{U}$.

Proposition 26.63.

$$V(G;\mathbf{y}) = \overline{U}\Big(G; z_{i,j} = \sum_{k=0}^{j}(-1)^{j-k}\binom{j}{k} y_k\Big).$$

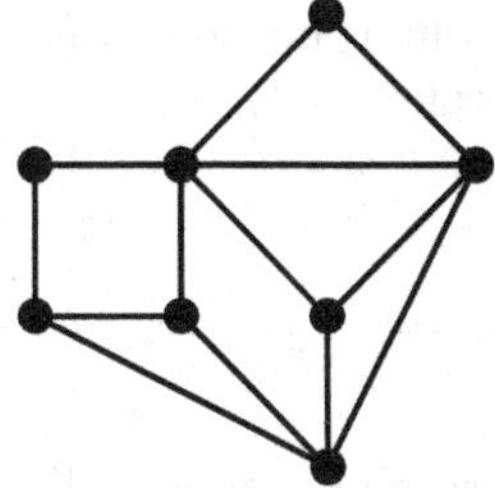
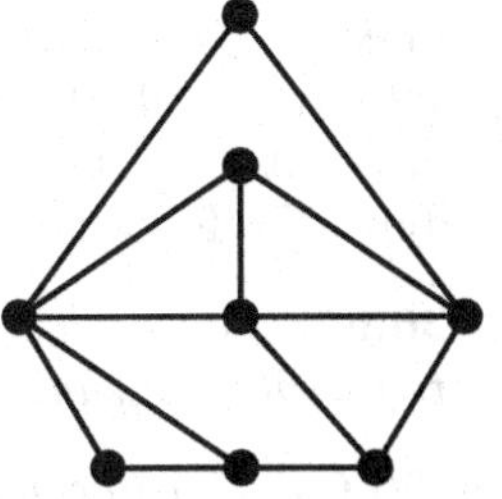

FIGURE 26.2: Two simple graphs with the same U-polynomial.

26.4 Graphs determined by their U-polynomial

We have already seen an example of two graphs with different U-polynomials, but the same Tutte polynomial. This raises the question of which pairs of non-isomorphic graphs have the same U-polynomials. Clearly an easy example can be constructed by taking two copies of a graph that is not vertex-transitive and attaching a loop to a suitably chosen vertex in two different ways. Markström [819] showed that the two graphs in Figure 26.2 are the simple graphs with the fewest vertices having the same U-polynomial and that any pair of simple graphs with fewer than ten vertices, having the same U-polynomial, also has the same strong U-polynomial.

Stanley mentioned in [1033] that it was not known whether there are two non-isomorphic trees with the same chromatic symmetric function. It follows from Theorems 26.48 and 26.51, and from (26.2) that this is equivalent to determining whether there are non-isomorphic trees with the same U-polynomial. The question has attracted a lot of attention, but so far progress has been surprisingly difficult. There are impressive computational results: in [821], it was reported that Tan verified that there are no non-isomorphic trees sharing the same U-polynomial with up to 23 vertices and, as reported in [889], this was extended to up to 25 vertices by Russell. As described in [351], this has been further extended to 29 vertices by Heil and Ji. Orellana and Scott have established several results on graphs determined by the chromatic symmetric function [889].

Definition 26.64. For $k \in \mathbb{N}$, the restricted U-polynomial U_k is defined as follows.

$$U_k(G;\mathbf{x},y) = \sum_{\substack{A\subseteq E \\ |E\setminus A|\leq k}} \mathbf{x}(\tau(\pi(A)))(y-1)^{|A|-r(A)}.$$

Thus if G is a tree, $U_k(G)$ records only the information from $U(G)$ corresponding to forests with up to $k+1$ components. Smith, Smith and Tian [1018] prove that for $k = 1, 2, 3, 4$, the smallest non-isomorphic trees with the same U_k-polynomial have 7, 10, 16 and 22 vertices respectively.

This work is extended by Aliste, De Mier and Zamora in [20], where the authors establish the following.

Theorem 26.65. *For each k, there exist arbitrarily many pairwise non-isomorphic trees with the same U_k-polynomial.*

Definition 26.66. A *caterpillar* is a tree with the property that the tree obtained by deleting all the leaves is a path.

We now present an important result of Loebl and Sereni [784], which answers Stanley's question in the affirmative when restricted to the class of caterpillars and constitutes the most substantial progress towards a positive answer to Stanley's question. Their work extends earlier results from [21] and [821]. The latter work also shows that whether or not a tree is a caterpillar may be determined from its U-polynomial.

Theorem 26.67. *No two non-isomorphic caterpillars share the same U-polynomial.*

26.5 Complexity issues

Consider the problem of evaluating U. This is clearly a hard problem in general, because U specializes to the Tutte polynomial and this was shown to be ♯P-hard to evaluate at most points, even for bipartite planar graphs [653, 1119] (see also Chapter 9). However, we shall see that there is a polynomial time algorithm to evaluate both U and $\overline{U}$ at any point, when the class of input graphs is restricted to having tree-width bounded by a constant k. (See [383] for the definition of tree-width.) This suggests that the complexities of evaluating U and T are similar. On the other hand, results from [885] show that evaluating or determining a coefficient of W is ♯P-hard even when the input graph is restricted to being a tree or a complete graph. Clearly there is little hope of finding large classes of graphs for which one might compute or evaluate W efficiently. We refer the reader to [509] for the necessary background on complexity theory. Let n, m and p denote the number of vertices, edges and the largest size of a set of mutually parallel edges of a graph, respectively. The following is the main result of [883].

Theorem 26.68. *For any $k \in \mathbb{N}$, there exists an algorithm $\mathcal{A}_k$ with the following properties.*

1. *The input is any graph G with tree-width at most k, and rationals $x_1 = p_1/q_1, \ldots, x_n = p_n/q_n$ and $y = p_0/q_0$, such that for all i, the integers p_i and q_i are coprime.*

2. The output is $U(G;\mathbf{x},y)$.

3. The algorithm requires $O(a_k n^{2k+3} + \log p)$ arithmetical operations, where a_k depends only on k, and overall running time

$$O(b_k(n^{2k+3} + \log p)(n^2 + m)r \log(r(n+m)) \log(\log(r(n+m)))),$$

where $r = \log(\max\{1 + |p_0|, \ldots, 1 + |p_n|, |q_0|, \ldots, |q_n|\})$.

We describe a key result used in the proof, giving a splitting formula for U, which may be useful elsewhere.

A *weighted partition* of a set X consists of a partition π of X into non-empty blocks, together with the assignment to each block of a nonnegative integer weight. Given two weighted partitions π_1 and π_2 of the same set, we define their *join* $\pi = \pi_1 \vee \pi_2$ as follows. The blocks are minimal sets such that if two elements are in the same block of either π_1 or π_2, then they are in the same block of π. The weight of each block B of π is the sum of the weights of all blocks B' in either π_1 or π_2 with $B' \subseteq B$.

Let π be a weighted partition of a set S of vertices. The *partial U-polynomial* $U(G, S, \pi; x, y)$ is defined by

$$U(G, S, \pi; \mathbf{x}, y) = \sum_{A \subseteq E: \pi_G(S,A) = \pi} x_{n_1} \cdots x_{n_k} (y-1)^{|A| - r(A)},$$

where the weighted partition $\pi_G(S, A)$ is formed from $\pi_G(A)$ by labelling each block B with $|B \setminus S|$ and deleting all the elements of $V \setminus S$ together with any resulting empty blocks, and $n_1, \ldots, n_k$ are the numbers of vertices in the connected components of $G|A$ having no vertices in S. The following is a splitting formula for U.

Proposition 26.69. *Let $G_1 = (V_1, E_1)$ and $G_2 = (V_2, E_2)$ be graphs with $E_1 \cap E_2 = \emptyset$. Let $S = V_1 \cap V_2$ and let $G = G_1 \cup G_2$. Then for all $\pi \in \Pi(S)$,*

$$U(G, S, \pi; \mathbf{x}, y) = \sum U(G_1, S, \pi_1; \mathbf{x}, y) U(G_2, S, \pi_2; \mathbf{x}, y)(y-1)^{\#\pi + |S| - \#\pi_1 - \#\pi_2},$$

where the summation is over all π_1, π_2 in $\Pi(S)$ such that $\pi_1 \vee \pi_2 = \pi$.

The full details of the algorithm are described in [882]. It is not difficult to extend these ideas to obtain a polynomial time algorithm to evaluate the strong U-polynomial. Working through the details of the algorithm in exactly the same way as in [882] leads to the following result.

Theorem 26.70. *For any $k \in \mathbb{N}$, there exists an algorithm $\mathcal{A}_k$ with the following properties.*

1. The input is any graph G with tree-width at most k, and rationals $z_{i,j} = p_{i,j}/q_{i,j}$ for $1 \leq i \leq n$ and $0 \leq j \leq m$ such that for all i and j, the integers $p_{i,j}$ and $q_{i,j}$ are coprime.

2. *The output is* $\overline{U}(G;\mathbf{z})$.

3. *The algorithm requires* $O(a_k n^{2k+3} m^{2k+2})$ *arithmetical operations, where* a_k *depends only on* k*, and overall running time*

$$O(b_k n^{2k+3} m^{2k+2}(n^2+m) r \log(r(n+m)) \log(\log(r(n+m)))),$$

where $r = \log(\max_{i,j}\{1+|p_{i,j}|, |q_{i,j}|\}$.

Thus the problem of evaluating the U-polynomial is not known to be fixed parameter tractable with respect to tree-width. In contrast there is an algorithm to evaluate the Tutte polynomial of a graph that is fixed parameter tractable with respect to tree-width [882] and indeed Courcelle, Makowsky and Rotics showed in [339] that evaluating any graph polynomial definable in monadic second order logic is fixed parameter tractable with respect to tree-width. Godlin, Katz and Makowsky [545] have shown that the U-polynomial cannot be defined using second order logic. Moving from the class of graphs of bounded tree-width to the class of graphs of bounded clique-width, we have the following result of Giménez, Hliněný and Noy [523].

Theorem 26.71. *For any* $k \in \mathbb{N}$*, there exists an algorithm* $\mathcal{A}_k$ *with the following properties.*

1. *The input is any graph* G *with clique-width at most* k *and a* k*-expression for* G.

2. *The output is* $U(G;\mathbf{x},y)$.

3. *The algorithm requires* $\exp(O(n^{1-1/(k+2)}))$ *operations.*

26.6 Related polynomials

We close this chapter with a brief description of other polynomials that are related to U, but which we do not have space to discuss in full. In the introduction, we mentioned Gebhard and Sagan's generalization of Stanley's symmetric chromatic function [515] which introduces a simple deletion–contraction relation at the cost that the variables no longer commute. In a recent paper, Markström [819] introduces an interesting family $\{C_q\}_{q\geq 2}$ of polynomials based on weighted homomorphisms to the complete graph on q vertices with a loop at each vertex. These polynomials have many specializations in common with U but their precise relationship with U is unclear. None of them determines U, but conversely the two graphs in Figure 26.2 have different C_q polynomials for $q \geq 3$. It is not known whether C_2 is a specialization of U.

There are a number of objects other than graphs for which one might try to define a polynomial analogous to the U-polynomial. For directed graphs the closest analogue of the Tutte polynomial is perhaps the cover polynomial [320]. Chow [316] has given a generalization of this that is analogous to Stanley's symmetric Tutte function, and D'Antona and Munarini [359] have defined the cycle-path indicator polynomial of a vertex-weighted digraph. (See Chapter 23 for a discussion of digraph polynomials.) Moving in another direction, Billera, Jia and Reiner [118] define a quasi-symmetric function of matroids and Derksen [374] defines several symmetric or quasi-symmetric functions of polymatroids, one of which specializes to Billera, Jia and Reiner's invariant. Forge and Zaslavsky [495] define the analogue of the Tutte polynomial for weighted gain graphs. This has many similarities with W and indeed when restricted to vertex-weighted graphs without gains, essentially reduces to W, but more generally includes many other specializations including Zaslavsky's analogue of the Tutte polynomial of gain graphs [1184].

26.7 Open problems

Most open problems involving the U-polynomial concern its distinctive power. It would be interesting to find new relationships with the other polynomials discussed in this book. Are there polynomials or collections of polynomials from this book that jointly distinguish the same graphs as the U-polynomial? In a similar vein one might ask how much redundancy is there in the definition of U. More precisely, is there a polynomial with the same distinctive power having significantly fewer variables? The relationship between U and certain specific polynomials is unclear. For example, we do not know of any non-isomorphic loopless graphs with the same U-polynomial but different strong U-polynomials, although we strongly suspect that they must exist. Similarly we do not know whether the polynomial C_2 from [819] is a specialization of U or whether the q-dichromate and U are equivalent.

Probably the most important outstanding question is to determine whether or not there are two non-isomorphic trees sharing the same U-polynomial. As we mentioned in Section 26.4, this is equivalent to Stanley's question from [1033].

As discussed in Section 26.6 analogues of the U-polynomial have been introduced for various objects generalizing graphs. One might hope to define such a polynomial for embedded graphs with weights on both vertices and faces so that its behavior under geometric duality extends that of the Tutte polynomial of a plane graph. The extension to embedded graphs cannot, however, be straightforward because deletion and contraction in an embedded graph can create new vertices and faces as well as merge them.

Finally we close with a complexity question that may be related to the distinctive power of the U-polynomial. It is curious that the exponent of n in the running time of the algorithms in both Theorems 26.68 and 26.70

grows linearly with the tree-width k. It seems unlikely that there is room for substantial improvement here, although this question is worthy of more consideration. We do not know of any other natural problem concerning the evaluation of a graph polynomial where this situation occurs and it would be interesting to understand why this happens.

27

Topological extensions of the Tutte polynomial

Sergei Chmutov

Synopsis

This chapter covers generalizations of the Tutte polynomial to graphs embedded in surfaces. The study of topological Tutte polynomials has seen recent rapid development, and we review the main polynomials in the area and their basic properties.

- Ribbon graphs.
- The Bollobás–Riordan polynomial.
- The Las Vergnas polynomial.
- The Krushkal polynomial.
- Topological Tutte polynomials through deletion–contraction.
- Topological Tutte polynomials through flows and tensions.
- Quasi-tree activities and polynomial expansions.

27.1 Introduction

This chapter covers recent work on topological extensions of the Tutte polynomial to graphs embedded in surfaces. There are several books devoted to graphs embedded in surfaces, [457, 587, 740, 857]. Mostly we consider cellular embeddings of graphs, which are equivalent to ribbon graphs.

DOI: 10.1201/9780429161612-27

In 1980 M. Las Vergnas [746] found a generalization of the Tutte polynomial to cellularly embedded graphs as an application of his matroid perspectives (see Chapter 28). Recently his polynomial was extended to graphs in pseudo-surfaces [447], a setting where the matroidal and topological deletion–contraction relations coincide.

In 2001 B. Bollobás and O. Riordan [160, 161], motivated by knot theory, introduced a different generalization of the Tutte polynomial for ribbon graphs, now generally called the Bollobás–Riordan polynomial. For non-planar graphs, there is no duality relation analogous to (2.20) for the Tutte polynomial but there is one for the Bollobás–Riordan polynomial. In [161], this was proved for a one-variable specialization. J. Ellis-Monaghan and I. Sarmiento [462] extended it to a two-variable relation. Independently the duality formula was rediscovered in [854].

In knot theory, Thistlethwaite's theorem relates the Jones polynomial and the Tutte polynomial of a corresponding planar graph as in Chapter 18. I. Pak suggested using the Bollobás–Riordan polynomial for Thistlethwaite-type theorems. This idea was first realized in [309] for a special class of (checkerboard colorable) virtual links. Then it was realized for classical links in [360], and for arbitrary virtual links in [310]. Formally all three theorems from [309, 360, 310] were different. They used different constructions of a ribbon graph from a link diagram and different substitutions in the Bollobás–Riordan polynomials of these graphs. An attempt to understand and unify these theorems led to the discovery of *partial duality* in [306] (called *generalized duality* there) and to a proof of an invariance of a certain specializations of the Bollobás–Riordan polynomial under it. Partial duality, which arose from graph polynomials, has proven to be an important construction in topological graph theory.

In 2011 V. Krushkal [724] found a four variable generalization of the Tutte polynomial. It can be reduced to the Bollobás–Riordan polynomial under a certain substitution. It turns out [52] that it also can be reduced to the Las Vergnas polynomial under a different substitution, which is noteworthy since the Las Vergnas and the Bollobás–Riordan polynomials are essentially independent of each other. Generalizing the classical spanning tree expansion of the Tutte polynomial, C. Butler found [255] an elegant quasi-tree expansion of the Krushkal polynomial. Figure 27.1 shows various relations among these polynomials.

Building upon work of T. Krajewski, I. Moffatt and A. Tanasa [718], and of I. Moffatt and B. Smith [856], S. Huggett and I. Moffatt in [855] constructed a family of four topological Tutte polynomials associated with different frameworks for embedded graphs, and showed how the Krushkal, Bollobás–Riordan, and Las Vergnas polynomials can be recovered from them by restricting the domains. These polynomials differ from the other topological Tutte polynomials in this chapter in that they have known deletion–contraction relations that apply to all edges types. These polynomials also have matroidal counterparts [718].

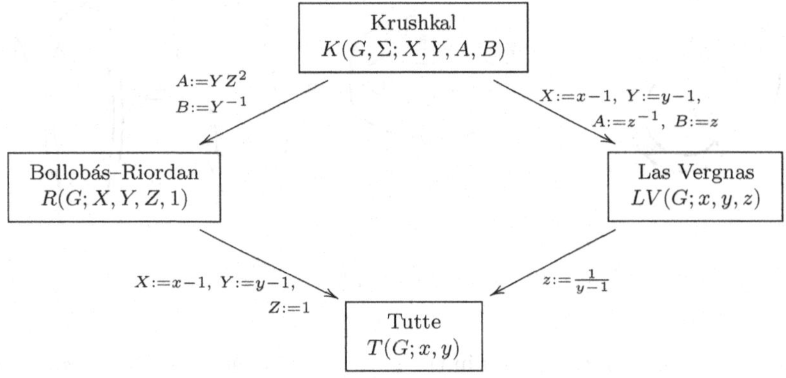

FIGURE 27.1: The relationships within a family of graph polynomials.

A further Tutte polynomial for embedded graphs was introduced in [560] by A. Goodall, T. Krajewski, G. Regts and L. Vena. This polynomial is of particular interest since it can be used to count nowhere-zero local flows and local tensions, just as the classical Tutte polynomial counts nowhere-zero flows and tensions.

The Tutte polynomial has also been extended to higher dimensional cell complexes. We do not pursue this connection here, but the interested reader may see [725], and also [63] for a relation to cellular spanning trees, [356] for arithmetic matroids, and [183, 1128] for the simplicial chromatic polynomial.

Other related topics that are beyond the scope of this chapter include the arrow generalization of the Bollobás–Riordan polynomial of [194], and its application to virtual knot theory; generalization of the polynomial invariants to graphs embedded in pseudo-surfaces [447]; and the relative Tutte polynomial [256, 380]. Some of these topics are discussed further in an extended version of this chapter [307].

27.2 Ribbon graphs

The topological objects we consider are graphs cellularly embedded in a surface.

Definition 27.1. A *cellular embedding* of a graph $G = (V, E)$ maps each vertex to a distinct point in the surface, and each edge to a curve in the surface whose end points are the points corresponding to the endpoints of the edge. Furthermore, none of these curves intersects any other curve (including itself) at any point other than an endpoint. Finally, each connected component

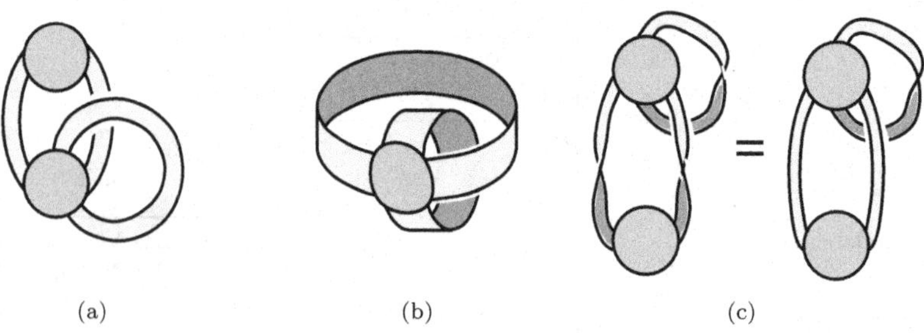

FIGURE 27.2: Three ribbon graphs.

of the complement of the graph in the surface, i.e. each *face*, is homeomorphic to a disc.

Considering a small regular neighborhood of a graph on a surface gives the equivalent definition of a ribbon graph.

Definition 27.2. A *ribbon graph* is a (not necessarily orientable) compact surface with boundary, decomposed into a number of closed topological discs of two types, *vertex-discs* and *edge-ribbons*, satisfying the following conditions: the discs of the same type are pairwise disjoint; the vertex-discs and the edge-ribbons intersect in disjoint line segments, each such line segment lies on the boundary of precisely one vertex and precisely one edge, and every edge contains exactly two such line segments.

We consider ribbon graphs up to a homeomorphism of the corresponding surfaces preserving the decomposition into vertex-discs and edge-ribbons. It is important to note that a ribbon graph is an abstract two-dimensional surface with boundary; its embedding into three-space shown in pictures is irrelevant. Figure 27.2 shows some examples of ribbon graphs.

Ribbon graphs are equivalent to cellularly embedded graphs. Gluing a disc to each boundary component of a ribbon graph G yields a closed surface without boundary. Then contract each vertex-disc in the surface to a point, followed by contracting each edge-ribbon in the surface to a simple curve between its endpoints. This yields a graph, $\widehat{G}$, called the *core* graph of G, cellularly embedded into the surface.

In addition to the usual graph parameters given in Chapter 1, we use the topological parameter $\mathrm{bc}(G)$, which denotes the number of connected components of the boundary of the surface of G.

A *spanning ribbon subgraph* F of a ribbon graph G is a ribbon graph consisting of all the vertices of G and a subset of the edges of G. Deletion for ribbon graphs is defined as expected: $G\backslash e$ is the ribbon graph obtained from G by removing the edge e. However, the definition of contraction requires some care. If u and v are the (not necessarily distinct) vertices incident with e, then G/e denotes the ribbon graph obtained as follows: consider the boundary component(s) of $e \cup \{u, v\}$ as curves on G. For each resulting curve, attach a disc (which will form a vertex of G/e) by identifying the boundary component of the disc with the curve. Delete e, u and v from the resulting complex, to get the ribbon graph G/e. For example, contracting a non-loop edge of the ribbon graph in Figure 27.2a, results in the ribbon graph in Figure 27.2b, but contracting the loop results in a path on two edges.

27.3 The Bollobás–Riordan polynomial

The Bollobás–Riordan polynomial was introduced in [160, 161].

Definition 27.3. For a ribbon graph G, the *Bollobás–Riordan polynomial* is

$$R(G; X, Y, Z, W) = \sum_{F \subseteq G} (X-1)^{r(G)-r(F)} Y^{n(F)} Z^{k(F)-\mathrm{bc}(F)+n(F)} W^{t(F)},$$

where the sum is over all spanning ribbon subgraphs F of G. The parameter $t(F)$ is equal to 0 for orientable ribbon subgraphs F, and 1 for non-orientable F. The polynomial takes values in the quotient ring $\mathbb{Z}[X, Y, Z, W]/(W^2 - W)$.

Note that the exponent $k(F) - \mathrm{bc}(F) + n(F)$ of the variable Z is equal to $2k(F) - \chi(\widetilde{F})$, where $\chi(\widetilde{F})$ is the *Euler characteristic* of the surface $\widetilde{F}$ obtained by gluing a disc to each boundary component of F.

For a planar ribbon graph G (i.e. when the surface G has genus zero), the variables Z and W do not appear in the Bollobás–Riordan polynomial $R(G; X, Y, Z, W)$. In this case, the Bollobás–Riordan polynomial is equal to the classical Tutte polynomial $T(\widehat{G}; x, y)$ of the core graph of G with the substitution $X = x$ and $Y = y - 1$:

$$R(G; x, y-1, Z, W) = T(\widehat{G}; x, y) .$$

Similarly, a specialization $Z = 1$ and $W = 1$ of the Bollobás–Riordan polynomial of an arbitrary ribbon graph G gives the Tutte polynomial of the core graph:

$$R(G; x, y-1, 1, 1) = T(\widehat{G}; x, y) .$$

Example 27.4. Table 27.1 shows the computation of $R(G; X, Y, Z, W)$ using Definition 27.3 for the right-most ribbon graph in Figure 27.2, giving $R(G; X, Y, Z, W) = Y + 1 + X + Y^2Z^2W + YZW + XYZW$.

Deletion–contraction relations hold for the Bollobás–Riordan polynomial for certain types of edges [161]:

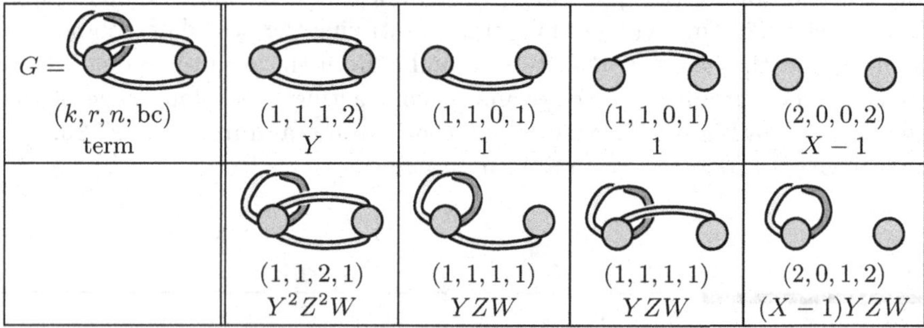

$G =$				
(k, r, n, bc) term	$(1,1,1,2)$ Y	$(1,1,0,1)$ 1	$(1,1,0,1)$ 1	$(2,0,0,2)$ $X-1$
	$(1,1,2,1)$ Y^2Z^2W	$(1,1,1,1)$ YZW	$(1,1,1,1)$ YZW	$(2,0,1,2)$ $(X-1)YZW$

TABLE 27.1: Computing $R(G; X, Y, Z, W)$.

Theorem 27.5. *The Bollobás–Riordan polynomial, $R(G) := R(G; X, Y, Z, W)$, satisfies the following relations.*

$$R(G) = \begin{cases} R(G/e) + R(G\backslash e) & \text{if } e \text{ is ordinary,} \\ X \cdot R(G/e) & \text{if } e \text{ is a bridge,} \\ (1+Y) \cdot R(G\backslash e) & \text{if } e \text{ is a trivial orientable loop,} \\ (1+YZW) \cdot R(G\backslash e) & \text{if } e \text{ is a trivial non-orientable loop,} \end{cases}$$

and $R(G_1 \sqcup G_2) = R(G_1) \cdot R(G_2)$, where $\sqcup$ is disjoint union.

Here a *trivial loop* is a loop such that deleting a chord in the vertex disk with an endpoint in each of the two intervals where the edge and vertex intersect, together with deleting the edge itself, separates the ribbon graph. An ordinary edge is one that is not a bridge or a loop.

The Bollobás–Riordan polynomial was generalized to a multivariable polynomial of weighted ribbon graphs in [854, 1121].

27.4 The Las Vergnas polynomial

M. Las Vergnas [746] arrived at his polynomials of graphs on surfaces through his general approach to matroid perspectives (see Chapter 28). It turns out that for a graph G, with edge set E, cellularly embedded on a surface, there is a geometric dual graph G^*. For this graph there is a dual matroid to its cycle matroid, $(M(G^*))^*$, which is the *bond matroid* of G^*. (Here $M(G)$, etc., refers to the cycle matroid of the core graph of G.) Then the natural bijection of the ground sets $(M(G^*))^* \to M(G)$ forms a matroid perspective according to [746]. As in Chapter 28, the Tutte polynomial of a matroid perspective

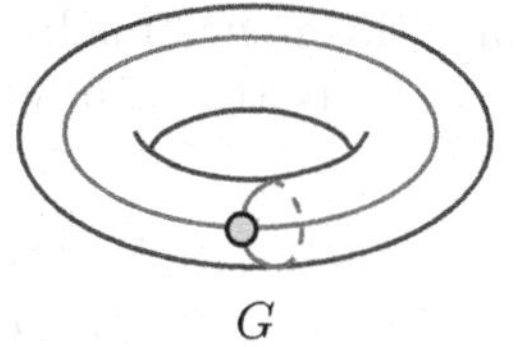

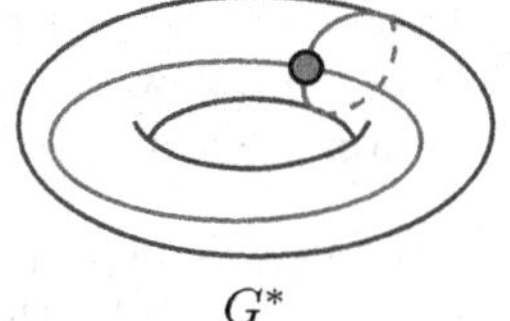

FIGURE 27.3: An embedded graph that is isomorphic to its dual.

$M \to M'$ on E is a polynomial $T(M, M'; x, y, z)$ in three variables defined by

$$T(M, M'; x, y, z) := \sum_{A \subseteq M} (x-1)^{r_{M'}(E)-r_{M'}(A)}(y-1)^{|A|-r_M(A)} z^{(r_M(E)-r_M(A))-(r_{M'}(E)-r_{M'}(A))},$$

where in the rank and nullity functions, we use the subscript to indicate at which matroid the function is considered in.

Definition 27.6. The *Las Vergnas polynomial* $LV(G; x, y, z)$ of G is the Tutte polynomial of the matroid perspective $(M(G^*))^* \to M(G)$:

$$LV(G; x, y, z) := T((M(G^*))^*, M(G); x, y, z).$$

If G is a plane graph, then the Whitney planarity criteria gives that the matroid $(M(G^*))^*$ is isomorphic to $M(G)$. So the matroid perspective is the identity, and then the Las Vergnas polynomial is equal to the classical Tutte polynomial of the core graph of G.

Example 27.7. Let G be a graph with one vertex and two loops embedded into a torus as shown in Figure 27.3. Then the dual graph G^* is isomorphic to G. In this case the bond matroid $M = (M(G^*))^*$ has rank 2, and the cycle matroid $M' = M(G)$ has rank 0. For any subset A of $E(G)$, we have $r_M(A) = |A|$, $n_M(A) = 0$, and $r_{M'}(A) = 0$. Thus

$$LV(G; x, y, z) = z^2 + 2z + 1 \; .$$

27.5 The Krushkal polynomial

A generalization of the previous two topological Tutte polynomials for graphs on surfaces was discovered by Krushkal in his research in topological quantum field theories and the algebraic and combinatorial properties of models of statistical mechanics [724]. Krushkal defined his polynomial for orientable surfaces only. Here we follow the exposition of C. Butler [255] who extended the definition to non-orientable surfaces.

Definition 27.8. Let G be a graph embedded in a surface Σ. The embedding does not have to be cellular and the surface Σ does not have to be orientable. Then the *Krushkal polynomial* is defined by

$$K(G,\Sigma;X,Y,A,B) := \sum_{F\subseteq G} X^{k(F)-k(G)}Y^{k(\Sigma\backslash F)-k(\Sigma)}A^{s(F)/2}B^{s^{\perp}(F)/2}.$$

Here, the sum runs over all spanning subgraphs F considered as ribbon graphs. These are the ribbon graphs that result from taking a small neighborhood in Σ of a spanning graph F of G, and deleting the complement of the neighborhood to yield a ribbon graph. The parameters $s(F)$ and $s^{\perp}(F)$ are defined by

$$s(F) := 2k(F) - \chi(\widetilde{F}), \quad \text{and} \quad s^{\perp}(F) := 2k(\Sigma\backslash F) - \chi(\widetilde{\Sigma\backslash F})$$

Here $\chi(\widetilde{F})$ (respectively $\chi(\widetilde{\Sigma\backslash F})$) is the Euler characteristic of the surface $\widetilde{F}$ (respectively $\widetilde{\Sigma\backslash F}$) obtained by gluing a disc to each boundary component of the ribbon graph F (respectively into the boundary components of $\Sigma\backslash F$).

Remark 27.9. In the orientable case Krushkal [724] indicated that that the parameters $s(F)$ and $s^{\perp}(F)$ have another interpretation in terms of the symplectic bilinear form on the vector space $H_1(\Sigma;\mathbb{R})$ given by the intersection number. For a given spanning subgraph F, let V be its image in the homology group:

$$H_1(\Sigma;\mathbb{R}) \supset \mathsf{V} := \mathrm{im}(\mathrm{H}_1(\mathrm{F};\mathbb{R}) \to \mathrm{H}_1(\Sigma;\mathbb{R}))\ .$$

For the subspace V we can define its orthogonal complement $\mathsf{V}^{\perp}$ in $H_1(\Sigma;\mathbb{R})$ with respect to the symplectic intersection form. Then

$$s(F) = \dim(\mathsf{V}/(\mathsf{V}\cap\mathsf{V}^{\perp}))\ , \qquad s^{\perp}(F) = \dim(\mathsf{V}^{\perp}/(\mathsf{V}\cap\mathsf{V}^{\perp}))\ .$$

Theorem 27.10. *Let G be a ribbon graph, or equivalently a cellular embedding of the core graph $\widehat{G}$ in the surface $\Sigma := \widetilde{G}$. Then the following hold.*

1. *The Tutte polynomial is a specialization of the Krushkal polynomial:*

$$T(G;x,y) = (y-1)^{s(\Sigma)/2}K(\widehat{G},\Sigma;x-1,y-1,y-1,(y-1)^{-1}).$$

2. *When $W=1$, the Bollobás–Riordan polynomial is a specialization of the Krushkal polynomial:*

$$R(G;X,Y,Z,1) = Y^{s(G)/2}K(\widehat{G},\Sigma;X-1,Y,YZ^2,Y^{-1}).$$

3. *The Las Vergnas polynomial is a specialization of the Krushkal polynomial:*

$$LV(G;x,y,z) = z^{s(G)/2}K(\widehat{G},\Sigma;x,y,z^{-1},z).$$

The first item of this theorem is from [724], the second from [724, 255], and the third from [52, 255].

For the following theorem, from [724, 255], a *separable loop* is one whose removal, together with its incident vertex, separates the surface Σ, and an ordinary edge is one that is neither a loop nor a bridge.

Theorem 27.11. *The Krushkal polynomial,* $K(G,\Sigma) := K(G,\Sigma;X,Y,A,B)$, *satisfies the following relations.*

$$K(G,\Sigma) = \begin{cases} K(G/e,\Sigma) + K(G\backslash e,\Sigma) & \text{if } e \text{ is ordinary,} \\ (1+X)\cdot K(G/e,\Sigma) & \text{if } e \text{ is a bridge,} \\ (1+Y)\cdot K(G\backslash e,\Sigma) & \text{if } e \text{ is a separable loop,} \end{cases}$$

and

$$K(G_1 \sqcup G_2, \Sigma_1 \sqcup \Sigma_2) = K(G_1,\Sigma_1)\cdot K(G_2,\Sigma_2),$$

where $\sqcup$ *is the disjoint union.*

The deletion–contraction properties for the Bollobás-Riordan, the Las Vergnas, and the Krushkal polynomials are not quite the same. The problem arises when deletion of an edge of a ribbon graph changes its genus. The genus might decrease by 1 with the removal of an edge as shown in Example 27.13. We refer the reader to [52] for more detailed discussion.

The classical relation, $T(G;x,y) = T(G^*;y,x)$, for the Tutte polynomials of dual plane graphs can be generalized to a relation for the Krushkal polynomial of arbitrary ribbon graphs, [255, 724].

Theorem 27.12. *Let* G *be a ribbon graph, with its core graph* $\widehat{G}$ *naturally embedded into the surface* $\Sigma = \widetilde{G}$, *with* $\widehat{G}^* \subset \Sigma$ *the (geometric) dual of* $\widehat{G}$ *in* Σ. *Then*

$$K(\widehat{G},\Sigma;X,Y,A,B) = K(\widehat{G}^*,\Sigma;Y,X,B,A)\ .$$

Example 27.13. This is an example of a calculation of the three polynomials $R(G;X,Y,Z,W)$, $K(G,\Sigma;X,Y,A,B)$, and $LV(G,x,y,z)$. Here G is a graph on the torus Σ shown in Figure 27.4, with two vertices and three edges a, b, and c, considered as a ribbon graph. The torus Σ is $\widetilde{G}$. The dual of G is G^*, which has one vertex and three loops. We use the same symbols a, b, c to denote the corresponding edges of G^*. To reduce clutter, we write here $R(G)$, $K(\widehat{G},\Sigma)$, and $LV(G)$ for $R(G;X,Y,Z,W)$, $K(\widehat{G},\Sigma;X,Y,A,B)$, and $LV(G,x,y,z)$.

The matroid $M' = M(G)$ is of rank 1, and for any nonempty subset F, $r_{M'}(F) = 1$. The cycle matroid $M(G^*)$ of the dual graph is of rank zero because G^* has only loops. So its dual $M = (M(G^*))^*$ has rank 3, all subsets F are independent and $r_M(F) = |F|$. Table 27.2 shows the values of various parameters and the contributions of all eight subsets $F \subseteq \{a,b,c\}$ to the three polynomials. Thus

$$K(\widehat{G},\Sigma) = 3 + 3B + XB + A, \qquad LV(G) = 3z + 3z^2 + (x-1)z^2 + 1,$$

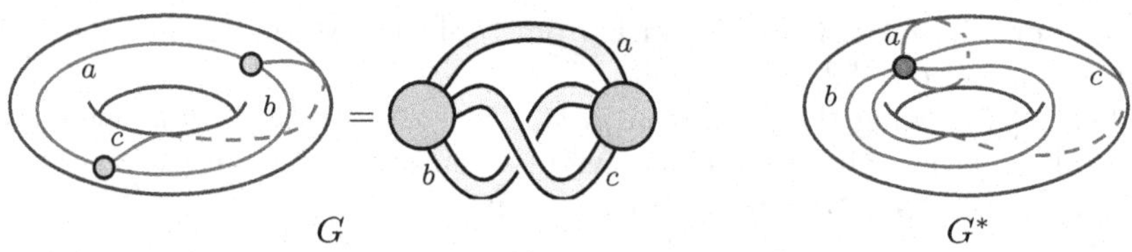

FIGURE 27.4: An embedded graph and its dual.

$$R(G) = 3Y + 3 + X + Y^2Z^2.$$

Now if we contract the edge c, the graph $\widehat{G}/c$ will still be cellularly embedded in the same torus Σ. The right four columns of Table 27.2 give the following polynomials:

$$K(\widehat{G}/c, \Sigma) = B{+}2{+}A, \qquad LV(G/c) = z^2{+}2z{+}1, \qquad R(G/c) = 1{+}2Y{+}Y^2Z^2.$$

Meanwhile if we delete the edge c, then

$$K(\widehat{G}\backslash c, \Sigma) = XB + 2B + 1\ .$$

But the graph $\widehat{G}\backslash c$ is not cellularly embedded into the torus Σ any more. Thus the Las Vergnas and the Bollobás–Riordan polynomials are not defined for it. The ribbon graph $G\backslash c$, after gluing discs to its two boundary components, results in the sphere $\widetilde{G\backslash c} = S^2$. Thus the graph $\widehat{G}\backslash c$ embeds cellularly into the sphere S^2. For this embedding we have

$$K(\widehat{G}\backslash c, S^2) = X{+}2{+}Y, \quad LV(G\backslash c) = (x{-}1){+}2{+}(y{-}1), \quad R(G\backslash c) = X{+}2{+}Y.$$

Therefore

$$K(\widehat{G}, \Sigma) = K(\widehat{G}\backslash c, \Sigma) + K(\widehat{G}/c, \Sigma) \qquad \text{and} \qquad R(G) = R(G\backslash c) + R(G/c)\ ,$$

but

$$K(\widehat{G}, \Sigma) \neq K(\widehat{G}\backslash c, S^2){+}K(\widehat{G}/c, \Sigma) \qquad \text{and} \qquad LV(G) \neq LV(G\backslash c){+}LV(G/c)\ .$$

However, deletion–contractions relations can be obtained by extending the domains of the polynomials, as described in the next section.

27.6 Polynomials from deletion–contraction relations

In [718], T. Krajewski, I. Moffatt, and A. Tanasa defined a Tutte polynomial of a Hopf algebra, and used this to define generalizations of the Bollobás–Riordan

F	$\emptyset$	$\{a\}$	$\{b\}$	$\{a,b\}$	$\{c\}$	$\{a,c\}$	$\{b,c\}$	$\{a,b,c\}$
$k(F)$	2	1	1	1	1	1	1	1
$k(\Sigma \setminus F) - k(\Sigma)$	0	0	0	0	0	0	0	0
$s(F)$	0	0	0	0	0	0	0	2
$s^{\perp}(F)$	2	2	2	0	2	0	0	0
$K(\widehat{G},\Sigma)$	XB	B	B	1	B	1	1	A
$r_M(F)$	0	1	1	2	1	2	2	3
$r_{M'}(F)$	0	1	1	1	1	1	1	1
$n_M(F)$	0	0	0	0	0	0	0	0
$LV(G)$	$(x-1)z^2$	z^2	z^2	z	z^2	z	z	1
$k(F)$	2	1	1	1	1	1	1	1
$n(F)$	0	0	0	1	0	1	1	2
$\mathrm{bc}(F)$	2	1	1	2	1	2	2	1
$R(G)$	X	1	1	Y	1	Y	Y	Y^2Z^2

TABLE 27.2: Computing the polynomials $R(G)$, $K(G,\Sigma)$, and $LV(G)$.

and Kruskal polynomials that have deletion–contraction relations that apply to all edge types (see Section 19.4). In [855], S. Huggett and I. Moffatt gave a purely combinatorial approach to these polynomials, and extended the construction to obtain a family of topological Tutte polynomials for four different notions of an embedded graph. There they considered four classes of embedded graph: graphs in pseudo-surface (i.e., surfaces with pinch-points), graphs in surfaces, graphs cellularly embedded in pseudo-surface, and graphs cellularly embedded in pseudo-surface. A Tutte polynomial was canonically defined for each of the four classes. The relationships among the four classes of embedded graph is reflected in the polynomials as indicated in Figure 27.5.

The most general polynomial in this family is that for graphs in pseudo-surfaces:

$$T_{ps}(G; w, x, y, z) := \sum_{A \subseteq E} w^{r_1(E)-r_1(A)} x^{r_2(E)-r_2(A)} y^{r_3(A)} z^{r_4(A)},$$

where

$$r_1(A) := r_G(A), \qquad r_3(A) := r_{G^*}(E) - r_{G^*}(E \backslash A),$$
$$r_2(A) := \rho(A) - r_G(A), \qquad r_4(A) := |A| + r_{G^*}(E \backslash A) - r_{G^*}(E) - \rho(A).$$

Here r_G is the rank function for the underlying abstract graph of G; r_{G^*} is the rank function for the underlying abstract graph of the dual G^* of the graph G in the pseudo-surface; and $\rho(A)$ is obtained by resolving all pinch-points to get a graph in a surface, forming a ribbon graph by taking a neighborhood of it, and computing the parameter $\frac{1}{2}(|A| + |V| - \mathrm{bc}(A))$, where $|V|$ is the number of vertices of the ribbon graph. The other three polynomials can be

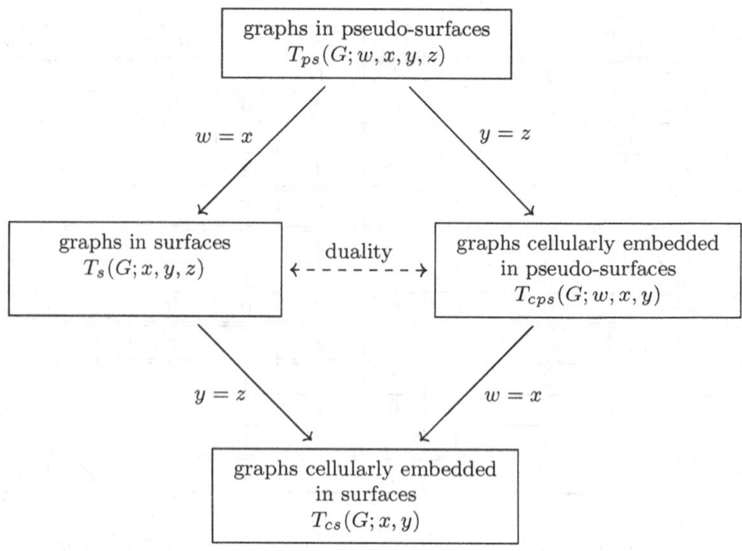

FIGURE 27.5: A family of topological Tutte polynomials.

obtained by specialization as indicated in Figure 27.5 (or see [855] for tidier forms).

Each of the polynomials in this family has a deletion–contraction relation in which the terminal forms consist of edgeless embedded graphs. The polynomials also satisfy duality identities, for example, $T_{ps}(G^*; w, x, y, z) = T_{ps}(G; y, z, w, x)$. Upon restricting the domains, the Krushkal, Bollobás–Riordan, and Las Vergnas polynomials can be recovered from them (see [855] for details).

In [856], I. Moffatt and B. Smith defined a Tutte polynomial for a triple (M, D, N) where M and N are matroids, and D is a delta-matroid. If this triple is read from a graph in a pseudo-surface, then this matroidal polynomial agrees with T_{ps}. Thus the polynomials in [856] can be be considered as the matroid versions on the polynomials in this section, as well as the Krushkal and Bollobás–Riordan polynomials.

27.7 Polynomials arising from flows and tensions

A. Goodall, T. Krajewski, G. Regts and L. Vena, in [560], introduced a topological Tutte polynomial, called the *surface Tutte polynomial*. This polynomial is of particular interest since it can be used to count nowhere-zero local flows and local tensions, just as the classical Tutte polynomial counts nowhere-zero flows and tensions (or proper colorings), as in Chapter 12.

Let $\mathbf{x} = (x, x_0, x_1, x_2, \ldots)$ and $\mathbf{y} = (y, y_0, y_1, y_2, \ldots)$ be two infinite sequences of variables, and G be a graph embedded in an orientable surface. Then the *surface Tutte polynomial* is

$$\mathcal{T}(G, \mathbf{x}, \mathbf{y}) := \sum_{A \subseteq E} x^{n^*(G/A)} y^{n(G \backslash A^c)} \prod_{\substack{\text{conn. cpts} \\ G_i \text{ of } G/A}} x_{g(G_i)} \prod_{\substack{\text{conn. cpts} \\ G_j \text{ of } G \backslash A^c}} y_{g(G_j)},$$

where $A^c = E \backslash A$, where $g(G)$ is the genus of an embedded graph G, and where $n^*(G) = r(G) + 2g(G)$.

The surface Tutte polynomial also satisfies a duality relation $\mathcal{T}(G, \mathbf{x}, \mathbf{y}) = \mathcal{T}(G^*, \mathbf{y}, \mathbf{x})$. Furthermore, the Krushkal polynomial, and therefore the Las Vergnas and Bollobás–Riordan polynomials, can be recovered from $\mathcal{T}(G, \mathbf{x}, \mathbf{y})$. It has deletion–contraction relations for certain edge types, but not all.

27.8 Quasi-trees

Quasi-trees, as well as a definition of activities relative to a spanning quasi-tree, were introduced and used for a *quasi-tree expansion* of the Bollobás–Riordan polynomial in [276], and generalized to non-orientable ribbon graphs in [1122] (and independently by E. Dewey in an unpublished manuscript).

Definition 27.14. A *quasi-tree* is a ribbon graph with one boundary component. We denote the set of all spanning quasi-trees of a given ribbon graph G by $\mathcal{Q}_G$.

Example 27.15. If G is a plane graph of genus zero, then a spanning quasi-tree is a spanning tree. But for non-plane graphs the set of spanning quasi-trees is bigger than the set of spanning trees. In Example 27.7, the whole graph consisting of a vertex and two loops on a torus is a quasi-tree but, of course, not a tree. Table 27.2 in Example 27.13 shows that the ribbon graph has four spanning quasi-trees on edge sets $\{a\}$, $\{b\}$, $\{c\}$, and $\{a, b, c\}$.

Let $\prec$ be a total order on edges $E(G)$ of a ribbon graph G, and Q be a spanning quasi-tree of G. Tracing the boundary component of Q yields a round trip passing the boundary arcs of each edge-ribbon twice. These two boundary arcs will intersect faces for edge in Q, and will intersect vertices for edges not in Q. This information can be encoded in a chord diagram $C_G(Q)$ consisting of a circle corresponding to the boundary of Q with edge arcs recorded in the cyclic order they appear on the boundary of Q, and with pairs corresponding to the same edge-ribbon connected by chords. Thus the set of chords inherits

FIGURE 27.6: Chord diagrams from quasi-trees.

the total order $\prec$. Figure 27.6b shows the chord diagram corresponding to the quasi-tree on $\{a, b, c\}$ in the graph from Example 27.13.

Definition 27.16. An edge is called *live* if the corresponding chord is smaller than any chord intersecting it relative to the order $\prec$. Otherwise, it is called *dead.*

For a plane graph G a spanning quasi-tree is a spanning tree and the notion of *live/dead* coincides with the classical Tutte's notion of *active/inactive.* But for higher genus the concepts are different.

Example 27.17. In Example 27.7, for the graph consisting of a single vertex and two loops, there are two spanning quasi-trees, one consisting of a vertex without edges and one the whole graph. Both of them have a chord diagram depicted in Figure 27.6a. In Example 27.13 there are four spanning quasi-trees. All of them have the same chord diagram, depicted in Figure 27.6b. In either case there is only one live edge, the smallest in the order $\prec$. It could be either *internal* or *external* depending on whether it belongs or does not belong to the spanning quasi-tree.

Definition 27.18. An edge of a non-orientable ribbon graph G can be classified as *orientable or non-orientable* relative to a spanning quasi-tree Q for G. Choose an orientation on the boundary circle ∂Q of Q and on the boundary of each edge-ribbon considered as a topological disc. Then, tracing ∂Q in the direction of the orientation we meet exactly two arcs of each-ribbon edge. If the orientations of these two arcs are coherent, either both along the tracing, or both against the tracing, then we call the edge *orientable* relative to Q. Otherwise we call it *non-orientable.*

The definition above is from [1122]. It is easy to see that the properties do not depend on the choice of the orientations. Essentially the chord diagram $C_G(Q)$ depicts a partial dual ribbon graph $G^{E(Q)}$ in the sense of [306] (see Section 1.2.16 for partial duals). All edges of $G^{E(Q)}$ are loops. For loops we can define orientability in the usual topological sense which coincides with our definition above.

Thus each edge has three pieces of additional information relative to a spanning tree Q: internal/external, live/dead, and orientable/non-orientable. This information is used in the quasi-tree expansion of the Krushkal polynomial.

Definition 27.19. With each spanning quasi-tree Q of a ribbon graph G we associate:

1. a spanning ribbon subgraph $F(Q)$ that results from deleting the (internally) live orientable edges of Q;

2. an abstract (not embedded) graph $G(Q)$ whose vertices are the connected components of $F(Q)$ and whose edges are the internally live orientable edges of Q.

We will also need a dual construction. For a ribbon graph G, regarded as a graph cellularly embedded into the surface $\Sigma = \widetilde{G}$, we consider the dual ribbon graph G^*. A spanning subgraph F for G determines a spanning subgraph $F^\star$ containing all edges of G^* which do not intersect edges of F. Note that $F^\star$ is not a dual ribbon graph for F: its edge set is rather the complement of the set of edges of G^* corresponding to the edges of F. The asterisk applied to an entire graph means the dual graph, but the star applied to a subgraph means the spanning subgraph given by this construction.

The spanning ribbon subgraphs F and $F^\star$ have common boundary and their gluing along this common boundary gives the whole surface Σ. In particular, for a spanning quasi-tree Q for G, the subgraph $Q^\star$ is a quasi-tree for G^*. Moreover, these quasi-trees have the same chord diagrams, $C_G(Q) = C_{G^*}(Q^\star)$. Also the natural bijection of edges of G and G^* leads to the total order $\prec^*$ on edges of G^* induced by $\prec$. Consequently the property of an edge of being live/dead relative to Q is mapped by the bijection to the same property relative to $Q^\star$. Also one may check that the property of an edge of being orientable/non-orientable relative to Q is preserved by the bijection to the same property relative to $Q^\star$. But its property of being internal/external is changed to the opposite. Now we can apply definition 27.19 to the quasi-tree of G^* just constructed to obtain the following.

- A spanning ribbon subgraph $F(Q^\star)$ that results from deleting the internally live orientable edges of $Q^\star$. They correspond to externally live orientable edges of Q.

- An abstract graph $G(Q^\star)$ that consists of vertices that are the connected components of $F(Q^\star)$ and edges that are the internally live orientable edges of $Q^\star$.

We can now state the quasi-tree expansion from [255].

Theorem 27.20. *For a ribbon graph G naturally embedded into the surface $\Sigma = \widetilde{G}$, the Krushkal polynomial has the following expansion over the set of quasi-trees.*

$$K(G, \Sigma; X, Y, A, B) = \sum_{Q \in \mathcal{Q}_G} A^{s(F(Q))/2} T(Q) \cdot B^{s(F(Q^\star))/2} T(Q^\star) \, ,$$

where $T(Q) = T(G(Q); X+1, A+1)$ *and* $T(Q^\star) = T(G(Q^\star); Y+1, B+1)$ *stand for the classical Tutte polynomial of the abstract graphs* $G(Q)$ *and* $G(Q^\star)$, *and the parameter* s *was defined in* (27.8).

Example 27.21. For the ribbon graph G on the torus Σ with a single vertex and two loops, longitude and meridian, in Example 27.7 we have two spanning quasi-trees, one consisting of a vertex only and the other the whole graph, as in Example 27.17. For Q being a vertex, both $F(Q)$ and $G(Q)$ consist of one vertex. So $s(F(Q)) = 0$ and $T(Q) = 1$. The dual quasi-tree $Q^\star = G^*$ is the whole graph in this case, and one of its loops is internally active. So $F(Q^\star)$ consists of one vertex with a loop attached to it. The abstract graph $G(Q^\star)$ is the same. Thus $s(F(Q^\star)) = 0$ and $T(Q^\star) = T(G(Q^\star); Y+1, B+1) = B+1$. Therefore, the contribution of the single vertex spanning quasi-tree is $B+1$. For the second quasi-tree Q, the whole graph, the situation is completely symmetrical. Its dual quasi-tree $Q^\star$ consists of a single vertex, and its contribution is $A+1$. So

$$K(\widehat{G}, \Sigma; X, Y, A, B) = A + 2 + B,$$

as seen in Example 27.13.

Example 27.22. For the graph G in Examples 27.13 and 27.17 with edges labeled a, b, and c there are four spanning quasi-trees $Q_{\{a\}}$, $Q_{\{b\}}$, $Q_{\{c\}}$, and $Q_{\{a,b,c\}}$. For each of them a is a live edge and b and c are dead relative to the total order $a \prec b \prec c$. Table 27.3 shows the ribbon graphs $F(Q)$, $F(Q^\star)$ and abstract graphs $G(Q)$, $G(Q^\star)$ and their contributions to the Krushkal polynomial for each of these spanning quasi-trees. So the Krushkal polynomial of the core graph with $\Sigma = \widetilde{G}$ is

$$K(\widehat{G}, \Sigma) = (X+1)B + (B+1) + (B+1) + (A+1) = XB + A + 3B + 3,$$

which coincides with the direct calculation in Example 27.13

Butler's Theorem 27.20 specializes to the quasi-tree expansion of the Bollobás–Riordan polynomial from [276] in the orientable case and from [1122] in the non-orientable case. Also it specializes to a quasi-tree expansion of the Las Vergnas polynomial, see [255].

Quasi-tree expansions for the polynomials in Section 27.6 were given in [855].

27.9 Open problems

The study of topological Tutte polynomials is still a fairly new field, and as such, many fundamental questions remain unanswered and areas unexplored. Perhaps the most pressing of these is in finding combinatorial interpretations of topological Tutte polynomials at points—what do topological Tutte

Q	$Q_{\{a\}}$	$Q_{\{b\}}$	$Q_{\{c\}}$	$Q_{\{a,b,c\}}$
$F(Q)$		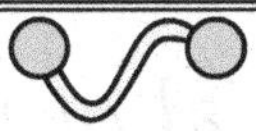	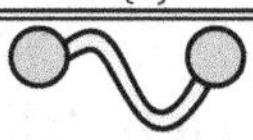	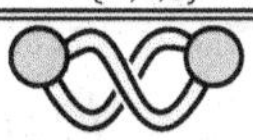
$A^{s(F(Q))/2}$	1	1	1	1
$\Gamma(Q)$				
$T(Q)$	$X+1$	1	1	$A+1$

$Q^{\star}$	$Q^{\star}_{\{a\}}$	$Q^{\star}_{\{b\}}$	$Q^{\star}_{\{c\}}$	$Q^{\star}_{\{a,b,c\}}$
$F(Q^{\star})$				
$B^{s(F(Q^{\star}))/2}$	B	1	1	1
$G(Q^{\star})$				
$T(Q^{\star})$	1	$B+1$	$B+1$	1

TABLE 27.3: Computing the Krushkal polynomial through quasi-trees.

polynomials count? When does a topological Tutte polynomial determine an embedded graph? What is the computational complexity of the polynomial, what are their easy points and hard points? How does this depend upon, say, genus? What about analogous polynomials for hypermaps and simplicial complexes?

28

The Tutte polynomial of matroid perspectives

Emeric Gioan

Synopsis

This chapter covers the Tutte polynomial of matroid perspectives. Matroid perspectives adapt the idea of morphisms to matroid theory. The Tutte polynomial of a matroid perspective is a three-variable generalization of the Tutte polynomial that shares many of its properties.

- Definition of matroid perspectives, or matroid strong maps.
- Tutte polynomial generalized to matroid perspectives and various expansions.
- Structural and enumerative properties in terms of subset activities when the ground set is linearly ordered.
- Extensions to matroid perspectives of classical counting results.
- Applications, evaluations, computational complexity, and related polynomials.

28.1 Introduction

This chapter covers the Tutte polynomial of matroid perspectives. Matroid perspectives (which are equivalent to strong maps) are a counterpart to linear maps and to graph homomorphisms. They were introduced to adapt the idea of a morphism to matroid theory. Their significance is that they allow access to richer structures in matroid theory, for example, in studying the relations

DOI: 10.1201/9780429161612-28

between the two matroid minors obtained by deleting and contracting the same subset.

We present the three-variable generalization of the Tutte polynomial for matroid perspectives. In this polynomial, two variables record information about each of the two matroids in the matroid perspective, and the third variable takes into account the differences in the matroids. It can be defined in several equivalent ways, allows generalizations of several properties of matroid polynomials, and is related to several other polynomials. We note that results involving matroid perspectives apply to matroids by considering the special case when the two matroids of the perspective are equal.

We also give results for particular classes, such as representable, binary, or graphical perspectives of matroids. Moreover, we examine properties that arise when the ground set of a matroid perspective is linearly ordered. We extend and refine several results about basis activities, yielding notably an important general unifying Tutte polynomial formula in five variables.

28.2 Matroid perspectives

We introduce *matroid perspectives*, also called in the literature *matroid strong maps*, or *matroid quotients*, or *morphisms of matroids* (up to unimportant variants). More details can be found in [729] or [894, Section 7.3].

Definition 28.1. Let M and M' be two matroids on the same ground set E. The pair (M, M') is a *matroid perspective*, or a *perspective* for short, denoted $M \to M'$, if there is a matroid N on $E \sqcup A$, for some set A, such that $M = N \backslash A$ and $M' = N/A$. Such a matroid N is called a *major* of the matroid perspective.

Proposition 28.2. *For M, M' matroids on E, the following are equivalent:*

1. *the pair (M, M') is a perspective;*
2. $r_{M'}(X) - r_{M'}(Y) \leq r_M(X) - r_M(Y)$, *for all $Y \subseteq X \subseteq E$;*
3. *every flat of M' is a flat of M;*
4. *every cocircuit of M' is a union of cocircuits of M;*
5. *every circuit of M is a union of circuits of M';*
6. *no circuit of M and cocircuit of M' intersect in exactly one element.*

Example 28.3. Let $G = (V, E)$ be a graph and $V = V_1 \sqcup V_2 \sqcup \cdots \sqcup V_k$ be a partition of V. Let $G' = (V', E)$ be the graph obtained from G by identifying vertices in each V_i, for $1 \leq i \leq k$. The relation between G and G' can be seen as a graph homomorphism $G \to G'$. In our setting, the cycle matroids of G

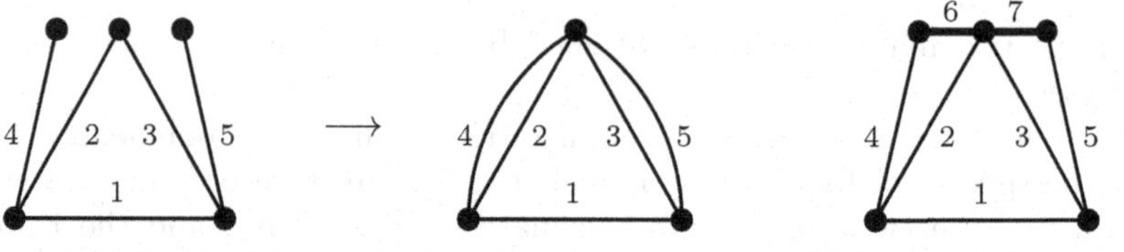

(a) Graphs giving a matroid perspective. (b) A graph giving a major.

FIGURE 28.1: Matroid perspective from a graph homomorphism, see Example 28.3.

and G' form a matroid perspective $M(G) \to M(G')$. For example, taking the cycle matroids of the graphs in Figure 28.1a results in a matroid perspective. A major is the cycle matroid of the graph in Figure 28.1b, where $A = \{6, 7\}$. Matroid perspectives arising in this way are called *graphic.*

Example 28.4. Let G be a graph embedded in a surface, let G^* be the geometric dual of G, and identify the edges on G and G^* in the natural way. Then $M^*(G^*) \to M(G)$ is a matroid perspective [440] (see Sections 27.4 and 28.8.7 for more on this example).

Example 28.5. Let V be a vector space, $u : \mathsf{V} \to \mathsf{V}$ be a linear transformation, and $E \subseteq \mathsf{V}$ be a finite set. The vector matroid M of E and the vector matroid M' of $\{u(e) : e \in E\}$ form a matroid perspective $M \to M'$ (using the natural bijection between the ground sets of M and M').

Example 28.6. If M is a matroid, then $M \to M$ is a matroid perspective. Hence, any result for matroid perspectives $M \to M'$ gives a result for matroids by setting $M = M'$. In fact, if $M \to M'$ is a matroid perspective and $r(M) = r(M')$, then $M = M'$. Trivial cases of matroid perspectives are also $M \to U_{0,|E|}$ and $U_{|E|,|E|} \to M'$. Additionally, if $M \to M'$ is a matroid perspective, then $M'^* \to M^*$ is also a matroid perspective.

28.3 The Tutte polynomial of a matroid perspective

The 3-variable generalization of the Tutte polynomial to matroid perspectives was introduced by M. Las Vergnas in [742, 743, 746]. This polynomial was further developed in the series of papers [470, 752, 754, 756], with the case of oriented matroids in [750, 755] (discussed in Chapter 31). This section follows [752].

Definition 28.7. Let $M \to M'$ be a matroid perspective on E. The *Tutte polynomial* of this matroid perspective is the polynomial in three variables defined by

$$T(M,M';x,y,z) = \sum_{A\subseteq E} (x-1)^{r(M')-r_{M'}(A)}(y-1)^{|A|-r_M(A)} z^{r(M)-r(M')-\left(r_M(A)-r_{M'}(A)\right)}.$$

Notation 28.8. For two matroids M, M' on E and $A \subseteq E$, we set

$$\begin{aligned} \mathrm{cr}_M(A) &:= r(M) - r_M(A),\\ \mathrm{nl}_M(A) &:= |A| - r_M(A),\\ \mathrm{rcd}_{M,M'}(A) &:= r(M) - r(M') - \big(r_M(A) - r_{M'}(A)\big). \end{aligned}$$

With this, Definition 28.7 can be written as

$$T(M,M';x,y,z) = \sum_{A\subseteq E} (x-1)^{\mathrm{cr}_{M'}(A)}(y-1)^{\mathrm{nl}_M(A)} z^{\mathrm{rcd}_{M,M'}(A)}.$$

The polynomial $T(M,M';x,y,z)$ is of degree $r(M')$ in x, degree $|E|-r(M)$ in y, and degree $r(M)-r(M')$ in z. The fact that the pair (M,M') forms a matroid perspective yields numerous interesting properties for this polynomial. The first is that its coefficients are nonnegative integers.

Sets that are independent in M and span M' are called *independent/spanning sets.* Their number is given by

$$T(M,M';1,1,1) = \#\{A\subseteq E : A \text{ independent in } M \text{ and spanning in } M'\}.$$

The following relations with the classical Tutte polynomial are direct.

Proposition 28.9. *For a matroid perspective $M \to M'$, the following relations hold.*

1. $T(M;x,y) = T(M,M;x,y,z)$.
2. $T(M;x,y) = T(M,M';x,y,x-1)$.
3. $T(M';x,y) = (y-1)^{r(M)-r(M')}T(M,M';x,y,\frac{1}{y-1})$.
4. $T(M;z+1,y) = T(M,U_{0,|E|};x,y,z)$.
5. $T(M';x,\frac{1}{z}+1) = z^{r(M')-|E|}\ T(U_{|E|,|E|},M';x,y,z)$.

In general, $T(M,M';x,y,z)$ is not determined by $T(M;x,y)$ and $T(M';x,y)$, but we do have the following special case.

Proposition 28.10. *Let $M \to M'$ be a matroid perspective. If $r(M) - r(M') = 1$, then*

$$T(M,M';x,y,z) = \frac{z(y-1)-1}{xy-x-y}T(M;x,y) + \frac{-z+x-1}{xy-x-y}T(M';x,y).$$

The polynomial $T(M, M'; x, y, z)$ has the following basic properties.

Proposition 28.11. *Let $M \to M'$ be a matroid perspective. Then*

$$T((M')^*, M^*; x, y, z) = z^{r(M)-r(M')}\, T(M, M'; y, x, \tfrac{1}{z}).$$

Proposition 28.12. *Let $M \to M'$ be a matroid perspective on E. Suppose that $E = E_1 \sqcup E_2$ is a partition of E such that $M = M|E_1 \oplus M|E_2$ and $M' = M'|E_1 \oplus M'|E_2$. Then*

$$T(M, M'; x, y, z) = T(M|E_1, M'|E_1; x, y, z) \cdot T(M|E_2, M'|E_2; x, y, z).$$

Linear relations among coefficients of the Tutte polynomial of a matroid perspective, generalizing those of [251] for matroids (see Section 4.9), are given in [746, 752]. Moreover, [746, 752] contains results involving the *Higgs factorization* of a matroid perspective (named after the notion of Higgs major [614], and consisting of a sequence of matroid perspectives whose majors have only one more element at each step). The main result is that the Tutte polynomial of the matroid perspective $M \to M'$ is computationally equivalent to the $r(M)-r(M')+1$ Tutte polynomials of the matroids of its Higgs factorization.

The evaluation $T(M, M'; 0, 0, 1)$ is of special interest because it has several interpretations. It can be seen as a generalization of the β-invariant (the coefficient $t_{1,0}$ of the Tutte polynomial), as shown below in a particular case of matroid perspectives (a result from [746]). It is also involved in connectivity properties (see Section 28.9, and [746, Section 8] for more details), in oriented matroid properties (see Sections 28.8.7 and 31.5), and in other formulas (see Section 28.16, and see also [752] for an expression in terms of β-invariants in a Higgs factorization).

Proposition 28.13. *Let M be a matroid on E, let $p \in E$ and let p_{loop} be a single loop labeled by p, so that $M \to M/p \oplus p_{\text{loop}}$ is a matroid perspective. Then*

$$T(M, M/p \oplus p_{\text{loop}}; 0, 0, 1) = 2\,\beta(M).$$

The complexity of computing the Tutte polynomial of graphic matroid perspectives, from [754], is given below. (See Chapter 9 for the computational complexity of the classical Tutte polynomial.)

Theorem 28.14. *The points (x, y, z) in the complex space $\mathbb{C}^3$ such that there is a polynomial time algorithm to evaluate $T(M, M'; x, y, z)$ for any graphic matroid perspective $M \to M'$ are the following.*

1. *All points $(t+1, \frac{1}{t}+1, t)$ for $t \in \mathbb{C}$.*

2. *The fifteen points of the form $(x, y, x-1)$, or $(x, y, \frac{1}{y-1})$ if $y \neq 1$, where (x, y) is one of the eight points from the list of Theorem 9.4: $\{(1,1), (-1,-1), (0,-1), (-1,0), (i,-i), (-i,i), (j,j^2), (j^2,j)\}$, where $j = e^{2\pi i/3}$.*

3. *The point* $(-1,-1,-1)$ *(see Theorem 28.31).*
4. *Eventually, by convention, the limit evaluation at* $(1,1,\infty)$.

The limit evaluation at $(1,1,\infty)$ above is dual to the evaluation at $(1,1,0)$.

28.4 Deletion–contraction, convolution, and the Möbius function

The Tutte polynomial of a matroid perspective can be defined through a deletion–contraction relation [746, 752].

Theorem 28.15. *Let $M \to M'$ be a matroid perspective on E. Its Tutte polynomial is uniquely defined by the following properties.*

1. *If $e \in E$ is neither a coloop nor a loop of M, then*
$$T(M,M';x,y,z) \;=\; T(M\backslash e, M'\backslash e;x,y,z) \;+\; T(M/e,M'/e;x,y,z).$$
2. *If $e \in E$ is a coloop of M' (and hence also a coloop of M), then*
$$T(M,M';x,y,z) \;=\; x\,T(M\backslash e, M'\backslash e;x,y,z).$$
3. *If $e \in E$ is a loop of M (and hence also a loop of M'), then*
$$T(M,M';x,y,z) \;=\; y\,T(M/e,M'/e;x,y,z).$$
4. *If $e \in E$ is a coloop of M and is not a coloop of M', then*
$$T(M,M';x,y,z) \;=\; z\,T(M\backslash e, M'\backslash e;x,y,z) \;+\; T(M/e,M'/e;x,y,z).$$
5. *For $U_{0,0} \to U_{0,0}$,*
$$T(U_{0,0},U_{0,0};x,y,z) \;=1.$$

An immediate consequence of Theorem 28.15 is that the coefficients of the Tutte polynomial of a matroid perspective are nonnegative integers.

Recall (see Definition 4.72) that the *lattice of flats* $\mathcal{L}(M)$ of a matroid M is the set of flats of M ordered by inclusion. This partially ordered set is known to be a geometric lattice, the Möbius function of which can be used to calculate the Tutte polynomial of M (see Section 4.7). We refer the reader to Definition 4.141 or [1182] for the definition of the Möbius function, and to Equation (4.9) or [247] for the relation with the Tutte polynomial in matroids. This property of the Tutte polynomial generalizes to matroid perspectives [752], as stated below.

Theorem 28.16. *Let $M \to M'$ be a matroid perspective. Then*

$$T(M,M';x,y,z) = \sum_{X,Y\in\mathcal{L}(M)} \mu_{\mathcal{L}(M)}(Y,X)\ (x-1)^{r(M')-r_{M'}(X)} \times y^{|Y|}\ (y-1)^{-r_M(X)}\ z^{r(M)-r(M')-\left(r_M(X)-r_{M'}(X)\right)}$$

where $\mu_{\mathcal{L}(M)}$ denotes the Möbius function of the lattice $\mathcal{L}(M)$.

Corollary 28.17. *If M is loopless, then*

1. $T(M,M';0,0,0) = (-1)^{r(M)} \sum_{\substack{X\in\mathcal{L}(M)\\ r_M(X)-r_{M'}(X)=r(M)-r(M')}} \mu_{\mathcal{L}(M)}(0,X),$

2. $T(M,M';0,0,1) = (-1)^{r(M')} \sum_{X\in\mathcal{L}(M)} \mu_{\mathcal{L}(M)}(0,X)(-1)^{r_M(X)-r_{M'}(X)}.$

Observe that $T(M,M';0,0,0)$ is not equal to 0 in general, in contrast with the fact that $T(M;0,0)=0$ for a matroid M.

The usual convolution formula for the Tutte polynomial of a matroid (see Theorem 4.150) generalizes to matroid perspectives. Several proofs in the setting of matroid perspectives have been given: first in [684] in terms of the equivalent linking polynomial (see Section 28.8.5); next in [685], constructively, by means of a decomposition of internal/external activities (see Section 28.5) with respect to cyclic flats, extending to matroid perspectives that of [469]; and recently in [718], algebraically, by a general formulation of the Tutte polynomial via a character in a Hopf algebra (see Section 28.8.6).

Theorem 28.18. *Let $M \to M'$ be a matroid perspective on E. Then*

$$T(M,M';x,y,z) = \sum_{F\subseteq E} T(M|F,M'|F;0,y,-1)\ T(M/F,M'/F;x,0,z).$$

The sum in Theorem 28.18 can be restricted to the cyclic flats F of $M \to M'$. A *cyclic flat* of a perspective $M \to M'$ is a flat of M and a union of circuits of M' (this extends the usual definition in matroids).

28.5 Subset activities in matroids and perspectives

In this section, we exploit a linear ordering of the ground set. Extending activities from matroid bases (given in Definition 4.133) to subsets yields a rich theory. First, as introduced in [746] and stated in Theorem 28.20 below, which is from [752], subset activities enable an expression of the Tutte polynomial of a matroid perspective in terms of activities of independent/spanning subsets, which extends the usual expression of the Tutte polynomial in terms of activities of bases. Beyond this, as addressed in [573] for matroids and in [756] for matroid perspectives, we consider four subset activities, related to a fundamental partition of the power set of the ground set into boolean intervals, and leading to multivariate expansions and partial derivatives of the Tutte polynomial (see Section 28.6). Subset activities can also be expressed in terms of closure operators and active orders for matroid bases; see [753] (and see Section 28.9). In what follows, we use notation from [756], and we denote by $\min(A)$ the smallest element of an ordered finite set A.

Definition 28.19. Let M be a matroid on a linearly ordered set E. For $A \subseteq E$, set

$$\begin{aligned} \mathrm{IA}_M(A) &:= \big\{e \in A : e = \min(C),\ C \text{ a cocircuit contained in } (E \setminus A) \cup \{e\}\big\}, \\ \mathrm{EA}_M(A) &:= \big\{e \in E \setminus A : e = \min(C),\ C \text{ a circuit contained in } A \cup \{e\}\big\}. \end{aligned}$$

Then the *internal* and *external activities* of A are, respectively,

$$\begin{aligned} \iota_M(A) &= |\mathrm{IA}_M(A)|, \text{ and} \\ \varepsilon_M(A) &= |\mathrm{EA}_M(A)|. \end{aligned}$$

By matroid duality, for $A \subseteq E$, we have $\mathrm{EA}_M(A) = \mathrm{IA}_{M^*}(E \setminus A)$ and $\varepsilon_M(A) = \iota_{M^*}(E \setminus A)$. If A is a basis of M, then the definitions of $\mathrm{IA}_M(A)$, $\iota_M(A)$, $\mathrm{EA}_M(A)$, and $\varepsilon_M(A)$ agree with the usual definition of basis activities.

Theorem 28.20. *For a matroid perspective $M \to M'$ on a linearly ordered set E,*

$$T(M, M'; x, y, z) = \sum_{\substack{B \subseteq E \\ \text{independent/spanning}}} x^{\iota_{M'}(B)}\, y^{\varepsilon_M(B)}\, z^{\mathrm{rcd}_{M,M'}(B)}.$$

The proof in [752] follows by applying the deletion–contraction relation to the greatest element of E. Observe that the number of independent/spanning subsets with given activities does not depend on the linear ordering of E, and their total number equals $T(M, M'; 1, 1, 1)$ as noted in Section 28.3.

Given a matroid M on a linearly ordered set E, a classical result [349] is that the set 2^E of subsets of E is partitioned into the *basis (boolean) intervals* $\big[B \setminus \mathrm{IA}_M(B), B \cup \mathrm{EA}_M(B)\big]$, where B ranges over all bases of M (see Section 5.4). This simple but fundamental result generalizes to matroid perspectives, as stated below. It is obtained in [756] using a more general combinatorial construction [361], which applies to both matroids and matroid perspectives (see also [131, 573] for other viewpoints in matroids, [571] for an independent generalization to greedoids, and [240] for an inverse property in matroids).

Proposition 28.21. *For a matroid perspective $M \to M'$ on a linearly ordered set E,*

$$2^E = \bigsqcup_{\substack{B \subseteq E \\ independent/spanning}} \Big[B \setminus \mathrm{IA}_{M'}(B), B \cup \mathrm{EA}_M(B)\Big].$$

Definition 28.22. Let M be a matroid on a linearly ordered set E. For $A \subseteq E$, let

$$P_M(A) := \big\{e \in E \setminus A : e = \min(C),\ C \text{ a cocircuit contained in } E \setminus A\big\}, \text{ and}$$
$$Q_M(A) := \big\{e \in A : e = \min(C),\ C \text{ a circuit contained in } A\big\}.$$

By matroid duality, for $A \subseteq E$, we have $Q_M(A) = P_{M^*}(E \setminus A)$. The cardinalities of these sets turn out to be well-known parameters, independent of the ordering of E:

$$\begin{aligned} |P_M(A)| &= \mathrm{cr}_M(A) &= r(M) - r_M(A), \\ |Q_M(A)| &= \mathrm{nl}_M(A) &= |A| - r_M(A). \end{aligned}$$

Proposition 28.23 below shows the consistency of the four subset activities for general subsets and for independent/spanning subsets, as related by Proposition 28.21. From all this we get the unifying Tutte polynomial formula given in Section 28.6. See [756, Prop. 4.4] for more details on Proposition 28.23, and see [528] for the last equation (which is missing in [756]).

Proposition 28.23. *Let $M \to M'$ be a matroid perspective on a linearly ordered set E. Let $B \subseteq E$ be independent in M and spanning in M'. Let A be in the basis boolean interval $[B \setminus \mathrm{IA}_{M'}(B), B \cup \mathrm{EA}_M(B)]$. Then*

$$\begin{aligned} \mathrm{IA}_{M'}(A) &= \mathrm{IA}_{M'}(B) \cap A, & P_{M'}(A) &= \mathrm{IA}_{M'}(B) \setminus A, \\ \mathrm{EA}_M(A) &= \mathrm{EA}_M(B) \setminus A, & Q_M(A) &= \mathrm{EA}_M(B) \cap A, \end{aligned}$$
$$\begin{aligned} \iota_{M'}(A) + \mathrm{cr}_{M'}(A) &= \iota_{M'}(B), \\ \varepsilon_M(A) + \mathrm{nl}_M(A) &= \varepsilon_M(B), \\ \mathrm{rcd}_{M,M'}(A) &= \mathrm{rcd}_{M,M'}(B). \end{aligned}$$

28.6 Five-variable expansion and partial derivatives

Building on Section 28.5, we obtain a rich four-variable, or five-variable, expansion formula for the Tutte polynomial of matroids, or matroid perspectives, respectively. (See also Chapter 5.) This result brings together Definition 28.7 and Theorem 28.20. Originally, the formula for matroids was given in [573] which introduced subset activities in a more general way. An alternative proof and the generalization to matroid perspectives was given in [756].

Theorem 28.24. *Let M be a matroid on a linearly ordered set E. Then*

$$T(M; x+u, y+v) = \sum_{A\subseteq E} x^{\iota_M(A)}\, u^{\mathrm{cr}_M(A)}\, y^{\varepsilon_M(A)}\, v^{\mathrm{nl}_M(A)}.$$

Theorem 28.25. *For a matroid perspective $M \to M'$ on a linearly ordered set E,*

$$T(M, M'; x+u, y+v, z) = \sum_{A\subseteq E} x^{\iota_{M'}(A)}\, u^{\mathrm{cr}_{M'}(A)}\, y^{\varepsilon_M(A)}\, v^{\mathrm{nl}_M(A)}\, z^{\mathrm{rcd}_{M,M'}(A)}.$$

The formula in Theorem 28.25 can be seen as the natural enumerative counterpart of the fundamental partition of Proposition 28.21. It can be proved directly by expanding each of $(x+u)^{\iota_{M'}(B)}$ and $(y+v)^{\epsilon_M(B)}$ in terms of the sets A with $B \setminus \mathrm{IA}_{M'}(B) \subseteq A \subseteq B \cup \mathrm{EA}_M(B)$ and using Proposition 28.23 with appropriate simple transformations (see [528] for details). The proof in [756] uses an equivalent formulation in terms of partial derivatives (see below).

A remarkable feature of this formula, for both matroids and matroid perspectives, is that it unifies the two celebrated Tutte polynomial formulas in terms of rank function (Definition 28.7) and in terms of activities (Theorem 28.20). In fact, numerous 2-variable expansions follow from Theorems 28.24 and 28.25 by specializing variables. The most noteworthy are obtained by setting some of x, u, y, or v to either 0 or ± 1, or replacing these by $x/2$ or $y/2$, and performing an appropriate change of variables. A list is given for matroids in [573] and completed in [756] for matroids and matroid perspectives, finally yielding nine expansions on different summands, including the two aforementioned ones. The other three main ones are given below. A detailed example is given in [756].

In what follows, we state formulas for matroid perspectives. The matroid results can be obtained by setting $M = M'$ (and thus also omitting variable z).

Corollary 28.26. *Under the hypotheses of Theorem 28.25,*

$$\begin{aligned} T(M, M'; x, y, z) &= \sum_{\substack{A\subseteq E \\ \textit{spanning in } M'}} x^{\iota_{M'}(A)} (y-1)^{\mathrm{nl}_M(A)} z^{\mathrm{rcd}_{M,M'}(A)} \\ &= \sum_{\substack{A\subseteq E \\ \textit{independent in } M}} (x-1)^{\mathrm{cr}_{M'}(A)} y^{\varepsilon_M(A)} z^{\mathrm{rcd}_{M,M'}(A)} \\ &= \sum_{A\subseteq E} \left(\frac{x}{2}\right)^{\mathrm{cr}_{M'}(A)+\iota_{M'}(A)} \left(\frac{y}{2}\right)^{\mathrm{nl}_M(A)+\varepsilon_M(A)} z^{\mathrm{rcd}_{M,M'}(A)}. \end{aligned}$$

By Taylor's theorem, the expansion formulas of Theorems 28.24 and 28.25 can be formulated as expressions of partial derivatives. They are given in [756] along with corollaries and examples.

Theorem 28.27. *Let $M \to M'$ be a matroid perspective on a linearly ordered set E, and p, q be nonnegative integers. Then*

$$\begin{aligned}\frac{\partial^{p+q}}{\partial x^p \partial y^q} T(M,M';x,y,z) &= p!\,q! \sum_{\substack{A\subseteq E\\ \mathrm{cr}_{M'}(A)=p\\ \mathrm{nl}_M(A)=q}} x^{\iota_{M'}(A)} y^{\varepsilon_M(A)} z^{\mathrm{rcd}_{M,M'}(A)}\\
&= p!\,q! \sum_{\substack{A\subseteq E\\ \iota_{M'}(A)=p\\ \mathrm{nl}_M(A)=q}} x^{\mathrm{cr}_{M'}(A)} y^{\varepsilon_M(A)} z^{\mathrm{rcd}_{M,M'}(A)}\\
&= p!\,q! \sum_{\substack{A\subseteq E\\ \mathrm{cr}_{M'}(A)=p\\ \varepsilon_M(A)=q}} x^{\iota_{M'}(A)} y^{\mathrm{nl}_M(A)} z^{\mathrm{rcd}_{M,M'}(A)}\\
&= p!\,q! \sum_{\substack{A\subseteq E\\ \iota_{M'}(A)=p\\ \varepsilon_M(A)=q}} x^{\mathrm{cr}_{M'}(A)} y^{\mathrm{nl}_M(A)} z^{\mathrm{rcd}_{M,M'}(A)}.\end{aligned}$$

A counterpart of Theorem 28.24 for oriented matroids is given in Section 31.9, along with a canonical bijection relating the two settings (derived from Section 31.8; see Figure 31.6).

28.7 Representable and binary cases

The results in this section may be found in [470]. From a general purely combinatorial identity (Theorem 28.28), by a formal algebraic/combinatorial transformation, it is possible to derive a powerful enumerative tool in terms of the Tutte polynomial. Theorem 28.29 formulates this in terms of any pair of matroids on the same set and representable over the same finite field, using Definition 28.7 for the polynomial T. This result can be seen as a far-reaching generalization of the well-known link between the Tutte polynomial and the generating function of weights of codewords of a linear code and its dual [649] (see Chapter 16).

Theorem 28.28. *Let $\mathcal{S} = (S_i)_{i\in I}$ and $\mathcal{T} = (T_j)_{j\in J}$ be two families of subsets of a set E. For $A \subseteq E$, denote $\mathcal{S}(A) = (S_i)_{i\in I,\, S_i\subseteq A}$ and $\mathcal{T}(A) = (T_j)_{j\in J,\, T_j\subseteq A}$. The following identity holds in $\mathbb{Z}[x]$:*

$$\sum_{\substack{i\in I,\ j\in J,\\ S_i\cap T_j=\emptyset}} x^{|S_i|}\,(1-x)^{|T_j|} = \sum_{A\subseteq E} |\mathcal{S}(A)|\;|\mathcal{T}(E\setminus A)|\; x^{|A|}\,(1-x)^{|E\setminus A|}.$$

Let $\mathbb{F}$ be a field and V be a subspace of $\mathbb{F}^E$. The support of $u \in \mathsf{V}$ is denoted $s(u)$. The matroid whose circuits are the inclusion-minimal supports of non-zero vectors of V is called the *support matroid* of V, denoted $\mathrm{Supp}(\mathsf{V})$. It is a classical result that a matroid M is $\mathbb{F}$-representable if and only if $M = \mathrm{Supp}(\mathsf{V})$ for some subspace V of $\mathbb{F}^E$. (See [894, Proposition 9.2.4] for details, and note that such a subspace V can thus be thought of as an alternative representation of M, which is sometimes named *Tutte representation* of M after [1093]. See also Section 16.2.) Then, we have $r(M) = |E| - \dim(\mathsf{V})$ and $M^* = \mathrm{Supp}(\mathsf{V}^\perp)$, where $\mathsf{V}^\perp$ is the orthogonal complement of V in $\mathbb{F}^E$, that is, $\mathsf{V}^\perp = \{v \in \mathbb{F}^E : \sum_{e\in E} u(e)v(e) = 0 \text{ for all } u \in \mathsf{V}\}$.

Theorem 28.29. *Let E be a finite set, and q be a prime power. Let V and V' be subspaces of $\mathbb{F}_q^E$. Set $M = \mathrm{Supp}(\mathsf{V})$ and $M' = \mathrm{Supp}(\mathsf{V}')$. Let k, k', l, l' be nonnegative integers. For $u = (u_1, u_2, \ldots, u_{k+l}) \in \mathsf{V}^k \times \mathsf{V}'^l$, set $s(u) = \bigcup_{1\le i\le k+l} s(u_i)$. The following identity holds in $\mathbb{Q}[x]$:*

$$\sum_{\substack{u\in \mathsf{V}^k\times \mathsf{V}'^l,\\ u'\in(\mathsf{V}^\perp)^{k'}\times(\mathsf{V}'^\perp)^{l'},\\ s(u)\cap s(u')=\emptyset}} x^{|s(u)|}(1-x)^{|s(u')|}$$

$$= q^{l\left(r(M)-r(M')\right)} x^{r(M)}(1-x)^{|E|-r(M)}$$

$$\cdot T\Big(M, M';\ q^{k'+l'}\frac{1-x}{x}+1,\ q^{k+l}\frac{x}{1-x}+1,\ q^{k'-l}\frac{1-x}{x}\Big).$$

The last two items in the following corollary should be compared with Theorems 12.27 and 12.24, respectively.

Corollary 28.30. *Let q be a prime power, V be a subspace of $\mathbb{F}_q^E$, and $M = \mathrm{Supp}(\mathsf{V})$. Then the following hold.*

1. *The number of k-tuples $u \in \mathsf{V}^k$ such that $s(u) = E$ is*

$$(-1)^{|E|-r(M)}T(M; 0, -q^k + 1).$$

2. $$\sum_{u\in\mathsf{V}} x^{|s(u)|} = x^{r(M)}(1-x)^{|E|-r(M)}\ T(M; \frac{1}{x}, q\frac{x}{1-x} + 1).$$

3. $$\sum_{\substack{u\in\mathsf{V},\ u'\in\mathsf{V}^\perp,\\ s(u)\cap s(u')=\emptyset}} x^{|s(u)|}(1-x)^{|s(u')|} = x^{r(M)}(1-x)^{|E|-r(M)}\ T(M; q\frac{1-x}{x} + 1, q\frac{x}{1-x} + 1).$$

4. $$\sum_{\substack{u\in\mathsf{V},\ u'\in\mathsf{V}^\perp,\\ s(u)\cap s(u')=\emptyset,\\ s(u)\cup s(u')=E}} (-1)^{|s(u)|} = (-1)^{|E|-r(M)}\ T(M; -q+1, -q+1).$$

The first item of the corollary is from [251, 350], the second from [579], and the final two, which are well-known in terms of linear codes, from [649].

We note that a more general theorem, whose statement we omit here, is given for matroids in [470]. For applications to both representable matroids and flows in a graph, the setting of representable matroids can be extended by replacing the field $\mathbb{F}$ with a unitary commutative ring $\mathfrak{R}$ and the subspace V with a regular submodule of $\mathfrak{R}^E$. As a corollary, one can derive a counting result of [349, 1092] (see also Section 12.3) for nowhere-zero flows in a graph or a regular matroid by means of a Tutte polynomial evaluation.

We now give two theorems from [470] for perspectives of binary matroids, and known corollaries of these theorems for matroids. The first, Theorem 28.31, was corrected in [754]. The second, Theorem 28.33, generalizes to perspectives of binary matroids a property proved for planar graphs in [751] and for binary matroids in [189] using isotropic systems. See Section 28.9 for related conjectures from [751].

A matroid perspective $M \to M'$ is called *binary* if M and M' are the support matroids of subspaces V and V' of $\mathbb{F}_2^E$, respectively, with $\mathsf{V} \subseteq \mathsf{V}'$. We denote by $\mathbf{1}_E$ the vector of $\mathbb{F}_2^E$ whose coordinates are all equal to 1.

Theorem 28.31. *Let $M \to M'$ be a binary matroid perspective such that $M = \mathrm{Supp}(\mathsf{V})$ and $M' = \mathrm{Supp}(\mathsf{V}')$ for binary subspaces $\mathsf{V} \subseteq \mathsf{V}' \subseteq \mathbb{F}_2^E$. Then*

$$T(M, M'; -1, -1, -1) = \begin{cases} 0 & \text{if } \mathbf{1}_E \notin \mathsf{V} + \mathsf{V}'^{\perp}, \\ (-1)^{|E| - \dim(\mathsf{V} \cap \mathsf{V}^{\perp})} 2^{\dim(\mathsf{V} \cap \mathsf{V}'^{\perp})} & \text{if } \mathbf{1}_E \in \mathsf{V} + \mathsf{V}'^{\perp}. \end{cases}$$

Corollary 28.32. *Let M be a binary matroid with $M = \mathrm{Supp}(\mathsf{V})$. Then $T(M; -1, -1) = (-1)^{|E| - \dim(\mathsf{V} \cap \mathsf{V}^{\perp})} 2^{\dim(\mathsf{V} \cap \mathsf{V}^{\perp})}$.*

Theorem 28.33. *For a binary matroid perspective $M \to M'$ with $M = \mathrm{Supp}(\mathsf{V})$ and $M' = \mathrm{Supp}(\mathsf{V}')$, the evaluation $T(M, M'; 3, 3, 1)$ is an integer multiple of $2^{\dim(\mathsf{V} \cap \mathsf{V}'^{\perp})}$. Moreover, if $\mathbf{1}_E \in \mathsf{V} + \mathsf{V}'^{\perp}$ then $2^{-\dim(\mathsf{V} \cap \mathsf{V}'^{\perp})} T(M, M'; 3, 3, 1)$ is an odd integer.*

Corollary 28.34. *Let M be a binary matroid with $M = \mathrm{Supp}(\mathsf{V})$. Then $2^{-\dim(\mathsf{V} \cap \mathsf{V}^{\perp})} T(M; 3, 3)$ is an odd integer.*

28.8 Brief accounts of related polynomials

28.8.1 Tutte polynomial of a set-pointed matroid

Because of the existence of a major for a matroid perspective (see Definition 28.1), an alternative equivalent setting for the Tutte polynomial of matroid perspectives is provided by *set-pointed matroids*, which are matroids with a distinguished subset of their ground sets. Hence, every result in the previous setting can be reformulated in this alternative setting (e.g., deletion–contraction relations, and a duality property). See [752, Section 3] for details.

Definition 28.35. Let M be a matroid on E, and $A \subseteq E$. The *Tutte polynomial of M pointed by A* is the polynomial in three variables defined by

$$T(M; A; x, y, z) = \sum_{X \subseteq E \setminus A} (x-1)^{r(M)-r_M(X \cup A)} (y-1)^{|X|-r_M(X)} z^{\left(r_M(X\cup A)-r_M(X)\right)}.$$

In the case where $A = \emptyset$, we have $T(M; \emptyset; x, y, z) = T(M; x, y)$, the usual Tutte polynomial. In the case where $A = \{e\}$ for $e \in E$, the polynomial $T(M; \{e\}; x, y, z)$ is equivalent to the Tutte polynomial of M pointed by e introduced in [251]. The equivalence between the polynomial of Definition 28.35 above and that of Definition 28.7 for matroid perspectives is given by the next result.

Proposition 28.36. *Let $M \to M'$ be a matroid perspective on E and let N be a major of $M \to M'$ on $E \sqcup A$. Then*

$$T(M, M'; x, y, z) = z^{-\left(r(N)-r(M)\right)}\, T(N; A; x, y, z).$$

Further results can be found in [752, Section 3], notably a "big" set-pointed Tutte polynomial in $|\mathcal{M}(A)|+2$ variables, where $\mathcal{M}(A)$ denotes the set of all matroids on A. Moreover, Tutte polynomials of set-pointed matroids can be used in certain situations to simplify the calculation of the Tutte polynomial in matroids by means of *normal subsets* of the matroid; see [752, Section 4].

28.8.2 Tutte polynomial of a matroid perspective sequence

As proposed in [742, 746, 752], a Tutte polynomial in $k+1$ variables can be associated with any sequence of k matroids on the same set, pairwise related by matroid perspectives (called a *matroid perspective sequence*). Details of this setting can be found in [273], including a relation with the set-pointed matroids of Section 28.8.1 and applications to electrical network theory.

28.8.3 Poincaré and coboundary polynomials

Theorem 28.16 provides a definition of the Tutte polynomial of a matroid perspective in terms of the Möbius function of the lattice of flats. From this, one can derive two polynomials in two variables for a matroid perspective that extend the Poincaré polynomial and the coboundary polynomial of a matroid [348, 349]. See [752, Equations 7.4 and 7.5] for some of the properties of these polynomials. Moreover, using the expression in terms of activities from Theorem 28.20, one can obtain an expression of the Poincaré polynomial of a matroid perspective in terms of subsets containing no broken-circuit of M. The coefficients of the chromatic polynomial have been expressed in terms of broken-circuits by Whitney. This result has been generalized to matroids in [241]. A similar result holds for matroid perspectives in terms of a generalized Poincaré polynomial, as detailed in [752, Theorem 8.3].

28.8.4 Polynomial in terms of activities of S-trees in a graph

A polynomial was introduced by Tutte [1099] for a graph G with a distinguished set of vertices S and a linearly ordered set of edges. This polynomial is the generating function of activities of S-trees of G, where an *S-tree* is a spanning forest of G with exactly $|S|$ components, each containing exactly one vertex of S. Let G' be the graph obtained by identifying all vertices in S to one vertex, and keeping all edges. Set $M = M(G)$ and $M' = M(G')$. Write $T(M, M'; x, y, z) = \sum_{0 \le k \le d} T_k(M, M'; x, y) z^k$. It turns out that Tutte's polynomial alluded to above is exactly $T_{|S|-1}(M, M'; x, y)$. (See [752, Section 8].)

28.8.5 Linking polynomial of two matroids

An alternative form of the 3-variable Tutte polynomial for matroid perspectives was considered as a 4-variable polynomial in [1135, 1136]. The *linking polynomial* is defined for any pair of matroids M and M' on E, by

$$Q(M, M'; x, y, u, v) = \sum_{X \subseteq E} (x-1)^{r_M(E) - r_M(X)} (y-1)^{|X| - r_M(X)} (u-1)^{r_{M'}(E) - r_{M'}(X)} (v-1)^{|X| - r_{M'}(X)}.$$

As pointed out in [1136], this polynomial is essentially equivalent to the Tutte polynomial of a matroid perspective. For a matroid perspective $M \to M'$, we have

$$T(M, M'; x, y, z) = Q(M, M'; z+1, y, 1 + \tfrac{x-1}{z}, 2).$$

Thus, these two polynomials share similar constructions and properties (e.g., evaluation interpretations, formulas in terms of deletion and contraction and activities); see [1135] where various results can be compared with [752]. This setting was used in [684] for a proof of the convolution formula (see Section 28.4).

28.8.6 Tutte polynomial of a Hopf algebra

In [718], a general construction is given of a Tutte polynomial associated with a Hopf algebra, provided that it is defined on a class of objects for which deletion and contraction make sense in the algebraic structure (see also [424] and Sections 19.4 and 30.3). This construction applies to various usual classes of objects such as matroids, ribbon graphs (or equivalently combinatorial maps), or matroid perspectives. Here, the key part of the definition of the Hopf algebra is its coproduct, which, in the case of matroid perspectives, is given by

$$\Delta(M \to M') = \sum_{A \subseteq E} (M|A \to M'|A) \otimes (M/A \to M'/A).$$

In this context, for a matroid perspective $M \to M'$ on E, the expression

$$x_1{}^{r(M')} y_2{}^{|E|-r(M)} x_3{}^{r(M)-r(M')} T\Big(M, M'; \tfrac{y_1}{x_1} + 1, \tfrac{x_2}{y_2} + 1, \tfrac{y_3}{x_3}\Big)$$

can be directly defined through an algebraic character in the associated Hopf algebra. This yields algebraic interpretations of various Tutte polynomial concepts, and notably a proof of the convolution formula of Section 28.4.

28.8.7 Las Vergnas polynomial of an embedded graph

Let G be a connected graph embedded in an orientable surface, and G^* be its dual embedded graph. Then, as in Example 28.4, $M^*(G^*) \to M(G)$ is a matroid perspective. Its Tutte polynomial

$$T\big(M^*(G^*), M(G); x, y, z\big)$$

is called the *Las Vergnas polynomial* of G. It was introduced in [746] (see also [747]), and later named and surveyed in [52, 447]. In [447], it is also extended to pseudo-surfaces, and is related to several objects and polynomials (ribbon graph polynomial, or Bollobás–Riordan polynomial, or the more general Krushkal polynomial), see Chapter 27 for details. A notable property, from [747], is that $T\big(M^*(G^*), M(G); 0, 0, 1\big)$ counts the number of pairs of consistent dual orientations for G and G^* such that both orientations are strongly connected; see Corollary 31.22 and Section 31.5 for the oriented-matroid version of this result.

28.8.8 Tutte polynomial of an oriented matroid perspective

An oriented matroid perspective $M \to M'$ is a matroid perspective involving two oriented matroids and a supplementary sign consistency condition. When the ground set is linearly ordered, the orientation-activity $o(M)$, respectively dual-orientation-activity $o^*(M)$, of the oriented matroid M is the number of smallest elements of positive circuits, respectively cocircuits, of M. Then, the main theorem of [750] states that

$$T(\underline{M}, \underline{M}'; x, y, 1) = \sum_{A \subseteq E} \left(\frac{x}{2}\right)^{o^*(-_A M')} \left(\frac{y}{2}\right)^{o(-_A M)},$$

where $\underline{M}$ and $\underline{M}'$ are the underlying matroids of M and M', respectively, and $-_A M$ denotes the reorientation of M with respect to A. See Section 31.5 for details.

28.9 Open problems

Generally, various properties of the Tutte polynomial of graphs or matroids can be explored for matroid perspectives. The reader may think of such questions by reading the present handbook. In particular, for graphs, matroid perspectives are related to graph homomorphisms (see Example 28.3), and this particular case may have interesting properties in terms of graph theory. Also, a natural question is to find combinatorial interpretations of special points of the Tutte polynomial of a matroid perspective in different settings.

Open Question 28.37. Characterize matroid perspectives with minimal $T(M, M'; 0, 0, 1)$.

Let $M \to M'$ be a matroid perspective on E. A *factor* of $M \to M'$ is a subset $A \subseteq E$ such that $M = M|A \oplus M\backslash A$ and $M' = M'|A \oplus M'\backslash A$. The matroid perspective is *connected* if its only non-empty factor is E. The following result reformulates a result given without proof in [746]. Note that $M|A = M'|A$ can be replaced with $r(M|A) = r(M'|A)$ in the statement below.

Proposition 28.38. *Let $M \to M'$ be a matroid perspective on E. Then, $T(M, M'; 0, 0, 1) = 0$ if and only if $M \to M'$ has a non-empty factor $A \subseteq E$ such that $M|A = M'|A$. Thus, $T(M, M'; 0, 0, 1) = 0$ implies that $M = M'$ (when $A = E$) or $M \to M'$ is not connected (when $A \subset E$).*

Beyond this result, it is claimed without proof in [746, Theorem 8.4] that, for a connected matroid perspective,

$$T(M, M'; 0, 0, 1) \geq 2^{r(M)-r(M')+1} - 2,$$

and that this bound is optimal. The gist of a conjecture in [746, Section 8] is to characterize connected matroid perspectives $M \to M'$ for which equality holds. It is also claimed in [746, Section 8] that such a matroid perspective has a binary major, and that both M and M' are series-parallel.

Open Question 28.39. Determine divisibility properties at some special evaluations.

Three conjectures of increasing difficulty were proposed in [751]. The first has been proved in [189] and is stated here as Corollary 28.34. The second is still open, and stated below as Conjecture 28.40. In 2008, Gordon Royle communicated to Las Vergnas that he has a counter-example to the third conjecture for binary matroids and graphs, that computations make it seem plausible for planar graphs and maybe larger graph classes, and that it has been tested for planar graphs up to 11 vertices. The conjecture is thus restricted and stated below for planar graphs as Conjecture 28.40. These results and conjectures suggest investigating similar properties in matroid perspectives.

Conjecture 28.40. Let M be a binary matroid with $M = \mathrm{Supp}(\mathsf{V})$. Let $d = \dim(\mathsf{V} \cap \mathsf{V}^{\perp})$. For all integer $p \in \mathbb{Z}$, the number 2^d divides $T(M; 4p-1, 4p-1)$ and the quotient is an odd integer.

Conjecture 28.41. Let G be a planar graph, and $M = M(G)$. Let d be defined as above. For $k = 0, \ldots, d$, the k-th derivative of $T(M; x, x)$ evaluated at $x = -1$ is an integer multiple of 2^{d-k}.

Open Question 28.42. Study active orders for independent/spanning subsets.

In [753], three orders are defined for bases of a matroid on a linearly ordered set (see also Section 5.9). They can be defined in terms of successive exchanges of elements between bases, or, equivalently, by comparisons of the extreme sets in the associated intervals (see Proposition 28.21). This viewpoint provides properties of subset activities in terms of closure operators. It was shown in [753] that these orders are lattices, and the author informally observed that the Möbius function of these lattices often turned out to be equal to zero. This property has been explained in [143] by a topological approach of the complex defined by these orders. Further properties of a topological flavor have been recently given in [39]. As mentioned in [753], these orders can be generalized to a matroid perspective $M \to M'$, as orders for independent/spanning subsets. A natural question is to study the topological/structural/enumerative properties of these active orders in the setting of matroid perspectives.

29

Tutte polynomials of hyperplane arrangements and the finite field method

Federico Ardila

Synopsis

This chapter discusses the Tutte polynomial of a hyperplane arrangement. This polynomial captures important enumerative, algebraic, and topological information about the arrangement. The chapter also describes the finite field method, a useful tool to compute Tutte polynomials of many graphs, matroids, and arrangements.

- Hyperplane arrangements and their complements.
- The characteristic and Tutte polynomials, and Tutte–Grothendieck invariants.
- Complements in $\mathbb{R}, \mathbb{C}, \mathbb{F}_q$: regions, Poincaré polynomial, number of points.
- Topological and algebraic invariants of arrangements.
- The finite field method.
- Multivariate and arithmetic Tutte polynomials.
- Zonotopes and toric arrangements.

The author was partially supported by NSF grants DMS-1600609 and DMS-1855610 and Simons Fellowship 613384.

29.1 Introduction

This chapter focuses on some of the most important results on Tutte polynomials of hyperplane arrangements. We show that many enumerative, algebraic,

DOI: 10.1201/9780429161612-29

geometric, and topological invariants of a hyperplane arrangement can be expressed in terms of its Tutte polynomial. We also show that, even if one is only interested in computing the Tutte polynomial of a graph or a matroid, the theory of hyperplane arrangements provides a powerful finite field method for this computation.

The presentation here is influenced by a 2002 graduate course on Hyperplane Arrangements by Richard Stanley at MIT, much of which became the survey [1039]. See [891] for a thorough introduction to more algebraic and topological aspects of the theory of hyperplane arrangements.

29.2 Hyperplane arrangements

Let $\mathbb{F}$ be a field and $\mathsf{V} = \mathbb{F}^d$ be a vector space over $\mathbb{F}$. Let V^* be the dual vector space, which consists of the linear maps or *functionals* from V to $\mathbb{F}$.

Definition 29.1. A *hyperplane arrangement* $\mathcal{A}$ is a collection of affine hyperplanes in V. For each hyperplane $H \in \mathcal{A}$, let $l_H \in \mathsf{V}^*$ be a nonzero linear functional and $b_H \in \mathbb{F}$ be a scalar such that

$$H = \{x \in \mathsf{V} \, : \, l_H(x) = b_H\}.$$

We say $\mathcal{A}$ is *central* if all hyperplanes have a common point; in the most natural examples, the origin is a common point. Furthermore, we say $\mathcal{A}$ is *essential* if the intersection of all hyperplanes is a point. Figure 29.1a shows an essential arrangement of four hyperplanes in $\mathbb{R}^3$.

A key object of study in the theory of hyperplane arrangements is the *complement* of an arrangement, $V(\mathcal{A}) := \mathsf{V} \setminus \left(\bigcup_{H \in \mathcal{A}} H\right)$.

29.2.1 Intersection posets and matroids

We define a *flat* of $\mathcal{A}$ to be an affine subspace obtained as an intersection of hyperplanes in $\mathcal{A}$. We often identify a flat F with the set of hyperplanes $\{H_1, \ldots, H_k\}$ of $\mathcal{A}$ containing it since clearly we have $F = H_1 \cap \cdots \cap H_k$.

Definition 29.2. The *intersection poset* $L_{\mathcal{A}}$ is the set of flats partially ordered by reverse inclusion of the flats (or inclusion of the sets of hyperplanes containing them). This is a ranked poset, with $r(F) = \dim(\mathsf{V}) - \dim(F)$.

If $\mathcal{A}$ is central, then $L_{\mathcal{A}}$ is a *geometric lattice* [350, 1038]. If $\mathcal{A}$ is not central, then $L_{\mathcal{A}}$ is only a *geometric meet semilattice* [1124]. The *rank* $r = r(\mathcal{A})$ of $\mathcal{A}$ is the height of $L_{\mathcal{A}}$. Figure 29.1 shows an arrangement and its intersection poset.

Every central hyperplane arrangement has an associated matroid as follows.

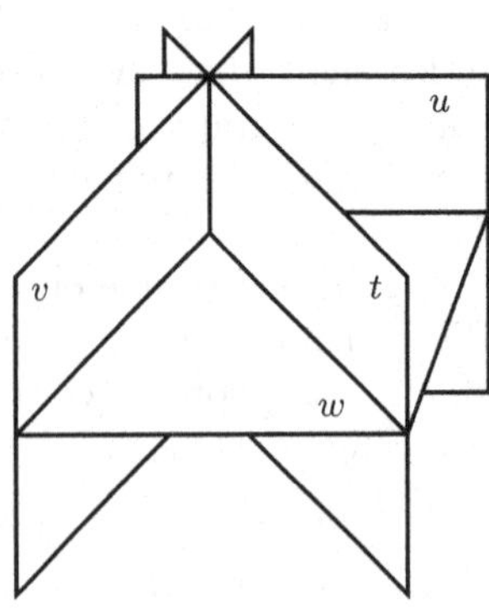

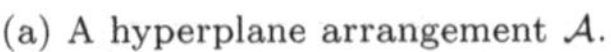

(a) A hyperplane arrangement $\mathcal{A}$.

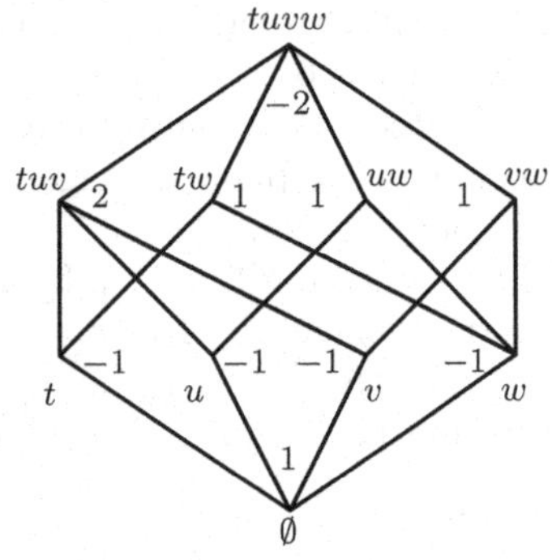

(b) Its intersection poset $L(\mathcal{A})$ and Möbius function.

FIGURE 29.1: A hyperplane arrangement $\mathcal{A}$ and its intersection poset $L(\mathcal{A})$ and Möbius function. Each flat of $\mathcal{A}$ is labeled by the list of hyperplanes containing it. The integers are explained in Example 29.9.

Definition 29.3. Let $\mathcal{A}$ be a hyperplane arrangement in a vector space V. If $\mathcal{A}$ is central, the *matroid* $M(\mathcal{A})$ of $\mathcal{A}$ is the matroid on the ground set $\mathcal{A}$ given by the rank function

$$r(\mathcal{B}) = \dim(V) - \dim\left(\bigcap \mathcal{B}\right) \qquad \text{for } \mathcal{B} \subseteq \mathcal{A}.$$

In general, the *semimatroid* of $\mathcal{A}$ is the collection of central subsets together with their ranks.

Semimatroids are equivalent to the *pointed matroids* of [250] (see [33]).

29.2.2 Deletion, contraction, centralization, essentialization

A common technique for inductive arguments is to choose a hyperplane H in an arrangement $\mathcal{A}$ and study how $\mathcal{A}$ behaves without H (in the deletion $\mathcal{A}\backslash H$) and how H interacts with the rest of $\mathcal{A}$ (in the contraction $\mathcal{A}/H$).

Definition 29.4. Let $\mathcal{A}$ be an arrangement in V and let H be a hyperplane in $\mathcal{A}$.

1. The *deletion*

$$\mathcal{A}\backslash H = \{A \in \mathcal{A} : A \neq H\}$$

is the arrangement in the same ambient space V consisting of the hyperplanes other than H.

2. The *contraction*

$$\mathcal{A}/H = \{A \cap H : A \in \mathcal{A}, A \neq H\}$$

is the arrangement in the new ambient space H consisting of the intersections of the other hyperplanes with H.

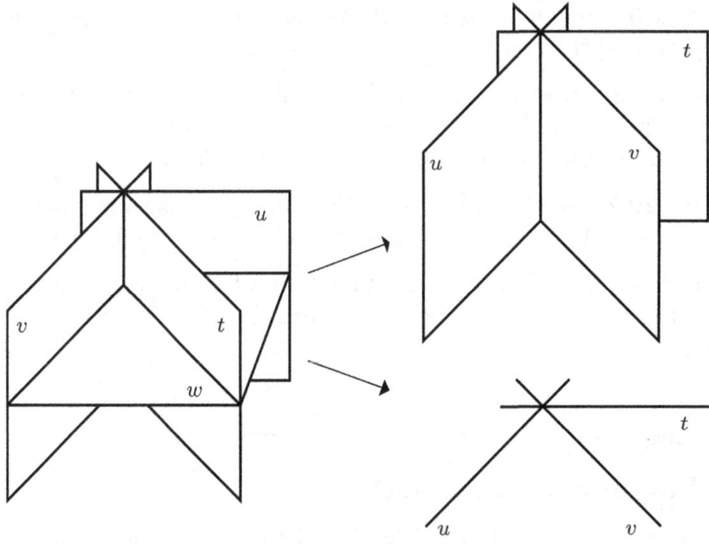

FIGURE 29.2: An arrangement $\mathcal{A}$ and its deletion $\mathcal{A}\backslash w$ (top) and contraction $\mathcal{A}/w$ (bottom).

Figure 29.2 shows an arrangement $\mathcal{A} = \{t, u, v, w\}$ in $\mathbb{R}^3$ together with the deletion $\mathcal{A}\backslash w$ and contraction $\mathcal{A}/w$.

Remark 29.5. It is somewhat inconvenient that hyperplane arrangements are not closed under contraction. For example, in Figure 29.2, the image of t in $(\mathcal{A}/u)/v$ is not a hyperplane, but the whole ambient space. We will circumvent this difficulty by considering arrangements where the full-dimensional ambient space is allowed as a degenerate "hyperplane". However, when we make statements about the complement $V(\mathcal{A})$, we will assume that $\mathcal{A}$ does not contain the degenerate hyperplane.

A more robust solution is to work more generally in the context of matroids [350, 894] for central arrangements, and pointed matroids [250] or semimatroids [33] for affine arrangements. However, to keep the presentation short and self-contained, we will not pursue this point of view.

We say a hyperplane H of an arrangement $\mathcal{A}$ in a vector space V is a *loop* if it is the degenerate hyperplane $H = \mathsf{V}$. We say it is a *coloop* if it intersects the rest of the arrangement transversally; that is, if $r(\mathcal{A}) = r(\mathcal{A}\backslash H) + 1$. For example, w is a coloop in the arrangement of Figure 29.2.

In some ways central arrangements are slightly better behaved than affine arrangements. We can *centralize* an affine arrangement $\mathcal{A}$ as follows.

Definition 29.6. The *centralization* or *cone* of a hyperplane arrangement $\mathcal{A}$ in $\mathbb{F}^d$ is the arrangement $c\mathcal{A}$ in $\mathbb{F}^{d+1}$ obtained by converting each hyperplane

$a_1x_1 + \cdots + a_dx_d = a$ in $\mathbb{F}^d$ into the hyperplane $a_1x_1 + \cdots + a_dx_d = ax_{d+1}$ in $\mathbb{F}^{d+1}$, and adding the new hyperplane $x_{d+1} = 0$.

Sometimes arrangements are "too central", in the sense that their intersection is a subspace $L \subset \mathsf{V}$ of positive dimension. In that case, there is little harm in modding out the arrangement by L, as follows.

Definition 29.7. The *essentialization* of a central arrangement $\mathcal{A}$ in V is the arrangement $\text{ess}(\mathcal{A}) = \{H/L : H \in \mathcal{A}\}$ in the quotient vector space V/L.

The resulting arrangement is essential. In most situations of interest, there is no important difference between $\mathcal{A}$ and $\text{ess}(\mathcal{A})$.

29.3 Polynomial invariants

Different choices of the ground field $\mathbb{F}$ lead to different questions about arrangements $\mathcal{A}$ and their complements $V(\mathcal{A})$. In many of these questions, a crucial role is played by two combinatorial polynomials which we now define.

29.3.1 The characteristic and Tutte polynomials

Definition 29.8. The *Möbius function* $\mu : L(\mathcal{A}) \to \mathbb{Z}$ of (the intersection poset of) an arrangement $\mathcal{A}$ is defined recursively by decreeing that for every flat $G \in L(\mathcal{A})$,

$$\sum_{F \leq G} \mu(F) = \begin{cases} 1 & \text{if } G \text{ is the minimum element of } L(\mathcal{A}), \\ 0 & \text{otherwise.} \end{cases} \tag{29.1}$$

The *characteristic polynomial* of $\mathcal{A}$ is

$$\chi(\mathcal{A}; q) = \sum_{F \in L_{\mathcal{A}}} \mu(F) q^{\dim(F)}.$$

Example 29.9. For the arrangement of Figure 29.1, the Möbius function is shown in face labels next to the intersection poset. The coefficients of the characteristic polynomial $\chi(\mathcal{A}; q) = q^3 - 4q^2 + 5q - 2$ are easily computed by adding the Möbius numbers on each level of $L_{\mathcal{A}}$.

Definition 29.10. The *Tutte polynomial* of an arrangement $\mathcal{A}$ in a vector space V is

$$T(\mathcal{A}; x, y) = \sum_{\substack{\mathcal{B} \subseteq \mathcal{A} \\ \mathcal{B} \text{ central}}} (x-1)^{r - r(\mathcal{B})} (y-1)^{|\mathcal{B}| - r(\mathcal{B})}, \tag{29.2}$$

where the sum is taken over all the central subarrangements $\mathcal{B}$ of $\mathcal{A}$, and where $r(\mathcal{B}) = \dim(\mathsf{V}) - \dim\left(\bigcap \mathcal{B}\right)$ and $r = r(\mathcal{A})$.

The Tutte polynomial was defined for graphs, matroids, and arrangements in [1092], [349], and [32] respectively. When $\mathcal{A}$ is central, the above definition coincides with the usual matroid-theoretic definition.

Example 29.11. For the arrangement of Figure 29.1, the Tutte polynomial of Definition (29.10) yields

$$T(\mathcal{A};x,y) = (x-1)^3+4(x-1)^2+6(x-1)+3+(x-1)(y-1)+(y-1) = x^3+x^2+xy.$$

The large amount of cancellation in the computation above is systematically explained by the following theorem. Let us fix a linear order on $\mathcal{A}$, and let $\mathcal{C}_{>H} = \{C \in \mathcal{C} : C > H\}$ for any subarrangement $\mathcal{C} \subseteq \mathcal{A}$ and hyperplane $H \in \mathcal{A}$. We define a *basis* of $\mathcal{A}$ to be a central subset of maximal rank r.

Let $\mathcal{B}$ be a basis. We say a hyperplane $H \notin \mathcal{B}$ is *externally active* with respect to $\mathcal{B}$ if $\mathcal{B} \cup H$ is central and $r(\mathcal{B}_{>H} \cup H) = r(\mathcal{B}_{>H})$, and we say $H \in \mathcal{B}$ is *internally active* if $r((\mathcal{B}-H) \cup \mathcal{A}_{<H}) = r(\mathcal{B}-H) = r-1$. Let $\mathrm{EA}(\mathcal{B})$ and $\mathrm{IA}(\mathcal{B})$ be the sets of externally and internally active elements with respect to $\mathcal{B}$, respectively. The following was shown in [33].

Theorem 29.12. *For any linear order on the hyperplanes of an arrangement $\mathcal{A}$, the Tutte polynomial of $\mathcal{A}$ is given by* $T(\mathcal{A};x,y) = \sum_{\mathcal{B} \text{ basis}} x^{|\mathrm{IA}(\mathcal{B})|}y^{|\mathrm{EA}(\mathcal{B})|}$.

In Example 29.11, the three monomials of $T(\mathcal{A};x,y)$ correspond to the bases tuw, tvw, uvw of $\mathcal{A}$. We invite the reader to choose a linear order for $\mathcal{A} = \{t,u,v,w\}$ and verify that these bases give the monomials x^3, x^2, and xy.

Although it is not obvious from its definition, the characteristic polynomial is a specialization of the Tutte polynomial, [32, 1159]:

Theorem 29.13 (Whitney's Theorem). *The characteristic polynomial and the Tutte polynomial of an arrangement of rank r in $\mathbb{F}^d$ are related by*

$$\chi(\mathcal{A};q) = (-1)^{r(\mathcal{A})}q^{d-r(\mathcal{A})}T(\mathcal{A};1-q,0).$$

This is part of a general phenomenon that we explore in the next section.

29.3.2 Tutte–Grothendieck invariants, recursion, and universality

The Tutte polynomial appears naturally in numerous different contexts, and provides the answer to many enumerative, algebraic, topological, and geometric questions. This is certainly true in the context of hyperplane arrangements; when we encounter a new quantity or polynomial associated to an arrangement, a good first question to ask is whether it is an evaluation of the Tutte polynomial.

The ubiquity of the Tutte polynomial is not accidental: this polynomial is universal among a large, important family of invariants of hyperplane arrangements, as we now make precise. Let $\mathfrak{R}$ be a ring, and let HypArr be the collection of all hyperplane arrangements over a field $\mathbb{F}$. As explained in Remark 29.5, we need to allow our arrangements to contain the ambient space as a degenerate hyperplane.

Definition 29.14. A function f : HypArr $\to \mathfrak{R}$ is a *generalized Tutte–Grothendieck invariant* if $f(\mathcal{A}_1) = f(\mathcal{A}_2)$ for any arrangements $\mathcal{A}_1$ and $\mathcal{A}_2$ with isomorphic semimatroids, and if for every arrangement $\mathcal{A}$ and every hyperplane $H \in \mathcal{A}$, we have

$$f(\mathcal{A}) = \begin{cases} af(\mathcal{A}\backslash H) + bf(\mathcal{A}/H) & \text{if } H \text{ is neither a loop nor a coloop,} \\ f(\mathcal{A}\backslash H)f(L) & \text{if } H \text{ is a loop,} \\ f(\mathcal{A}/H)f(C) & \text{if } H \text{ is a coloop,} \end{cases} \tag{29.3}$$

for some non-zero constants $a, b \in \mathfrak{R}$. Here $f(L)$ and $f(C)$ denote the (necessarily well-defined) function of a single loop L and a single coloop C, respectively. We say $f(\mathcal{A})$ is a *Tutte–Grothendieck invariant* when $a = b = 1$.

As in the case of graphs and matroids [349, 1092], the Tutte polynomial satisfies a universality property for hyperplane arrangements and for semimatroids [32, 33].

Theorem 29.15. *The Tutte polynomial is a universal Tutte–Grothendieck invariant for* HypArr*, namely,*

1. *The Tutte polynomial $T(\mathcal{A}; x, y)$ satisfies Equation (29.3) with $a = b = 1$, $f(C) = x$, and $f(L) = y$.*

2. *Any generalized Tutte–Grothendieck invariant is a function of the Tutte polynomial. Explicitly, if f satisfies Equation (29.3), then*

$$f(\mathcal{A}) = a^{n-r}\, b^{r(\mathcal{A})}\, T\left(\mathcal{A}; \frac{f(C)}{b}, \frac{f(L)}{a}\right).$$

where n is the number of elements and r is the rank of $\mathcal{A}$.

Part 1 of this theorem implies that the Tutte polynomial can also be defined alternatively by the recursion (29.3) with $a = b = 1$, $f(C) = x$, and $f(L) = y$. In Part 2 we do not need to assume that a and b are invertible; when we multiply by $a^{n-r}\, b^r$, all denominators cancel.

29.4 Topological and algebraic invariants

As is the case with graphs and matroids, many important invariants of a hyperplane arrangement are generalized Tutte–Grothendieck invariants, and hence are evaluations of the Tutte polynomial. In this section we collect a few selected results of this flavor.

29.4.1 Topological invariants of arrangements

Theorem 29.16. *The characteristic polynomial $\chi(\mathcal{A};x)$ contains the following information about the complement $V(\mathcal{A})$ of a hyperplane arrangement $\mathcal{A}$.*

1. *Let $\mathcal{A}$ be a hyperplane arrangement in $\mathbb{R}^d$. Let $\mathcal{A}$ separate its complement $V(\mathcal{A})$ into $a(\mathcal{A})$ connected components or* regions. *Let $b(\mathcal{A})$ be the number of bounded regions of the essentialization* ess$(\mathcal{A})$. *Then*

$$a(\mathcal{A}) = (-1)^d\chi(\mathcal{A};-1), \text{ and } \qquad b(\mathcal{A}) = (-1)^{r(\mathcal{A})}\chi(\mathcal{A};1).$$

2. *Let $\mathcal{A}$ be a hyperplane arrangement in $\mathbb{C}^d$. The integral cohomology ring of the complement $V(\mathcal{A})$ has Poincaré polynomial*

$$\sum_{k\geq 0} \operatorname{rank} H^k(V(\mathcal{A}),\mathbb{Z})q^k = (-q)^d\chi\left(\mathcal{A};\frac{-1}{q}\right).$$

3. *Let $\mathcal{A}$ be a hyperplane arrangement in $\mathbb{F}_q^d$ where $\mathbb{F}_q$ is the finite field of q elements. The complement $V(\mathcal{A})$ has size*

$$|V(\mathcal{A})| = \chi(\mathcal{A};q).$$

The first item of Theorem 29.16 is from [1181], the second from [576, 890], and the third from [54, 350]. The next theorem is from [247].

Theorem 29.17. *Let $\mathcal{A}$ be a central arrangement in $\mathbb{R}^d$.*

1. *Consider an affine hyperplane H which is in general position with respect to $\mathcal{A}$. Then the number of regions of $\mathcal{A}$ which have a bounded (and nonempty) intersection with H equals $T(\mathcal{A};1,0)$, the absolute value of the last coefficient of $\chi(\mathcal{A};q)$. In particular, this number is independent of H.*

2. *Add to $\mathcal{A}$ an affine hyperplane H' which is a parallel translation of one of the hyerplanes $H \in \mathcal{A}$. The number of bounded regions of $\mathcal{A}\cup H'$ is the* β-invariant *of $\mathcal{A}$, which is the coefficient of x^1y^0 and of x^0y^1 in $T(\mathcal{A};x,y)$. In particular, this number is independent of H.*

One very important algebraic topological invariant of a complex arrangement $\mathcal{A}$ is the cohomology ring $H^*(V(\mathcal{A}),\mathbb{Z})$ of its complement, known as the *Orlik–Solomon algebra* of $\mathcal{A}$ (see Chapter 30). It has the following combinatorial presentation from [890].

Theorem 29.18. *Let $\mathcal{A}$ be a central arrangement in $\mathbb{C}^d$. Let E be the exterior algebra with generators e_H for each $H \in \mathcal{A}$. For each ordered set of hyperplanes $S = \{H_1,\dots,H_k\}$ let $e_S = e_{H_1}\wedge\cdots\wedge e_{H_k}$ and let $\partial e_S = \sum_{j=1}^k(-1)^{j-1}e_{S-H_j}$. Say that S is* dependent *if $\dim(\cap S) > d-|S|$, or equivalently, if $l_{H_1},\dots,l_{H_k}$ are linearly dependent. Then*

$$H^*(V(\mathcal{A}),\mathbb{Z}) \cong E/\left\langle \partial e_S \,:\, S\subseteq\mathcal{A} \text{ is dependent}\right\rangle.$$

Another important invariant of a complex arrangement $\mathcal{A}$ is the cohomology ring $H^*(W(\mathcal{A}), \mathbb{Z})$ of the *wonderful compactification* $W(\mathcal{A})$ of the complement $V(\mathcal{A})$ constructed by De Concini and Procesi [363]. It also has an elegant combinatorial presentation, from [485]:

Theorem 29.19. *Let $\mathcal{A}$ be a central arrangement in $\mathbb{C}^d$ and $W(\mathcal{A})$ be the maximal wonderful compactification of its complement. Then*

$$H^*(W(\mathcal{A}), \mathbb{Z}) \cong S_{\mathcal{A}}/(I_{\mathcal{A}} + J_{\mathcal{A}}) \tag{29.4}$$

where

$$S_{\mathcal{A}} = \mathbb{Z}[x_F : \emptyset \subsetneq F \subsetneq \mathcal{A} \text{ is a flat of } \mathcal{A}],$$
$$I_{\mathcal{A}} = \langle x_{F_1}x_{F_2} : F_1, F_2 \text{ are incomparable proper flats}\rangle, \text{ and}$$
$$J_{\mathcal{A}} = \left\langle \sum_{F \ni i} x_F - \sum_{F \ni j} x_F : i \neq j \text{ in } \mathcal{A} \right\rangle.$$

One may use (29.4) as the definition of the *Chow ring* of any matroid. This ring is one of the crucial ingredients in the solution to the following central problem in matroid theory, due to Rota, Heron, and Welsh in the early 1970s [612, 971, 1137].

Theorem 29.20. *The characteristic polynomial of a matroid*

$$\chi_M(q) = q^n - a_{n-1}q^{n-1} + a_{n-2}q^{n-2} - \cdots + (-1)^n a_n q^0$$

has coefficients with alternating signs, so $a_i \geq 0$ for all i. Furthermore, the a_i's are unimodal and log-concave, that is,

$$a_1 \leq a_2 \leq \cdots \leq a_{i-1} \leq a_i \geq a_{i+1} \geq \cdots \geq a_n \qquad \text{for some } i, \text{ and}$$
$$a_{j-1}a_{j+1} \leq a_j^2 \qquad \text{for all } j.$$

Theorem 29.20 was proved by Huh [629] for hyperplane arrangement in fields of characteristic 0, by Huh and Katz [631] for arbitrary hyperplane arrangements, and by Adiprasito, Huh, and Katz for arbitrary matroids [3].

29.4.2 Algebras from arrangements

There are other natural algebras related to the Tutte polynomial of an arrangement arising in commutative algebra, hyperplane arrangements, box splines, and index theory; we discuss a few. Other important algebras associated to a general matroid are discussed in Chapter 30.

Throughout this section we assume our arrangement $\mathcal{A}$ is central. For each hyperplane H in a hyperplane arrangement $\mathcal{A}$ in $\mathbb{F}^d$ let l_H be a linear functional in $(\mathbb{F}^d)^*$ such that H is given by the equation $l_H(x) = 0$.

Our first example is a family of graded vector spaces $C_{\mathcal{A},k}$ associated to an arrangement $\mathcal{A}$. For $k = 0, -1, -2$, they arose in the theory of splines [357, 366, 623] as the spaces of solutions to certain systems of differential equations.

Theorem 29.21. *Let $\mathcal{A}$ be an arrangement of rank r in dimension d over a field of characteristic zero.*

1. *Let $C_{\mathcal{A},0} = \text{span}\{\prod_{H\in\mathcal{B}} l_H : \mathcal{B} \subseteq \mathcal{A}\}$. This is a subspace of a polynomial ring in d variables, graded by degree. Its dimension is $T(\mathcal{A}; 2, 1)$ and its Hilbert polynomial is*

$$\text{Hilb}(C_{\mathcal{A},0}; q) := \sum_{j\geq 0} \dim(C_{\mathcal{A},0})_j \, q^j = q^{d-r} T\left(\mathcal{A}; 1+q, \frac{1}{q}\right).$$

2. *More generally, let $C_{\mathcal{A},k}$ be the vector space of polynomial functions such that the restriction of f to any line h has degree at most $\rho_{\mathcal{A}}(h) + k$, where $\rho_{\mathcal{A}}(h)$ is the number of hyperplanes of $\mathcal{A}$ not containing h. (It is not obvious, but true, that this definition of $C_{\mathcal{A},0}$ matches the one above.) We have*

$$\text{Hilb}(C_{\mathcal{A},-1}; q) = q^{d-r} T\left(\mathcal{A}; 1, \frac{1}{q}\right), \qquad \text{Hilb}(C_{\mathcal{A},-2}; q) = q^{d-r} T\left(\mathcal{A}; 0, \frac{1}{q}\right)$$

and similar formulas hold for any $k \geq -2$.

The first item of Theorem 29.21 is from [1125]. The different parts of the second item, in various levels of generality, are shown in [31, 42, 357, 366, 623, 924, 925]. Another example, from [208, 931, 1060], that arises in several contexts is the following.

Theorem 29.22. *Let $\mathcal{A}$ be an arrangement in dimension d over a field of characteristic zero. Let $R(\mathcal{A})$ be the vector space of rational functions whose poles are in $\mathcal{A}$. It is the $\mathbb{F}$-algebra of rational functions generated by $\{1/l_H : H \in \mathcal{A}\}$, and we grade it so that $\deg(1/l_H) = 1$. Then*

$$\text{Hilb}(R(\mathcal{A}); q) = \frac{q^d}{(1-q)^d} T\left(\mathcal{A}; \frac{1}{q}, 0\right).$$

29.5 The finite field method

Even in the special case of fields of characteristic zero, it is also quite useful to consider hyperplane arrangements over the finite field $\mathbb{F}_q$ of q elements, where q is a prime power. The following variant of the Tutte polynomial plays an important role.

Definition 29.23. The *coboundary polynomial* $\overline{\chi}(\mathcal{A}; X, Y)$ is the following simple transformation of the Tutte polynomial:

$$\overline{\chi}(\mathcal{A}; X, Y) = (Y-1)^r T\left(\mathcal{A}; \frac{X+Y-1}{Y-1}, Y\right). \tag{29.5}$$

A simple change of variables allows us to recover $T(\mathcal{A}; x, y)$ from $\overline{\chi}(\mathcal{A}; X, Y)$.

Let $\mathcal{A}$ be an arrangement over a field of characteristic zero. We say $\mathcal{A}$ is a $\mathbb{Q}$-*arrangement* if its defining equations have rational coefficients. For any power q of a large enough prime, the equations of $\mathcal{A}$ also define a hyperplane arrangement $\mathcal{A}_q$ over $\mathbb{F}_q$. We say that $\mathcal{A}$ *reduces correctly* over $\mathbb{F}_q$ if the intersection posets of $\mathcal{A}$ and $\mathcal{A}_q$ are isomorphic, and hence $\mathcal{A}$ and $\mathcal{A}_q$ have the same Tutte polynomial.

Theorem 29.24 (Finite Field Method).

1. *Let $\mathcal{A}$ be a hyperplane arrangement of rank r in $\mathbb{F}_q^d$. For each point $p \in \mathbb{F}_q^d$ let $h(p)$ be the number of hyperplanes of $\mathcal{A}$ containing p. Then*

$$\sum_{p \in \mathbb{F}_q^d} t^{h(p)} = q^{d-r}\overline{\chi}(\mathcal{A}; q, t). \tag{29.6}$$

2. *Let $\mathcal{A}$ be a $\mathbb{Q}$-arrangement over a field of characteristic zero. For any power q of a large enough prime, $\mathcal{A}$ reduces correctly over $\mathbb{F}_q$, and the coboundary polynomial $\overline{\chi}(\mathcal{A}; q, t) = \overline{\chi}(\mathcal{A}_q; q, t)$ of $\mathcal{A}$ may be computed using Equation (29.6).*

Theorem 29.24, which is from [32, 350, 579, 1146], is one of the most effective methods for computing Tutte polynomials of a hyperplane arrangement $\mathcal{A}$. It reduces the computation of $T(\mathcal{A}; x, y)$ to an enumerative problem over finite fields, which can sometimes be solved [32, 54]. This method also works for any graph or any matroid realizable over $\mathbb{Q}$, since they can be regarded as hyperplane arrangements as well. Let us illustrate this with two simple examples.

Example 29.25. If we think of the arrangement of Figure 29.1 as a subset of the ambient space $\mathbb{F}_q^3$, a careful enumeration gives

$$\overline{\chi}(\mathcal{A}; q, t) = t^4 + (q-1)t^3 + 3(q-1)t^2 + (4q^2 - 9q + 5)t + (q^3 - 4q^2 + 5q - 2),$$

in agreement with Example 29.11.

Example 29.26. We can show that the *coordinate arrangement* $\mathcal{H}_n$, consisting of the n coordinate hyperplanes in $\mathbb{C}^n$, has coboundary polynomial

$$\overline{\chi}(\mathcal{H}_n; X, Y) = (X + Y - 1)^n.$$

This follows by regarding $\mathcal{H}_n$ as an arrangement over $\mathbb{F}_q$ for a power q of a large prime. By (29.6), we need to count the points in $\mathbb{F}_q^n$ that are on exactly k of the n hyperplanes $x_i = 0$ for $1 \leq i \leq n$. To choose such a point, we may first choose which k hyperplanes it is on, and then choose its remaining $n - k$ non-zero coordinates independently, for a total of $\binom{n}{k}(q-1)^{n-k}$ choices. It then follows that

$$\overline{\chi}(\mathcal{H}_n; q, t) = \sum_{p \in \mathbb{F}_q^n} t^{h(p)} = \sum_{k=0}^{n} \binom{n}{k} (q-1)^{n-k} t^k = (q + t - 1)^n.$$

29.6 A catalog of characteristic and Tutte polynomials

Computing Tutte polynomials is extremely difficult in general, as explained in Chapter 9. However, the computation is possible in some cases. We now survey some of the most interesting examples; see [843] for others. Some of these formulas are best expressed in terms of the coboundary polynomial $\overline{\chi}(\mathcal{A}; X, Y)$, which is equivalent to the Tutte polynomial $T(\mathcal{A}; x, y)$ by Equation (29.5). Almost all of them are most easily proved using the finite field method of Theorems 29.16 and 29.24.

1. If the characteristic of $\mathbb{F}$ is 0, a *sufficiently generic* central arrangement in $\mathbb{F}^d$ is one in which the intersection of any m hyperplanes has codimension m for $1 \leq m \leq d$. If $\mathcal{A}_{n,d}$ is such a sufficiently generic arrangement of n hyperplanes in $\mathbb{F}^d$, then

$$T(\mathcal{A}_{n,d}; x, y) = \sum_{i=1}^{d} \binom{n-i-1}{n-d-1} x^i + \sum_{j=1}^{n-d} \binom{n-j-1}{d-1} y^j.$$

2. A graph G on n vertices gives rise to the *graphical arrangement* $\mathcal{A}_G$ in $\mathbb{F}^n$ which has a hyperplane $x_i = x_j$ for every edge ij of G. If $\mathbb{F} = \mathbb{R}$, the regions of $\mathcal{A}_G$ are in bijection with the orientations of the edges of G that form no directed cycles.

   By the finite field method, the characteristic polynomial $\chi(\mathcal{A}_G; q)$ is equal to the *chromatic polynomial* $\chi(G; q)$, which counts the vertex colorings of G with q colors such that no edge joins two vertices of the same color. This gives a proof that $\chi(G; q)$ is indeed polynomial in q. Similarly,

$$q^{n-r}\overline{\chi}(\mathcal{A}_G; q, t) = \sum_{f:[n]\to[q]} t^{h(f)}$$

   where we sum over all vertex colorings f of G with q colors, and $h(f)$ is the number of edges of G whose ends have the same color in f.

   An important special case is the graphical arrangement for the complete graph K_n, consisting of the $\binom{n}{2}$ hyperplanes in $\mathbb{R}^n$ given by equations $x_i = x_j$ for $1 \leq i < j \leq n$. This is known as the *braid arrangement* or the *type A Coxeter arrangement*, discussed further in the next item.

3. Root systems are arguably the most important vector configurations; these highly symmetric arrangements are fundamental in many branches of mathematics. For the definition and properties, see for example [633]; we

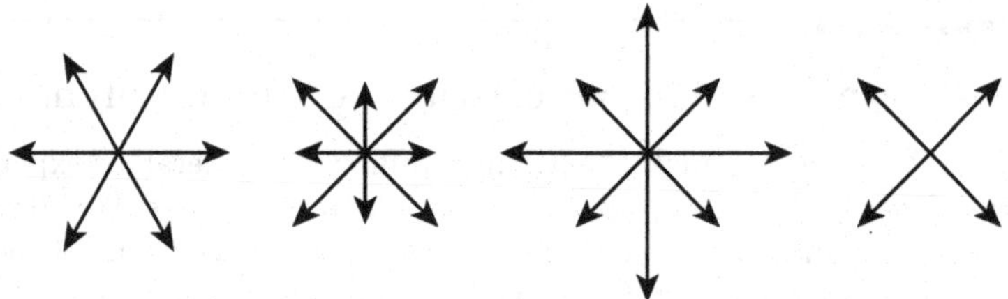

FIGURE 29.3: The root systems A_2, B_2, C_2, and D_2, respectively.

focus on the four infinite families of *classical root systems*:

$$\begin{aligned}
A_{n-1} &= \{e_i - e_j, : 1 \leq i < j \leq n\}, \\
B_n &= \{e_i - e_j, e_i + e_j : 1 \leq i < j \leq n\} \cup \{e_i : 1 \leq i \leq n\}, \\
C_n &= \{e_i - e_j, e_i + e_j : 1 \leq i < j \leq n\} \cup \{2e_i : 1 \leq i \leq n\}, \textit{and} \\
D_n &= \{e_i - e_j, e_i + e_j : 1 \leq i < j \leq n\},
\end{aligned}$$

regarded as linear functionals in $(\mathbb{F}^n)^*$, where $e_1, \ldots, e_n$ is the standard basis. Figure 29.3 illustrates the two-dimensional examples. Aside from the infinite families, there are five exceptional root systems: E_6, E_7, E_8, F_4, G_2. See [365, 366] for details.

The classical root systems lead to the *Coxeter arrangements* $\mathcal{A}_{n-1}, \mathcal{BC}_n$, and $\mathcal{D}_n$ of hyperplanes determined by the roots. For example, the Coxeter arrangement $\mathcal{A}_{n-1}$ is the braid arrangement in $\mathbb{R}^n$. Note that B_n and C_n lead to the same arrangement $\mathcal{BC}_n$.

The characteristic polynomials of Coxeter arrangements are very elegant:

$$\begin{aligned}
\overline{\chi}(A_{n-1}; q) &= q(q-1)(q-2)\cdots(q-n+1), \\
\overline{\chi}(BC_n; q) &= (q-1)(q-3)\cdots(q-2n+3)(q-2n+1), \\
\overline{\chi}(D_n; q) &= (q-1)(q-3)\cdots(q-2n+3)(q-n+1).
\end{aligned}$$

Similar expressions hold for the exceptional root systems. There are several conceptual explanations for the factorization of these polynomials into linear forms; see [34, Section 1.7.4] or [978] for references. In view of Theorem 29.16.1, when $\mathbb{F} = \mathbb{R}$ these formulas lead to

$$a(A_{n-1}) = n!, \qquad a(BC_n) = 2^n n!, \qquad a(D_n) = 2^{n-1} n!,$$

consistent with the general fact that the regions of a Coxeter arrangement are in bijection with the elements of the corresponding Coxeter group.

To compute the Tutte polynomials of the classical Coxeter arrangements, let the *deformed exponential function* be $F(\alpha, \beta) := \sum_{n \geq 0} \alpha^n \beta^{\binom{n}{2}}/n!$. The

Tutte generating functions of A and $\Phi \in \{B, C, D\}$, defined as

$$\begin{aligned} T_A(X,Y,Z) &:= 1 + X \sum_{n \geq 1} \overline{\chi}(A_{n-1}; X, Y) \frac{Z^n}{n!}, \\ T_\Phi(X,Y,Z) &:= \sum_{n \geq 0} \overline{\chi}(\Phi_n; X, Y) \frac{Z^n}{n!}, \end{aligned}$$

are given by

$$\begin{aligned} T_A(X,Y,Z) &= F(Z,Y)^X, \\ T_{BC}(X,Y,Z) &= F(2Z,Y)^{(X-1)/2} F(YZ, Y^2), \\ T_D(X,Y,Z) &= F(2Z,Y)^{(X-1)/2} F(Z, Y^2). \end{aligned}$$

Here we are following the convention that if C and $A = 1 + B$ are formal power series such that the constant coefficient of A is 1, then $A^C := e^{C \log(1+B)}$. This is well defined because the formal power series for e^D and $\log(1 + D)$ are well defined for any D with constant coefficient equal to 0. The Tutte polynomials of the exceptional root systems are computed in [365, 366].

4. The formula in the previous item for the root systems of type A gives the Tutte polynomials of the complete graphs; they are due to Tutte [1095]. The coboundary polynomials of the *complete bipartite graphs* $K_{m,n}$ are given by

$$1 + X \sum_{\substack{m,n \geq 0 \\ (m,n) \neq (0,0)}} \overline{\chi}(K_{m,n}; X, Y) \frac{Z_1^m}{m!} \frac{Z_2^n}{n!} = \left(\sum_{m,n \geq 0} Y^{mn} \frac{Z_1^m}{m!} \frac{Z_2^n}{n!} \right)^X.$$

5. Let p be a prime power and consider the arrangement $\mathcal{A}(p, n)$ of *all* linear hyperplanes in $\mathbb{F}_p^n$. The characteristic polynomial is

$$\chi(\mathcal{A}(p,n); q) = (q-1)(q-p)(q-p^2) \cdots (q - p^{n-1}).$$

The *p-exponential generating function* of the coboundary polynomials is

$$\sum_{n \geq 0} \overline{\chi}(\mathcal{A}(p,n); X, Y) \frac{u^n}{(p;p)_n} = \frac{(u;p)_\infty}{(Xu;p)_\infty} \sum_{n \geq 0} Y^{1+p+\cdots+p^{n-1}} \frac{u^n}{(p;p)_n},$$

where we define $(a;p)_n = (1-a)(1-pa) \cdots (1-p^{n-1}a)$ for $n \in \mathbb{N}$ and $(a;p)_\infty = (1-a)(1-pa)(1-p^2a) \cdots$.

6. The *threshold arrangement* $\mathcal{T}_n$ in $\mathbb{F}^n$ consists of the $\binom{n}{2}$ hyperplanes $x_i + x_j = 0$ for $1 \leq i < j \leq n$. We have

$$\sum_{n \geq 0} \overline{\chi}(\mathcal{T}_n; X, Y) \frac{Z^n}{n!} = \left(\sum_{r,s \geq 0} \frac{Y^{rs} Z^{r+s}}{r!s!} \right)^{(X-1)/2} \left(\sum_{n \geq 0} \frac{Y^{\binom{n}{2}} Z^n}{n!} \right).$$

FIGURE 29.4: The braid, Catalan, and Shi arrangements $\mathcal{A}_2, \mathrm{Cat}_2$, and Shi_2.

When $\mathbb{F} = \mathbb{R}$, the regions of $\mathcal{T}_n$ are in bijection with the *threshold graphs* on $[n]$. These are the graphs for which there exist vertex weights $w(i)$ for $1 \leq i \leq n$ and a *threshold* w such that edge ij is present in the graph if and only if $w(i) + w(j) > w$. They have many interesting properties and applications; see [808].

7. If $\mathcal{A}^{(k)}$ is the arrangement obtained from $\mathcal{A}$ by replacing each hyperplane by k copies of itself, then

$$T(\mathcal{A}^{(k)}; x, y) = (y^{k-1}+\cdots+y^2+y+1)^r \, T\left(\mathcal{A}; \frac{y^{k-1}+\cdots+y^2+y+x}{y^{k-1}+\cdots+y^2+y+1}, y^k\right).$$

For arrangements with integer coefficients, this formula follows readily from the finite field method: notice that a point p which is on m hyperplanes of $\mathcal{A}$ is on km hyperplanes of $\mathcal{A}^{(k)}$, and this implies that $\overline{\chi}(\mathcal{A}^{(k)}; X, Y) = \overline{\chi}(\mathcal{A}; X, Y^k)$. For a generalization, see Theorem 29.28.

8. There are many interesting *deformations of the braid arrangement*, obtained by considering hyperplanes of the form $x_i - x_j = a$ for various constants a. Two particularly elegant ones are the *Catalan* and *Shi* arrangements:

$$\begin{aligned} \mathrm{Cat}_{n-1} &: \quad x_i - x_j \in \{-1, 0, 1\} \qquad (1 \leq i < j \leq n), \textit{and} \\ \mathrm{Shi}_{n-1} &: \quad x_i - x_j \in \{0, 1\} \qquad (1 \leq i < j \leq n). \end{aligned}$$

The left panel of Figure 29.4 shows the arrangement $\mathcal{A}_2$ consisting of the planes $x_1 = x_2$, $x_2 = x_3$, and $x_1 = x_3$ in $\mathbb{R}^3$. Since all planes contain the line $x_1 = x_2 = x_3$, we quotient by it, obtaining a two-dimensional picture. The other panels show the Catalan and Shi arrangements.

When $\mathbb{F} = \mathbb{R}$, we have the simple formulas

$$\begin{aligned} a(\mathrm{Cat}_{n-1}) &= n!C_n, & a(\mathrm{Shi}_{n-1}) &= (n+1)^{n-1}, \\ b(\mathrm{Cat}_{n-1}) &= n!C_{n-1}, & b(\mathrm{Shi}_{n-1}) &= (n-1)^{n-1}, \end{aligned}$$

where $C_n = \frac{1}{n+1}\binom{2n}{n}$ is the n-th *Catalan number*, which famously has hundreds of different combinatorial interpretations [1041]. The number $(n+1)^{n-1}$ also has many combinatorial interpretations of interest; *parking functions* are particularly relevant [1038]. We have

$$\begin{aligned}\chi(\mathrm{Cat}_{n-1}; q) &= q(q-n-1)(q-n-2)\cdots(q-2n+1),\\ \chi(\mathrm{Shi}_{n-1}; q) &= q(q-n)^{n-1}.\end{aligned}$$

There are substantially more complicated formulas for the Tutte polynomials of the Catalan and Shi arrangements [32], but it is not known whether they can be used to compute these polynomials efficiently.

In the above list, Item 1 is from [247], Item 2 from [247], Item 3 from [32, 1095], Item 4 from [822], Item 5 from [75, 867], Item 6 from [32], Item 7 from [247], and Item 8 is from [32, 926].

29.7 Multivariate and arithmetic Tutte polynomials

We now discuss two useful variants of the Tutte polynomial.

29.7.1 The multivariate Tutte polynomial

The first variant, introduced in [32, 1023], is a refinement of the ordinary Tutte polynomial which is inspired by statistical mechanics. See Chapters 20, 24, and 26 .

Definition 29.27. The *multivariate Tutte polynomial* of a hyperplane arrangement $\mathcal{A}$ is

$$\widetilde{Z}(\mathcal{A}; q, \mathbf{w}) = \sum_{\substack{\mathcal{B}\subseteq\mathcal{A}\\ \mathcal{B}\text{ central}}} q^{-r(\mathcal{B})} \prod_{e\in\mathcal{B}} w_e,$$

where q and $(w_e)_{e\in\mathcal{B}}$ are indeterminates.

When $\mathcal{A} = \mathcal{A}_G$ is a graphical arrangement, $\widetilde{Z}(\mathcal{A}; q, \mathbf{w})$ is equal to the partition function of the q-state Potts model on G; see also Chapter 20 and [1023]. Note that if we set $w_e = w$ for all e in $\mathcal{A}$, then we have $\widetilde{Z}(\mathcal{A}; q, \mathbf{w}) = (w/q)^{r(\mathcal{A})} T(\mathcal{A}; \frac{q}{w} + 1, w+1)$, which is simply a transformation of the Tutte polynomial. The following theorem is from [42].

Theorem 29.28. *For a vector* $\mathbf{a} \in \mathbb{N}^n$, *let* $\mathcal{A}(\mathbf{a})$ *be the arrangement* $\mathcal{A}$ *where each hyperplane* e *is replaced by* a_e *copies of* e.

1. *The Tutte polynomial of* $\mathcal{A}(\mathbf{a})$ *is*

$$T(\mathcal{A}(\mathbf{a}); x, y) = (x-1)^{r(\mathrm{supp}(\mathbf{a}))} \widetilde{Z}\left(\mathcal{A}; (x-1)(y-1), y^{a_1-1}, \ldots, y^{a_n-1}\right).$$

2. *The generating function for the Tutte polynomials of* all *the arrangements* $\mathcal{A}(\mathbf{a})$ *is essentially equivalent to the multivariate Tutte polynomial:*

$$\sum_{\mathbf{a}\in\mathbb{N}^n} \frac{T(\mathcal{A}(\mathbf{a});x,y)}{(x-1)^{r(\mathrm{supp}(\mathbf{a}))}} w_1^{a_1}\cdots w_n^{a_n}$$
$$= \frac{1}{\prod_{i=1}^n (1-w_i)} \widetilde{Z}\left(\mathcal{A};(x-1)(y-1);\frac{(y-1)w_1}{1-yw_1},\ldots,\frac{(y-1)w_n}{1-yw_n}\right).$$

Here $\mathrm{supp}(\mathbf{a})$ *denotes the set of hyperplanes e for which* $a_e > 0$.

There is also an algebraic manifestation of the multivariate Tutte polynomial: the multigraded Hilbert series of the *zonotopal Cox ring* of $\mathcal{A}$ is an evaluation of the multivariate Tutte polynomial of $\mathcal{A}$ [42, 1048].

29.7.2 The arithmetic Tutte polynomial

The second variant takes arithmetic into account, and is defined for vector arrangements, or for arrangements of subtori of codimension 1 inside a torus. See [853] for further details.

Definition 29.29. For a collection $A \subseteq \mathbb{Z}^d$ of integer vectors, the *arithmetic Tutte polynomial* is

$$T_{\mathrm{arith}}(A;x,y) = \sum_{B\subseteq A} m(B)(x-1)^{r(A)-r(B)}(y-1)^{|B|-r(B)},$$

where, for each $B \subseteq A$, the *multiplicity* $m(B)$ is the index of $\mathbb{Z}B$ as a sublattice of $(\mathrm{span}\, B) \cap \mathbb{Z}^d$. If we use the vectors in B as the columns of a matrix, then $m(B)$ equals the greatest common divisor of the minors of full rank. The *arithmetic characteristic polynomial* of A is $(-1)^{r(A)} q^{d-r(A)} T_{\mathrm{arith}}(A;1-q,0)$.

There is also a multivariate arithmetic Tutte polynomial, see [195].

The next theorem shows that $T_{\mathrm{arith}}(A;x,y)$ encodes information about the zonotope of A (see [366, 1032]).

Theorem 29.30. *Let* $A \subseteq \mathbb{Z}^d$ *be a set of integer vectors and let the* zonotope *of* A *be the Minkowski sum of the vectors in* A, *that is*

$$\mathrm{Zono}(A) := \left\{\sum_{a\in A} \lambda_a a \,:\, 0 \le \lambda_a \le 1 \textit{ for } a \in A\right\}.$$

1. *The volume of the zonotope* $\mathrm{Zono}(A)$ *is* $T_{\mathrm{arith}}(A;1,1)$.
2. *The zonotope* $\mathrm{Zono}(A)$ *contains* $T_{\mathrm{arith}}(A;2,1)$ *lattice points,* $T_{\mathrm{arith}}(A;0,1)$ *of which are in its interior.*

3. *The Ehrhart polynomial of the zonotope* Zono(A), *which counts the lattice points in the dilation* qZono(A) *for* $q \in \mathbb{N}$, *equals* $q^{r(A)}T_{\text{arith}}(A; 1+\frac{1}{q}, 1)$.

The arithmetic Tutte polynomial is also intimately related to the geometry of toric arrangements, as follows. Let the torus $T = \text{Hom}(\mathbb{Z}^d, G)$ be the group of homomorphisms from $\mathbb{Z}^d$ to a multiplicative group G, such as the unit circle $\mathbb{S}^1$ or $\mathbb{F}^* = \mathbb{F}\backslash\{0\}$ for a field $\mathbb{F}$. The collection A determines a *toric arrangement* in T, consisting of the codimension 1 subtori for each vector $a \in A$:

$$T_a := \{t \in T \,:\, t(a) = 1\} \subset T.$$

For instance $a = (2, -3, 5)$ gives the torus $x^2y^{-3}z^5 = 1$.

The following results are the toric analogs of Theorems 29.16 and 29.24 about hyperplane arrangements.

Theorem 29.31. *Let* $A \subset \mathbb{Z}^d$ *and* $T = \text{Hom}(\mathbb{Z}^d, G)$ *for a group* G. *Consider the* toric arrangement *of* A *and its* complement, *namely,*

$$\mathcal{T}(A) = \{T_a \,:\, a \in A\}, \quad \text{and} \quad R(A) = T \setminus \bigcup_{a \in \mathcal{T}(A)} T_a.$$

1. *If* $G = \mathbb{S}^1$, *the number of regions of* $R(A)$ *in the torus* $(\mathbb{S}^1)^d$ *is equal to* $T_{\text{arith}}(A; 1, 0)$.

2. *If* $G = \mathbb{C}^*$, *the Poincaré polynomial of* $R(A)$ *is equal to* $q^r T_{\text{arith}}(A; 2+\frac{1}{q}, 0)$.

3. *If* $G = \mathbb{F}_{q+1}^*$ *where* $q+1$ *is a prime power, then the number of elements of* $R(A)$ *is* $(-1)^r q^{d-r} T_{\text{arith}}(A; 1-q, 0)$, *the* arithmetic characteristic polynomial. *Furthermore,*

$$\sum_{p \in (\mathbb{F}_{q+1}^*)^d} t^{h(p)} = (t-1)^r q^{d-r} T_{\text{arith}}\left(A; \frac{q+t-1}{t-1}, t\right),$$

where $h(p)$ *is the number of hypertori of* $\mathcal{T}(A)$ *that* p *lies on.*

The first item of the theorem is from [443, 853], the second from [364, 366, 853], and the third from [38, 195].

As with ordinary Tutte polynomials, the third item of Theorem 29.31 may be used as a finite field method to compute arithmetic Tutte polynomials for some vector configurations and toric arrangements. At the moment there are very few results along these lines.

An important family that is well understood is the family of classical root systems, whose geometric properties motivate much of the theory of toric arrangements. Formulas for the arithmetic Tutte and characteristic polynomials of the classical root systems A_n, B_n, C_n, and D_n are given in [38]. Most of them resemble the formulas for the ordinary Tutte polynomials of the hyperplane arrangements $\mathcal{A}_n, \mathcal{BC}_n$, and $\mathcal{D}_n$ mentioned earlier. However, as should be expected, more subtle arithmetic issues arise—especially in type A. For related results on non-integral zonotopes related to the classical root systems, see [35, 43].

29.8 Open problems

There are many other conjectural inequalities for sequences related to matroids, and to their Tutte polynomial in particular. For a summary of some of them, see [1127]. It seems likely that the solution of such conjectures would be related to algebraic, geometric, and topological properties of arrangements and matroids, similar to those highlighted in Section 29.4.

Section 29.6 gives formulas for the Tutte polynomials of many families of arrangements. Similarly, it would be interesting to find formulas for the arithmetic Tutte polynomials of other families of integer vector configurations and toric arrangements.

30

Some algebraic structures related to the Tutte polynomial

Michael J. Falk • Joseph P.S. Kung

Synopsis

There are many algebraic structures connected with the Tutte polynomial or its specializations. In this chapter we discuss two such structures, focusing on the Orlik–Solomon algebra of a simple matroid.

- The Orlik–Solomon algebra.
- An algebraic interpretation of the characteristic polynomial.
- Hyperplane arrangements.
- The restriction-contraction coalgebra of a matroid.

30.1 Introduction

In addition to playing a central role in graph and matroid theory, Tutte polynomials and related polynomials have also played key roles in other areas. We begin with a description of the Orlik–Solomon algebra of a matroid M. This is a quotient of an exterior algebra by an ideal determined by the circuits of M. The dimension of the subspace of elements of a given grade is a coefficient of the characteristic polynomial. We end with a short account of coalgebras associated with graphs and matroids. Throughout this chapter, we will usually abbreviate the set $\{a, b, \dots, d\}$ by $ab \dots d$; for example, 123 is the set $\{1, 2, 3\}$.

DOI: 10.1201/9780429161612-30

30.2 Orlik–Solomon algebras

Often is it useful to realize a sequence of nonnegative integers, or coefficients of a polynomial or formal power series, as the sequence of dimensions of the graded pieces of a graded algebra. The Orlik–Solomon algebra accomplishes this for the coefficients of the characteristic polynomial of a loopless matroid. Indeed, its Hilbert series is, after a simple algebraic transformation, the characteristic polynomial. The Orlik–Solomon algebra first arose as a presentation of an algebra of differential forms associated with a complex hyperplane arrangement, which is in turn isomorphic to the cohomology ring of the complement of the union of the hyperplanes.

We shall assume a basic knowledge of the exterior algebra of a vector space. See, for example, the books [741, 803] for background on this topic.

30.2.1 Broken-circuits

Let M be a rank-r loopless matroid on the set E of size n with elements labeled by the integers $1, 2, \ldots, n$. A *broken-circuit* is a subset of E of the form $C \setminus m$ where C is a circuit of M and m is the minimum element in C. A subset $I \subseteq E$ is a *no-broken-circuit-* or *nbc-set* if I has no subset that is a broken circuit. An nbc-set cannot have a circuit as a subset and hence is independent. In addition, as a subset of an nbc-set is nbc, the collection of nbc-sets forms a simplicial complex, the *broken-circuit complex* of M (relative to the labeling).

The following simple observation, made by Whitney in [1154], underlies the combinatorics of nbc-sets.

Lemma 30.1. *Let Y be a nonempty flat of M and let a be the element of Y with the smallest label. Let $I \subset Y$ be a subset not containing a. Then I is an nbc-set if and only if $I \cup a$ is an nbc-set.*

No-broken-circuit sets give a counting interpretation for the Möbius function (See Section 4.7). The *characteristic polynomial* $\chi(M;\lambda)$ and *Whitney numbers of the first kind* $w_k(M)$ are defined by

$$\chi(M;\lambda) := \sum_{X \in L(M)} \mu(\emptyset, X)\lambda^{r-r(X)} = \sum_{k=0}^{r} (-1)^k w_k(M)\lambda^{r-k},$$

where the first sum ranges over all flats X in the lattice $L(M)$ of flats of M, and μ is the Möbius function. (See also Definitions 4.136 and 4.141.) The following result is due to Rota [970].

Theorem 30.2. *Let M be a rank-r loopless matroid and X be a flat of M. Then $(-1)^{r(X)}\mu(\emptyset, X)$ equals the number of nbc-sets with closure X. In particular, $w_k(M)$ equals the number of nbc-sets of size k.*

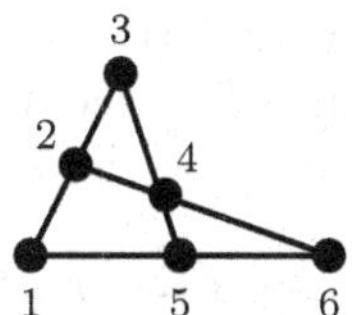

FIGURE 30.1: The matroid K.

Example 30.3. Let K be the rank-3 simple matroid on the set 123456 with 3-element circuits $123, 156, 246, 345$ shown in Figure 30.1. The matroid K is the cycle matroid of the complete graph K_4 on 4 vertices and the matroid of the Coxeter hyperplane arrangement of type A_3 (see Chapter 29). Its nbc-sets are the empty set $\emptyset$; all 1-element subsets, $1, 2, 3, 4, 5, 6$; the 2-element subsets

$$12, 13, 24, 25, 34, 35, 15, 16, 14, 25, 36;$$

and the 3-element subsets

$$124, 125, 126, 134, 135, 136.$$

As predicted by Theorem 30.2, $\chi(K; \lambda) = \lambda^3 - 6\lambda^2 + 11\lambda - 6$.

30.2.2 Exterior and graded algebras

Let $\mathbb{F}$ be a field of characteristic not equal to 2, let E be a finite set labeled by the integers $1, 2, \ldots, n$, and let $\Lambda(E)$ be the exterior algebra of the $|E|$-dimensional vector space $\mathbb{F}^E$ spanned by the standard basis e_i, for $i \in E$. If S is a sequence $(i_1, i_2, \ldots, i_k)$ with terms in E, then the *exterior product* e_S is defined by

$$e_S = e_{i_1} \wedge e_{i_2} \wedge \cdots \wedge e_{i_k}.$$

Note that $e_\emptyset = 1$. The 2^n exterior products e_S, where S is an increasing sequence, form a basis for $\Lambda(E)$.

For $0 \leq k \leq n$, let $\Lambda_k(E)$ be the subspace (of dimension $\binom{n}{k}$)) spanned by the exterior products e_S, where S is a length-k sequence. If $a \in \Lambda_k(E)$ for some k, then a is said to be *homogeneous*. If $a \neq 0$, it is assigned the *grade* k and we write $|a| = k$. Under this grading, $\Lambda(E)$ forms a graded algebra, in the sense that, as a vector space,

$$\Lambda(E) = \bigoplus_{k=0}^{n} \Lambda_k(E),$$

and the product of an element in $\Lambda_j(E)$ and an element in $\Lambda_k(E)$ is an element in $\Lambda_{j+k}(E)$. Multiplication in $\Lambda(E)$ is *graded-commutative:* if a and b are homogeneous, then

$$a \wedge b = (-1)^{|a||b|} b \wedge a.$$

Let M be a matroid on E with lattice of flats $L(M)$. Then M defines another grading on $\Lambda(E)$. If X is a flat, let $\Lambda_X(E)$ be the subspace of $\Lambda(E)$ spanned by the exterior products e_S such that the closure $\mathrm{cl}(S)$ equals X. Then

$$\Lambda(E) = \bigoplus_{X \in L(M)} \Lambda_X(E)$$

and the product of an element in $\Lambda_X(E)$ and an element in $\Lambda_Y(E)$ is an element in $\Lambda_{X \vee Y}(E)$, where $X \vee Y$ is the join $\mathrm{cl}(X \cup Y)$. If $a \in \Lambda_X(E)$ for some flat X, we say that a is *M-homogeneous* and has *M-grade* X.

The Orlik–Solomon algebra of M, which is defined in the next section, is the quotient of $\Lambda(E)$ by a homogeneous ideal. An ideal I in $\Lambda(E)$ is *homogeneous* (respectively, *M-homogeneous*) if I is generated by homogeneous (respectively, M-homogeneous) elements. The following easy proposition summarizes the underlying algebra; in the statement $\Lambda(E)$ can be replaced with an arbitrary graded ring.

Proposition 30.4. *Let I be a homogeneous ideal in $\Lambda(E)$. Then the quotient $\Lambda(E)/I$ is graded, with*

$$(\Lambda(E)/I)_k = \Lambda_k(E)/(I \cap \Lambda_k(E)).$$

An analogous assertion holds for M-homogeneous ideals.

The *boundary operator* $\partial : \Lambda(E) \to \Lambda(E)$ is defined on the basis elements by $\partial 1 = 0$ and, if $k > 0$ and $S = (i_1, i_2, \ldots, i_k)$,

$$\partial e_S = \sum_{j-1}^{k} (-1)^{j-1} e_{S \backslash i_j},$$

where $S\backslash i_j$ is the length-$(k-1)$ sequence obtained by deleting the j-th term i_j from S. The definition is then extended by linearity to all of $\Lambda(E)$. We note three elementary properties.

Lemma 30.5.

1. $\partial\partial = 0$.

2. $\partial(e_1 \wedge e_2 \wedge \cdots \wedge e_k) = (e_2 - e_1) \wedge (e_3 - e_1) \wedge \cdots \wedge (e_k - e_1)$.

3. ∂ *is a* graded derivation: *for homogeneous elements a and b,*

$$\partial(a \wedge b) = (\partial a) \wedge b + (-1)^{|a|} a \wedge (\partial b).$$

30.2.3 Orlik–Solomon algebras defined

We are now ready to define Orlik–Solomon algebras [890].

Definition 30.6. Let M be a rank-r matroid on the set E labeled by $1, 2, \ldots, n$ and $I(M)$ be the ideal of $\Lambda(E)$ generated by the set

$$\{\partial e_C : C \text{ is a circuit of } M\}.$$

The *Orlik–Solomon algebra* $\mathsf{A}(M)$ of M is the quotient algebra defined by

$$\mathsf{A}(M) = \Lambda(E)/I(M).$$

We will denote by a_S the image of e_S in $\mathsf{A}(M)$.

Example 30.7. Continuing Example 30.3, the ideal $I(K)$ is generated by seven homogeneous elements. Four are from the circuits 123, 156, 246, 345 of size 3:

$$e_{23} - e_{13} + e_{12},\ e_{56} - e_{16} + e_{15},\ e_{46} - e_{26} + e_{24},\ e_{45} - e_{35} + e_{34};$$

and three from the circuits 1245, 1346, 2356 of size 4 :

$$e_{245} - e_{145} + e_{125} - e_{124},\ e_{346} - e_{146} + e_{136} - e_{134},\ e_{356} - e_{256} + e_{236} - e_{235}.$$

We note some easy consequences of Definition 30.6. If M has a loop i, then $\partial e_i = 1$ and $1 \in I(M)$, also $\mathsf{A}(M) = 0$. Thus, we may restrict our attention to loopless matroids. If i and j are parallel elements in a loopless matroid, then $\{i, j\}$ is a circuit, $\partial(e_i \wedge e_j) = e_j - e_i \in I(M)$, and $a_i = a_j$ in $\mathsf{A}(M)$. Thus, $\mathsf{A}(M)$ is naturally isomorphic to the Orlik–Solomon algebra of a simplification of M, so we may as well work with simple matroids or equivalently, as Orlik and Solomon did in [890], geometric lattices.

It is immediate from the definition that the ideal $I(M)$ is homogeneous. Moreover, if C is a circuit and $i \in C$, then $\mathrm{cl}(C \setminus i) = \mathrm{cl}(C)$, hence $I(M)$ is M-homogeneous as well. By Proposition 30.4, it then follows that $\mathsf{A}(M)$ is graded and M-graded. A non-zero homogeneous or M-homogeneous element has the same grade or M-grade in $\mathsf{A}(M)$ it had in $\Lambda(E)$. It is easy to see that an element a_S is non-zero in $\mathsf{A}(M)$ only if S is independent, so all the non-zero elements in $\mathsf{A}_M(X)$ have grade k, where $k = r(X)$.

Proposition 30.8. $\mathsf{A}_k(M) = \bigoplus_{\substack{X \in L(M) \\ r(X)=k}} \mathsf{A}_X(M).$

Example 30.9. Concluding Examples 30.3 and 30.7, let X be the flat 123 in K. Then $\mathsf{A}_X(K)$ is the quotient of the 4-dimensional vector space spanned by $e_{123}, e_{12}, e_{13}, e_{23}$ with the linear relations $e_{123} = 0$ and $e_{23} = e_{13} - e_{12}$. Thus $\mathsf{A}_X(K)$ is a 2-dimensional space with basis e_{12}, e_{13}. Now let $E = 123456$. Then $\mathsf{A}_E(K)$ is the quotient of a 20-dimensional space (spanned by e_S, where $S \subseteq 123456$, and $|S| = 3$) modulo linear relations implied algebraically by relations given in Example 30.7. The theory of nbc-monomials, developed in the next subsection, will show that $\mathsf{A}_E(K)$ has dimension 6.

30.2.4 An nbc-basis

An *nbc-monomial* is an element in $\mathsf{A}(M)$ of the form a_J, where J is the sequence obtained by putting an nbc-set in increasing order. The following result can be found in [131, 654].

Theorem 30.10. *Let M be a simple rank-r matroid on the set E labeled by $1, 2, \ldots, n$ and let X be a flat of M. Then*

1. *The nbc-monomials a_S, where $\mathrm{cl}(S) = X$, form a basis for $\mathsf{A}_X(M)$.*

2. *The nbc-monomials a_S, where $|S| = k$, form a basis for $\mathsf{A}_k(M)$.*

In particular, the nbc-monomials form a basis for $\mathsf{A}(M)$.

The first statement implies the second, by Proposition 30.8; the two are proved simultaneously by induction on k, using Lemma 30.1 and Proposition 30.8.

Remark 30.11. It follows from Theorem 30.10 that the set

$$\{\partial e_C : C \text{ is a circuit of } M\}$$

forms a (not necessarily reduced) Gröbner basis for the ideal $I(M)$, for the graded lexicographic term order (see [384]).

The *Hilbert series* $H(\mathsf{R}; t)$ of a graded algebra $\mathsf{R} = \bigoplus_{k \geq 0} \mathsf{R}_k$ is the formal power series

$$H(\mathsf{R}; t) = \sum_{k \geq 0} \dim(\mathsf{R}_k)\, t^k.$$

Theorems 30.2 and 30.10 imply the following corollary.

Corollary 30.12. *Let M be a rank-r simple matroid and Y a flat of M. Then*

$$\dim \mathsf{A}_Y(M) = (-1)^{r(Y)} \mu(\emptyset, Y)$$

and

$$H(\mathsf{A}(M); t) = \sum_{X \in L(M)} (-1)^{r(X)} \mu(\emptyset, X) t^{r(X)} = (-t)^r \chi(M; -1/t).$$

Remark 30.13. For a rank-r matroid M,

$$t^r T(M; 1 + 1/t, 0) = \begin{cases} 0 \text{ if } M \text{ has a loop,} \\ (-t)^r \chi(\mathrm{si}(M); -1/t) \text{ if } M \text{ is loopless;} \end{cases}$$

where $\mathrm{si}(M)$ is a simplification of M. Then for any rank-r matroid M,

$$H(\mathsf{A}(M); t) = t^r T(M; 1 + 1/t, 0)$$

and $t^{-r} H(\mathsf{A}(M); t)$ is an evaluation of the Tutte polynomial.

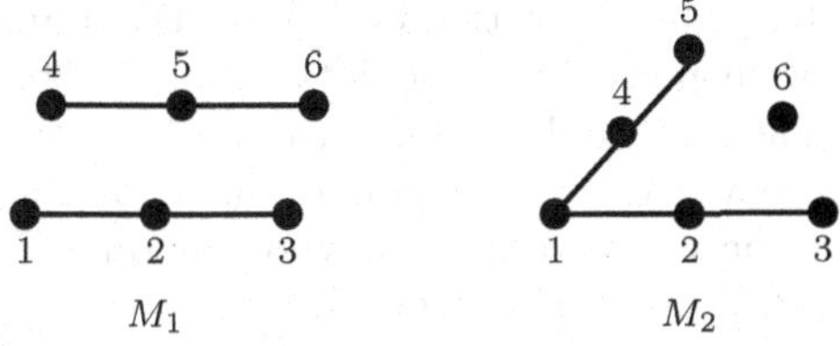

FIGURE 30.2: Two matroids M_1 and M_2 with isomorphic Orlik–Solomon algebras.

30.2.5 A-equivalence

To what extent does the algebra $\mathsf{A}(M)$ determine the (simple) matroid M? The following lemma might seem to answer the question.

Lemma 30.14. *Let $S \subseteq E$. Then S is independent in M if and only if $a_S \neq 0$ in $\mathsf{A}(M)$.*

However, if $\mathsf{A}(M)$ is only known as an algebra (up to isomorphism), then one may not be able to identify the elements e_S. The following example, due to Rose and Terao [891, Example 3.77], shows that this situation may occur.

Example 30.15. Let $E = 123456$. Let M_1 be the simple rank-3 matroid on E with 3-element circuits 123 and 456 and M_2 the simple rank-3 matroid with 3-element circuits 123 and 345 depicted in Figure 30.2. Then $\mathsf{A}(M_1)$ is isomorphic to $\mathsf{A}(M_2)$.

An explicit isomorphism $\mathsf{A}(M_1) \to \mathsf{A}(M_2)$ is induced by the isomorphism $\Phi : \Lambda(E) \to \Lambda(E)$ given by

$$\begin{array}{lll} e_1 \mapsto e_1, & e_2 \mapsto e_2, & e_3 \mapsto e_3, \\ e_4 \mapsto e_3 - e_5 + e_6, & e_5 \mapsto e_4 - e_5 + e_6, & e_6 \mapsto e_6, \end{array}$$

on $\Lambda_1(E)$. An easy calculation (with the help of Lemma 30.5) shows that Φ maps the ideal $I(M_1)$ into the ideal $I(M_2)$ and hence Φ gives a homomorphism between the Orlik–Solomon algebras. Since the two matroids have the same characteristic polynomial $\lambda^3 - 6\lambda^2 + 13\lambda - 8$, Φ is an isomorphism by Corollary 30.12.

The matroids of Example 30.15 have the same Tutte polynomial. But each is a truncation of a (disconnected) rank-4 matroid, and those matroids have isomorphic Orlik–Solomon algebras and different Tutte polynomials [467]. In fact, the general construction of [467] yields many other examples. In particular, for any simple matroid M and positive integer m, there are m extensions of M having isomorphic Orlik–Solomon algebras but pairwise distinct Tutte polynomials. The Orlik–Solomon algebra determines the Tutte polynomial for a broad class of rank-three simple (paving) matroids [471, Corollary 3.18], including those having no rank-two flats of size bigger than three and those having a real realization [329, 472]. Whether the statement holds for arbitrary simple rank-three (complex-realizable) matroids remains an open question.

In the other direction, we have the following example from [471].

Example 30.16. Let P_1 (respectively, P_2) be the rank-3 sparse paving matroid on 123457 with non-trivial copoints $123, 145, 356, 476$ (respectively, $123, 145, 356, 176$). Then P_1 and P_2 have the same Tutte polynomial, but $\mathsf{A}(P_1) \ncong \mathsf{A}(P_2)$. Moreover there is a pair of supersolvable rank-3 (paving) matroids with the same characteristic polynomial (and Tutte polynomial), but non-isomorphic Orlik–Solomon algebras.

From these examples, one sees that the answer to the question at the start of this subsection is far from clear. Example 30.16 suggests the following question. Can Orlik–Solomon algebras distinguish non-isomorphic projective planes of the same order or Dowling matroids with the same rank based on non-isomorphic groups of the same order?

30.2.6 Deletion and contraction

Let M be a simple matroid on E. Let $e \in E$ and let $M' = M\backslash e$ and $M'' = M/e$. Both M' and M'' have ground set $E \setminus e$. Any circuit of M' is a circuit of M, so the inclusion of $E \setminus e$ into E induces an algebra homomorphism $\mathsf{A}(M') \to \mathsf{A}(M)$. The map of exterior algebras $\Lambda(E) \to \Lambda(E \setminus e)$ defined by

$$e_S \longmapsto \begin{cases} e_{S\backslash e} \text{ if } e \in S, \\ 0 \text{ if } e \notin S \end{cases}$$

maps $I(M)$ into $I(M'')$, so it induces a homogeneous linear map $\rho\colon \mathsf{A}(M) \to \mathsf{A}(M'')$ of degree -1.

The following proposition is from [890].

Proposition 30.17. *For each nonnegative integer p, the sequence*

$$0 \longrightarrow \mathsf{A}(M')_p \longrightarrow \mathsf{A}(M)_p \xrightarrow{\rho} \mathsf{A}(M'')_{p-1} \longrightarrow 0$$

is exact and splits.

The exact sequence of Proposition 30.17 expresses the Tutte–Grothendieck invariant of Remark 30.13 in algebraic form.

30.2.7 Topology of complex hyperplane arrangements

The motivation behind Orlik–Solomon algebras lies in the topology of complex hyperplane arrangements. This section describes the main result, here Theorem 30.20, of [890] (see also [384, 891]). Note that additional background on hyperplane arrangements can be found in Chapter 29.

A *hyperplane* H in $\mathbb{C}^r$ is the kernel of a nonzero linear form $\alpha_H\colon \mathbb{C}^r \to \mathbb{C}$, that is

$$H = \{x \in \mathbb{C}^r : \alpha_H(x) = c_1x_1 + \cdots + c_rx_r = 0\}$$

for a nonzero vector $(c_1, \ldots, c_r)$ of complex numbers, determined by H up to a nonzero scalar multiple. If $\mathcal{A}$ is an *arrangement* (that is, a finite set) of distinct hyperplanes, then linear dependence of the linear forms α_H, for $H \in \mathcal{A}$, defines a simple matroid M on the set $\mathcal{A}$ that does not depend on the choice of defining forms α_H. The matroid M has rank r if and only if $\bigcap_{H \in \mathcal{A}} H = 0$ and we shall assume that is the case. The *complement* X of $\mathcal{A}$ is defined by

$$X = \mathbb{C}^r \setminus \bigcup_{H \in \mathcal{A}} H.$$

Complements of arrangements provide models for the motion of non-colliding particles and braids, and play an important role in singularity theory. Of special interest is the cohomology algebra $\mathsf{H}^{\bullet}(X, \mathbb{C})$ of X. This is a topological invariant of X, in the sense that homeomorphic spaces have isomorphic cohomology algebras. Building upon work of Arnol'd [45] and Brieskorn [201], Orlik and Solomon [890] proved the cohomology algebra of the complement of an arbitrary complex hyperplane arrangement is isomorphic to the Orlik–Solomon algebra of the associated matroid.

To begin with, $\mathsf{A}(M)$ is isomorphic to an algebra of holomorphic differential forms on X associated with $\mathcal{A}$. See [1085] for the general theory of differential forms and cohomology. Since X is an open subset of $\mathbb{C}^r$, differential forms on X have an elementary description: a holomorphic differential p-form on X is a holomorphic function ω from X to the degree p part of the exterior algebra on the dual vector space of $\mathbb{C}^r$. Denoting the standard dual basis of $\mathbb{C}^r$ by $dz_1, \ldots, dz_r$, a differential p-form ω can be written

$$\omega = \sum_J f_J(z) dz_J$$

where the sum is over strictly increasing p-tuples $J = (j_1, \ldots, j_p)$,

$$dz_J = dz_{j_1} \wedge \cdots \wedge dz_{j_p},$$

and f_J is holomorphic on X for each J. A 0-form is just a holomorphic function $f \colon X \to \mathbb{C}$. There is a natural product of differential forms coming from the product in the exterior algebra, making the set of all holomorphic differential forms into a graded algebra. If $f(z)$ is a 0-form, its differential is defined by

$$df = \sum_{i=1}^{r} \frac{\partial f}{\partial z_i} dz_i.$$

This function d extends to a unique graded derivation of differential forms of degree $+1$, satisfying $dd = 0$.

For $H \in \mathcal{A}$ with defining linear form α_H, define the 1-form ω_H by

$$\omega_H = \frac{d\alpha_H}{\alpha_H}.$$

Note that ω_H does not depend on the choice of defining linear form α_H for H, since any two differ by a nonzero constant factor. The ω_H are called logarithmic 1-forms because, locally on X, ω_H is equal to $d\log(\alpha_H)$, where $\log(\alpha_H)$ is defined using a locally-defined branch of the complex natural logarithm function. The function $\log(\alpha_H)$ itself is multi-valued on X but it has a single-valued differential. Let $\mathsf{R}(\mathcal{A})$ be the algebra of differential forms generated by 1 and $\{\omega_H : H \in \mathcal{A}\}$.

Proposition 30.18. *There is a surjective homomorphism of graded algebras* $\mathsf{A}(M) \to \mathsf{R}(\mathcal{A})$ *mapping* a_H *to* ω_H.

The connection with cohomology is via integration of differential forms: if ω is a holomorphic p-form on X and σ is a singular p-simplex in X, then $\int_\sigma \omega$ is a well-defined complex number. Because ω_H is a *closed* form, that is $d\omega_H = 0$, this pairing induces a well-defined mapping from $\mathsf{R}(\mathcal{A})$ to the dual space $\mathsf{H}^\bullet(X, \mathbb{C}) = \mathrm{Hom}_\mathbb{Z}(\mathsf{H}_\bullet(X, \mathbb{Z}), \mathbb{C})$ of the singular homology $\mathsf{H}_\bullet(X, \mathbb{Z})$ of X. Thus we have a sequence of maps $\mathsf{A}(M) \to \mathsf{R}(\mathcal{A}) \to \mathsf{H}^\bullet(X, \mathbb{C})$. (The vector space $\mathsf{H}^\bullet(X, \mathbb{C})$ is isomorphic to the singular cohomology ring of X with complex coefficients, and with this identification the second map is a homomorphism of graded rings.)

Using deletion and contraction, combined with Proposition 30.18, this implies each of the two maps is an isomorphism [473]. The fact that the algebra of forms $\mathsf{R}(\mathcal{H})$ is isomorphic to $H^\bullet(X, \mathbb{C})$ has important topological consequences (see [384]).

Let $H \in \mathcal{A}$, corresponding to $e \in E$. The *deletion* of $\mathcal{A}$ relative to H is the arrangement $\mathcal{A} \backslash H$ in $\mathbb{C}^r$, while the *contraction* of $\mathcal{A}$ relative to H, denoted $\mathcal{A}/H$, is the arrangement $\{H \cap K : K \in \mathcal{A}'\}$ in the vector space $H \cong \mathbb{C}^{r-1}$. The matroids of these arrangements are $M \backslash e$ and a simplification of M/e, respectively. Denote the complements of $\mathcal{A}'$ and $\mathcal{A}''$ by X' and X'', respectively. Note $X \subseteq X'$ and $X' \setminus X = X''$. One obtains the following sequence from the long exact sequence of the pair (X', X) in homology, identifying the relative homology $\mathsf{H}_{p+1}(X', X)$ with $\mathsf{H}_{p-1}(X'', \mathbb{Z})$ using excision, the tubular neighborhood theorem, and the Künneth formula.

Lemma 30.19. *There is an exact sequence of vector spaces*

$$\cdots \to \mathsf{H}^p(X', \mathbb{C}) \to \mathsf{H}^p(X, \mathbb{C}) \to \mathsf{H}^{p-1}(X'', \mathbb{C}) \to \mathsf{H}^{p+1}(X', \mathbb{C}) \to \cdots.$$

The main result of this section is the following.

Theorem 30.20. *Let* $\mathcal{A}$ *be a complex hyperplane arrangement with matroid* M. *Then* $\mathsf{A}(M)$, $\mathsf{R}(\mathcal{A})$, *and* $\mathsf{H}^\bullet(X, \mathbb{C})$ *are isomorphic as graded algebras.*

A proof of the theorem runs along the following lines. That the composite $\mathsf{A}(M) \to \mathsf{R}(\mathcal{A}) \to \mathsf{H}^\bullet(X, \mathbb{C})$ is an isomorphism can be shown by induction on $|\mathcal{A}|$. By Proposition 30.18 this is sufficient. The case $|\mathcal{H}| = 1$ is treated in

Example 30.21. For $|\mathcal{A}| > 1$ fix $H \in \mathcal{A}$ corresponding to $e \in E$ and form the associated deletion and contraction of $\mathcal{H}$ as above. The maps in Lemma 30.19 make a commutative diagram:

$$\begin{array}{ccccccccc} 0 & \longrightarrow & \mathsf{A}(M\backslash e)_p & \longrightarrow & \mathsf{A}(M)_p & \longrightarrow & \mathsf{A}(M/e)_{p-1} & \longrightarrow & 0 \\ & & \downarrow & & \downarrow & & \downarrow & & \\ \cdots & \longrightarrow & \mathsf{H}^p(X',\mathbb{C}) & \longrightarrow & \mathsf{H}^p(X,\mathbb{C}) & \longrightarrow & \mathsf{H}^{p-1}(X'',\mathbb{C}) & \longrightarrow & \cdots . \end{array}$$

The rows are exact and the map $\mathsf{A}(M)_p \to \mathsf{A}(M/e)_{p-1}$ has a section, by Propositions 30.17 and 30.19. By the inductive hypothesis the first and third vertical maps are isomorphisms. Then the lower long exact sequence breaks into short exact sequences, and the five lemma, or an elementary diagram-chase, shows the map in the middle is an isomorphism, establishing the inductive step.

Example 30.21. In case $r = 1$, the arrangement $\mathcal{A}$ consists of the single hyperplane H given $z = 0$ in $\mathbb{C}$. The matroid M is the isthmus and the Orlik–Solomon algebra $\mathsf{A}(M)$ is isomorphic to $\mathbb{C}^2$ with basis $\{1, a_H\}$. The complement $X = \mathbb{C} \setminus \{0\}$ is a connected open subset of $\mathbb{C}$ with the homotopy type of the unit circle. The integral homology $\mathsf{H}_\bullet(X, \mathbb{Z})$ has rank two, with basis the homology classes of the singular 0-simplex 1 and a singular 1-simplex τ parametrizing the unit circle. The holomorphic 1-form ω_H is $\frac{dz}{z}$, and $\mathsf{R}(\mathcal{A})$ is isomorphic to $\mathbb{C}^2$, with basis $\{1, \omega_H\}$. Since $\int_\tau \omega_H = 2\pi\sqrt{-1} \neq 0$, the integration map $\mathsf{R}(\mathcal{A}) \to \mathsf{H}^\bullet(X, \mathbb{C})$ sends $\{1, \omega_H\}$ to a basis of $H^\bullet(X, \mathbb{C})$, and is indeed is an isomorphism of graded algebras.

Corollary 30.22. *Let $\mathcal{A}$ be a complex arrangement with complement X and underlying matroid M. Then*

$$H(\mathsf{H}^\bullet(X); t) = (-t)^r \chi(M; -1/t).$$

Corollary 30.23. *The underlying matroid M determines the cohomology ring of the complement of the complex hyperplane arrangement $\mathcal{A}$.*

The preceding corollaries have motivated much of the research in the topology of arrangements. Detailed accounts of the theory can be found in [891]. In [986], Schechtman and Varchenko gave a combinatorial model for the dual $\mathrm{Hom}_{\mathbb{C}}(\mathsf{A}(M), \mathbb{C})$ of the Orlik–Solomon algebra of a complex hyperplane arrangement, called the *flag complex*, as a quotient of the vector space with basis the set of chains in the lattice of flats. Equipped with a remarkable bilinear form, the flag complex becomes an important tool in the study of local systems over the complement X, and in representation theory and mathematical physics. The theory of flag complexes is developed for matroids in [248].

30.3 Coalgebras associated with matroids

The Tutte polynomial satisfies identities which are manifestations of an underlying coalgebra structure. We give a brief informal account of a coalgebra constructed from matroids (see, for example, [1, 942, 1050] for additional background).

Coalgebras are opposites of algebras. Instead of a (bilinear) multiplication $\mathsf{R} \otimes \mathsf{R} \to \mathsf{R}$, with $(x, y) \mapsto x \circ y$, in an algebra R "merging" two elements into one, a *combinatorial coalgebra* C is defined by a *comultiplication*

$$\mathbf{\Delta} : \mathsf{C} \to \mathsf{C} \otimes \mathsf{C} : \quad x \mapsto \sum x' \otimes x'',$$

where the sum ranges over all decompositions of x into two parts x' and x''. Combinatorial coalgebras may have additional structure. A *bialgebra* is a coalgebra with a multiplication $\circ$ compatible with the comultiplication, and a *Hopf algebra* is a bialgebra with an *antipode*, a comultiplication analog of the inverse. The notion of a Hopf algebra originated with the study of cohomology rings of topological groups: the group multiplication induces a comultiplication in cohomology, and inversion in the group defines an antipode, endowing the cohomology ring with a Hopf algebra structure.

The *restriction-contraction coalgebra* of a minor-closed class $\mathcal{C}$ of matroids is defined by the comultiplication

$$\mathbf{\Delta}(M) = \sum_{A \subseteq E} M|A \otimes M/A$$

on the vector space of formal linear combinations of isomorphism-classes of matroids in $\mathcal{C}$. Usually, one imposes a multiplication $\circ$, defined by

$$M \circ N = M \oplus N,$$

to form a bialgebra. By a method of Schmitt [988], restriction-contraction bialgebras have antipodes and are Hopf algebras. An explicit cancellation-free formula for the antipode was reported in [4]. Crapo and Schmitt [344] have shown that the restriction-contraction bialgebra of freedom matroids is generated freely as a bialgebra by a loop and a coloop. Also, Section 19.4.1 describes how Hopf algebras can be used to construct Tutte polynomials of various objects.

Many natural invariants of matroids are compatible with the restriction-contraction coalgebra. For example, the nullity–corank polynomial of a rank-r matroid M on the set E, defined by

$$R_{\mathrm{nc}}(M; x, \lambda) := \sum_{A \subseteq E} x^{|A| - r(A)} \lambda^{r(M) - r(A)},$$

(cf. Definition 4.122) satisfies the convolution–multiplication identity [735]

$$R_{\mathrm{nc}}(M; xy, \lambda\xi) = \sum_{A \subseteq E} \lambda^{r-r(A)}(-y)^{|A|-r(A)} R_{\mathrm{nc}}(M|A; -x, -\lambda) R_{\mathrm{nc}}(M/A; y, \xi). \tag{30.1}$$

This identity converts multiplication of variables into a convolution and is a typical example of compatibility. The corank–subset or multivariate Tutte polynomial, the $\mathcal{G}$-invariant, and the F-invariant satisfy similar identities (see [118, 374, 735]). Intuitively, the reason is that these invariants can be expressed as sums over subsets or chains of subsets and such sums can be decomposed according to the restriction-contraction comultiplication. The convolution formula of [469] and [704] is a special case of (30.1)—see [373] for a generalization (cf. Theorems 4.150, 5.8, 12.24, 19.7, 26.32, and 28.18).

The Hopf algebra QSym of quasisymmetric functions (see, for example, [799]) is used in a fundamental way in the study of the F- and $\mathcal{G}$-invariants (see Chapter 32). Indeed, $\mathcal{G}$ defines a Hopf algebra homomorphism from the restriction-contraction coalgebra on all matroids into QSym [374]. In addition, from the F-invariant, Luoto has constructed a "matroid-friendly" basis for QSym [800].

Coalgebra-compatibility may involve other binary operations and coalgebras. Let $\chi(G; x)$ be the chromatic polynomial of the graph G with vertex set V. If $U \subseteq V$, let $G[U]$ be the *induced* subgraph on the vertex set U (with all the edges in G having both endpoints in U). Tutte [1095] observed that

$$\chi(G; x+y) = \sum_{U \subseteq V} \chi(G[U]; x)\chi(G[V \setminus U]; y).$$

Thus, the chromatic polynomial (under addition of variables) is compatible with the "Boolean coalgebra" defined by the comultiplication

$$\boldsymbol{\Delta}(G) = \sum_{U \subseteq V} G[U] \otimes G[V \backslash U].$$

30.4 Open problems

1. Does the Orlik–Solomon algebra of a rank-three matroid determine its Tutte polynomial? Equivalently, does the Orlik–Solomon algebra of a matroid determine the multiset of cardinalities of its rank-two flats? The analogous statement for higher-rank matroids is false (see [467]).

2. Under what conditions on M is the Orlik–Solomon ideal of M generated by its elements of degree two? See [136] for the supersolvable case.

3. Can the Orlik–Solomon algebra distinguish non-isomorphic projective planes of the same order?

4. Can the Orlik–Solomon algebra distinguish Dowling geometries of the same rank on non-isomorphic groups of the same order?

31

The Tutte polynomial of oriented matroids

Emeric Gioan

Synopsis

This chapter covers properties of the Tutte polynomial for oriented matroids. It covers the basics of oriented matroids, as well as their interactions with the Tutte polynomial.

- Oriented matroids, interpretations and translations in hyperplane arrangements and in directed graphs.
- The Tutte polynomial in terms of orientation-activities, generalization to oriented-matroid perspectives, and a 4-variable expansion.
- Geometric interpretations of the β-invariant, of the other coefficients, and of particular evaluations.
- Expression of the Tutte polynomial of a matroid in terms of active filtrations/partitions and β-invariants of minors.
- Activity-preserving bijections between bases/subsets/no-broken-circuit-subsets and activity-classes/reorientations/regions.
- Circuit/cocircuit reversal classes in directed graphs and regular matroids.

31.1 Introduction

While a matroid captures the linear dependence relations of given real vectors, an oriented matroid captures, in addition, the convexity relations among these vectors. Alternatively, a matroid captures the lattice of intersections of given real hyperplanes, while an oriented matroid captures, in addition, the

DOI: 10.1201/9780429161612-31

relative positions of these intersections. The theories of matroids and oriented matroids are related but distinct: an oriented matroid has an underlying matroid, but not every matroid is orientable, and distinct oriented matroids may have the same underlying matroid (e.g., the uniform ones). However, the Tutte polynomial of an oriented matroid is that of its underlying matroid. At the most general level, oriented matroids are both combinatorial and topological objects. In rank three, they are equivalent to pseudo-line arrangements. Results about oriented matroids apply to real hyperplane arrangements and thus to directed graphs. Oriented matroids offer a geometric understanding of the relations between different graph orientations. We treat some results and interpretations of the Tutte polynomial in these settings. More specifically, we investigate oriented matroids on a linearly ordered ground set (or edge set, for directed graphs) through the notion of orientation-activities. This yields counting results involving the coefficients, or particular evaluations, or general formulas, of the Tutte polynomial; enumerative counterparts of constructions such as special partitions of the ground set; decompositions into bounded regions; activity-preserving bijections between basis/subsets and reorientations; and two types of equivalence classes for reorientations.

31.2 Oriented matroids and related structures

Oriented matroid theory arose from work from the 1970s on graph theory and linear programming [138], and on polytope theory and topological representations in terms of pseudo-sphere arrangements [491]. Numerous equivalent combinatorial axiomatizations for oriented matroids, together with an equivalence with a topological object, make these structures very rich and natural. In this section, we summarize the oriented matroid theory needed to follow this chapter and to apply it to directed graphs and real hyperplane arrangements. See [133] for further background.

Notation 31.1. A *signed subset* A of a finite set E is a subset $A \subseteq E$ in which each element of A is equipped with a *sign* in $\{+,-\}$. It is partitioned into a *positive part* and a *negative part*: $A = A^+ \sqcup A^-$. Exchanging the two parts in A yields the *opposite* signed subset $-A$. A signed subset with no negative element is called *positive*, and with no positive element is called *negative*. Throughout the chapter, we write, for example, 123 for the set $\{1,2,3\}$. When a set is signed, we write negative elements with a bar. For example, $A = 1\overline{23}$ denotes that $A^+ = \{1\}$ and $A^- = \{2,3\}$.

31.2.1 Directed graphs

Let $\vec{G} = (V,E)$ be a digraph, with underlying undirected graph G. Let C be the edge set of a cycle of G, recalling that the edge sets of cycles of G are

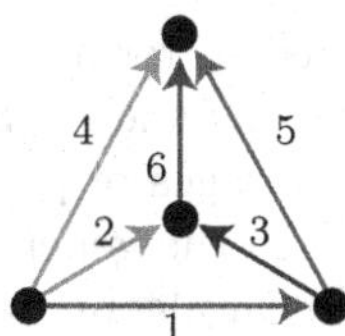

FIGURE 31.1: An orientation of K_4, serving as a running example throughout the chapter.

non-comparable under the partial ordering of set inclusion. Fix a direction in which to traverse the cycle and then assign to each edge in C the sign + or − according to whether its direction in $\vec{G}$ matches, or disagrees with, the fixed direction.

The resulting signed sets C (for all cycles of G) form the set of *circuits* of an oriented matroid denoted by $M(\vec{G})$ (which satisfies the axiomatics from Definition 31.4).

A cycle of G thus provides two opposite circuits of $M(\vec{G})$ (depending on the chosen fixed direction). Directed cycles of $\vec{G}$ thus yield positive (and negative) circuits of $M(\vec{G})$.

Next, consider the edge set C of a minimal edge cut of G, and a direction for this cut, e.g., from A to B if $V = A \sqcup B$ is the partition corresponding to the cut. Then $e \in C$ gets the sign + if its direction matches the direction of the cut in $\vec{G}$, and the sign − otherwise. This gives a *cocircuit* of the oriented matroid $M(\vec{G})$, and its opposite. A directed cut yields a positive cocircuit (and its opposite).

Note that the opposite of a circuit, or cocircuit, of $M(\vec{G})$ is also a circuit, or cocircuit, of $M(\vec{G})$, respectively. Omitting the signs from the circuits and cocircuits of $M(\vec{G})$ yields the circuits and cocircuits of the cycle matroid of G.

Example 31.2. Consider the digraph in Figure 31.1. The circuits of its oriented matroid are $1\bar{2}3$, $1\bar{2}5\bar{6}$, $13\bar{4}6$, $1\bar{4}5$, $2\overline{34}5$, $2\bar{4}6$, $3\bar{5}6$, and their opposites. The cocircuits are 124, $12\overline{56}$, $1\bar{3}46$, $1\overline{35}$, 2345, $23\bar{6}$, 456, and their opposites. Observe that the digraph is acyclic so there are no positive circuits, and every element belongs to a positive cocircuit.

31.2.2 Real vector sets and real hyperplane arrangements

Let E be a finite set of vectors spanning a real vector space V, with the convention that the same vector can be repeated under different labels in E. The minimal linearly dependent subsets of E are the circuits of the matroid on E. Adding the signs of the non-zero scalars used in the dependence relations yields the *circuits* of the oriented matroid M on E (the same dependence relation yields two opposite circuits). In geometric terms, the +/− partition of the circuit is the unique partition such that the convex hulls of the two parts intersect.

Cocircuits of M can be defined in terms of subspaces of V, or, as we do here, in terms of a hyperplane arrangement. (See Chapter 29 for hyperplane arrangements.) This has the advantages of providing an all-in-one geometric representation and of being close to the general oriented matroid topological representation. From duality in linear algebra, a non-zero vector $h \in E$ defines a linear form. We consider its kernel, the hyperplane $\{x \in \mathsf{V} : h \cdot x = 0\}$, as well as the positive and negative half-spaces, $\{x \in \mathsf{V} : h \cdot x > 0\}$ and $\{x \in \mathsf{V} : h \cdot x < 0\}$. We obtain a *signed real central hyperplane arrangement*, which we may simply call an *arrangement*, or a *representation*, of M. Then, each $x \in \mathsf{V}$ yields a signed subset C of E in the following way: for every $h \in E$, we have that $h \in C^+$ if $h \cdot x > 0$, that $h \in C^-$ if $h \cdot x < 0$, and that $h \notin C$ if $h \cdot x = 0$. Informally, the signed subset C gives the position of the vector x with respect to the hyperplanes defined by the linear forms $h \in E$ (if $h \in E$ is the zero vector, then $h \notin C$ and h is called a *loop*).

The signed subsets obtained in this way thus locate the faces of all dimensions in the cell decomposition defined by the hyperplanes. In this chapter, we focus on the faces of minimal and maximal dimension. The signed subsets corresponding to 1-dimensional faces are the *cocircuits* of M, and the signed subsets corresponding to *regions* (or faces of maximal dimension) are called *topes* of M.

When the intersection of all positive half-spaces is a region, the signed subset whose positive part equals E is a tope, and the cocircuits corresponding to the boundary of the region are positive. In this case, M is called *acyclic*. Cocircuits and topes are illustrated in the left and right part of Figure 31.2.

The matroid defined by E captures the lattice of intersections of the hyperplanes (that is, the linear-dependence relations of the initial vectors), whereas the oriented matroid captures, in addition, the relative positions of these intersections (that is, the convexity relations).

Example 31.3. Starting from a digraph, each edge (u, v) yields a vector $x_v - x_u$ (for independent variables x_v indexed by vertices v of the graph), and thus a signed hyperplane. The arrangement in Figure 31.2 is obtained in this way from the digraph of Figure 31.1. Thus, the cocircuits are the same as the cocircuits listed in Example 31.2, and the left part of Figure 31.2 provides a geometric representation of cocircuits (that is, minimal cuts) of the initial digraph.

In more detail, Figure 31.2 shows a representation of a 3-dimensional arrangement on $E = \{1, \ldots, 6\}$. Each 2-dimensional hyperplane is represented by its intersection with a half-sphere. The half-sphere is delimited by the intersection of 1 and a central sphere. Only half of the arrangement is represented: the opposite half-sphere is symmetric. Each hyperplane has a positive and a negative side, indicated by the gray region which belongs to the positive side of each hyperplane. In the left figure, the cocircuits are written next to the vertices representing 1-dimensional faces. This arrangement represents the same rank-3 oriented matroid as the digraph of Figure 31.1. Observe the positive cocircuits 124, 2345, and 456, corresponding to the vertices of the boundary

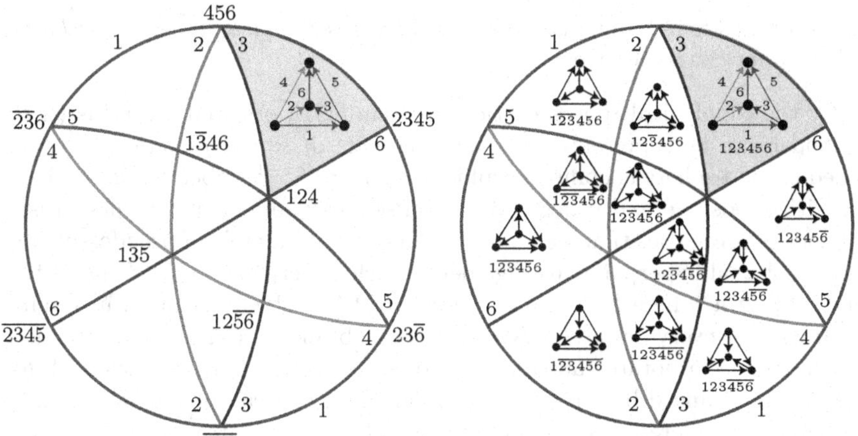

FIGURE 31.2: Representation of a 3-dimensional arrangement on $E = \{1, \ldots, 6\}$. The figure is detailed in Example 31.3.

of the region, and to directed cuts of the digraph. The right figure shows the canonical bijections between: (1) regions of the arrangement; (2) topes of the oriented matroid (signed subsets written inside regions); (3) acyclic reorientations of the oriented matroid (negative elements of the tope, with a bar, are reoriented with respect to the gray region); and (4) acyclic orientations of the graph (drawn inside regions)—see Section 31.2.4.

31.2.3 Combinatorial axioms and topological pseudo-spheres

In this section, we briefly present general oriented matroids. However, reading this is not required to follow this chapter, and the best support for intuition is given by real hyperplane arrangements in Section 31.2.2.

There are about twenty equivalent combinatorial axiom systems for oriented matroids, involving cryptomorphic objects that determine the whole structure. Although we do not use it in this chapter, we mention an axiom system that applies both to circuits and to cocircuits (besides this, circuits and cocircuits are related by an orthogonality property which we omit).

Definition 31.4. An *oriented matroid* M on a finite set E is determined by a set $\mathcal{C}$ of signed subsets of E called *cocircuits* such that:

1. $\emptyset \notin \mathcal{C}$;
2. for all $X \in \mathcal{C}$, we have $-X \in \mathcal{C}$;
3. for all $X, Y \in \mathcal{C}$, if $X^+ \cup X^- \subseteq Y^+ \cup Y^-$, then $X = Y$ or $X = -Y$;
4. for all $X, Y \in \mathcal{C}$, with $X \neq Y$, and $e \in X^+ \cap Y^-$, there exists $Z \in \mathcal{C}$ such that $Z^+ \subseteq (X^+ \cup Y^+) \setminus e$ and $Z^- \subseteq (X^- \cup Y^-) \setminus e$.

The second to fourth axioms are referred to as *symmetry*, *non-comparability*, and *elimination*, respectively.

By the topological representation theorem [491], oriented matroids (with no loops) can be represented by *arrangements of signed pseudo-spheres.* A signed pseudo-sphere is homeomorphic to a sphere S^{d-1} embedded in a sphere S^d, with a positive and a negative side. Pseudo-sphere arrangements satisfy topological properties that we omit. The first usual example of pseudo-spheres is given by real hyperplane arrangements: each hyperplane in $\mathbb{R}^{d-1}$ meets the sphere S^d in a sphere S^{d-1}, as in Figure 31.2 (see the description in Example 31.3). Not every oriented matroid can be obtained from real hyperplanes. In comparison, a notable advantage of oriented matroids is that the cell decomposition induced by a pseudo-sphere arrangement can be combinatorially characterized, whereas the one induced by a hyperplane arrangement cannot. The second usual example is in rank 3, where pseudo-spheres are equivalent to classical *pseudo-lines* (each is the image of a line under an homeomorphism of the affine plane, and pseudo-lines must cross each other exactly once).

The faces of the cell decomposition of the sphere defined by the pseudo-spheres define the sets of *cocircuits* and *topes* of M, in a similar way to that for real hyperplane arrangements in Section 31.2.2. These objects can also be defined from one another, by purely combinatorial means. In practice, one often switches between the complementary combinatorial and topological viewpoints. As in the real case, we will call a signed pseudo-sphere arrangement an *arrangement*, or a *representation*, of M.

31.2.4 Oriented matroids concepts and terminology

Given the set of cocircuits of an oriented matroid M, the underlying unsigned subsets form the set of cocircuits of a matroid, called the *underlying matroid of* M, and denoted by $\underline{M}$. Not every matroid is the underlying matroid of an oriented matroid (a matroid can be non-orientable), and various unrelated oriented matroids may have the same underlying matroid (uniform matroids are trivial while the theory of uniform oriented matroids is complicated).

An (oriented) matroid M on E is called *ordered* when E is linearly ordered.

Let M be an oriented matroid on E, and $A \subseteq E$. The oriented matroid $-_A M$ is obtained from M by reversing the signs of elements in A in every circuit, cocircuit, and tope of M. Obviously, it has the same underlying matroid as M. The set of all $-_A M$ for $A \subseteq E$ is called the set of *reorientations* of M. This set is isomorphic to 2^E, that is, we distinguish between $-_A M$ and $-_{E\setminus A} M$ as reorientations of M, even if those two oriented matroids are equal (by symmetry of circuits or cocircuits). If $M = M(\vec{G})$ is defined by a digraph $\vec{G}$ with edge set E, then the $2^{|E|}$ reorientations of M are in canonical bijection with the $2^{|E|}$ orientations of the underlying graph G. Adapting the notation above, for $A \subseteq E$, we write $-_A \vec{G}$

and we have $M(-_A\vec{G}) = -_A M(\vec{G})$. In a representation, the $2^{|E|}$ reorientations are in canonical bijection with the $2^{|E|}$ ways of signing the arrangement.

The notion of *minor* in an oriented matroid is exactly the same as in a matroid, with naturally induced signs. In the case of a digraph, it is consistent with the usual definition of a minor of a graph. A *flat* is a flat of the underlying matroid. In a representation, a flat is represented by the intersection of its elements. A *basis* is a basis of the underlying matroid.

An oriented matroid M has a *dual* oriented matroid denoted by M^*. Circuits and cocircuits of M^* are the cocircuits and circuits of M, respectively. Thus the dual of the underlying matroid is the underlying matroid of the dual. Oriented matroid duality is consistent with planar graph duality and linear algebra orthogonality.

In particular, *loops* and *coloops* of M are loops and coloops of $\underline{M}$, and they are coloops and loops of M^*, respectively.

In an oriented matroid, every element belongs either to a positive circuit or to a positive cocircuit, but not both. This result generalizes the well-known fact that in a digraph every edge belongs to either a directed cycle or a directed cut, but no both. In the real case, this result is equivalent to Farkás lemma on linear inequalities (see [133, Corollary 3.4.6]).

Definition 31.5. An oriented matroid is *acyclic* if any one of the following equivalent conditions holds:

1. all its elements belong to a positive cocircuit,
2. no element belongs to a positive circuit,
3. the signed subset whose positive part equals E is a tope,
4. the intersection of all positive half-spheres is a region (in a representation).

As illustrated in Figure 31.2 (see the description in Example 31.3), acyclic reorientations are in canonical bijection with regions of a representation (topologically), and with topes (combinatorially). Thus, in what follows, we may use the term *regions* for topes or acyclic reorientations. In the graph case, $M(\vec{G})$ is acyclic if and only if $\vec{G}$ is acyclic. Dually, an oriented matroid is *totally cyclic* if its dual is acyclic. One can replace each condition in Definition 31.5 with its dual formulation (for the third/fourth, use a tope/representation of the dual). In the graph case, when G is connected, $M(\vec{G})$ is totally cyclic if and only if $\vec{G}$ is strongly connected.

The Tutte polynomial of the oriented matroid M is that of its underlying matroid $\underline{M}$, and $t_{i,j}$ denotes the coefficient of $x^i y^j$ in this Tutte polynomial, that is, $T(\underline{M}; x, y) = \sum_{i,j} t_{i,j} x^i y^j$.

31.3 Tutte polynomial in terms of orientation-activities

We consider here the main results relating the Tutte polynomial with oriented structures. This relies upon a linear ordering of the ground set and was introduced in [750]. We use $\min(A)$ to denote the smallest element of an ordered finite set A.

Definition 31.6. Let M be an ordered oriented matroid on E. The set of *(orientation-)active elements* of M is

$$O(M) := \{\min(C) : \ C \text{ positive circuit of } M\},$$

and the set of *dual (orientation-)active elements* of M is

$$O^*(M) := O(M^*) = \{\min(C) : \ C \text{ positive cocircuit of } M\}.$$

The *(orientation-)activity* and the *dual (orientation-)activity* of M are

$$o(M) = |O(M)| \qquad \text{and} \qquad o^*(M) = o(M^*) = |O^*(M)|.$$

By the properties of oriented matroids above, we have $O(M) \cap O^*(M) = \emptyset$. By definition, we have $o(M) = 0$ if and only if M is acyclic; also, $o^*(M) = 0$ if and only if M is totally cyclic. Moreover, $O^*(M)$ is contained in the minimal basis, and, dually, $O(M)$ is contained in the complement of the maximal basis.

In a topological representation, dual orientation-active elements of an acyclic reorientation, which corresponds to a region, are given by the smallest elements of positive cocircuits, which correspond to vertices of the boundary of the region. In this way, dual orientation-active elements situate the regions with respect to the sequence of nested flats induced by the minimal basis. Note that dual orientation-active elements of regions are thus independent of the signature of the arrangement. To make this precise, the (lexicographic) minimal basis $b_1 < \ldots < b_r$ induces the sequence of nested flats $\mathrm{cl}(\{b_1\}) \subset \mathrm{cl}(\{b_1, b_2\}) \subset \cdots \subset \mathrm{cl}(\{b_1, \ldots, b_{r-1}\}) \subset E$ of M, which, in the representation, corresponds to the sequence of nested intersections $b_1 \supset b_1 \cap b_2 \supset \cdots \supset b_1 \cap b_2 \cap \ldots b_r$. A cocircuit whose representation does not belong to $b_1 \cap \ldots \cap b_i$, where i is the smallest possible and $1 \leq i \leq r$, is a cocircuit whose smallest element is b_i. See Figure 31.3 and Example 31.7.

Example 31.7. Consider Figure 31.3. In each region, we write the set of its dual active elements. The left part continues the left part of Figure 31.2. We use the ordering $1 < \ldots < 6$ (and we keep the smallest elements of cocircuits). The right part shows a (non-graphical) arrangement on 13 elements and its minimal basis $1 < 2 < 3$ (the rest of the ordering is not used). In the left part of Figure 31.3, dual active elements of regions depend on their positions with respect to the flats represented by $1 \cap 2 \cap 3 \subset 1$. The gray region M touches both flats, and we have $O^*(M) = \{1, 2, 4\}$. The region $-_6M$ touches 1 but not $1 \cap 2 \cap 3$, hence $O^*(-_6M) = \{1, 2\}$. The region $-_3M$ does not touch 1 but touches $1 \cap 2 \cap 3$, hence $O^*(-_3M) = \{1, 4\}$.

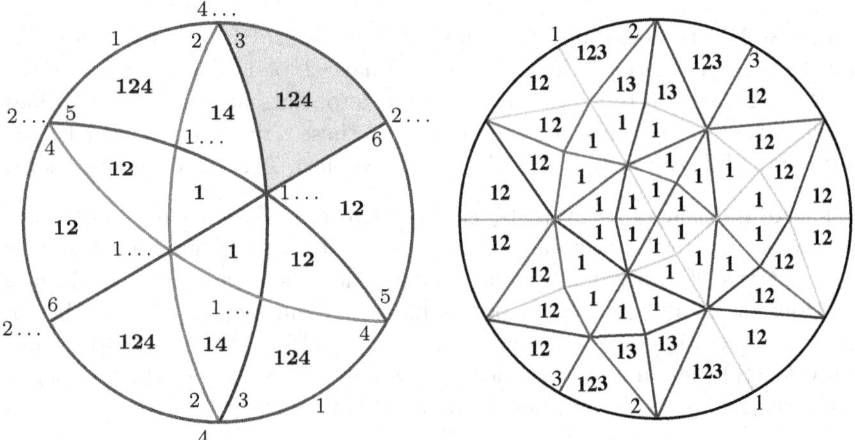

FIGURE 31.3: Dual active elements of regions; see details in Example 31.7.

Theorem 31.8. *Let M be an ordered oriented matroid on E. Then*

$$T(\underline{M}; x, y) = \sum_{A \subseteq E} \left(\frac{x}{2}\right)^{o^*(-_A M)} \left(\frac{y}{2}\right)^{o(-_A M)}.$$

Theorem 31.8 is the central result of this chapter. It asserts that, for an oriented matroid M, the number of reorientations with given activities does not depend on the linear ordering. Moreover, it depends only on the underlying matroid (not on the way this matroid is oriented). Furthermore, it can be expressed in terms of a Tutte polynomial coefficient, as follows.

Corollary 31.9. *Let M be an ordered oriented matroid. Then*

$$2^{i+j}\, t_{i,j} = \#\,\{\textit{reorientations of } M \textit{ with dual activity } i \textit{ and activity } j\}.$$

A direct application of Definition 31.6 and Theorem 31.8 is the following.

Corollary 31.10. *Let M be an oriented matroid (or a real hyperplane arrangement), and let G be a connected graph. Then*

1. $T(\underline{M}; 2, 0) = \#\{\textit{acyclic reorientations of } M\} = \#\{\textit{regions of } M\}$,
2. $T(G; 2, 0) = \#\{\textit{acyclic orientations of } G\}$,
3. $T(\underline{M}; 0, 2) = \#\{\textit{totally cyclic reorientations of } M\} = \#\{\textit{regions of } M^*\}$,
4. $T(G; 0, 2) = \#\{\textit{strongly connected orientations of } G\}$.

Example 31.11. Continuing Example 31.7, in Figure 31.3 (which represents half of the acyclic orientations of K_4), we have $2^2 \times 1 = 4$, $2^1 \times 3 = 6$, or $2^0 \times 2 = 2$, regions touching two, one, or zero, flats among $1 \cap 2 \cap 3$ and 1, respectively. According to Theorem 31.8, these values correspond to the coefficients of $T(K_4; x, 0) = x^3 + 3x^2 + 2x$. We have $T(K_4; 2, 0) = 24$ regions.

The evaluation of T at $(2, 0)$ counts, in order of increasing generality: the number of acyclic orientations of a graph [1027]; the number of acyclic reorientations of a regular matroid [246]; twice the number of non-Radon partitions of a finite set of real points [253] or (equivalently) the number of regions of a real hyperplane arrangement [1166, 1181] (see also [270]) or also (equivalently) the number of vertices of a zonotope [580, 1181]; the number of acyclic reorientations of an oriented matroid [742, 743].

31.4 Counting bounded regions and bipolar orientations

The β-invariant of M is that of its underlying matroid $\underline{M}$ (see Section 3.3.2). It is the coefficient of x in $T(\underline{M}; x, y)$:

$$\beta(\underline{M}) = t_{1,0}.$$

The bounded reorientations addressed in this section have been counted in hyperplane arrangements in [1181], and in oriented matroids in [743, 750]. We combine these approaches, along with terminology from [534, 537, 538, 530]; see also [580] for related results in terms of Whitney numbers.

Definition 31.12. Let M be an oriented matroid on E and $p \in E$. The oriented matroid M is called *bounded* with respect to p, or *p-bounded*, if either $|E| = 1$ and p is a coloop (so $M = U_{1,1}$), or M is acyclic and $-_pM$ is totally cyclic. Equivalently, M is p-bounded if and only if every element of E belongs to a positive cocircuit of M and every positive cocircuit of M contains p.

For a digraph $\vec{G}$ and an edge p of $\vec{G}$, the oriented matroid $M(\vec{G})$ is p-bounded if and only if $\vec{G}$ is *bipolar* with respect to p, that is, $\vec{G}$ is acyclic and has a unique source and a unique sink, which are the endpoints of p.

In a representation, p-bounded reorientations of M canonically correspond to regions that do not meet the element p. We can also call them *p-bounded regions* of M. Starting from an affine hyperplane arrangement, one can derive a central hyperplane arrangement in the natural way, and add a hyperplane p representing the hyperplane at infinity for the initial arrangement. In this setting, p-bounded regions exactly correspond to bounded regions of the initial arrangement in the usual sense (which explains the term "bounded").

Assume that the ground set E of M is linearly ordered with $p = \min(E)$. Then, M is p-bounded if and only if $o(M) = 0$ and $o^*(M) = 1$. This follows directly from Definitions 31.6 and 31.12. Then, Theorem 31.8 yields the following well-known result, a striking consequence of which is that the number of p-bounded reorientations does not depend on p.

Corollary 31.13. *For an element p of an oriented matroid M, or an edge p of a graph G, we have*

1. $2\,\beta(\underline{M}) = \#\{$*$p$-bounded reorientations of M*$\}$,
2. $\beta(\underline{M}) = \#\{$*$p$-bounded regions of M on the positive side of p*$\}$,
3. $2\,\beta(G) = \#\{$*bipolar orientations of G with respect to p*$\}$.

Example 31.14. Continuing Example 31.7, there are $\beta(K_4) = 2$ bounded regions with respect to 1 on the positive side of 1. Their only dual active element is 1. They correspond to the reorientations $-_{35}M$ and $-_{356}M$, and to the two bipolar orientations of K_4 with respect to 1 with fixed orientation for 1.

Remark 31.15. Assume $|E| > 1$. The involution $M \mapsto -_pM^*$ provides a canonical bijection between p-bounded regions of M and p-bounded regions of M^* (by Definition 31.4). In terms of activities, the involution $M \mapsto -_pM$ provides a canonical bijection between reorientations of M with activity 0 and dual activity 1, and reorientations of M with activity 1 and dual activity 0. Furthermore, it is known that, assuming $|E| > 1$, we have $\beta(\underline{M}) = t_{1,0} = t_{0,1} = \beta(\underline{M}^*)$. Hence, the setting and counting result above yield several dual formulations.

In a graph G, a *cyclic-bipolar orientation* with respect to p is a strongly connected orientation where each directed cycle contains p. This corresponds to having a p-bounded dual. By Remark 31.15, if G has at least two edges,

$$2\,\beta(G) = \#\{\text{cyclic-bipolar orientations of } G \text{ with respect to } p\}.$$

31.5 Generalizations to oriented matroid perspectives

This section is independent of the remainder of the chapter, and uses concepts on matroid perspectives from Chapter 28. Two, or more, signed subsets are called *conformal* (to each other) if any element belonging to any two of the subsets has the same sign in these two subsets.

Definition 31.16. Two oriented matroids M and M' on the same ground set form an *oriented matroid perspective* (also called *strong map* or *quotient map*) $M \to M'$ if any of the following equivalent properties hold:

1. any cocircuit of M' is a union of conformal cocircuits of M,

2. any circuit of M is a union of conformal circuits of M',
3. no circuit of M and cocircuit of M' with a non-empty intersection are conformal.

Note that the underlying matroids of an oriented matroid perspective $M \to M'$ form a matroid perspective $\underline{M} \to \underline{M}'$ (see Proposition 28.2). In contrast with matroid perspectives, for which a major always exists, oriented matroid perspectives are not equivalent in general to an extension followed by a contraction, as shown in [961]. Obviously, for an oriented matroid M, we have that $M \to M$ is an oriented matroid perspective. Also, if $M \to M'$ is an oriented matroid perspective, then $M'^* \to M^*$ is also an oriented matroid perspective. See [133, Section 7.7] for more information.

The Tutte polynomial of a matroid perspective is addressed in Chapter 28. With this, Theorem 31.8 generalizes to oriented matroid perspectives as follows.

Theorem 31.17. *Let $M \to M'$ be an oriented matroid perspective on a linearly ordered set E. Then*

$$T(\underline{M}, \underline{M}'; x, y, 1) = \sum_{A \subseteq E} \left(\frac{x}{2}\right)^{o^*(-_A M')} \left(\frac{y}{2}\right)^{o(-_A M)}.$$

We highlight a notable evaluation from [743] in Theorem 31.17, and special choices of oriented matroid perspectives, allowing us to retrieve results from Corollary 31.10 and 31.4. See also Section 28.3 for more on this evaluation.

Corollary 31.18. *Let $M \to M'$ be an oriented matroid perspective. Then*

$$T(\underline{M}, \underline{M}'; 0, 0, 1) = \#\{A \subseteq E : -_A M \text{ is acyclic and } -_A M' \text{ is totally cyclic}\}.$$

Remark 31.19. Consider the oriented matroid perspective $M \to U_{0,|E|}$. Since every element of $U_{0,|E|}$ is a loop, every reorientation of this oriented matroid is totally cyclic. Hence, by Corollary 31.18, $T(\underline{M}, U_{0,|E|}; 0, 0, 1)$ counts the number of acyclic reorientations of M, and equals $T(\underline{M}; 2, 0)$ (by properties seen in Section 28.3). Observe that Corollary 31.18 thus generalizes Corollary 31.10 in a self-dual way.

Remark 31.20. Consider the oriented matroid perspective $M \to M'$, where $M' = M/p \oplus p_{\text{loop}}$, with $p \in E$ and p_{loop} is a loop in M' labeled by p. Adapting Definition 31.12, M is p-bounded if and only if M is acyclic and M/p is totally cyclic. So, by Proposition 28.13 and Corollary 31.18, $T(\underline{M}, \underline{M}'; 0, 0, 1) = 2\beta(\underline{M})$ counts the number of p-bounded regions of M, which is Corollary 31.13. In a graph G, the perspective amounts to identifying the endpoints of p: bipolar orientations of G with respect to p yield strongly connected orientations of G/p.

The following generalization of Remark 31.20 is given in [330, Statement (b) p.154] in equivalent terms. More results of this flavor can be found in [330].

Corollary 31.21. *Let M be an oriented matroid and let X be a flat of M. Let X_{loop} be the uniform matroid with only loops, labelled by elements of X. Then $T(\underline{M}, \underline{M}/X \oplus X_{loop}; 0, 0, 1)$ counts the number of acyclic reorientations of M such that no complement of a positive cocircuit contains X (that is, in a representation, the number of regions that do not meet the intersection of elements of X).*

Let G be a connected graph embedded in an orientable surface, and G^* be its geometric dual. Then $M^*(G^*) \to M(G)$ is a matroid perspective, whose Tutte polynomial is called the *Las Vergnas polynomial* (see Chapter 27). When G and G^* are oriented with consistent orientations, we get an oriented matroid perspective, and Corollary 31.18 gives the following.

Corollary 31.22. *For a connected graph G embedded in an orientable surface, $T\big(M^*(G^*), M(G); 0, 0, 1\big)$ counts the number of consistent orientations of G and its geometric dual G^* such that both orientations are strongly connected.*

31.6 Activity classes and active partitions

Activity classes are canonical equivalence classes on the set of reorientations, built from active partitions (or active filtrations) of an ordered oriented matroid. They are enumerated by the Tutte polynomial, and they are fundamental with respect to various results: a decomposition into bounded regions or bipolar digraphs (see below), a Tutte polynomial formula in terms of filtrations and β-invariants of minors (Section 31.7), an activity-preserving bijection with bases (Section 31.8), and a Tutte polynomial formula in terms of four refined orientation-activities (Section 31.9). They were introduced and developed in [525] (building on an initial construction in [749]); see also [528, 531, 537, 538, 530] for more reformulations, examples and results. They are also related to graph constructions (see the end of this section).

Definition 31.23. Let M be an ordered oriented matroid on E with i dual orientation-active elements $a_1 < \ldots < a_i$ and j orientation-active elements $a'_1 < \ldots < a'_j$. The *active filtration* of M is the sequence of subsets

$$\emptyset = F'_j \subset \cdots \subset F'_0 = F_c = F_0 \subset \cdots \subset F_i = E$$

defined as follows. The subset F_c is the union of all positive circuits, or, equivalently, $E \setminus F_c$ is the union of all positive cocircuits. For every $0 \le k \le j-1$,

$$F'_k = \bigcup_{\substack{C \text{ positive circuit} \\ \min(C) \ge a'_{k+1}}} C.$$

Moreover, $F'_j = \emptyset$, $F_i = E$, and, dually, for every $0 \leq k \leq i-1$,

$$F_k = E \setminus \bigcup_{\substack{C \text{ positive cocircuit} \\ \min(C) \geq a_{k+1}}} C.$$

The *active partition* of M is the partition induced by successive differences:

$$E = (F'_{j-1} \setminus F'_j) \sqcup \cdots \sqcup (F'_0 \setminus F'_1) \sqcup (F_1 \setminus F_0) \sqcup \cdots \sqcup (F_i \setminus F_{i-1}),$$

with $\min(F'_{k-1} \setminus F'_k) = a'_k$ for $1 \leq k \leq j$, $\min(F_k \setminus F_{k-1}) = a_k$ for $1 \leq k \leq i$.

The *active minors* of M are the $i+j$ minors

$$M'_k = (M|F'_{k-1})/F'_k, \quad 1 \leq k \leq j, \text{ and } M_k = (M|F_k)/F_{k-1}, \quad 1 \leq k \leq i.$$

Proposition 31.24. *With the notation of Definition 31.23,*

1. *for $1 \leq k \leq i$, the active minor $(M|F_k)/F_{k-1}$ is bounded with respect to a_k (bipolar in the graph case);*

2. *for $1 \leq k \leq j$, the active minor $(M|F'_{k-1})/F'_k$ has a bounded dual with respect to a'_k (cyclic-bipolar in the graph case).*

Proposition 31.25. *Let M be an ordered oriented matroid with $o^*(M) = i$ and $o(M) = j$. The 2^{i+j} reorientations $-_A M$, where A is an arbitrary union of subsets in the active partition of M, have the same active filtration and active partition as M (and hence the same active and dual active elements).*

Definition 31.26. The *activity class* of M is the set of 2^{i+j} reorientations of M considered in Proposition 31.25.

By extension to all reorientations of M, we obtain an equivalence relation over all reorientations of M.

Geometrically, the active filtration of a region is given by the intersections of its boundary with the flats induced by the minimal basis. This refines the interpretation of dual activities seen in Section 31.3. See the illustrations of Definitions 31.23 and 31.26 given in Figures 31.4 and 31.5.

Corollary 31.27. *Activity classes of reorientations of an ordered oriented matroid partition its set of reorientations, and*

$$t_{i,j} = \#\{\textit{activity classes of reorientations with dual activity } i \textit{ and activity } j\}.$$

This corollary is obtained by combining Definition 31.26 and Theorem 31.8. Beyond this, each activity class is naturally isomorphic to a boolean lattice, and we get a partition of 2^E into boolean lattices. Note that, in Sections 31.8 and 31.9, this partition is bijectively related to the partition of 2^E into basis intervals (see Chapter 5 and Proposition 28.21). Furthermore, consider M as a reference reorientation, and say that an element has *fixed orientation* if it is not reoriented with respect to M. There obviously exists one and only one reorientation with fixed orientation for active elements and dual active elements in each activity class of reorientations. Hence, we obtain the following corollary. (See also Tables 31.1 and 31.2 for various derived enumerations using Tutte polynomial evaluations.)

Corollary 31.28. *For an ordered oriented matroid M on E,*

$$\begin{aligned} t_{i,j} &= \#\{\textit{reorientations of } M \textit{ with dual activity } i \textit{ and activity } j \\ &\quad \textit{and fixed orientation for active elements and dual active elements}\}. \end{aligned}$$

Example 31.29. An example of active partitions and of Corollaries 31.27 and 31.28 can be found in Figure 31.5. In each region of the left part, we write its active partition. This refines Figure 31.3 since dual active elements are the smallest elements of the parts. In the right part, there are $T(\underline{M}; 1, 0) = 6$ gray regions. They show the unique representative in each activity class, with fixed orientation for dual active elements, with respect to the oriented matroid M represented by the darker region (Corollary 31.28). They also correspond to acyclic graph orientations with a fixed unique source, which is the source endpoint of edge 1, as one can check in Figure 31.2 (Proposition 31.30).

The following proposition is from [530], and its corollary from [580].

Proposition 31.30. *Let G be a connected graph on a linearly ordered set of edges. Let s be a vertex of G. Let T be the lexicographically smallest spanning tree of G. Assume that the edges increase in the ordering along any path of T beginning at s. Then there exists one and only one orientation with unique source s in each activity class of acyclic orientations of G.*

Corollary 31.31. *For a connected graph G and a vertex s of G,*

$$T(G; 1, 0) = \#\{\textit{acyclic orientations of } G \textit{ with unique source } s\}.$$

Remark 31.32. As shown in [530, Section 7], active partitions generalize the components of an acyclic orientation of a graph, defined for certain linear orderings of the vertex set, which appear in different combinatorial contexts [271, 757, 1120] (see details in [757]). Here, they are defined for any ordering of the edge-set of a graph, any type of orientation, and general oriented matroids. This also extends to oriented matroid perspectives [528].

31.7 Filtrations in matroids and β-invariants of minors

By Proposition 31.24, the active filtration yields a decomposition of an ordered oriented matroid (or digraph) into bounded and dually bounded minors (or

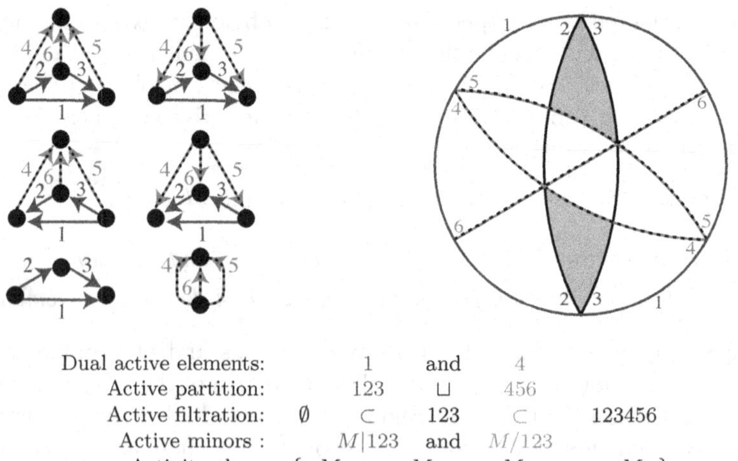

Dual active elements: 1 and 4
Active partition: 123 ⊔ 456
Active filtration: ∅ ⊂ 123 ⊂ 123456
Active minors : M|123 and M/123
Activity class: { M, −123M, −456M, −123456M }

FIGURE 31.4: Details of an activity class formed by the four orientations of K_4, or by the two gray regions and their opposite (here, M is represented by any of these regions).

bipolar and cyclic-bipolar minors). When all reorientations are considered, it is possible to describe globally all the involved filtrations (see below), and the involved minor reorientations are counted by the β-invariant (see Section 31.4). It is also possible to refine Proposition 31.24 with a uniqueness result, and derive a structural decomposition of the set of all reorientations, whose enumerative counterpart is the formula below. More generally, this formula applies to ordered matroids, as it can be also derived from a similar decomposition based on active partitions/filtrations for matroid bases, that we omit (briefly, each basis is decomposed into bases of minors with internal/external activities equal to 1/0 or 0/1, refining the internal/external partition of [469]). In this way, the formula below refines at the same time the formulas in terms of basis activities (Theorem 28.20) and in terms of orientation activities (Theorem 31.17), as well as the well-known convolution formula, Equation (4.150), obtained by using only the subsets F_c below. All these constructions were given in [525]; see also [536, 537, 538] for more details in graphs, matroids, or oriented matroids, respectively.

Definition 31.33. Let M be a matroid on a linearly ordered set E. A *filtration* of E is a sequence $(F'_j, \ldots, F'_0, F_c, F_0, \ldots, F_i)$ of subsets of E such that

1. $\emptyset = F'_j \subset \cdots \subset F'_0 = F_c = F_0 \subset \cdots \subset F_i = E$;
2. the sequence $\min(F_k \setminus F_{k-1})$, $1 \leq k \leq i$ is increasing with k;
3. the sequence $\min(F'_{k-1} \setminus F'_k)$, $1 \leq k \leq j$, is increasing with k.

The sequence is a *connected filtration* of M if, in addition

4. for $1 \leq k \leq i$, the minor $(M|F_k)/F_{k-1}$ is connected and is not $U_{0,1}$;

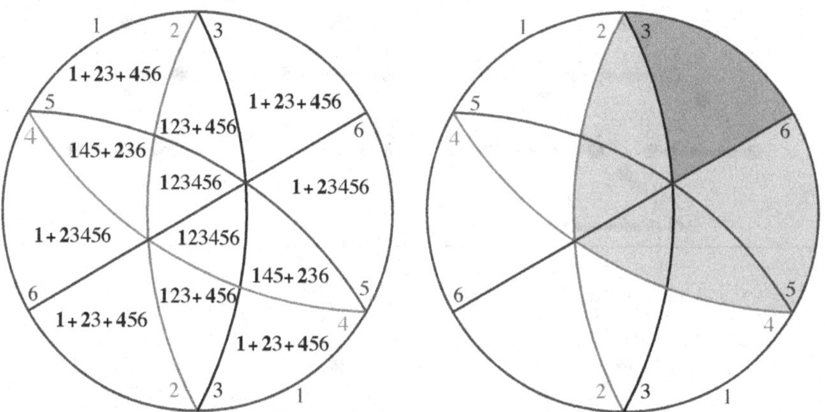

FIGURE 31.5: Active partitions, and representatives of activity classes. See Example 31.29.

5. for $1 \leq k \leq j$, the minor $(M|F'_{k-1})/F'_k$ is connected and is not $U_{1,1}$.

Recall that for a loopless graph G with at least 3 vertices, the cycle matroid $M(G)$ is connected if and only if G is 2-connected.

Theorem 31.34. *Let M be a matroid on a linearly ordered set E. Then*

$$T(M;x,y) = \sum \Big(\prod_{1 \leq k \leq i} \beta((M|F_k)/F_{k-1}) \Big) \Big(\prod_{1 \leq k \leq j} \beta^*((M|F'_{k-1})/F'_k) \Big) x^i y^j,$$

where $\beta^* = \beta$ for matroids with more than one element, $\beta^*(U_{1,1}) = \beta(U_{0,1}) = 0$ and $\beta^*(U_{0,1}) = \beta(U_{1,1}) = 1$, and where the sum can be equivalently:

1. *over all connected filtrations $(F'_j, \ldots, F'_0, F_c, F_0, \ldots, F_i)$ of M; or*
2. *over all filtrations $(F'_j, \ldots, F'_0, F_c, F_0, \ldots, F_i)$ of E.*

The two ranges of summation yield the same sum since $\beta(M) \neq 0$ if and only if M is not $U_{0,1}$ and is connected. Hence, non-zero terms in the second sum correspond to connected filtrations.

31.8 The active basis and the canonical active bijection

Given an ordered oriented matroid M (or a digraph on a linearly ordered set of edges), one can exhibit, in a canonical way, a particular basis (or spanning tree), called *the active basis* of M and denoted by $\alpha(M)$, with important properties. This is the central concept of a three-level construction called the *active bijection.* Each level provides a structural understanding of the relation between reorientations on one side and bases on the other side. From the

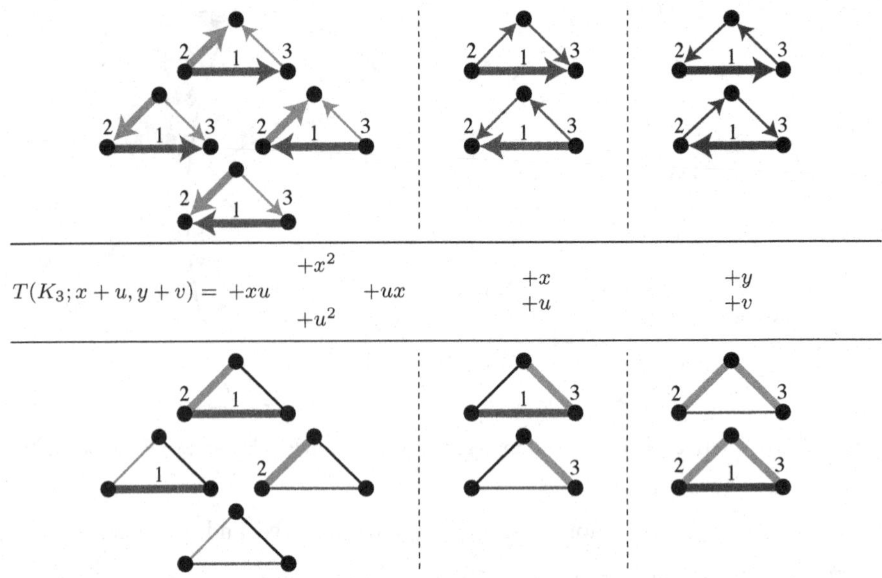

FIGURE 31.6: Active bijection illustrated on the graph K_3. See Example 31.35 for details.

enumerative viewpoint, each level provides a bijective interpretation of Tutte polynomial properties. The first level involves activities in the bounded case. The second level involves general activities. These two levels yield the canonical bijection presented in this section. In terms of arrangements (or graphs), it is intrinsic to the unsigned (or undirected) ordered structure. The third level involves refined activities and is derived from the above in the next section. These constructions were given in [525] (they are different from the correspondence proposed without proof for graphs in [749], which was abandoned as it was non-canonical and presumably non-valid beyond graphs). See also [534, 536, 537], [538, 530], or [529, 531], for more reformulations, details, and results in oriented matroids, graphs, or other special cases, respectively.

Example 31.35. Figure 31.6 illustrates the ideas we discuss below. The figure shows the active bijection for the graph K_3. The layout reflects the bijections. We have $T(K_3; x, y) = x^2 + x + y$. Each monomial in this expression corresponds to an activity class in the top part and to a basis in the top row of the bottom part associated by the canonical active bijection. Each basis yields a boolean interval of subsets (bold edges). Orientations in the top part and subsets in the bottom part are associated by the refined active bijection (with respect to the orientation displayed first in the top row), consistently with the monomials of the expanded four-variable formula (written between the top and the bottom parts of the figure).

Definition 31.36. For any ordered oriented matroid M on E, the *active basis* of M, denoted by $\alpha(M)$, is determined by the following properties.

1. *Fully optimal basis*: if M is bounded with respect to $p = \min(E)$, then $\alpha(M)$ is the unique basis B of M, such that:

   (a) for all $b \in B \setminus \{p\}$, the signs of b and $\min(C^*(b, B))$ are opposite in $C^*(b, B)$;

   (b) for all $e \in E \setminus B$, the signs of e and $\min(C(e, B))$ are opposite in $C(e, B)$.

2. *Duality:* $\alpha(M) = E \setminus \alpha(M^*)$.

3. *Decomposition of activities:* if the active filtration of M is

$$\emptyset = F'_j \subset \cdots \subset F'_0 = F_c = F_0 \subset \cdots \subset F_i = E,$$

   then

$$\alpha(M) = \bigsqcup_{1 \leq i \leq i} \alpha\big((M|F_k)/F_{k-1}\big) \sqcup \bigsqcup_{1 \leq i \leq j} \alpha\big((M|F'_{k-1})/F'_k\big).$$

Observe that the consistency of Definition 31.36 is given by Proposition 31.24 (the active minors of M are all bounded or dually bounded with respect to their smallest element). Observe also that, trivially, $\alpha(M) = \alpha(-_E M)$, so all reorientations of M in the same activity class have the same image under α.

Remark 31.37. The third property above can be replaced with the following: $\alpha(M) = \alpha(M|F) \sqcup \alpha(M/F)$ where F is the union of all positive circuits C with $\min(C) \geq a$ for $a = \max(O(M))$, or also (equivalently) for any $a \in E$.

Theorem 31.38. *Let M be an ordered oriented matroid on E.*

1. *The subset $\alpha(M)$ is well-defined, it is a basis of M, and it satisfies*

$$\mathrm{IA}_{\underline{M}}(\alpha(M)) \;=\; O^*(M), \qquad \mathrm{EA}_{\underline{M}}(\alpha(M)) \;=\; O(M).$$

2. *The 2^{i+j} reorientations of M in the same activity class with dual activity i and activity j have the same active basis $\alpha(M)$.*

3. *The mapping $-_A M \mapsto \alpha(-_A M)$ for $A \subseteq E$ provides a bijection between all activity classes of reorientations of M and all bases of M.*

The *uniactive bijection* of M is the restricted bijection between (pairs of opposite) p-bounded regions and bases with internal/external activities equal to 1/0. It thus gives a bijective interpretation of $\beta(\underline{M}) = t_{1,0}$. This restriction of Theorem 31.38 (equivalent to the existence and uniqueness of the fully optimal basis) is a difficult key result with various interpretations (it is proved in [534]). The inverse bijection (from bases to p-bounded regions) can be computed by a single pass over E (just reorient or not the elements one by one

Activity classes	Bases	#
all	all	$T(\underline{M};1,1)$
acyclic	internal	$T(\underline{M};1,0)$
totally cyclic	external	$T(\underline{M};0,1)$
bounded	internal with internal activity 1	$\beta(\underline{M}) = t_{1,0}$
dually bounded	external with external activity 1	$\beta^*(\underline{M}) = t_{0,1}$

TABLE 31.1: Canonical active bijection between activity classes of reorientations and bases. Here internal, respectively external, refers to a basis with external, respectively internal, activity 0.

following the ordering so that the sign criterion is satisfied), but the direct computation is complicated. Notably it contains a pseudo/real linear programming problem and further elaborations (hence the term *fully optimal*). It also involves the *active duality property*, $\alpha(M) = \big(E\setminus\alpha(-_pM^*)\big)\setminus\{p'\}\cup\{p\}$ where $p' = \min(E \setminus \{p\})$ (when $|E| > 1$), which strengthens linear programming duality and which highlights the consistency of the construction with the involutions seen in Remark 31.15. See [534, 530]. See also [537, 538] for summaries and [529, 535] for special cases.

The *canonical active bijection* of M is the bijection provided by Theorem 31.38. Most importantly, it depends only on the reorientation class of the oriented matroid, that is, on the non-signed arrangement in terms of a representation, or on the underlying undirected graph in the graph case (i.e., the basis $\alpha(-_AM)$ depends only on $-_AM$, not on M and $A \subseteq E$). We get a bijective transformation between Theorem 31.8 and the usual formula:

$$T(\underline{M};x,y) = \sum_{A\subseteq E} \left(\frac{x}{2}\right)^{o^*(-_AM)} \left(\frac{y}{2}\right)^{o(-_AM)} = \sum_{B \text{ basis}} x^{|\mathrm{IA}_{\underline{M}}(B)|} y^{|\mathrm{EA}_{\underline{M}}(B)|}.$$

In addition to preserving activities, this bijection is consistent with the decompositions of activities alluded to in Section 31.7: it actually transforms the active partition of M into the active partition of the basis $\alpha(M)$. The inverse construction, from bases to reorientations, can be built by a single pass over E (by reorienting or not each element one by one). See [525]; see [537, 538] for more details. See Table 31.1 for restrictions.

Remark 31.39. Various noteworthy bijections can be derived in particular cases: between permutations and increasing trees (in the case of the complete graph, or the braid arrangement, or the Coxeter arrangement A_n [531]); between signed permutations and signed increasing trees (in the case of the hyperoctahedral arrangement, or the Coxeter arrangement B_n [531]); between bounded regions and linear programming optimal vertices (in uniform oriented matroids or general position arrangements [529]); between unique source acyclic orientations and spanning trees with external activity 0 (in the case of graphs for suitable edge orderings [530], as addressed in Proposition 31.30).

31.9 Reorientations/subsets bijection, refined orientation-activities

In this section, we formally rephrase a definition and discussion given in [525]. From Theorem 31.38 and the classical partition of 2^E into basis intervals (see Section 5.4), we get various corollaries and refinements (see [537, 538] for details).

Definition 31.40. For an ordered oriented matroid M on E and $A \subseteq E$, set

$$\alpha_M(A) := \alpha(-_A M) \setminus \Big(O^*(-_A M) \cap A\Big) \cup \Big(O(-_A M) \cap A\Big).$$

Corollary 31.41. *Let M be an ordered oriented matroid on E.*

1. *The mapping $A \mapsto \alpha_M(A)$ for $A \subseteq E$ can be viewed as a bijection between all reorientations $-_A M$ of M and all subsets of E.*

2. *This bijection maps activity classes of reorientations of M onto intervals of type $[B \setminus \mathrm{IA}_{\underline{M}}(B), B \cup \mathrm{EA}_{\underline{M}}(B)]$ for bases B of M, and consists of boolean lattice isomorphisms for these restrictions.*

The *refined active bijection* with respect to (the reference reorientation) M is the bijection provided by Corollary 31.41. See Figure 31.6 for an example. It is important that, in contrast with the canonical one, this bijection is induced by the choice of a reference reorientation. By Theorem 31.38, we have a bijection between activity classes and basis intervals (see also Chapter 5 and Proposition 28.21). The role of the reference orientation is simply to "break the symmetry" and define an isomorphism for each such pair of boolean lattices. In detail, consider the unique representative of an activity class with fixed orientation with respect to M for active and dual active elements (that is, $-_A M$ such that $O^*(-_A M) \cap A = O(-_A M) \cap A = \emptyset$; see above Corollary 31.28). It is mapped onto its active basis by α_M. Then, the other elements of the activity class are mapped onto the other elements of the basis interval in the obvious induced way. Formally, combining Corollary 31.41 with subset activities of Proposition 28.23, we obtain the following.

Corollary 31.42. *For an ordered oriented matroid M on E, and $A \subseteq E$,*

$$\begin{aligned} \mathrm{IA}_{\underline{M}}(\alpha_M(A)) &= O^*(-_A M) \setminus A, & P_M(\alpha_M(A)) &= O^*(-_A M) \cap A, \\ \mathrm{EA}_{\underline{M}}(\alpha_M(A)) &= O(-_A M) \setminus A, & Q_M(\alpha_M(A)) &= O(-_A M) \cap A. \end{aligned}$$

Definition 31.43. For an ordered oriented matroid M on E, we define the four *refined orientation-activities* of M as follows, for $A \subseteq E$:

$$\begin{aligned} \theta^*_M(A) &= |O^*(-_A M) \setminus A|, & \bar{\theta}^*_M(A) &= |O^*(-_A M) \cap A|, \\ \theta_M(A) &= |O(-_A M) \setminus A|, & \bar{\theta}_M(A) &= |O(-_A M) \cap A|. \end{aligned}$$

Recall the following Tutte polynomial expansion in terms of subset activities (see Chapter 5; see also Theorem 28.24), where $\underline{M}$ is an ordered matroid on E:

$$T(\underline{M}; x+u, y+v) = \sum_{A\subseteq E} x^{|\mathrm{IA}_{\underline{M}}(A)|}\, u^{r(M)-r_M(A)}\, y^{|\mathrm{EA}_{\underline{M}}(A)|}\, v^{|A|-r_M(A)}.$$

The Tutte polynomial expansion below (Corollary 31.44) is the counterpart of the expansion above in terms of refined orientation-activities. It can be deduced from the latter by the bijection of Corollary 31.41 and Corollary 31.42. It can also be directly proved, without using the machinery of the active bijection, by using Proposition 31.25 to expand the formula of Theorem 31.8 in an appropriate way (similarly to proving Theorem 28.24 from Proposition 28.21; see [528, 538, 537]). This formula is extended to oriented matroid perspectives in [528, 755].

Corollary 31.44. *For an ordered oriented matroid M on E,*

$$T(\underline{M}; x+u, y+v) \;=\; \sum_{A\subseteq E} x^{\theta^*_M(A)}\; u^{\bar\theta^*_M(A)}\; y^{\theta_M(A)}\; v^{\bar\theta_M(A)}.$$

From Corollary 31.44, one can derive various formulas by specializing variables, in the same way as for subset activities in Section 28.6. Each formula given in Section 28.6 (in matroids) gives a formula in terms of refined orientation activities as above. We give some of them below (see also [528, 755]).

Corollary 31.45. *For an ordered oriented matroid M on E, and two nonnegative integers p, q, it follows that*

1. $T(\underline{M}; x, y) = \sum_{A\subseteq E} (x-1)^{\theta^*_M(A)} (y-1)^{\theta_M(A)} = \sum_{\substack{A\subseteq E\\ \bar\theta^*_M(A)=\bar\theta_M(A)=0}} x^{\theta^*_M(A)} y^{\theta_M(A)}$,

2. $\dfrac{\partial^{p+q}}{\partial x^p \partial y^q} T(\underline{M}; x, y) = p!q! \sum_{\substack{A\subseteq E\\ \bar\theta^*_M(A)=p,\bar\theta_M(A)=q}} x^{\theta^*_M(A)} y^{\theta_M(A)}$,

3. $T(\underline{M}; 2, 0) = \sum_{\substack{A\subseteq E\\ \bar\theta^*_M(A)=\theta_M(A)=\bar\theta_M(A)=0}} 2^{\theta^*_M(A)} = \sum_{\substack{A\subseteq E\\ \theta^*_M(A)=\theta_M(A)=\bar\theta_M(A)=0}} 2^{\bar\theta^*_M(A)}$.

Table 31.2 shows some noteworthy restrictions of Corollary 31.41. Each one of them corresponds to special values of activities in Corollary 31.42.

31.10 Counting circuit/cocircuit reversal classes

We consider equivalence classes on the set of reorientations, which are enumerated by the Tutte polynomial in regular oriented matroids. These classes

Reorientations	Subsets	#
all	all	$T(\underline{M};2,2)$
acyclic	no-broken-circuit subsets	$T(\underline{M};2,0)$
totally cyclic	supersets of external bases	$T(\underline{M};0,2)$
acyclic and dual-active-fixed	internal bases	$T(\underline{M};1,0)$
totally cyclic and active-fixed	external bases	$T(\underline{M};0,1)$
active-fixed	independent subsets	$T(\underline{M};2,1)$
dual-active-fixed	spanning subsets	$T(\underline{M};1,2)$
active-fixed and dual-active-fixed	bases	$T(\underline{M};1,1)$

TABLE 31.2: Refined active bijection between reorientations and subsets. Call a reorientation with fixed orientation for active (respectively dual active) elements with respect to the reference reorientation M *active-fixed* (respectively *dual-active-fixed*).

have been introduced and studied in [526] for graphs (where they have special interpretations), and in [527] for oriented matroids (reproved and completed in [539] by means of a relation with activity classes). Regular matroids can always be oriented (similarly to graphs, two oriented matroids with the same underlying regular matroid are reorientations of each other). They can be characterized as orientable binary matroids (that is, orientable matroids with no $U_{2,4}$ minor).

Definition 31.46. Let M be an oriented matroid on E. Write $M \sim -_C M$ if C is a positive circuit of M. Applying the same rule to reorientations $-_A M$ for $A \subseteq E$ (i.e., writing $-_A M \sim -_{C \triangle A} M$ when C is a positive circuit of $-_A M$) and taking the transitive closure of the relation, we define the *circuit-reversal equivalence relation* for the set of reorientations of M. Dually, using positive cocircuits instead of circuits, we define the *cocircuit-reversal equivalence relation.* Taking the transitive closure of the union of the two above relations, we define the *circuit-cocircuit-reversal equivalence relation.*

For a graph G, we obtain equivalence relations for the set of orientations of G, which amounts to reversing the directions of edges of successive directed cycles/cuts. Alternative definitions for these classes in digraphs can be given (e.g., see [297]). The *indegree-sequence* of an orientation $\vec{G}$ of G is the sequence of numbers of incoming edges at each vertex of G (in a fixed order). Some of the equivalence classes above are bijectively related with indegree-sequences of G, consistently with the theorem below. (The results about indegree-sequences can be rephrased in terms of score vectors.)

Theorem 31.47. *Let M be a regular oriented matroid. Let G be a connected graph with a given vertex s. The results below involving G have to be read assuming that $\underline{M} = M(G)$.*

$$\begin{aligned} T(\underline{M};1,1) &= \#\{\textit{circuit-cocircuit-reversal classes of } M \ (\textit{or } G)\} \\ &= \#\{\textit{indegree-sequences of orientations of } G \textit{ such} \end{aligned}$$

$$\begin{aligned}
&\qquad\qquad \textit{that } s \textit{ is reachable from any other vertex}\},\\
T(\underline{M};1,0) &= \#\{\textit{acyclic cocircuit-reversal classes of } M \ (\textit{or } G)\}\\
&= \#\{\textit{acyclic orientations of } G \textit{ with unique sink at } s\}\\
&= \#\{\textit{indegree-sequences of acyclic orientations of } G\\
&\qquad\qquad \textit{with unique sink at } s\},\\
T(\underline{M};1,2) &= \#\{\textit{cocircuit-reversal classes of } M \ (\textit{or } G)\}\\
&= \#\{\textit{orientations of } G \textit{ such that } s \textit{ is reachable}\\
&\qquad\qquad \textit{from any other vertex}\},\\
T(\underline{M};0,1) &= \#\{\textit{totally cyclic circuit-reversal classes of } M\\
&\quad (\textit{or strongly connected circuit-reversal classes of } G)\}\\
&= \#\{\textit{indegree-sequences of strongly connected}\\
&\qquad\qquad \textit{orientations of } G\},\\
T(\underline{M};2,1) &= \#\{\textit{circuit-reversal classes of } M \ (\textit{or } G)\}\\
&= \#\{\textit{indegree-sequences of all orientations of } G\}
\end{aligned}$$

Conversely, let M be a non-regular oriented matroid. Then, the equalities above involving $T(\underline{M};1,1)$, $T(\underline{M};1,2)$ and $T(\underline{M};2,1)$, fail: the left-hand is larger than the right hand (and the same result holds for the equalities involving $T(\underline{M};1,0)$ and $T(\underline{M};0,1)$ under the assumption that M has no loop nor coloop).

Theorem 31.47 combines statements from [526, 527] (enumerations) and [539] (converse). This theorem thus yields a characterization of regular matroids among orientable matroids. In graphs, the second and the last equalities can also be found in [580] and in [247, ?], respectively.

Remark 31.48. For any integer k, the number of acyclic cocircuit-reversal classes of a uniform oriented matroid $U_{2,k}$ equals 1, or 2, if k is even, or odd, respectively (see [527], and see [539] for using a $U_{2,4}$ minor in the non-regular case).

Remark 31.49. As discussed in [526], circuit-cocircuit reversals generalize in a self-dual way the dynamics of the chip firing game (see Chapter 17 for background on the chip firing game) at a certain level and share enumerative properties with this system.

Remark 31.50. For a linear ordering of the ground set of M when M is regular (respectively, a suitable ordering of the edge set of G), acyclic reorientations with fixed orientation for dual active elements (respectively, acyclic orientations with a unique given sink), are at the same time representative of acyclic cocircuit-reversal classes and of acyclic activity classes (see the discussion above Corollary 31.28, Proposition 31.30, and [539] for details).

31.11 Open problems

Any Tutte polynomial property or formula available for matroids applies in particular to oriented matroids. Hence a general natural question is to search for interpretations or refinements arising when the structure is oriented, notably in terms of a topological representation. For instance, the Merino–Welsh conjecture (see Section 3.8) can be formulated in oriented matroids with a topological flavor, since $T(\underline{M}; 2, 0)$ is the number of regions of a pseudo-sphere arrangement and $T(\underline{M}; 0, 2)$ the number of regions of the dual pseudo-sphere arrangement (see Section 31.3), while $T(\underline{M}; 1, 1)$ is the number of pseudo-simplices (matroid bases) in these two arrangements.

Beyond this, a rather general problem is to address matroid perspective properties, or to generalize oriented matroid properties, in the context of oriented matroid perspectives. An interesting feature of an oriented matroid perspective $M \to M'$ in terms of the Tutte polynomial is that the two oriented matroids are addressed with dual viewpoints: positive circuits of M and positive cocircuits of M' play the main role (see Section 31.5). The generalization to oriented matroid perspectives of the active bijection and related constructions is work in progress (for example, see [528] on active partitions), as well as deletion/contraction and linear programming constructions in oriented matroids (see [538] for graphs, and [537] for further references).

An interesting special case of oriented matroids is given by regular matroids (that is, binary orientable matroids). This class is thus a good candidate to generalize, or dualize, directed graph properties in oriented matroids, notably Tutte polynomial properties (for example, see Section 31.10).

On another note, one can wonder if there are any properties of Tutte polynomials that are particular to the oriented case. Not every Tutte polynomial of a matroid is the Tutte polynomial of an oriented matroid. For example, the Fano plane is Tutte unique (see Section 6.4.3) and non-orientable, so no oriented matroid has the same Tutte polynomial as the Fano plane. The question is then whether the Tutte polynomials that arise from oriented matroids have some special purely algebraic or enumerative properties that distinguish them.

32

Valuative invariants on matroid basis polytopes

Michael J. Falk • Joseph P.S. Kung

Synopsis

The Tutte polynomial is a valuative invariant on matroid basis polytopes. We give an introduction to such invariants, focusing on the $\mathcal{G}$-invariant, which is the universal valuative invariant, and its relation to the Tutte polynomial.

- Valuative functions on matroid base polytopes.
- The $\mathcal{G}$-invariant and its specialization to the Tutte polynomial.

32.1 Introduction

As the universal invariant of matroids under the deletion–contraction recursion, the Tutte polynomial seems to occupy an unassailable position in enumerative matroid theory. However, there are other decompositions, equally fundamental, such as decompositions of matroid basis polytopes. Here, the universal invariant is the $\mathcal{G}$-invariant. The Tutte polynomial turns out to be a specialization of the $\mathcal{G}$-invariant so that the $\mathcal{G}$-invariant offers a wider environment in which the Tutte polynomial lives. This chapter covers a new and evolving area of research, and our account is brief and exploratory.

DOI: 10.1201/9780429161612-32

32.2 Valuative functions on matroid base polytopes

32.2.1 Subdivisions of base polytopes

Let E be a finite set, and let $\mathbb{R}^E$ be the $|E|$-dimensional real vector space with coordinates labeled by the set E, so that the standard basis vectors e_s, for $s \in E$, form a basis. The *(matroid) base polytope* $Q(M)$ of a matroid M on the set E is the convex polytope in $\mathbb{R}^E$ obtained by taking the convex closure of indicator vectors of bases of M, that is,

$$Q(M) = \mathrm{conv}\Big\{ \sum_{b \in B} e_b : B \text{ is a basis of } M \Big\}.$$

A *(base polytope) subdivision* is a decomposition

$$Q(M) = Q(M_1) \cup Q(M_2) \cup \cdots \cup Q(M_k),$$

where

1. $Q(M_1), Q(M_2), \ldots, Q(M_k)$ are all base polytopes of matroids $M_1, M_2, \ldots, M_k$ on the set E; and
2. for every subset $I \subseteq \{1, 2, \ldots, k\}$ with two or more elements, the intersection $\bigcap_{i \in I} Q(M_i)$ is a proper face of the polytopes $Q(M_i)$, for $i \in I$.

32.2.2 Valuative invariants

A function v defined on matroid base polytopes is *invariant* if it depends only on the isomorphism class of the matroid. It is *valuative* if for every subdivision $Q(M) = Q(M_1) \cup Q(M_2) \cup \cdots \cup Q(M_k)$, it satisfies the inclusion-exclusion identity:

$$v(Q(M)) = \sum_{\substack{I \subseteq \{1,2,\ldots,k\} \\ I \neq \emptyset}} (-1)^{|I|-1} v\Big(\bigcap_{i \in I} Q(M_i)\Big).$$

For example, any valuation of polytopes, such as the volume or a quermassintegral, gives a valuative invariant on base polytopes. (A formula for the volume of a base polytope can be found in [36]. Examples of valuative functions which are not invariant can be found in [40].)

32.2.3 The $\mathcal{G}$-invariant

In [374], Derksen introduced the $\mathcal{G}$-invariant. It can be defined in the following way. Let M be a rank-r matroid on the set $\{1, \ldots, n\}$. Let π be a permutation of $\{1, \ldots, n\}$. The *rank sequence* $\underline{r}(\pi) = (r_1, \ldots, r_n)$ of π is the sequence defined by $r_1 = r(\{\pi(1)\})$, where $r(A)$ denotes the rank of the set A in M, and, for $j \geq 2$,

$$r_j = r(\{\pi(1), \pi(2), \ldots, \pi(j)\}) - r(\{\pi(1), \pi(2), \ldots, \pi(j-1)\}).$$

For matroids, $r_j = 0$ or 1, there are exactly r 1's, and the set $B(\underline{r})$ of elements $\pi(j)$ where $r_j = 1$ is a basis of M.

Definition 32.1. Let $[\underline{r}]$ be a variable or formal symbol, one for each $(0,1)$-sequence $\underline{r}$. The *$\mathcal{G}$-invariant* and its coefficients $g_{\underline{r}}$ are defined by

$$\mathcal{G}(M) = \sum_{\pi} [\underline{r}(\pi)] = \sum_{\underline{r}} g_{\underline{r}}(M)[\underline{r}]$$

where the first sum ranges over all $n!$ permutations of E.

Derksen showed that the $\mathcal{G}$-invariant is a valuative invariant. Note that if 1^r0^{n-r} is the sequence beginning with r 1's followed by $n-r$ 0's, then

$$g_{1^r0^{n-r}}(M) = r!(n-r)!b(M),$$

where $b(M)$ is the number of bases in M.

Remark 32.2. Derksen defines $\mathcal{G}(M)$ with $[\underline{r}]$ an element of "a convenient basis" of the algebra of quasisymmetric functions. However, the only property used is that basis elements are linearly independent and hence can be assigned values independently. It seems simpler to define $[\underline{r}]$ as a formal symbol, as we do here, to which one may assign any interpretation, including a quasisymmetric function.

A *specialization* of the $\mathcal{G}$-invariant with values in an abelian group A is a function assigning a value in A to each symbol $[\underline{r}]$. As was noted by Derksen, the formula for the rank function of the dual M^* implies that the rank sequence of π in M^* can be obtained by switching 0's and 1's in the rank sequence of π^{rev} in M, where π^{rev} is the permutation defined by $\pi^{\text{rev}}(j) = \pi(n-j)$. Thus we have the following.

Proposition 32.3. *$\mathcal{G}(M^*)$ is the specialization of $\mathcal{G}(M)$ given by replacing the symbol $[\underline{r}]$ by the symbol $[\underline{r}^*]$, where $\underline{r}^*$ is the sequence obtained by reversing $\underline{r}$ and switching 0's and 1's.*

The fundamental theorem in this area, due to Derksen and Fink [375], says that the $\mathcal{G}$-invariant encompasses all valuative invariants.

Theorem 32.4. *The $\mathcal{G}$-invariant is a universal valuative invariant on base polytopes, in the sense that every valuative invariant on base polytopes is a specialization of the $\mathcal{G}$-invariant.*

Derksen [374] gave an explicit specialization of the $\mathcal{G}$-invariant to the Tutte polynomial using quasisymmetric functions. Hence, the Tutte polynomial is a valuative invariant. Derksen's specialization is restated in the following lemma.

Lemma 32.5. *The assignment*

$$[r_1 r_2 \dots r_n] \mapsto \sum_{m=0}^{n} \frac{x^{r-\mathrm{wt}(r_1 r_2 \dots r_m)} y^{m-\mathrm{wt}(r_1 r_2 \dots r_m)}}{m!(n-m)!},$$

where the Hamming weight $\mathrm{wt}(r_1 r_2 \dots r_m)$ *is the number of* 1*'s in the initial segment* $r_1 \cdots r_m$ *of the sequence* $r_1 \cdots r_m r_{m+1} \cdots r_n$*, sends the* $\mathcal{G}$*-invariant to the Tutte polynomial.*

32.2.4 Paving matroids

Calculations of $\mathcal{G}$-invariants are, in general, time-consuming. However, for paving matroids, which have a simple structure, explicit formulas can be written down. Recall from Section 4.6 that a rank-r matroid P on the set E is *paving* if all circuits have r or $r+1$ elements. A copoint X (or rank-$(r-1)$ flat) is *trivial* if $|X| = r-1$ and *non-trivial* if $|X| \geq r$. A paving matroid is *sparse* if all non-trivial copoints have size r.

Let P be a paving matroid on $\{1, 2, \dots, n\}$. Since every subset of size $r-1$ is independent, the rank sequence of a permutation π starts with $(r-1)$ 1's. The remaining 1 occurs in position i, where

$$i = 1 + \max\{j : \{\pi(1), \pi(2), \dots, \pi(j-1)\} \subseteq X\}.$$

Thus, if X a trivial copoint, the last 1 occurs in position $i = r$, and in all cases $r \leq i \leq |X| + 1$. Hence we have the following.

Proposition 32.6. *For a rank-r paving matroid P on n elements,*

$$\begin{aligned}\mathcal{G}(P) = &\sum_{X \text{ trivial}} (r-1)!(n-r+1)![1^r 0^{n-r}] \\ &+ \sum_{X \text{ non-trivial}} \Big(\sum_{i=r}^{|X|+1} \frac{|X|!}{(|X|-i+1)!}(n-|X|)(n-i)![1^{r-1}0^{i-r}10^{n-i}] \Big),\end{aligned}$$

where $1^{r-1}0^{i-r}10^{n-i}$ *is the sequence with* 1*'s in the leading* $r-1$ *positions and the remaining* 1 *in the* i*-th position.*

Example 32.7. Since all rank-3 simple matroids are paving, we have explicit formulas for their $\mathcal{G}$-invariants. For example, let P be the paving matroid on the set $\{1, 2, 3, 4, 5, 6\}$, with non-trivial copoints $\{1, 2, 3, 4\}$, $\{4, 5, 6\}$ and trivial copoints $\{1, 5\}$, $\{1, 6\}$, $\{2, 5\}$, $\{2, 6\}$, $\{3, 5\}$, $\{3, 6\}$. Geometrically, P is the rank-3 matroid consisting of a 4-point line $\{1, 2, 3, 4\}$ and a 3-point line $\{4, 5, 6\}$ intersecting at the point 4. Then

$$\mathcal{G}(P) = 48[110010] + 132[110100] + 540[111000].$$

For sparse paving matroids, there is an even simpler formula derived by counting bases. If M is a rank-r sparse paving matroid on n elements with α non-trivial copoints, then M has $\binom{n}{r} - \alpha$ bases. Hence we have the following.

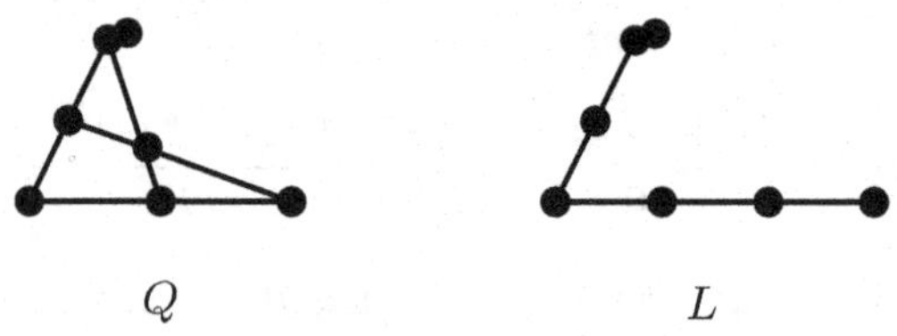

FIGURE 32.1: The matroids Q and L.

Proposition 32.8. *If M is a rank-r sparse paving matroid on n elements with α non-trivial copoints, then*

$$\mathcal{G}(M) = \left(\binom{n}{r} - \alpha\right) r!(n-r)![1^r 0^{n-r}] + \alpha r!(n-r)![1^{r-1}010^{n-r-1}].$$

As an example of Proposition 32.8,

$$\mathcal{G}(M(K_4)) = 16 \cdot 36[111000] + 4 \cdot 36[110100].$$

The next proposition is an immediate consequence of the explicit formulas.

Proposition 32.9.

1. *The $\mathcal{G}$-invariant of a paving matroid P depends only on the rank, the number of elements, and the multiset*

$$\{|X| : X \text{ is a non-trivial copoint in } P\}.$$

2. *The $\mathcal{G}$-invariant of a sparse paving matroid depends only on the rank, the number of elements, and the number of bases.*

The analogue of Proposition 32.9 holds for the Tutte polynomial [245]. Thus, for paving matroids, the $\mathcal{G}$-invariant and the Tutte polynomial contain exactly the same information. As a consequence, the Orlik–Solomon algebra (see Chapter 30) $\mathsf{A}(M)$ determines $\mathcal{G}(M)$ for rank-3 simple matroids M having a real realization, but not conversely.

Example 32.10. Consider the two matroids Q and L in Figure 32.1 given in [247, p. 133]. They are the smallest pair of non-paving matroids with the same Tutte polynomial. Then $\mathcal{G}(Q)$ equals

$$48[1010100] + 192[1011000] + 240[1100100] + 1104[1101000] + 3456[111000],$$

and $\mathcal{G}(L)$ equals

$$24[1010100] + 216[1011000] + 264[1100100] + 1080[1101000] + 3456[111000].$$

32.2.5 The specialization of $\mathcal{G}$ to T

We discuss some consequences of the fact that the Tutte polynomial is a specialization of the $\mathcal{G}$-invariant. Let $\mathcal{G}(n,r)$ be the vector space of dimension $\binom{n}{r}$ consisting of formal linear combinations of the symbols $[\underline{r}]$ where $\underline{r}$ ranges over all length-n $(0,1)$-sequence with r 1's with coefficients in $\mathbb{Q}$ (or a field of characteristic zero). We shall construct a natural basis for this vector space.

Definition 32.11. Let $\underline{r}$ be the length-n $(0,1)$-sequence with 1's in positions $b_1, b_2, \ldots, b_r$, where $b_1 < b_2 < \cdots < b_r$. The *freedom matroid* $F(\underline{r})$ defined by the $(0,1)$-sequence $\underline{r}$ is the rank-r matroid on the set $1, 2, \ldots, n$ in which

1. $1, 2, \ldots, b_1 - 1$ are loops;
2. if $1 \leq j \leq r-1$, then b_j is an isthmus in the restriction of $F(\underline{r})$ to $\{b_1, \ldots, b_j\}$, and $b_j, b_j + 1, \ldots, b_{j+1} - 1$ are freely positioned in $\mathrm{cl}(\{b_1, b_2, \ldots, b_j\})$; and
3. $b_r, b_r + 1, \ldots, n$ are freely positioned in the entire matroid.

Freedom matroids were first studied by Crapo [346]; they have been rediscovered many times and are also known as nested, counting, or Schubert matroids. Combinatorial formulas for the Tutte polynomials of freedom matroids can be found in [170].

The matroid $F(\underline{r})$ can be represented by a generic matrix in lower reduced row-echelon form having pivots in columns $b_1, b_2, \ldots, b_r$. In particular, freedom matroids are the matroids associated with Schubert cells in the complex Grassmannian.

We shall need a partial order on $(0,1)$-sequences. If $\underline{r}$ and $\underline{s}$ are two length-n $(0,1)$-sequences with the same number of 1's, then $\underline{s} \trianglerighteq \underline{r}$ if for every index j,

$$s_1 + s_2 + \cdots + s_j \geq r_1 + r_2 + \cdots + r_j,$$

in other words, reading from the left, there are always at least as many 1's in $\underline{s}$ as there are in $\underline{r}$. This order has maximum $1^r 0^{n-r}$ and minimum $0^{n-r} 1^r$ for each r. It is easy to see that the symbol $[\underline{s}]$ occurs with non-zero coefficient in $\mathcal{G}(F(\underline{r}))$ only if $\underline{s} \trianglerighteq \underline{r}$. Hence, the system of equations

$$\sum_{\underline{s}} g_{\underline{s}}(F(\underline{r}))[\underline{s}] = \mathcal{G}(F(\underline{r}))$$

is triangular with non-zero diagonal entries. We can invert the system and write a symbol as a linear combination of $\mathcal{G}$-invariants of freedom matroids. This yields the following theorem from [375].

Theorem 32.12. *The $\mathcal{G}$-invariants $\mathcal{G}(F(\underline{r}))$ of freedom matroids form a basis for $\mathcal{G}(n,r)$.*

Since the specialization of $\mathcal{G}$ to T maps $\mathcal{G}(F(\underline{r}))$ to $T(F(\underline{r}))$, we have the following corollary of Theorem 32.12 (see [736]).

Corollary 32.13. *The Tutte polynomial of a rank-r matroid on n elements is a linear combination of the Tutte polynomials of freedom matroids $T(F(\underline{r}))$, where $\underline{r}$ ranges over all length-n $(0,1)$-sequences with r 1's.*

Example 32.14. For a rank-r sparse paving matroid M on n elements with α non-trivial copoints (all of size r),

$$T(M) \quad = \quad -(\alpha-1)T(F(1^r0^{n-r})) + \alpha T(F(1^{r-1}010^{n-r-1})).$$

Note that $F(1^r0^{n-r})$ is the uniform matroid $U_{r,n}$ and $F(1^{r-1}010^{n-r-1})$ is the paving matroid with one non-trivial r-element copoint.

The assignment $\mathcal{G}(F(\underline{r}))$ to $T(F(\underline{r}))$ gives a linear transformation Sp from $\mathcal{G}(n,r)$ to the vector space $\mathbb{Q}[x,y]$ of polynomials in two variables x and y with coefficients in the rational numbers $\mathbb{Q}$ or any field of characteristic zero. The image is the subspace $\mathcal{T}(n,r)$ of $\mathbb{Q}[x,y]$ spanned by Tutte polynomials of rank-r matroids on n elements. Such Tutte polynomials are linear combinations of monomials $(x-1)^i(y-1)^j$, where $0 \le i \le r$ and $0 \le j \le n-r$. Hence an upper bound on the dimension of $\mathcal{T}(n,r)$ is $(r+1)(n-r+1)$. Thus, when n is sufficiently large compared to r, Sp is not an injection because $\dim \mathcal{G}(n,r) = \binom{n}{r}$. A specific example of a *syzygy* of Sp, that is, an element in its kernel, is given by Example 32.10. The linear combination

$$[1010100] - [1011000] - [1100100] + [1101000]$$

is in the kernel because $T(Q) = T(L)$. It converts to the linear relation

$$\begin{aligned} T(F(1010100)) - T(F(1011000)) - T(F(1100100)) \\ +2T(F(1101000)) - T(F(1110000)) = 0 \end{aligned}$$

on Tutte polynomials of freedom matroids. A solution to the *syzygy problems,* to find an explicit generating or spanning set for (i) the kernel of Sp, and (ii) the linear relations among Tutte polynomials of freedom matroids, can be found in [736]. This solution shows that $\dim \mathcal{T}(n,r) = r(n-r)+1$. We remark that linear relations among (arbitrary) Tutte polynomials generalize Tutte-equivalence, which are linear relations between two Tutte polynomials.

32.2.6 The F-invariant

A precursor and motivator of the $\mathcal{G}$-invariant is the F-invariant defined by Billera, Jia, and Reiner [118]. Our description of the F-invariant assumes an acquaintance with the greedy algorithm axiomatization (Edmonds [441]; see, for example, [894, p. 55]).

Let M be a matroid on E. A function $f : E \to \{1,2,3,\ldots\}$ is *M-generic* if the minimum

$$\min\Big\{\sum_{b\in B} f(b) : B \text{ is a basis of } M\Big\}$$

is achieved by exactly one basis. Let $x_1, x_2, x_3, \ldots$ be variables and define the F*-invariant* $F(M)$ to be formal power series given by

$$F(M) = \sum_{f} \prod_{i \in E} x_{f(i)},$$

where the sum ranges over all M-generic functions. The power series $F(M)$ is a quasisymmetric function. The definition of the F-invariant inspired that of $\mathcal{G}$. The specialization of $\mathcal{G}$ to F is given in [374]. The precise relationship of F with the Tutte polynomial remains unresolved.

32.3 Open problems

The $\mathcal{G}$-invariant was discovered by Derksen [374] in 2010. Although much fundamental work has been done since then, the study of the combinatorial aspects, especially the interpretation of its specializations, is open, uncharted territory, rather like the area of Tutte polynomials in the 1960s and 1970s. The challenge is not to settle open problems, but to find the problems themselves.

33

Non-matroidal generalizations

Gary Gordon • Elizabeth McMahon

Synopsis

This chapter covers several generalizations of the Tutte polynomial to combinatorial structures that are not matroids, including structures arising from trees, rooted graphs, posets, finite subsets of points in Euclidean space. In each case, the generalized Tutte polynomial encodes meaningful combinatorial information about the object under consideration.

- Defining a Tutte polynomial for a set with a rank function.
- Computing the Tutte polynomial using a deletion–contraction recursion.
- Application to greedoids, specialized to rooted graphs and digraphs.
- Application to antimatroids, specialized to trees, rooted trees and posets.
- The β-invariant for antimatroids, specialized to chordal graphs and finite subsets of Euclidean space.

33.1 Introduction

Chapter 4 of this handbook discusses the generalization of the Tutte polynomial to matroids; many of the standard results that hold for the Tutte polynomial of a graph generalize to matroids. We will see that many of these results can be further generalized to interesting combinatorial structures that are not matroids. These non-matroidal generalizations arise from trees, rooted graphs, posets, finite subsets of points in Euclidean space, and more. In each of these cases, the generalized Tutte polynomial encodes meaningful combinatorial information about the object under consideration.

DOI: 10.1201/9780429161612-33

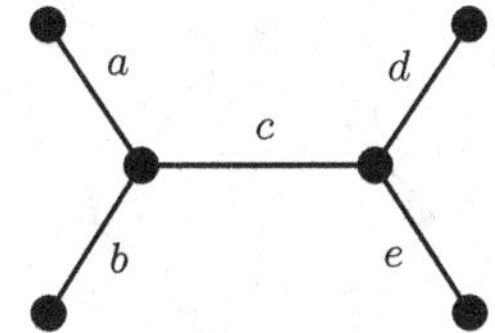

FIGURE 33.1: The tree for Example 33.3.

Recall that the Tutte polynomial of a matroid can be defined as a *corank-nullity* generating function. This definition depends on a *rank function* $r : 2^S \to \mathbb{Z}$, where S is the ground set of the matroid.

Definition 33.1. Let S be a finite (ground) set with a rank function $r : 2^S \to \mathbb{Z}$. Then the Tutte polynomial of the pair $G = (S, r)$ is

$$T(G; x, y) = \sum_{A \subseteq S} (x-1)^{r(S)-r(A)}(y-1)^{|A|-r(A)}.$$

Remark 33.2. The rank functions considered in this chapter will be reasonably "well-behaved". In particular, these functions will satisfy the following *greedoid* rank conditions:

(r1) $r(\emptyset) = 0$,

(r2) $r(A) \leq r(A \cup \{p\})$,

(r3) $r(A) \leq |A|$,

(r4) If $r(A) = r(A \cup p_1) = r(A \cup p_2)$, then $r(A \cup \{p_1, p_2\}) = r(A)$,

known as the *normalization, increasing, subcardinality, local semimodularity* conditions, respectively. (More information about greedoids appears in Section 33.2.)

Example 33.3. Consider the tree T of Figure 33.1. A subset of the edges is said to be *feasible* if it is the complement of a subtree. (This makes the tree a *pruning (or shelling) antimatroid*, a topic we will return to in Section 33.3.) The rank of a subset is then the size of the largest feasible set contained in that subset. For instance, $r(\{a, b, c\}) = 3$ since this set is the complement of a subtree (namely, $\{d, e\}$), while $r(\{b, c\}) = 1$ since $\{b\}$ is the largest subtree complement contained in $\{b, c\}$.

Checking the rank of each of the 32 subsets gives us the Tutte polynomial directly from Definition 33.1:

$$T(G; x, y) = x^5y - x^4y - 2x^3y + 2x^3 + 2x^2y + xy - x - y.$$

Unlike the case of graphs and matroids, this generalized Tutte polynomial can have negative coefficients. But many of the evaluations for this polynomial remain meaningful:

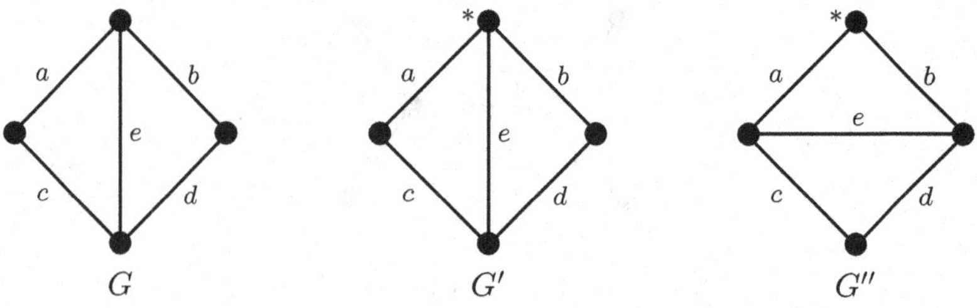

FIGURE 33.2: One unrooted and two rooted graphs for Example 33.4.

1. $T(G;1,1) = 1$. In general, $T(G;1,1)$ is the number of generalized bases, i.e., sets satisfying $r(A) = |A| = r(S)$. In this case, the entire tree is the only basis, agreeing with the matroid interpretation.
2. $T(G;1,2) = 1$. In general, $T(G;1,2)$ is the number of generalized spanning sets, i.e., sets containing a basis. When T is a tree, the entire tree is the only spanning set, again agreeing with the matroid interpretation.
3. $T(G;2,1) = 23$. In general, $T(G;2,1)$ is the number of feasible sets. When T is a tree, this equals the total number of subtrees of the tree, because there is a one-to-one correspondence between the feasible sets and subtrees. (The empty set is considered a subtree.) For a matroid, this evaluation is the number of independent sets.
4. $T(G;2,2) = 32$. In general, $T(G;2,2)$ is always 2^n, where n is the size of the ground set S.

As a comparison with the classical graph or matroid Tutte polynomial, the matroid $M(T)$ associated with a tree T on n edges is the free matroid $U_{n,n}$, so the *classical* Tutte polynomial of T is just x^n.

We give another comparison between the (classical) Tutte polynomial and the generalized version in the next example.

Example 33.4. Consider the three graphs G, G' and G'' of Figure 33.2. The graph G is an ordinary, unrooted graph, so the Tutte polynomial is the classical one. On the other hand, G' and G'' are rooted graphs, where a certain vertex has been distinguished (marked with a $*$ in the Figure 33.2). We can define a rank function on a rooted graph as follows: the rank of a subset A of edges is the size of the largest rooted subtree contained in A (where rooted subtrees contain the same root as the graph).

Each graph in the Figure 33.2 has rank 3 with a ground set of size 5. Definition 33.1 gives:

$$\begin{aligned} T(G;x,y) &= x^3 + 2x^2 + 2xy + x + y^2 + y, \\ T(G';x,y) &= x^3y^2 - 3x^2y^2 + 2x^2y + x^2 + 3xy^2 - 2xy + 3x + 3y, \\ T(G'';x,y) &= x^3y^3 - 3x^2y^3 + 2x^2y + 3xy^3 - 3xy + 4x - y^3 + y^2 + 4y. \end{aligned}$$

As in Example 33.3, G' and G'' give greedoid Tutte polynomials (defined in the next section) with negative coefficients. But, also as above, we can interpret the standard evaluations of each polynomial.

1. $T(G;1,1) = T(G';1,1) = T(G'';1,1) = 8$. In general this evaluation gives the number of spanning trees for each graph (rooted or not).

2. $T(G;1,2) = T(G';1,2) = T(G'';1,2) = 14$. In general this evaluation gives the number of spanning sets, i.e., sets that contain a spanning tree. Since rooted spanning trees coincide with spanning trees, these numbers are the same for all three graphs.

3. $T(G;2,1) = 24, T(G';2,1) = 19$, and $T(G'';2,1) = 16$. For the unrooted graph G, the $x = 2, y = 1$ evaluation counts the number of forests, i.e., acyclic subsets of edges (independent sets in the graphic matroid). For the rooted graphs, we get the number of rooted subtrees (where the empty set is counted as a subtree). Since subtrees are also forests, we will always have $T(G;2,1) \geq T(H;2,1)$ when H is is a rooted graph obtained from G by choosing some root vertex.

4. $T(G;2,2) = T(G';2,2) = T(G'';2,2) = 2^5$. In general this evaluation gives 2^n where n is the size of the ground set.

33.2 Greedoids and the Tutte polynomial

Greedoids were defined by Korte and Lovász in the 1980s [707] as a generalization of matroids. Intuitively, greedoids are structures where a version of Prim's greedy algorithm (for rooted graphs) produces a minimum weight spanning tree. Greedoids include several interesting combinatorial objects that have no meaningful matroid structure. See [137, 708] for much more information about greedoids. Greedoids are usually defined in terms of the *feasible* sets, as follows.

Definition 33.5. A *greedoid* G is a pair $(S, \mathcal{F})$ where S is a finite set and $\mathcal{F} \subseteq 2^S$ is a family of subsets, called the *feasible* sets, satisfying:

(F1) $\emptyset \in \mathcal{F}$.

(F2) If $F \in \mathcal{F}$, then $F \setminus x \in \mathcal{F}$ for some $x \in F$.

(F3) If $F_1, F_2 \in \mathcal{F}$ with $|F_1| < |F_2|$, then there is some $x \in F_2 \setminus F_1$ such that $F_1 \cup x \in \mathcal{F}$.

The axioms are called the *non-triviality*, *accessibility*, and *augmentation* axioms, respectively.

The feasible sets of a greedoid are analogous to the independent sets of a matroid. Indeed, the properties **(F1)**, **(F2)** and **(F3)** are all satisfied by the independent sets of a matroid.

If G is a greedoid, we define the rank function via feasible sets.

Definition 33.6. Let G be a greedoid with ground set S and feasible sets $\mathcal{F}$. Then the *rank function* $r : 2^S \to \mathbb{Z}^+ \cup \{0\}$ is defined by

$$r(A) = \max_{F \in \mathcal{F}} \{|F| : F \subseteq A\}.$$

It is straightforward to prove that this rank function satisfies properties **(r1)**, **(r2)**, **(r3)**, and **(r4)** of Remark 33.2. Definition 33.1 now gives us a Tutte polynomial for greedoids. As with matroids, we can define the operations of deletion and contraction for greedoids. The feasible sets of the deletion $G \backslash e$ are just the feasible sets of G that avoid e. For contraction, assume $\{e\} \in \mathcal{F}$, i.e., e is a feasible singleton. Then the feasible sets of the contraction G/e are those feasible sets of G that include e, with e then removed. (We must have $\{e\}$ feasible to guarantee $\emptyset$ is feasible in G/e.) It is an easy exercise to show that both $G \backslash e$ and G/e are always greedoids when $\{e\}$ is feasible.

Proposition 33.7. *Let G be a greedoid on the ground set S with feasible sets $\mathcal{F}$, and let $e \in S$. Then deletion and contraction may be defined consistently as follows.*

1. *Deletion: Let $\mathcal{F}_d = \{F \in \mathcal{F} : e \notin F\}$. Then $\mathcal{F}_d$ satisfies the feasible set properties (F1), (F2) and (F3) with the ground set $S \setminus e$.*

2. *Contraction: Suppose $\{e\} \in \mathcal{F}$, and let $\mathcal{F}_c = \{F \setminus e : F \in \mathcal{F}, e \in F\}$. Then $\mathcal{F}_c$ satisfies the feasible set properties (F1), (F2) and (F3) with the ground set $S \setminus e$.*

33.2.1 Basic properties of the greedoid Tutte polynomial

The greedoid version of the Tutte polynomial satisfies many familiar properties and evaluations. These include a deletion–contraction recursion, a direct sum property, a notion of (weak) duality and various identities involving the coefficients of the polynomial.

For the fundamental deletion–contraction recursion, the loss of the unit rank increase property introduces a correction term.

Theorem 33.8. *Let G be a greedoid with greedoid Tutte polynomial $T(G; x, y)$, and let e be a feasible singleton. Then*

$$T(G; x, y) = T(G/e; x, y) + (x-1)^{r(G)-r(G \backslash e)} T(G \backslash e; x, y).$$

The $(x-1)^{r(G)-r(G \backslash e)}$ term does not appear in the deletion–contraction recursion for the classical (matroid) Tutte polynomial because $r(G) = r(G \backslash e)$

for all non-coloops e. (Ordinarily, the matroid deletion–contraction recursion is not used on loops and coloops.)

Let G_1 and G_2 be greedoids on disjoint ground sets S_1 and S_2, respectively. Then the *direct sum* $G_1 \oplus G_2$ is a greedoid on the ground set $S_1 \cup S_2$ with feasible sets $F_1 \cup F_2$, where F_i is feasible in G_i. The proof of the next result is straightforward.

Proposition 33.9. *Let $G_1 \oplus G_2$ be the direct sum of greedoids G_1 and G_2. Then $T(G_1 \oplus G_2; x, y) = T(G_1; x, y) \cdot T(G_2; x, y)$.*

As with matroids, a special case of this proposition describes the behavior of $T(G; x, y)$ for loops and coloops. In a greedoid G, an element e is a *greedoid loop* if it is in no feasible set, and e is a *coloop* (or an isthmus) if $r(A \cup e) = r(A) + 1$ for all A not containing e.

Corollary 33.10. *Let G be a greedoid.*

1. *If e is a coloop in G, then $T(G; x, y) = x \cdot T(G/e; x, y)$.*
2. *If e is a loop in G, then $T(G; x, y) = y \cdot T(G\backslash e; x, y)$.*

Some evaluations and coefficients of the greedoid Tutte polynomial have combinatorial interpretations. A *basis* of a greedoid is a maximal feasible set, and a *spanning set* is a set that contains a basis. As is the case with matroids, all greedoid bases have the same cardinality.

Proposition 33.11. *Let $T(G; x, y)$ be the greedoid Tutte polynomial of the greedoid G.*

1. *$T(G; 2, 2) = 2^{|E|}$.*
2. *$T(G; 2, 1)$ is the number of feasible sets.*
3. *$T(G; 1, 2)$ is the number of spanning sets.*
4. *$T(G; 1, 1)$ is the number of bases.*

The coefficients of the classical matroid Tutte polynomial satisfy certain linear relations. Write $T(M; x, y) = \sum_{i,j \geq 0} t_{i,j} x^i y^j$. If the ground set has at least two elements, then it is easy to show that $t_{1,0} = t_{0,1}$. This (common) coefficient is Crapo's *β-invariant* [347], and is usually written $\beta(M)$ (see Section 3.3.2). We study a generalization of this invariant to greedoids and antimatroids in Section 33.4.

More generally, Brylawski proved that the coefficients of the matroid Tutte polynomial satisfy a series of linear relations [251] (see Section 3.3). Assume $r(M) = r$ and $|S| = n$. Then for all $0 \leq k < n$,

$$\sum_{i=0}^{k} \sum_{j=0}^{k-i} (-1)^j \binom{k-i}{j} t_{i,j} = 0.$$

These linear relations hold for the greedoid version of the polynomial, too.

Theorem 33.12. *Let G be a greedoid with $|S| = n$ and greedoid Tutte polynomial $T(G; x, y) = \sum_{i,j\in\mathbb{Z}} t_{i,j}x^i y^j$. Then*

1. *For all $0 \le k < n$,*
$$\sum_{i=0}^{k}\sum_{j=0}^{k-i}(-1)^j\binom{k-i}{j}t_{i,j} = 0.$$

2. *For $k = n$,*
$$\sum_{i=0}^{n}\sum_{j=0}^{n-i}(-1)^j\binom{n-i}{j}t_{i,j} = (-1)^{n-r(S)}.$$

Finally, there is a notion of weak duality for greedoids; this is explored in [567]. Given a greedoid G with rank function r, define the dual object G^* via the rank function:
$$r^*(A) := r(S \setminus A) + |A| - r(S).$$
The proof of the next result is immediate.

Proposition 33.13. *Let G be a greedoid with rank function r and define the dual object G^* on the same ground set with rank function $r^*(A)$ given above. Then $T(G^*; x, y) = T(G; y, x)$.*

Unfortunately, G^* is not a greedoid in general since $r^*(A)$ may be negative. But $T(G^*; x, y)$ remains a polynomial in x and y.

We conclude this section by noting that a one-variable greedoid polynomial $\lambda_G(t)$ was defined in [137] using *basis activities.* The two-variable polynomial $T(G; x, y)$ considered here generalizes this one-variable polynomial.

Proposition 33.14. *Let G be a greedoid with one-variable polynomial $\lambda_G(t)$. Then $\lambda_G(t) = T(G; 1, t)$.*

Propositions 33.8–33.11, and 33.14 are from [570], Theorem 33.12 is from [568].

33.2.2 Rooted graphs and digraphs

Let $G = (V, E)$ be a rooted graph or digraph, with the root vertex denoted $*$. When G is undirected, we say $A \subseteq E$ is *feasible* if A is a rooted subtree. When G is a digraph, the feasible subsets are the rooted *arborescences*, i.e., rooted trees where all the edges are directed away from the root (that is, every vertex in the subtree is uniquely accessible from the root). In both cases, this defines a greedoid structure on the rooted graph or digraph, usually called the *branching greedoid.*

We remark that the greedoid structure (in particular, the list of feasible sets) for a rooted graph or digraph with no greedoid loops and no isolated vertices uniquely determines the rooted graph [834]. Thus, if G_1 and G_2 are rooted graphs (or digraphs) with no greedoid loops whose associated greedoids are isomorphic, then G_1 and G_2 are isomorphic as rooted graphs or digraphs.

33.2.2.1 Rooted graphs

What additional graphical information is encoded in the greedoid Tutte polynomial? We list several graph invariants of interest that can be recovered from $T(G; x, y)$ for the rooted graph G.

The number of loops in the graph or edges not reachable from the root by any path (these are the greedoid loops) can be recovered from $T(G; x, y)$. It is clear that if G has k greedoid loops, then $y^k \mid T(G; x, y)$. The converse is also true [834]:

Proposition 33.15. *Suppose $T(G; x, y)$ factors as $f(x, y)g(y)$ where the degree of g is as large as possible. Then $g(y) = y^a$, where a is the number of greedoid loops in G.*

The number of acyclic orientations with the root as the unique source is $T(G; 1, 0)$. This invariant is determined by the evaluation $T(G; 1, 0)$ for the (classical) Tutte polynomial of an unrooted graph G [580]. The same evaluation works for the greedoid version of the polynomial [572].

Proposition 33.16. *Let G be a rooted graph with greedoid Tutte polynomial $T(G; x, y)$. Then the number of acyclic orientations of G with the root as the unique source is given by $T(G; 1, 0)$.*

For example, in Example 33.4, the rooted graphs G' and G'' both have $T(G'; 1, 0) = T(G''; 1, 0) = 4$. The reader can find the four relevant acyclic orientations.

Note that $T(G; 1, 0) = 4$ for the unrooted graph in Example 33.4. In fact, the rooted and unrooted versions of the polynomial always agree when evaluated at $x = 1$, $y = 0$. This follows from the remarkable result of Greene and Zaslavsky [580] that the number of acyclic orientations of a graph with a unique source is independent of the vertex chosen to be the source.

Corollary 33.17. *Suppose G^r is a rooted graph with underlying unrooted graph G. Then the classical Tutte polynomial $T(G; 1, 0)$ equals the greedoid Tutte polynomial $T(G^r; 1, 0)$.*

The number of minimum tree covers of a rooted digraph can be recovered from $T(G; x, 0)$, [572]. Define a set of edges H in a rooted graph to be a *tree cover* if H is the edge set of a rooted subtree, and every edge of G is adjacent to some edge of H. A *minimum tree cover* is a tree cover of minimum size.

Proposition 33.18. *Suppose G is a connected rooted graph with r vertices (not including the root) and with greedoid Tutte polynomial $T(G; x, y)$. Further assume G has no greedoid loops. Let d be the degree of $T(G; x, 0)$. Then the size of the minimum tree cover is $r - d$.*

Returning to the two rooted graphs G' and G'' in Example 33.4, we have

$$T(G'; x, 0) = x^2 + 3x \quad \text{and} \quad T(G''; x, 0) = 4x.$$

We have $r = 3$ for both G' and G'', so the proposition tells us that the minimum size tree cover for G' consists of a single edge, while the minimum size tree cover for G'' consists of two edges.

Remark 33.19. Corollary 4.7 of [572] as it is stated is incorrect. In that paper, it is claimed that the coefficient of the leading term of $T(G; x, 0)$ counts the number of minimum tree covers, In fact, this coefficient only gives an upper bound on that number.

33.2.2.2 Rooted digraphs

When G is a rooted digraph, there is a different interpretation for the largest power of y that divides the Tutte polynomial [571].

Theorem 33.20. *Let G be a rooted digraph with no greedoid loops and with greedoid Tutte polynomial $T(G; x, y)$. Suppose $T(G; x, y) = f(x, y)g(y)$ where the degree of g is as large as possible. Then $g(y) = y^a$, where a is the minimum number of edges that need to be removed from G to leave a spanning, acyclic digraph with no greedoid loops.*

The special case when $g(y)$ is non-trivial, giving a result from [132, 834], is worth noting:

Corollary 33.21. *Let G be a rooted digraph with no greedoid loops. Then G has a directed cycle if and only if $T(G; x, 0) = 0$.*

Example 33.22. Let G be the rooted digraph in Figure 33.3. Note that there are no greedoid loops, and exactly one directed cycle: cfd. Then the greedoid Tutte polynomial $T(G; x, y)$ is:

$$T(G; x, y) = x^4y^4 - 4x^3y^4 + 2x^3y^3 + 6x^2y^4 - 6x^2y^3 + 3x^2y^2 - 4xy^4 + 6xy^3 - 4xy^2 + 2xy + y^4 - 2y^3 + 2y^2.$$

Note that $y \mid T(G; x, y)$, but $y^2 \nmid T(G; x, y)$. By Theorem 33.20, we can find an edge whose removal eliminates the directed cycle and creates no loops. Removing any of c, d or f breaks the only directed cycle, but only the removal or d or f guarantees the loopless condition.

We point out that deleting c turns f into a greedoid loop, and we have

$$T(G\backslash c) = y(x^3y^2 - 2x^2y^2 + x^2y + xy^2 - xy + x + y).$$

If we delete d or f, no greedoid loops are created. We find

$$\begin{aligned} T(G\backslash d) &= T(G\backslash f) \\ &= x^4y^3 - 4x^3y^3 + 2x^3y^2 + 6x^2y^3 - 5x^2y^2 + 2x^2y \\ &\quad -4xy^3 + 4xy^2 - 2xy + x + y^3 - y^2 + y. \end{aligned}$$

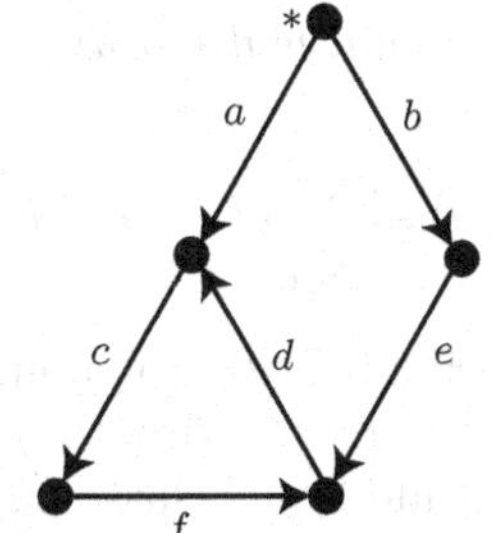

FIGURE 33.3: A rooted digraph.

The reader can check that $G\backslash d$ and $G\backslash f$ are not isomorphic as rooted digraphs. This gives us an example of two non-isomorphic rooted digraphs with the same greedoid Tutte polynomial. As a consequence we have the following.

1. Since every vertex in $G\backslash d$ has degree two, but $G\backslash f$ has a vertex of degree one, the polynomial cannot distinguish the degree sequence of the digraph.
2. The edges a, b and c are in every basis for $G\backslash d$, but only b and c are in every basis for $G\backslash f$. Thus, the polynomial cannot distinguish the number of edges in every basis.

Finally, we note that Tedford [1054] has generalized Theorem 33.20 to *rooted mixed graphs*, i.e., rooted graphs that include directed and undirected edges.

33.3 Antimatroids

Antimatroids are a class of *full greedoids* (meaning the ground set S is a feasible set) whose feasible sets are closed under taking unions. See [137] for a treatment of antimatroids as a class of greedoids, or [436], where antimatroids are used as a model for generalized convexity. Antimatroids have been rediscovered several times in the literature [859].

Definition 33.23. Let $G = (S, \mathcal{F})$ be a greedoid with rank function r given by Definition 33.6 and feasible sets $\mathcal{F}$. Then G is an *antimatroid* if $r(S) = |S|$ and $F_1 \cup F_2 \in \mathcal{F}$ whenever $F_1, F_2 \in \mathcal{F}$.

If G is an antimatroid on the ground set S, define $C \subseteq S$ to be *convex* if its complement $S \setminus C$ is feasible. For C convex, let $\text{ex}(C)$ be the *extreme* points of C (i.e., $p \in \text{ex}(C)$ if $p \in C$ but $p \notin \overline{C \setminus p}$, where $\overline{A}$ is the *convex closure* of A, the smallest convex set containing A). We write $\text{nex}(C) := C \setminus \text{ex}(C)$ for the *non-extreme* points of C. The set of non-extreme points $\text{nex}(C)$ has combinatorial interpretations for antimatroids [564].

Theorem 33.24. *Let G be an antimatroid and let $\mathcal{C}$ denote its collection of convex sets. Then*

$$T(G;x,y) = \sum_{C\in\mathcal{C}} (x-1)^{|C|} y^{|\mathrm{nex}(C)|}.$$

We now examine the greedoid Tutte polynomial in more detail for three classes of antimatroids: rooted trees, (unrooted) trees, and posets. (In Section 33.4, we will consider two additional classes of antimatroids in the context of the β-invariant: chordal graphs and finite subsets of Euclidean space.)

33.3.1 Rooted trees

Let G be a rooted tree with edge set S, i.e., a tree with a distinguished vertex, the *root*. There are two complementary ways to define an antimatroid on the ground set S. The empty set is considered a rooted tree throughout.

Definition 33.25. Let G be a rooted tree with edge set S. The *branching antimatroid*, G_b, and *pruning antimatroid*, G_p, are both antimatroids on S with feasible sets given by:

1. for the branching antimatroid G_b, a set $F \subseteq S$ is feasible if the edges in F form a rooted subtree; and

2. for the pruning antimatroid G_p, a set $F \subseteq S$ is feasible if the edges in $S \setminus F$ form a rooted subtree.

We can use Theorem 33.24 to get direct combinatorial interpretations for these two Tutte polynomials (a result from [295]). For a rooted subtree C, define the *boundary* $b(C)$ of C to be the set of edges of $S \setminus C$ that are incident to some vertex of C, and let $l(C)$ be the set of edges of C that are leaves (not including the root).

Proposition 33.26. *Let G_b and G_p be the branching and pruning antimatroids associated to a rooted tree G. Let $\mathcal{R}$ be the collection of all rooted subtrees of G. Then the greedoid Tutte polynomials of G_b and G_p are given by:*

1. $T(G_b;x,y) = \sum_{C\in\mathcal{R}} (x-1)^{|S|-|C|} y^{|S|-|C|-|b(C)|},$

2. $T(G_p;x,y) = \sum_{C\in\mathcal{R}} (x-1)^{|C|} y^{|C|-|l(C)|}.$

Example 33.27. Consider the rooted tree of Figure 33.4. Then, for instance, the rooted subtree $C = \{a,c,d\}$ has 3 edges, with $b(C) = \{b,e\}$ and $l(C) = \{c,d\}$. This subtree contributes $(x-1)^2$ to the branching antimatroid polynomial, and $(x-1)^3 y$ to the pruning antimatroid polynomial.

The reader can check the branching Tutte polynomial is

$$T(G_b;x,y) = x^5y^3 - 4x^4y^3 + x^4 + 6x^3y^3 - 4x^2y^3 + xy^3,$$

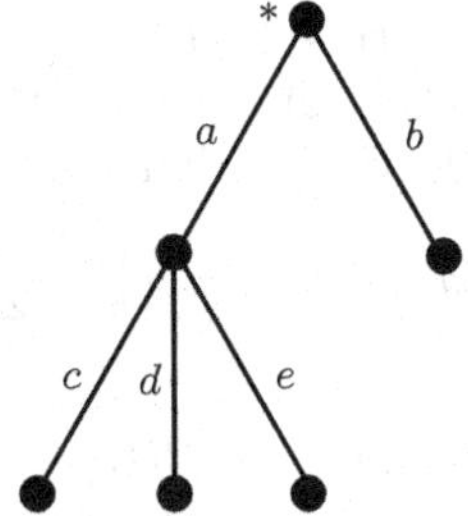

FIGURE 33.4: A rooted tree with the root labeled $*$.

and the pruning Tutte polynomial is given by

$$T(G_p; x, y) = x^5y - x^4y - x^2y + x^2 + xy.$$

The greedoid Tutte polynomials of the branching and pruning antimatroids are complete invariants for rooted trees, as shown in [570] and [295], respectively.

Theorem 33.28. *Let G and G' be rooted trees.*

1. *$T(G_b; x, y) = T(G'_b; x, y)$ if and only if G and G' are isomorphic as rooted trees.*

2. *$T(G_p; x, y) = T(G'_p; x, y)$ if and only if G and G' are isomorphic as rooted trees.*

An immediate corollary, from [295], of Theorem 33.28 is the following purely graph theoretic result.

Corollary 33.29. *Let G be a rooted tree, and let $\mathcal{R}$ be the collection of all rooted subtrees of G. Then the following hold.*

1. *G can be uniquely reconstructable from the list of all ordered pairs $(|C|, |b(C)|)_{C\in\mathcal{R}}$.*

2. *G can be reconstructed from the list of all ordered pairs $(|C|, |l(C)|)_{C\in\mathcal{R}}$.*

The proofs of both parts of Theorem 33.28 use algebraic techniques. It would be of interest to find a purely combinatorial proof. We also remark that Tedford [1055] defines a 3-variable version of $T(G; x, y)$ that generalizes Theorem 33.28 to rooted graphs with at most one cycle.

33.3.1.1 Connecting leaf and boundary data

We now give an application of the Tutte polynomial approach to rooted trees. The main result, given in Theorem 33.31, is an explicit connection between the boundary data and the leaf data for the rooted subtrees of a given rooted tree. We emphasize that our connection will be purely combinatorial, with the Tutte polynomials serving as an intermediate step in proving the formulas.

Definition 33.30. Let G be a tree, let $l_{i,j}$ be the number of rooted subtrees of G with i edges and j leaf edges, and $m_{i,j}$ be the number of rooted subtrees of G with i edges and j boundary edges. Then define two generating functions:

1. the *subtree-leaf polynomial* by $L(G;u,v) = \sum_{i,j\geq 0} l_{i,j}u^iv^j$, and

2. the *subtree-boundary polynomial* by $B(G;u,v) = \sum_{i,j\geq 0} m_{i,j}u^iv^j$.

The generating functions $L(G;u,v)$ and $B(G;u,v)$ for the rooted tree in Figure 33.4 are

$$\begin{aligned} L(G;u,v) &= u^5v^4 + 4u^4v^3 + 6u^3v^2 + 3u^2v + u^2v^2 + 2uv + 1, \\ B(G;u,v) &= u^5 + 4u^4v + 6u^3v^2 + 4u^2v^3 + uv^4 + uv + v^2. \end{aligned}$$

Our goal is purely graph-theoretic: connect these two invariants. By Corollary 33.29, we know we can use either $L(G;u,v)$ or $B(G;u,v)$ to reconstruct a rooted tree. Thus, for example, if we are given $L(G)$, we can compute $B(G)$ by first reconstructing the rooted tree, then computing $B(G)$ directly from the rooted tree. But this approach uses a rather cumbersome tree reconstruction procedure (which relies on factoring associated Tutte polynomials and induction) as an intermediate step. We seek a direct connection, which is the main result of this section.

Theorem 33.31. *Let G be a rooted tree with $L(G;u,v)$ and $B(G;u,v)$ as above. Then*

1. $L(G;u,v) = B(G;u,uv-u+1)$,

2. $B(G;u,v) = L\left(G;u,\dfrac{u+v-1}{u}\right)$.

The formulas in Theorem 33.31 are new. Our proof of Theorem 33.31 relies on two lemmas. The first lemma connects the Tutte polynomial descriptions from Proposition 33.26 to the generating functions $L(G;u,v)$ and $B(G;u,v)$. We leave the straightforward proofs of both of these lemmas to the interested reader.

Lemma 33.32. *Let G be a rooted tree with n edges, and let G_b and G_p be the associated branching and pruning antimatroid structures on the edges of G. Then*

1. $L(G;u,v) = T\left(G_p;uv+1,\frac{1}{v}\right)$,

2. $T(G_p;x,y) = L\left(G;xy-y,\frac{1}{y}\right)$,

3. $B(G;u,v) = u^nT\left(G_b;\frac{u+v}{u},\frac{1}{v}\right)$,

4. $T(G_b;x,y) = ((x-1)y)^n B\left(G; \frac{1}{(x-1)y}, \frac{1}{y}\right)$.

The next lemma connects the branching and pruning Tutte polynomials. The proof follows immediately from Theorem 33.41(3) in Section 33.3.3 when we view the pruning and branching antimatroids of a rooted tree as dual posets.

Lemma 33.33. *Let G be a rooted tree with n edges, and let G_b and G_p be the respective branching and pruning antimatroids associated to G. Then*

$$T(G_p;x,y) = (xy-y)^n T\left(G_b; \frac{x}{xy-y}, \frac{1}{x+y-xy}\right).$$

We remark that the formula in Lemma 33.33 is an involution—it remains valid if we swap G_b and G_p. It would be of interest to find a purely combinatorial explanation of the relation between $L(G)$ and $B(G)$.

33.3.2 Trees

Let G be a tree with edge set E. The *pruning antimatroid* $G = (E, r)$ is defined on the set of edges of the tree, where the feasible sets are the complements of subtrees of G. This structure is completely analogous to the pruning antimatroid for rooted trees. (Unfortunately, there is no analogous way to define an antimatroid structure for an unrooted tree that is based on a branching rank function.) We now interpret Theorem 33.24 for trees.

Proposition 33.34. *Let G be the pruning antimatroid associated with a tree. Let $\mathcal{S}$ be the collection of all subtrees, and let $l(C)$ be the set of leaves of the subtree C. Then* $T(G;x,y) = \sum_{C \in \mathcal{R}} (x-1)^{|C|} y^{|C|-|l(C)|}$.

Does the Tutte polynomial uniquely determine the tree? Unlike the situation for rooted trees, the answer is no.

Example 33.35. Let G and G' be the two non-isomorphic trees of Figure 33.5. The number of subtrees with i edges and j leaves for all i and j is the same for these two trees (see Table 33.1).

A procedure for creating more counterexamples is given in [445]. Although the non-rooted tree version of Theorem 33.28(2) is false, the Tutte polynomial still determines several interesting invariants of the tree.

Proposition 33.36. *Suppose G and G' are trees with the same pruning Tutte polynomial $T(G;x,y) = T(G';x,y)$. Then the following hold.*

1. *G and G' have the same degree sequence.*
2. *G and G' have the same number of paths of length k, for all $k \geq 1$.*

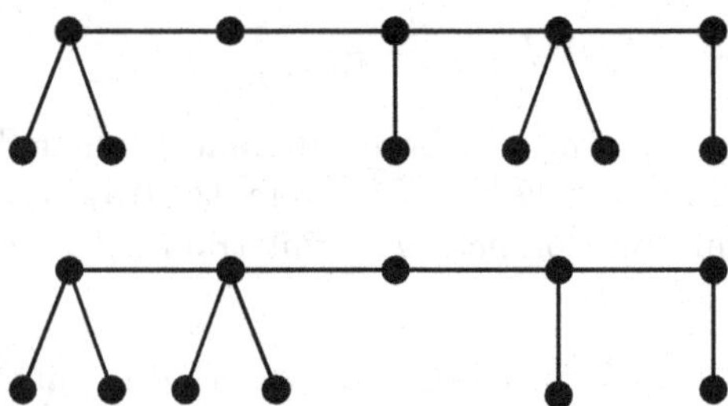

FIGURE 33.5: Two trees with the same number of subtrees with i edges and j leaves for all $i, j \geq 1$.

		Number of edges i of subtree									
		1	2	3	4	5	6	7	8	9	10
No. of leaves j	1	10	0	0	0	0	0	0	0	0	0
	2	0	14	13	9	7	2	0	0	0	0
	3	0	0	6	14	15	18	7	0	0	0
	4	0	0	0	1	6	9	17	9	0	0
	5	0	0	0	0	0	1	2	7	5	0
	6	0	0	0	0	0	0	0	0	1	1

TABLE 33.1: The entry in row j and column i is the number of subtrees with i edges and j leaves for both of the trees of Figure 33.5.

3. *$\lambda(G) = \lambda(G')$ where $\lambda(H)$ is the size of the largest matching in the graph H.*

4. *$\mu(G) = \mu(G')$ where $\mu(H)$ is the size of the largest independent set of vertices in the graph H.*

Proposition 33.36 is from [569]. Part 1 of the proposition tells us we can determine the degree sequence of a tree from its antimatroid Tutte polynomial. In this case, if a tree G has n edges and d_i vertices of degree i, then we can write the generating function $\sum_{i\geq 1} d_i z^i$ in terms of the antimatroid Tutte polynomial (see [569]) as follows:

$$\sum_{i\geq 1} d_i z^i = T(G; z, 0) + nz + 1.$$

Although trees are not determined by their antimatroid Tutte polynomials, it is still possible that large classes of trees are uniquely reconstructable from the polynomial. (For instance, it is trivial to reconstruct stars or paths from their polynomials.) It would be of interest to know other subclasses of trees that are uniquely determined by their antimatroid Tutte polynomials.

Finally, another invariant for trees is Stanley's *chromatic symmetric function* (see Definition 26.45), a multivariate generalization of the chromatic polynomial. Martin, Morin and Wagner [821] proved that Stanley's polynomial is stronger than the generalized Tutte polynomial of Definition 33.1 considered here. They give examples of two trees with the same generalized Tutte polynomial, but different chromatic symmetric functions. It is an open question if the chromatic symmetric polynomial of a tree uniquely determines the tree (see Chapter 26).

33.3.3 Posets

There are several ways to give a poset P an antimatroid structure. We use *order ideals* to define the *poset antimatroid.* A set I is an *order ideal* (or *down-set*) if $x \in I$ and $y \leq x$ implies $y \in I$. Dually, a subset F is an *order filter* (or *up-set*) if $x \in F$ and $y \geq x$ implies $y \in F$. The feasible sets of the poset antimatroid are the order ideals of the poset.

If A is an antichain of P (i.e., a subset in which each pair of different elements is incomparable), we let $I(A)$ be the order ideal generated by A, so

$$I(A) = \{x \in P : x \leq y \text{ for some } y \in A\};$$

and $I^*(A)$ the order filter generated by A, so

$$I^*(A) = \{x \in P : x \geq y \text{ for some } y \in A\}.$$

Then Theorem 33.24 takes the following form for poset antimatroids.

Proposition 33.37. *Let P be the poset antimatroid with feasible sets corresponding to the order ideals of P. Let $\mathcal{A}$ be the collection of all antichains of P. Then*

$$T(P;x,y) = \sum_{A \in \mathcal{A}} (x-1)^{|I^*(A)|} y^{|I^*(A)|-|A|}.$$

Let $a_{i,j}$ be the number of antichains A of size i with $|I^*(A)| = j$. As before, it is straightforward to find an algebraic connection between the Tutte polynomial and the antichain-order filter generating function $f(P;u,v) = \sum_{i,j\geq 0} a_{i,j}u^i v^j$. The following result is from [564].

Proposition 33.38. *Let P be a poset, $a_{i,j}$ be the number of antichains A of size i with $|I^*(A)| = j$, and let $f(P;u,v) = \sum_{i,j\geq 0} a_{i,j}u^i v^j$. Then*

1. $f(P;u,v) = T\left(P; uv+1, \frac{1}{u}\right)$,
2. $T(P;x,y) = f\left(P; \frac{1}{y}, xy-y\right)$.

The Tutte polynomial does not determine the poset. The smallest counterexample is given in the next example.

FIGURE 33.6: Two non-isomorphic posets with the same Tutte polynomial.

Example 33.39. The two posets in Figure 33.6 have the same Tutte polynomial. One quick way to see this is to list the ordered pairs $(|A|, |I^*(A)|)$ for all antichains A. For both posets, we have

$$f(P_1) = f(P_2) = u^2v^4 + u^2v^3 + u^2v^2 + uv^3 + uv^2 + 2uv + 1,$$

so

$$T(P_1) = T(P_2) = x^4y^2 - 3x^3y^2 + x^3y + 3x^2y^2 - 2x^2y + x^2 - xy^2 + xy.$$

While the Tutte polynomial does not determine the poset, the polynomial does distinguish interesting subclasses of posets. Before proceeding, we first define three familiar operations on posets.

Definition 33.40. Let P and Q be disjoint posets.

1. The *direct sum* of P and Q, written $P + Q$, is the poset on $P \cup Q$ where $x \leq y$ in $P + Q$ if
   (a) $x, y \in P$ and $x \leq y$ in P, or
   (b) $x, y \in Q$ and $x \leq y$ in Q.
2. The *ordinal sum* of P and Q, written $P \oplus Q$, is the poset on $P \cup Q$ where $x \leq y$ in $P \oplus Q$ if
   (a) $x, y \in P$ and $x \leq y$ in P,
   (b) $x, y \in Q$ and $x \leq y$ in Q, or
   (c) $x \in P$ and $y \in Q$.
3. The *dual* poset P^* has $x \leq y$ in P^* precisely when $y \leq x$ in P.

Thus, the order diagram for $P+Q$ is formed by placing the order diagrams for P and Q side by side, and the order diagram for $P \oplus Q$ is formed by placing the order diagram for P beneath the diagram for Q, then making all of the maximal elements of P less than all of the minimal elements of Q. The order diagram for P^* is obtained by rotating the diagram for P by 180°.

The next theorem gives the Tutte polynomial of the sum, the ordinal sum and the dual. Parts 1 and 2 are proven in [564] and Part 3 appears in [565].

Theorem 33.41. *Let P and Q be posets. Then*

1. $T(P+Q;x,y) = T(P;x,y)T(Q;x,y)$,
2. $T(P \oplus Q;x,y) = T(Q;x,y) + ((x-1)y)^{|Q|}(T(P;x,y)-1)$,
3. $T(P^*;x,y) = (xy-y)^{|P|}T\left(P;\dfrac{x}{xy-y},\dfrac{1}{x+y-xy}\right)$.

Note that the formula for duals is an involution: $T((P^*)^*) = T(P)$. When T is a rooted tree, there is an induced poset structure on the edges, and the associated poset is isomorphic (as an antimatroid) to the branching antimatroid polynomial (where edges adjacent to the root are minimal elements of the poset). Then Theorem 33.41 (3) immediately implies Lemma 33.33, which was an important step in proving the formulas relating the subtree-leaf polynomial and the subtree-boundary polynomial (Theorem 33.31).

Posets that can be obtained from the 1-element poset **1** by repeatedly applying the two operations of direct sum and ordinal sum give the class of *series-parallel posets.* Note that series-parallel posets are also closed under duality.

Conjecture 33.42. Let P and Q be series-parallel posets. Then $T(P) = T(Q)$ if and only if P and Q are isomorphic.

This conjecture has been verified for the following proper subclass of series-parallel posets (see [565]). Let $\mathbf{n} = \underbrace{\mathbf{1}+\mathbf{1}+\cdots+\mathbf{1}}_{n}$. Now define a subclass $\mathcal{P}$ of the class of series-parallel posets as follows:

1. The one element poset **1** is in $\mathcal{P}$.
2. If $P \in \mathcal{P}$, then so is $P \oplus \mathbf{n}$ for all n.
3. If $P, Q \in \mathcal{P}$, then so is $P + Q$.
4. If $P \in \mathcal{P}$, then so is P^*.

Theorem 33.43. *Let $P, Q \in \mathcal{P}$. Then $T(P) = T(Q)$ if and only if P and Q are isomorphic posets.*

33.4 The β-invariant

For a matroid M, the *β-invariant* $\beta(M)$ is the coefficient of x in the Tutte polynomial $T(M;x,y)$ (see Section 3.3.2). We use this same definition for greedoids.

Definition 33.44. Let G be a greedoid with greedoid Tutte polynomial $T(G;x,y) = \sum_{i,j\geq 0} t_{i,j}x^i y^j$. Then the *$\beta$-invariant* of G is $\beta(G) = t_{1,0}$.

When $|S| > 1$, we note that the coefficient of x and the coefficient of y are equal, i.e., $t_{1,0} = t_{0,1}$. (The proof follows immediately by induction, using the deletion–contraction recursion of Theorem 33.8.) Crapo proved that when M is a matroid on more than one point, $\beta(M) > 0$ if and only if M is connected [347].

The following subset expansion holds for all greedoids [566].

Proposition 33.45. *Let G be a greedoid on the ground set S. Then*

$$\beta(G) = (-1)^{r(S)} \sum_{A\subseteq S} (-1)^{|A|} r(A).$$

For an antimatroid G, we can find an expansion for $\beta(G)$ in terms of the subclass of the convex sets, called the *free convex* sets (or, more simply, the *free sets*) of the antimatroid.

Definition 33.46. Let G be an antimatroid on the ground set S, with $C \subseteq S$. Then C is *free* if every subset of C is convex.

The following was shown in [566].

Theorem 33.47. *Let G be an antimatroid, and f_i denote the number of free sets of size i. Then*

$$\beta(G) = \sum_{C \text{ free}} (-1)^{|C|-1}|C| = \sum_{i\geq 0} (-1)^{i-1} f_i.$$

We will apply Theorem 33.47 to two classes of antimatroids: chordal graphs and finite subsets of Euclidean space. In each case, we will be interested in combinatorial interpretations of our results.

33.4.1 Chordal graphs

A graph G is *chordal* if every cycle of length greater than 3 has a chord. When G is a chordal graph, a vertex is *simplicial* if its neighbors form a clique. The ground set of the *simplicial shelling antimatroid* is the collection of vertices of G. An ordered set $\{v_1, v_2, \ldots, v_k\}$ is *shellable* if v_1 is simplicial, and v_2 is simplicial in $G\backslash v_1$, and v_3 is simplicial in $G\backslash\{v_1, v_2\}$, and so on. The subsets of vertices that can be shelled in this way form the feasible sets of an antimatroid on G. As before, the convex sets are defined to be the complements of the feasible sets, and it is straightforward to see that a set of vertices is free convex if it induces a clique in G.

A graph is a *block* if it contains no cut–vertices, i.e., every pair of edges is in some cycle. The next result, from [566], gives a formula that allows us to interpret the β-invariant combinatorially.

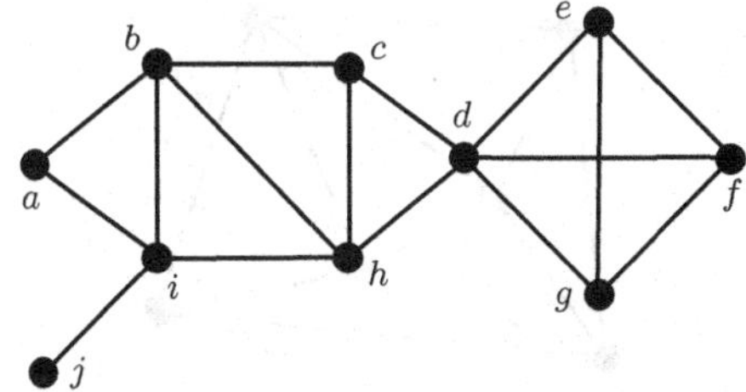

FIGURE 33.7: A chordal graph G with $\beta(G) = 2$.

Theorem 33.48. *Let G be a chordal graph with b blocks, and let f_i be the number of cliques of size i. Then*

$$\sum_{i=0}^{n}(-1)^i i f_i = b - 1.$$

Note that $\sum_{i=0}^{n}(-1)^i i f_i = -\beta(G)$ which gives a cleaner combinatorial interpretation.

Example 33.49. Consider the chordal graph of Figure 33.7. Note that G has three blocks. Counting the number of cliques of various sizes gives $-\beta(G) = -10 + 2 \cdot 16 - 3 \cdot 8 + 1 \cdot 4 = 2$, which is one less than the number of blocks.

33.4.2 Finite point sets in Euclidean space

Finite subsets of $\mathbb{R}^n$ are the prototypical examples of convex geometries, or, dually, antimatroids. A subset C of a finite set S of points in $\mathbb{R}^n$ is *convex* if $C = \overline{C} \cap S$, where $\overline{C}$ is the convex hull of C in $\mathbb{R}^n$. As usual, the complements of the convex sets are the feasible sets of the antimatroid.

A point $p \in S$ is in the *interior* of S if there is a positive real number r so that the ball of radius r centered at p is completely contained in the convex hull of S. Then the β-invariant counts the number of these interior points, with a sign indicating the parity of the dimension of the set S. This theorem is the main result of [437].

Theorem 33.50. *Let S be a finite subset of $\mathbb{R}^n$, and let $\text{int}(S)$ be the set of points in the interior of S. Then*

$$(-1)^{n-1}\beta(S) = \sum_{i=0}^{n}(-1)^i i f_i = (-1)^{n-1}|\text{int}(S)|.$$

Example 33.51. The set S of six points in the plane in Figure 33.8 has two interior points, e and f. (The point set is triangulated to help visualize the convex sets.) Then each point is free, and every pair of points is also a free set. There are 15 free sets of size three (empty triangles in the configuration),

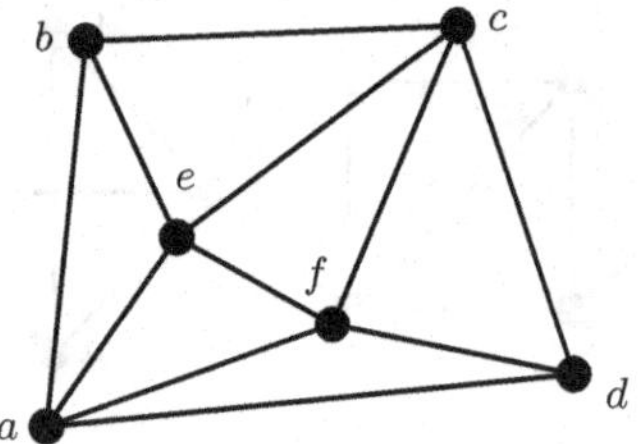

FIGURE 33.8: A finite subset of the plane. $\beta(G) = 2$, the number of interior points.

six empty quadrilaterals and one empty pentagon. Putting this together gives us

$$\beta(G) = 6 \cdot 1 - 15 \cdot 2 + 15 \cdot 3 - 6 \cdot 4 + 1 \cdot 5 = 2,$$

the number of interior points.

The formula in Theorem 33.50 was first proven in $\mathbb{R}^2$ in [9], where the generalization to $\mathbb{R}^n$ was conjectured. Edelman and Reiner [437] applied techniques from combinatorial topology to prove this theorem. Klain also [692] gave a proof of Theorem 33.50 (using valuation theory), and Pinchasi, Radoičić, and Sharir [921] found an elementary proof of the theorem. We emphasize that, although the β-invariant provided motivation for Theorems 33.48 and 33.50, these results are purely combinatorial.

33.5 Open problems

We list several problems for future investigation.

1. For a matroid M with Tutte polynomial $T(M; x, y)$, the highest power of x that can be factored out of the polynomial is the number of coloops (or isthmuses) in the matroid. This is false for greedoids: If G is a greedoid on the ground set $\{a, b, c\}$ with feasible sets $\{\emptyset, a, b, ac, bc, abc\}$, then $T(G; x, y) = x(x^2y - 2xy + x + y)$. This greedoid has no coloops, however.

   But is still may true that the highest factor of x dividing $T(G; x, y)$ has a combinatorial interpretation for some classes of greedoids. In particular, we conjecture the following:

   Conjecture 33.52. Let G be a rooted graph. Then the highest power of x dividing $T(G; x, y)$ is the number of leaves incident to the root.

2. Find a large subclass of trees that are uniquely determined by their antimatroid Tutte polynomials, as defined in Section 33.3.2.

3. Prove Conjecture 33.42 for series-parallel posets from Section 33.3.3.

4. More generally, find other classes of combinatorial objects (greedoids or not) that are Tutte-unique using the Tutte polynomial of Definition 33.1.

Part VI

History

34

The history of Tutte–Whitney polynomials

Graham Farr

Synopsis

This chapter covers the history of Tutte–Whitney polynomials, mainly through commentary on the early papers on the topic. It focuses particularly on the following:

- Notation and definitions used in the seminal papers.
- Whitney, early 1930s: his papers and thesis on his rank generating function.
- Tutte, 1947–1967: his papers and thesis on the Tutte polynomial.
- Potts, 1952: his papers and thesis on the Potts model partition function.
- Zykov's early work on related polynomials, 1949 and 1964.
- Further generalizations, abstractions, interpretations of evaluations, and links to other fields, in the late 1960s and early 1970s.
- Unexpectedly early appearances of many graph polynomial properties.

34.1 Introduction

The Tutte polynomial and the Whitney rank generating function have their origins in papers by Hassler Whitney in 1932 and William (Bill) Tutte in 1947 and 1954. Their close relative, the Potts model partition function, was introduced by Renfrey Potts in a 1952 paper. Each of these works generalizes earlier functions that count structures associated with a graph. (For this

DOI: 10.1201/9780429161612-34

chapter, ten foundational papers of Tutte, Potts and Whitney have special reference keys that can be found in Section 34.2.)

The bivariate polynomials of Whitney and Tutte were introduced as generalizations of the chromatic polynomial, a univariate polynomial introduced by Birkhoff [121] to count colorings of graphs (see Chapter 11). Whitney and Tutte came from different directions. Whitney's rank generating function [W32d] built on his edge-subgraph expansion [W32a] for the chromatic polynomial. Tutte started with the deletion–contraction relation, and sought a generalization that embraced other functions of graphs that satisfy a deletion–contraction relation, such as the number of spanning trees [T47]. The work of Whitney and Tutte introduced these polynomials as a topic in combinatorics.

A separate stream of research, in statistical mechanics, began with the paper by Ising in 1925, introducing the Ising model partition function [635]. This work was generalized by Ashkin and Teller [51] and Potts [P52]. The Potts model partition function contains, in a formal sense, the same combinatorial information as the Whitney rank generating function and Tutte polynomial. The two streams of research were united by Essam, Fortuin and Kasteleyn in the early 1970s, [468, 498].

Incidentally, we find that various concepts and results turn out to be older than is generally recognized. These include: chromatic polynomials of clique-separable graphs; flow polynomials; Tutte equivalence; the question of which polynomials are Tutte polynomials; a graph-theoretical instance of the exponential formula in enumerative combinatorics; recipe theorems; Tutte polynomials of matroids; the question of the computational complexity of Tutte polynomials; clique polynomials; and a convolution-type identity for the Tutte polynomial.

This chapter describes the history of Tutte–Whitney polynomials. This is done mainly through commentary on the early papers on the topic.

In the next section, we list these classic works, and set the scene for the detailed commentary later. In Section 34.3, we give the notation and key definitions we will need throughout this chapter. Note that for historical reasons, this notation may not always agree with that used elsewhere in the handbook. In Section 34.4, we turn to the notation used in those early papers. Then, in Sections 34.5–34.12, we give our commentary on each of the works we consider. This is followed in Section 34.13 with a short discussion of the work of Zykov, a Russian mathematician who did some early work on these and other polynomials in parallel with Tutte. In Section 34.14, we consider a few highlights of Tutte–Whitney polynomial research over the decade or so after the period covered by the included papers, i.e., from the late 1960s to late 1970s. We finish with some closing remarks in Section 34.15.

34.2 The classic papers

The early papers on the topic by Whitney, Tutte and Potts are the source material for the story of the birth of the eponymous polynomials. These seminal papers are well worth reading, and not only to see the introduction and early development of the polynomials. They give a window into how some things were done in the past, in the world of mathematics. They are rich in mathematical insight, and contain results that, even now, are not as well known as they should be. They reveal the thinking of the greatest minds in their fields.

It is not necessary to read our commentary before reading the papers, and in fact, reading at least some of each paper first will help. The commentary expands on several aspects of the papers: notation or terminology that may seem strange to a modern reader; some connections with other works of that time; some subsequent developments; some suggestions that may be helpful in understanding some definitions and proofs; and some remarks on alternative approaches. It does not purport to be comprehensive, since to say everything that could now be said about these papers would be to discuss a large portion of enumerative graph theory. But hopefully it will see the reader through most points in the papers that may bring perplexity or curiosity.

The publications discussed in detail in this chapter are as follows. We use special reference numbers [W32a, W32c, W32d, W33a, T47, T48, T54, T67, P52] for these, so they are clearly distinguished from other citations.

Foundational papers

[P51] R. B. Potts. *The mathematical investigation of some cooperative phenomena.* D.Phil. dissertation, University of Oxford, 1951.

[P52] R. B. Potts. Some generalized order-disorder transformations. *Proc. Cambridge Philos. Soc.*, 48:106–109, 1952.

[T47] W. T. Tutte. A ring in graph theory. *Proc. Cambridge Philos. Soc.*, 43:26–40, 1947.

[T48] W. T. Tutte. *An algebraic theory of graphs.* PhD thesis, University of Cambridge, 1948.

[T54] W. T. Tutte. A contribution to the theory of chromatic polynomials. *Canadian J. Math.*, 6:80–91, 1954.

[T67] W. T. Tutte. On dichromatic polynomials. *J. Combinatorial Theory*, 2:301–320, 1967.

[W32a] H. Whitney. A logical expansion in mathematics. *Bull. Amer. Math. Soc.*, 38(8):572–579, 1932.

[W32c] H. Whitney, *The coloring of graphs*, PhD thesis, Harvard University, 1932.

[W32d] H. Whitney. The coloring of graphs. *Ann. of Math. (2)*, 33(4):688–718, 1932.

[W33a] H. Whitney. A set of topological invariants for graphs. *Amer. J. Math.*, 55(1-4):231–235, 1933.

Other sources of commentary

Other sources of commentary on some of these papers include:

- Tutte's own commentary on [T47, T54, T67], in [1101];
- some of Tutte's papers:
  - [1098], which briefly redoes parts of [T47];
  - [1099], which redoes much of [T47] and part of [T54];
  - [1102], which redoes [T47], emphasizing Section 5 and Section 7;
- Tutte's textbook exposition in [1104, Ch. IX] and personal account in [1108];
- Biggs, Lloyd and Wilson [115, pp. 180–185]: short commentary on [W32a], with an extract from the paper;
- Biggs's discussions in [104, 105, 106, 107, 111], which are very illuminating on the main results of [T67] and aspects of [W32d] (for the latter, see especially [106]);
- Kung's commentary on [T47] in [728, Section IV.1];
- Kung's commentary on [W32d] in [734];
- the author's two-page overview in [479, Section 3.3].

The research reported in the papers by Whitney was done while he was a PhD student. Tutte's 1947 paper predated his PhD thesis (1948); some ideas from that thesis were developed further in his 1954 paper. Potts's paper too reported some of his PhD work. So it can be said that the foundations of Tutte–Whitney polynomial theory were laid by student researchers.

Whitney "had become interested in the four-color problem in the late 1920s, while studying physics in Germany" [115, p. 148]. His PhD, at Harvard (1932), was supervised by G. D. Birkhoff, who had introduced the chromatic polynomial in 1912 and done much foundational work on it [121, 121, 122, 123, 124] (see also the extract and discussion in [115, pp. 166–170,180,185]).

Tutte's supervisor was Shaun Wylie, a topologist whom he had known at Bletchley Park. Their respective supervisors were not co-authors of papers by these PhD students, and it seems that the advisors were not closely involved in their students' research.

The styles of Whitney and Tutte were very different. Whitney was an effective communicator of the key ideas in proofs, but did not lay out the proofs as formally and rigorously as would be expected today. Tutte was formal and precise, often terse, restrained, and dry, with occasional touches of whimsy.

34.3 Preliminaries

In this section, we give some notation and terminology used throughout this chapter, and state some basic definitions and results about chromatic, Tutte, and related polynomials that are used in the papers we discuss. This enables us to discuss the papers, with their varied notation and terminology, in a reasonably consistent way. We briefly attribute the origin of each of these polynomials, but this gives a very simplified version of the history. The full story is given in later sections.

Throughout, $G = (V, E)$ is a graph with n vertices, m edges, $k(G)$ components, and rank $r(G) = r(E) = n - k(G)$. The *components* of $X \subseteq E$ are the components of $G[X]$, the induced subgraph on X. Note that the number of components of the graph G equals the number of components of the set $E(G)$ if and only if G has no vertices of degree zero. The function $k(X)$ thus gives the number of components of X plus the number of vertices of degree zero in G.

The *chromatic polynomial* $\chi(G; x)$ is described in Chapter 11. It was introduced by Birkhoff in 1912 [121]. Its deletion–contraction relation appears in [123, p. 93], and is a corollary of the deletion–contraction relation for the coefficients of the Whitney rank generating function due to Foster and reported by Whitney [W32d]. It extends to cycle matroids of graphs, with a small adjustment: $\chi(G; x) = x^{k(G)}\chi(M(G); x)$. Its coefficients alternate in sign [W32d].

Whitney showed in [W32a] that

$$\chi(G; x) = \sum_{X \subseteq E} (-1)^{|X|} x^{n-r(X)}. \tag{34.1}$$

This is the seed from which his rank generating function grew.

The sum (34.1) is easily rewritten in a few different ways (see Equations (11.1) and (11.4)).

$$\chi(G;x) = \sum_{X\subseteq E}(-1)^{|X|}x^{n-r(X)} = \sum_{X\subseteq E}(-1)^{|X|}x^{k(X)} \tag{34.2}$$

$$= x^{k(G)}\sum_{X\subseteq E}(-1)^{|X|}x^{r(E)-r(X)} \tag{34.3}$$

$$= \sum_{i=0}^{n-1}\left(\sum_{j=0}^{m-i}(-1)^{i+j}m_{i,j}\right)x^{n-i}. \tag{34.4}$$

where $m_{ij} = |\{X\subseteq E : r(X) = i, \text{ and } |X| - r(X) = j\}|$. The expression using coefficients m_{ij}, due to Whitney [W32a], reveals the chromatic polynomial as a specialization of a bivariate polynomial with coefficients m_{ij}, which Whitney introduced in [1151] and [W32d] and which we define below.

The third major contribution of [W32a] to chromatic polynomials was the interpretation of the coefficients of $\chi(G;x)$ in terms of broken-cycles; see Theorem 11.22.

Flows, tensions, and flow polynomials were discussed in Chapter 12. Colorings (and "improper" colorings) are analogous to electrical potentials, tensions are analogous to potential differences, and flows are analogous to electrical currents. The analogy can be made precise by letting flows and tensions be real-valued and thinking of edges as wires of unit resistance. Flows are dual to tensions in the same way as currents are dual to potential differences.

Definition 34.1. The *tension polynomial* $\theta(G;x)$ gives the number of nowhere-zero A-tensions of G whenever $x = |\mathsf{A}|$, where A is an additively written abelian group.

The tension polynomial is related to the chromatic polynomial by $\chi(G;x) = x^{k(G)}\theta(G;x)$ and $\theta(G;x) = \chi(M(G);x)$. Tensions and tension polynomials were introduced by Tutte [T54]. The flow polynomial itself, as essentially the dual of the chromatic polynomial, is due to Whitney [W33a], but the interpretation in terms of flows is due to Tutte [T47, T54].

Lastly, we define the bivariate polynomials that are our main focus.

Definition 34.2. The *Whitney rank generating function* [W32d] is

$$W(G;x,y) = \sum_{X\subseteq E} x^{r(E)-r(X)}y^{|X|-r(X)} = \sum_{i=0}^{r(E)}\sum_{j=0}^{|E|-r(E)} r_{ij}x^iy^j \tag{34.5}$$

where $r_{ij} = |\{X\subseteq E : r(E) - r(X) = i, \text{ and } |X| - r(X) = j\}|$.

If G^* is the dual of G (which will be another graph if G is planar, and a non-graphic but cographic matroid otherwise), then

$$W(G^*;x,y) = W(G;y,x). \tag{34.6}$$

This was observed for planar graphs in [W32d]. $W(G;x,y)$ has sometimes been called the *corank–nullity polynomial*, but we avoid this term because it can be confusing. The term "corank", here meaning $r(E) - r(X)$, is also used for $r^*(X)$.

It is sometimes convenient to introduce the variable $q := xy$ and let

$$S(G;q,y) = W(G;qy^{-1},y) = \sum_{X \subseteq E} q^{r(E)-r(X)} y^{|X|-r(E)}. \tag{34.7}$$

The *Tutte polynomial* [T54] may be defined in any of three ways, which we give in order of increasing mathematical complexity (which happens also to be the reverse of their chronological order).

Definition 34.3.

1.

$$T(G;x,y) = W(G;x-1,y-1). \tag{34.8}$$

2.

$$T(G;x,y) = \begin{cases} 1, & \text{if } E(G) = \emptyset; \\ x\,T(G/e;x,y), & \text{if } e \text{ is a bridge;} \\ y\,T(G\backslash e;x,y), & \text{if } e \text{ is a loop;} \\ T(G\backslash e;x,y) + T(G/e;x,y), & \text{otherwise.} \end{cases} \tag{34.9}$$

3.

$$T(G;x,y) = \sum_{T} x^{|\mathrm{IA}(T)|} y^{|\mathrm{EA}(T)|}, \tag{34.10}$$

where the sum is over all bases of G (which are spanning trees if G is connected and are spanning forests containing one spanning tree for each component of G otherwise) and, for a given linear order $<$ on $E(G)$, we write $|\mathrm{IA}(T)|$ (respectively, $|\mathrm{EA}(T)|$) for the number of edges e in T (respectively, $E(G) \setminus T$) that are the greatest, under $<$, in the unique cocycle (respectively, cycle) contained in $(E(G) \setminus T) \cup \{e\}$ (respectively, $T \cup \{e\}$). (For further explanation of this activities-based definition, see Section 2.2.3.)

Graphs (or matroids) G and H are *Tutte equivalent* if $W(G;x,y) = W(H;x,y)$, or equivalently, $T(G;x,y) = T(H;x,y)$ (see Chapter 6). This notion is due to Whitney, appearing in [W32d, p. 689] and [W33a, p. 235]. Tutte's term for this was *codichromatic* [1097].

Prior to Whitney's work in the 1930s, there were three graph invariants in the literature that would later be shown to be specializations of the polynomials of Whitney and Tutte. These were:

- the chromatic polynomial $\chi(G;x)$, due to Birkhoff [121], mostly studied just for planar graphs;
- the number of spanning trees of a graph, studied by Kirchhoff [691];

- the partition function of the Ising model in statistical mechanics [635], with the same interaction energy on all edges and no external magnetic field.

We next meet them in the work of Whitney (Section 34.5, Section 34.6), Tutte (Section 34.8, Section 34.10) and Potts (Section 34.12) respectively.

34.4 Notation in the papers

A table of the notation and terminology used in the papers under discussion can be found at the end of this chapter (pp. 663–668). The first column lists some notation and terms used by Whitney [W32a, W32c, W32d, W33a], with the terms from the first three unless otherwise indicated. The second column lists notation and terms used by Tutte [T47, T48, T54, T67], with the terms from [T47] unless otherwise indicated. The third column lists a modern equivalent. Blank table entries, and terms that do not appear in the table at all, mean that the particular notation/term either has its modern meaning or is not used at all.

The modern terms listed in the third column are not always a precise mathematical translation from Whitney or Tutte. They are meant to convey the essence of the term for the modern reader who is interested in the contributions of the paper to graph theory. Tutte's early work in particular is often embedded in a formalism derived from algebraic and combinatorial topology which, while beautiful and very precisely expressed, can be unnecessarily ornate. This may distract the modern reader and make the paper's audience narrower than it needs to be. For example, Tutte's "linear graphs" and "cubical networks" [T47] are incarnated as complexes whose constituent cells are points and curves, but we translate them as "graphs" and "cubic graphs" respectively. The modern graph theorist loses nothing from such slightly "forgetful" translations.

The topological terminology used by Tutte may be found, for example, in Veblen's classic text [1115].[1] Chapter I, on "Linear Graphs", is an elegant introduction to graph theory in the language of complexes, cells, boundaries and so on, and emphasizes the use of linear algebra.[2] This text was very influential in topology and seems to have influenced Tutte too.[3] Much of

[1]Veblen shared the same PhD supervisor, E. Moore, as G. D. Birkhoff, at the University of Chicago in the first decade of the twentieth century. Veblen was about four years ahead, in age and PhD dates.

[2]Veblen's view of the state of development of graph theory, even then, is interesting: "There is a rather elaborate theory of linear graphs in existence ..." [1115, p. 9].

[3]Tutte comments on the influence of the topological texts by Veblen [1115] and Seifert and Threlfall [995] in [1108, p. 46]. The second of these (used at Cambridge when Tutte was an undergraduate) does not mention graphs explicitly, but they may be discerned as special

Tutte's early work may be viewed as a very advanced development of the linear algebraic approach to graph theory inspired by combinatorial topology.

In [W32a, W32c, W32d, W33a], the following terms all have their usual meaning: graph (assumed undirected), vertex, isolated vertex, connected, forest, planar, and non-separable. Whitney's circuits are our cycles, and his broken-circuits are here broken-cycles.

34.5 Whitney, 1932: A logical expansion in mathematics

This short (8pp) paper is an accessible exposition of the inclusion-exclusion principle with a sample of its applications in mathematics. The principle, which goes back to the eighteenth century [1045], is described by Whitney as "...well known to logicians. It should be better known to mathematicians also ...". A proof of the principle is presented in Section 2, with the inductive step only going from two sets to three sets, with the remark that it "obviously holds if the numbers two and three are replaced by i and $i+1$"! The principle is extended in Section 3 (to non-integer-valued functions of sets), referring to related work in [1157]. Example applications are to number theory (divisibility, Section 4), probability (derangements, Section 5), and graph coloring (Section 6). This latter section proves the edge-subgraph expansion (state model) for the chromatic polynomial, which we gave in Section 34.3 as (34.1)–(34.3), in a couple of closely related forms. The concluding Section 7 gives the interpretation of the coefficients of the chromatic polynomial, as signed numbers of subgraphs of each possible size containing no broken-cycle; see Theorem 11.22.

Whitney gives a dual result in a later paper, [W33a], which we discuss in Section 34.7.

The paper [W32a] does not mention the Whitney rank generating function. The only hint of it is in the brief appearance of m_{ij}, defined as the number of sets of edges of G of rank i and nullity j, in the last few lines of p. 577, where he proves the expression for $\chi(G;x)$ we gave in (34.4). This number array is a simple reflection of the coefficient array (r_{ij}) for the Whitney rank generating function, since $m_{i,j} = r_{r(E)-i,j}$.

Whitney credited his edge-subgraph expansion [W32a, Section 6] to Birkhoff [121], mentioning that his own discovery of it was independent. But he later realized that this attribution was incorrect: in footnote ‡ on p. 232 of

---

cases in some of the discussion of complexes; see, e.g., Sections 21–23. All mathematicians—and mathematics students, from senior undergraduates onwards—are encouraged to read the beautiful and accessible Chapter 1 of [995], "Illustrative material" (although it does not bear on the subject of this article).

Another topological text known to Tutte was Lefschetz [760], which introduces "linear graphs" as one-dimensional complexes (p. 93). In [T47], Tutte cites [760] in support of the observation that the number of λ-flows on a graph G is $\lambda^{|E|-r(E)}$.

[W33a], he observes that his m_{ik} are not the same as Birkhoff's symbols (i,k) from [121], although he had supposed they were in [W32a, Section 6].

Incidentally, Whitney's papers [W32a] and [W32d] appear to be the first to describe the chromatic polynomial for general graphs, rather than just for planar graphs as done previously by Birkhoff [121, 122] (who did it in terms of map coloring, where faces are colored rather than vertices). This step by Whitney was noted briefly by Tutte [T54, p. 81]; see also [123, p. 86].

An extract from this paper, with short commentary, is given in [115, pp. 180–185].

34.6 Whitney, 1932: The coloring of graphs

This paper is essentially Whitney's PhD thesis (Harvard, 1932). Some of the results were announced in January 1931 [1151]. The paper is a treatise on the numbers m_{ij} mentioned briefly in [W32a]. It may be regarded as the first paper on the Whitney rank generating function. Curiously, Whitney does not give this function explicitly as a polynomial in this paper or thesis. Nonetheless, he develops a lot of theory about it in terms of its coefficients, and he does discuss in Section 11 the relationship between the coefficients m_{ij} of a general bivariate power series $\sum_{i,j} m_{ij}x^iy^j$ and those of its logarithm $F(x,y) = \sum_{i,j} f_{ij}x^iy^j$. The clear purpose of this discussion is to apply it to the case when the m_{ij} are the coefficients of the rank generating function (indeed, this is the only section of the paper where m_{ij} may stand for anything else), which he does for the rest of the paper. So he had the rank generating function, as a bivariate polynomial, in mind, even though he happened to present it as a coefficient array. It is only in his summary, [1151, p. 123][4], that he writes of the m_{ij} for a graph as the coefficients of a polynomial.

The rank generating function is thus an "accumulation point" of the theory in [W32d], in that Whitney put the foundations of its theory in place and came "arbitrarily close" to defining it explicitly, only giving the "limit point" itself in an earlier summary, [1151], that has not been read or cited nearly as much as the paper. As such, it is fitting to attach Whitney's name to the rank generating function, as Welsh did in [1138, Ch. 15], and as we do here. On the naming of the Tutte polynomial, Tutte wrote that "This may be unfair to Hassler Whitney who knew and used analogous coefficients without bothering to affix them to two variables" [1107, p. 8].

We describe Whitney's results in terms of $W(G;x,y)$, rather than the array (m_{ij}), when that would be more natural to the modern reader. These descriptions should be read with the foregoing remarks in mind.

[4]In those days, the *Proc. Nat. Acad. Sci. U.S.A.* was "an effective means for the rapid publication of first announcements of scientific discoveries", as described in the foreword of [1164].

Whitney observes the relationship between the Whitney rank generating function of a planar graph and its dual (see Equation (34.6)), in the introduction to [W32d], in the first paragraph after his outline of the paper. He does not extend $W(G; x, y)$ to arbitrary matroids[5], so does not extend this duality observation to arbitrary graphs (whose duals are cographic matroids, but are not graphic if the original graph is nonplanar). However, he immediately introduces the class of graphs G—let us call them *Whitney graphs*—for which the dual $M(G)^*$ of the cycle matroid $M(G)$ of G is Tutte-equivalent to some other graph G', i.e., $W(M(G)^*; x, y) = W(G'; x, y)$. This includes the class of planar graphs (for which we have $G' = G^*$), and Whitney conjectured in January 1931 [1151] that there were no others. But in [W32d] he gives a counterexample, found on 13 May 1931 by R. M. Foster, which is nonplanar but *Tutte self-dual* in that it is Tutte equivalent to its dual matroid. Whitney asks whether or not every Whitney graph is four-colorable, noting that this is a strengthening of the (then) four color conjecture.

Part I (Sections 2–4) of [W32d] is something of a warm-up for the main results, which come in Parts II–IV. Both here and in [W32a], Whitney shows how to obtain the chromatic polynomial, $\chi(G; x)$, from the rank generating function [W32d, Section 1, (1.1)]. He illustrates the broken-cycle viewpoint [W32a, Section 7] by using it to show that the coefficients of $\chi(G; x)$ alternate in sign and the lowest-order term has degree $n - r(G)$ [W32d, Section 2]. He then illustrates some elementary observations on the m_{ij} by using them to show that $\chi(G; 1) = 0$ and to compute the chromatic polynomial of cycle graphs, $\chi(C_n; x) = (x - 1)^n + (-1)^n(x - 1)$.

In [W32d, Section 4], the inclusion-exclusion principle is applied in two different ways to the coefficients of the chromatic polynomial, using the broken-cycle viewpoint. Recall that the size a_i of the coefficient of x^{n-i} in $\chi(G; x)$ is the number of subsets of size i that contain no broken-cycle (see Theorem 11.22). These subsets are precisely those that do *not* contain *some* broken-cycle, and we can use this as a starting point for inclusion-exclusion. On the other hand, for a subset of edges *not* to contain a broken-cycle, its *complement* must contain *some* set of edges that meets *every* broken-cycle. This too launches us into inclusion-exclusion; the sets used, and the numbers added and subtracted, differ from those of the first approach, but the end result is the same. These two inclusion-exclusion approaches are complementary in a formal sense; see, for example, [1040, Section 2.1]. Most of the time, when using inclusion-exclusion, we settle on whichever viewpoint is more convenient—using sets, or their complements. Whitney had already described the formal connection between these two viewpoints in [1157]. The two approaches described above are instances of his second and third "normal forms", respectively. He uses notation α_k and β_k for the contributions to the respective inclusion-exclusion sums from sets of k edges; this notation is used both for the special broken-cycle situation case here, and for the general

[5]Indeed, his seminal paper on matroids was not published until three years later [1159].

treatment in [1157, Section 10] where he gives expressions for the α_k in terms of the β_k and vice versa.

We now come to Parts II–IV and the main results of the paper: determination of $W(G;x,y)$ from the non-separable subgraphs of G (Part II); the relationship between coefficients of generating functions $F_a(x,y)$, associated with objects a of some type, and their exponentials $M_a(x,y) = \exp F_a(x,y)$, when F_a behaves additively with respect to addition of the underlying objects a (Part III); and an expression for $W(G;x,y)$ as an exponential of another generating function whose coefficients are linear in numbers of non-separable graphs of the various isomorphism types (Part IV), an early relative of the exponential formula in enumerative combinatorics.

The first main result in the paper, developed in Part II (Sections 5–8), is that each coefficient m_{ij} can be expressed as a polynomial in the numbers $N_t = N_t(G)$ of non-separable subgraphs of G of each possible isomorphism type t. The polynomial itself depends only on i, j, and not on G; the dependence of m_{ij} on G is *only* via the numbers $N_t(G)$. Calculating the m_{ij} using this method compares favorably with using the definition directly, since the former only uses non-separable subgraphs of G, while the latter uses *all* subgraphs of G that have no isolated vertices.

The proof is based on a clever application of (again) inclusion-exclusion. Two graphs are *equivalent* if there is a bijection between their respective sets of blocks such that pairs of blocks that correspond (under the bijection) are isomorphic. In other words, detaching each graph into a disjoint union of its blocks gives two isomorphic graphs. (Equivalence in this sense implies Tutte equivalence, since $W(G;x,y)$ is multiplicative over blocks, but the converse is not true in general.) To begin with, subgraphs of G of rank i and nullity j are partitioned according to their equivalence class. The subgraphs in each equivalence class are counted separately, and then combined in an appropriately weighted sum to give m_{ij}. Inclusion-exclusion appears in the counting within an equivalence class.

Our task here is to count subgraphs of rank i, nullity j, and with a specified multiset of (isomorphism types of) blocks $H_1, \ldots, H_q$, i.e., the subgraphs are equivalent to a graph H whose q blocks are isomorphic to $H_1, \ldots, H_q$.

Whitney showed in [1155, Theorem 17] that a specified set of non-separable graphs is indeed the set of blocks of a graph H if and only if no subset of two or more of them forms a *circuit of graphs* in H. Here, a *circuit of graphs* is either two non-separable graphs with at least two vertices (and possibly some edges) in common, or is formed from some set of $c \geq 3$ non-separable graphs by an ordering $B_1, \ldots, B_c$ of the blocks such that, for each $i \in \{1, \ldots, c\}$, there is a vertex v_i shared by B_i and B_{i+1} (with wrap-around of subscripts), all the v_i are distinct, and no two of the blocks share any other vertices (or edges).

Our requirement is that that *no subset* of two or more of $H_1, \ldots, H_q$ is realized as a circuit of graphs in G. This forms a starting point for inclusion-exclusion, in which each *set of subsets* $\mathcal{B}$ of two or more of the H_i defines one of the summands. This inclusion-exclusion is being done "one level higher" than the inclusion-exclusion arguments used earlier in the paper: there, the summands were determined just by sets of edges, rather than (as we have here) *sets of sets* of edges.

A summand in this inclusion-exclusion has the form $(-1)^{|\mathcal{B}|}n_{\mathcal{H}}(\mathcal{B})$, where

- $\mathcal{H} := \{H_1, \ldots, H_q\}$ is the set of isomorphism types of blocks under consideration,
- $\mathcal{B}$ is a set of subsets of $\mathcal{H}$ each containing at least two of the H_i, and
- $n_{\mathcal{H}}(\mathcal{B})$ is the number of ways of realizing the graphs $H_i \in \mathcal{H}$ as subgraphs of G such that, for every block-set $B \in \mathcal{B}$, the particular blocks H_i in B are realized as a circuit of graphs.

In the last point, the *realizations* of graphs H_i and H_j as subgraphs of G may overlap, provided H_i and H_j do not belong to the same $B \in \mathcal{B}$.

An important issue in the proof is dependence among the realizations of the H_i as subgraphs of G. For a start, if two H_i belong to a single part of $\mathcal{B}$, then that means they must be realized as members of a circuit of graphs, so the realizations are not independent. Also, circuits of graphs can "join up" to make more complicated structures, extending the dependence further. This is addressed in [W32d, Section 7, pp. 697–698].

Over Sections 6–7 of [W32d], it is shown that $n_{\mathcal{H}}(\mathcal{B})$ is a polynomial in the $N_t(G)$, and hence that each m_{ij} itself is a polynomial (with no constant term) in the $N_t(G)$, and that the dependence on G is entirely captured by these numbers.

Especially significant, in the polynomial expressions for m_{ij}, are the linear terms. These arise in the inclusion-exclusion when $\mathcal{B}$ has the property that all isomorphism types in $\mathcal{H}$ are mutually dependent in the sense described two paragraphs previously. The linear terms are characterized in [W32d, Theorem 8A].

In the rest of Section 8 of [W32d], Whitney analyses the set of pairs (p, q) such that a non-separable graph of rank p and nullity q contributes the linear terms of m_{ij}. Lemmas A–F are established, giving bounds for the set of these (p, q).

In Part III (Sections 9–11), Whitney determines the relationship between the coefficients of a general univariate (Sections 9–10) or bivariate (Section 11) generating function F and its exponential $M := e^F$. Proofs are given for the univariate case; this is used solely to introduce the reader to the bivariate case, which is then treated more briefly and without proofs. He notes in passing that the treatment can be extended to more variables. We will only refer to the bivariate case.

Whitney describes this relationship using expressions for the coefficients f_{ij} of $F(x, y)$ in terms of the coefficients m_{ij} of $M(x, y) = e^{F(x,y)}$, and vice versa (without giving any interpretation to these coefficients). In these expressions, the only linear term of f_{ij} is m_{ij}, and the only linear term in m_{ij} is f_{ij}.

The nonlinear transformation that maps the f_{ij} array to the m_{ij} array is denoted by T; it just describes, at the level of the coefficients, the effect of exponentiation of $F(x, y)$ to obtain $M(x, y)$.

Each of these generating functions is taken to be associated with objects that can be added in some way. Whitney has in mind the set of graphs under disjoint union ("sum"), but the discussion in these sections is abstract and general.

The generating functions $F(x, y)$ behave additively: the generating function of a sum of objects equals the sum of the generating functions of the objects, and so their coefficients behave in the same way. Whitney shows that this occurs if and only if the coefficients of $M(x, y)$ combine according to a convolution expression [W32d, Theorem 11B].

Note the use of m_{ij} to denote the coefficients of $M(x, y)$. It is no coincidence that this is the same notation used earlier in the paper for the coefficients of the rank generating function. In Part IV, Whitney specializes to the case where the m_{ij} pertain to graphs and have their earlier interpretation. Once he does that, he no longer refers directly to any bivariate polynomial, and so bypasses explicit definition of $W(G)$. But by then it is clear that he has such a polynomial in mind. The lack of *explicit* definition is more a quirk of his approach to the exposition than any real failure to define it.

Part IV (Sections 12–17) makes very elegant use of the machinery set up in Parts II and III together with the notion of a *generalized graph*, namely, a function mapping isomorphism types of non-separable graphs to integers which is nonzero only finitely many times. The main result is that, when all the m_{ij} and f_{ij} are expressed in terms of the numbers N_t of non-separable graphs (as was done for m_{ij} in Part II, and for f_{ij} using T^{-1} in Theorem 12C), f_{ij} is just the linear terms of m_{ij} (Theorem 15A).[6] So f_{ij} is a certain integer linear combination of some finite number of the N_t, and m_{ij} can be obtained from it using the expression given by T.

The historical significance of this result goes beyond its contribution to the theory of Tutte–Whitney polynomials. In [W32c, Theorem 8A], it is shown that the linear terms of m_{ij} are (apart from their integer coefficients) numbers of ways of forming non-separable subgraphs of G from blocks of a graph of rank i and nullity j by combining the blocks so that certain circuits of graphs are formed. Now that we recognize these linear terms as the f_{ij}, we have an interpretation of each f_{ij} in terms of numbers of non-separable subgraphs of G whose isomorphism type is drawn from a fixed finite menu. The transformation from the f_{ij} to the m_{ij}—which, as we have seen, is just the exponentiation $M(x, y) = e^{F(x,y)}$—gives numbers of *all* subgraphs (separable or not) of the required rank and nullity. This relationship between the combinatorial interpretations of the coefficients of $F(x, y)$ and those of $e^{F(x,y)}$ may be regarded

[6] This is not to be confused with our remarks a few paragraphs ago, on page 635, on linear terms in the relations involving the f_{ij} and the m_{ij} in [W32d, part II]. Those earlier remarks apply when these coefficients are expressed in terms of *each other*. Our current point is about expressing them in terms of the N_t.

as a close relative of the exponential formula in enumerative combinatorics (see, e.g., [1038, Section 5.1]). There are some differences, most obviously that $F(x, y)$ and $M(x, y)$ here are ordinary generating functions, not exponential generating functions. Also, the combinatorial interpretation of the f_{ij} is not as crisp as in classical applications of the exponential formula, and the coefficients here pertain to substructures (in this case, subgraphs) of a specific fixed finite structure (in our case, a graph G), rather than *all* structures of a given order (without thinking of them as contained within something else). Nonetheless, Whitney's achievement here is significant and deserves better recognition in the history of the formula. It came well before the early application of the formula to the relationships between numbers of graphs and connected graphs developed since the 1950s (see [600]).

In Section 17, Whitney discusses inequalities for the m_{ij} that must be satisfied for all graphs. He gives a few, but remarks on the "great difficulty" of finding them all. Here he is implicitly asking a question he asks explicitly (in coefficient language) in [1151, p. 124]: which polynomials are $W(G; x, y)$ for some graph G? This is one of Crapo's problems in [342] (Problem 3, see also the related Problem 4), and remains open.

In Part V (Sections 18–19), Whitney applies his results to calculation of his polynomial expressions for his coefficients, for $i \leq 5$, and then applies these expressions to certain planar graphs. He mentions Birkhoff's result [121] that, to prove the four color conjecture (as it then was), it is sufficient to prove it for 5-connected nearly-6-connected triangulations. (Here, a graph is *nearly 6-connected* if, for any $U \subseteq V(G)$ of size 6, $G \backslash U$ has a singleton component.) The graph of this type with fewest vertices is the icosahedron, so Whitney demonstrates the utility of his methods by using them to compute the first six coefficients of its chromatic polynomial. He then gives the whole polynomial, "which was calculated by direct means". This is no mean feat. Computers had not yet been built, and Birkhoff wrote that "a considerable computation would be necessary to determine" this particular chromatic polynomial [121].

Whitney closes with a "Note added in proof", stating the deletion–contraction relation for $W(G; x, y)$, again in terms of the m_{ij}, and attributing it to R. M. Foster.[7]

There are some important results in Whitney's PhD thesis not included in this paper. In Section 5 (p. 10, (5.1)) of the thesis, Whitney uses a simple counting argument to prove that

$$(|E| - i - j)\, m_{ij}(G) = \sum_{e \in E} m_{ij}(G \backslash e). \tag{34.11}$$

This is mentioned briefly in [1151], but with an error in the prefactor on the left.

[7]Unfortunately, Whitney misspells Foster's name in this note. The work of Foster on these numbers seems to have remained unpublished by him. There is a glimpse of it in [500] where he lists all multigraphs, up to 2-isomorphism, of rank i and nullity j for $i, j \leq 3$, but he does not there develop the theory of m_{ij}.

Affixing variables to (34.11) and summing each side over i, j, aiming for a statement about polynomials, gives Tutte's result from [T67, (23)], which we give later in Equation (34.19). In this case, Whitney's result has all the combinatorial content of Tutte's, three decades earlier, albeit expressed in terms of coefficients rather than polynomials.

In Section 14 (pp. 28–29) of the thesis, Whitney proves that if G consists of two graphs G_1 and G_2 that overlap in a clique K_r, so that the removal of this clique would disconnect G, then

$$\chi(G;x) = \frac{\chi(G_1;x)\chi(G_2;x)}{\chi(K_r;x)}. \tag{34.12}$$

This is sometimes credited to Zykov [1191], and Read [945] is also sometimes given as a reference. It was proved for $r = 3$ by Birkhoff [122, pp. 86–87], who also does it for $r \leq 3$ in [123, pp. 91–92]. Those papers were confined to planar graphs, though there is a sentence at the end of [122] indicating some extendability to other surfaces. It is now known that *chromatic factorizations* of this type, (34.12), can occur even when G is not clique-separable in the given way and is not even chromatically equivalent to such a graph (see Subsection 11.3.5 and [861, 862]).

34.7 Whitney, 1933: Topological invariants for graphs

This paper is another of the many published by Whitney in the early 1930s, and seems to have attracted less attention than most of the others. Yet it is notable for, in effect, introducing the flow polynomial as the dual of the chromatic polynomial (see Chapter 12), and for another brief mention (in different language) of Tutte equivalence. Again, Whitney works mainly with coefficients, denoting the flow polynomial's coefficients by p_i and obtaining them as alternating sums of his array m_{ij} down its columns (instead of rows, as for the chromatic polynomial), in effect showing that the flow polynomial is a partial evaluation of $W(G; x, y)$. In keeping with his approach in [W32d], he refrains from explicitly giving a polynomial with these coefficients for general graphs, although he surely had that viewpoint in mind. He only uses a polynomial in the planar case, for which he shows that the polynomial is the chromatic polynomial of the dual graph. Furthermore, he does not give a combinatorial interpretation of the *values* of the flow polynomial, only of its coefficients, which he expresses in terms of broken-cutsets. In effect, he defines the flow polynomial without flows!

Whitney shows that the flow polynomial is multiplicative over blocks (presenting it in terms of convolution of coefficient sequences) and that it is identically zero for any graph with a bridge. He also shows that it is *topologically invariant*, i.e., invariant under edge subdivision. This implies that if two graphs

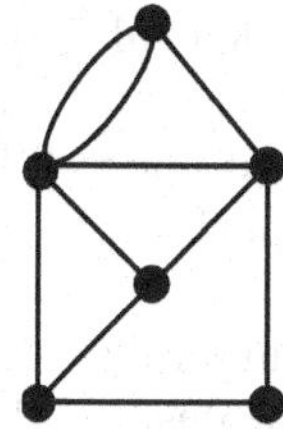
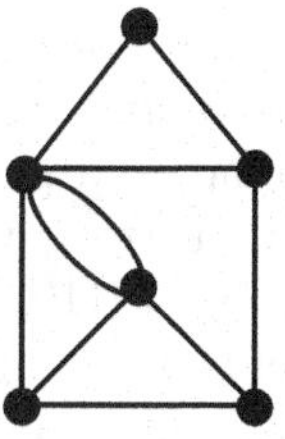

FIGURE 34.1: The Gray graphs.

are homeomorphic (i.e., both can be obtained by sequences of subdivisions from some third graph) then they have the same flow polynomial. Whitney shows that the converse does not necessarily hold, by giving two graphs with the same flow polynomial that are not homeomorphic.

He closes the paper with another early appearance of what we now call Tutte equivalence. He asks whether Tutte-equivalent graphs must necessarily be 2-isomorphic. The answer was soon shown to be negative, by M. Gray, using two Tutte-equivalent 2-connected planar six-vertex ten-edge multigraphs with non-isomorphic cycle matroids (see Figure 34.1). The graphs are given in [1097]. Tutte's source was R. M. Foster (private communication, no date given). Tutte places Gray's counterexample "forty years ago"—which sounds like it was within a year or so of the publication of the conjecture in [W33a]. Tutte goes on to describe non-isomorphic 5-connected Tutte equivalent graphs.

34.8 Tutte, 1947: A ring in graph theory

This paper by Tutte is arguably the most important in the history of Tutte–Whitney polynomials. The polynomials are introduced explicitly, and many of the themes developed in the paper remain central to research in the field. Quite apart from the specific results proved, it makes the major conceptual contribution of placing reduction relations, such as deletion–contraction, at the heart of the topic; these had been literally only a "note added in proof" in Whitney's work [W32d]. In this paper, Tutte also identifies concrete combinatorial objects, other than colorings, that are counted by some evaluations of the polynomials. Moreover, he introduces universal invariants with an associated "recipe theorem".

There are some superficial differences with subsequent work. The paper [T47] works entirely in the realm of graphs with no mention of matroids. There are some technical differences between the polynomials introduced in this paper and those that later acquired Tutte's and Whitney's names. The paper is not easy to read, partly by virtue of its sheer originality but mostly due to Tutte's compact style and use of topological language. Nonetheless, it is beautiful and deep, and richly repays the effort of reading it. It is often cited but not often read, and parts of it are still not well known.

Many years later, Tutte gave a short exposition of some of the theory developed in this paper in [1102].

A function F defined on graphs and taking values in an abelian group is a *W-function* if it is invariant under isomorphism and, for all G and any non-loop edge $e \in E(G)$,

$$F(G) = F(G\backslash e) + F(G/e). \tag{34.13}$$

If it is also multiplicative over components (with its range a commutative ring rather than just an abelian group), i.e.,

$$F(G_1 \sqcup G_2) = F(G_1)F(G_2), \tag{34.14}$$

then it is a *V-function.* These definitions are from [T47, (4)–(6), pp. 26–27]. Tutte's definition of a V-function was the inspiration for the later definition of a Tutte–Grothendieck invariant, as we discuss in Section 34.14.1.

As examples of V-functions, Tutte gives (with an appropriate sign) numbers of colorings and nowhere-zero flows. As an example of a W-function, he gives the number of spanning trees. This fails to be a V-function due to the technicality of being 0 for disconnected graphs, but the number of spanning *forests* (i.e., bases of the cycle matroid) is easily seen to be a V-function.

The reader familiar with Tutte polynomials will notice some differences between their definition and (34.13)–(34.14) above. Firstly, the deletion–contraction rule (34.13) contains no rule for loops. Secondly, bridges are treated the same as other edges. Thirdly, (34.14) only specifies multiplicativity over components, whereas Tutte–Whitney polynomials are multiplicative over blocks. In fact, Tutte–Whitney polynomials in general are not W- or V-functions, and W- or V-functions are not, in general, obtainable from Tutte–Whitney polynomials. But the relationship between them is very close. A simple variant of the Whitney rank generating function, adjusted to take account of the number of components of a graph, turns out to be a V-function. This adjustment is just multiplication by $x^{k(G)}$, and corresponds exactly to the adjustment needed to convert the chromatic polynomial of the cycle matroid of a graph to the usual chromatic polynomial of the graph itself. Similarly, multiplying the Tutte polynomial $T(G; x, y)$ by $(x-1)^{k(G)}$ makes it a V-function.

It is natural to ask, under what circumstances is $W(G; x, y)$ itself a W-function? The only thing to check is the deletion–contraction relation, [T47, (5)]. We know that $W(G; x, y)$ satisfies this for any edge that is neither a loop nor a bridge. To be a W-function, the relation must also hold for bridges. It is routine to show that this holds if and only if $x = 1$. In this case, $W(G; x, y)$ is also a V-function, since it is multiplicative over blocks, and hence over components. Similarly, $T(G; x, y)$ is a W-function if and only if $x = 2$, and then it too is a V-function.

This multiplication of $W(G;x,y)$ by $x^{k(G)}$, giving a V-function, has the effect of converting a matroid polynomial (i.e., one which extends immediately to matroids, since it only depends on the sizes and ranks of the subsets of $E(G)$) to a polynomial that is restricted to graphs because of its dependence on $k(G)$. So, some generality is lost in focusing on V-functions. It seems that Tutte did not have matroids in mind in this paper. It was received by the journal on 10 April 1946; Tutte does mention the extension of his polynomials to (representable) matroids in his thesis [T48], which was submitted in 1948. Even then, he apparently was not aware of other work on matroids as his thesis does not cite Whitney's 1935 paper [1159]. His later papers [T54] and [T67] on these polynomials also consider only graphs, though many of the ideas are extendable to matroids, as was done, for example, by Crapo [349] and Brylawski [251].[8]

We now give an overview of the rest of [T47]. Tutte considers *graphic forms*, which are formal integer linear combinations of isomorphism classes of graphs with only finitely many nonzero coefficients.[9] They have a ring structure under formal addition and multiplication (with the latter corresponding to disjoint union of graphs), and he calls this ring B.

Consider the deletion–contraction relation (34.13), recast as a relation among graphic forms, in which any graph may be replaced by a linear combination of two graphs in any of three ways for each non-loop edge e: $G \to G\backslash e + G/e$, or $G\backslash e \to G - G/e$, or $G/e \to G - G\backslash e$. Starting with any graph, deletion–contraction can be applied repeatedly to give a sequence of graphic forms, leading eventually to a graphic form whose summands are all disjoint unions of bouquets (where a *bouquet* has a single vertex and some nonnegative integer number of loops). Tutte defines an *elementary graph* to be a bouquet with $r \geq 0$ loops. He denotes it by y_r and its isomorphism class by $\mathbf{y}_r$. So the deletion–contraction process can transform the isomorphism class of any graph to a multivariate polynomial expression in the $\mathbf{y}_r$. Indeed, it can do the same for any graphic form at all. He constructs the ring R, obtained from the ring B of graphic forms by factoring out by the deletion–contraction relation.[10] In other words, two graphic forms are equivalent if one can be transformed to the other by repeated application of the deletion–contraction relation.

To describe this construction more formally, much as Tutte does: for any graph G and any non-loop edge $e \in E(G)$, construct a graphic form $G - (G\backslash e) - (G/e)$ (which Tutte calls a *W-form*) from these three isomorphism classes; take the ideal W generated by all these three-term graphic forms; and

[8]Tutte made many major contributions to matroid theory, going well beyond its role as a setting for the polynomials we discuss here. See [477].

[9]Graphic forms may be viewed as an extension of Whitney's generalized graphs, the difference being that the formal summands of the latter are restricted to isomorphism classes of non-separable graphs. The two authors however use their structures in different ways.

[10]Although rings are denoted $\mathfrak{R}$ in this handbook, here we use R for consistency with the original sources.

form the ring $R := B/W$ (or $B - W$, in [T47]) whose members are the cosets modulo W.

Each member of R—being an equivalence class of graphic forms under deletion–contraction, i.e., a coset of W in B—contains exactly one graphic form consisting entirely of elementary graphs. So every graph gives, under this equivalence, a unique graphic form of this elementary type. This graphic form, considered as a multivariate polynomial in the $\mathbf{y}_r$, is an analogue of the Tutte polynomial, and indeed specializes to it using appropriate substitutions and an algebraic adjustment of the kind described earlier. The ring R is the starting point for the theory of *Tutte–Grothendieck rings* in [251]; see also [728, Section IV.1].

Tutte proves that a function F is a W-function if and only if it is a group-homomorphism on R, and it is a V-function if and only if it is a ring-homomorphism on R (with its range being an abelian group or commutative ring, respectively). This is Theorem I, which may be viewed as the first "recipe theorem". It does not state the required substitutions explicitly, but they may be derived from the proof. Theorem V states that the ring R is isomorphic to the ring of all polynomials in $\mathbb{Z}[z_0, z_1, \ldots]$ with no constant term, where $(z_i : i \geq 0)$ is an infinite sequence of independent variables. Not every such polynomial can be obtained by repeated deletion–contraction from a graph; many come from other graphic forms. Then Theorem VI shows that any sequence of leafless connected graphs $(x_i : i \geq 0)$ with x_i having nullity i can be used as the "fundamental" graphs in terms of which any other graph (or graphic form) can be uniquely expressed. In other words, any member (coset) of R has a unique element that is a polynomial in the x_i. The bouquet sequence $(y_i : i \geq 0)$ mentioned above is just one such sequence.

In [T47, Section 3], Tutte introduces the V-function $Z \in \mathbb{Z}[z_0, z_1, \ldots]$, defined for any graph by

$$Z(G) = \sum_{X \subseteq E} \prod_{Y \in \{\text{components of X}\}} z_{|Y|-r(Y)}\,. \tag{34.15}$$

Caution: note that in this chapter, $Z(G)$ should not be confused with the multivariate Tutte polynomial $Z(G; q, \mathbf{v})$ that appears elsewhere in the handbook, nor with the partition function also denoted $Z(G)$ that appears in Section 34.12.

The $Z(G)$ in (34.15) turns out to be a universal invariant for the class of V-functions, in that any V-function can be obtained from it by appropriate substitutions for the variables z_i. (Throughout the paper, substitutions for variables are described in terms of homomorphisms of abelian groups or commutative rings.) Tutte does not quite state this as a theorem, but it follows from his proof of Theorem V (in particular, the ring isomorphism h), in Section 4, using Theorems I and III. In fact, the most important part of Theorem V is the specific ring isomorphism h in its proof, which respects the graph-theoretic interpretations of the two rings. (The mere assertion of an isomorphism between R and R_0 would in itself be a routine observation, using

Theorem II and its Corollary.) More recently, Bollobás, Pebody and Riordan [156] noted that, for the class of functions required to satisfy (34.13) only for edges which are neither bridges nor loops (i.e., unlike V-functions, they do not have to satisfy (34.13) for bridges), Z still serves as the universal invariant.

The V-function Z is an analogue of the Whitney rank generating function and specializes to it using an appropriate substitution together with the usual algebraic adjustment, as we discuss below. The isomorphism h gives, for any graph G, a coordinate transformation linking its equivalent graphic form in elementary graphs (an adjusted multivariate Tutte polynomial, in variables y_r) and the universal V-function $Z(G)$ (an adjusted multivariate Whitney rank generating function, in variables z_r). This isomorphism subsumes the simple coordinate translation relating the Tutte polynomial and Whitney rank generating function, (34.8). We return to this point in about a page.

The rest of Section 4 is devoted to Theorem VI, which describes conditions under which an infinite family of graphs $(x_i)_{i\geq 0}$ may serve as the elementary graphs, instead of the bouquets $(y_i)_{i\geq 0}$.

In [T47, Section 5], Tutte considers W- and V-functions that are invariant under edge subdivision. Such invariance imposes another relation, as well as deletion–contraction, and factoring out by this relation (i.e., by the ideal N generated by its corresponding graphic forms) gives another commutative ring R/N. Topologically invariant W- and V-functions are characterized in terms of homomorphisms to R/N (Theorem VII). In his later commentary on this work [1101, p. 53], Tutte writes: "It now seems to me that the most important result [of Section 5] is that a V-function is topologically invariant if $V(y_0) = -1$. This should have been more clearly stated. It is the result to be carried over to the following Sections". He later noted that the condition $V(y_0) = -1$ is also necessary, for any non-identically-zero V-function; see [1098, (2.1)] and [1099, Section 2].

It is a short step from Tutte's results to the observation that the V-function $Z(G)$ is topologically invariant if and only if $z_0 = -1$. We return to this shortly.

In [T47, Section 6], Tutte substitutes $z_i \mapsto tz^i$, where t and z are two new indeterminates, into $Z(G)$ and obtains the polynomial he calls $Q(G;t,z)$. This is nothing other than the Whitney rank generating function adjusted by the factor $t^{k(G)}$ which ensures it is a V-function:

$$Q(G;t,z) = t^{k(G)}W(G;t,z).$$

He remarks that, for any fixed a, b, the coefficient of $t^a z^b$ (hence, each m_{ij} from [W32d]) is a W-function.

It now follows from Tutte's previous section, [T47, Section 5], that $Q(G;t,z)$ is topologically invariant if and only if $t = -1$, and hence that $W(G;x,y)$ is topologically invariant if and only if $x = -1$.[11] This last observation tells us that the only topologically invariant evaluations of Tutte–

[11] As mentioned earlier, $W(G;x,y)$ is not a W-function, but we can still speak of its topological invariance or otherwise.

Whitney polynomials are evaluations of the flow polynomial. (One direction is clear, since flows are essentially unaffected by subdivisions.)

With the specialization of $Z(G)$ to $Q(G;t,z)$ in mind, the coordinate translation (34.8) can be discerned as a special case of the theory in [T47]. Theorem IV gave the universal V-function Z of a bouquet y_r, and under $z_i \mapsto tz^i$ this reduces to $Z(y_r) = Q(y_r;t,z) = t(z+1)^r$. For a single loop, we have $Q(y_1;t,z) = t(z+1)$. For a bridge, deletion–contraction shows that Q is just $t(t+1)$. The factor t is just the usual adjustment with $k(G)$ in its exponent. Ignoring this factor, we see loops and bridges yielding factors of $z+1$ and $t+1$ respectively. From the proof of Theorem V (and using Theorem IV), it can be seen that h maps a loop y_1 to $Z(y_1)$, i.e., to $Q(y_1;t,z) = t(z+1)$.

Continuing in [T47, Section 6], Tutte observes in Theorem IX that duality of plane graphs essentially interchanges the variables—(34.6) in our Section 34.3—which we saw earlier that Whitney had shown implicitly using the coefficients.

Tutte goes on to introduce polynomials that extend the chromatic and flow polynomials by counting the number of color assignments or flows, respectively, with given numbers of "bad" edges, i.e., edges that are monochromatic (having identically colored endpoints) or have zero flow, respectively. For colorings, let $\chi_b(G;x)$ be the number of x-assignments with b monochromatic edges (so $\chi_0(G;x)$ is the chromatic polynomial). Then the *coboundary polynomial* (using Crapo's term [348]) is $\sum_{k=0}^{m} \chi_k(G;x)s^k$, and Tutte proves in [T47, Theorem X] that this is (apart from an elementary factor) a coordinate transformation of the Whitney rank generating function:

$$\sum_{b=0}^{m} \chi_b(G;x)s^b = (s-1)^{r(G)}x^{k(G)}W(G;x/(s-1), s-1). \tag{34.16}$$

Tutte observes that Whitney's sum over sets of edges, given as (34.1) above, is a special case of this. The coboundary polynomial is also discussed in [247, Section 6.3] and [1139, pp. 63–64].

Tutte then does the same for the flow polynomial. Let $\phi_b(G;x)$ be the number of x-flows where the number of edges with zero flow is b. Tutte's Theorem XI is that

$$\sum_{b=0}^{m} \phi_b(G;x)s^b = (s-1)^{m-r(G)}W(G;s-1, x/(s-1)). \tag{34.17}$$

Tutte deduces from this that $\phi_b(G;x)$ depends only on the size x of the abelian group of flow values, not on other details of the group[12], and observes that the signed flow polynomial $(-1)^{m-r(G)}\phi_0(G;x)$ is a topologically invariant V-function. He explains the relationship between flows of a plane graph and proper colorings of its dual, and gives an identity [T47, (41)] (obtained by

[12]This property fails in some more general contexts, e.g., signed graphs [84].

combining (34.16) and (34.17) above) relating all the χ_b and all the ϕ_b, for a given G.

The paper concludes with Section 7, on some invariants of graphs in which every vertex has degree two or three. Let us call such graphs *topologically cubic* (not Tutte's term). The term "node" is reserved for a vertex of degree three; an "arc" is essentially a path between nodes whose internal vertices have degree two in the graph. In this way, the graph is viewed as a subdivision of a cubic graph. Throughout, nodeless (or "loose") edges are permitted. A nodeless edge in a topologically cubic graph may be pictured as a loop-shaped edge that is disjoint from every vertex, and forms a component in its own right. If such an edge is subdivided, then it becomes a conventional cycle. But nodeless edges, and any vertices or edges belonging to cycles formed by subdividing them, do not belong to arcs as defined here. Any topologically cubic graph H yields a cubic graph G whose vertices are the nodes of H, and whose edges are the arcs of H together with a nodeless edge for each nodeless component of H. Tutte's term *cubical network* may be taken to refer to a cubic graph with nodeless edges allowed. He has the cubical network embodied in a 1-complex, but that need not concern us. In his 1980 exposition of this work [1102, Sections 3–4], he focuses only on cubic graphs (potentially with nodeless edges) without mention of 1-complexes and only a glance at edge subdivision.

Tutte is interested in isomorphism invariants of these topologically cubic graphs that are also topologically invariant, so that subdivision does not change them. So, as far as these invariants are concerned, a single cubic graph may be taken to represent all topologically cubic graphs obtained by subdivisions from it.

Tutte applies the deletion–contraction relation to two different cubic graphs that are the same on one side of a 4-cut[13] and have two adjacent vertices on the other side, the graphs differing only in the way those two vertices meet the cut. This yields a four-term relation involving cubic graphs (Theorem XII). Topologically invariant W-functions of topologically cubic graphs that satisfy this four-term relation are called F-functions, or ϕ-functions in [1102]. Tutte shows that every topologically invariant W-function, restricted to topologically cubic graphs, is also an F-function (Theorem XII). Conversely, any F-function gives a unique topologically invariant W-function, or V-function, on topologically cubic graphs (Theorem XIV and its Corollary, with the V-function case requiring the F-function to be multiplicative over components). These converses make beautiful use of the theory developed earlier in the paper, including the results on possible families of elementary graphs. A family of elementary graphs for F-functions is introduced; these can be obtained from paths of odd length by adding a loop at each end and duplicating every second path edge, ensuring 3-regularity (see Theorem XIII).

Tutte's first example of an F-function is the trivial $x^{n/2}$, where n is the number of vertices of a cubic graph (or the number of vertices of degree three

[13] i.e., a set of four edges forming a member of the cutset space.

in a topologically cubic graph). He returns to this simple case at the very end of the paper, using his theory to give a simple proof that, up to a constant factor, this is the only F-function satisfying another binary linear recurrence on cubic graphs [T47, (45)]. This is worth a look; it is brief and understated, but quite a surprise, and is yet another example of the great depth of insight contained in just a few lines of Tutte's writing.

His next example of an F-function is very interesting: $(-1)^{n/2}\pi_k(G)$, where G is cubic and $\pi_k(G)$ is the number of subsets $X \subseteq E(G)$ such that (i) every vertex of G meets an edge in X, and (ii) the number of vertices of G that meet an odd number of edges in X is $2k$ (noting that this number is necessarily even).

Particular interest attaches to the case $k = 0$, since $\pi_0(G)$ is the number of X such that each vertex meets X twice.[14] Such an X is just a 2-factor in G, and its complement in $E(G)$ is a 1-factor, or perfect matching, of G.[15] So the (appropriately signed) number of perfect matchings of a cubic graph is an F-function, and hence (a restriction to cubic graphs of) a topologically invariant V-function, and hence can be obtained from an evaluation of the flow polynomial of the cubic graph.[16] This is an impressive connection between matching theory and the theory of colorings and flows. In general, the number of perfect matchings of a graph G cannot be obtained from the Tutte polynomial. Instead, it can be found from the independence polynomial [593]—or its close relative, the stability polynomial [478, 605]—of the line graph $L(G)$. It is also an evaluation of the Oxley–Whittle polynomial [898, 899] of G. The Oxley–Whittle polynomial is the analogue of the Whitney rank generating function for graphic 2-polymatroids, and the stability polynomial of a graph is just the chromatic polynomial of its graphic 2-polymatroid. So, counting perfect matchings is not so far, in spirit, from Tutte polynomials. But Tutte's work here shows that, for cubic graphs, the connection is even closer.

The appearance of $(-1)^{n/2}\pi_0(G)$ as a restriction to cubic graphs of a V-function on general graphs raises the question of the graph-theoretic meaning of that V-function for non-cubic graphs. Tutte considers this in [1102, Section 5] and the paper almost finishes by leaving the question open, with a "hope" of resolving it. He then adds a "Postscript" section, [1102, Section 6], giving such an interpretation, as a certain sum over some Eulerian subgraphs.

Tutte then defines $D(G;x)$ to be the generating polynomial for the $\pi_k(G)$, and observes that $(-1)^{n/2}D(G;x)$ is also an F-function, and one that is multiplicative over components.

His last example of an F-function is remarkable: the parity of the number

[14]This quantity $\pi_0(G)$ is also denoted by $f(G)$ in [1102, Section 4].

[15]Tutte says in [1106] that "the theory of 1-factors was only developed in those days for the cubic case".

[16]For the F-function recurrence to work, each nodeless edge must be allowed to belong, or not belong, to any perfect matching, thereby doubling the number of perfect matchings of the graph without that nodeless edge. This makes sense, since such edges have no effect on the degree of any vertex in the subgraph these edges form.

of Hamiltonian cycles of a cubic graph, with the F-function now taking values in $\mathbb{Z}_2$. He gives no further comment on this.

Matchings and Hamiltonian cycles developed into major interests of Tutte's, and are generally seen as quite distinct from his work on colorings, flows, and graph polynomials. So it is interesting to see these topics being touched on even in this foundational paper on graph polynomials.

Tutte was very pleased with his work on F-functions, but noted that it did not generate a lot of research by others [1106].

34.9 Tutte's PhD thesis

Tutte's PhD thesis was titled *An Algebraic Theory of Graphs* [T48]. The selection of topics is not typical of the field called "algebraic graph theory" today. Nonetheless, the title is apt, reflecting Tutte's aim of bringing much of graph theory into the kind of linear algebraic framework known today under the name of "representable matroids". He develops this theory for over four hundred pages, culminating in his famous excluded minor characterizations of binary, graphic, and cographic matroids. Only Chapter V (the shortest in the thesis), "Chromatic functions", concerns us here.

In Tutte's PhD thesis [T48, p. 144], he uses the term *dichromate* and the symbol χ for what we now know as the Whitney rank generating function of a matroid M. *Be careful!* Tutte uses both the term "dichromate" and the notation $\chi(M; x, y)$ for different things elsewhere: see above for his more usual usage. Outside his thesis, Tutte seems consistent in his use of "dichromatic polynomial" and "dichromate" for $W(G; x, y)$ (essentially) and $T(G; x, y)$, respectively. Welsh [1134] notes the confusion surrounding the two terms, and recalls (from 1973) that Tutte "...seemed prone to refer to the dichromate in conversation while his written work concentrated more on the dichromatic polynomial".

Tutte was fond of word play, and his choice of the name "dichromate" was surely a play on words. As a chemistry graduate with two publications [930, 1049] and a Masters degree [1089] in the field, he would have been well aware of the chemical meaning of the term. But, as far as I know, there is no evidence of any deeper reason for borrowing this term from chemistry: dichromate is not involved in his chemistry papers or thesis.

Care is also needed in interpreting the equation [T48, p. 144, equation (4)] Tutte gives for χ. That equation is, in Tutte's notation,

$$\chi(N; x, y) = \sum_{W} x^{\rho(p(N;W))} y^{\rho(c(N^*;M-W))},$$

where N is the (representable) matroid, M is its ground set, $W \subseteq M$, and $\rho = r$ is rank. The notations p and c may *seem* to refer to restriction and contraction, respectively, *to* a subset of the ground set (i.e., to deletion and contraction, respectively, *of* that set's complement). That is how the notation is used for graphs (pp. 11–12), and his definitions for *nets* (p. 45)—which correspond to representable matroids—seem to read that way too. However, Tutte's view of representable matroids is dual to the usual one: see his definitions of the terms "cycle" and "cycle space" for nets on p. 40, and the discussion in [1138, Section 9.4]. So, in the context of modern terminology for representable matroids, we must regard the meanings of p and c as being switched, relative to Tutte's use of that notation for graphs. When this is done, it is seen that [T48, p. 144, equation (4)] gives exactly the Whitney rank generating function.

Outside of Tutte's thesis, extension of Tutte–Whitney polynomials to matroids does not occur in print until Crapo's beautiful 1969 paper [349].

Tutte establishes a "recipe theorem" for $W(M;x,y)$ in Theorem IV (p. 150) of Section V.2 of the thesis (returning now to our usual notation, rather than Tutte's, with M again standing for a representable matroid).

His PhD thesis also sees the Tutte polynomial start to emerge in its usual form, although, as we saw in Section 34.8, it is already discernible in [T47]. Theorem V (p. 152) of Section V.2 introduces, without name or notation, the polynomial in s and t obtained from the Whitney rank generating function by the coordinate translation $s = 1 + x$, $t = 1 + y$, and states that that polynomial has nonnegative coefficients. The modern reader may see this translation coming a couple of pages earlier, when $1 + x$ and $1 + y$ are seen to be the values of χ on a bridge and a loop, respectively. Tutte does not yet relate this polynomial to internal and external activities.

Tutte finds some interesting evaluations of $W(M;x,y)$. In Section V.3, he shows that $W(M;0,0)$ is the number of bases (which he calls the *complexity* of M, in keeping with his use of that term for the number of spanning trees of a graph in [220][17]) (p. 153). For Section V.4, let M be represented by a matrix A over $\mathbb{Q}$, and define $Z(M,q)$ to be the number of all-nonzero vectors over $\mathbb{Z}_q$ that can be obtained from all-integer members of the rowspace of A by reducing each of their entries modulo q. He proves that $Z(M,q) = (-1)^{r(M)}W(M;-q,-1)$. For the special case of graphs, Tutte relates this quantity to the chromatic polynomial in Section VII.10. There, the objects being counted are essentially just nowhere-zero q-tensions, though Tutte does not give the usual definition of tensions using a fixed but arbitrary orientation of the graph.

[17]There are several different meanings of the term "complexity of a graph" in the literature. See the discussion at the start of [875, Section 1].

34.10 Tutte, 1954: A contribution to the theory of chromatic polynomials

The main contributions of this paper are to develop the theory of flows and flow polynomials, and to introduce the Tutte polynomial $T(G; x, y)$ as well as proving that it is well-defined. Tutte calls $T(G; x, y)$ the *dichromate* (as distinct from the Whitney rank generating function, which he calls the *dichromatic polynomial*). As in [T47], here in [T54] Tutte also uses terminology and notation drawn from algebraic topology.

Curiously, the paper does not mention the Whitney rank generating function $W(G; x, y)$, or even cite [T47]. The connection between the polynomials,

$$T(G; x, y) = W(G; x - 1, y - 1), \tag{34.18}$$

was established in [T47] at a more general level, as we discussed in Section 34.8; it was not stated in this simple form until [T67]. It appears that Tutte just decided to maintain a very tight focus for [T54], and postponed further discussion of the connection with $W(G; x, y)$ until his next paper on the topic.

This paper, [T54], begins by considering the tension and flow polynomials $\theta(G; x)$ and $\phi(G; x)$. Tutte does not name these polynomials, but does use the notation $\theta(G; x)$ and $\phi(G; x)$ for them. We follow Tutte's notation in this chapter, but note that the flow polynomial is $F(G; x)$ elsewhere in the handbook. The flow polynomial had previously appeared in [W33a] (without flows), of which Tutte seems unaware, and in [T47, Section 6]. Tutte states without proof the duality between the two, that $\theta(G^*; x) = \phi(G; x)$ where G is plane and G^* is its dual. He proves that $\phi(G; x) > 0 \Rightarrow \phi(G; x + 1) > 0$ for all $x \in \mathbb{N}$, and proposes two very influential conjectures, the second a strengthening of the first:

- There is a positive integer k such that every bridgeless graph has a k-flow. (This was proved by Jaeger [645] with $k = 8$. Seymour then reduced k to 6 in [999].)
- Every bridgeless graph has a 5-flow. (This is still open. It is best possible, in the sense that not every bridgeless graph has a 4-flow. The Petersen graph is such a case, as Tutte shows in [T54, p. 83] using the fact that the Petersen graph is not 3-edge-colorable [919].)

Tutte defines the *dichromate* (Tutte polynomial) in [T54, Section 3], denoted $\chi(G, x, y)$, using the third and most complex definition, (34.10), in Definition 34.3. In some ways, this is puzzling to the modern reader. The definition uses the order $<$ on $E(G)$, and the exponents in (34.10) certainly depend on the ordering, so the fact that the Tutte polynomial is independent of this ordering seems surprising and mysterious. The second (recursive) definition, (34.9), captures the essence of the Tutte polynomial, and this appears to have been the starting point for Tutte's research on the topic [79, p. 50]. In Tutte's later commentary, he preferred the first and simplest definition, (34.8), for the more straightforward proofs it yields [1101, p. 155], and followed that route in [1104, Section IX.6].

Edge order comes into the second definition too: on the face of it, the definition depends on the order in which the edges are deleted/contracted. But deletion and contraction can be shown to commute, so the order of edges makes no difference. In fact, it turns out that the edge order used for the third (activities-based) definition is also the deletion–contraction order, and independence of edge order for this definition is also based only on the commutativity of deletion and contraction. These observations can be used to give a more natural proof of the main theorem of this paper [T54, Section 3]. This more down-to-earth perspective is not apparent in [T54].

This may make the reader wonder how Tutte arrived at the internal/external activities approach of (34.10), and whether the way he presents these results reflects the order in which he conceived them.[18] On this question, it is worth reading Tutte's own account of his journey in [1108, p. 57].[19] It reflects the great significance he attached to the number of spanning trees in a graph, ever since his first paper [220]. He observed that $T(G;x,y) = W(G;x-1,y-1)$ satisfies the standard recurrences and was typically simpler than $W(G;x,y)$ in having fewer terms, with smaller coefficients. He noticed that the sum of the coefficients equals the number of spanning trees. So he looked for a sum over spanning trees. He was led to impose a linear order on the edges, since he noticed that symmetry among spanning trees was not reflected in the polynomial. He then had to find combinatorial interpretations, in terms of spanning tree and edge order, of the exponents, and prove the "curious theorem" that the resulting polynomial does not depend on edge order.[20] This is a formidable achievement, though the approach alluded to in the previous paragraph seems more intuitive for the modern reader and makes the theorem less surprising (but no less interesting).

The recursive relations Tutte gives for $T(G;x,y)$ are

$$\begin{aligned} T(G;x,y) &= x^{\#\text{bridges}}y^{\#\text{loops}}, && \text{if } G \text{ consists only of bridges and loops;}\\ T(G;x,y) &= T(G\backslash e;x,y) + T(G/e;x,y), && \\ & && \text{if } e \in E(G) \text{ is neither a bridge nor a loop;}\\ T(G;x,y) &= \prod_{i=1}^{c} T(G_i;x,y), && \text{if } G \text{ has components } G_1,\ldots,G_c. \end{aligned}$$

[18] And it is, of course, very common for a write-up to bear little relation to the journey of discovery.

[19] Tutte's account there is consistent with one he gave to Ruth Bari and which she summarized in [79, p. 50]. Bari notes that, in particular, Tutte was *not* trying to generalize Whitney's broken-cycle theorem.

[20] About a quarter of a century later, Tutte's view had not changed: "very curious" [1101, p. 156].

$$T(G;x,y) = \prod_{i=1}^{b} T(H_i;x,y), \qquad \text{if } G \text{ has blocks } H_1,\ldots,H_b.$$

The third of these is used to extend the definition of $T(G;x,y)$ to disconnected graphs. The other three are proved using spanning trees, judicious choices of edge order, and our third definition, (34.10), in Section 34.3.

Tutte combines these results with the deletion–contraction results he gives earlier for the tension and flow polynomials to obtain expressions for both those polynomials as partial evaluations of $T(G;x,y)$.

Spanning trees are a strong theme of this paper. Tutte not only uses them to define $T(G;x,y)$ and prove that it is independent of the edge order, but describes this result in the paper's abstract as "a new result about spanning trees". After establishing that the tension and flow polynomials can be obtained from $T(G;x,y)$, by appropriate evaluations on the lines $y = 0$ and $x = 0$ respectively, he remarks that the number of spanning trees can be obtained by putting $(x,y) = (1,1)$. His comment on this, in the abstract, is that "the theory of spanning trees now links the theory of graph-colorings to that of electrical networks". See our remarks on colorings, tensions and flows in Section 34.3.

Tutte closes the paper with the remark

> "[The number of spanning trees of a graph] has a simple expression as a determinant, and its properties are well known. Perhaps some of them will suggest new properties of the dichromate and hence of chromatic polynomials."

This reflects the inspiration that Tutte drew from the properties of numbers of spanning trees, since his earliest work. He used it in "squaring the square" [220], and it was the first member of his collection of deletion–contraction invariants [1106]; it was abstraction from this collection that led him to this polynomial [1098, 1108]. In more recent times, it has become clear that the number of spanning trees is quite atypical of Tutte polynomial evaluations, in that it is one of very few points where evaluating $T(G;x,y)$ can be done in polynomial time [653] (see Chapter 9). This suggests that its mathematical theory can be expected to be richer and more tractable than that for other evaluations (including, for example, colorings with more than two colors).

Much of the theory of this paper was extended to matroids by Crapo [349], including tensions, flows and the activities-based approach.

34.11 Tutte, 1967: On dichromatic polynomials

This paper comes across as a mixture of tying up loose ends from previous work (Sections 2 and 5), a miscellany of important new results with short proofs (Sections 3–5, and 9), and a new theory that occupies half the paper (Sections 6–8).

An early highlight is the following identity, which is most neatly expressed in terms of $Q(G;x,y) = x^{k(G)}W(G;x,y)$:

$$\sum_{W \subseteq V} Q(G[W];x_1,y)Q(G[V \setminus W];x_2,y) = Q(G;x_1+x_2,y).$$

This may be regarded as the first "convolution identity" for the Tutte polynomial. See [735] for a recent treatment of the theory of such identities, and [994] for multivariate Tutte polynomial convolution identities. In some respects, Tutte's identity is atypical of most later ones, since it is based on a sum over sets of vertices rather than edges, and does not use minors.

The recursive definition of $Q(G;x,y)$ (deletion–contraction and multiplicativity over components, with bouquets as the base case [T67, (18)–(20)]) is compared with the analogous results for $T(G;x,y)$ in [T54, (18)–(20)][21] to point out that (in our notation) $T(G;x,y) = W(G;x-1,y-1)$. As mentioned above, Tutte knew this much earlier, but had not published the fact.

Later, [T67, (23)], he establishes one of the first results involving derivatives of these polynomials.

$$\sum_{e \in E} Q(G\backslash e;x,y) = (|E|-|V|)Q(G;x,y) + x\frac{\partial}{\partial x}Q(G;x,y) - y\frac{\partial}{\partial y}Q(G;x,y). \tag{34.19}$$

As discussed in our Section 34.6, this result is in Whitney's thesis [W32c, Section 5, (5.1)], in a different form.

Sections 6–8 form the most substantial part of the paper. Tutte shows how to express $Q(G;x,y)$ as a product over all the non-separable subgraphs of G. This is a far more sophisticated result than mere multiplicativity over blocks [T54, p. 90]. It is not at all obvious that a useful multiplicative expression for $Q(G;x,y)$ exists at all, when G is non-separable; nor is it obvious that there is *any* useful expression—of multiplicative or any other form—just in terms of the non-separable subgraphs (rather than *all* subgraphs, as in the definition of $Q(G;x,y)$).[22]

Tutte's account of the genesis of this work, in [1108, pp. 107–109], is worth reading. It started as an attempt to find a multiplicative recursion for the chromatic polynomial, in the hope that it might be useful for investigating chromatic roots.

The material of Sections 6–8 can be challenging. Indeed, it does not seem to have been easy for Tutte: he later recounted how his first submitted paper on the topic, a "decade or two" earlier (and with the more restricted setting

[21] The equation numbers for the corresponding results in the two papers are the same, coincidentally.

[22] Incidentally, Tutte's expression (34.10) may be viewed as a result of diametrically opposite kind, being a sum over subgraphs that are as separable as possible, rather than a product over non-separable subgraphs.

of chromatic polynomials), had an error in a crucial proof and was rejected [1101, p. 526].

Tutte reworked this material in a later paper [1100], using a more abstract setting and somewhat more direct proofs. Even so, it is hard to see the intuition behind it and connections with other results.

Norman Biggs has given a beautiful algebraic treatment that encompasses this work by Tutte, Whitney's theorem expressing coefficients of $W(G;x,y)$ in terms of numbers of non-separable subgraphs each type [W32d], and some parallel work in the statistical physics community [65]. See [105] for the details. There is also a shorter exposition in [106], and an earlier one in [110, Chapter 4] (which, incidentally, shows a larger portion of the matrices at the heart of this theory). See also [111]. For the material in [T67, Sections 6–8], Biggs gives an elegant simplified treatment in [104] (without discussing there the connections with the aforementioned work of Whitney) and in [107, Chapters 11 and 12].

We make some remarks that may help the intuition of the reader of Sections 6–8 and demystify some of the results. For a more detailed analysis, see the work of Biggs.

Suppose we seek an expression for $W(G;x,y)$ as a *product* of factors, with one factor for each subset $X \subseteq E$. Then $\log W(G;x,y)$ must be a sum, over edge subsets, of the logarithms of those factors. Using Möbius inversion, we can express each of these logarithms of factors—one for each X—as a sum, over all $Y \subseteq X$, of terms of the form $\pm \log W(G[Y];x,y)$. Putting these summands in the form $\log(W(G[Y];x,y)^{\pm 1})$, and taking exponentials to remove the logarithms and go back to products, we find that our original factors of $W(G;x,y)$ are themselves products of $W(G[Y];x,y)^{\pm 1}$.

$$W(G;x,y) = \prod_{X \subseteq E} \left(\prod_{Y \subseteq X} W(G[Y];x,y)^{(-1)^{|E \setminus Y|}} \right). \qquad (34.20)$$

Thus we have an expression for $W(G;x,y)$ as a product over sets X of edges. But the factors are computationally complex (in contrast to the simple monomial summands of $W(G;x,y)$), and furthermore, one of them is $W(G;x,y)$ itself! (This comes from $Y = X = E$.)

The expression is in fact simpler than it seems. Routine counting arguments show that the factor corresponding to any separable subgraph—i.e., when $G[X]$ is separable—is 1, and so may be ignored in the product. So we have a product over non-separable subgraphs, which are usually much fewer in number.

In fact, all our arguments so far can be carried through for sums and products over subsets of vertices, rather than edges, with each vertex subset contributing a term, or factor, relating to its induced subgraph. The arguments have the same structure, with only some notational change. We obtain a product over non-separable induced subgraphs.

The arguments given use very few properties of $W(G;x,y)$ other than multiplicativity over blocks. So we can develop analogous expressions for many other functions on graphs. In fact, for the chromatic polynomial—not quite multiplicative over blocks, but multiplicative over components and factorizing according to (34.12) for graphs with a separating clique—we can use essentially the same arguments to develop a product over induced subgraphs that are not only non-separable, but have no separating clique (or, in other words, graphs with a separating clique contribute factor 1). Again, this uses few properties of the chromatic polynomial except this chromatic factorization property.

This approach reduces the number of factors considerably. It remains to address the circularity presented by the appearance of $W(G;x,y)$ as one of the factors in its own product expansion. This depends on a "Theorem of Vanishing Coefficients", due to Tutte. The theorem had been known in some form to statistical physicists, but Tutte gave the first rigorous proof and was the first to do it for $W(G;x,y)$ (see [107, p. 90] and [1108, p. 101]).

This theorem is most neatly described using $S(G;q,y)$ rather than $W(G;x,y)$, where $q = xy$ and recalling (34.7). In particular,

$$S(G;q,y)\cdot(q^{-1}y)^{r(E)} = \frac{W(G;x,y)}{x^{r(E)}} = \sum_{X\subseteq E} y^{|X|}q^{-r(X)}.$$

Although the argument we now give spans several lines, the strategy is straightforward, and each step is natural at the point where it is made: start with the logarithm of the above expression for $S(G;q,y)\cdot(q^{-1}y)^{r(E)}$; express it as $\log(1+z)$, for appropriate z, to enable power series expansion; do the power series expansion; expand each power-of-a-sum into a sum-of-products; tidy up a little; and interchange the order of summation. We have

$$\begin{aligned}
&\log(S(G;q,y)\cdot(q^{-1}y)^{r(E)})\\
&= \log\sum_{X\subseteq E} y^{|X|}q^{-r(X)}\\
&= \log\left(1+\sum_{X\subseteq E\,:\,X\neq\emptyset} y^{|X|}q^{-r(X)}\right)\\
&= \sum_{i=1}^{\infty}\frac{(-1)^{i+1}}{i}\left(\sum_{X\subseteq E\,:\,X\neq\emptyset} y^{|X|}q^{-r(X)}\right)^{i}\\
&\qquad\qquad \text{(using the power series } \log(1+z)=\textstyle\sum_{i=1}^{\infty}(-1)^{i+1}z^i/i\text{)}\\
&= \sum_{i=1}^{\infty}\frac{(-1)^{i+1}}{i}\sum_{\substack{(X_1,\ldots,X_i)\,:\\ \forall j\,:\,X_j\subseteq E,\,X_j\neq\emptyset}}\prod_{j=1}^{i} y^{|X_j|}q^{-r(X_j)}\\
&= \sum_{i=1}^{\infty}\frac{(-1)^{i+1}}{i}\sum_{\substack{(X_1,\ldots,X_i)\,:\\ \forall j\,:\,X_j\subseteq E,\,X_j\neq\emptyset}} y^{\sum_{j=1}^{i}|X_j|}q^{-\sum_{j=1}^{i}r(X_j)}
\end{aligned}$$

$$= \sum_{X\subseteq E\,:\,X\neq\emptyset} \sum_{i=1}^{\infty} \frac{(-1)^{i+1}}{i} \sum_{\substack{(X_1,\ldots,X_i)\,:\\ \forall j\,:\,X_j\subseteq E,\ X_j\neq\emptyset;\\ \bigcup_{j=1}^{i} X_j = X}} y^{\sum_{j=1}^{i}|X_j|} q^{-\sum_{j=1}^{i} r(X_j)}$$

(grouping the sequences $(X_1, \ldots, X_i)$ according to their union)

$$= \sum_{X\subseteq E\,:\,X\neq\emptyset} \left(y^{|X|} q^{-r(X)} + \sum_{i=2}^{\infty} \frac{(-1)^{i+1}}{i} \sum_{\substack{(X_1,\ldots,X_i)\,:\\ \forall j\,:\,X_j\subseteq E,\ X_j\neq\emptyset;\\ \bigcup_{j=1}^{i} X_j = X}} y^{\sum_{j=1}^{i}|X_j|} q^{-\sum_{j=1}^{i} r(X_j)} \right).$$

Therefore

$$S(G;q,y)\cdot(q^{-1}y)^{r(E)} = \prod_{X\subseteq E\,:\,X\neq\emptyset} \exp\left(y^{|X|} q^{-r(X)} + \sum_{i=2}^{\infty} \frac{(-1)^{i+1}}{i} \sum_{\substack{(X_1,\ldots,X_i):\\ \forall j:X_j\subseteq E,\ X_j\neq\emptyset;\\ \bigcup_{j=1}^{i} X_j=X}} y^{\sum_{j=1}^{i}|X_j|} q^{-\sum_{j=1}^{i} r(X_j)} \right). \quad (34.21)$$

The exponent of q in the innermost sum is $\leq -r(X)$, by submodularity. From the definition of $S(G;q,y)$ in (34.7), we know that the exponents of q in the left-hand side of (34.21) are nonpositive and range from 0 down to $-r(E)$. Now consider the contribution from the $X = E$ factor in (34.21). If q varies and y is fixed, we find that the $X = E$ factor has the form $1 + O(q^{-r(E)})$ (using $\exp z = 1 + z +$ higher order terms). Observe the gap, in powers of q, between the constant term and the term in $q^{-r(E)}$; typically a very long gap, with $r(E) - 1$ "missing" terms (the "vanished" coefficients). If we write the product of all the other factors (i.e., for $X \neq E$) as a general sum, $\sum_{l=0}^{\infty} a_l q^{-l}$, where the coefficients a_l depend only on y, then we have

$$\begin{aligned} S(G;q,y)\cdot(q^{-1}y)^{r(E)} &= \left(\sum_{l=0}^{\infty} a_l q^{-l}\right)\cdot\left(1 + O(q^{-r(E)})\right) \\ &= \left(\sum_{l=0}^{r(E)-1} a_l q^{-l}\right) + O(q^{-r(E)}). \end{aligned}$$

Notice the effect of the long gap just mentioned: it means that the coefficients of the first $r(E)$ powers of q, namely $1, q^{-1}, q^{-2}, \ldots, q^{-r(E)+1}$ are unchanged by multiplication by the $X = E$ factor, $1 + O(q^{-r(E)})$. Thus, the first $r(E)$ terms in the expansion in powers of q, can be obtained without using that last factor. This *almost* removes the circularity referred to earlier; the only remaining unknown in $S(G;q,y)\cdot(q^{-1}y)^{r(E)}$ is the coefficient of $q^{-r(E)}$. But this can be read from the definition, (34.7). We see that the coefficient of $q^{-r(E)}$ in $S(G;q,y)\cdot(q^{-1}y)^{r(E)}$ is just $y^{|E|}$.

We therefore have a recursive method for calculating $S(G;q,y)\cdot(q^{-1}y)^{r(E)}$, and hence $W(G;x,y)$, which requires only its value on proper subgraphs, and furthermore, we can restrict attention to proper induced non-separable subgraphs.

In his later paper on this topic [1100], Tutte applies the result to reconstruction, showing that both the Tutte polynomial and the number of Hamiltonian cycles of a graph are reconstructible (in that each of these can be determined for G once it is known for every unlabelled graph obtained from G by deleting a single vertex).

The final section (Section 9) applies some of the theory of Section 8 to chromatic roots. Tutte shows that, for every graph G and every positive integer k, the sum of the k-th powers of the chromatic roots of G is a linear combination, with coefficients independent of G, of the numbers of non-separable subgraphs of each possible type.

34.12 Potts, 1952: Some generalized order-disorder transformations

Whitney and Tutte created modern graph theory as a branch of pure mathematics. Their graph polynomials were studied from a pure mathematician's perspective other than very occasional mentions of links with electrical circuit theory. Separately and in parallel, statistical physicists studied partition functions of certain models on graphs, beginning with Ising [635] and Ashkin and Teller [51]. Then Renfrey Potts introduced a model which generalized that of Ising and the simpler models of Ashkin and Teller. Remarkably, its partition function turns out to be equivalent to the Tutte polynomial, though it was not until nearly two decades later that this connection was formalized and published (see Section 34.14.2).

Pott's doctoral supervisor was Cyril Domb, who was only around five years older and whose own PhD (Cambridge, 1949) was supervised by the astronomer Fred Hoyle. Potts was one of six DPhil students of Domb's during the latter's Oxford years (1949–52) [390]. Potts's short paper was published in 1952 [P52] and summarizes some results from his doctoral thesis [P51]. The paper was received by the journal in July 1951, shortly after the thesis was submitted. Potts's paper and thesis used the language of statistical mechanics. We briefly give some background to help the modern reader from outside that field (see also Chapter 20).

Let Θ be a finite set of size r. This is a set of available *states* or *colors* that may be assigned to vertices. Let $G = (V, E)$ be a graph. In [P51, P52], the graphs are always regular lattices and their vertices are called *units*.[23] A *configuration* of G is a function $\theta : V \to \Theta$. We write θ_v for $\theta(v)$; Potts calls it the *configuration of* v, meaning the *state* or *color* of v.[24]

The *interaction energy* along an edge is determined by the states of its ends, and is given by some function $J : \Theta \times \Theta \to \mathbb{R}$. Define $x : \Theta \times \Theta \to \mathbb{R}$ by $x(\theta_1, \theta_2) = e^{-J(\theta_1,\theta_2)/(kT)}$. Physically, T is temperature and k is Boltzmann's constant, though these details need not concern us.

The *partition function* of G is

$$Z(G) = \sum_{\theta:V\to\Theta} \prod_{uv\in E} x(\theta_u, \theta_v) = \sum_{\theta:V\to\Theta} \exp\left(\sum_{uv\in E} -J(\theta_u, \theta_v)/(kT)\right).$$

Again, this partition function $Z(G)$ should not be confused with the $Z(G)$ of Equation (34.15). However, with appropriate choices of θ and x, this $Z(G)$ does agree with the partition functions given in Chapter 20.

The *Ising model* [635] has two states. The interaction energy on an edge has magnitude J and sign determined by whether or not the states of the ends are the same. In [389], Domb recounts suggesting to Potts a generalization of the Ising model to $r \geq 3$ states[25] represented as two-dimensional vectors all of the same length and with their directions equally spaced. Interaction energy is determined by the scalar product of the vectors. This model became known sometimes as the *Potts vector model* or the *$\mathbb{Z}_r$ clock model.* The particular result they sought did not work out for $r > 4$, but Potts found that a different generalization of the $r = 2, 3$ cases was more amenable. This involved just two interaction energies: one for identical states, and one for different states. This became known as the *Potts model.* Potts notes that the $r = 4$ case had been studied earlier by Ashkin and Teller [51], as a special case of what became known as the Ashkin–Teller model, but Potts was the first to describe and study it for general r. It is identical to the Potts vector model for $r = 3$ (and of course for $r = 2$, when both are equivalent to the Ising model), but the two models differ for $r \geq 4$.

In the special case of the Potts model, $x(\theta_1, \theta_2)$ is determined solely by whether the states are the same or different.

$$x(\theta_1, \theta_2) = \begin{cases} x_0, & \text{if } \theta_1 = \theta_2, \\ x_1, & \text{if } \theta_1 \neq \theta_2. \end{cases}$$

In this case,

$$Z(G) = \sum_{\theta:V\to\Theta} \prod_{uv\in E} x_0^{|\{uv\in E:\theta_u=\theta_v\}|} x_1^{|\{uv\in E:\theta_u\neq\theta_v\}|}.$$

[23]This is common in papers in statistical mechanics, where the graphs used are very special but some of the definitions and ideas readily extend at least to planar graphs and often to general graphs.

[24]In [P52], the term *species* is also used for the states in the Potts model, but not for the Potts vector model.

[25]The number r of states is more often denoted by q nowadays.

The thesis [P51] considers a range of lattice graphs, including the one-dimensional lattice (path), square lattice (called the Ising lattice in [P51]), rectangular lattice (which has the same graph as the square lattice, but we use $x(\theta_1, \theta_2)$ only for vertical edges; for horizontal edges, we use another two-valued function $w(\theta_1, \theta_2)$ instead), triangular lattices, cubic lattices, and square/rectangular lattice plus diagonals (which is nonplanar; the infinite version is 8-regular). In each case, the real interest is in limits as one or more of the dimensions goes to infinity, so that details of edges at boundaries (e.g., flat versus cylindrical lattices) are not important. The paper [P52] just considers an $m \times \infty$ square lattice with m rows in which the j-th vertex in the top row has, as an extra neighbor, the $(j + 1)$-th vertex in the bottom row, so forming a *screw* (following [722]), and thus the vertical edges actually form an infinite helical path. Potts considers the partition function of the subgraph induced by the first i vertices of this helical path, broken down according to contributions from each possible configuration of the last m of these i vertices. This is expressed in matrix form; each successive vertex on the helical path corresponds to multiplication by a sparse "characteristic matrix" whose nonzero entries are described in the second displayed equation of [P52], just below the middle of the first page.

The paper [P52] is mostly about using certain "inversion transformations" (related to duality) to determine transition points. This is done for the Potts model for $r \geq 3$ in [P52, Section 2]. Potts credits Ashkin and Teller [51] for the transition point (as well as for the model itself) in the $r = 4$ case. Potts treats the Potts vector model for $r = 4$ in [P52, Section 3] but could not go to higher r for that model. For a later discussion of some of this theory from a graph-theoretic perspective, see [110, esp. Section 3.5].

Potts does not write out the partition function for his model explicitly in this paper, but it was not necessary to do so, since the partition function was so foundational that any statistical physicist reading the paper would have been able to write it down immediately on seeing the Potts model defined. Partition functions are described in the thesis [P51], but again are not themselves the main object of study. The material of the paper [P52] is mainly based on Section 3.7 of the thesis.

34.13 Zykov

Alexander A. Zykov is usually regarded as the first graph theorist in the former Soviet Union. In a 1949 paper [1191]—"the first Soviet paper devoted entirely to graph theory" [1088]—he develops some algebraically-flavored graph theory including some graph polynomials. Like Tutte, he introduced graphs as linear complexes. For our purposes, the main points of interest are: a polynomial for counting color partitions, very closely related to the chromatic polynomial;

addition-identification and deletion–contraction relations for it; the theorem on the chromatic polynomial of the join of two graphs (obtained by adding all possible edges between them), and the use of diagrammatic expressions, essentially integer linear combinations of diagrams of graphs in which each graph stands for its corresponding polynomial.[26] Incidentally, it also introduced what we now call the clique polynomial. Zykov considers the clique polynomial again briefly in [1193]. After these papers, the clique polynomial does not appear again in the literature for many years, maybe not until [482].[27] Zykov's terminology is nonstandard, for example "rank" and "dimension" refer to chromatic number and clique number respectively.

At a conference in 1963 [1193], Zykov presented a very general theory of recursively defined graph invariants. This theory makes fleeting contact with the Tutte polynomial, which is a special case of one of his recursive invariants [1193, Section 5]. The clique polynomial puts in another appearance [1193, Section 7].

In a problem session at the same conference [1192], Zykov appears to have been the first to ask if the Tutte polynomial can be computed in polynomial time. (The question was asked later by Crapo [342, Problem 6].) This is paraphrasing Zykov's question, since the term "polynomial time" came later. He asks for a method that improves on brute force[28] in the same way in which efficient methods for the determinant (e.g., Gaussian elimination) improve on direct calculation from the definition. This is now considered very unlikely, due to the NP-hardness of graph coloring and ♯P-completeness of counting colorings (see, e.g., [509]). If the Tutte polynomial is computable in polynomial time, then P = NP. Even computation of a single evaluation of $T(G; x, y)$, for an arbitrary input graph G, is ♯P-hard except on the curve $xy = 1$ and at a small number of points (x, y) [653] (see Chapter 9).

34.14 Some other work

Tutte kept working on these polynomials, and continued to initiate new and unexpected directions for research on them: "golden roots" of chromatic polynomials, beginning with [96]; "dichromatic sums" (related to his work on enumerating rooted planar maps), beginning with [1096]; Tutte equivalence (a topic that had lain dormant and unnoticed in Whitney's work) [1097]; and applications to reconstruction [1100]. He also wrote several expository articles on them: [1098, 1099, 1105, 1107], [1104, Ch. 9] and [1108, esp. Ch. 5].

[26] These expressions may be regarded as examples of Tutte's graphic forms, though Zykov uses them somewhat differently and does not develop their theory.

[27] Polynomials for independent sets appear earlier [605]. See our remarks on p. 646 for connections between independent set polynomials and Tutte–Whitney polynomials.

[28] I.e., looping over all X, determining $|X|$ and $r(X)$ for each.

For the rest of this section, we briefly review some other work, from the late 1960s to the late 1970s, that is important in the history of Tutte–Whitney polynomials.

34.14.1 Generalization and abstraction

The late 1960s and early 1970s saw some generalization and abstraction of Tutte–Whitney polynomial theory that placed it in a much wider mathematical context and revealed the depth and foresight of Tutte's earliest work on the topic.

Crapo's landmark paper of 1969 [349] was based on his PhD thesis (1964) supervised by Gian-Carlo Rota at M.I.T.. This paper gives the name *Tutte polynomial* to $T(G;x,y)$, and introduces the theory of Tutte polynomials for matroids. The treatment is geometric, with matroids called "combinatorial pregeometries". Aspects of Tutte polynomial theory Crapo extended to matroids include the deletion–contraction relation, duality, activities, tensions and flows, coboundary polynomials, and the relationship $T(M;x,y) = W(M;x-1,y-1)$ with the Whitney rank generating function. Connections are made with Rota's theory of incidence algebras, Möbius inversion, and the characteristic or Poincaré polynomial of a finite lattice [970]. See also Crapo's survey in [342]. Other early discussions of Tutte polynomials of matroids are in [1016, 1017].

Brylawski [251, 252] discerns in Tutte's "ring in graph theory" [T47] a relative of the Grothendieck ring of K-theory.[29] To express this idea, he develops in [252] a theory of the *Tutte bidecomposition*, under which a matroid M may be decomposed into the formal sum $M\backslash e + M/e$ (where $e \in E(M)$), and if M is a disjoint union (i.e., direct sum) of M_1 and M_2, it may also be decomposed into the formal product M_1M_2. In so doing, he generalizes the core of [T47] to matroids (geometrically viewed), expanding on the algebraic aspects (and without Tutte's topological language). Instead of Tutte's notion of V-functions, we now have *Tutte(-Grothendieck) invariants*, which are like V-functions except (i) they apply to matroids, not just graphs; (ii) the deletion–contraction relation need not hold for bridges; and (iii) they must be multiplicative over blocks, not just over components. In [251], Brylawski gives a "reaxiomatization" of the Grothendieck ring, calling it the *Tutte-Grothendieck ring* and noting that Tutte's work was earlier. The theory covers other settings with their own Tutte–Grothendieck rings, including some involving vector spaces, matrices, and groups. These papers report some of the work of Brylawski's PhD (Dartmouth College, 1970), also supervised by Rota. Kung [728, Section IV.1] gives an excellent discussion of the connection with K-theory.

Extending Tutte–Whitney polynomials to more general *objects* (e.g., from

[29] Brylawski credits the initial insight for these papers to G.-C. Rota's Hedrick Lectures at the Mathematical Association of America's Summer Meeting in Toronto in 1967.

graphs to matroids) is just one way of generalizing them; see the discussion of different types of generalization in [479, Section 3.4]. Another is to extend the bivariate *polynomials* by introducing more variables (see Chapter 24). Tutte's multivariate polynomials in [T47] are cases in point (although he specializes to bivariate polynomials, rather than generalizing from them). In statistical mechanics, it is common to give each edge its own variable; see for example [468, 498]. This approach was taken up in combinatorics by Traldi [1072], Kung [732] and others; see [479, pp. 38–39]. Sokal [1021, 1023] has promoted its use in proving new results and finding shorter proofs of some existing ones.

34.14.2 Evaluations and connections

The same period of time saw an expansion in results linking Tutte–Whitney polynomials to other invariants for a variety of combinatorial structures. Connections were made with statistical mechanics, network reliability, and coding theory. Significant new evaluations of the polynomial were discovered.

The two streams of research on the polynomials—combinatorial (Whitney, Tutte) and statistical mechanical (Potts)—seem to have joined properly with the work of Essam, Fortuin and Kasteleyn [468, 498]. They gave the transformation linking the Potts model partition function with Tutte–Whitney polynomials.[30] Other points of contact between the streams occurred in [1056].

Later, the weight enumerator of a linear code [807] and the all-terminal network reliability polynomial [1114], were brought into the Tutte–Whitney family by Greene [579] and Oxley and Welsh [904], respectively. In the following decade, the Jones polynomial of an alternating link [668] was linked to Tutte–Whitney polynomials [1061]. Significant new specializations were found within graph theory as well, including the number of acyclic orientations of a graph [1027] and the Martin polynomial [823]. See Chapters 13–16, 18, and 31.

34.15 Closing remarks

Research on Tutte–Whitney polynomials has grown enormously since the work reviewed here. This is not just due to the riches the polynomials contain and the work of the mathematicians who have discovered these riches. It is also due to the expository writing by many of the field's leading researchers. This has brought the topic to a much wider mathematical audience, stimulated further progress, and made connections with other fields.

We have already mentioned Tutte's own significant body of expository

[30]The Ashkin–Teller model partition function cannot be obtained from the Tutte polynomial. But, in the symmetric case, even this can be obtained from a Tutte–Whitney-type polynomial of more general combinatorial objects [476].

writing on the topic. Since the 1970s, Dominic Welsh has been particularly influential in promoting Tutte–Whitney polynomials, and often with an eye for their history. His book chapter [1138, Ch. 15] and book [1139] were especially influential, and he wrote many survey articles (see, e.g., the list in [479]). In Section 34.11 we touched on the work of Biggs. His book [107], first published in 1974, not only includes a beautiful and erudite treatment of Tutte–Whitney polynomials, but its whole selection of topics ranges very closely over Tutte's early interests. Some other significant expositions are by Bollobás [152], Brylawski and Oxley [247], Crapo [342], Ellis-Monaghan and Merino [454], Godsil and Royle [546, Ch. 15], Kung [728, Ch. IV], [734], and Sokal [1023]. The contribution of these and other expository writers to the development of the field is sometimes underestimated but nonetheless enormous.

Whitney	Tutte	Modern
	rational integer	integer
	G	additive abelian group
	H	commutative ring
	n_m [T67]	$(n)_m = n(n-1)\cdots(n-m+1)$, falling factorial
	L	a graph $G = (V, E)$
	L^*, $\mathbf{L}$	isomorphism class of the graph
	congruent	isomorphic
	equivalent	blocks pairwise isomorphic
	topologically invariant	invariant under edge subdivision
	graphic form	formal integer linear combination of isomorphism classes, with only finitely many coefficients nonzero
	B	ring of graphic forms
	0-cell	vertex
arc	1-cell	edge
	arc	edge in a cubic graph; edge whose endpoints have degree three
	A	some specific edge e
	link	non-loop edge
	isthmus	bridge, coloop
V	$\alpha_0(L)$	n, $\|V\|$, number of vertices
E	$\alpha_1(L)$	m, $\|E\|$, number of edges

Continued

Whitney	Tutte	Modern
R	$\alpha_1(L) - p_1(L)$	$r(G)$, rank of G
$N := E - R$	$p_1(L)$, [T47] cycle-rank $r(G)$ [T67]	$\|E\| - r(G)$, nullity, cyclomatic number
P	$p_0(L)$	$k(G)$, number of components
T		number of faces (for plane G)
	node; triple vertex	vertex of degree three; vertex in a cubic graph
	cubical network	cubic graph (thought of as representing a class of graphs, with all vertex degrees two or three, related by edge subdivisions and their reversals)
	path [T54]	walk
chain	simple path [T54]	path
	circular path [T54]	circuit
k-circuit		circuit of size k
connected piece		component
component	cyclic element, block [T67]	block
sum, $G_1 + G_2$	product, L_1L_2	disjoint union
subgraph		subset of edges of G (usually without isolated vertices)
	subgraph; [T47] spanning subgraph, [T54] $G : S$ [T67]	(V, S) where $S \subseteq E(G)$

Continued

Whitney	Tutte	Modern
	$G \cdot S$ [T67]	induced subgraph $G[S]$
	subgraph [T54]	usual meaning
	vertex-generated subgraph [T67]	induced subgraph
	$E(W)$ [T67]	set of edges with both endpoints in W
	$G[W]$; [T54] $G(W)$ [T67]	subgraph of G induced by $W \subseteq V$
	subtree	spanning tree
	$C(G)$ [T54]	number of spanning trees (called the *complexity* in [220])
	y_r	bouquet; graph with one vertex and r loops
	L'_A	$G\backslash e$
	L''_A	G/e
	$p(G; M(G) - \{A\})$ [T48]	$G\backslash e$
	$c(G; M(G) - \{A\})$ [T48]	G/e
	$p(G; M(G) - X)$ [T48]	$G\backslash X$
	$c(G; M(G) - X)$ [T48]	G/X
	$M(G)$ [T48]	edge set E of graph G
	net, N [T48]	representable matroid, with its representation (but note: representation is dual to the usual form)
	$M(N)$ [T48]	ground set, E

Continued

Whitney	Tutte		Modern
	$c(N; M(N) - \{A\})$	[T48]	$G\backslash e$
	$p(N; M(N) - \{A\})$	[T48]	G/e
λ	n		number of colors, x
	Q_n	[T54]	a set of n colors, typically $\mathbb{Z}_n$
	I_n	[T67]	$\{1, 2, \ldots, n\}$, used as a set of n colors
$M(\lambda)$	$J(L; \lambda, 0)$; $P(G, n)$		$\chi(G; x)$, chromatic polynomial
	$\theta(G, n)$	[T54]	tension polynomial; chromatic/characteristic polynomial of cycle matroid
	α-coloring; 0-chain	[T47] [T54]	x-assignment, $f : V \to Q_x$, i.e., an "x-coloring" that may be proper or improper
	f	[T47]	an x-assignment
	F_n	[T67]	the set of all n-assignments
	f-monochromatic edge	[T67]	edge whose endpoints have the same color under f
	$\phi(f)$	[T47]	number of monochromatic ("bad") edges, under assignment f
	$e(P)$	[T67]	number of monochromatic ("bad") edges, under color-partition P
	$J(L; \lambda, \phi)$		number of λ-assignments with ϕ monochromatic edges
	1-chain	[T54]	a function $f : E \to Q_x$, where usually Q_x is an abelian group of size x, typically $\mathbb{Z}_x$

Continued

Whitney	Tutte	Modern
	coboundary [T54]	tension
	color-coboundary [T54]	nowhere-zero tension
	1-cycle; β-coloring	flow
	color-cycle [T54]	nowhere-zero flow
	λ	number of values a flow may take
	g_G	number of flows, with flow values in the subscript group
	$\psi(g_G)$	number of edges with zero flow
$M'(\lambda)$ [W33a]	$\phi(G,n)$ [T54]	flow polynomial
	$E(L;G,0)$	number of nowhere-zero G-flows. In terms of $\|G\|$, this is the flow polynomial.
	$E(L;G,\psi)$	number of G-flows with ψ zero-flow edges
m_i		coefficient of x^{n-i} in $\chi(G;x)$
m_i', p_i [W33a]		coefficient of flow polynomial
m_{ij}		coefficient of $x^{r(G)-i}y^j$ in $W(G;x,y)$
	$Q(L;t,z)$; [T54] dichromatic polynomial [T67]	$x^{k(G)}W(G;x,y)$ ($t=x$, $z=y$)
	$R(G;t,z)$, rank polynomial [T67]	$x^{r(G)}W(G;x^{-1},y)$
	dichromate, $\chi(M;x,y)$ [T48]	$x^{r(M)}y^{r^*(M)}W(M;x^{-1},y^{-1})$ where M is a (representable) matroid
	dichromate, $\chi(G,x,y)$ [T54]	Tutte polynomial, $T(G;x,y)$

Continued

Whitney	Tutte	Modern
N_i		number of non-separable subgraphs of G of a particular isomorphism class
	$\theta(G, F)$, where F is an isomorphism class [T67]	number of non-separable *induced* subgraphs of G of a particular isomorphism class
logical expansion		inclusion-exclusion

Acknowledgments

The editors would like to express their immense gratitude to all of the contributors and anonymous referees. This project could not have happened without their time and effort.

Federico Ardila was partially supported by NSF grants DMS-1600609 and DMS-1855610 and Simons Fellowship 613384.

Spencer Backman would like to thank Seth Chaiken and Sam Hopkins for comments.

Jason Brown acknowledges the support of the Natural Sciences and Engineering Research Council of Canada (NSERC), RGPIN-2018-05227.

Jin-Yi Cai was supported in part by the National Science Foundation under grant CCF-1714275.

Michael Falk and Joseph Kung thank Graham Denham and Vic Reiner for their comments and suggestions on Chapter 30; and Joseph Bonin, Alex Fink, and Vic Reiner for their comments and suggestions on Chapter 32.

Graham Farr is grateful to many colleagues for comments on earlier versions of this chapter, including Norman Biggs, Thomas Britz, Henry Crapo, Joseph Kung, Dillon Mayhew, James Oxley, Carrie Rutherford, Robert Shrock, Alan Sokal, Dominic Welsh, Geoff Whittle. He also thanks Joseph Kung for pointing out some references and the link with K-theory, and Donald McNaughton for comments on Tutte's chemistry papers. He was supported in part by Australian Research Council Discovery Grants DP110100957 and DP130100300.

Emeric Gioan dedicates his chapters to Michel Las Vergnas.

Geoffrey Grimmett was supported in part by the Engineering and Physical Sciences Research Council under grant EP/I03372X/1.

Tomer Kotek and Johann A. Makowsky note that parts of their chapter were prepared while funded by the Simons Institute for the Theory of computing. Both authors while attending "Logical Structures in Computation" in Fall 2016, and the second author also while attending "Counting Complexity and Phase Transitions" in Spring 2016.

Anna de Mier was supported by the Spanish Ministerio de Economía y Competitividad project MTM2017-82166-P.

Bibliography

[1] E. Abe. *Hopf Algebras*, volume 74 of *Cambridge Tracts in Mathematics.* Cambridge University Press, Cambridge–New York, 1980.

[2] C. C. Adams. *The Knot Book.* American Mathematical Society, Providence, RI, 2004.

[3] K. Adiprasito, J. Huh, and E. Katz. Hodge theory for combinatorial geometries. *Ann. of Math.*, 188(2):381–452, 2018.

[4] M. Aguiar and F. Ardila. Hopf monoids and generalized permutahedra. arXiv:1709.07504, 2017.

[5] M. Aguiar, N. Bergeron, and F. Sottile. Combinatorial Hopf algebras and generalized Dehn-Somerville relations. *Copos. Math.*, 142:1–30, 2006.

[6] D. Aharonov and I. Arad. The BQP-hardness of approximating the Jones polynomial. *New J. Phys.*, 13(3):035019, 2011.

[7] D. Aharonov, I. Arad, E. Eban, and Z. Landau. Polynomial quantum algorithms for additive approximations of the Potts model and other points of the Tutte plane. arXiv:quant-ph/0702008, 2007.

[8] D. Aharonov, V. Jones, and Z. Landau. A polynomial quantum algorithm for approximating the Jones polynomial. *Algorithmica*, 55(3):395–421, 2008.

[9] C. Ahrens, G. Gordon, and E. W. McMahon. Convexity and the beta invariant. *Discrete Comput. Geom.*, 22(3):411–424, 1999.

[10] M. Aigner. The Penrose polynomial of a plane graph. *Math. Ann.*, 307(2):173–189, 1997.

[11] M. Aigner. *A course in enumeration*, volume 238 of *Graduate Texts in Mathematics.* Springer, Berlin, 2007.

[12] M. Aigner and H. Mielke. The Penrose polynomial of binary matroids. *Monatsh. Math.*, 131(1):1–13, 2000.

[13] M. Aigner and H. van der Holst. Interlace polynomials. *Linear Algebra Appl.*, 377:11–30, 2004.

[14] M. Aizenman. Geometric analysis of φ^4 fields and Ising models. In *Mathematical Problems in Theoretical Physics (Berlin, 1981)*, volume 153 of *Lecture Notes in Phys.*, pages 37–46. Springer-Verlag, Berlin, 1982.

[15] M. Aizenman, D. J. Barsky, and R. Fernández. The phase transition in a general class of Ising-type models is sharp. *J. Stat. Phys.*, 47:343–374, 1987.

[16] M. Aizenman, H. Duminil-Copin, and V. Sidoravicius. Random currents and continuity of Ising model's spontaneous magnetization. *Comm. Math. Phys.*, 334:719–742, 2015.

[17] M. Aizenman and R. Fernández. On the critical behavior of the magnetization in high-dimensional Ising models. *J. Stat. Phys.*, 44:393–454, 1986.

[18] F. Alex and D. E. Speyer. K-classes for matroids and equivariant localization. *Duke Math. J.*, 161(14):2699–2723, 2012.

[19] J. W. Alexander and G. B. Briggs. On types of knotted curves. *Ann. of Math*, 28(1-4):562–586, 1926/27.

[20] J. Aliste-Prieto, A. de Mier, and J. Zamora. On trees with the same restricted U-polynomial and the Prouhet-Tarry-Escott problem. *Discrete Math.*, 340(6):1435–1441, 2017.

[21] J. Aliste-Prieto and J. Zamora. Proper caterpillars are distinguished by their chromatic symmetric function. *Discrete Math.*, 315:158–164, 2014.

[22] N. Alon, A. Frieze, and D. J. A. Welsh. Polynomial time randomised approximation schemes for Tutte-Grothendieck invariants: the dense case. *Random Struct. Algor.*, 6:459–478, 1995.

[23] R. Anderson, L. Lovász, P. Shor, J. Spencer, É. Tardos, and S. Winograd. Disks, balls, and walls: analysis of a combinatorial game. *Amer. Math. Monthly*, 96(6):481–493, 1989.

[24] A. Andrzejak. Splitting formulas for Tutte polynomials. *J. Combin. Theory Ser. B*, 70(2):346–366, 1997.

[25] A. Andrzejak. An algorithm for the Tutte polynomials of graphs of bounded treewidth. *Discrete Math.*, 190(1-3):39–54, 1998.

[26] R. M. Ankney and J. E. Bonin. Characterizations of $\mathrm{PG}(n-1,q)\setminus \mathrm{PG}(k-1,q)$ by numerical and polynomial invariants. *Adv. Appl. Math.*, 28(3-4):287–301, 2002.

[27] J. D. Annan. *The Complexity of Counting Problems.* D.Phil. Dissertation, University of Oxford, 1994.

[28] J. D. Annan. A randomised approximation algorithm for counting the number of forests in dense graphs. *Combin. Probab. Comput.*, 3:273–283, 1994.

[29] R. Anstee, J. H. Przytycki, and D. Rolfsen. Knot polynomials and generalized mutation. *Topology Appl.*, 32:237–249, 1989.

[30] F. Ardila. The Catalan matroid. *J. Combin. Theory Ser. A*, 104(1):49–62, 2003.

[31] F. Ardila. *Enumerative and Algebraic Aspects of Matroids and Hyperplane Arrangements.* PhD Thesis, Massachusetts Institute of Technology, 2003.

[32] F. Ardila. Computing the Tutte polynomial of a hyperplane arrangement. *Pacific J. Math.*, 230(1):1–26, 2007.

[33] F. Ardila. Semimatroids and their Tutte polynomials. *Rev. Colombiana Mat.*, 41(1):39–66, 2007.

[34] F. Ardila. Algebraic and geometric methods in enumerative combinatorics. In *Handbook of Enumerative Combinatorics*, CRC Press Ser. Discrete Math. Appl., pages 3–172. CRC, Boca Raton, FL, 2015.

[35] F. Ardila, M. Beck, and J. McWhirter. The arithmetic of coxeter permutahedra. *Rev. Acad. Colomb. Cienc. Ex. Fis. Nat.*, 44(173):1152–1166, octubre-diciembre de 2020.

[36] F. Ardila, C. Benedetti, and J. Doker. Matroid polytopes and their volumes. *Discrete Comput. Geom.*, 43(4):841–854, 2010.

[37] F. Ardila and A. Boocher. The closure of a linear space in a product of lines. *J. Algebraic Combin.*, 43(1):199–235, 2016.

[38] F. Ardila, F. Castillo, and M. Henley. The arithmetic Tutte polynomials of the classical root systems. *Int. Math. Res. Not. IMRN*, (12):3830–3877, 2015.

[39] F. Ardila, F. Castillo, and J. A. Samper. The topology of the external activity complex of a matroid. *Electron. J. Combin.*, 23(3):P3.8, 2016.

[40] F. Ardila, A. Fink, and F. Rincón. Valuations for matroid polytope subdivisions. *Canad. J. Math.*, 62(6):1228–1245, 2010.

[41] F. Ardila and C. J. Klivans. The Bergman complex of a matroid and phylogenetic trees. *J. Combin. Theory Ser. B*, 96(1):38–49, 2006.

[42] F. Ardila and A. Postnikov. Combinatorics and geometry of power ideals. *Trans. Amer. Math. Soc.*, 362(8):4357–4384, 2010.

[43] F. Ardila, M. Supina, and A. R. Vindas-Meléndez. The equivariant Ehrhart theory of the permutahedron. *Proc. Amer. Math. Soc.*, 148(12):5091–5107, 2020.

[44] M. Arenas, M. Munoz, and C. Riveros. Descriptive complexity for counting complexity classes. In *Logic in Computer Science (LICS), 2017 32nd Annual ACM/IEEE Symposium on*, pages 1–12. IEEE, 2017.

[45] V. I. Arnol'd. The cohomology ring of the group of dyed braids. *Mat. Zametki*, 5:227–231, 1969.

[46] J. L. Arocha and B. Llano. Mean value for the matching and dominating polynomial. *Discuss. Math. Graph Theory*, 20(1):57–69, 2000.

[47] S. Arora and B. Barak. *Computational Complexity: A Modern Approach.* Cambridge University Press, Cambridge, 2009.

[48] R. Arratia, B. Bollobás, and G. B. Sorkin. The interlace polynomial: a new graph polynomial. In *Proceedings of the Eleventh Annual ACM-SIAM Symposium on Discrete Algorithms (San Francisco, CA, 2000)*, pages 237–245. ACM, New York, 2000.

[49] R. Arratia, B. Bollobás, and G. B. Sorkin. The interlace polynomial of a graph. *J. Combin. Theory Ser. B*, 92(2):199–233, 2004.

[50] R. Arratia, B. Bollobás, and G. B. Sorkin. A two-variable interlace polynomial. *Combinatorica*, 24(4):567–584, 2004.

[51] J. Ashkin and E. Teller. Statistics of two-dimensional lattices with four components. *Phys. Rev.*, 64:178–184, 1943.

[52] R. Askanazi, S. Chmutov, C. Estill, J. Michel, and P. Stollenwerk. Polynomial invariants of graphs on surfaces. *Quantum Topology*, 3:77–90, 2013.

[53] E. F. Assmus, Jr. and H. F. Mattson, Jr. New 5-designs. *J. Combin. Theory*, 6:122–151, 1969.

[54] C. A. Athanasiadis. Characteristic polynomials of subspace arrangements and finite fields. *Adv. Math.*, 122(2):193–233, 1996.

[55] I. Averbouch, B. Godlin, and J. A. Makowsky. An extension of the bivariate chromatic polynomial. *European J. Combin.*, 31(1):1–17, 2010.

[56] I. Averbouch and J. Makowsky. The complexity of multivariate matching polynomials. Preprint, 2007.

[57] J. Awan and O. Bernardi. Tutte polynomials for directed graphs. *J. Combin. Theory Ser. B*, 140:192–247, 2020.

[58] J. Azarija. Tutte polynomials and a stronger version of the Akiyama-Harary problem. *Graphs Combin.*, 31(5):1155–1161, 2015.

[59] S. Backman. Partial graph orientations and the Tutte polynomial. *Adv. in Appl. Math.*, 94:103–119, 2018.

[60] S. Backman and S. Hopkins. Fourientations and the Tutte polynomial. *Res. Math. Sci.*, 4:Paper No. 18, 2017.

[61] S. Backman, S. Hopkins, and L. Traldi. Fourientation activities and the Tutte polynomial. *European J. Combin.*, 67:40–60, 2018.

[62] S. Backman and M. Lenz. A convolution formula for Tutte polynomials of arithmetic matroids and other combinatorial structures. *Sém. Lothar. Combin.*, 78B:Art. 4, 2017.

[63] C. Bajo, B. Burdick, and S. Chmutov. Evaluations of the Tutte-Krushkal-Renardy polynomial for cell complexes. *J. Combin. Theory Ser. A*, 123:186–201, 2014.

[64] P. Bak, C. Tang, and K. Wiesenfeld. Self-organized criticality. *Phys. Rev. A (3)*, 38(1):364–374, 1988.

[65] G. A. Baker, Jr. Linked-cluster expansion for the graph-vertex coloration problem. *J. Combin. Theory Ser. B*, 10:217–231, 1971.

[66] M. Baker and S. Norine. Riemann-Roch and Abel-Jacobi theory on a finite graph. *Adv. Math.*, 215(2):766–788, 2007.

[67] M. Baker and F. Shokrieh. Chip-firing games, potential theory on graphs, and spanning trees. *J. Combin. Theory Ser. A*, 120(1):164–182, 2013.

[68] P. N. Balister, B. Bollobás, O. M. Riordan, and A. D. Scott. Alternating knot diagrams, Euler circuits and the interlace polynomial. *European J. Combin.*, 22(1):1–4, 2001.

[69] M. O. Ball, C. J. Colbourn, and J. S. Provan. Network reliability. In *Network models*, volume 7 of *Handbooks Oper. Res. Management Sci.*, pages 673–762. North-Holland, Amsterdam, 1995.

[70] M. O. Ball and G. L. Nemhauser. Matroids and a reliability analysis problem. *Math. Oper. Res.*, 4(2):132–143, 1979.

[71] M. O. Ball and J. S. Provan. Bounds on the reliability polynomial for shellable independence systems. *SIAM J. Algebraic Discrete Methods*, 3(2):166–181, 1982.

[72] M. O. Ball and J. S. Provan. Calculating bounds on reachability and connectedness in stochastic networks. *Networks*, 13(2):253–278, 1983.

[73] X. W. Bao and X. E. Chen. Chromaticity of the graph $\theta(a, b, c, d, e)$. *J. Xinjiang Univ. Natur. Sci.*, 11(3):19–22, 1994.

[74] V. Baranovsky and R. Sazdanović. Graph homology and graph configuration spaces. *J. Homotopy Relat. Struct.*, 7:1–13, 2012.

[75] M. Barany and V. Reiner. The Tutte polynomial of a finite projective space, unpublished, 2005.

[76] A. Barg. The matroid of supports of a linear code. *Appl. Algebra Engrg. Comm. Comput.*, 8(2):165–172, 1997.

[77] A. Barg. On some polynomials related to weight enumerators of linear codes. *SIAM J. Discrete Math.*, 15(2):155–164, 2002.

[78] A. Barg and W. Park. On linear ordered codes. *Mosc. Math. J.*, 15(4):679–702, 2015.

[79] R. A. Bari. Chromatic polynomials and the internal and external activities of Tutte. In *Graph Theory and Related Topics (Proc. Conf., Univ. Waterloo, Waterloo, Ont., 1977)*, pages 41–52. Academic Press, New York-London, 1979.

[80] R. A. Bari and D. W. Hall. Chromatic polynomials and Whitney's broken circuits. *J. Graph Theory*, 1(3):269–275, 1977.

[81] J. E. Bartels, J. Mount, and D. J. A. Welsh. The polytope of win vectors. *Ann. Comb.*, 1(1):1–15, 1997.

[82] J. E. Bartels and D. J. A. Welsh. The Markov chain of colourings. In *Integer Programming and Combinatorial Optimization (Copenhagen, 1995)*, volume 920 of *Lecture Notes in Comput. Sci.*, pages 373–387. Springer, Berlin, 1995.

[83] L. Beaudin, J. Ellis-Monaghan, G. Pangborn, and R. Shrock. A little statistical mechanics for the graph theorist. *Discrete Math.*, 310(13-14):2037–2053, 2010.

[84] M. Beck and T. Zaslavsky. The number of nowhere-zero flows on graphs and signed graphs. *J. Combin. Theory Ser. B*, 96(6):901–918, 2006.

[85] A. Bedini and J. L. Jacobsen. A tree-decomposed transfer matrix for computing exact Potts model partition functions for arbitrary graphs, with applications to planar graph colourings. *J. Phys. A: Math. Theor.*, 43(38):385001, 2010.

[86] V. Beffara, H. Duminil-Copin, and S. Smirnov. On the critical parameters of the $q \geq 4$ random-cluster model on isoradial graphs. *J. Phys. A*, 48:484003, 2015.

[87] J. S. Beissinger. On external activity and inversions in trees. *J. Combin. Theory Ser. B*, 33(1):87–92, 1982.

[88] D. Bénard, A. Bouchet, and A. Duchamp. On the Martin and Tutte polynomials. Unpublished manuscript, 1997.

[89] C. Benedetti and N. Bergeron. Cancelation free formula for the antipode of linearized Hopf monoid. arXiv:1611.01657, 2016.

[90] M. K. Bennett, K. P. Bogart, and J. E. Bonin. The geometry of Dowling lattices. *Adv. Math.*, 103(2):131–161, 1994.

[91] B. Benson, D. Chakrabarty, and P. Tetali. G-parking functions, acyclic orientations and spanning trees. *Discrete Math.*, 310(8):1340–1353, 2010.

[92] S. Beraha. *Infinite Non-Trivial Families of Maps and Chromials.* PhD Thesis, Johns Hopkins University, 1975.

[93] S. Beraha, J. Kahane, and N. J. Weiss. Limits of chromatic zeros of some families of maps. *J. Combin. Theory Ser. B*, 28(1):52–65, 1980.

[94] J. van den Berg and H. Kesten. Inequalities with applications to percolation and reliability. *J. Appl. Probab.*, 22:556–569, 1985.

[95] G. Berman. The dichromate and orientations of a graph. *Canad. J. Math.*, 29(5):947–956, 1977.

[96] G. Berman and W. T. Tutte. The golden root of a chromatic polynomial. *J. Combin. Theory*, 6:301–302, 1969.

[97] K. A. Berman. Bicycles and spanning trees. *SIAM J. Algebraic Discrete Methods*, 7(1):1–12, 1986.

[98] K. A. Berman. Balanced vertex weightings and arborescences. *J. Combin. Theory Ser. B*, 42(3):274–301, 1987.

[99] O. Bernardi. A characterization of the Tutte polynomial via combinatorial embeddings. *Ann. Comb.*, 12(2), 2008.

[100] E. Bernstein and U. Vazirani. Quantum complexity theory. *SIAM J. Comput.*, 26(5):1411–1473, 1997.

[101] T. Beth, M. Hain, G. Sagerer, and N. Schäfer. Materialien zur Codierungstheorie. II. *Arbeitsber. Inst. Math. Masch. Datenverarb. (Inform.)*, 12(10), 1979/80.

[102] T. Beth and V. Strehl. Materialien zur Codierungstheorie. *Arbeitsber. Inst. Math. Masch. Datenverarb. (Inform.)*, 11(14), 1978.

[103] A. Beutelspacher and U. Rosenbaum. *Projective Geometry: From Foundations to Applications.* Cambridge University Press, Cambridge, 1998.

[104] N. Biggs. Expansions of the chromatic polynomial. *Discrete Math.*, 6:105–113, 1973.

[105] N. Biggs. On cluster expansions in graph theory and physics. *Quart. J. Math. Oxford Ser.*, 29(114):159–173, 1978.

[106] N. Biggs. On the algebra of graph types. In *Graph Theory and Related Topics (Proc. Conf., Univ. Waterloo, Waterloo, Ont., 1977)*, pages 81–89. Academic Press, New York-London, 1979.

[107] N. Biggs. *Algebraic Graph Theory.* Cambridge Mathematical Library. Cambridge University Press, Cambridge, second edition, 1993.

[108] N. Biggs. A matrix method for chromatic polynomials. *J. Combin. Theory Ser. B*, 82(1):19–29, 2001.

[109] N. Biggs. Tutte polynomials of bracelets. *J. Algebraic Combin.*, 32(3):389–398, 2010.

[110] N. L. Biggs. *Interaction Models.* Cambridge University Press, Cambridge-New York-Melbourne, 1977.

[111] N. L. Biggs. Resonance and reconstruction. In *Surveys in combinatorics (Proc. Seventh British Combinatorial Conf., Cambridge, 1979)*, volume 38 of *London Math. Soc. Lecture Note Ser.*, pages 1–21. Cambridge University Press, Cambridge-New York, 1979.

[112] N. L. Biggs. Chip-firing and the critical group of a graph. Technical Report LSE-CDAM-96-03, Centre for Discrete and Applicable Mathematics, London School of Economics, London, 1996.

[113] N. L. Biggs. Chip-firing and the critical group of a graph. *J. Algebraic Combin.*, 9(1):25–45, 1999.

[114] N. L. Biggs, R. M. Damerell, and D. A. Sands. Recursive families of graphs. *J. Combin. Theory Ser. B*, 12:123–131, 1972.

[115] N. L. Biggs, E. K. Lloyd, and R. J. Wilson. *Graph Theory: 1736–1936.* Clarendon Press, Oxford, 1976.

[116] N. L. Biggs and P. Winkler. Chip-firing and the chromatic polynomial. Technical Report LSE-CDAM-97-03, Centre for Discrete and Applicable Mathematics, London School of Economics, London, 1997.

[117] L. J. Billera. Polyhedral theory and commutative algebra. In *Mathematical Programming: The State of the Art (Bonn, 1982)*, pages 57–77. Springer, Berlin-New York, 1983.

[118] L. J. Billera, N. Jia, and V. Reiner. A quasisymmetric function for matroids. *European J. Combin.*, 30(8):1727–1757, 2009.

[119] G. Birkhoff. Abstract linear dependence and lattices. *Amer. J. Math.*, 57(4):800–804, 1935.

[120] G. D. Birkhoff. A determinant formula for the number of ways of coloring a map. *Ann. of Math.*, 14(1-4):42–46, 1912/13.

[121] G. D. Birkhoff. The reducibility of maps. *Amer. J. Math.*, 35(2):115–128, 1913.

[122] G. D. Birkhoff. On the number of ways of colouring a map. *Proc. Edinburgh Math. Soc.*, 2(2):83–91, 1930.

[123] G. D. Birkhoff. On the polynomial expressions for the number of ways of coloring a map. *Ann. Scuola Norm.-Sci.*, 3(1):85–103, 1934.

[124] G. D. Birkhoff and D. C. Lewis. Chromatic polynomials. *Trans. Amer. Math. Soc.*, 60:355–451, 1946.

[125] Z. W. Birnbaum, J. D. Esary, and S. C. Saunders. Multicomponent systems and structures and their reliability. *Technometrics*, 3:55–77, 1961.

[126] M. Biskup, C. Borgs, J. T. Chayes, and R. Kotecký. Gibbs states of graphical representations of the Potts model with external fields. *J. Math. Phys.*, 41:1170–1210, 2000.

[127] A. Björklund, T. Husfeldt, P. Kaski, and M. Koivisto. Computing the Tutte polynomial in vertex-exponential time. In *Foundations of Computer Science, 2008. FOCS'08. IEEE 49th Annual IEEE Symposium on*, pages 677–686. IEEE, 2008.

[128] J. E. Björnberg. Vanishing critical magnetization in the quantum Ising model. *Comm. Math. Phys.*, 337:879–907, 2015.

[129] A. Björner. Shellable and Cohen-Macaulay partially ordered sets. *Trans. Amer. Math. Soc.*, 260(1):159–183, 1980.

[130] A. Björner. Some matroid inequalities. *Discrete Math.*, 31(1):101–103, 1980.

[131] A. Björner. The homology and shellability of matroids and geometric lattices. In *Matroid Applications*, volume 40 of *Encyclopedia Math. Appl.*, pages 226–283. Cambridge University Press, Cambridge, 1992.

[132] A. Björner, B. Korte, and L. Lovász. Homotopy properties of greedoids. *Adv. in Appl. Math.*, 6(4):447–494, 1985.

[133] A. Björner, M. Las Vergnas, B. Sturmfels, N. White, and G. M. Ziegler. *Oriented Matroids*, volume 46 of *Encyclopedia of Mathematics and its Applications.* Cambridge University Press, Cambridge, second edition, 1999.

[134] A. Björner and L. Lovász. Chip-firing games on directed graphs. *J. Algebraic Combin.*, 1(4):305–328, 1992.

[135] A. Björner, L. Lovász, and P. W. Shor. Chip-firing games on graphs. *European J. Combin.*, 12(4):283–291, 1991.

[136] A. Björner and G. M. Ziegler. Broken circuit complexes: factorizations and generalizations. *J. Combin. Theory Ser. B*, 51(1):96–126, 1991.

[137] A. Björner and G. M. Ziegler. Introduction to greedoids. In *Matroid Applications*, volume 40 of *Encyclopedia Math. Appl.*, pages 284–357. Cambridge University Press, Cambridge, 1992.

[138] R. G. Bland and M. Las Vergnas. Orientability of matroids. *J. Combin. Theory Ser. B*, 24(1):94–123, 1978.

[139] M. Bläser and H. Dell. Complexity of the cover polynomial. In *Automata, Languages and Programming*, volume 4596 of *Lecture Notes in Comput. Sci.*, pages 801–812. Springer, Berlin, 2007.

[140] M. Bläser, H. Dell, and M. Fouz. Complexity and approximability of the cover polynomial. *Comput. Complexity*, 21:359–419, 2012.

[141] M. Bläser, H. Dell, and J. A. Makowsky. Complexity of the Bollobás-Riordan polynomial—exceptional points and uniform reductions. *Theory Comput. Syst.*, 46(4):690–706, 2010.

[142] M. Bläser and C. Hoffmann. On the complexity of the interlace polynomial. In *STACS 2008: 25th International Symposium on Theoretical Aspects of Computer Science*, volume 1 of *LIPIcs. Leibniz Int. Proc. Inform.*, pages 97–108. Schloss Dagstuhl. Leibniz-Zent. Inform., Wadern, 2008.

[143] R. Blok and B. E. Sagan. Topological properties of activity orders for matroid bases. *J. Combin. Theory Ser. B*, 94(1):101–116, 2005.

[144] L. Blum, F. Cucker, M. Shub, and S. Smale. *Complexity and Real Computation.* Springer-Verlag, New York, 1998.

[145] L. Blum, M. Shub, and S. Smale. On a theory of computation and complexity over the real numbers: NP-completeness, recursive functions and universal machines. *Bull. Amer. Math. Soc. (N.S.)*, 21(1):1–46, 1989.

[146] H. L. Bodlaender and T. Kloks. Efficient and constructive algorithms for the pathwidth and treewidth of graphs. *J. Algorithms*, 21(2):358–402, 1996.

[147] F. T. Boesch, A. Satyanarayana, and C. L. Suffel. Least reliable networks and the reliability domination. *IEEE Trans. Communications*, 38:2004–2009, 1990.

[148] F. T. Boesch, A. Satyanarayana, and C. L. Suffel. Some alternate characterizations of reliability domination. *Prob. Eng. Inf. Sci.*, 4:257–276, 1990.

[149] C. Bogner and S. Weinzierl. Feynman graph polynomials. *Int. J. Mod. Phys.*, A25:2585–2618, 2010.

[150] A. Bohn. Chromatic roots as algebraic integers. In *FPSAC 2012 (24th International Conference on Formal Power Series and Algebraic Combinatorics, Nagoya, Japan)*, DMTCS proceedings, pages 539–550, 2012.

[151] M. Boij, J. C. Migliore, R. M. Miró-Roig, U. Nagel, and F. Zanello. On the shape of a pure O-sequence. *Mem. Amer. Math. Soc.*, 218(1024), 2012.

[152] B. Bollobás. *Modern graph theory*, volume 184 of *Graduate Texts in Mathematics*. Springer-Verlag, New York, 1998.

[153] B. Bollobás. Evaluations of the circuit partition polynomial. *J. Combin. Theory Ser. B*, 85(2):261–268, 2002.

[154] B. Bollobás and S. Eldridge. Packings of graphs and applications to computational complexity. *J. Combin. Theory Ser. B*, 25(2):105–124, 1978.

[155] B. Bollobás, G. Grimmett, and S. Janson. The random-cluster model on the complete graph. *Probab. Theory Related Fields*, 104:283–317, 1996.

[156] B. Bollobás, L. Pebody, and O. Riordan. Contraction-deletion invariants for graphs. *J. Combin. Theory Ser. B*, 80(2):320–345, 2000.

[157] B. Bollobás, L. Pebody, and D. Weinreich. A state space definition of the HOMFLY invariant. In *Contemporary Combinatorics*, volume 10 of *Bolyai Soc. Math. Stud.*, pages 139–184. János Bolyai Math. Soc., Budapest, 2002.

[158] B. Bollobás and O. Riordan. A Tutte polynomial for coloured graphs. *Combin. Probab. Comput.*, 8(1-2):45–93, 1999.

[159] B. Bollobás and O. Riordan. Polychromatic polynomials. *Discrete Math.*, 219(1-3):1–7, 2000.

[160] B. Bollobás and O. Riordan. A polynomial invariant of graphs on orientable surfaces. *Proc. London Math. Soc.*, 83(3):513–531, 2001.

[161] B. Bollobás and O. Riordan. A polynomial of graphs on surfaces. *Math. Ann.*, 323(1):81–96, 2002.

[162] J. A. Bondy and U. S. R. Murty. *Graph Theory with applications.* American Elsevier Publishing Co., Inc., New York, 1976.

[163] J. A. Bondy and U. S. R. Murty. *Graph Theory*, volume 244 of *Graduate Texts in Mathematics.* Springer, New York, 2008.

[164] J. E. Bonin. Matroids with no $(q+2)$-point-line minors. *Adv. in Appl. Math.*, 17(4):460–476, 1996.

[165] J. E. Bonin. Strongly inequivalent representations and Tutte polynomials of matroids. *Algebra Universalis*, 49(3):289–303, 2003.

[166] J. E. Bonin and A. de Mier. T-uniqueness of some families of k-chordal matroids. *Adv. in Appl. Math.*, 32(1-2):10–30, 2004.

[167] J. E. Bonin and A. de Mier. Tutte polynomials of generalized parallel connections. *Adv. in Appl. Math.*, 32(1-2):31–43, 2004.

[168] J. E. Bonin and A. de Mier. Lattice path matroids: structural properties. *European J. Combin.*, 27(5):701–738, 2006.

[169] J. E. Bonin and A. de Mier. The lattice of cyclic flats of a matroid. *Ann. Comb.*, 12(2):155–170, 2008.

[170] J. E. Bonin, A. de Mier, and M. Noy. Lattice path matroids: enumerative aspects and Tutte polynomials. *J. Combin. Theory Ser. A*, 104(1):63–94, 2003.

[171] J. E. Bonin and J. P. S. Kung. The $\mathcal{G}$-invariant and catenary data of a matroid. *Adv. in Appl. Math.*, 94:39–70, 2018.

[172] J. E. Bonin and W. P. Miller. Characterizing combinatorial geometries by numerical invariants. *European J. Combin.*, 20(8):713–724, 1999.

[173] J. E. Bonin and H. Qin. Tutte polynomials of q-cones. *Discrete Math.*, 232(1-3):95–103, 2001.

[174] J. Borcea, P. Brändén, and T. M. Liggett. Negative dependence and the geometry of polynomials. *J. Amer. Math. Soc.*, 22:521–567, 2009.

[175] M. Bordewich. *The Complexity of Counting and Randomised Approximation.* PhD Thesis, New College, Oxford University, 2003.

[176] M. Bordewich. Approximating the number of acyclic orientations for a class of sparse graphs. *Combin. Probab. Comput.*, 13:1–16, 2004.

[177] M. Bordewich, M. Freedman, L. Lovász, and D. Welsh. Approximate counting and quantum computation. *Combin. Probab. Comput.*, 14:737–754, 2005.

[178] M. Bordewich, C. Greenhill, and V. Patel. Mixing of the Glauber dynamics for the ferromagnetic Potts model. *Random Struct. Algor.*, 48(1):21–52, 2016.

[179] C. Borgs, J. Chayes, A. Frieze, J. H. Kim, P. Tetali, E. Vigoda, and V. H. Vu. Torpid mixing of some Monte Carlo Markov chain algorithms in statistical physics. In *Foundations of Computer Science, 1999. 40th Annual Symposium on*, pages 218–229, 1999.

[180] O. V. Borodin and I. G. Dmitriev. Characterization of chromatically rigid polynomials. *Sibirsk. Mat. Zh.*, 32(1):22–27, 219, 1991.

[181] P. Borwein. *Computational Excursions in Analysis and Number Theory.* CMS Books in Mathematics/Ouvrages de Mathématiques de la SMC, 10. Springer-Verlag, New York, 2002.

[182] R. C. Bose and R. C. Burton. A characterization of flat spaces in a finite geometry and the uniqueness of the Hamming and the MacDonald codes. *J. Combin. Theory*, 1:96–104, 1966.

[183] R. Bott. Reflections of the theme of the posters. In *Topological Methods in Modern Mathematics: A Symposium in Honor of John Milnor's Sixtieth Birthday*, pages 125–135. Goldberg and Phillips, 1993.

[184] A. Bouchet. Isotropic systems. *European J. Combin.*, 8(3):231–244, 1987.

[185] A. Bouchet. Graphic presentations of isotropic systems. *J. Combin. Theory Ser. B*, 45(1):58–76, 1988.

[186] A. Bouchet. Representability of $\triangle$-matroids. In *Combinatorics (Eger, 1987)*, volume 52 of *Colloq. Math. Soc. János Bolyai*, pages 167–182. North-Holland, Amsterdam, 1988.

[187] A. Bouchet. Transforming trees by successive local complementations. *J. Graph Theory*, 12(2):195–207, 1988.

[188] A. Bouchet. κ-transformations, local complementations and switching. In *Cycles and Rays (Montreal, PQ, 1987)*, volume 301 of *NATO Adv. Sci. Inst. Ser. C Math. Phys. Sci.*, pages 41–50. Kluwer Acad. Publ., Dordrecht, 1990.

[189] A. Bouchet. Tutte-Martin polynomials and orienting vectors of isotropic systems. *Graphs Combin.*, 7(3):235–252, 1991.

[190] A. Bouchet. Multimatroids. I. Coverings by independent sets. *SIAM J. Discrete Math.*, 10(4):626–646, 1997.

[191] A. Bouchet. Graph polynomials derived from Tutte-Martin polynomials. *Discrete Math.*, 302(1-3):32–38, 2005.

[192] A. Bouchet and A. Duchamp. Representability of $\triangle$-matroids over GF(2). *Linear Algebra Appl.*, 146:67–78, 1991.

[193] A. Bouchet and B. Jackson. Parity systems and the delta-matroid intersection problem. *Electron. J. Combin.*, 7:R14, 2000.

[194] R. Bradford, C. Butler, and S. Chmutov. Arrow ribbon graphs. *J. Knot Theory Ramif.*, 21(13): 1240002, 2012.

[195] P. Brändén and L. Moci. The multivariate arithmetic Tutte polynomial. *Trans. Amer. Math. Soc.*, 366(10):5523–5540, 2014.

[196] K. Braun, M. Kretz, B. Walter, and M. Walter. Die chromatischen Polynome unterringfreier Graphen. *Manuscripta Math.*, 14:223–234, 1974.

[197] C. Brennan, T. Mansour, and E. Mphako-Banda. Tutte polynomials of wheels via generating functions. *Bull. Iranian Math. Soc.*, 39(5):881–891, 2013.

[198] F. Brenti. Expansions of chromatic polynomials and log-concavity. *Trans. Amer. Math. Soc.*, 332(2):729–756, 1992.

[199] F. Brenti, G. F. Royle, and D. G. Wagner. Location of zeros of chromatic and related polynomials of graphs. *Canad. J. Math.*, 46(1):55–80, 1994.

[200] F. Breuer and R. Sanyal. Ehrhart theory, modular flow reciprocity, and the Tutte polynomial. *Math. Z.*, 270(1-2):1–18, 2012.

[201] E. Brieskorn. Sur les groupes de tresses [d'après V. I. Arnol′d]. In *Séminaire Bourbaki, 24ème année (1971/1972), Exp. No. 401*, pages 21–44. Lecture Notes in Math., Vol. 317. Springer, Berlin, 1973.

[202] R. Brijder and H. J. Hoogeboom. Quaternary bicycle matroids and the Penrose polynomial for delta-matroids. arXiv:1210.7718, 2012.

[203] R. Brijder and H. J. Hoogeboom. The group structure of pivot and loop complementation on graphs and set systems. *European J. Combin.*, 32(8):1353–1367, 2011.

[204] R. Brijder and H. J. Hoogeboom. Nullity invariance for pivot and the interlace polynomial. *Linear Algebra Appl.*, 435(2):277–288, 2011.

[205] R. Brijder and H. J. Hoogeboom. Nullity and loop complementation for delta-matroids. *SIAM J. Discrete Math.*, 27(1):492–506, 2013.

[206] R. Brijder and H. J. Hoogeboom. Interlace polynomials for multimatroids and delta-matroids. *European J. Combin.*, 40:142–167, 2014.

[207] G. Brinkmann and B. D. McKay. Fast generation of planar graphs. *MATCH Commun. Math. Comput. Chem.*, 58(2):323–357, 2007.

[208] M. Brion and M. Vergne. Arrangement of hyperplanes. I. Rational functions and Jeffrey-Kirwan residue. *Ann. Sci. École Norm. Sup. (4)*, 32(5):715–741, 1999.

[209] D. Britz, T. Britz, K. Shiromoto, and H. Sørensen. The higher weight enumerators of the doubly-even, self-dual [48, 24, 12] code. *IEEE Trans. Inform. Theory*, 53(7):2567–2571, 2007.

[210] T. Britz. MacWilliams identities and matroid polynomials. *Electron. J. Combin.*, 9(1):R19, 2002.

[211] T. Britz. Extensions of the critical theorem. *Discrete Math.*, 305(1-3):55–73, 2005.

[212] T. Britz. Higher support matroids. *Discrete Math.*, 307(17-18):2300–2308, 2007.

[213] T. Britz. Code enumerators and Tutte polynomials. *IEEE Trans. Inform. Theory*, 56(9):4350–4358, 2010.

[214] T. Britz, T. Johnsen, D. Mayhew, and K. Shiromoto. Wei-type duality theorems for matroids. *Des. Codes Cryptogr.*, 62(3):331–341, 2012.

[215] T. Britz, G. Royle, and K. Shiromoto. Designs from matroids. *SIAM J. Discrete Math.*, 23(2):1082–1099, 2009.

[216] T. Britz and C. G. Rutherford. Covering radii are not matroid invariants. *Discrete Math.*, 296(1):117–120, 2005.

[217] T. Britz and K. Shiromoto. Designs from subcode supports of linear codes. *Des. Codes Cryptogr.*, 46(2):175–189, 2008.

[218] T. Britz and K. Shiromoto. A MacWilliams type identity for matroids. *Discrete Math.*, 308(20):4551–4559, 2008.

[219] T. Britz, K. Shiromoto, and T. Westerbäck. Demi-matroids from codes over finite Frobenius rings. *Des. Codes Cryptogr.*, 75(1):97–107, 2015.

[220] R. L. Brooks, C. A. B. Smith, A. H. Stone, and W. T. Tutte. The dissection of rectangles into squares. *Duke Math. J.*, 7:312–340, 1940.

[221] F. Brown. The Massless higher-loop two-point function. *Commun. Math. Phys.*, 287:925–958, 2009.

[222] J. Brown and D. Cox. Inflection points of reliability polynomials are dense in [0,1]. *Networks*, 67(4):266–269, 2016.

[223] J. Brown and A. Erey. New bounds for chromatic polynomials and chromatic roots. *Discrete Math.*, 338(11):1938–1946, 2015.

[224] J. Brown and L. Mol. On the roots of all-terminal reliability polynomials. *Discrete Math.*, 340(6):1287–1299, 2017.

[225] J. I. Brown. Subdivisions and chromatic roots. *J. Combin. Theory Ser. B*, 76(2):201–204, 1999.

[226] J. I. Brown and C. J. Colbourn. Roots of the reliability polynomial. *SIAM J. Discrete Math.*, 5(4):571–585, 1992.

[227] J. I. Brown and C. J. Colbourn. Non-Stanley bounds for network reliability. *J. Algebraic Combin.*, 5(1):13–36, 1996.

[228] J. I. Brown, C. J. Colbourn, and J. S. Devitt. Network transformations and bounding network reliability. *Networks*, 23(1):1–17, 1993.

[229] J. I. Brown, C. J. Colbourn, and R. J. Nowakowski. Chip firing and all-terminal network reliability bounds. *Discrete Optim.*, 6(4):436–445, 2009.

[230] J. I. Brown, C. J. Colbourn, and D. G. Wagner. Cohen-Macaulay rings in network reliability. *SIAM J. Discrete Math.*, 9(3):377–392, 1996.

[231] J. I. Brown and D. Cox. The closure of the set of roots of strongly connected reliability polynomials is the entire complex plane. *Discrete Math.*, 309(16):5043–5047, 2009.

[232] J. I. Brown and D. Cox. Nonexistence of optimal graphs for all terminal reliability. *Networks*, 63(2):146–153, 2014.

[233] J. I. Brown, D. Cox, and R. Ehrenborg. The average reliability of a graph. *Discrete Appl. Math.*, 177:19–33, 2014.

[234] J. I. Brown and K. Dilcher. On the roots of strongly connected reliability polynomials. *Networks*, 54(2):108–116, 2009.

[235] J. I. Brown, C. Hickman, A. D. Sokal, and D. G. Wagner. On the chromatic roots of generalized theta graphs. *J. Combin. Theory Ser. B*, 83(2):272–297, 2001.

[236] J. I. Brown and C. A. Hickman. On chromatic roots with negative real part. *Ars Combin.*, 63:211–221, 2002.

[237] J. I. Brown, Y. Koç, and R. Kooij. Inflection points for network reliailbity. *Telecomm. Syst. J.*, 56:79–84, 2014.

[238] J. I. Brown and X. Li. Uniformly optimal digraphs for strongly connected reliability. *Networks*, 49(2):145–151, 2007.

[239] R. A. Brualdi, J. S. Graves, and K. M. Lawrence. Codes with a poset metric. *Discrete Math.*, 147(1-3):57–72, 1995.

[240] J. Brunat, A. Guedes de Oliveira, and M. Noy. Partitions of a finite boolean lattice into intervals. *European J. Combin.*, 30:1801–1809, 2009.

[241] T. Brylawski. The broken-circuit complex. *Trans. Amer. Math. Soc.*, 234(2):417–433, 1977.

[242] T. Brylawski. Intersection theory for embeddings of matroids into uniform geometries. *Stud. Appl. Math.*, 61(3):211–244, 1979.

[243] T. Brylawski. Hyperplane reconstruction of the Tutte polynomial of a geometric lattice. *Discrete Math.*, 35:25–38, 1981.

[244] T. Brylawski. Intersection theory for graphs. *J. Combin. Theory Ser. B*, 30(2):233–246, 1981.

[245] T. Brylawski. The Tutte polynomial. I. General theory. In *Matroid Theory and Its Applications*, pages 125–275. Liguori, Naples, 1982.

[246] T. Brylawski and D. Lucas. Uniquely representable combinatorial geometries. In B. Segre, editor, *Teorie Combinatorie (volume 1)*, pages 83–108. Accademia Nazionale dei Lincei, Roma, 1976.

[247] T. Brylawski and J. Oxley. The Tutte polynomial and its applications. In *Matroid Applications*, volume 40 of *Encyclopedia Math. Appl.*, pages 123–225. Cambridge University Press, Cambridge, 1992.

[248] T. Brylawski and A. Varchenko. The determinant formula for a matroid bilinear form. *Adv. Math.*, 129(1):1–24, 1997.

[249] T. H. Brylawski. *The Tutte-Grothendiek Ring.* PhD Thesis, Dartmouth College, 1970.

[250] T. H. Brylawski. A combinatorial model for series-parallel networks. *Trans. Amer. Math. Soc.*, 154:1–22, 1971.

[251] T. H. Brylawski. A decomposition for combinatorial geometries. *Trans. Amer. Math. Soc.*, 171:235–282, 1972.

[252] T. H. Brylawski. The Tutte-Grothendieck ring. *Algebra Universalis*, 2:375–388, 1972.

[253] T. H. Brylawski. A combinatorial perspective on the Radon convexity theorem. *Geom. Dedicata*, 5:459–466, 1976.

[254] P. Bürgisser. *Completeness and Reduction in Algebraic Complexity Theory*, volume 7 of *Algorithms and Computation in Mathematics.* Springer-Verlag, Berlin, 2000.

[255] C. Butler. A quasi-tree expansion of the Krushkal polynomial. *Adv. in Appl. Math.*, 94:3–22, 2018.

[256] C. Butler and S. Chmutov. Bollobás-Riordan and relative Tutte polynomials. *Arnold Math. J.*, 1(3):283–298, 2015.

[257] J.-Y. Cai and X. Chen. *Complexity Dichotomies for Counting Problems.* Cambridge University Press, Cambridge, 2018.

[258] J.-Y. Cai, H. Guo, and T. Williams. A complete dichotomy rises from the capture of vanishing signatures. *SIAM J. Comput.*, 45(5):1671–1728, 2016.

[259] J.-Y. Cai, H. Guo, and T. Williams. The complexity of counting edge colorings and a dichotomy for some higher domain Holant problems. *Res. Math. Sci.*, 3:Paper No. 18, 77, 2016.

[260] J.-Y. Cai, P. Lu, and M. Xia. Holographic reduction, interpolation and hardness. *Comput. Complexity*, 21(4):573–604, 2012.

[261] N. Calkin, C. Merino, S. Noble, and M. Noy. Improved bounds for the number of forests and acyclic orientations in the square lattice. *Electron. J. Combin.*, 10:R4, 2003.

[262] P. Cameron. Cycle index, weight enumerator, and Tutte polynomial. *Electron. J. Combin.*, 9(1):N2, 2002.

[263] P. Cameron. Permutation codes. *European J. Combin.*, 31(2):482–490, 2010.

[264] P. Cameron, B. Jackson, and J. Rudd. Orbit-counting polynomials for graphs and codes. *Discrete Math.*, 308(5-6):920–930, 2008.

[265] P. J. Cameron. *Permutation Groups*, volume 45 of *London Mathematical Society Student Texts.* Cambridge University Press, Cambridge, 1999.

[266] P. J. Cameron and D. G. Fon-Der-Flaass. Bases for permutation groups and matroids. *European J. Combin.*, 16(6):537–544, 1995.

[267] P. J. Cameron and K. K. Kayibi. Orbital chromatic and flow roots. *Combin. Probab. Comput.*, 16(3):401–407, 2007.

[268] P. J. Cameron and K. Morgan. Algebraic properties of chromatic roots. *Electron. J. Combin.*, 24(1):P1.21, 2017.

[269] J. Carter, S. Kamada, and M. Saito. Stable equivalence of knots on surfaces and virtual knot cobordisms. *J. Knot Theory Ramif.*, 11:311–322, 2002.

[270] P. Cartier. Les arrangements d'hyperplans: un chapitre de géométrie combinatoire. *Séminaire N. Bourbaki*, exp. no 561:1–22, 1980–1981.

[271] P. Cartier and D. Foata. *Problèmes combinatoires de commutation et réarrangements*. Springer-Verlag, Berlin, 1969.

[272] U. A. Celmins. *On Cubic Graphs That Do Not Have an Edge-3-Colouring*. PhD Thesis, University of Waterloo, 1985.

[273] S. Chaiken. The Tutte polynomial of a ported matroid. *J. Combin. Theory Ser. B*, 46(1):96–117, 1989.

[274] A. Champanerkar and I. Kofman. On mutation and Khovanov homology. *Commun. Contemp. Math.*, 10(suppl. 1):973–992, 2008.

[275] A. Champanerkar and I. Kofman. A survey on the Turaev genus of knots. *Acta Math. Vietnam*, 39(4):497–514, 2014.

[276] A. Champanerkar, I. Kofman, and N. Stoltzfus. Quasi-tree expansion for the Bollobás-Riordan-Tutte polynomial. *Bull. London Math. Soc.*, 43(5)(5):972–984, 2011.

[277] S. H. Chan. Abelian sandpile model and Biggs-Merino polynomial for directed graphs. *J. Combin. Theory Ser. A*, 154:145–171, 2018.

[278] T. Chan, A. Grant, and T. Britz. Quasi-uniform codes and their applications. *IEEE Trans. Inform. Theory*, 59(12):7915–7926, 2013.

[279] H. Chang, J. Ma, and Y.-N. Yeh. Tutte polynomials and G-parking functions. *Adv. in Appl. Math.*, 44(3):231–242, 2010.

[280] S.-C. Chang, J. Salas, and R. Shrock. Exact Potts model partition functions for strips of the square lattice. *J. Stat. Phys.*, 107(5-6):1207–1253, 2002.

[281] S.-C. Chang and R. Shrock. Exact Potts model partition function on strips of the triangular lattice. *Phys. A*, 286(1-2):189–238, 2000.

[282] S.-C. Chang and R. Shrock. Exact Potts model partition functions on wider arbitrary-length strips of the square lattice. *Phys. A*, 296(1-2):234–288, 2001.

[283] S.-C. Chang and R. Shrock. Zeros of Jones polynomials for families of knots and links. *Phys. A*, 301(1-4):196–218, 2001.

[284] S.-C. Chang and R. Shrock. Tutte polynomials and related asymptotic limiting functions for recursive families of graphs. *Adv. in Appl. Math.*, 32(1-2):44–87, 2004.

[285] S.-C. Chang and R. Shrock. Exact Potts model partition functions for strips of the honeycomb lattice. *J. Stat. Phys.*, 130(5):1011–1024, 2008.

[286] C.-Y. Chao, N. Z. Li, and S. J. Xu. On q-trees. *J. Graph Theory*, 10(1):129–136, 1986.

[287] C.-Y. Chao and G. A. Novacky, Jr. On maximally saturated graphs. *Discrete Math.*, 41(2):139–143, 1982.

[288] C.-Y. Chao and E. G. Whitehead, Jr. On chromatic equivalence of graphs. In *Theory and Applications of Graphs (Proc. Internat. Conf., Western Mich. Univ., Kalamazoo, Mich., 1976)*, pages 121–131. Lecture Notes in Math., Volume 642. Springer, Berlin, 1978.

[289] C.-Y. Chao and E. G. Whitehead, Jr. Chromatically unique graphs. *Discrete Math.*, 27(2):171–177, 1979.

[290] C.-Y. Chao and L. C. Zhao. Chromatic polynomials of a family of graphs. *Ars Combin.*, 15:111–129, 1983.

[291] M. K. Chari. Steiner complexes, matroid ports, and shellability. *J. Combin. Theory Ser. B*, 59(1):41–68, 1993.

[292] M. K. Chari. Matroid inequalities. *Discrete Math.*, 147(1-3):283–286, 1995.

[293] M. K. Chari. Two decompositions in topological combinatorics with applications to matroid complexes. *Trans. Amer. Math. Soc.*, 349(10):3925–3943, 1997.

[294] M. K. Chari and C. J. Colbourn. Reliability polynomials: a survey. *J. Combin. Inform. System Sci.*, 22(3-4):177–193, 1997.

[295] S. Chaudhary and G. Gordon. Tutte polynomials for trees. *J. Graph Theory*, 15(3):317–331, 1991.

[296] B. Chen. Dual complementary polynomials of graphs and combinatorial-geometric interpretation on the values of Tutte polynomial at positive integers. *European J. Combin.*, 36:206–230, 2014.

[297] B. Chen, A. Yang, and T. Zhang. A bijection for eulerian-equivalence classes of totally cyclic orientations. *Graphs Combin.*, 24:519–530, 2008.

[298] G. L. Chia. A note on chromatic uniqueness of graphs. *J. Graph Theory*, 10(4):541–543, 1986.

[299] G. L. Chia. The chromaticity of wheels with a missing spoke. *Discrete Math.*, 82(2):209–212, 1990.

[300] G. L. Chia. The chromaticity of wheels with a missing spoke. II. *Discrete Math.*, 148(1-3):305–310, 1996.

[301] G. L. Chia. A bibliography on chromatic polynomials. *Discrete Math.*, 172(1-3):175–191, 1997.

[302] G. L. Chia. On the chromatic equivalence class of graphs. *Discrete Math.*, 178(1-3):15–23, 1998.

[303] G.-L. Chia, B.-H. Goh, and K.-M. Koh. The chromaticity of some families of complete tripartite graphs. *Sci. Ser. A Math. Sci. (N.S.)*, 2:27–37, 1988.

[304] G. L. Chia and C.-K. Ho. Chromatic equivalence classes of complete tripartite graphs. *Discrete Math.*, 309(1):134–143, 2009.

[305] M. Chmutov, S. Chmutov, and Y. Rong. Knight move in chromatic cohomology. *European J. Combin.*, 29(1):311–321, 2008.

[306] S. Chmutov. Generalized duality for graphs on surfaces and the signed Bollobás-Riordan polynomial. *J. Combin. Theory Ser. B*, 99(3):617–638, 2009.

[307] S. Chmutov. Topological Tutte polynomial. arXiv:1708.08132, 2017.

[308] S. Chmutov, S. Duzhin, and Y. Mostovoy. *Introduction to Vassiliev Knot Invariants.* Cambridge University Press, 2012.

[309] S. Chmutov and I. Pak. The Kauffman bracket of virtual links and the Bollobás-Riordan polynomial. *Mosc. Math. J.*, 7(3):409–418, 2007.

[310] S. Chmutov and J. Voltz. Thistlethwaite's theorem for virtual links. *J. Knot Theory Ramif.*, 17(10):1189–1198, 2008.

[311] S. V. Chmutov, S. V. Duzhin, and S. K. Lando. Vassiliev knot invariants. I. Introduction. In *Singularities and Bifurcations*, volume 21 of *Adv. Soviet Math.*, pages 117–126. Amer. Math. Soc., Providence, RI, 1994.

[312] S. V. Chmutov, S. V. Duzhin, and S. K. Lando. Vassiliev knot invariants. II. Intersection graph conjecture for trees. In *Singularities and Bifurcations*, volume 21 of *Adv. Soviet Math.*, pages 127–134. Amer. Math. Soc., Providence, RI, 1994.

[313] S. V. Chmutov, S. V. Duzhin, and S. K. Lando. Vassiliev knot invariants. III. Forest algebra and weighted graphs. In *Singularities and Bifurcations*, volume 21 of *Adv. Soviet Math.*, pages 135–145. Amer. Math. Soc., Providence, RI, 1994.

[314] S. V. Chmutov and S. K. Lando. Mutant knots and intersection graphs. *Algebr. Geom. Topol.*, 7:1579–1598, 2007.

[315] T. Chow. A short proof of the rook reciprocity theorem. *Electron. J. Combin.*, 3:R10, 1996.

[316] T. Y. Chow. The path-cycle symmetric function of a digraph. *Adv. Math.*, 118(1):71–98, 1996.

[317] M. Chudnovsky and P. Seymour. The roots of the independence polynomial of a clawfree graph. *J. Combin. Theory Ser. B*, 97(3):350–357, 2007.

[318] C. Chun, I. Moffatt, S. D. Noble, and R. Rueckriemen. On the interplay between embedded graphs and delta-matroids. *Proc. London Math. Soc.*, 118(3):675–700, 2019.

[319] F. Chung and R. Graham. The matrix cover polynomial. *J. Comb.*, 7(2-3):375–412, 2016.

[320] F. R. K. Chung and R. L. Graham. On the cover polynomial of a digraph. *J. Combin. Theory Ser. B*, 65(2):273–290, 1995.

[321] M. Cohn and A. Lempel. Cycle decomposition by disjoint transpositions. *J. Combin. Theory Ser. A*, 13:83–89, 1972.

[322] C. Colbourn, J. Provan, and D. Vertigan. The complexity of computing the Tutte polynomial on transversal matroids. *Combinatorica*, 15(1):1–10, 1995.

[323] C. J. Colbourn. *The Combinatorics of Network Reliability.* International Series of Monographs on Computer Science. The Clarendon Press, Oxford University Press, New York, 1987.

[324] C. J. Colbourn. Some open problems on reliability polynomials. *Congr. Numer.*, 93:187–202, 1993.

[325] C. J. Colbourn, M. S. Keranen, and D. L. Kreher. f-vectors of pure complexes and pure multicomplexes of rank three. *Discrete Math.*, 320:26–39, 2014.

[326] C. J. Colbourn and W. R. Pulleyblank. Matroid Steiner problems, the Tutte polynomial and network reliability. *J. Combin. Theory Ser. B*, 47(1):20–31, 1989.

[327] A. Connes and M. Marcolli. *Noncommutative Geometry, Quantum Fields and Motives*, volume 55 of *American Mathematical Society Colloquium Publications.* American Mathematical Society, Providence, RI; Hindustan Book Agency, New Delhi, 2008.

[328] J. Conway. An enumeration of knots and links, and some of their algebraic properties. In J. Leech, editor, *Computational Problems in Abstract Algebra*, pages 329–358. Pergamon, 1969.

[329] R. Cordovil and D. Forge. A note on Tutte polynomials and Orlik-Solomon algebras. *European J. Combin.*, 24(8):1081–1087, 2003.

[330] R. Cordovil, M. Las Vergnas, and A. Mandel. Euler's relation, Möbius functions, and matroid identities. *Geom. Dedicata*, 12(2):147–162, 1982.

[331] R. Cori. *Un code pour les graphes planaires et ses applications.* Société Mathématique de France, Paris, 1975.

[332] R. Cori and Y. Le Borgne. The sand-pile model and Tutte polynomials. *Adv. in Appl. Math.*, 30(1-2):44–52, 2003.

[333] R. Cori, D. Rossin, and B. Salvy. Polynomial ideals for sandpiles and their Gröbner bases. *Theoret. Comput. Sci.*, 276(1-2):1–15, 2002.

[334] H. Corrales and C. E. Valencia. On the critical ideals of graphs. *Linear Algebra Appl.*, 439(12):3870–3892, 2013.

[335] S. Corry. Genus bounds for harmonic group actions on finite graphs. *Int. Math. Res. Not. IMRN*, 2011(19):4515–4533, 2011.

[336] S. Corry and D. Perkinson. *Divisors and Sandpiles.* American Mathematical Society, Providence, RI, 2018.

[337] B. Courcelle. The monadic second-order logic of graphs. I. recognizable sets of finite graphs. *Inform. and Comput.*, 85(1):12–75, 1990.

[338] B. Courcelle. A multivariate interlace polynomial and its computation for graphs of bounded clique-width. *Electron. J. Combin.*, 15(1):R69, 2008.

[339] B. Courcelle, J. A. Makowsky, and U. Rotics. On the fixed parameter complexity of graph enumeration problems definable in monadic second-order logic. *Discrete Appl. Math.*, 108(1-2):23–52, 2001.

[340] B. Courcelle and S. Olariu. Upper bounds to the clique width of graphs. *Discrete Appl. Math.*, 101(1):77–114, 2000.

[341] J. Courtiel. *Combinatorics of the Tutte Polynomial and Planar Maps.* PhD Thesis, Université de Bordeaux, 2014.

[342] H. Crapo. Numerical invariants of geometries and graphs. In *Proceedings of the Twenty-Fifth Summer Meeting of the Canadian Mathematical Congress (Lakehead Univ., Thunder Bay, Ont., 1971)*, pages 42–54, 1971.

[343] H. Crapo and W. Schmitt. The free product of matroids. *European J. Combin.*, 26(7):1060–1065, 2005.

[344] H. Crapo and W. Schmitt. A free subalgebra of the algebra of matroids. *European J. Combin.*, 26(7):1066–1085, 2005.

[345] H. Crapo and W. Schmitt. A unique factorization theorem for matroids. *J. Combin. Theory Ser. A*, 112(2):222–249, 2005.

[346] H. H. Crapo. Single-element extensions of matroids. *J. Res. Nat. Bur. Standards Sect. B*, 69B:55–65, 1965.

[347] H. H. Crapo. A higher invariant for matroids. *J. Combin. Theory*, 2:406–417, 1967.

[348] H. H. Crapo. Möbius inversion in lattices. *Arch. Math. (Basel)*, 19:595–607 (1969), 1968.

[349] H. H. Crapo. The Tutte polynomial. *Aequationes Math.*, 3:211–229, 1969.

[350] H. H. Crapo and G.-C. Rota. *On the Foundations of Combinatorial Theory: Combinatorial Geometries.* The M.I.T. Press, Cambridge, Mass.-London, preliminary edition, 1970.

[351] L. Crew and S. Spirkl. A deletion-contraction relation for the chromatic symmetric function. *European J. Combin.*, 89:103143, 20, 2020.

[352] P. Cromwell. *Knots and Links.* Cambridge University Press, Cambridge, 2004.

[353] P. Curie. Propriétés magnétiques des corps à diverses températures. *Ann. Chim. Phys.*, séries 7:289–405, 1895.

[354] R. Curticapean. Block interpolation: a framework for tight exponential-time counting complexity. In *Automata, Languages, and Programming. Part I*, volume 9134 of *Lecture Notes in Comput. Sci.*, pages 380–392. Springer, Heidelberg, 2015.

[355] M. Cygan, F. V. Fomin, Ł. u. Kowalik, D. Lokshtanov, D. Marx, M. Pilipczuk, M. Pilipczuk, and S. Saurabh. *Parameterized Algorithms.* Springer, Cham, 2015.

[356] M. D'Adderio and L. Moci. Arithmetic matroids, the Tutte polynomial and toric arrangements. *Adv. Math.*, 232:335–367, 2013.

[357] W. Dahmen and C. A. Micchelli. On the local linear independence of translates of a box spline. *Studia Math.*, 82(3):243–263, 1985.

[358] Z. Dancso and A. Licata. Odd Khovanov homology for hyperplane arrangements. *J. Algebra*, 436:102–144, 2015.

[359] O. M. D'Antona and E. Munarini. The cycle-path indicator polynomial of a digraph. *Adv. in Appl. Math.*, 25(1):41–56, 2000.

[360] O. T. Dasbach, D. Futer, E. Kalfagianni, X.-S. Lin, and N. W. Stoltzfus. The Jones polynomial and graphs on surfaces. *J. Combin. Theory Ser. B*, 98:384–399, 2008.

[361] J. E. Dawson. A construction for a family of sets and its application to matroids. In *Combinatorial Mathematics, VIII (Geelong, 1980)*, volume 884 of *Lecture Notes in Math.*, pages 136–147. Springer, Berlin-New York, 1981.

[362] J. E. Dawson. A collection of sets related to the Tutte polynomial of a matroid. In *Graph Theory, Singapore 1983*, volume 1073 of *Lecture Notes in Math.*, pages 193–204. Springer, Berlin, 1984.

[363] C. De Concini and C. Procesi. Wonderful models of subspace arrangements. *Selecta Math.*, 1(3):459–494, 1995.

[364] C. De Concini and C. Procesi. On the geometry of toric arrangements. *Transform. Groups*, 10(3-4):387–422, 2005.

[365] C. De Concini and C. Procesi. The zonotope of a root system. *Transform. Groups*, 13(3-4):507–526, 2008.

[366] C. De Concini and C. Procesi. *Topics in Hyperplane Arrangements, Polytopes and Box-Splines.* Universitext. Springer-Verlag New York, 2010.

[367] P. de la Harpe and F. Jaeger. Chromatic invariants for finite graphs: theme and polynomial variations. *Linear Algebra Appl.*, 226/228:687–722, 1995.

[368] P. de la Harpe and V. F. R. Jones. Graph invariants related to statistical mechanical models: examples and problems. *J. Combin. Theory Ser. B*, 57(2):207–227, 1993.

[369] J. A. De Loera, Y. Kemper, and S. Klee. h-vectors of small matroid complexes. *Electron. J. Combin.*, 19(1):P14, 2012.

[370] H. Dell, T. Husfeldt, D. Marx, N. Taslaman, and M. Wahlen. Exponential time complexity of the permanent and the Tutte polynomial. *ACM Trans. Algorithms*, 10(4):21:1–21:32, 2014.

[371] P. Delsarte. An algebraic approach to the association schemes of coding theory. *Philips Res. Rep. Suppl.*, (10), 1973.

[372] P. Delsarte. Four fundamental parameters of a code and their combinatorial significance. *Inf. Control*, 23:407–438, 1973.

[373] G. Denham. The combinatorial Laplacian of the Tutte complex. *J. Algebra*, 242(1):160–175, 2001.

[374] H. Derksen. Symmetric and quasi-symmetric functions associated to polymatroids. *J. Algebraic Combin.*, 30(1):43–86, 2009.

[375] H. Derksen and A. Fink. Valuative invariants for polymatroids. *Adv. Math.*, 225(4):1840–1892, 2010.

[376] M. DeVos, J. Nešetřil, and A. Raspaud. On edge-maps whose inverse preserves flows or tensions. In *Graph Theory in Paris*, Trends Math., pages 109–138. Birkhäuser, Basel, 2007.

[377] D. Dhar. Self-organized critical state of sandpile automaton models. *Phys. Rev. Lett.*, 64(14):1613–1616, 1990.

[378] D. Dhar and S. N. Majumdar. Abelian sandpile model on the Bethe lattice. *J. Phys. A*, 23(19):4333–4350, 1990.

[379] Y. Diao and G. Hetyei. Relative Tutte polynomials for coloured graphs and virtual knot theory. *Combin. Probab. Comput.*, 19(3):343–369, 2010.

[380] Y. Diao and G. Hetyei. Relative Tutte polynomials of tensor products of coloured graphs. *Combin. Probab. Comput.*, 22(6):801–828, 2013.

[381] Y. Diao, G. Hetyei, and K. Hinson. Tutte polynomials of tensor products of signed graphs and their applications in knot theory. *J. Knot Theory Ramif.*, 18(1):561–589, 2009.

[382] Y. Diao, G. Hetyei, and K. Hinson. A Tutte-style proof of Brylawski's tensor product formula. *European J. Combin.*, 32(6):775–781, 2011.

[383] R. Diestel. *Graph Theory*, volume 173 of *Graduate Texts in Mathematics.* Springer, Berlin, fifth edition, 2017.

[384] A. Dimca and S. Yuzvinsky. Lectures on Orlik-Solomon algebras. In *Arrangements, Local Systems and Singularities*, volume 283 of *Progr. Math.*, pages 83–110. Birkhäuser Verlag, Basel, 2010.

[385] G. A. Dirac. On rigid circuit graphs. *Abh. Math. Sem. Univ. Hamburg*, 25:71–76, 1961.

[386] I. G. Dmitriev. Weakly cyclic graphs with integral chromatic spectra. *Metody Diskret. Analiz.*, 34:3–7, 100, 1980.

[387] I. G. Dmitriev. Characterization of a class of k-trees. *Metody Diskret. Analiz.*, 38:9–18, 1982.

[388] K. Dohmen, A. Pönitz, and P. Tittmann. A new two-variable generalization of the chromatic polynomial. *Discrete Math. Theor. Comput. Sci.*, 6(1):69–89, 2003.

[389] C. Domb. Configurational studies of the Potts models. *J. Phys. A: Math. Nuc. Gen.*, 7(11):1335–1348, 1974.

[390] C. Domb. Some reminiscences about my early career. *Phys. A*, 168(1):1–21, 1990.

[391] F. Dong. Uniqueness of the chromatic polynomial of a generalized wheel graph (Chinese). *J. Math. Res. Exposition*, 10(3):447–454, 1990.

[392] F. Dong. A survey on the study of real zeros of flow polynomials. *J. Graph Theory*, 92(4):361–376, 2019.

[393] F. M. Dong. On chromatic uniqueness of two infinite families of graphs. *J. Graph Theory*, 17(3):387–392, 1993.

[394] F. M. Dong. Proof of a chromatic polynomial conjecture. *J. Combin. Theory Ser. B*, 78(1):35–44, 2000.

[395] F. M. Dong and B. Jackson. A zero-free interval for chromatic polynomials of nearly 3-connected plane graphs. *SIAM J. Discrete Math.*, 25(3):1103–1118, 2011.

[396] F. M. Dong and K. M. Koh. On the structure and chromaticity of graphs in which any two colour classes induce a tree. *Discrete Math.*, 176(1-3):97–113, 1997.

[397] F. M. Dong and K. M. Koh. Non-chordal graphs having integral-root chromatic polynomials. *Bull. Inst. Combin. Appl.*, 22:67–77, 1998.

[398] F. M. Dong and K. M. Koh. On graphs having no chromatic zeros in $(1,2)$. *SIAM J. Discrete Math.*, 20(3):799–810, 2006.

[399] F. M. Dong and K. M. Koh. Bounds for the real zeros of chromatic polynomials. *Combin. Probab. Comput.*, 17(6):749–759, 2008.

[400] F. M. Dong and K. M. Koh. Domination numbers and zeros of chromatic polynomials. *Discrete Math.*, 308(10):1930–1940, 2008.

[401] F. M. Dong and K. M. Koh. A maximal zero-free interval for chromatic polynomials of bipartite planar graphs. *Discrete Math.*, 308(11):2285–2287, 2008.

[402] F. M. Dong and K. M. Koh. The 3-connectivity of a graph and the multiplicity of zero "2" of its chromatic polynomial. *J. Graph Theory*, 70(3):262–283, 2012.

[403] F. M. Dong, K. M. Koh, and C. A. Soh. Divisibility of certain coefficients of the chromatic polynomials. *Discrete Math.*, 275(1-3):311–317, 2004.

[404] F. M. Dong, K. M. Koh, and K. L. Teo. *Chromatic Polynomials and Chromaticity of Graphs.* World Scientific Publishing Co. Pte. Ltd., Hackensack, NJ, 2005.

[405] F. M. Dong, K. M. Koh, K. L. Teo, C. H. C. Little, and M. D. Hendy. Sharp bounds for the number of 3-independent partitions and the chromaticity of bipartite graphs. *J. Graph Theory*, 37(1):48–77, 2001.

[406] F. M. Dong and Y. P. Liu. All wheels with two missing consecutive spokes are chromatically unique. *Discrete Math.*, 184(1-3):71–85, 1998.

[407] F. M. Dong, Y. P. Liu, and K. M. Koh. The chromaticity of odd wheels with a missing spoke. *New Zealand J. Math.*, 26(1):31–44, 1997.

[408] F. M. Dong, K. L. Teo, K. M. Koh, and M. D. Hendy. Non-chordal graphs having integral-root chromatic polynomials. II. *Discrete Math.*, 245(1-3):247–253, 2002.

[409] F. M. Dong, K. L. Teo, C. H. C. Little, M. Hendy, and K. M. Koh. Chromatically unique multibridge graphs. *Electron. J. Combin.*, 11(1):R12, 2004.

[410] F. M. Dong, K. L. Teo, C. H. C. Little, and M. D. Hendy. Chromaticity of some families of dense graphs. *Discrete Math.*, 258(1-3):303–321, 2002.

[411] A. Donno and D. Iacono. The Tutte polynomial of the Sierpiński and Hanoi graphs. *Adv. Geom.*, 13(4):663–694, 2013.

[412] T. Došlić. Planar polycyclic graphs and their Tutte polynomials. *J. Math. Chem.*, 51(6):1599–1607, 2013.

[413] T. Dowling. Codes, packings and the critical problem. In *Atti del Convegno di Geometria Combinatoria e sue Applicazioni (Univ. Perugia, Perugia, 1970)*, pages 209–224. Ist. Mat., Univ. Perugia, Perugia, 1971.

[414] T. A. Dowling. A class of geometric lattices based on finite groups. *J. Combin. Theory Ser. B*, 14:61–86, 1973.

[415] T. A. Dowling. A q-analog of the partition lattice. In *A Survey of Combinatorial Theory (Proc. Internat. Sympos., Colorado State Univ., Ft Collins, Colo., 1971)*, pages 101–115. North-Holland, Amsterdam, 1973.

[416] T. A. Dowling and R. M. Wilson. The slimmest geometric lattices. *Trans. Amer. Math. Soc.*, 196:203–215, 1974.

[417] R. G. Downey and M. R. Fellows. *Parameterized Complexity*. Monographs in Computer Science. Springer-Verlag, New York, 1999.

[418] R. G. Downey and M. R. Fellows. *Fundamentals of Parameterized Complexity*. Texts in Computer Science. Springer, London, 2013.

[419] J. Draisma, D. C. Gijswijt, L. Lovász, G. Regts, and A. Schrijver. Characterizing partition functions of the vertex model. *J. Algebra*, 350:197–206, 2012.

[420] Q. Du. Chromaticity of the complements of paths and cycles. *Discrete Math.*, 162(1-3):109–125, 1996.

[421] Y. Duan, H. Wu, and Q. Yu. On chromatic and flow polynomial unique graphs. *Discrete Appl. Math.*, 156(12):2300–2309, 2008.

[422] Y. Duan, H. Wu, and Q. Yu. On Tutte polynomial uniqueness of twisted wheels. *Discrete Math.*, 309(4):926–936, 2009.

[423] G. Duchamp, N. Hoang-Nghia, T. Krajewski, and A. Tanasa. Recipe theorem for the Tutte polynomial for matroids, renormalization group-like approach. *Adv. in Appl. Math.*, 51:345–358, 2013.

[424] C. Dupont, A. Fink, and L. Moci. Universal Tutte characters via combinatorial coalgebras. *Algebr. Comb.*, 1(5):603–651, 2018.

[425] I. Duursma. Extremal weight enumerators and ultraspherical polynomials. *Discrete Math.*, 268(1-3):103–127, 2003.

[426] I. Duursma. Combinatorics of the two-variable zeta function. In *Finite Fields and Applications*, volume 2948 of *Lecture Notes in Comput. Sci.*, pages 109–136. Springer, Berlin, 2004.

[427] A. M. Duval, C. J. Klivans, and J. L. Martin. Critical groups of simplicial complexes. *Ann. Comb.*, 17(1):53–70, 2013.

[428] S. Duzhin and M. Shkolnikov. Bipartite knots. *Fund. Math.*, 225(1):95–102, 2014.

[429] M. Dworkin. Factorization of the cover polynomial. *J. Combin. Theory Ser. B*, 71:17–53, 1997.

[430] H. A. Dye and L. H. Kauffman. Virtual crossing number and the arrow polynomial. *J. Knot Theory Ramif.*, 18(10):1335–1357, 2009.

[431] M. Dyer, L. A. Goldberg, C. Greenhill, and M. R. Jerrum. On the relative complexity of approximate counting problems. *Algorithmica*, 38(3):471–500, 2003.

[432] M. Dyer and C. Greenhill. The complexity of counting graph homomorphisms. *Random Struct. Algor.*, 17(3-4):260–289, 2000.

[433] M. Eastwood and S. Huggett. Euler characteristics and chromatic polynomials. *European J. Combin.*, 28:1553–1560, 2007.

[434] H.-D. Ebbinghaus, J. Flum, and W. Thomas. *Mathematical Logic*. Undergraduate Texts in Mathematics. Springer-Verlag, New York, second edition, 1994.

[435] J. N. Eberhardt. Computing the Tutte polynomial of a matroid from its lattice of cyclic flats. *Electron. J. Combin.*, 21(3):P3.47, 2014.

[436] P. H. Edelman and R. E. Jamison. The theory of convex geometries. *Geom. Dedicata*, 19(3):247–270, 1985.

[437] P. H. Edelman and V. Reiner. Counting the interior points of a point configuration. *Discrete Comput. Geom.*, 23(1):1–13, 2000.

[438] J. Edmonds. Lehman's switching game and a theorem of Tutte and Nash-Williams. *J. Res. Nat. Bur. Standards Sect. B*, 69B:73–77, 1965.

[439] J. Edmonds. Minimum partition of a matroid into independent subsets. *J. Res. Nat. Bur. Standards Sect. B*, 69:67–72, 1965.

[440] J. Edmonds. On the surface duality of linear graphs. *J. Res. Nat. Bur. Stand.*, 69:121–123, 1965.

[441] J. Edmonds. Matroids and the greedy algorithm. *Math. Programming*, 1:127–136, 1971.

[442] K. Edwards. The complexity of colouring problems on dense graphs. *Theoret. Comput. Sci.*, 43:337–343, 1986.

[443] R. Ehrenborg, M. Readdy, and M. Slone. Affine and toric hyperplane arrangements. *Discrete Comput. Geom.*, 41(4):481–512, 2009.

[444] B. Eisenberg. Characterization of a tree by means of coefficients of the chromatic polynomial. *Trans. New York Acad. Sci.*, 34(2):146–153, 1972.

[445] D. Eisenstat and G. Gordon. Non-isomorphic caterpillars with identical subtree data. *Discrete Math.*, 306(8-9):827–830, 2006.

[446] S. Eliahou, L. H. Kauffman, and M. B. Thistlethwaite. Infinite families of links with trivial Jones polynomial. *Topology*, 42(1):155–169, 2003.

[447] J. Ellis-Monaghan and I. Moffatt. The Las Vergnas polynomial for embedded graphs. *European J. Combin.*, 50:97–114, 2015.

[448] J. A. Ellis-Monaghan. New results for the Martin polynomial. *J. Combin. Theory Ser. B*, 74(2):326–352, 1998.

[449] J. A. Ellis-Monaghan. Differentiating the Martin polynomial. *Congr. Numer.*, 142:173–183, 2000.

[450] J. A. Ellis-Monaghan. Exploring the Tutte-Martin connection. *Discrete Math.*, 281(1-3):173–187, 2004.

[451] J. A. Ellis-Monaghan. Identities for circuit partition polynomials, with applications to the Tutte polynomial. *Adv. in Appl. Math.*, 32(1-2):188–197, 2004.

[452] J. A. Ellis-Monaghan, A. J. Goodall, J. A. Makowsky, and I. Moffatt. Graph Polynomials: Towards a Comparative Theory (Dagstuhl Seminar 16241). *Dagstuhl Reports*, 6(6):26–48, 2016.

[453] J. A. Ellis-Monaghan, L. H. Kauffman, and I. Moffatt. Edge colourings and topological graph polynomials. *Australas. J. Combin.*, 72:290–305, 2018.

[454] J. A. Ellis-Monaghan and C. Merino. Graph polynomials and their applications I: The Tutte polynomial. In *Structural Analysis of Complex Networks*, pages 219–255. Birkhäuser/Springer, New York, 2011.

[455] J. A. Ellis-Monaghan and I. Moffatt. The Tutte-Potts connection in the presence of an external magnetic field. *Adv. in Appl. Math.*, 47(4):772–782, 2011.

[456] J. A. Ellis-Monaghan and I. Moffatt. Twisted duality for embedded graphs. *Trans. Amer. Math. Soc.*, 364(3):1529–1569, 2012.

[457] J. A. Ellis-Monaghan and I. Moffatt. *Graphs on Surfaces: Dualities, Polynomials, and Knots.* Springer-Briefs in Mathematics. Springer, New York, 2013.

[458] J. A. Ellis-Monaghan and I. Moffatt. A Penrose polynomial for embedded graphs. *European J. Combin.*, 34(2):424–445, 2013.

[459] J. A. Ellis-Monaghan and I. Moffatt. A note on recognizing an old friend in a new place: list coloring and the zero-temperature Potts model. *Ann. Inst. Henri Poincaré D*, 1(4):429–442, 2014.

[460] J. A. Ellis-Monaghan and I. Sarmiento. Generalized transition polynomials. *Congr. Numer.*, 155:57–69, 2002.

[461] J. A. Ellis-Monaghan and I. Sarmiento. Distance hereditary graphs and the interlace polynomial. *Combin. Probab. Comput.*, 16(6):947–973, 2007.

[462] J. A. Ellis-Monaghan and I. Sarmiento. A recipe theorem for the topological Tutte polynomial of Bollobás and Riordan. *European J. Combin.*, 32(6):782–794, 2011.

[463] J. A. Ellis-Monaghan and L. Traldi. Parametrized Tutte polynomials of graphs and matroids. *Combin. Probab. Comput.*, 15(6):835–854, 2006.

[464] B. Ellzey. A directed graph generalization of chromatic quasisymmetric functions. arXiv:1709.00454, 2017.

[465] E. Engeler. Algorithmic properties of structures. *Theory Comput. Syst.*, 1(2):183–195, 1967.

[466] M. Epple. Geometric aspects in the development of knot theory. In *History of Topology*, pages 301–357. North-Holland, Amsterdam, 1999.

[467] C. J. Eschenbrenner and M. J. Falk. Orlik-Solomon algebras and Tutte polynomials. *J. Algebraic Combin.*, 10(2):189–199, 1999.

[468] J. W. Essam. Graph theory and statistical physics. *Discrete Math.*, 1(1):83–112, 1971/72.

[469] G. Etienne and M. Las Vergnas. External and internal elements of a matroid basis. *Discrete Math.*, 179(1-3):111–119, 1998.

[470] G. Etienne and M. Las Vergnas. The Tutte polynomial of a morphism of matroids. III. Vectorial matroids. *Adv. in Appl. Math.*, 32(1-2):198–211, 2004.

[471] M. J. Falk. Arrangements and cohomology. *Ann. Comb.*, 1(2):135–157, 1997.

[472] M. J. Falk and S. Yuzvinsky. Multinets, resonance varieties, and pencils of plane curves. *Compos. Math.*, 143(4):1069–1088, 2007.

[473] M. J. Falk. *Geometry and Topology of Hyperplane Arrangements.* PhD Thesis, The University of Wisconsin, Madison, 1983.

[474] G. Farr. A generalization of the Whitney rank generating function. *Math. Proc. Cambridge Philos. Soc.*, 113(2):267–280, 1993.

[475] G. Farr. Some results on generalised Whitney functions. *Adv. in Appl. Math.*, 32(1-2):239–262, 2004.

[476] G. Farr. On the Ashkin-Teller model and Tutte-Whitney functions. *Combin. Probab. Comput.*, 16(2):251–260, 2007.

[477] G. Farr and J. Oxley. The contributions of W. T. Tutte to matroid theory. In *2017 MATRIX Annals*, volume 2 of *MATRIX Book Ser.*, pages 343–361. Springer, Cham, 2019.

[478] G. E. Farr. A correlation inequality involving stable set and chromatic polynomials. *J. Combin. Theory Ser. B*, 58(1):14–21, 1993.

[479] G. E. Farr. Tutte-Whitney polynomials: some history and generalizations. In *Combinatorics, Complexity, and Chance*, volume 34 of *Oxford Lecture Ser. Math. Appl.*, pages 28–52. Oxford University Press, Oxford, 2007.

[480] G. E. Farr. Minors for alternating dimaps. *Q. J. Math.*, 69(1):285–320, 2018.

[481] E. J. Farrell. On chromatic coefficients. *Discrete Math.*, 29(3):257–264, 1980.

[482] E. J. Farrell. On a class of polynomials associated with the cliques in a graph and its applications. *Internat. J. Math. Math. Sci.*, 12(1):77–84, 1989.

[483] G. Fath-Tabar, Z. Gholam-Rezaei, and A. R. Ashrafi. On the Tutte polynomial of benzenoid chains. *Iran. J. Math. Chem.*, 3(2):113–119, 2012.

[484] T. Feder and M. Mihail. Balanced matroids. In *24th ACM Symposum on the Theory of Computing*, pages 26–38, 1992.

[485] E. M. Feichtner and S. Yuzvinsky. Chow rings of toric varieties defined by atomic lattices. *Invent. Math.*, 155(3):515–536, 2004.

[486] R. Fernández and A. Procacci. Regions without complex zeros for chromatic polynomials on graphs with bounded degree. *Combin. Probab. Comput.*, 17(2):225–238, 2008.

[487] A. Fink, D. E. Speyer, and A. Woo. A Gröebner basis for the graph of the reciprocal plane. *J. Commut. Algebra.*, 12(1):77–86 (Spring 2020).

[488] H. Fleischner. *Eulerian graphs and related topics. Part 1. Volume 1*, volume 45 of *Annals of Discrete Mathematics*. North-Holland Publishing Co., Amsterdam, 1990.

[489] H. Fleischner. *Eulerian Graphs and Related Topics. Part 1. Volume 2*, volume 50 of *Annals of Discrete Mathematics*. North-Holland Publishing Co., Amsterdam, 1991.

[490] J. Flum and M. Grohe. *Parameterized Complexity Theory*. Texts in Theoretical Computer Science. An EATCS Series. Springer-Verlag, Berlin, 2006.

[491] J. Folkman and J. Lawrence. Oriented matroids. *J. Combin. Theory Ser. B*, 25:199–236, 1978.

[492] F. Fomin, P. Golovach, D. Lokshtanov, and S. Saurabh. Algorithmic lower bounds for problems parameterized with clique-width. In M. Charikar, editor, *Proceedings of the Twenty-First Annual ACM-SIAM Symposium on Discrete Algorithms, SODA 2010, Austin, Texas, USA, January 17-19, 2010*, pages 493–502. SIAM, 2010.

[493] F. Fomin, P. Golovach, D. Lokshtanov, and S. Saurabh. Intractability of clique-width parameterizations. *SIAM J. Comput.*, 39(5):1941–1956, 2010.

[494] F. V. Fomin and D. Kratsch. *Exact Exponential Algorithms*. Texts in Theoretical Computer Science. An EATCS Series. Springer, Heidelberg, 2010.

[495] D. Forge and T. Zaslavsky. Lattice points in orthotopes and a huge polynomial Tutte invariant of weighted gain graphs. *J. Combin. Theory Ser. B*, 118:186–227, 2016.

[496] C. M. Fortuin. On the random-cluster model. II. The percolation model. *Physica*, 58:393–418, 1972.

[497] C. M. Fortuin. On the random-cluster model. III. The simple random-cluster model. *Physica*, 59:545–570, 1972.

[498] C. M. Fortuin and P. W. Kasteleyn. On the random-cluster model. I. Introduction and relation to other models. *Physica*, 57:536–564, 1972.

[499] C. M. Fortuin, P. W. Kasteleyn, and J. Ginibre. Correlation inequalities on some partially ordered sets. *Comm. Math. Phys.*, 22:89–103, 1971.

[500] R. Foster. Geometrical circuits of electrical networks. *Trans. Amer. Inst. Elect. Engrs.*, 51:309–317, 1932.

[501] M. Freedman, A. Kitaev, M. Larsen, and Z. Wang. Topological quantum computation. *Bull. Amer. Math. Soc. (N.S.)*, 40(1):31–38, 2002.

[502] M. Freedman, L. Lovász, and A. Schrijver. Reflection positivity, rank connectivity, and homomorphism of graphs. *J. Amer. Math. Soc.*, 20(1):37–51, 2007.

[503] P. Freyd, D. Yetter, J. Hoste, W. B. R. Lickorish, K. Millett, and A. Ocneanu. A new polynomial invariant of knots and links. *Bull. Amer. Math. Soc. (N.S.)*, 12(2):239–246, 1985.

[504] H. Friedman. Algorithmic procedures, generalized Turing algorithms, and elementary recursion theory. In *Logic Colloquium '69 (Proc. Summer School and Colloq., Manchester, 1969)*, pages 361–389, 1971.

[505] H. Friedman and R. Mansfield. Algorithmic procedures. *Trans. Amer. Math. Soc.*, 332(1):297–312, 1992.

[506] A. Frieze and E. Vigoda. A survey on the use of Markov chains to randomly sample colorings. In *Combinatorics, Complexity and Chance.* Oxford University Press, 2007.

[507] A. Gabrielov. Abelian avalanches and Tutte polynomials. *Phys. A*, 195(1-2):253–274, 1993.

[508] A. Gabrielov. Avalanches, sandpiles and Tutte decomposition. In *The Gelfand Mathematical Seminars, 1990–1992*, pages 19–26. Birkhäuser Boston, Boston, MA, 1993.

[509] M. R. Garey and D. S. Johnson. *Computers and Intractability.* W. H. Freeman and Co., San Francisco, Calif., 1979.

[510] D. Garijo, M. E. Gegúndez, A. Márquez, M. P. Revuelta, and F. Sagols. Computing the Tutte polynomial of Archimedean tilings. *Appl. Math. Comput.*, 242:842–855, 2014.

[511] D. Garijo, A. J. Goodall, and J. Nešetřil. Distinguishing graphs by their left and right homomorphism profiles. *European J. Combin.*, 32(7):1025–1053, 2011.

[512] D. Garijo, A. J. Goodall, and J. Nešetřil. Polynomial graph invariants from homomorphism numbers. *Discrete Math.*, 339(4):1315–1328, 2016.

[513] D. Garijo, A. Márquez, and M. P. Revuelta. Tutte uniqueness of locally grid graphs. *Ars Combin.*, 92:377–396, 2009.

[514] D. Garijo, A. Márquez, and M. P. Revuelta. Hexagonal tilings: Tutte uniqueness. *Ars Combin.*, 96:41–63, 2010.

[515] D. D. Gebhard and B. E. Sagan. A chromatic symmetric function in noncommuting variables. *J. Algebraic Combin.*, 13(3):227–255, 2001.

[516] J. Geelen. Some open problems on excluding a uniform matroid. *Adv. in Appl. Math.*, 41(4):628–637, 2008.

[517] J. F. Geelen. A generalization of Tutte's characterization of totally unimodular matrices. *J. Combin. Theory Ser. B*, 70(1):101–117, 1997.

[518] T. W. Geldon. *Computing the Tutte Polynomial of Hyperplane Arrangements.* PhD Thesis, The University of Texas at Austin, 2009.

[519] I. M. Gessel. Generalized rook polynomials and orthogonal polynomials. In D. Stanton, editor, *q-Series and Partitions*, pages 159–176. Springer, 1989.

[520] I. M. Gessel and B. E. Sagan. The Tutte polynomial of a graph, depth-first search, and simplicial complex partitions. *Electron. J. Combin.*, 3(2):R9, 1996.

[521] B. R. Gillespie. *The Generalized External Order, and Applications to Zonotopal Algebra.* PhD Thesis, UC Berkeley, 2018.

[522] O. Giménez. *Polinomi de Tutte Dels Grafs Outerplanars.* Bachelor thesis, Universitat Politècnica de Catalunya, 2002.

[523] O. Giménez, P. Hliněný, and M. Noy. Computing the Tutte polynomial on graphs of bounded clique-width. *SIAM J. Discrete Math.*, 20(4):932–946, 2006.

[524] O. Giménez and M. Noy. On the complexity of computing the Tutte polynomial of bicircular matroids. *Combin. Probab. Comput.*, 15(3):385–395, 2006.

[525] E. Gioan. *Correspondance naturelle entre bases et réorientations des matroïdes orientés.* PhD Thesis, University of Bordeaux 1, 2002.

[526] E. Gioan. Enumerating degree sequences in digraphs and a cycle-cocycle reversing system. *European J. Combin.*, 28(4):1351–1366, 2007.

[527] E. Gioan. Circuit-cocircuit reversing systems in regular matroids. *Ann. Comb.*, 12:171–182, 2008.

[528] E. Gioan. On Tutte polynomial expansion formulas in perspectives of matroids and oriented matroids. *Discrete Math.*, to appear.

[529] E. Gioan and M. Las Vergnas. Bases, reorientations, and linear programming, in uniform and rank-3 oriented matroids. *Adv. in Appl. Math.*, 32(1-2):212–238, 2004.

[530] E. Gioan and M. Las Vergnas. Activity preserving bijections between spanning trees and orientations in graphs. *Discrete Math.*, 298(1-3):169–188, 2005.

[531] E. Gioan and M. Las Vergnas. The active bijection between regions and simplices in supersolvable arrangements of hyperplanes. *Electron. J. Combin.*, 11(2):R30, 2006.

[532] E. Gioan and M. Las Vergnas. Fully optimal bases and the active bijection in graphs, hyperplane arrangements, and oriented matroids. *Electron. Notes Discrete Math.*, 29:365–371, 2007.

[533] E. Gioan and M. Las Vergnas. On the evaluation at (j, j^2) of the Tutte polynomial of a ternary matroid. *J. Algebraic Combin.*, 25(1):1–6, 2007.

[534] E. Gioan and M. Las Vergnas. The active bijection in graphs, hyperplane arrangements, and oriented matroids. I. The fully optimal basis of a bounded region. *European J. Combin.*, 30(8):1868–1886, 2009.

[535] E. Gioan and M. Las Vergnas. A linear programming construction of fully optimal bases in graphs and hyperplane arrangements. *Electron. Notes Discrete Math.*, 34:307–311, 2009.

[536] E. Gioan and M. Las Vergnas. The active bijection - 2.a - Decomposition of activities for matroid bases, and Tutte polynomial of a matroid in terms of beta invariants of minors. arXiv:1807.06516, 2018.

[537] E. Gioan and M. Las Vergnas. The active bijection - 2.b - Decomposition of activities for oriented matroids, and general definitions of the active bijection. arXiv:1807.06578, 2018.

[538] E. Gioan and M. Las Vergnas. The active bijection for graphs. *Adv. in Appl. Math.*, 104:165–236, 2019.

[539] E. Gioan and C. H. Yuen. On the number of circuit-cocircuit reversal classes of an oriented matroid. *Discrete Math.*, 342(4):1056–1059, 2019.

[540] R. E. Giudici. Some new families of chromatically unique graphs. In *Analysis, Geometry, and Probability (Valparaíso, 1981)*, volume 96 of *Lecture Notes in Pure and Appl. Math.*, pages 147–158. Dekker, New York, 1985.

[541] R. E. Giudici and M. Lopez. Chromatic uniqueness of $\overline{sK_n}$. *Report No.85-03, de mat.y ciencia de la Comp. Simon Bolivar*, 1985.

[542] D. B. Glass. Critical groups of graphs with dihedral actions II. *European J. Combin.*, 61:25–46, 2017.

[543] D. B. Glass and C. Merino. Critical groups of graphs with dihedral actions. *European J. Combin.*, 39:95–112, 2014.

[544] L. A. Goddyn, M. Tarsi, and C.-Q. Zhang. On (k,d)-colorings and fractional nowhere-zero flows. *J. Graph Theory*, 28(3):155–161, 1998.

[545] B. Godlin, E. Katz, and J. A. Makowsky. Graph polynomials: from recursive definitions to subset expansion formulas. *J. Logic Comput.*, 22(2):237–265, 2012.

[546] C. Godsil and G. Royle. *Algebraic Graph Theory*, volume 207 of *Graduate Texts in Mathematics*. Springer-Verlag, New York, 2001.

[547] L. Goldberg and M. Jerrum. Approximating the partition function of the ferromagnetic Potts model. *J. ACM*, 59(5):Article 25, 2012.

[548] L. Goldberg and M. Jerrum. The complexity of computing the sign of the Tutte polynomial. *SIAM J. Comput.*, 43(6):1921–1952, 2014.

[549] L. Goldberg and M. R. Jerrum. Approximating the Tutte polynomial of a binary matroid and other related combinatorial polynomials. *J. Comput. Syst. Sci.*, 79(1):68–78, 2013.

[550] L. A. Goldberg, M. Jalsenius, R. Martin, and M. Paterson. Improved mixing bounds for the anti-ferromagnetic Potts model on Z2. *LMS J. Comput. Math.*, 9:1–20, 2006.

[551] L. A. Goldberg and M. R. Jerrum. Inapproximability of the Tutte polynomial. *Inform. and Comput.*, 206(7):908–929, 2008.

[552] L. A. Goldberg and M. R. Jerrum. Inapproximability of the Tutte polynomial of a planar graph. *Comput. Complexity*, 21(4):605–642, 2012.

[553] J. Goldman and J. Haglund. Generalized rook polynomials. *J. Combin. Theory Ser. A*, 91:509–530, 2000.

[554] J. R. Goldman, J. T. Joichi, and D. E. White. Rook theory I. Rook equivalence of Ferrers boards. *Proc. Amer. Math. Soc.*, 52:485–492, 1975.

[555] J. R. Goldman, J. T. Joichi, and D. E. White. Rook theory III. Rook polynomials and the chromatic structure of graphs. *J. Combin. Theory Ser. B*, 25:135–142, 1978.

[556] O. Goldreich. *Computational Complexity: A Conceptual Perspective.* Cambridge University Press, Cambridge–New York, 2008.

[557] H. Gong and X. Jin. Potts model partition functions on two families of fractal lattices. *Phys. A*, 414:143–153, 2014.

[558] A. J. Goodall. Parity, Eulerian subgraphs and the Tutte polynomial. *J. Combin. Theory Ser. B*, 98(3):599–628, 2008.

[559] A. J. Goodall, A. de Mier, S. D. Noble, and M. Noy. The Tutte polynomial characterizes simple outerplanar graphs. *Combin. Probab. Comput.*, 20(4):609–616, 2011.

[560] A. J. Goodall, T. Krajewski, G. Regts, and L. Vena. A Tutte polynomial for maps. *Combin. Probab. Comput.*, 27(6):913–945, 2018.

[561] A. J. Goodall, J. Nešetřil, and P. Ossona de Mendez. Strongly polynomial sequences as interpretations. *J. Appl. Log.*, 18:129–149, 2016.

[562] A. J. Goodall, G. Regts, and L. Vena. Matroid invariants and counting graph homomorphisms. *Linear Algebra Appl.*, 494:263–273, 2016.

[563] R. Goodman and N. R. Wallach. *Symmetry, Representations, and Invariants*, volume 255 of *Graduate Texts in Mathematics.* Springer, Dordrecht, 2009.

[564] G. Gordon. A Tutte polynomial for partially ordered sets. *J. Combin. Theory Ser. B*, 59(1):132–155, 1993.

[565] G. Gordon. Series-parallel posets and the Tutte polynomial. *Discrete Math.*, 158(1-3):63–75, 1996.

[566] G. Gordon. A β invariant for greedoids and antimatroids. *Electron. J. Combin.*, 4(1):R13, 1997.

[567] G. Gordon. On Brylawski's generalized duality. *Math. Comput. Sci.*, 6(2):135–146, 2012.

[568] G. Gordon. Linear relations for a generalized Tutte polynomial. *Electron. J. Combin.*, 22(1):P1.79, 2015.

[569] G. Gordon, E. McDonnell, D. Orloff, and N. Yung. On the Tutte polynomial of a tree. *Congr. Numer.*, 108:141–151, 1995.

[570] G. Gordon and E. McMahon. A greedoid polynomial which distinguishes rooted arborescences. *Proc. Amer. Math. Soc.*, 107(2):287–298, 1989.

[571] G. Gordon and E. McMahon. Interval partitions and activities for the greedoid Tutte polynomial. *Adv. Appl. Math.*, 18(1):33–49, 1997.

[572] G. Gordon and E. McMahon. A characteristic polynomial for rooted graphs and rooted digraphs. *Discrete Math.*, 232(1-3):19–33, 2001.

[573] G. Gordon and L. Traldi. Generalized activities and the Tutte polynomial. *Discrete Math.*, 85(2):167–176, 1990.

[574] G. Gordon and L. Traldi. Polynomials for directed graphs. *Congr. Numer.*, 94:187–201, 1993.

[575] G. Gordon and L. Traldi. Addendum to "Polynomials for directed graphs". *Congr. Numer.*, 100:5–6, 1994.

[576] M. Goresky and R. MacPherson. *Stratified Morse Theory*, volume 14 of *Ergebnisse der Mathematik und ihrer Grenzgebiete (3) [Results in Mathematics and Related Areas (3)]*. Springer-Verlag, Berlin, 1988.

[577] C. Graves and D. Milan. Reliability polynomials having arbitrarily many inflection points. *Networks*, 64(1):1–5, 2014.

[578] C. Greene. Some partitions associated with a partially ordered set. *J. Combin. Theory Ser. A*, 20(1):69–79, 1976.

[579] C. Greene. Weight enumeration and the geometry of linear codes. *Studies in Appl. Math.*, 55(2):119–128, 1976.

[580] C. Greene and T. Zaslavsky. On the interpretation of Whitney numbers through arrangements of hyperplanes, zonotopes, non-Radon partitions, and orientations of graphs. *Trans. Amer. Math. Soc.*, 280(1):97–126, 1983.

[581] D. L. Greenwell. Reconstructing graphs. *Proc. Amer. Math. Soc.*, 30:431–433, 1971.

[582] G. R. Grimmett. *Percolation*, volume 321 of *Grundlehren der Mathematischen Wissenschaften*. Springer-Verlag, Berlin, second edition, 1999.

[583] G. R. Grimmett. *The Random-Cluster Model*, volume 333 of *Grundlehren der Mathematischen Wissenschaften*. Springer-Verlag, Berlin, 2006.

[584] G. R. Grimmett. Flows and ferromagnets. In *Combinatorics, Complexity, and Chance*, volume 34 of *Oxford Lecture Ser. Math. Appl.*, pages 130–143. Oxford University Press, Oxford, 2007.

[585] G. R. Grimmett. *Probability on Graphs*, volume 1 of *Institute of Mathematical Statistics Textbooks*. Cambridge University Press, Cambridge, 2010.

[586] G. R. Grimmett and S. N. Winkler. Negative association in uniform forests and connected graphs. *Random Struct. Algor.*, 24:444–460, 2004.

[587] J. Gross and T. Tucker. *Topological Graph Theory*. Wiley, 1987.

[588] H. Grötzsch. Zur Theorie der diskreten Gebilde. VII. Ein Dreifarbensatz für dreikreisfreie Netze auf der Kugel. *Wiss. Z. Martin-Luther-Univ. Halle-Wittenberg Math.-Natur. Reihe*, 8:109–120, 1958/59.

[589] B. Grünbaum. Acyclic colorings of planar graphs. *Israel J. Math.*, 14:390–408, 1973.

[590] J. Gruska. *Quantum Computing*, volume 2005 of *Advanced Topics in Computer Science Series*. McGraw-Hill London, 1999.

[591] H. Guo and T. Williams. The complexity of planar Boolean #CSP with complex weights. In *Automata, Languages, and Programming. Part I*, volume 7965 of *Lecture Notes in Comput. Sci.*, pages 516–527. Springer, Heidelberg, 2013.

[592] Z.-Y. Guo and E. G. Whitehead, Jr. Chromaticity of a family of K_4 homeomorphs. *Discrete Math.*, 172(1-3):53–58, 1997.

[593] I. Gutman and F. Harary. Generalizations of the matching polynomial. *Utilitas Math.*, 24:97–106, 1983.

[594] J. Guzmán and C. Klivans. Chip-firing and energy minimization on M-matrices. *J. Combin. Theory Ser. A*, 132:14–31, 2015.

[595] J. Guzmán and C. Klivans. Chip firing on general invertible matrices. *SIAM J. Discrete Math.*, 30(2):1115–1127, 2016.

[596] G. Haggard, D. J. Pearce, and G. Royle. Computing Tutte polynomials. *ACM Trans. Math. Software*, 37(3):Art. 24, 17, 2010.

[597] J. Haglund. Rook theory and hypergeometric series. *Adv. in Appl. Math.*, 17:408–459, 1996.

[598] A. Hall. *The Tutte Polynomial Formula for the Class of Twisted Wheel Graphs*. Bachelor Thesis, University of Misssissippi, 2014.

[599] B. T. Han. Chromaticity of q_k-trees. *Acta Math. Appl. Sinica*, 11(4):457–467, 1988.

[600] F. Harary and E. M. Palmer. *Graphical Enumeration*. Academic Press, New York-London, 1973.

[601] D. Harvey and G. Royle. Chromatic roots at 2 and the beraha number b_{10}. *J. Graph Theory*, 95(3):445–456, 2020.

[602] D. Hausmann and B. Korte. Algorithmic versus axiomatic definitions of matroids. *Math. Programming Stud.*, (14):98–111, 1981.

[603] B. R. Heap. Random matrices and graphs. *Numer. Math.*, 8:114–122, 1966.

[604] P. J. Heawood. Map-colour theorem. *Proc. London Math. Soc.*, 51:161–175, 1949.

[605] T. Helgason. Aspects of the theory of hypermatroids. In *Hypergraph Seminar (Proc. First Working Sem., Ohio State Univ., Columbus, Ohio, 1972; dedicated to Arnold Ross)*, pages 191–213. Lecture Notes in Math., Volume 411, 1974.

[606] P. Hell and J. Nešetřil. *Graphs and Homomorphisms*, volume 28 of *Oxford Lecture Series in Mathematics and Its Applications*. Oxford University Press, Oxford, 2004.

[607] T. Helleseth, T. Kløve, V. I. Levenshtein, and Ø. Ytrehus. Bounds on the minimum support weights. *IEEE Trans. Inform. Theory*, 41(2):432–440, 1995.

[608] T. Helleseth, T. Kløve, and J. Mykkeltveit. The weight distribution of irreducible cyclic codes with block length $n_1((q^l - 1)/N)$. *Discrete Math.*, 18(2):179–211, 1977.

[609] T. Helleseth, T. Kløve, and Ø. Ytrehus. Generalized Hamming weights of linear codes. *IEEE Trans. Inform. Theory*, 38(3):1133–1140, 1992.

[610] L. Helme-Guizon and Y. Rong. A categorification for the chromatic polynomial. *Algebr. Geom. Topol.*, 5:1365–1388, 2005.

[611] S. Hernández and F. Luca. Integer roots chromatic polynomials of non-chordal graphs and the Prouhet-Tarry-Escott problem. *Graphs Combin.*, 21(3):319–323, 2005.

[612] A. P. Heron. Matroid polynomials. In *Combinatorics (Proc. Conf. Combinatorial Math., Math. Inst., Oxford, 1972)*, pages 164–202, 1972.

[613] T. Hibi. What can be said about pure *O*-sequences? *J. Combin. Theory Ser. A*, 50(2):319–322, 1989.

[614] D. A. Higgs. Strong maps of geometries. *J. Combin. Theory*, 5:185–191, 1968.

[615] A. J. W. Hilton. The number of spanning trees of labeled wheels, fans and baskets. In *Combinatorics (Proc. Conf. Combinatorial Math., Math. Inst., Oxford, 1972)*, pages 203–206. Inst. Math. Appl., Southend-on-Sea, 1972.

[616] P. Hliněný. The Tutte polynomial for matroids of bounded branch-width. *Combin. Probab. Comput.*, 15(03):397–409, 2006.

[617] P. Hliněný, S. Oum, D. Seese, and G. Gottlob. Width parameters beyond tree-width and their applications. *Comput. J.*, 51(3):326–362, 2008.

[618] C. Hoffmann. *Computational Complexity of Graph Polynomials.* PhD Thesis, Naturwissenschaftlich-Technische Fakultät der Universität des Saarlandes, Saarbrücken, Germany, 2010.

[619] C. Hoffmann. A most general edge elimination polynomial–thickening of edges. *Fundam. Inform.*, 98.4:373–378, 2010.

[620] S. G. Hoggar. Chromatic polynomials and logarithmic concavity. *J. Combin. Theory Ser. B*, 16:248–254, 1974.

[621] R. Holley. Remarks on the FKG inequalities. *Comm. Math. Phys.*, 36:227–231, 1974.

[622] A. E. Holroyd, L. Levine, K. Mészáros, Y. Peres, J. Propp, and D. B. Wilson. Chip-firing and rotor-routing on directed graphs. In *In and Out of Equilibrium. 2*, volume 60 of *Progr. Probab.*, pages 331–364. Birkhäuser, Basel, 2008.

[623] O. Holtz and A. Ron. Zonotopal algebra. *Adv. Math.*, 227(2):847–894, 2011.

[624] Y. Hong. A note on the coefficients of chromatic polynomials. *J. East China Norm. Univ. Natur. Sci. Ed.*, 4:33–35, 1984.

[625] S. Hopkins and D. Perkinson. Bigraphical arrangements. *Trans. Amer. Math. Soc.*, 368(1):709–725, 2016.

[626] S. Huang and P. Lu. A dichotomy for real weighted Holant problems. *Comput. Complexity*, 25(1):255–304, 2016.

[627] S. Huggett. On tangles and matroids. *J. Knot Theory Ramif.*, 14(7):1–11, 2005.

[628] D. R. Hughes and F. C. Piper. *Projective Planes.* Springer-Verlag, New York, 1973.

[629] J. Huh. Milnor numbers of projective hypersurfaces and the chromatic polynomial of graphs. *J. Amer. Math. Soc.*, 25(3):907–927, 2012.

[630] J. Huh. h-vectors of matroids and logarithmic concavity. *Adv. Math.*, 270:49–59, 2015.

[631] J. Huh and E. Katz. Log-concavity of characteristic polynomials and the Bergman fan of matroids. *Math. Ann.*, 354(3):1103–1116, 2012.

[632] J. Huh and B. Wang. Enumeration of points, lines, planes, etc. *Acta Math.*, 218(2):297–317, 2017.

[633] J. E. Humphreys. *Reflection Groups and Coxeter Groups*, volume 29 of *Cambridge Studies in Advanced Mathematics*. Cambridge University Press, Cambridge, 1990.

[634] A. B. Huseby. Domination theory and the Crapo β-invariant. *Networks*, 19(1):135–149, 1989.

[635] E. Ising. Beitrag zur Theorie des Ferromagnetismus. *Zeitschrift für Physik*, 31:253–258, 1925.

[636] C. Itzykson and J.-B. Zuber. *Quantum Field Theory*. Dover Publications, 1980.

[637] B. Jackson. A zero-free interval for chromatic polynomials of graphs. *Combin. Probab. Comput.*, 2(3):325–336, 1993.

[638] B. Jackson. Zeros of chromatic and flow polynomials of graphs. *J. Geom.*, 76(1-2):95–109, 2003.

[639] B. Jackson. An inequality for Tutte polynomials. *Combinatorica*, 30(1):69–81, 2010.

[640] B. Jackson. Chromatic polynomials. In *Topics in chromatic graph theory*, volume 156 of *Encyclopedia Math. Appl.*, pages 56–72. Cambridge University Press, Cambridge, 2015.

[641] B. Jackson, A. Procacci, and A. D. Sokal. Complex zero-free regions at large $|q|$ for multivariate Tutte polynomials (alias Potts-model partition functions) with general complex edge weights. *J. Combin. Theory Ser. B*, 103(1):21–45, 2013.

[642] B. Jackson and A. D. Sokal. Zero-free regions for multivariate Tutte polynomials (alias Potts-model partition functions) of graphs and matroids. *J. Combin. Theory Ser. B*, 99(6):869–903, 2009.

[643] J. L. Jacobsen and J. Salas. Is the five-flow conjecture almost false? *J. Combin. Theory Ser. B*, 103(4):532–565, 2013.

[644] F. Jaeger. On nowhere-zero flows in multigraphs. In *Proceedings of the Fifth British Combinatorial Conference (Univ. Aberdeen, Aberdeen, 1975)*, pages 373–378. Congressus Numerantium, No. XV. Utilitas Math., Winnipeg, Man., 1976.

[645] F. Jaeger. Flows and generalized coloring theorems in graphs. *J. Combin. Theory Ser. B*, 26(2):205–216, 1979.

[646] F. Jaeger. On some algebraic properties of graphs. In *Progress in Graph Theory (Waterloo, Ont., 1982)*, pages 347–366. Academic Press, Toronto, ON, 1984.

[647] F. Jaeger. Nowhere-zero flow problems. In *Selected Topics in Graph Theory, 3*, pages 71–95. Academic Press, San Diego, CA, 1988.

[648] F. Jaeger. Tutte polynomials and link polynomials. *Proc. Amer. Math. Soc.*, 103(2):647–654, 1988.

[649] F. Jaeger. On Tutte polynomials of matroids representable over GF(q). *European J. Combin.*, 10(3):247–255, 1989.

[650] F. Jaeger. Tutte polynomials and bicycle dimension of ternary matroids. *Proc. Amer. Math. Soc.*, 107(1):17–25, 1989.

[651] F. Jaeger. On transition polynomials of 4-regular graphs. In *Cycles and Rays (Montreal, PQ, 1987)*, volume 301 of *NATO Adv. Sci. Inst. Ser. C Math. Phys. Sci.*, pages 123–150. Kluwer Acad. Publ., Dordrecht, 1990.

[652] F. Jaeger, N. Linial, C. Payan, and M. Tarsi. Group connectivity of graphs—a nonhomogeneous analogue of nowhere-zero flow properties. *J. Combin. Theory Ser. B*, 56(2):165–182, 1992.

[653] F. Jaeger, D. L. Vertigan, and D. J. A. Welsh. On the computational complexity of the Jones and Tutte polynomials. *Math. Proc. Cambridge Philos. Soc.*, 108(1):35–53, 1990.

[654] M. Jambu and H. Terao. Arrangements of hyperplanes and broken circuits. In *Singularities (Iowa City, IA, 1986)*, volume 90 of *Contemp. Math.*, pages 147–162. Amer. Math. Soc., Providence, RI, 1989.

[655] S. Janson, T. Łuczak, and A. Rucinski. *Random Graphs.* Wiley-Interscience Series in Discrete Mathematics and Optimization. Wiley-Interscience, New York, 2000.

[656] E. Jasso-Hernandez and Y. Rong. A categorification for the Tutte polynomial. *Algebr. Geom. Topol.*, 6:2031–2049, 2006.

[657] P. Jensen and B. Korte. Complexity of matroid property algorithms. *SIAM J. Comput.*, 11(1):184–190, 1982.

[658] T. R. Jensen and B. Toft. *Graph Coloring Problems.* Wiley-Interscience Series in Discrete Mathematics and Optimization. John Wiley & Sons, Inc., New York, 1995.

[659] M. Jerrum. *On the Complexity of Evaluating Multivariate Polynomials.* PhD Thesis, University of Edinburgh, 1981.

[660] M. Jerrum. Two remarks concerning balanced matroids. *Combinatorica*, 26(6):733–742, 2006.

[661] M. R. Jerrum. A very simple algorithm for estimating the number of k-colourings of a low degree graph. *Random Struct. Algor.*, 7(2):157–165, 1995.

[662] M. R. Jerrum and A. Sinclair. Polynomial-time approximation algorithms for the Ising model. *SIAM J. Comput.*, 22:1087–1116, 1993.

[663] M. R. Jerrum, L. G. Valiant, and V. V. Vazirani. Random generation of combinatorial structures from a uniform distribution. *Theoret. Comput. Sci.*, 43:169–188, 1986.

[664] X. Jin. Tutte polynomials for two recursive families of graphs. *J. Math. Study*, 39(4):345–353, 2006.

[665] X. Jin and F. Zhang. The HOMFLY and dichromatic polynomials. *Proc. Amer. Math. Soc.*, 140(4):1459–1472, 2012.

[666] D. S. Johnson. A catalog of complexity classes. In *Handbook of theoretical computer science, Volume A*, pages 67–161. Elsevier, Amsterdam, 1990.

[667] D. S. Johnson and M. A. Trick, editors. *Cliques, coloring, and satisfiability*, volume 26 of *DIMACS Series in Discrete Mathematics and Theoretical Computer Science.* American Mathematical Society, Providence, RI, 1996.

[668] V. F. R. Jones. A polynomial invariant for knots via von Neumann algebras. *Bull. Amer. Math. Soc. (N.S.)*, 12(1):103–111, 1985.

[669] P. Jössang and A. Jössang. Monsieur C. S. M. Pouillet, de l'Académie, qui découvrit le point "de Curie" en ... 1832. *Science Tribune*, 1997.

[670] R. Jurrius. Relations between Möbius and coboundary polynomials. *Math. Comput. Sci.*, 6(2):109–120, 2012.

[671] R. Jurrius and R. Pellikaan. Truncation formulas for invariant polynomials of matroids and geometric lattices. *Math. Comput. Sci.*, 6(2):121–133, 2012.

[672] R. Jurrius and R. Pellikaan. Defining the q-analogue of a matroid. *Electron. J. Combin.*, 25(3):P3.2, 2018.

[673] N. Kaplan. MacWilliams identities for m-tuple weight enumerators. *SIAM J. Discrete Math.*, 28(1):428–444, 2014.

[674] D. R. Karger. A randomized fully polynomial time approximation scheme for the all-terminal network reliability problem. *SIAM J. Comput.*, 29(2):492–514, 1999.

[675] D. R. Karger and R. P. Tai. Implementing a fully polynomial time approximation scheme for all terminal network reliability. In *Proceedings of the Eighth Annual ACM-SIAM Symposium on Discrete Algorithms (New Orleans, LA, 1997)*, pages 334–343. ACM, New York, 1997.

[676] R. Karp and M. Luby. Monte Carlo algorithms for the planar multiterminal network reliability problem. *J. Complexity*, 1:45–64, 1985.

[677] R. Karp, M. Luby, and N. Madras. Monte Carlo approximation algorithms for enumeration problems. *J. Algorithms*, 10:429–448, 1989.

[678] P. Kasteleyn. The statistics of dimers on a lattice. *Physica*, 27:1209–1225, 1961.

[679] G. Katona. A theorem of finite sets. In *Theory of Graphs (Proc. Colloq., Tihany, 1966)*, pages 187–207. Academic Press, New York, 1968.

[680] L. Kauffman. A Tutte polynomial for signed graphs. *Discrete Appl. Math.*, 25:105–127, 1989.

[681] L. H. Kauffman. State models and the Jones polynomial. *Topology*, 26(3):395–407, 1987.

[682] L. H. Kauffman. Virtual knot theory. *European J. Combin.*, 20:663–691, 1999.

[683] L. H. Kauffman. Introduction to virtual knot theory. *J. Knot Theory Ramif.*, 21(13):1240007, 37, 2012.

[684] K. Kayibi. A decomposition theorem for the linking polynomial of two matroids. *Discrete Math.*, 308:583–596, 2008.

[685] K. Kayibi and S. Pirzada. On the activities of p-basis of matroid perspectives. *Discrete Math.*, 339:1629–1639, 2016.

[686] K. K. Kayibi. *On Some Extensions of the Tutte Polynomial.* PhD Thesis, University of Oxford, 2002.

[687] P. Kelly. *On Isometric Transformations.* PhD thesis, University of Wisconsin, 1942.

[688] A. K. Kelmans. On graphs with randomly deleted edges. *Acta Math. Acad. Sci. Hungar.*, 37(1-3):77–88, 1981.

[689] M. Khovanov. A categorification of the Jones polynomial. *Duke Math. J.*, 101:359–426, 2000.

[690] P. Kilpatrick. *Tutte's First Colour-Cycle Conjecture.* Master's Thesis, University of Cape Town, 1975.

[691] G. Kirchoff. Über die Auflösung der Gleichungen auf welche man bei der Untersuchung der Linearen Verteilung Galvanischer Ströme Geführt wird. *Poggendorg's Ann. Phys. Chem.*, 72:497–508, 1847. (Translation in *IRE Trans. Circuit Theory* C-5 (1958), 4–7).

[692] D. A. Klain. An Euler relation for valuations on polytopes. *Adv. Math.*, 147(1):1–34, 1999.

[693] M. Klazar, M. Loebl, and I. Moffatt. The Potts model and chromatic functions of graphs. *Ann. Inst. Henri Poincaré D*, 1(1):47–60, 2014.

[694] D. J. Kleitman and K. J. Winston. Forests and score vectors. *Combinatorica*, 1(1):49–54, 1981.

[695] C. J. Klivans. *The Mathematics of Chip-Firing.* Discrete Math. Appl. (Boca Raton). CRC Press, Boca Raton, FL, 2019.

[696] T. Kløve. Support weight distribution of linear codes. *Discrete Math.*, 106/107:311–316, 1992.

[697] D. E. Knuth. The asymptotic number of geometries. *J. Combin. Theory Ser. A*, 16:398–400, 1974.

[698] M. Kochol. An equivalent version of the 3-flow conjecture. *J. Combin. Theory Ser. B*, 83(2):258–261, 2001.

[699] M. Kochol. Polynomials associated with nowhere-zero flows. *J. Combin. Theory Ser. B*, 84(2):260–269, 2002.

[700] M. Kochol. Reduction of the 5-flow conjecture to cyclically 6-edge-connected snarks. *J. Combin. Theory Ser. B*, 90(1):139–145, 2004.

[701] M. Kochol. Restrictions on smallest counterexamples to the 5-flow conjecture. *Combinatorica*, 26(1):83–89, 2006.

[702] M. Kochol. Smallest counterexample to the 5-flow conjecture has girth at least eleven. *J. Combin. Theory Ser. B*, 100(4):381–389, 2010.

[703] A. G. Konheim and B. Weiss. An occupancy discipline and applications. *SIAM J. Appl. Math.*, 14(6):1266–1274, 1966.

[704] W. Kook, V. Reiner, and D. Stanton. A convolution formula for the Tutte polynomial. *J. Combin. Theory Ser. B*, 76(2):297–300, 1999.

[705] M. Korn and I. Pak. Combinatorial evaluations of the Tutte polynomial. Preprint, 2003.

[706] M. Korn and I. Pak. Tilings of rectangles with T-tetrominoes. *Theoret. Comput. Sci.*, 319(1-3):3–27, 2004.

[707] B. Korte and L. Lovász. Mathematical structures underlying greedy algorithms. In *Fundamentals of Computation Theory (Szeged, 1981)*, volume 117 of *Lecture Notes in Comput. Sci.*, pages 205–209. Springer, Berlin-New York, 1981.

[708] B. Korte, L. Lovász, and R. Schrader. *Greedoids*, volume 4 of *Algorithms and Combinatorics*. Springer-Verlag, Berlin, 1991.

[709] D. Kostić and C. H. Yan. Multiparking functions, graph searching, and the Tutte polynomial. *Adv. in Appl. Math.*, 40(1):73–97, 2008.

[710] T. Kotek. On the reconstruction of graph invariants. In *European Conference on Combinatorics, Graph Theory and Applications (EuroComb 2009)*, volume 34 of *Electron. Notes Discrete Math.*, pages 375–379. Elsevier Sci. B. V., Amsterdam, 2009.

[711] T. Kotek, J. Makowsky, and E. Ravve. A computational framework for the study of partition functions and graph polynomials. In *2012 14th International Symposium on Symbolic and Numeric Algorithms for Scientific Computing*, pages 365–368. IEEE, 2012.

[712] T. Kotek and J. A. Makowsky. Connection matrices and the definability of graph parameters. *Log. Methods Comput. Sci.*, 10(4):4:1, 33, 2014.

[713] T. Kotek, J. A. Makowsky, and E. V. Ravve. A computational framework for the study of partition functions and graph polynomials. In *Proceedings of the 12th Asian Logic Conference*, pages 210–230. World Sci. Publ., Hackensack, NJ, 2013.

[714] T. Kotek, J. A. Makowsky, and B. Zilber. On counting generalized colorings. In *Computer Science Logic*, volume 5213 of *Lecture Notes in Comput. Sci.*, pages 339–353. Springer, Berlin, 2008.

[715] A. Kotzig. Eulerian lines in finite 4-valent graphs and their transformations. In *Theory of Graphs (Proc. Colloq., Tihany, 1966)*, pages 219–230. Academic Press, New York, 1968.

[716] D. N. Kozlov. *Combinatorial algebraic topology*, volume 21 of *Algorithms and Computation in Mathematics*. Springer, Berlin, 2008.

[717] T. Krajewski and P. Martinetti. Wilsonian renormalization, differential equations and Hopf algebras. *Contemp. Math.*, 539:187, 2011.

[718] T. Krajewski, I. Moffatt, and A. Tanasa. Hopf algebras and Tutte polynomials. *Adv. in Appl. Math.*, 95:271–330, 2018.

[719] T. Krajewski, V. Rivasseau, and A. Tanasa. Combinatorial Hopf algebraic description of the multiscale renormalization in quantum field theory. *Sém. Lothar. Combin.*, B70c, 2014.

[720] T. Krajewski, V. Rivasseau, A. Tanasa, and Z. Wang. Topological graph polynomials and quantum field theory. I. Heat kernel theories. *J. Noncommut. Geom.*, 4(1):29–82, 2010.

[721] D. Král. Decomposition width of matroids. *Discrete Appl. Math.*, 160(6):913–923, 2012.

[722] H. A. Kramers and G. H. Wannier. Statistics of the two-dimensional ferromagnet. I. *Phys. Rev.*, 60:252–262, 1941.

[723] G. Kreweras. Une famille de polynômes ayant plusieurs propriétés énumeratives. *Period. Math. Hungar.*, 11(4):309–320, 1980.

[724] V. Krushkal. Graphs, links, and duality on surfaces. *Combin. Probab. Comput.*, 20:267–287, 2011.

[725] V. Krushkal and D. Renardy. A polynomial invariant and duality for triangulations. *Electron. J. Combin.*, 21(3):P3.42, 2014.

[726] J. B. Kruskal. The number of simplices in a complex. In *Mathematical Optimization Techniques*, pages 251–278. University of California Press, Berkeley, California, 1963.

[727] J. S. Kuhl. The Tutte polynomial and the generalized Petersen graph. *Australas. J. Combin.*, 40:87–97, 2008.

[728] J. P. S. Kung. *A Source Book in Matroid Theory.* Birkhäuser Boston, Inc., Boston, MA, 1986.

[729] J. P. S. Kung. Strong maps. In N. White, editor, *Theory of Matroids*, volume 26 of *Encyclopedia of Mathematics and Its Applications.* Cambridge University Press, 1986.

[730] J. P. S. Kung. Extremal matroid theory. In *Graph Structure Theory (Seattle, WA, 1991)*, volume 147 of *Contemp. Math.*, pages 21–61. Amer. Math. Soc., Providence, RI, 1993.

[731] J. P. S. Kung. Critical problems. In *Matroid Theory (Seattle, WA, 1995)*, volume 197 of *Contemp. Math.*, pages 1–127. Amer. Math. Soc., Providence, RI, 1996.

[732] J. P. S. Kung. Twelve views of matroid theory. In *Combinatorial & Computational Mathematics (Pohang, 2000)*, pages 56–96. World Sci. Publ., River Edge, NJ, 2001.

[733] J. P. S. Kung. Curious characterizations of projective and affine geometries. *Adv. in Appl. Math.*, 28(3-4):523–543, 2002.

[734] J. P. S. Kung. Preface: Old and new perspectives on the Tutte polynomial. *Ann. Comb.*, 12(2):133–137, 2008.

[735] J. P. S. Kung. Convolution-multiplication identities for Tutte polynomials of graphs and matroids. *J. Combin. Theory Ser. B*, 100(6):617–624, 2010.

[736] J. P. S. Kung. Syzygies on Tutte polynomials of freedom matroids. *Ann. Comb.*, 21(4):605–628, 2017.

[737] J. P. S. Kung and G. F. Royle. Graphs whose flow polynomials have only integral roots. *European J. Combin.*, 32(6):831–840, 2011.

[738] G. Kuperberg. How hard is it to approximate the Jones polynomial? *Theory Comput.*, 11(6):183–219, 2015.

[739] M. Lackenby. A polynomial upper bound on Reidemeister moves. *Ann. of Math.*, 182(2):491–564, 2015.

[740] S. K. Lando and A. K. Zvonkin. *Graphs on Surfaces and Their Applications*, volume 141 of *Encyclopaedia of Mathematical Sciences*. Springer-Verlag, Berlin, 2004.

[741] S. Lang. *Algebra*, volume 211 of *Graduate Texts in Mathematics*. Springer-Verlag, New York, third edition, 2002.

[742] M. Las Vergnas. Extensions normales d'un matroïde, polynôme de Tutte d'un morphisme. *C.R. Acad Sci. Paris, sér. A*, 280(22):1479–1482, 1975.

[743] M. Las Vergnas. Acyclic and totally cyclic orientations of combinatorial geometries. *Discrete Math.*, 20:51–61, 1977/78.

[744] M. Las Vergnas. On Eulerian partitions of graphs. In *Graph Theory and Combinatorics (Proc. Conf., Open Univ., Milton Keynes, 1978)*, volume 34 of *Res. Notes in Math.*, pages 62–75. Pitman, Boston, Mass.-London, 1979.

[745] M. Las Vergnas. Convexity in oriented matroids. *J. Combin. Theory Ser. B*, 29(2):231–243, 1980.

[746] M. Las Vergnas. On the Tutte polynomial of a morphism of matroids. *Ann. Discrete Math.*, 8:7–20, 1980.

[747] M. Las Vergnas. Eulerian circuits of 4-valent graphs imbedded in surfaces. In L. Lovász and V. Sós, editors, *Algebraic Methods in Graph Theory*, pages 451–478. North Holland, 1981.

[748] M. Las Vergnas. Le polynôme de Martin d'un graphe eulérien. In *Combinatorial Mathematics (Marseille-Luminy, 1981)*, volume 75 of *North-Holland Math. Stud.*, pages 397–411. North-Holland, Amsterdam, 1983.

[749] M. Las Vergnas. A correspondence between spanning trees and orientations in graphs. In *Graph Theory and Combinatorics (Cambridge, 1983)*, pages 233–238. Academic Press, London, 1984.

[750] M. Las Vergnas. The Tutte polynomial of a morphism of matroids II. Activities of orientations. In J. Bondy and U. Murty, editors, *Progress in Graph Theory*, pages 367–380. Academic Press, Toronto, Canada, 1984.

[751] M. Las Vergnas. On the evaluation at $(3,3)$ of the Tutte polynomial of a graph. *J. Combin. Theory Ser. B*, 45(3):367–372, 1988.

[752] M. Las Vergnas. The Tutte polynomial of a morphism of matroids. I. Set-pointed matroids and matroid perspectives. *Ann. Inst. Fourier (Grenoble)*, 49(3):973–1015, 1999.

[753] M. Las Vergnas. Active orders for matroid bases. *European J. Combin.*, 22:709–721, 2001.

[754] M. Las Vergnas. The Tutte polynomial of a morphism of matroids IV. Computational complexity. *Port. Math. (N.S.)*, 64(3):303–309, 2007.

[755] M. Las Vergnas. The Tutte polynomial of a morphism of matroids 6. A multi-faceted counting formula for hyperplane regions and acyclic orientations. arXiv:1205.5424, 2012.

[756] M. Las Vergnas. The Tutte polynomial of a morphism of matroids—5. Derivatives as generating functions of Tutte activities. *European J. Combin.*, 34(8):1390–1405, 2013.

[757] B. Lass. Orientations acycliques et le polynôme chromatique. *European J. Combin.*, 22:1101–1123, 2001.

[758] B. Lass. Variations sur le thème $E + \overline{E} = XY$. *Adv. in Appl. Math.*, 29:215–242, 2002.

[759] F. Lazebnik. The maximum number of colorings of graphs of given order and size: a survey. *Discrete Math.*, 342(10):2783–2791, 2019.

[760] S. Lefschetz. *Algebraic Topology.* American Mathematical Society Colloquium Publications, v. 27. American Mathematical Society, New York, 1942.

[761] M. Lenz. The f-vector of a representable-matroid complex is log-concave. *Adv. in Appl. Math.*, 51(5):543–545, 2013.

[762] W. Lenz. Beiträge zum Verständnis der magnetischen Eigenschaften in festen Körpern. *Physikalische Zeitschrift*, 21:613–615, 1920.

[763] L. Levine. Parallel chip-firing on the complete graph: devil's staircase and Poincaré rotation number. *Ergodic Theory Dynam. Systems*, 31(3):891–910, 2011.

[764] L. Levine and J. Propp. What is ... a sandpile? *Notices Amer. Math. Soc.*, 57(8):976–979, 2010.

[765] N. Z. Li and R. Y. Liu. The chromaticity of the complete t-partite graph $K(1, p_2, \cdots, p_t)$. *J. Xinjiang Univ. Natur. Sci.*, 7(3):95–96, 1990.

[766] W. Li. Almost every K_4 homeomorph is chromatically unique. *Ars Combin.*, 23:13–35, 1987.

[767] W. M. Li. Some new results on chromatic uniqueness of K_4 homeomorphs. *Math. Appl. (Wuhan)*, 4(3):43–47, 1991.

[768] W. X. Li and F. Tian. Some problems concerning the chromatic polynomials of graphs. *Acta Math. Sinica*, 21(3):223–230, 1978.

[769] X. Li and X. Wei. The chromatic uniqueness of a family of 5-bridge graphs (Chinese). *J. Qinghai Normal Univ.*, 2:12–17, 2001.

[770] Y. Liao, A. Fang, and Y. Hou. The Tutte polynomial of an infinite family of outerplanar, small-world and self-similar graphs. *Phys. A*, 392(19):4584–4593, 2013.

[771] W. B. R. Lickorish. Polynomials for links. *Bull. London Math. Soc.*, 20(6):558–588, 1988.

[772] W. B. R. Lickorish. *An Introduction to Knot Theory*, volume 175 of *Graduate Texts in Mathematics*. Springer-Verlag, New York, 1997.

[773] E. Lieb. Residual entropy of square ice. *Phys. Rev.*, 162:162–172, 1967.

[774] J. Lieberum. Skein modules of links in cylinders over surfaces. *Internat. J. Math. Math. Sci.*, 32(9):515–554, 2002.

[775] C. H. C. Little. The parity of the number of 1-factors of a graph. *Discrete Math.*, 2:179–181, 1972.

[776] C. I. Liu and Y. Chow. Enumeration of connected spanning subgraphs of a planar graph. *Acta Math. Hungar.*, 41(1-2):27–36, 1983.

[777] R. Liu. A new method to find chromatic polynomial of graph and its applications. *Kexue Tongbao*, 32:1508–1509, 1987.

[778] R. Liu, H. Zhao, and C. Ye. A complete solution to a conjecture on chromatic uniqueness of complete tripartite graphs. *Discrete Math.*, 289(1-3):175–179, 2004.

[779] R. Y. Liu. Chromatic uniqueness of $K_n - E(kP_s \cup rP_t)$. *J. Systems Sci. Math. Sci.*, 12(3):207–214, 1992.

[780] M. Loebl. Chromatic polynomial, q-binomial counting and colored Jones function. *Adv. Math.*, 211(2):546–565, 2007.

[781] M. Loebl. *Discrete Mathematics in Statistical Physics.* Advanced Lectures in Mathematics. Vieweg + Teubner, Wiesbaden, 2010.

[782] M. Loebl, J. Matoušek, and O. Pangrác. Triangles in random graphs. *Discrete Math.*, 289(1-3):181–185, 2004.

[783] M. Loebl and I. Moffatt. The chromatic polynomial of fatgraphs and its categorification. *Adv. Math.*, 217(4):1558–1587, 2008.

[784] M. Loebl and J.-S. Sereni. Isomorphism of weighted trees and Stanley's isomorphism conjecture for caterpillars. *Ann. Inst. Henri Poincaré D*, 6(3):357–384, 2019.

[785] B. Loerinc. Chromatic uniqueness of the generalized θ-graph. *Discrete Math.*, 23(3):313–316, 1978.

[786] D. J. Lorenzini. Arithmetical graphs. *Math. Ann.*, 285(3):481–501, 1989.

[787] D. J. Lorenzini. A finite group attached to the Laplacian of a graph. *Discrete Math.*, 91(3):277–282, 1991.

[788] M. Lotz and J. Makowsky. On the algebraic complexity of some families of coloured Tutte polynomials. *Adv. in Appl. Math.*, 32(1):327–349, 2004.

[789] L. Lovász. Operations with structures. *Acta Math. Acad. Sci. Hungar.*, 18:321–328, 1967.

[790] L. Lovász. On the cancellation law among finite relational structures. *Period. Math. Hungar.*, 1(2):145–156, 1971.

[791] L. Lovász. Kneser's conjecture, chromatic number, and homotopy. *J. Combin. Theory Ser. A*, 25(3):319–324, 1978.

[792] L. Lovász. *Combinatorial Problems and Exercises.* North-Holland Publishing Co., Amsterdam, second edition, 1993.

[793] L. Lovász. The rank of connection matrices and the dimension of graph algebras. *European J. Combin.*, 27(6):962–970, 2006.

[794] L. Lovász. Connection matrices. In *Combinatorics, complexity, and chance*, volume 34 of *Oxford Lecture Ser. Math. Appl.*, pages 179–190. Oxford University Press, Oxford, 2007.

[795] L. Lovász. *Large Networks and Graph Limits*, volume 60 of *American Mathematical Society Colloquium Publications*. American Mathematical Society, Providence, RI, 2012.

[796] L. Lovász and B. Szegedy. Contractors and connectors of graph algebras. *J. Graph Theory*, 60(1):11–30, 2009.

[797] L. M. Lovász, C. Thomassen, Y. Wu, and C.-Q. Zhang. Nowhere-zero 3-flows and modulo k-orientations. *J. Combin. Theory Ser. B*, 103(5):587–598, 2013.

[798] P. H. Lundow and K. Markström. Broken-cycle-free subgraphs and the log-concavity conjecture for chromatic polynomials. *Experiment. Math.*, 15(3):343–353, 2006.

[799] K. Luoto, S. Mykytiuk, and S. van Willigenburg. *An Introduction to Quasisymmetric Schur Functions*. SpringerBriefs in Mathematics. Springer, New York, 2013.

[800] K. W. Luoto. A matroid-friendly basis for the quasisymmetric functions. *J. Combin. Theory Ser. A*, 115(5):777–798, 2008.

[801] K. Luse and Y. Rong. A categorification for the Penrose polynomial. *J. Knot Theory Ramif.*, 20(1):141–157, 2011.

[802] J. Ma and Y.-N. Yeh. Combinatorial interpretations for $T_G(1, -1)$. *J. Graph Theory*, 69(3):341–348, 2012.

[803] S. Mac Lane and G. Birkhoff. *Algebra*. Chelsea Publishing Co., New York, third edition, 1988.

[804] I. G. Macdonald. *Symmetric Functions and Hall Polynomials*. Oxford Mathematical Monographs. The Clarendon Press, Oxford University Press, New York, second edition, 1995.

[805] L. Mach and T. Toufar. Amalgam width of matroids. In *Parameterized and Exact Computation*, volume 8246 of *Lecture Notes in Comput. Sci.*, pages 268–280. Springer, Cham, 2013.

[806] F. MacWilliams and N. Sloane. *The Theory of Error-Correcting Codes. I.* North-Holland Publishing Co., Amsterdam-New York-Oxford, 1977.

[807] J. MacWilliams. A theorem on the distribution of weights in a systematic code. *Bell System Tech. J.*, 42:79–94, 1963.

[808] N. V. R. Mahadev and U. N. Peled. *Threshold Graphs and Related Topics*, volume 56 of *Annals of Discrete Mathematics*. North-Holland Publishing Co., Amsterdam, 1995.

[809] S. Majumdar and D. Dhar. Equivalence between the abelian sandpile model and the q→0 limit of the Potts model. *Phys. A*, 185(1):129–145, 1992.

[810] J. Makowsky and J. Mariño. The parametrized complexity of knot polynomials. *J. Comput. Syst. Sci.*, 67(4):742–756, 2003.

[811] J. A. Makowsky. Coloured Tutte polynomials and Kauffman brackets for graphs of bounded tree width. *Discrete Appl. Math.*, 145(2):276–290, 2005.

[812] J. A. Makowsky and K. Meer. On the complexity of combinatorial and metafinite generating functions of graph properties in the computational model of Blum, Shub and Smale. In *Computer Science Logic (Fischbachau, 2000)*, volume 1862 of *Lecture Notes in Comput. Sci.*, pages 399–410. Springer, Berlin, 2000.

[813] J. A. Makowsky, E. V. Ravve, and N. K. Blanchard. On the location of roots of graph polynomials. *European J. Combin.*, 41:1–19, 2014.

[814] J. A. Makowsky, U. Rotics, I. Averbouch, and B. Godlin. Computing graph polynomials on graphs of bounded clique-width. In *Graph-Theoretic Concepts in Computer Science*, volume 4271 of *Lecture Notes in Comput. Sci.*, pages 191–204. Springer, Berlin, 2006.

[815] V. Manturov. *Knot Theory.* Chapman & Hall/CRC, Boca Raton, FL, 2004.

[816] G. A. Margulis. Probabilistic characteristics of graphs with large connectivity. *Problemy Peredači Informacii*, 10(2):101–108, 1974.

[817] I. L. Markov and Y. Shi. Simulating quantum computation by contracting tensor networks. *SIAM J. Comput.*, 38(3):963–981, 2008.

[818] K. Markström. Closure properties and negatively associated measures violating the van den Berg–Kesten inequality. *Electron. Commun. Probab.*, 15:449–456, 2010.

[819] K. Markström. From the Ising and Potts models to the general graph homomorphism polynomial. In *Graph Polynomials*, Discrete Math. Appl. (Boca Raton), pages 123–138. CRC Press, Boca Raton, FL, 2017.

[820] A. Márquez, A. de Mier, M. Noy, and M. P. Revuelta. Locally grid graphs: classification and Tutte uniqueness. *Discrete Math.*, 266(1-3):327–352, 2003.

[821] J. L. Martin, M. Morin, and J. D. Wagner. On distinguishing trees by their chromatic symmetric functions. *J. Combin. Theory Ser. A*, 115(2):237–253, 2008.

[822] J. L. Martin and V. Reiner. Cyclotomic and simplicial matroids. *Israel J. Math.*, 150:229–240, 2005.

[823] P. Martin. Anneau de Tutte-Grothendieck associé aux dénombrements eulériens dans les graphes 4-réguliers planaires. *Cahiers Centre Études Recherche Opér.*, 15:343–349, 1973.

[824] P. Martin. *Enumérations eulériennes dans le multigraphs et invariants de Tutte-Grothendieck.* PhD Thesis, Grenoble, 1977.

[825] P. Martin. Remarkable valuation of the dichromatic polynomial of planar multigraphs. *J. Combin. Theory Ser. B*, 24(3):318–324, 1978.

[826] J. H. Mason. Matroids: unimodal conjectures and Motzkin's theorem. In *Combinatorics (Proc. Conf. Combinatorial Math., Math. Inst., Oxford, 1972)*, pages 207–220. Inst. Math. Appl., Southend-on-Sea, 1972.

[827] S. B. Maurer. Matrix generalizations of some theorems on trees, cycles and cocycles in graphs. *SIAM J. Appl. Math.*, 30(1):143–148, 1976.

[828] D. Mayhew. Matroid complexity and nonsuccinct descriptions. *SIAM J. Discrete Math.*, 22(2):455–466, 2008.

[829] D. Mayhew, M. Newman, D. Welsh, and G. Whittle. On the asymptotic proportion of connected matroids. *European J. Combin.*, 32:882–890, 2011.

[830] D. Mayhew and G. F. Royle. Matroids with nine elements. *J. Combin. Theory Ser. B*, 98(2):415–431, 2008.

[831] G. Mazzuoccolo and E. Steffen. Nowhere-zero 5-flows on cubic graphs with oddness 4. *J. Graph Theory*, 85(2):363–371, 2017.

[832] L. M. McDonald and I. Moffatt. On the Potts model partition function in an external field. *J. Stat. Phys.*, 146(6):1288–1302, 2012.

[833] B. D. McKay and A. Piperno. Practical graph isomorphism, II. *J. Symbolic Comput.*, 60:94–112, 2014.

[834] E. W. McMahon. On the greedoid polynomial for rooted graphs and rooted digraphs. *J. Graph Theory*, 17(3):433–442, 1993.

[835] K. Meer. Counting problems over the reals. *Theoret. Comput. Sci.*, 242(1-2):41–58, 2000.

[836] W. Menasco and M. Thistlethwaite. The classification of alternating links. *Ann. of Math.*, 138(1):113–171, 1993.

[837] W. W. Menasco and M. B. Thistlethwaite. The Tait flyping conjecture. *Bull. Amer. Math. Soc. (N.S.)*, 25(2):403–412, 1991.

[838] C. Merino. The chip firing game and matroid complexes. In *Discrete Models: Combinatorics, Computation, and Geometry (Paris, 2001)*, Discrete Math. Theor. Comput. Sci. Proc., AA, pages 245–255. Maison Inform. Math. Discrèt. (MIMD), Paris, 2001.

[839] C. Merino. The chip-firing game. *Discrete Math.*, 302(1-3):188–210, 2005.

[840] C. Merino, A. de Mier, and M. Noy. Irreducibility of the Tutte polynomial of a connected matroid. *J. Combin. Theory Ser. B*, 83(2):298–304, 2001.

[841] C. Merino and S. D. Noble. The equivalence of two graph polynomials and a symmetric function. *Combin. Probab. Comput.*, 18(4):601–615, 2009.

[842] C. Merino, S. D. Noble, M. Ramírez-Ibáñez, and R. Villarroel-Flores. On the structure of the h-vector of a paving matroid. *European J. Combin.*, 33(8):1787–1799, 2012.

[843] C. Merino, M. Ramírez-Ibáñez, and G. Rodríguez-Sánchez. The Tutte polynomial of some matroids. *Int. J. Comb.*, 2012:Article ID 430859.

[844] C. Merino and D. J. A. Welsh. Forests, colorings and acyclic orientations of the square lattice. *Ann. Comb.*, 3(2-4):417–429, 1999.

[845] C. Merino López. Chip firing and the Tutte polynomial. *Ann. Comb.*, 1(3):253–259, 1997.

[846] A. de Mier. *Graphs and Matroids Determined by Their Tutte Polynomials.* PhD Thesis, Universitat Politècnica de Catalunya, 2003.

[847] A. de Mier and M. Noy. On graphs determined by their Tutte polynomials. *Graphs Combin.*, 20(1):105–119, 2004.

[848] A. de Mier and M. Noy. On matroids determined by their Tutte polynomials. *Discrete Math.*, 302(1-3):52–76, 2005.

[849] A. de Mier and M. Noy. Tutte uniqueness of line graphs. *Discrete Math.*, 301(1):57–65, 2005.

[850] W. P. Miller. Techniques in matroid reconstruction. *Discrete Math.*, 170(1-3):173–183, 1997.

[851] H. Mine. Reliability of physical systems. *IRE Trans. Circuit Theory*, CT-6:138–151, 1959.

[852] Y. Miyazawa. A link invariant dominating the HOMFLY and the Kauffman polynomials. *J. Knot Theory Ramif.*, 19(11):1507–1533, 2010.

[853] L. Moci. A Tutte polynomial for toric arrangements. *Trans. Amer. Math. Soc.*, 364(2):1067–1088, 2012.

[854] I. Moffatt. Knot invariants and the Bollobás-Riordan polynomial of embedded graphs. *European J. Combin.*, 29:95–107, 2008.

[855] I. Moffatt and S. Huggett. Types of embedded graphs and their Tutte polynomials. *Math. Proc. Cambridge Philos. Soc.*, 169(2):255–297, 2020.

[856] I. Moffatt and B. Smith. Matroidal frameworks for topological Tutte polynomials. *J. Combin. Theory Ser. B*, 133:1–31, 2018.

[857] B. Mohar and C. Thomassen. *Graphs on Surfaces.* The Johns Hopkins University Press, 2001.

[858] M. Monagan. A new edge selection heuristic for computing the Tutte polynomial of an undirected graph. In *24th International Conference on Formal Power Series and Algebraic Combinatorics (FPSAC 2012)*, Discrete Math. Theor. Comput. Sci. Proc., AR, pages 839–850. Assoc. Discrete Math. Theor. Comput. Sci., Nancy, 2012.

[859] B. Monjardet. A use for frequently rediscovering a concept. *Order*, 1(4):415–417, 1985.

[860] E. F. Moore and C. E. Shannon. Reliable circuits using less reliable relays. I, II. *J. Franklin Inst.*, 262:191–208, 281–297, 1956.

[861] K. Morgan and G. Farr. Certificates of factorisation for a class of triangle-free graphs. *Electron. J. Combin.*, 16(1):R75, 2009.

[862] K. Morgan and G. Farr. Certificates of factorisation for chromatic polynomials. *Electron. J. Combin.*, 16(1):Research Paper 74, 2009.

[863] K. Morgan and G. Farr. Non-bipartite chromatic factors. *Discrete Math.*, 312(6):1166–1170, 2012.

[864] H. R. Morton and P. R. Cromwell. Distinguishing mutants by knot polynomials. *J. Knot Theory Ramif.*, 5(2):225–238, 1996.

[865] J. Morton and J. Turner. Computing the Tutte polynomial of lattice path matroids using determinantal circuits. *Theoret. Comput. Sci.*, 598(150–156), 2015.

[866] F. Moskowitz. The analysis of redundancy networks. *AIEE Trans. Commun. Electron.*, 39:627–632, 1958.

[867] E. G. Mphako. Tutte polynomials of perfect matroid designs. *Combin. Probab. Comput.*, 9(4):363–367, 2000.

[868] E. G. Mphako. The component numbers of links from graphs. *Proc. Edinb. Math. Soc.*, 45(3):723–730, 2002.

[869] E. Mphako-Banda. Tutte polynomials of flower graphs. *Bull. Iranian Math. Soc.*, 35(2):179–190, 279, 2009.

[870] K. Murasugi. Jones polynomials and classical conjectures in knot theory. *Topology*, 26(2):187–194, 1987.

[871] K. Murasugi. Jones polynomials and classical conjectures in knot theory. II. *Math. Proc. Cambridge Philos. Soc.*, 102(2):317–318, 1987.

[872] G. Musiker. The critical groups of a family of graphs and elliptic curves over finite fields. *J. Algebraic Combin.*, 30(2):255–276, 2009.

[873] W. Myrvold, K. H. Cheung, L. B. Page, and J. E. Perry. Uniformly most reliable networks do not always exist. *Networks*, 21(4):417–419, 1991.

[874] J. Nederlof. Fast polynomial-space algorithms using inclusion-exclusion: Improving on Steiner tree and related problems. *Algorithmica*, 65:868–884, 2013.

[875] D. L. Neel and M. E. Orrison. The linear complexity of a graph. *Electron. J. Combin.*, 13(1):R9, 2006.

[876] S. Negami. Polynomial invariants of graphs. *Trans. Amer. Math. Soc.*, 299(2):601–622, 1987.

[877] J. Nešetřil and R. Šámal. Tension continuous maps—their structure and applications. *European J. Combin.*, 33(6):1207–1225, 2012.

[878] C. M. Newman and D. L. Stein. *Spin Glasses and Complexity.* Princeton University Press, Princeton, N.J., 2013.

[879] B. L. Ng and F. Dong. The chromatic equivalence class of $K_{1,n,n+2}$. *Discrete Math.*, 338(5):674–687, 2015.

[880] M. A. Nielsen and I. L. Chuang. *Quantum Computation and Quantum Information: 10th Anniversary Edition.* Cambridge University Press, Cambridge, 2010.

[881] A. Nijenhuis and H. S. Wilf. *Combinatorial Algorithms.* Academic Press, Inc., New York-London, 1978.

[882] S. D. Noble. Evaluating the Tutte polynomial for graphs of bounded tree-width. *Combin. Probab. Comput.*, 7(3):307–321, 1998.

[883] S. D. Noble. Evaluating a weighted graph polynomial for graphs of bounded tree-width. *Electron. J. Combin.*, 16(1):R64, 2009.

[884] S. D. Noble and G. F. Royle. The Merino-Welsh conjecture holds for series-parallel graphs. *European J. Combin.*, 38:24–35, 2014.

[885] S. D. Noble and D. J. A. Welsh. A weighted graph polynomial from chromatic invariants of knots. *Ann. Inst. Fourier (Grenoble)*, 49(3):1057–1087, 1999.

[886] M. Noy. Graphs determined by polynomial invariants. *Theoret. Comput. Sci.*, 307(2):365–384, 2003.

[887] S. Oh. Generalized permutohedra, h-vectors of cotransversal matroids and pure O-sequences. *Electron. J. Combin.*, 20(3):P14, 2013.

[888] S. Ok and T. J. Perrett. Density of real zeros of the Tutte polynomial. *Combin. Probab. Comput.*, 27(3):398–410, 2018.

[889] R. Orellana and G. Scott. Graphs with equal chromatic symmetric functions. *Discrete Math.*, 320:1–14, 2014.

[890] P. Orlik and L. Solomon. Combinatorics and topology of complements of hyperplanes. *Invent. Math.*, 56(2):167–189, 1980.

[891] P. Orlik and H. Terao. *Arrangements of Hyperplanes*, volume 300 of *Grundlehren der Mathematischen Wissenschaften [Fundamental Principles of Mathematical Sciences]*. Springer-Verlag, Berlin, 1992.

[892] S. Oum and P. Seymour. Approximating clique-width and branch-width. *J. Combin. Theory Ser. B*, 96(4):514–528, 2006.

[893] J. Oxley. Graphs and series-parallel networks. In *Theory of Matroids*, volume 26 of *Encyclopedia Math. Appl.*, pages 97–126. Cambridge University Press, Cambridge, 1986.

[894] J. Oxley. *Matroid theory*, volume 21 of *Oxford Graduate Texts in Mathematics*. Oxford University Press, Oxford, second edition, 2011.

[895] J. Oxley, K. Prendergast, and D. Row. Matroids whose ground sets are domains of functions. *J. Austral. Math. Soc. Ser. A*, 32(3):380–387, 1982.

[896] J. Oxley, D. Vertigan, and G. Whittle. On inequivalent representations of matroids over finite fields. *J. Combin. Theory Ser. B*, 67(2):325–343, 1996.

[897] J. Oxley and D. Welsh. Chromatic, flow and reliability polynomials: the complexity of their coefficients. *Combin. Probab. Comput.*, 11(4):403–426, 2002.

[898] J. Oxley and G. Whittle. A characterization of Tutte invariants of 2-polymatroids. *J. Comb. Theory, Ser. B*, 59(2):210–244, 1993.

[899] J. Oxley and G. Whittle. Tutte invariants for 2-polymatroids. In *Graph Structure Theory (Seattle, WA, 1991)*, volume 147 of *Contemp. Math.*, pages 9–19. Amer. Math. Soc., Providence, RI, 1993.

[900] J. Oxley and G. Whittle. On the non-uniqueness of q-cones of matroids. *Discrete Math.*, 218(1-3):271–275, 2000.

[901] J. Oxley. On Crapo's beta invariant for matroids. *Stud. Appl. Math.*, 66(3):267–277, 1982.

[902] J. Oxley. On minor-minimally-connected matroids. *Discrete Math.*, 51(1):63–72, 1984.

[903] J. Oxley. A note on Negami's polynomial invariants for graphs. *Discrete Math.*, 76(3):279–281, 1989.

[904] J. Oxley and D. J. A. Welsh. The Tutte polynomial and percolation. In *Graph Theory and Related Topics (Proc. Conf., Univ. Waterloo, Waterloo, Ont., 1977)*, pages 329–339. Academic Press, New York-London, 1979.

[905] J. Oxley and D. J. A. Welsh. Tutte polynomials computable in polynomial time. *Discrete Math.*, 109(1-3):185–192, 1992.

[906] C. H. Papadimitriou. *Computational Complexity.* Addison-Wesley, 1994.

[907] L. Pauling. The structure and entropy of ice and of other crystals with some randomness of atomic arrangement. *J. Am. Chem. Soc.*, 57:2680–2684, 1935.

[908] L. Pauling. *The Nature of the Chemical Bond.* Cornell University Press, Ithaca, New York, 1960.

[909] D. J. Pearce, G. Haggard, and G. Royle. Edge-selection heuristics for computing Tutte Polynomials. In *Proceedings of the Computing: The Australasian Theory Symposium (CATS)*, pages 153–162, 2009.

[910] R. Peierls. On Ising's model of ferromagnetism. *Proc. Cambridge Philos. Soc.*, 36:477–481, 1936.

[911] A. Pellionisz and R. Llinas. Tensor network theory of the metaorganization of functional geometries in the central nervous system. *Neuroscience*, 16(2):245–273, 1985.

[912] R. Pendavingh. On the evaluation at $(-i, i)$ of the Tutte polynomial of a binary matroid. *J. Algebraic Combin.*, 39(1):141–152, 2014.

[913] R. A. Pendavingh and S. H. M. van Zwam. Skew partial fields, multilinear representations of matroids, and a matrix tree theorem. *Adv. in Appl. Math.*, 50(1):201–227, 2013.

[914] Y. Peng. On the chromatic coefficients of a graph and chromatic uniqueness of certain n-partition graphs. In *Combinatorics, Graph Theory, Algorithms and Applications (Beijing, 1993)*, pages 307–316, River Edge, NJ, 1993. World Scientific.

[915] Y.-H. Peng. On the chromatic coefficients of a bipartite graph. *Ars Combin.*, 34:107–117, 1992.

[916] R. Penrose. Applications of negative dimensional tensors. In *Combinatorial Mathematics and Its Applications (Proc. Conf., Oxford, 1969)*, pages 221–244. Academic Press, London, 1971.

[917] K. Perrot and T. V. Pham. Chip-firing game and a partial Tutte polynomial for Eulerian digraphs. *Electron. J. Combin.*, 23(1):P1.57, 2016.

[918] M. E. Peskin and D. V. Schroeder. *An Introduction to Quantum Field Theory.* Addison-Wesley Publishing Company, Advanced Book Program, Reading, MA, 1995.

[919] J. Petersen. Sur le théorème de Tait. *L'Intermédiaire de Mathématiciens*, 5:225–227, 1898.

[920] M. J. Piff and D. J. A. Welsh. On the vector representation of matroids. *J. London Math. Soc.*, 2:284–288, 1970.

[921] R. Pinchasi, R. Radoičić, and M. Sharir. On empty convex polygons in a planar point set. *J. Combin. Theory Ser. A*, 113(3):385–419, 2006.

[922] W. Plesken and T. Bächler. Counting polynomials for linear codes, hyperplane arrangements, and matroids. *Doc. Math.*, 19:285–312, 2014.

[923] J. Polchinski. Renormalization and Effective Lagrangians. *Nucl. Phys. B*, 231:269–295, 1984.

[924] A. Postnikov and B. Shapiro. Trees, parking functions, syzygies, and deformations of monomial ideals. *Trans. Amer. Math. Soc.*, 356(8):3109–3142, 2004.

[925] A. Postnikov, B. Shapiro, and M. Shapiro. Algebras of curvature forms on homogeneous manifolds. In *Differential Topology, Infinite-Dimensional Lie Algebras, and Applications*, volume 194 of *Amer. Math. Soc. Transl. Ser. 2*, pages 227–235. Amer. Math. Soc., Providence, RI, 1999.

[926] A. Postnikov and R. P. Stanley. Deformations of Coxeter hyperplane arrangements. *J. Combin. Theory Ser. A*, 91(1-2):544–597, 2000.

[927] R. B. Potts. *The Mathematical Investigation of Some Cooperative Phenomena.* D.Phil. Dissertation, Univeristy of Oxford, 1951.

[928] R. B. Potts. Some generalized order-disorder transformations. *Proc. Cambridge Philos. Soc.*, 48:106–109, 1952.

[929] M. Pouillet. *Eléments de Physique Expérimentale et de Météorologie.* Béchet Jeune, Paris, 1832.

[930] W. Price and W. Tutte. The absorption spectra of ethylene, deutero-ethylene and some alkyl-substituted ethylenes in the vacuum ultra-violet. *Proc. Roy. Soc. Ser. A*, 174:207–220, 1940.

[931] N. Proudfoot and D. Speyer. A broken circuit ring. *Beiträge Algebra Geom.*, 47(1):161–166, 2006.

[932] J. S. Provan. *Decompositions, Shellings, and Diameters of Simplicial Complexes and Convex Polyhedra.* PhD Thesis, Cornell University, 1977.

[933] J. S. Provan. The complexity of reliability computations in planar and acyclic graphs. *SIAM J. Comput.*, 15(3):694–702, 1986.

[934] J. S. Provan and M. O. Ball. The complexity of counting cuts and of computing the probability that a graph is connected. *SIAM J. Comput.*, 12(4):777–788, 1983.

[935] J. S. Provan and L. J. Billera. Decompositions of simplicial complexes related to diameters of convex polyhedra. *Math. Oper. Res.*, 5(4):576–594, 1980.

[936] J. H. Przytycki. Search for different links with the same Jones' type polynomials: ideas from graph theory and statistical mechanics. In *Panoramas of Mathematics (Warsaw, 1992/1994)*, volume 34 of *Banach Center Publ.*, pages 121–148. Polish Acad. Sci. Inst. Math., Warsaw, 1995.

[937] J. H. Przytycki. Graphs and links. arXiv:math/0601227, 2006.

[938] J. H. Przytycki. History of knot theory. arXiv:math/0703096, 2007.

[939] J. H. Przytycki and P. Traczyk. Invariants of links of Conway type. *Kobe J. Math.*, 4(2):115–139, 1988.

[940] H. Qin. Connected matroids with symmetric Tutte polynomials. *Combin. Probab. Comput.*, 10(2):179–186, 2001.

[941] H. Qin. Complete principal truncations of Dowling lattices. *Adv. in Appl. Math.*, 32(1-2):364–379, 2004.

[942] D. E. Radford. *Hopf Algebras*, volume 49 of *Series on Knots and Everything.* World Scientific Publishing Co. Pte. Ltd., Hackensack, NJ, 2012.

[943] A. Ramanathan and C. J. Colbourn. Counting almost minimum cutsets with reliability applications. *Math. Programming*, 39(3):253–261, 1987.

[944] R. Read. MR0354428 (50 #6906). Mathematical Reviews reviewer's remarks, 1975.

[945] R. C. Read. An introduction to chromatic polynomials. *J. Combin. Theory*, 4:52–71, 1968.

[946] R. C. Read. A large family of chromatic polynomials. In *Proceedings of the Third Caribbean Conference on Combinatorics and Computing (Bridgetown, 1981)*, pages 23–41. Univ. West Indies, Cave Hill Campus, Barbados, 1981.

[947] R. C. Read. Connectivity and chromatic uniqueness. *Ars Combin.*, 23:209–218, 1987.

[948] R. C. Read. A note on the chromatic uniqueness of W_{10}. *Discrete Math.*, 69(3):317, 1988.

[949] R. C. Read. Chain polynomials of graphs. *Discrete Math.*, 265(1-3):213–235, 2003.

[950] R. C. Read and W. T. Tutte. Chromatic polynomials. In *Selected Topics in Graph Theory, 3*, pages 15–42. Academic Press, San Diego, CA, 1988.

[951] R. C. Read and E. G. Whitehead, Jr. Chromatic polynomials of homeomorphism classes of graphs. *Discrete Math.*, 204(1-3):337–356, 1999.

[952] R. C. Read and E. G. Whitehead, Jr. A note on chain lengths and the Tutte polynomial. *Discrete Math.*, 308(10):1826–1829, 2008.

[953] F. Redig. Mathematical aspects of the abelian sandpile model. In *Mathematical Statistical Physics*, pages 657–729. Elsevier B. V., Amsterdam, 2006.

[954] G. Regts. *Graph Parameters and Invariants of the Orthogonal Group.* PhD Thesis, Universiteit van Amsterdam, 2013.

[955] G. Regts. Edge-reflection positivity and weighted graph homomorphisms. *J. Combin. Theory Ser. A*, 129:80–92, 2015.

[956] G. Regts and B. Sevenster. Graph parameters from symplectic group invariants. *J. Combin. Theory Ser. B*, 122:844–868, 2017.

[957] K. Reidemeister. Elementare Begründung der Knotentheorie. *Abh. Math. Sem. Univ. Hamburg*, 5(1):24–32, 1927.

[958] D. Reimer. Proof of the van den Berg–Kesten conjecture. *Combin. Probab. Comput.*, 9:27–32, 2000.

[959] V. Reiner. An interpretation for the Tutte polynomial. *European J. Combin.*, 20(2):149–161, 1999.

[960] H. Ren and S. Zhang. Chromatic uniqueness of a family of k_4-homeomorphs. *J. Qinghai Normal Univ. (Natural Sc.)*, 2:9–11, 2001.

[961] J. Richter-Gebert. Oriented matroids with few mutations. *Discrete Comput. Geom.*, 10:251–269, 1993.

[962] J. Riordan. Ballots and trees. *J. Combin. Theory*, 6:408–411, 1969.

[963] V. Rivasseau. *From Perturbative to Constructive Renormalization.* Princeton University Press, 1992.

[964] N. Robertson and P. Seymour. Graph minors. X. obstructions to tree-decomposition. *J. Combin. Theory Ser. B*, 52(2):153–190, 1991.

[965] N. Robertson and P. Seymour. Graph minors. XX. Wagner's conjecture. *J. Combin. Theory Ser. B*, 92:325–357, 2004.

[966] N. Robertson, P. Seymour, and R. Thomas. Tutte's edge-colouring conjecture. *J. Combin. Theory Ser. B*, 70(1):166–183, 1997.

[967] G. Robinson and D. Welsh. The computational complexity of matroid properties. *Math. Proc. Cambridge Philos. Soc.*, 87(1):29–45, 1980.

[968] A. Rosengren and B. Lindström. A combinatorial series expansion for the Ising model. *European J. Combin.*, 8(3):317–323, 1987.

[969] P. Rosenstiehl and R. C. Read. On the principal edge tripartition of a graph. *Ann. Discrete Math.*, 3:195–226, 1978.

[970] G.-C. Rota. On the foundations of combinatorial theory. I. Theory of Möbius functions. *Z. Wahrscheinlichkeitstheorie und Verw. Gebiete*, 2:340–368, 1964.

[971] G.-C. Rota. Combinatorial theory, old and new. In *Actes du Congrès International des Mathématiciens (Nice, 1970), Tome 3*, pages 229–233. Gauthier-Villars, Paris, 1971.

[972] G. Royle. Planar triangulations with real chromatic roots arbitrarily close to 4. *Ann. Comb.*, 12(2):195–210, 2008.

[973] G. Royle and A. D. Sokal. The Brown-Colbourn conjecture on zeros of reliability polynomials is false. *J. Combin. Theory Ser. B*, 91(2):345–360, 2004.

[974] G. F. Royle. Graphs with chromatic roots in the interval $(1, 2)$. *Electron. J. Combin.*, 14(1):N18, 2007.

[975] J. Rudd. Tutte polynomials for counting and classifying orbits. *Discrete Math.*, 310(2):206–222, 2010.

[976] D. Ruelle. *Thermodynamic Formalism.* Cambridge University Press, Cambridge, second edition, 2004.

[977] J. K. S. Beraha and R. Reid. b_7 and b_{10} are limit points of chromatic zeros. *Notices Amer. Math. Soc.*, 20:45, 1973.

[978] B. E. Sagan. Why the characteristic polynomial factors. *Bull. Amer. Math. Soc. (N.S.)*, 36(2):113–133, 1999.

[979] J. Salas and A. D. Sokal. Absence of phase transition for antiferromagnetic Potts models via the dobrushin uniqueness theorem. *J. Stat. Phys.*, 86(3-4):551–579, 1997.

[980] J. Salas and A. D. Sokal. Transfer matrices and partition-function zeros for antiferromagnetic Potts models. I. General theory and square-lattice chromatic polynomial. *J. Stat. Phys.*, 104(3-4):609–699, 2001.

[981] J. Salas and A. D. Sokal. Transfer matrices and partition-function zeros for antiferromagnetic Potts models VI. Square lattice with extra-vertex boundary conditions. *J. Stat. Phys.*, 144(5):1028–1122, 2011.

[982] P. M. Salzberg, M. A. López, and R. E. Giudici. On the chromatic uniqueness of bipartite graphs. *Discrete Math.*, 58(3):285–294, 1986.

[983] I. Sarmiento. A characterisation of jointless Dowling geometries. *Discrete Math.*, 197/198:713–731, 1999.

[984] I. Sarmiento. The polychromate and a chord diagram polynomial. *Ann. Comb.*, 4(2):227–236, 2000.

[985] A. Satyanarayana and Z. Khalil. On an invariant of graphs and the reliability polynomial. *SIAM J. Algebraic Discrete Methods*, 7(3):399–403, 1986.

[986] V. V. Schechtman and A. N. Varchenko. Arrangements of hyperplanes and Lie algebra homology. *Invent. Math.*, 106(1):139–194, 1991.

[987] E. R. Scheinerman and D. H. Ullman. *Fractional Graph Theory.* Wiley-Interscience Series in Discrete Mathematics and Optimization. John Wiley & Sons, Inc., New York, 1997.

[988] W. R. Schmitt. Incidence Hopf algebras. *J. Pure Appl. Algebra*, 96(3):299–330, 1994.

[989] A. Schrijver. Tait's flyping conjecture for well-connected links. *J. Combin. Theory Ser. B*, 58(1):65–146, 1993.

[990] A. Schrijver. Characterizing partition functions of the spin model by rank growth. *Indag. Math. (N.S.)*, 24(4):1018–1023, 2013.

[991] W. Schwärzler. Being Hamiltonian is not a Tutte invariant. *Discrete Math.*, 91(1):87–89, 1991.

[992] W. Schwärzler and D. J. A. Welsh. Knots, matroids and the Ising model. *Math. Proc. Cambridge Philos. Soc.*, 113(1):107–139, 1993.

[993] J. Schweig. On the h-vector of a lattice path matroid. *Electron. J. Combin.*, 17(1):N3, 2010.

[994] A. D. Scott and A. D. Sokal. Some variants of the exponential formula, with application to the multivariate Tutte polynomial (alias Potts model). *Sém. Lothar. Combin.*, 61A:Art. B61Ae, 33, 2009/11.

[995] H. Seifert and W. Threlfall. *Seifert and Threlfall: A Textbook of Topology*, volume 89 of *Pure and Applied Mathematics*. Academic Press, Inc. [Harcourt Brace Jovanovich, Publishers], New York-London, 1980.

[996] K. Sekine, H. Imai, and S. Tani. Computing the Tutte polynomial of a graph of moderate size. In *Algorithms and Computations (Cairns, 1995)*, volume 1004 of *Lecture Notes in Comput. Sci.*, pages 224–233. Springer, Berlin, 1995.

[997] K. Sekine and C. Q. Zhang. Decomposition of the flow polynomial. *Graphs Combin.*, 13(2):189–196, 1997.

[998] P. Seymour. Two chromatic polynomial conjectures. *J. Combin. Theory Ser. B*, 70(1):184–196, 1997.

[999] P. D. Seymour. Nowhere-zero 6-flows. *J. Combin. Theory Ser. B*, 30(2):130–135, 1981.

[1000] J. Shareshian and M. L. Wachs. Chromatic quasisymmetric functions. *Adv. Math.*, 295:497–551, 2016.

[1001] D. R. Shier. *Network Reliability and Algebraic Structures*. Oxford Science Publications. The Clarendon Press, Oxford University Press, New York, 1991.

[1002] K. Shiromoto. A new MacWilliams type identity for linear codes. *Hokkaido Math. J.*, 25(3):651–656, 1996.

[1003] K. Shiromoto. The weight enumerator of linear codes over GF(q^m) having generator matrix over GF(q). *Des. Codes Cryptogr.*, 16(1):87–92, 1999.

[1004] K. Shiromoto. Codes with the rank metric and matroids. *Des. Codes Cryptogr.*, 87(8):1765–1776, 2019.

[1005] K. Shoda. *Large Families of Matroids with the Same Tutte Polynomial.* PhD Thesis, The George Washington University, 2012.

[1006] R. Shrock. Exact Potts model partition functions on ladder graphs. *Phys. A*, 283(3-4):388–446, 2000.

[1007] R. Shrock. Exact Potts/Tutte polynomials for polygon chain graphs. *J. Phys. A*, 44(14):145002, 14, 2011.

[1008] J. M. S. Simoes-Pereira. On subgraphs as matroid cells. *Math. Z.*, 127:315–322, 1972.

[1009] B. Simon. Correlation inequalities and the decay of correlations in ferromagnets. *Comm. Math. Phys.*, 77:111–126, 1980.

[1010] J. Simonis. MacWilliams identities and coordinate partitions. *Linear Algebra Appl.*, 216:81–91, 1995.

[1011] J. Simonis and A. Ashikhmin. Almost affine codes. *Des. Codes Cryptogr.*, 14(2):179–197, 1998.

[1012] A. Sinclair. *Algorithms for Random Generation and Counting: A Markov Chain Approach.* Birkhauser, 1992.

[1013] M. Sipser. *Introduction to the Theory of Computation.* PWS Publishing Company, Boston, 1997.

[1014] A. N. Skorobogatov. Linear codes, strata of Grassmannians, and the problems of Segre. In *Coding Theory and Algebraic Geometry (Luminy, 1991)*, volume 1518 of *Lecture Notes in Math.*, pages 210–223. Springer, Berlin, 1992.

[1015] S. Smirnov. Conformal invariance in random cluster models. I. Holomorphic fermions in the Ising model. *Ann. of Math.*, 172(2):1435–1467, 2010.

[1016] C. A. B. Smith. Map colourings and linear mappings. In *Combinatorial Mathematics and Its Applications (Proc. Conf., Oxford, 1969)*, pages 259–283. Academic Press, London, 1971.

[1017] C. A. B. Smith. On Tutte's dichromate polynomial. *Ann. Discrete Math.*, 3:247–257, 1978.

[1018] I. Smith, Z. Smith, and P. Tian. Symmetric chromatic polynomial of trees. arXiv:1505.01889, 2015.

[1019] S. Snook. *Matroids, Complexity and Computation.* PhD Thesis, Victoria University of Wellington, 2013.

[1020] A. D. Sokal. Chromatic polynomials, Potts models and all that. *Phys. A*, 279(324–332), 2000.

[1021] A. D. Sokal. Bounds on the complex zeros of (di)chromatic polynomials and Potts-model partition functions. *Combin. Probab. Comput.*, 10(1):41–77, 2001.

[1022] A. D. Sokal. Chromatic roots are dense in the whole complex plane. *Combin. Probab. Comput.*, 13(2):221–261, 2004.

[1023] A. D. Sokal. The multivariate Tutte polynomial (alias Potts model) for graphs and matroids. In *Surveys in Combinatorics 2005*, volume 327 of *London Math. Soc. Lecture Note Ser.*, pages 173–226. Cambridge University Press, Cambridge, 2005.

[1024] J. Spencer. Balancing games. *J. Combin. Theory Ser. B*, 23(1):68–74, 1977.

[1025] E. Sperner. Ein Satzüber Untermengen einer endlichen Menge. *Math. Z.*, 27(1):544–548, 1928.

[1026] R. P. Stanley. Supersolvable lattices. *Algebra Universalis*, 2:197–217, 1972.

[1027] R. P. Stanley. Acyclic orientations of graphs. *Discrete Math.*, 5:171–178, 1973.

[1028] R. P. Stanley. Cohen-Macaulay complexes. In *Higher Combinatorics (Proc. NATO Advanced Study Inst., Berlin, 1976)*, pages 51–62. NATO Adv. Study Inst. Ser., Ser. C: Math. and Phys. Sci., 31. Reidel, Dordrecht, 1977.

[1029] R. P. Stanley. Hilbert functions of graded algebras. *Adv. Math.*, 28(1):57–83, 1978.

[1030] R. P. Stanley. Decompositions of rational convex polytopes. *Ann. Discrete Math.*, 6:333–342, 1980.

[1031] R. P. Stanley. Log-concave and unimodal sequences in algebra, combinatorics, and geometry. In *Graph Theory and Its Applications: East and West (Jinan, 1986)*, volume 576 of *Ann. New York Acad. Sci.*, pages 500–535. New York Acad. Sci., New York, 1989.

[1032] R. P. Stanley. A zonotope associated with graphical degree sequences. In *Applied Geometry and Discrete Mathematics*, volume 4 of *DIMACS Ser. Discrete Math. Theoret. Comput. Sci.*, pages 555–570. Amer. Math. Soc., Providence, RI, 1991.

[1033] R. P. Stanley. A symmetric function generalization of the chromatic polynomial of a graph. *Adv. Math.*, 111(1):166–194, 1995.

[1034] R. P. Stanley. *Combinatorics and Commutative Algebra*, volume 41 of *Progress in Mathematics*. Birkhäuser Boston, Inc., Boston, MA, second edition, 1996.

[1035] R. P. Stanley. Parking functions and noncrossing partitions. *Electron. J. Combin.*, 4(2):P20, 1997.

[1036] R. P. Stanley. Graph colorings and related symmetric functions: ideas and applications: a description of results, interesting applications, & notable open problems. *Discrete Math.*, 193(1-3):267–286, 1998.

[1037] R. P. Stanley. Hyperplane arrangements, parking functions and tree inversions. In *Mathematical Essays in Honor of Gian-Carlo Rota (Cambridge, MA, 1996)*, volume 161 of *Progr. Math.*, pages 359–375. Birkhäuser Boston, Boston, MA, 1998.

[1038] R. P. Stanley. *Enumerative Combinatorics. Volume 2*, volume 62 of *Cambridge Studies in Advanced Mathematics*. Cambridge University Press, Cambridge, 1999.

[1039] R. P. Stanley. An introduction to hyperplane arrangements. In *Geometric Combinatorics*, volume 13 of *IAS/Park City Math. Ser.*, pages 389–496. Amer. Math. Soc., Providence, RI, 2007.

[1040] R. P. Stanley. *Enumerative Combinatorics. Volume 1*, volume 49 of *Cambridge Studies in Advanced Mathematics*. Cambridge University Press, Cambridge, second edition, 2012.

[1041] R. P. Stanley. *Catalan Numbers*. Cambridge University Press, Cambridge, 2015.

[1042] R. P. Stanley and J. R. Stembridge. On immanants of Jacobi–Trudi matrices and permutations with restricted position. *J. Combin. Theory Ser. A*, 62:261–279, 1993.

[1043] D. Stark. The edge correlation of random forests. *Ann. Comb.*, 15:529–539, 2011.

[1044] E. Steffen. Tutte's 5-flow conjecture for highly cyclically connected cubic graphs. *Discrete Math.*, 310(3):385–389, 2010.

[1045] P. Stein. A brief history of enumeration. In G.-C. Rota, editor, *Science and Computers: A Volume Dedicated to Nicholas Metropolis*, pages 169–206. Academic Press, Orlando, 1986.

[1046] L. Stockmeyer. The polynomial-time hierarchy. *Theoret. Comput. Sci.*, 3(1):1–22, 1976.

[1047] M. Stošić. New categorifications of the chromatic and dichromatic polynomials for graphs. *Fund. Math.*, 190:231–243, 2006.

[1048] B. Sturmfels and X. Zhiqiang. Sagbi bases of Cox-Nagata rings. *J. Eur. Math. Soc.*, 12:429–459, 2010.

[1049] G. B. B. M. Sutherland and W. T. Tutte. Absorption of polymolecular films in the infra-red. *Nature*, 144(3651):707–707, 1939.

[1050] M. E. Sweedler. *Hopf Algebras*. Mathematics Lecture Note Series. W. A. Benjamin, Inc., New York, 1969.

[1051] B. Szegedy. Edge coloring models and reflection positivity. *J. Amer. Math. Soc.*, 20(4):969–988, 2007.

[1052] P. G. Tait. On knots. *Trans. Roy. Soc. Edinburgh*, 28:145–190, 1876-7.

[1053] P. G. Tait. On Knots I, II, and III. In *Scientific Papers*, volume 1, pages 270–347. Cambridge University Press, 1898.

[1054] S. J. Tedford. A characteristic polynomial for rooted mixed graphs. *Discrete Math.*, 304(1-3):121–127, 2005.

[1055] S. J. Tedford. A Tutte polynomial which distinguishes rooted unicyclic graphs. *European J. Combin.*, 30(2):555–569, 2009.

[1056] H. N. V. Temperley and E. H. Lieb. Relations between the "percolation" and "colouring" problem and other graph-theoretical problems associated with regular planar lattices: some exact results for the "percolation" problem. *Proc. Roy. Soc. London Ser. A*, 322(1549):251–280, 1971.

[1057] C. P. Teo and K. M. Koh. The chromaticity of complete bipartite graphs with at most one edge deleted. *J. Graph Theory*, 14(1):89–99, 1990.

[1058] C. P. Teo and K. M. Koh. The number of shortest cycles and the chromatic uniqueness of a graph. *J. Graph Theory*, 16(1):7–15, 1992.

[1059] C. P. Teo and K. M. Koh. On chromatic uniqueness of uniform subdivisions of graphs. *Discrete Math.*, 128(1-3):327–335, 1994.

[1060] H. Terao. Algebras generated by reciprocals of linear forms. *J. Algebra*, 250(2):549–558, 2002.

[1061] M. B. Thistlethwaite. A spanning tree expansion of the Jones polynomial. *Topology*, 26(3):297–309, 1987.

[1062] C. Thomassen. Tilings of the torus and the Klein bottle and vertex-transitive graphs on a fixed surface. *Trans. Amer. Math. Soc.*, 323(2):605–635, 1991.

[1063] C. Thomassen. On the number of Hamiltonian cycles in bipartite graphs. *Combin. Probab. Comput.*, 5(4):437–442, 1996.

[1064] C. Thomassen. The zero-free intervals for chromatic polynomials of graphs. *Combin. Probab. Comput.*, 6(4):497–506, 1997.

[1065] C. Thomassen. Chromatic roots and Hamiltonian paths. *J. Combin. Theory Ser. B*, 80(2):218–224, 2000.

[1066] C. Thomassen. The weak 3-flow conjecture and the weak circular flow conjecture. *J. Combin. Theory Ser. B*, 102(2):521–529, 2012.

[1067] C. Thomassen. Group flow, complex flow, unit vector flow, and the $(2+\epsilon)$-flow conjecture. *J. Combin. Theory Ser. B*, 108:81–91, 2014.

[1068] P. Tittmann, I. Averbouch, and J. A. Makowsky. The enumeration of vertex induced subgraphs with respect to the number of components. *European J. Combin.*, 32(7):954–974, 2011.

[1069] I. Tomescu. Some extremal results concerning the number of graph and hypergraph colorings. In *Proc. Combinatorics and Graph Theory*, volume 25, pages 187–194. Banach Center Publ., 1989.

[1070] I. Tomescu. Maximal chromatic polynomials of connected planar graphs. *J. Graph Theory*, 14(1):101–110, 1990.

[1071] I. Tomescu. Maximum chromatic polynomials of 2-connected graphs. *J. Graph Theory*, 18(4):329–336, 1994.

[1072] L. Traldi. A dichromatic polynomial for weighted graphs and link polynomials. *Proc. Amer. Math. Soc.*, 106(1):279–286, 1989.

[1073] L. Traldi. Generalized activities and K-terminal reliability. *Discrete Math.*, 96(2):131–149, 1991.

[1074] L. Traldi. Chain polynomials and Tutte polynomials. *Discrete Math.*, 248(1-3):279–282, 2002.

[1075] L. Traldi. Parallel connections and coloured Tutte polynomials. *Discrete Math.*, 290(2-3):291–299, 2005.

[1076] L. Traldi. On the colored Tutte polynomial of a graph of bounded treewidth. *Discrete Appl. Math.*, 154(6):1032–1036, 2006.

[1077] L. Traldi. Binary nullity, Euler circuits and interlace polynomials. *European J. Combin.*, 32(6):944–950, 2011.

[1078] L. Traldi. On the interlace polynomials. *J. Combin. Theory Ser. B*, 103(1):184–208, 2013.

[1079] L. Traldi. Interlacement in 4-regular graphs: a new approach using nonsymmetric matrices. *Contrib. Discrete Math.*, 9(1):85–97, 2014.

[1080] L. Traldi. Binary matroids and local complementation. *European J. Combin.*, 45:21–40, 2015.

[1081] L. Traldi. The transition matroid of a 4-regular graph: an introduction. *European J. Combin.*, 50:180–207, 2015.

[1082] M. Trinks. The covered components polynomial: a new representation of the edge elimination polynomial. *Electron. J. Combin.*, 19(1):P50, 2012.

[1083] M. Trinks. *Graph Polynomials and Their Representations.* PhD Thesis, Technische Universität Bergakademie Freiburg, 2012.

[1084] M. A. Tsfasman and S. G. Vlăduţ. Geometric approach to higher weights. *IEEE Trans. Inform. Theory*, 41(6, part 1):1564–1588, 1995.

[1085] L. W. Tu. *An Introduction to Manifolds.* Universitext. Springer, New York, second edition, 2011.

[1086] A. W. Tucker. A combinatorial equivalence of matrices. In *Proc. Sympos. Appl. Math., Volume 10*, pages 129–140. American Mathematical Society, Providence, R.I., 1960.

[1087] V. Turaev. A simple proof of the Murasugi and Kauffman theorems on alternating links. *Enseign. Math.*, 33(34):203–225, 1987.

[1088] J. Turner and W. H. Kautz. A survey of progress in graph theory in the Soviet Union. *Ann. New York Acad. Sci.*, 175:385–390, 1970.

[1089] W. T. Tutte. *The Ultra-violet and Infra-red Spectra of Some Simple Molecules.* Master's thesis, University of Cambridge, 1940.

[1090] W. T. Tutte. A ring in graph theory. *Proc. Cambridge Philos. Soc.*, 43:26–40, 1947.

[1091] W. T. Tutte. *An Algebraic Theory of Graphs.* PhD Thesis, University of Cambridge, 1948.

[1092] W. T. Tutte. A contribution to the theory of chromatic polynomials. *Canad. J. Math.*, 6:80–91, 1954.

[1093] W. T. Tutte. Lectures on matroids. *J. Res. Nat. Bur. Standards Sect. B*, 69B:1–47, 1965.

[1094] W. T. Tutte. On the algebraic theory of graph colorings. *J. Combin. Theory*, 1:15–50, 1966.

[1095] W. T. Tutte. On dichromatic polynominals. *J. Combin. Theory*, 2:301–320, 1967.

[1096] W. T. Tutte. Dichromatic sums for rooted planar maps. In *Combinatorics (Proc. Sympos. Pure Math., Volume XIX, Univ. California, Los Angeles, Calif., 1968)*, pages 235–245, 1971.

[1097] W. T. Tutte. Codichromatic graphs. *J. Combin. Theory Ser. B*, 16:168–174, 1974.

[1098] W. T. Tutte. Some polynomials associated with graphs. In *Combinatorics (Proc. British Combinatorial Conf., Univ. Coll. Wales, Aberystwyth, 1973)*, pages 161–167. London Math. Soc. Lecture Note Ser., No. 13, 1974.

[1099] W. T. Tutte. The dichromatic polynomial. In *Proc. Fifth British Combinatorial Conference (Aberdeen 1975), Utilitas Math., Winnipeg 1976*, pages 605–635, 1976.

[1100] W. T. Tutte. All the king's horses. A guide to reconstruction. In *Graph Theory and Related Topics (Proc. Conf., Univ. Waterloo, Waterloo, Ont., 1977)*, pages 15–33. Academic Press, New York-London, 1979.

[1101] W. T. Tutte. *Selected Papers of W. T. Tutte Volumes I & II.* Charles Babbage Research Center, Winnipeg, 1979.

[1102] W. T. Tutte. 1-factors and polynomials. *European J. Combin.*, 1(1):77–87, 1980.

[1103] W. T. Tutte. Rotors in graph theory. *Ann. Discrete Math.*, 6:343–347, 1980.

[1104] W. T. Tutte. *Graph Theory*, volume 21 of *Encyclopedia of Mathematics and its Applications.* Addison-Wesley Publishing Company, Advanced Book Program, Reading, MA, 1984.

[1105] W. T. Tutte. Polynomials. In *Graph Structure Theory (Seattle, WA, 1991)*, volume 147 of *Contemp. Math.*, pages 1–8. Amer. Math. Soc., Providence, RI, 1993.

[1106] W. T. Tutte. Sixty years in the nets. Lecture at Fields Institute to mark receipt of the 2001 CRM-Fields Prize, 25 October 2001., 2001, www.fields.utoronto.ca/audio/01-02/CRM-Fields/tutte/.

[1107] W. T. Tutte. Graph-polynomials. *Adv. in Appl. Math.*, 32(1-2):5–9, 2004.

[1108] W. T. Tutte. *Graph Theory as I Have Known It.* Oxford Lecture Series in Mathematics and its Applications. Oxford University Press, Oxford, 2012.

[1109] P. Vaderlind. Chromaticity of triangulated graphs. *J. Graph Theory*, 12(2):245–248, 1988.

[1110] L. G. Valiant. The complexity of enumeration and reliability problems. *SIAM J. Comput.*, 8(3):410–421, 1979.

[1111] L. G. Valiant. Reducibility by algebraic projections. In *Logic and Algorithmic (Zurich, 1980)*, volume 30 of *Monograph. Enseign. Math.*, pages 365–380. Univ. Genève, Geneva, 1982.

[1112] L. G. Valiant. Holographic algorithms. *SIAM J. Comput.*, 37(5):1565–1594, 2008.

[1113] L. G. Valiant and V. V. Vazirani. NP is as easy as detecting unique solutions. *Theoret. Comput. Sci.*, 47(1):85–93, 1986.

[1114] R. Van Slyke and H. Frank. Network reliability analysis. I. *Networks*, 1:279–290, 1971/72.

[1115] O. Veblen. *Analysis Situs.* Colloquium publications (American Mathematical Society), v. 5, pt. 2. Amer. Math. Soc., second edition, 1931.

[1116] D. Vertigan. Bicycle dimension and special points of the Tutte polynomial. *J. Combin. Theory Ser. B*, 74(2):378–396, 1998.

[1117] D. Vertigan. Latroids and their representation by codes over modules. *Trans. Amer. Math. Soc.*, 356(10):3841–3868, 2004.

[1118] D. Vertigan. The computational complexity of Tutte invariants for planar graphs. *SIAM J. Comput.*, 35(3):690–712, 2005.

[1119] D. Vertigan and D. J. A. Welsh. The computational complexity of the Tutte plane: the bipartite case. *Combin. Probab. Comput.*, 1:181–187, 1992.

[1120] X. Viennot. Heaps of pieces, I : basic definitions and combinatorial lemmas. In *Lect. Notes in Math. (Combinatoire énumérative, Proc. Colloq., Montréal Can.)*, volume 1234, pages 321–350, 1986.

[1121] F. Vignes-Tourneret. The multivariate signed Bollobás-Riordan polynomial. *Discrete Math.*, 309(20):5968–5981, 2009.

[1122] F. Vignes-Tourneret. Non-orientable quasi-trees for the Bollobás-Riordan polynomial. *European J. Combin.*, 32(4):510–532, 2011.

[1123] E. Vigoda. Improved bounds for sampling colorings. *J. Math. Phys.*, 41(3):1555–1569, 2000.

[1124] M. L. Wachs and J. W. Walker. On geometric semilattices. *Order*, 2(4):367–385, 1986.

[1125] D. G. Wagner. Algebras related to matroids represented in characteristic zero. *European J. Combin.*, 20(7):701–711, 1999.

[1126] D. G. Wagner. Zeros of reliability polynomials and f-vectors of matroids. *Combin. Probab. Comput.*, 9(2):167–190, 2000.

[1127] D. G. Wagner. Negatively correlated random variables and Mason's conjecture for independent sets in matroids. *Ann. Comb.*, 12(2):211–239, 2008.

[1128] Z. Wang. On Bott polynomials. *J. Knot Theory Ramif.*, 3(4):537–546, 1994.

[1129] T. Wanner. On the chromaticity of certain subgraphs of a q-tree. *J. Graph Theory*, 13(5):597–605, 1989.

[1130] S. Wehrli. Khovanov homology and Conway mutation. arXiv:math/0301312.

[1131] V. K. Wei. Generalized Hamming weights for linear codes. *IEEE Trans. Inform. Theory*, 37(5):1412–1418, 1991.

[1132] S. Weiberg. *The Quantum Theory of Fields.* Cambridge University Press, 2005.

[1133] D. Welsh. The Tutte polynomial. *Random Struct. Algor.*, 15(3-4):210–228, 1999.

[1134] D. Welsh. W. T. (Bill) Tutte—in memoriam. *Adv. in Appl. Math.*, 32(1-2):3–4, 2004.

[1135] D. Welsh and K. Kayibi. A linking polynomial of two matroids. *Adv. in Appl. Math.*, 32:391–419, 2004.

[1136] D. Welsh and K. Kayibi. Corrigendum to "A linking polynomial of two matroids". *Adv. in Appl. Math.*, 49:78–79, 2012.

[1137] D. J. A. Welsh. Combinatorial problems in matroid theory. In *Combinatorial Mathematics and Its Applications (Proc. Conf., Oxford, 1969)*, pages 291–306. Academic Press, London, 1971.

[1138] D. J. A. Welsh. *Matroid Theory.* Academic Press [Harcourt Brace Jovanovich, Publishers], London-New York, 1976.

[1139] D. J. A. Welsh. *Complexity: Knots, Colourings and Counting*, volume 186 of *London Mathematical Society Lecture Note Series.* Cambridge University Press, Cambridge, 1993.

[1140] D. J. A. Welsh. Randomised approximation in the Tutte plane. *Combin. Probab. Comput.*, 3(1):137–143, 1994.

[1141] D. J. A. Welsh. Randomised approximation schemes for Tutte-Grothendieck invariants. In D. Aldous, P. Diaconis, J. Spencer, and J. Steele, editors, *Discrete Probability and Algorithms*, volume 72 of *IMA Volume Math. Appl.*, pages 133–148, 1995.

[1142] D. J. A. Welsh. Counting colourings and flows in random graphs. In *Combinatorics, Paul Erdős is Eighty, Volume 2 (Keszthely, 1993)*, volume 2 of *Bolyai Soc. Math. Stud.*, pages 491–505. János Bolyai Math. Soc., Budapest, 1996.

[1143] D. J. A. Welsh. The Tutte polynomial. Statistical physics methods in discrete probability, combinatorics, and theoretical computer science. *Random Struct. Algor.*, 15:210–228, 1999.

[1144] D. J. A. Welsh. Graph polynomials: some questions. Talk at the One-Day Combinatorics Colloquium, University of Reading, 2005.

[1145] D. J. A. Welsh and C. Merino. The Potts model and the Tutte polynomial. *J. Math. Phys.*, 41(3):1127–1152, 2000.

[1146] D. J. A. Welsh and G. P. Whittle. Arrangements, channel assignments, and associated polynomials. *Adv. in Appl. Math.*, 23(4):375–406, 1999.

[1147] D. B. West. *Introduction to Graph Theory.* Prentice Hall, Inc., Upper Saddle River, NJ, 1996.

[1148] J. A. White. On multivariate chromatic polynomials of hypergraphs and hyperedge elimination. *Electron. J. Combin.*, 18(1):P160, 2011.

[1149] E. G. Whitehead, Jr. and L. C. Zhao. Chromatic uniqueness and equivalence of K_4 homeomorphs. *J. Graph Theory*, 8(3):355–364, 1984.

[1150] E. G. Whitehead, Jr. and L. C. Zhao. Cutpoints and the chromatic polynomial. *J. Graph Theory*, 8(3):371–377, 1984.

[1151] H. Whitney. The coloring of graphs. *Proc. Nat. Acad. Sci. U.S.A.*, 17:122–125, 1931.

[1152] H. Whitney. The coloring of graphs. *Ann. of Math.*, 33(4):688–718, 1932.

[1153] H. Whitney. *The Coloring of Graphs.* PhD Thesis, Harvard University, 1932.

[1154] H. Whitney. A logical expansion in mathematics. *Bull. Amer. Math. Soc.*, 38(8):572–579, 1932.

[1155] H. Whitney. Non-separable and planar graphs. *Trans. Amer. Math. Soc.*, 34(2):339–362, 1932.

[1156] H. Whitney. 2-Isomorphic Graphs. *Amer. J. Math.*, 55(1-4):245–254, 1933.

[1157] H. Whitney. Characteristic functions and the algebra of logic. *Ann. of Math. (2)*, 34(3):405–414, 1933.

[1158] H. Whitney. A set of topological invariants for graphs. *Amer. J. Math.*, 55(1-4):231–235, 1933.

[1159] H. Whitney. On the abstract properties of linear dependence. *Amer. J. Math.*, 57(3):509–533, 1935.

[1160] G. Whittle. Dowling group geometries and the critical problem. *J. Combin. Theory Ser. B*, 47(1):80–92, 1989.

[1161] G. Whittle. q-lifts of tangential k-blocks. *J. London Math. Soc.*, 39(1):9–15, 1989.

[1162] G. Whittle. Characteristic polynomials of weighted lattices. *Adv. Math.*, 99(2):125–151, 1993.

[1163] B. Widom and J. S. Rowlinson. New model for the study of liquid–vapor phase transitions. *J Chem. Phys.*, 52(4):1670–1684, 1970.

[1164] E. B. Wilson. *History of the Proceedings of the National Academy of Sciences, 1914–1963.* Washington, National Academy of Sciences, 1966.

[1165] S. E. Wilson. Operators over regular maps. *Pacific J. Math.*, 81(2):559–568, 1979.

[1166] R. Winder. Partitions of N-space by hyperplanes. *SIAM J. Appl. Math.*, 14:811–818, 1966.

[1167] P. Wocjan and J. Yard. The Jones polynomial: Quantum algorithms and applications in quantum complexity theory. *Quantum Inf. Comput.*, 8(1,2):147–180, 2008.

[1168] J. A. Wood. Foundations of linear codes defined over finite modules: the extension theorem and the MacWilliams identities. In *Codes Over Rings*, volume 6 of *Ser. Coding Theory Cryptol.*, pages 124–190. World Sci. Publ., Hackensack, NJ, 2009.

[1169] J. A. Wood. Applications of finite Frobenius rings to the foundations of algebraic coding theory. In *Proceedings of the 44th Symposium on Ring Theory and Representation Theory*, pages 223–245. Symp. Ring Theory Represent. Theory Organ. Comm., Nagoya, 2012.

[1170] D. R. Woodall. Zeros of chromatic polynomials. In *Combinatorial Surveys (Proc. Sixth British Combinatorial Conf., Royal Holloway College, Egham, 1977)*, pages 199–223. Academic Press, London, 1977.

[1171] D. R. Woodall. A zero-free interval for chromatic polynomials. *Discrete Math.*, 101(1-3):333–341, 1992.

[1172] D. R. Woodall. Tutte polynomial expansions for 2-separable graphs. *Discrete Math.*, 247(1-3):201–213, 2002.

[1173] F.-Y. Wu. Potts model and graph theory. *J. Stat. Phys.*, 52:99–112, 1988.

[1174] S. Xu. Complete-graph-basis and the chromaticity of graphs with more edges. *J. Shanghai Teachers College*, 3:18–23, 1983.

[1175] S. Xu. Some notes on chromatic uniqueness of graphs. *J. Shanghai Teach. Univ., Nat. Sci. Ed.*, 2:10–12, 1987.

[1176] S. Xu. Chromaticity of chordal graphs. *Graphs Combin.*, 13(3):287–294, 1997.

[1177] S. J. Xu and N. Z. Li. The chromaticity of wheels. *Discrete Math.*, 51(2):207–212, 1984.

[1178] S. J. Xu, J. J. Liu, and Y.-H. Peng. The chromaticity of s-bridge graphs and related graphs. *Discrete Math.*, 135(1-3):349–358, 1994.

[1179] D. N. Yetter. On graph invariants given by linear recurrence relations. *J. Combin. Theory Ser. B*, 48(1):6–18, 1990.

[1180] K. S. Yow, G. Farr, and K. Morgan. Tutte invariants for alternating dimaps. arXiv:1803.05539, 2018.

[1181] T. Zaslavsky. Facing up to arrangements: face-count formulas for partitions of space by hyperplanes. *Mem. Amer. Math. Soc.*, (154), 1975.

[1182] T. Zaslavsky. The Möbius function and the characteristic polynomial. In *Combinatorial Geometries*, volume 29 of *Encyclopedia Math. Appl.*, pages 114–138. Cambridge University Press, Cambridge, 1987.

[1183] T. Zaslavsky. Strong Tutte functions of matroids and graphs. *Trans. Amer. Math. Soc.*, 334(1):317–347, 1992.

[1184] T. Zaslavsky. Biased graphs. III. Chromatic and dichromatic invariants. *J. Combin. Theory Ser. B*, 64(1):17–88, 1995.

[1185] F. Zhang, editor. *The Schur Complement and Its Applications*, volume 4 of *Numerical Methods and Algorithms*. Springer-Verlag, New York, 2005.

[1186] S. Zhang. The chromatic uniqueness of a kind of k_4-homeomorphs (Chinese). *J. Qinghai Normal Univ. (Natural Sc.)*, 1:12–16, 2000.

[1187] H. Zhao, X. Li, R. Liu, and C. Ye. The chromaticity of certain complete multipartite graphs. *Graphs Combin.*, 20(3):423–434, 2004.

[1188] H. Zhao, X. Li, S. Zhang, and R. Liu. On the minimum real roots of the σ-polynomials and chromatic uniqueness of graphs. *Discrete Math.*, 281(1-3):277–294, 2004.

[1189] H. Zou. On the chromatic uniqueness of complete tripartite graphs $k(n_1, n_2, n_3)$ (Chinese). *J. Sys. Sc. and Math. Sc.*, 20(2):181–186, 2000.

[1190] H. Zou and Y. Shi. On the chromaticity of the complete tripartite graph $k(n-k, n, n)$ (Chinese). *J. Shanghai Teachers Univ. (Natural Sc.)*, 28(1):15–22, 1999.

[1191] A. A. Zykov. On some properties of linear complexes. *Mat. Sbornik N.S.*, 24(66):163–188, 1949. (Translation available in *Amer. Math. Soc. Translation*, 1952(79):33, 1952).

[1192] A. A. Zykov. Problem 31. In *Theory of Graphs and Its Applications*, Proceedings of the Symposium held in Smolenice in June 1963, page 165. Publishing House of the Czechoslovak Academy of Sciences, Prague, 1964.

[1193] A. A. Zykov. Recursively calculable functions of graphs. In *Theory of Graphs and Its Applications*, Proceedings of the Symposium held in Smolenice in June 1963, pages 99–105. Publishing House of the Czechoslovak Academy of Sciences, Prague, 1964.

Selected evaluations and interpretations

$T(G;0,0)$

- $T(G;0,0)$ is 1 if G is edgeless and is 0 otherwise.

$T(G;0,1)$

- $T(G;0,1)$, for G planar, is the number of totally cyclic orientations containing no clockwise cycle. (See p. 31.)
- $T(G;0,1)$, for G connected, is the number of indegree-sequences of strongly connected orientations. (See p. 588.)
- $T(G;0,1)$, for G the underlying graph of a digraph with a fixed ordering on its edges, is the number of totally cyclic reorientations such that in each cycle the lowest edge is not reoriented. (See [580].)
- $T(G;0,1)$, where γ a linear circle orientation of G, is the number of totally cyclic orientations such that each cycle is oriented as prescribed by γ. (See [580].)
- $T(\underline{M};0,1)$, for M a regular oriented matroid, is the number of totally cyclic circuit-reversal classes. (See p. 588.)

$T(G;0,2)$

- $T(G;0,2)$ is the number of totally cyclic orientations. (See pp. 31, 573.)
- $T(\underline{M};0,2)$, for M an oriented matroid or real hyperplane arrangement, is the number of totally cyclic reorientations of M and the number of regions of M^*. (See p. 573.)

$T(G; \frac{1}{2}, 1)$

- $2^{|V|-1}T(G; \frac{1}{2}, 1)$, for G connected, is the number of relatively bounded regions of a generic bigraphical arrangement. (See [625].)

$T(G; \frac{1}{2}, 3)$

- $2^{|V|-1}T(G; \frac{1}{2}, 3)$, for G connected, is the number of strongly connected partial orientations. (See [59].)

$T(G; 1, 0)$

- $T(G; 1, 0)$, for G connected, is the number of acyclic orientations in which some fixed vertex is the only source. (See p. 30.)
- $T(G; 1, 0)$, for G connected with fixed vertex s, is the number of indegree-sequences of acyclic orientations of G with unique sink at s. (See p. 588.)
- $T(\underline{M}; 1, 0)$, for M a regular oriented matroid, is the number of acyclic cocircuit-reversal classes. (See p. 588.)
- $T(M; 1, 0)$, where M is the simple matroid from a non-central hyperplane arrangement, is the number of bounded regions. (See [247, 1181].)

$T(G; 1, \frac{1}{2})$

- $2^{n(G)}T(G; 1, \frac{1}{2})$ is the number of initially connected acyclic suborientations (or partial orientations). (See [520].)

$T(G;1,1/(1-p))$

- $(1-p)^{n(G)}p^{r(G)}T(G;1,1/(1-p)) = \mathrm{Rel}(G;p)$, for $p \in [0,1]$ and G connected, which is the probability that a random subgraph is connected. (See p. 35, and p. 82 for a matroid version.)

$T(G;1,1)$

- $T(G;1,1)$, for G connected, is the number of spanning trees of G. (See p. 29.)
- $T(G;1,1)$, for G connected with fixed vertex s, is the number of indegree-sequences of orientations such that s is reachable from any other vertex. (See p. 588.)
- $T(M;1,1)$, for M a matroid, is the number of bases of M. (See p. 81.)
- $T(\underline{M};1,1)$, for M a regular oriented matroid, is the number of circuit-cocircuit-reversal classes. (See p. 588.)

$T(G;1,2)$

- $T(G;1,2)$, for G connected, is the number of connected spanning subgraphs of G. (See p. 29.)
- $T(G;1,2)$, for G connected with fixed vertex s, is the number of orientations such that s is reachable from any other vertex. (See p. 588.)
- $T(M;1,2)$ for a matroid M, is the number of spanning sets. (See p. 82.)
- $T(\underline{M};1,2)$, for M a regular oriented matroid, is the number of cocircuit-reversal classes. (See p. 588.)

$T(G;1,3)$

- $2^{r(G)}T(G;1,3)$, for G connected, is the number of partial orientations modulo cut reversals. (See [59].)

$T(G; \frac{3}{2}, 1)$

- $2^{r(G)}T(G; \frac{3}{2}, 1)$, for G connected, is the number of regions of a generic bigraphical arrangement. (See [625].)

$T(G; 2, 0)$

- $T(G; 2, 0)$ is the number of acyclic orientations. (See pp. 29, 573.)
- $T(G; 2, 0)$ is the number spanning subgraphs that contain no broken cycles with respect to any edge order. (Follows from Theorem 11.22.)
- $T(\underline{M}; 2, 0)$, for M an oriented matroid or real hyperplane arrangement, is the number of acyclic reorientations and the number of regions of M. (See p. 573.)

$T(G; 2, 1)$

- $T(G; 2, 1)$, for G connected, is the number of spanning forests. (See p. 29.)
- $T(G; 2, 1)$, for G connected, is the number of score vectors. (See p. 31, 587.)
- $T(M; 2, 1)$, for a matroid M, is the number of independent sets. (See p. 82.)
- $T(\underline{M}; 2, 1)$, for M a regular oriented matroid, is the number of circuit-reversal classes. (See p. 588.)

$T(G; 2, 2)$

- $T(G; 2, 2) = 2^{e(G)}$ is the number of spanning subgraphs. (See p. 29.)
- $T(M; 2, 2) = 2^{|E|}$, where M a matroid on E. (See p. 82.)

$T(G;3,\frac{1}{2})$

- $2^{n(G)}T(G;3,\frac{1}{2})$ for G connected is the number of acyclic suborientations (or partial orientations). (See [520].)

$T(G;3,1)$

- $2^{n(G)}T(G;3,1)$ for G connected is the number of partial orientations modulo cycle reversals. (See [59].)

$T(G;3,3)$

- $T(G;3,3) = q_G 2^{s(G)-1}$, for G plane, where s is the number of crossing circuits (anti-circuits) of G_m and q_G is some integer. (See p. 38.)
- $T(G;3,3) = \sum_{k\geq 0} 2^{k-1} e_k(G_m)$, for G plane, where $e_k(G_m)$ is the number of Eulerian orientations of G_m with exactly k saddle vertices. (See p. 281.)
- $2\,T(G;3,3) = \mathrm{Holant}_2(G_m; f)$, for G plane and connected, for a suitable quaternary function f. (See p. 401.)
- $2\,T(G;3,3)$, for G plane and connected, is the number of claw coverings of a graph associated with G . (See p. 281.)
- $2T(G;3,3)$, for G an $m\times n$ grid, is the number tilings of a $4m\times 4n$ rectangle with T-tetrominoes. (See pp. 32, 281.)

$T(G;k,0)$

- $n^{k(G)}T(G;n+1,0)$, for $n \in \mathbb{N}$, is the number of pairs (σ,ρ), where σ is an n-coloring and ρ is any orientation of G such that if $u \to v$ is in the orientation ρ, then $\sigma(u) > \sigma(v)$. (Follows from Theorem 11.30, see also p. 30.)

$T(G;k,k)$

- $T(G;1+n,1+n) = (\frac{1}{n})^{k(G)} \sum_\phi 2^{\mu(\phi)}$, for $n \in \mathbb{N}$, for G planar and $n \in \mathbb{N}$. Here the sum is over all edge n-colorings ϕ of $\vec{G}_m$ such that each set of monochromatic edges forms an Eulerian digraph, and where $\mu(\phi)$ is the number of monochromatic vertices. (See p. 281.)

- $nT(G;n+1,n+1) = \text{Holant}_n(G_m;f)$, for G plane and connected and $n \in \mathbb{N}$, for a suitable quaternary function f. (See p. 402.)

$T(G;k,j)$

- $T(G;p+1,q+1) = \sum 2^{|\ker(f)-\text{supp}(g)|}$, for $p,q \in \mathbb{N}$ and some orientation of G. Here the sum is over all pairs (f,g) with $\text{supp}(g) \subseteq \ker(f)$ where f is a p-tension, g a q-flow, $\ker(f)$ is the set of edges with zero-tension, and $\text{supp}(g)$ is the set of edges with nonzero flow. (Follows from Theorem 12.27, see [296, 959].)

- $T(G;p+1,q+1)$ for $p,q \in \mathbb{N}$ is the number of triples (f,g,ρ) where f is a $\mathbb{Z}_p$-tension, g is a $\mathbb{Z}_q$-flow of an orientation of G such that $\text{supp}(g) \subseteq \ker(f)$, and ρ is a reorientation on $\ker(f) \setminus \text{supp}(g)$. Here $\ker(f)$ is the set of edges with zero-tension, and $\text{supp}(g)$ is the set of edges with nonzero flow. (See [200] and also [296].)

- $nT(G;p+1,q+1) = \sum_\phi 2^{\alpha(L)}$, for G plane and connected, and $p,q \in \mathbb{N}$ with $\text{lcm}(p,q) = n$. Here the sum is over all edge n-colorings of the canonically checkerboard colored medial graph G_m such that at each vertex either all incident edges have the same color; or the two edges on one gray face have color a, the two edges on the other gray face have color b, and $a \equiv b \mod \frac{n}{p}$; or or the two edges on one white face color a, the two edges on the other white face have color b, and $a \equiv b \mod \frac{n}{q}$. Moreover $\alpha(\phi)$ is the number of monochromatic vertices. (See [705].)

- $nT(G;p+1,q+1) = \text{Holant}_n(G_m;f)$, for G plane and connected $p,q \in \mathbb{N}$ with $\text{lcm}(p,q) = n$, for a suitable quaternary function f. (See p. 402.)

- $pT(G;p+1,q+1) = \sum_W 2^{\alpha(W)}$, for G a grid graph and $p,q \in \mathbb{N}$ with $\text{lcm}(p,q) = n$. Here the sum is over "Wang tilings" of an associated graph and $\alpha(W)$ is the number of monochromatic tiles. (See [705].)

$T(G; \frac{k}{2}, \frac{j}{2})$

- $T(G; \frac{k}{2}, \frac{j}{2})$, for G connected and $0 \leq k, j \leq 4$, or $j = 3$ and $1 \leq k \leq 3$, or $k = 3$ and $1 \leq j \leq 3$, counts the numbers of fourientations subject to various conditions. A *fourientation* of a graph is a choice for each edge of the graph whether to orient that edge in either direction, leave it unoriented, or biorient it. (See [60].)

$T(G; 0, -2)$

- $T(G; 0, -2)$, for G a 4-regular graph, is the number of ice configurations of G. (See p. 31.)
- $T(G; 0, -2) = (-1)^{|V|+|E|+k(G)}\,\mathrm{Holant}_2(G; g)$, for G a 4-regular graph, for a suitable quaternary function g. (See p. 403.)

$T(G; 0, -3)$

- $(-1)^{n(G)} T(G; 0, -3)$, for G cubic, is the number of proper edge 3-colorings. (Follows from discussion on p. 264.)

$T(G; 0, -k)$

- $(-1)^{n(G)} T(G; 0, 1-k) = F(G; k)$, for $k \in \mathbb{N}$, where $F(G; k)$ is the number of nowhere-zero A-flows where $|\mathsf{A}| = k$. (See p. 34.)

$T(G; 1, -1)$

- $T(G; 1, -1)$ is the number of "even-left spanning trees" of G. (See [802].)

$T(G;-1,0)$

- $(-1)^{r(G)}2^{n(G)}T(G;-1,0) = \sum(-1)^{|A|}$, where the sum is over all $A \subseteq E(G)$ such that A induces an even subgraph. (See [558].)

$T(G;-1,-1)$

- $T(G;-1,-1) = (-1)^{|E|}(-2)^{\dim(B)}$ where B is the bicycle space of G. (See p. 31, 262, 281.)
- $T(G;-1,-1) = (-1)^{e(G)}(-2)^{s-1}$, for G plane, where s is the number of crossing circuits (or anti-circuits) of G_m. (See p. 38, 281, 361.)

$T(G;-3,0)$

- $(-4)^{n(G)}T(G;-3,0) = \sum(-3)^{|E|-|A\cup B|}$, where the sum is over all $A, B \subseteq E(G)$ that induce even subgraphs. (See [558].)

$T(G;-k,0)$

- $(-1)^{r(G)}\lambda^{k(G)}T(G;1-k,0) = \chi(G;k)$, for $k \in \mathbb{N}$, where $\chi(G;k)$ is the number of proper k-colorings of G. (See p. 34.)

$T(G;-k,-k)$

- $T(G;1-n,1-n) = (\frac{-1}{n})^{k(G)}\sum_{D_n(\vec{G}_m)}(-1)^{\sum_{i=1}^{n}k(D_i)}$, for G planar and $n \in \mathbb{N}$. Here $D_n(\vec{G}_m)$ is the set of all ordered partitions $(D_1,\ldots,D_n)$ of $E(\vec{G}_m)$ in such that G restricted to each D_i is 2-regular and consistently oriented. (See p. 282.)

$T(G; -k, -j)$

- $T(G; 1-p, 1-q) = (-1)^{r(G)} p^{-k(G)} \sum -(\mathbf{x}, \mathbf{y})(-1)^{|\mathrm{supp}(\mathbf{y})|}$, for $p, q \in \mathbb{N}$, where $(\mathbf{x}, \mathbf{y})$ are pairs in which $\mathbf{x}$ is a vertex p-coloring, $\mathbf{y}$ is q-flow, and each edge contains non-zero flow if and only if it is colored improperly. Here $|\mathrm{supp}(\mathbf{y})|$ is the number of improperly colored edges in $\mathbf{x}$. (Follows from Theorem 12.27, see [959] and [296] for tension-flow form.)
- $T(G; -k, -j)$, for (k, j) any of $(-3, 0)$, $(-2, -\frac{1}{3})$, $(-2, \frac{1}{3})$, $(-1, -1)$, $(-1, 0)$, $(-\frac{1}{2}, -\frac{1}{3})$, $(0, -3)$, $(0, -1)$, has been expressed in terms of correlations of events involving Eulerian graphs. (See [558].)

$T(G; i, -i)$

- $|T(G; i, -i)|$ is $\sqrt{2}^{b(G)}$ if every bicycle has size a multiple of 4, and 0 otherwise, where $b(G)$ is the dimension of the bicycle space of G. (See p. 263.)

$T(G; -i, i)$

- $T(M; -i, i) = e^{(\pi i/4)(\sigma(\tilde{q}_V) + |E| - 3r(E))} \sqrt{2}^{\dim(V \cap V^{\perp})}$, for $M = M(V)$ and V a subspace of $\mathbb{F}_2^E$, if the support of x is congruent to zero modulo 4 for all $x \in V \cap V^{\perp}$, and is zero otherwise. See [912] for the definition of $\sigma(\tilde{q}_V)$. (See p. 263 and [912].)

$T(G; e^{2\pi i/3}, e^{4\pi i/3})$

- $T(G; e^{2\pi i/3}, e^{4\pi i/3}) = (-1)^{q(G)} (e^{(2\pi i/3)(|E| + n(G))}) (i\sqrt{3})^{b_3(G)}$ where $b_3(G)$ is the dimension of the space of $\mathbb{Z}_3$-bicycles of G, and $q(G)$ is the number of vectors with support size congruent to 2 modulo 3 in any orthogonal basis for the space of $\mathbb{Z}_3$-flows of G. (See p. 263.)

Symbols

General mathematical notation

$\#X$ or $\lvert X\rvert$	cardinality of a set X
$\sqcup$	disjoint union for sets
$[n]$	$\{1, 2, \ldots, n\}$
$\binom{n}{k}$	binomial coefficient
$\left[{m \atop k}\right]_q$	Gaussian coefficient
A	abelian group (usually)
A^c	complement of a set A
A^T	transpose of a matrix A
$A * X$	principal pivot transform of a matrix
$\mathbb{C}$	complex numbers
$\mathbb{F}$	field
$\mathbb{F}_q$	finite field of q elements
G	group
I_r	$r \times r$ identity matrix
$\mathbb{N}$	natural numbers, $\{1, 2, 3, \ldots\}$
$\mathbb{N}_0$	nonnegative integers, $\{0, 1, 2, 3, \ldots\}$
$O(\cdot)$	big O asymptotic notation
O_q	complex orthogonal group in dimension q
$\mathbb{Q}$	field of rational numbers
$\mathfrak{R}$	ring
$\mathcal{R}(A)$	row space of the matrix A
$\mathbb{R}$	real numbers
V	vector space
$V(n, \mathbb{F})$	n-dimensional vector space over $\mathbb{F}$
$V(n, q)$	n-dimensional vector space over $\mathbb{F}_q$
$\mathsf{V}^{\perp}$	orthogonal subspace of V
$X \triangle Y$	symmetric difference of sets X and Y
$\mathbb{Z}$	integers
$\mathbb{Z}_k$	integers modulo k

Graph theoretic notation

$\alpha(G, k)$	no. of ways to partition $V(G)$ into k independent sets of G
$\delta(G)$	minimum degree of G
$\Delta(G)$	maximum degree of G
$\Delta_2(G)$	second largest degree of G
$\Delta(G)$	Laplacian matrix of G

DOI: 10.1201/9780429161612-34

$\Delta^v(G)$	reduced Laplacian matrix of G at vertex v
$\mathbf{\Delta}(G)$	coproduct of G
$\epsilon(G)$	mean size set of a broken-cycle-free spanning subgraph of G
$\eta_{i,j}$	operation adding edges between vertices labeled i and j
$\rho_{i\to j}$	operation renaming the label i of any vertex to label j
$\chi(G)$	chromatic number
$\chi_c(G)$	circular chromatic number
$\chi_f(G)$	fractional chromatic number
$\omega(G)$	clique number of G
$\omega^+(U)$	edges directed out of $U \subseteq V$ in G^ω
$\omega^-(U)$	edges directed into $U \subseteq V$ in G^ω
$\mathrm{bw}(G)$	branch-width of G
$\mathrm{cw}(G)$	clique-width of G
$\deg(v)$	degree of vertex v
$e(A)$	number of edges of $(V(G), A)$ of G, equals $\|A\|$
$e(G)$	number of edges G
$E(G)$	edge set of G
$\mathrm{EA}(F)$	set of externally active elements
$G = (V, E)$	graph G with vertex set V and edge set E
$\vec{G}$	directed graph
$G\backslash e$	G delete e
G/e	G contract e
G/v	G contract vertex v
$G\backslash v$	G delete vertex v
$G\|A$	spanning subgraph $(V(G), A)$
$G[X]$	subgraph of G induced by X for $X \subseteq E(G) \cup V(G)$
$G * v$	local complement of a graph
$G * e$	pivot of a graph
G^*	dual of G
$\overline{G}$	complement of G
$G * H$	one-point join of G and H
$G \sqcup H$	disjoint union of G and H
$G_1 \sqcup_\ell G_2$	gluing product of ℓ-labeled graphs G_1 and G_2
$G \times H$	Cartesian product of G and H
$G + H$	complete join of graphs G and H
G^ω	graph G with orientation ω
$G \sim H$	chromatically equivalent graphs
$G \sim_h H$	adjointly equivalent graphs
$\mathcal{G}[G]$	set of graphs H with $H \sim G$
$h(G, \lambda)$	adjoint polynomial of G
$\mathrm{Hom}(G, H)$	set of homomorphisms from G to H
$\mathrm{hom}(G, H)$	number of homomorphisms from G to H
$\mathrm{IA}(F)$	set of internally active elements
$I(C)$	interlace graph
$i_G(H)$	number of induced subgraphs in G isomorphic to H
$k(A)$	number of components of spanning subgraph $(V(G), A)$
$k(G)$	number of components of G
$\bar{k}(A)$	number of components $(V(G), A)$ that are not isolated vertices
$L(G)$	line graph of graph G

$M(f,\ell)$	ℓ-th connection matrix of graph invariant f
$n(A)$	nullity of spanning subgraph $(V(G),A)$
$n(G)$	nullity of G
$N_G(v)$	neighborhood
$\bar{N}_G(v)$	closed neighbourhood
$r(A)$	rank of spanning subgraph $(V(G),A)$
$r(G)$	rank of G
$\mathrm{rw}(G)$	rank-width of a graph G
$s_G(H)$	number of subgraphs in G which are isomorphic to H
$\mathrm{tw}(G)$	tree-width of a graph G
$v(A)$	number of vertices of $(V(G),A)$ of G, equals $\lvert V\rvert$
$v(G)$	number of vertices of G
$v_1 \parallel v_2$	parallel reduction operation
$v_1 \bowtie_q v_2$	series reduction operation
$V(G)$	vertex set of G

Graph families

$\Gamma(A,B)$	Cayley graph on A defined by B
$\theta(a_1,a_2,\cdots,a_k)$	generalized theta-graph
$\theta_s(a)$	s-theta graph
$B_{m,n}$	Cartesian product of P_m and P_n
C_m	cycle of length m
K_m	complete graph on m vertices
$K_q^{x,y}$	q-state Potts model graph
$KG_{q,r}$	Kneser graph
$J_{q,r}$	Johnson graph
$J_{q,r,D}$	generalized Johnson graph
N_r	edgeless graph on r vertices
P_m	path of length m
W_r	r-spoke wheel

Matroid theoretic notation

$-_AM$	reorientation of M with respect to A
$\alpha(M)$	active basis of an ordered oriented matroid M
α_M	refined active bijection with respect to M
$\beta(M)$	β-invariant of M
$\varepsilon_M(B)$	external activity of B in M
$\theta_M(A)$	$\lvert O(-_AM)\setminus A\rvert$
$\bar{\theta}_M(A)$	$\lvert O(-_AM)\cap A\rvert$
$\theta^*_M(A)$	$\lvert O^*(-_AM)\setminus A\rvert$
$\bar{\theta}^*_M(A)$	$\lvert O^*(-_AM)\cap A\rvert$
$\iota_M(B)$	internal activity of B in M
$\mu(X,Y)$	Möbius function of (X,Y)
$\mathsf{A}(M)$	The Orlik–Solomon algebra of M
$\mathrm{aw}(M)$	amalgam width of M
$\mathcal{B}(M)$	set of bases of M

$\mathrm{Bi}(G)$	bicircular matroid of G
$\mathrm{cl}_M(X)$	closure of a set X
$\mathrm{cr}_M(A)$	$r(M) - r_M(A)$, equals $\lvert P_M(A)\rvert$
$C(e, B)$	fundamental circuit of the e with respect to the basis B
$C^*(e, B)$	fundamental cocircuit of the e with respect to the basis B
$\mathcal{C}(M)$	set of circuits of M
$\mathrm{dw}(M)$	decomposition width of M
$\mathrm{EA}_M(B)$	set of externally active elements of B in M
$\mathcal{I}(M)$	set of independent sets of M
$\mathrm{IA}_M(B)$	set of internally active elements of B in M
$\mathcal{L}(M)$	lattice of flats of M
M^*	dual of M
$\underline{M}$	underlying matroid of the oriented matroid M
$M\backslash X$	deletion of the set X from M
M/X	contraction of the set X from M
$M\lvert X$	restriction of M to the set X
$M.X$	contraction of M to the set X
$M(G)$	cycle matroid of G
$M^*(G)$	bond matroid of G
$M + e$	free extension
$M \times e$	free coextension
$M \oplus N$	direct sum of M and N
$M \oplus_2 N$	2-sum of M and N
$M \otimes N$	tensor product of M and N
$M * X$	twist of a delta-matroid
$M \to M'$	matroid perspective
$\mathcal{M}_G$	delta-matroid of graph
$M \,\square\, N$	free product of M and N
$\mathrm{nl}_M(A)$	$\lvert A\rvert - r_M(A)$, equals $\lvert Q_M(A)\rvert$
$N \times e$	free coextension
$o(M)$	orientation-activity of M
$O(M)$	set of orientation-active elements of M
$o^*(M)$	dual orientation-activity of M
$O^*(M)$	set of dual-orientation-active elements of M
$P_M(A)$	$\mathrm{IA}_M(B) \setminus A$ where B basis associated with A
$Q_M(A)$	$\mathrm{EA}_M(B) \cap A$ where B basis associated with A
$r_M(X)$	rank of X in M
$\mathrm{rcd}_{M,M'}(A)$	$r(M) - r(M') - \big(r_M(A) - r_{M'}(A)\big)$
$s_k(G)$	k-stretch of M
$\mathrm{si}(M)$	simplification of M
$t_k(G)$	k-thickening M
$\mathrm{Tr}(M)$	truncation
$\mathrm{Tr}^k(M)$	k-fold truncation
$w_i(M)$	Whitney number of the first kind
$W_i(M)$	Whitney number of the second kind

Matroid families

$AG(r,q)$	affine geometry
M_n	Catalan matroid
$PG(r,q)$	projective geometry of rank $r+1$ over $\mathbb{F}_q$
$Q_r(G)$	Dowling geometry
$\mathcal{T}$	class of transversal matroids
$U_{r,n}$	uniform matroid of rank r on an n-element set
$\mathcal{W}_r$	cycle matroid of W_r
$\mathcal{W}^r$	rank-r whirl

Polynomials

$\xi(G;x,y,z)$	Averbouch–Godlin–Makowsky polynomial
$\Xi(D;\mathbf{x},\mathbf{y})$	path-cycle symmetric function
$\chi(G;\lambda)$	chromatic polynomial
$\chi(M;x)$	characteristic polynomial
$\overline{\chi}(M;x,y)$	coboundary polynomial
$B(G;u,v)$	subtree-boundary polynomial
$B(D;q,y,z)$	B-polynomial
$C(D;x,y)$	cover polynomial
$\hat{C}(D;x,y)$	geometric cover polynomial
$\mathrm{Ch}(G;\omega,\mathbf{w})$	chain polynomial
$\mathrm{Chg}(G;\omega)$	chain generating function
$f(G;t,x,y)$	a Negami polynomial
$f^*(G;t,x,y)$	a Negami polynomial
$F(G;x)$	flow polynomial
$F_{\mathbb{Z}}(G;x)$	integral flow polynomial
$\mathcal{G}(M)$	$\mathcal{G}$-invariant
$\mathrm{Holant}_q(G;\mathcal{F})$	Holant
$I(G;x)$	independence polynomial
$\mathrm{Ind}(D)$	cycle-path indicator polynomial
$j(\vec{G};x)$	circuit partition polynomial
$J(G;x)$	circuit partition polynomial
$K(G,\Sigma;X,Y,A,B)$	Krushkal polynomial
$L(G;u,v)$	subtree-leaf polynomial
$LV(G;x,y,z)$	Las Vergnas polynomial
$m(\vec{G};x)$	Martin polynomial
$M(G;x)$	Martin polynomial
$N(G;u,\mathbf{a},\mathbf{b})$	Negami's polynomial
$p_h(G)$	partition function of edge coloring model h
$P(G;\lambda)$	Penrose Polynomial
$P_c(G,q;y)$	critical configuration polynomial
$\mathrm{Prob}_{\mathrm{k}}(G;u,\mathbf{p})$	reliability polynomial with variable edge probabilities
$q(G;W,t)$	transition polynomial
$q(G;y)$	interlace polynomial of graph
$Q(G;y)$	global interlace polynomial of graph
$Q(G;x,y)$	Tittmann–Averbouch–Makowsky polynomial
$Q(G;(\boldsymbol{\alpha},\boldsymbol{\beta},\boldsymbol{\gamma}),t)$	topological transition polynomial

$Q(G; W, y)$	weighted interlace polynomial
$r(D; x, y)$	cycle rook polynomial
$R(G; X, Y, Z, W)$	Bollobás–Riordan polynomial
$\mathrm{Rel}(G; p)$	reliability polynomial
$\mathrm{Sh}(G; \omega, \mathbf{w})$	sheaf polynomial
t_{ij} or $t_{i,j}$	coefficient of $x^i y^j$ in $T(G; x, y)$
$T(G; x, y)$	Tutte polynomial
$T(M, M'; x, y, z)$	Tutte polynomial of a matroid perspective
$T_{\mathrm{par}}(G; x, y, \mathbf{a}, \mathbf{b})$	standard doubly parametrized Tutte polynomial
$T_{\mathrm{gen}}(G; x, y, \mathbf{a}, \mathbf{b})$	doubly-multivariate Tutte polynomial
$U(G; \bar{x}, y)$	Noble's weighted graph polynomial
$W(G; x, y)$	Whitney rank generating function
$X(G; \mathbf{x})$	Stanley's chromatic symmetric function
$Z(G; u, v)$	dichromatic polynomial
$Z(G; q, \mathbf{w})$	multivariate Tutte polynomial
$Z_{\mathrm{gen}}(G; q, \mathbf{v})$	multivariate dichromatic polynomial
$Z_{\mathrm{I}}(G)$	Ising partition function
$Z_{\mathrm{P}}(G)$	Potts partition function
$Z_{\mathrm{par}}(G; q, \mathbf{v})$	parameterized dichromatic polynomial
$Z_{\mathrm{RC}}(G)$	random-cluster partition function

Complexity and approximation

♯BIS	number of independent sets in a bipartite graph
♯ETH	exponential-time hypothesis
♯P	polynomial decision problems
♯$\mathrm{P}_{\mathcal{R}}$	analogue of ♯P to the BSS model
♯SAT	number of satisfying assignments for a Boolean formula
♯W[1]	parameterized counting problems in Wedge hierarchy
BPP	bounded-error probabilistic polynomial time
FP	polynomial-time computable functions
FPRAS	fully polynomial randomized approximation schemes
FPT	fixed-parameter tractable computable decision problems
MCMC	Markov chain Monte Carlo
NP	nondeterministic poly.-time computable decision problems
$\mathrm{NP}_{\mathcal{R}}$	analogue of NP to the BSS model
P	polynomial-time computable decision problems
$\mathrm{P}_{\mathcal{R}}$	analogue of P to the BSS model
RAS	randomized approximation scheme
RP	randomized polynomial time
VP	analogue of P to arithmetic complexity
VNP	analogue of NP to arithmetic complexity

Hyperplane arrangements

$\chi(\mathcal{A}; x)$	characteristic polynomial
$\overline{\chi}(\mathcal{A}; X, Y)$	coboundary polynomial
$a(\mathcal{A})$	number of regions of a real arrangement

$\mathcal{A}$	hyperplane arrangement
$b(\mathcal{A})$	number of bounded regions
$L(\mathcal{A})$	intersection poset
$R(A)$	complement of $\mathcal{T}(A)$
$T(\mathcal{A};x,y)$	Tutte polynomial
$\mathcal{T}(A)$	hypertorus arrangement
$T_{\mathrm{arith}}(A;x,y)$	arithmetic Tutte polynomial
$\mathcal{V}(\mathcal{A})$	complement of $\mathcal{A}$ in V
$W(\mathcal{A})$	wonderful compactification
$\mathrm{Zono}(A)$	zonotope
$\widetilde{Z}(\mathcal{A};q,\mathbf{w})$	multivariate Tutte polynomial

Knot theory

$\langle D\rangle$	Kauffman bracket
$D(G)$	link diagram of G
$\mathcal{D}(G)$	Jaeger link diagram of G
$H^{i,j}(G)$	chromatic cohomology
$P(L)$	HOMFLYPT polynomial
$V(L)$	Jones polynomial
$w(D)$	writhe
$X(L)$	Kauffman polynomial

Index

Printed in the United States
by Baker & Taylor Publisher Services

Printed in the United States
by Baker & Taylor Publisher Services